# Lecture Notes in Computer Science     11072

Commenced Publication in 1973
Founding and Former Series Editors:
Gerhard Goos, Juris Hartmanis, and Jan van Leeuwen

More information about this series at http://www.springer.com/series/7412

Alejandro F. Frangi · Julia A. Schnabel
Christos Davatzikos · Carlos Alberola-López
Gabor Fichtinger (Eds.)

# Medical Image Computing and Computer Assisted Intervention – MICCAI 2018

21st International Conference
Granada, Spain, September 16–20, 2018
Proceedings, Part III

 Springer

*Editors*
Alejandro F. Frangi (ID)
University of Leeds
Leeds
UK

Carlos Alberola-López (ID)
Universidad de Valladolid
Valladolid
Spain

Julia A. Schnabel
King's College London
London
UK

Gabor Fichtinger
Queen's University
Kingston, ON
Canada

Christos Davatzikos (ID)
University of Pennsylvania
Philadelphia, PA
USA

ISSN 0302-9743           ISSN 1611-3349   (electronic)
Lecture Notes in Computer Science
ISBN 978-3-030-00930-4        ISBN 978-3-030-00931-1   (eBook)
https://doi.org/10.1007/978-3-030-00931-1

Library of Congress Control Number: 2018909526

LNCS Sublibrary: SL6 – Image Processing, Computer Vision, Pattern Recognition, and Graphics

This Springer imprint is published by the registered company Springer Nature Switzerland AG
The registered company address is: Gewerbestrasse 11, 6330 Cham, Switzerland

# Preface

We are very pleased to present the conference proceedings for the 21st International Conference on Medical Image Computing and Computer Assisted Intervention (MICCAI), which was successfully held at the Granada Conference Center, September 16–20, 2018 in Granada, Spain.

The conference also featured 40 workshops, 14 tutorials, and ten challenges held on September 16 or 20. For the first time, we had events co-located or endorsed by other societies. The two-day Visual Computing in Biology and Medicine (VCBM) Workshop partnered with EUROGRAPHICS[1], the one-day Biomedical Workshop Biomedical Information Processing and Analysis: A Latin American perspective partnered with SIPAIM[2], and the one-day MICCAI Workshop on Computational Diffusion on MRI was endorsed by ISMRM[3]. This year, at the time of writing this preface, the MICCAI 2018 conference had over 1,400 firm registrations for the main conference featuring the most recent work in the fields of:

– Reconstruction and Image Quality
– Machine Learning and Statistical Analysis
– Registration and Image Guidance
– Optical and Histology Applications
– Cardiac, Chest and Abdominal Applications
– fMRI and Diffusion Imaging
– Neuroimaging
– Computer-Assisted Intervention
– Segmentation

This was the largest MICCAI conference to date, with, for the first time, four volumes of *Lecture Notes in Computer Science* (LNCS) proceedings for the main conference, selected after a thorough double-blind peer-review process organized in several phases as further described below. Following the example set by the previous program chairs of MICCAI 2017, we employed the Conference Managing Toolkit (CMT)[4] for paper submissions and double-blind peer-reviews, the Toronto Paper Matching System (TPMS)[5] for automatic paper assignment to area chairs and reviewers, and Researcher.CC[6] to handle conflicts between authors, area chairs, and reviewers.

---

[1]  https://www.eg.org.

[2]  http://www.sipaim.org/.

[3]  https://www.ismrm.org/.

[4]  https://cmt.research.microsoft.com.

[5]  http://torontopapermatching.org.

[6]  http://researcher.cc.

In total, a record 1,068 full submissions (ca. 33% more than the previous year) were received and sent out to peer-review, from 1,335 original intentions to submit. Of those submissions, 80% were considered as pure Medical Image Computing (MIC), 14% as pure Computer-Assisted Intervention (CAI), and 6% as MICCAI papers that fitted into both MIC and CAI areas. The MICCAI 2018 Program Committee (PC) had a total of 58 area chairs, with 45% from Europe, 43% from the Americas, 9% from Australasia, and 3% from the Middle East. We maintained an excellent gender balance with 43% women scientists on the PC.

Using TPMS scoring and CMT, each area chair was assigned between 18 and 20 manuscripts using TPMS, for each of which they suggested 9–15 potential reviewers. Subsequently, 600 invited reviewers were asked to bid for the manuscripts they had been suggested for. Final reviewer allocations via CMT took PC suggestions, reviewer bidding, and TPMS scores into account, allocating 5–6 papers per reviewer. Based on the double-blind reviews, 173 papers (16%) were directly accepted and 314 papers (30%) were directly rejected – these decisions were confirmed by the handling area chair. The remaining 579 papers (54%) were invited for rebuttal. Two further area chairs were added using CMT and TPMS scores to each of these remaining manuscripts, who then independently scored these to accept or reject, based on the reviews, rebuttal, and manuscript, resulting in clear paper decisions using majority voting: 199 further manuscripts were accepted, and 380 rejected.

The overall manuscript acceptance rate was 34.9%. Two PC teleconferences were held on May 14, 2018, in two different time zones to confirm the final results and collect PC feedback on the peer-review process (with over 74% PC attendance rate). For the MICCAI 2018 proceedings, the 372 accepted papers[7] have been organized in four volumes as follows:

- Volume LNCS 11070 includes: Image Quality and Artefacts (15 manuscripts), Image Reconstruction Methods (31), Machine Learning in Medical Imaging (22), Statistical Analysis for Medical Imaging (10), and Image Registration Methods (21)
- Volume LNCS 11071 includes: Optical and Histology Applications (46); and Cardiac, Chest, and Abdominal Applications (59)
- Volume LNCS 11072 includes: fMRI and Diffusion Imaging (45); Neuroimaging and Brain Segmentation (37)
- Volume LNCS 11073 includes: Computer-Assisted Intervention (39) grouped into image-guided interventions and surgery; surgical planning, simulation and work flow analysis; and visualization and augmented reality; and Image Segmentation Methods (47) grouped into general segmentation methods; multi-organ segmentation; abdominal, cardiac, chest, and other segmentation applications.

We would like to thank everyone who contributed greatly to the success of MICCAI 2018 and the quality of its proceedings. These include the MICCAI Society, for support and insightful comments; and our sponsors for financial support and their presence on site. We are especially grateful to all members of the Program Committee for their diligent work in the reviewer assignments and final paper selection, as well as the 600

---

[7] One paper was withdrawn.

reviewers for their support during the entire process. Finally, and most importantly, we thank all authors, co-authors, students, and supervisors, for submitting and presenting their high-quality work which made MICCAI 2018 a greatly enjoyable, informative, and successful event. We are especially indebted to those reviewers and PC members who helped us resolve last-minute missing reviews at a very short notice.

We are looking forward to seeing you in Shenzhen, China, at MICCAI 2019!

August 2018

Julia A. Schnabel
Christos Davatzikos
Gabor Fichtinger
Alejandro F. Frangi
Carlos Alberola-López
Alberto Gomez Herrero
Spyridon Bakas
Antonio R. Porras

# Organization

## Organizing Committee

### General Chair and Program Co-chair

Alejandro F. Frangi      University of Leeds, UK

### General Co-chair

Carlos Alberola-López      Universidad de Valladolid, Spain

### Associate to General Chairs

Antonio R. Porras      Children's National Medical Center, Washington D.C., USA

### Program Chair

Julia A. Schnabel      King's College London, UK

### Program Co-chairs

Christos Davatzikos      University of Pennsylvania, USA
Gabor Fichtinger      Queen's University, Canada

### Associates to Program Chairs

Spyridon Bakas      University of Pennsylvania, USA
Alberto Gomez Herrero      King's College London, UK

### Tutorial and Educational Chair

Anne Martel      University of Toronto, Canada

### Tutorial and Educational Co-chairs

Miguel González-Ballester      Universitat Pompeu Fabra, Spain
Marius Linguraru      Children's National Medical Center, Washington D.C., USA
Kensaku Mori      Nagoya University, Japan
Carl-Fredrik Westin      Harvard Medical School, USA

### Workshop and Challenge Chair

Danail Stoyanov      University College London, UK

**Workshop and Challenge Co-chairs**

Hervé Delingette            Inria, France
Lena Maier-Hein            German Cancer Research Center, Germany
Zeike A. Taylor            University of Leeds, UK

**Keynote Lecture Chair**

Josien Pluim            TU Eindhoven, The Netherlands

**Keynote Lecture Co-chairs**

Matthias Harders            ETH Zurich, Switzerland
Septimiu Salcudean            The University of British Columbia, Canada

**Corporate Affairs Chair**

Terry Peters            Western University, Canada

**Corporate Affairs Co-chairs**

Hayit Greenspan            Tel Aviv University, Israel
Despina Kontos            University of Pennsylvania, USA
Guy Shechter            Philips, USA

**Student Activities Facilitator**

Demian Wasserman            Inria, France

**Student Activities Co-facilitator**

Karim Lekadir            Universitat Pompeu-Fabra, Spain

**Communications Officer**

Pedro Lopes            University of Leeds, UK

**Conference Management**

DEKON Group

# Program Committee

Ali Gooya            University of Sheffield, UK
Amber Simpson            Memorial Sloan Kettering Cancer Center, USA
Andrew King            King's College London, UK
Bennett Landman            Vanderbilt University, USA
Bernhard Kainz            Imperial College London, UK
Burak Acar            Bogazici University, Turkey

| | |
|---|---|
| Sotirios Tsaftaris | University of Edinburgh, UK |
| Stamatia Giannarou | Imperial College London, UK |
| Stefanie Speidel | National Center for Tumor Diseases (NCT) Dresden, Germany |
| Stefanie Demirci | Technical University of Munich, Germany |
| Tammy Riklin Raviv | Ben-Gurion University, Israel |
| Tanveer Syeda-Mahmood | IBM Research, USA |
| Ulas Bagci | University of Central Florida, USA |
| Vamsi Ithapu | University of Wisconsin-Madison, USA |
| Yanwu Xu | Baidu Inc., China |

## Scientific Review Committee

| | |
|---|---|
| Amir Abdi | Martin Benning |
| Ehsan Adeli | Aïcha BenTaieb |
| Iman Aganj | Ruth Bergman |
| Ola Ahmad | Alessandro Bevilacqua |
| Amr Ahmed | Ryoma Bise |
| Shazia Akbar | Isabelle Bloch |
| Alireza Akhondi-asl | Sebastian Bodenstedt |
| Saad Ullah Akram | Hrvoje Bogunovic |
| Amir Alansary | Gerda Bortsova |
| Shadi Albarqouni | Sylvain Bouix |
| Luis Alvarez | Felix Bragman |
| Deepak Anand | Christopher Bridge |
| Elsa Angelini | Tom Brosch |
| Rahman Attar | Aurelien Bustin |
| Chloé Audigier | Irène Buvat |
| Angelica Aviles-Rivero | Cesar Caballero-Gaudes |
| Ruqayya Awan | Ryan Cabeen |
| Suyash Awate | Nathan Cahill |
| Dogu Baran Aydogan | Jinzheng Cai |
| Shekoofeh Azizi | Weidong Cai |
| Katja Bühler | Tian Cao |
| Junjie Bai | Valentina Carapella |
| Wenjia Bai | M. Jorge Cardoso |
| Daniel Balfour | Daniel Castro |
| Walid Barhoumi | Daniel Coelho de Castro |
| Sarah Barman | Philippe C. Cattin |
| Michael Barrow | Juan Cerrolaza |
| Deepti Bathula | Suheyla Cetin Karayumak |
| Christian F. Baumgartner | Matthieu Chabanas |
| Pierre-Louis Bazin | Jayasree Chakraborty |
| Delaram Behnami | Rudrasis Chakraborty |
| Erik Bekkers | Rajib Chakravorty |
| Rami Ben-Ari | Vimal Chandran |

Catie Chang
Pierre Chatelain
Akshay Chaudhari
Antong Chen
Chao Chen
Geng Chen
Hao Chen
Jianxu Chen
Jingyun Chen
Min Chen
Xin Chen
Yang Chen
Yuncong Chen
Jiezhi Cheng
Jun Cheng
Veronika Cheplygina
Farida Cheriet
Minqi Chong
Daan Christiaens
Serkan Cimen
Francesco Ciompi
Cedric Clouchoux
James Clough
Dana Cobzas
Noel Codella
Toby Collins
Olivier Commowick
Sailesh Conjeti
Pierre-Henri Conze
Tessa Cook
Timothy Cootes
Pierrick Coupé
Alessandro Crimi
Adrian Dalca
Sune Darkner
Dhritiman Das
Johan Debayle
Farah Deeba
Silvana Dellepiane
Adrien Depeursinge
Maria Deprez
Christian Desrosiers
Blake Dewey
Jwala Dhamala
Qi Dou
Karen Drukker

Lei Du
Lixin Duan
Florian Dubost
Nicolas Duchateau
James Duncan
Luc Duong
Nicha Dvornek
Oleh Dzyubachyk
Zach Eaton-Rosen
Mehran Ebrahimi
Matthias J. Ehrhardt
Ahmet Ekin
Ayman El-Baz
Randy Ellis
Mohammed Elmogy
Marius Erdt
Guray Erus
Marco Esposito
Joset Etzel
Jingfan Fan
Yong Fan
Aly Farag
Mohsen Farzi
Anahita Fathi Kazerooni
Hamid Fehri
Xinyang Feng
Olena Filatova
James Fishbaugh
Tom Fletcher
Germain Forestier
Denis Fortun
Alfred Franz
Muhammad Moazam Fraz
Wolfgang Freysinger
Jurgen Fripp
Huazhu Fu
Yang Fu
Bernhard Fuerst
Gareth Funka-Lea
Isabel Funke
Jan Funke
Francesca Galassi
Linlin Gao
Mingchen Gao
Yue Gao
Zhifan Gao

Utpal Garain
Mona Garvin
Aimilia Gastounioti
Romane Gauriau
Bao Ge
Sandesh Ghimire
Ali Gholipour
Rémi Giraud
Ben Glocker
Ehsan Golkar
Polina Golland
Yuanhao Gong
German Gonzalez
Pietro Gori
Alejandro Granados
Sasa Grbic
Enrico Grisan
Andrey Gritsenko
Abhijit Guha Roy
Yanrong Guo
Yong Guo
Vikash Gupta
Benjamin Gutierrez Becker
Séverine Habert
Ilker Hacihaliloglu
Stathis Hadjidemetriou
Ghassan Hamarneh
Adam Harrison
Grant Haskins
Charles Hatt
Tiancheng He
Mehdi Hedjazi Moghari
Tobias Heimann
Christoph Hennersperger
Alfredo Hernandez
Monica Hernandez
Moises Hernandez Fernandez
Carlos Hernandez-Matas
Matthew Holden
Yi Hong
Nicolas Honnorat
Benjamin Hou
Yipeng Hu
Heng Huang
Junzhou Huang
Weilin Huang

Xiaolei Huang
Yawen Huang
Henkjan Huisman
Yuankai Huo
Sarfaraz Hussein
Jana Hutter
Seong Jae Hwang
Atsushi Imiya
Amir Jamaludin
Faraz Janan
Uditha Jarayathne
Xi Jiang
Jieqing Jiao
Dakai Jin
Yueming Jin
Bano Jordan
Anand Joshi
Shantanu Joshi
Leo Joskowicz
Christoph Jud
Siva Teja Kakileti
Jayashree Kalpathy-Cramer
Ali Kamen
Neerav Karani
Anees Kazi
Eric Kerfoot
Erwan Kerrien
Farzad Khalvati
Hassan Khan
Bishesh Khanal
Ron Kikinis
Hyo-Eun Kim
Hyunwoo Kim
Jinman Kim
Minjeong Kim
Benjamin Kimia
Kivanc Kose
Julia Krüger
Pavitra Krishnaswamy
Frithjof Kruggel
Elizabeth Krupinski
Sofia Ira Ktena
Arjan Kuijper
Ashnil Kumar
Neeraj Kumar
Punithakumar Kumaradevan

Manuela Kunz
Jin Tae Kwak
Alexander Ladikos
Rodney Lalonde
Pablo Lamata
Catherine Laporte
Carole Lartizien
Toni Lassila
Andras Lasso
Matthieu Le
Maria J. Ledesma-Carbayo
Hansang Lee
Jong-Hwan Lee
Soochahn Lee
Etienne Léger
Beatrice Lentes
Wee Kheng Leow
Nikolas Lessmann
Annan Li
Gang Li
Ruoyu Li
Wenqi Li
Xiang Li
Yuanwei Li
Chunfeng Lian
Jianming Liang
Hongen Liao
Ruizhi Liao
Roxane Licandro
Lanfen Lin
Claudia Lindner
Cristian Linte
Feng Liu
Hui Liu
Jianfei Liu
Jundong Liu
Kefei Liu
Mingxia Liu
Sidong Liu
Marco Lorenzi
Xiongbiao Luo
Jinglei Lv
Ilwoo Lyu
Omar M. Rijal
Pablo Márquez Neila
Henning Müller

Kai Ma
Khushhall Chandra Mahajan
Dwarikanath Mahapatra
Andreas Maier
Klaus H. Maier-Hein
Sokratis Makrogiannis
Grégoire Malandain
Anand Malpani
Jose Manjon
Tommaso Mansi
Awais Mansoor
Anne Martel
Diana Mateus
Arnaldo Mayer
Jamie McClelland
Stephen McKenna
Ronak Mehta
Raphael Meier
Qier Meng
Yu Meng
Bjoern Menze
Liang Mi
Shun Miao
Abhishek Midya
Zhe Min
Rashika Mishra
Marc Modat
Norliza Mohd Noor
Mehdi Moradi
Rodrigo Moreno
Kensaku Mori
Aliasghar Mortazi
Peter Mountney
Arrate Muñoz-Barrutia
Anirban Mukhopadhyay
Arya Nabavi
Layan Nahlawi
Ana Ineyda Namburete
Valery Naranjo
Peter Neher
Hannes Nickisch
Dong Nie
Lipeng Ning
Jack Noble
Vincent Noblet
Alexey Novikov

Ilkay Oksuz
Ozan Oktay
John Onofrey
Eliza Orasanu
Felipe Orihuela-Espina
Jose Orlando
Yusuf Osmanlioglu
David Owen
Cristina Oyarzun Laura
Jose-Antonio Pérez-Carrasco
Danielle Pace
J. Blas Pagador
Akshay Pai
Xenophon Papademetris
Bartlomiej Papiez
Toufiq Parag
Magdalini Paschali
Angshuman Paul
Christian Payer
Jialin Peng
Tingying Peng
Xavier Pennec
Sérgio Pereira
Mehran Pesteie
Loic Peter
Igor Peterlik
Simon Pezold
Micha Pfeifer
Dzung Pham
Renzo Phellan
Pramod Pisharady
Josien Pluim
Kilian Pohl
Jean-Baptiste Poline
Alison Pouch
Prateek Prasanna
Philip Pratt
Raphael Prevost
Esther Puyol Anton
Yuchuan Qiao
Gwénolé Quellec
Pradeep Reddy Raamana
Julia Rackerseder
Hedyeh Rafii-Tari
Mehdi Rahim
Kashif Rajpoot

Parnesh Raniga
Yogesh Rathi
Saima Rathore
Nishant Ravikumar
Shan E. Ahmed Raza
Islem Rekik
Beatriz Remeseiro
Markus Rempfler
Mauricio Reyes
Constantino Reyes-Aldasoro
Nicola Rieke
Laurent Risser
Leticia Rittner
Yong Man Ro
Emma Robinson
Rafael Rodrigues
Marc-Michel Rohé
Robert Rohling
Karl Rohr
Plantefeve Rosalie
Holger Roth
Su Ruan
Danny Ruijters
Juan Ruiz-Alzola
Mert Sabuncu
Frank Sachse
Farhang Sahba
Septimiu Salcudean
Gerard Sanroma
Emine Saritas
Imari Sato
Alexander Schlaefer
Jerome Schmid
Caitlin Schneider
Jessica Schrouff
Thomas Schultz
Suman Sedai
Biswa Sengupta
Ortal Senouf
Maxime Sermesant
Carmen Serrano
Amit Sethi
Muhammad Shaban
Reuben Shamir
Yeqin Shao
Li Shen

Bibo Shi
Kuangyu Shi
Hoo-Chang Shin
Russell Shinohara
Viviana Siless
Carlos A. Silva
Matthew Sinclair
Vivek Singh
Korsuk Sirinukunwattana
Ihor Smal
Michal Sofka
Jure Sokolic
Hessam Sokooti
Ahmed Soliman
Stefan Sommer
Diego Sona
Yang Song
Aristeidis Sotiras
Jamshid Sourati
Rachel Sparks
Ziga Spiclin
Lawrence Staib
Ralf Stauder
Darko Stern
Colin Studholme
Martin Styner
Heung-Il Suk
Jian Sun
Xu Sun
Kyunghyun Sung
Nima Tajbakhsh
Sylvain Takerkart
Chaowei Tan
Jeremy Tan
Mingkui Tan
Hui Tang
Min Tang
Youbao Tang
Yuxing Tang
Christine Tanner
Qian Tao
Giacomo Tarroni
Zeike Taylor
Kim Han Thung
Yanmei Tie
Daniel Toth

Nicolas Toussaint
Jocelyne Troccaz
Tomasz Trzcinski
Ahmet Tuysuzoglu
Andru Twinanda
Carole Twining
Eranga Ukwatta
Mathias Unberath
Tamas Ungi
Martin Urschler
Maria Vakalopoulou
Vanya Valindria
Koen Van Leemput
Hien Van Nguyen
Gijs van Tulder
S. Swaroop Vedula
Harini Veeraraghavan
Miguel Vega
Anant Vemuri
Gopalkrishna Veni
Archana Venkataraman
François-Xavier Vialard
Pierre-Frederic Villard
Satish Viswanath
Wolf-Dieter Vogl
Ingmar Voigt
Tomaz Vrtovec
Bo Wang
Guotai Wang
Jiazhuo Wang
Liansheng Wang
Manning Wang
Sheng Wang
Yalin Wang
Zhe Wang
Simon Warfield
Chong-Yaw Wee
Juergen Weese
Benzheng Wei
Wolfgang Wein
William Wells
Rene Werner
Daniel Wesierski
Matthias Wilms
Adam Wittek
Jelmer Wolterink

Ken C. L. Wong
Jonghye Woo
Pengxiang Wu
Tobias Wuerfl
Yong Xia
Yiming Xiao
Weidi Xie
Yuanpu Xie
Fangxu Xing
Fuyong Xing
Tao Xiong
Daguang Xu
Yan Xu
Zheng Xu
Zhoubing Xu
Ziyue Xu
Wufeng Xue
Jingwen Yan
Ke Yan
Yuguang Yan
Zhennan Yan
Dong Yang
Guang Yang
Xiao Yang
Xin Yang
Jianhua Yao
Jiawen Yao
Xiaohui Yao
Chuyang Ye
Menglong Ye
Jingru Yi
Jinhua Yu
Lequan Yu
Weimin Yu
Yixuan Yuan
Evangelia Zacharaki
Ernesto Zacur

Guillaume Zahnd
Marco Zenati
Ke Zeng
Oliver Zettinig
Daoqiang Zhang
Fan Zhang
Han Zhang
Heye Zhang
Jiong Zhang
Jun Zhang
Lichi Zhang
Lin Zhang
Ling Zhang
Mingli Zhang
Pin Zhang
Shu Zhang
Tong Zhang
Yong Zhang
Yunyan Zhang
Zizhao Zhang
Qingyu Zhao
Shijie Zhao
Yitian Zhao
Guoyan Zheng
Yalin Zheng
Yinqiang Zheng
Zichun Zhong
Luping Zhou
Zhiguo Zhou
Dajiang Zhu
Wentao Zhu
Xiaofeng Zhu
Xiahai Zhuang
Aneeq Zia
Veronika Zimmer
Majd Zreik
Reyer Zwiggelaar

## Mentorship Program (Mentors)

Stephen Aylward        Kitware Inc., USA
Christian Barillot      IRISA/CNRS/University of Rennes, France
Kayhan Batmanghelich    University of Pittsburgh/Carnegie Mellon University,
                          USA
Christos Bergeles       King's College London, UK

# Sponsors and Funders

## Platinum Sponsors

- NVIDIA Inc.
- Siemens Healthineers GmbH

## Gold Sponsors

- Guangzhou Shiyuan Electronics Co. Ltd.
- Subtle Medical Inc.

## Silver Sponsors

- Arterys Inc.
- Claron Technology Inc.
- ImSight Inc.
- ImFusion GmbH
- Medtronic Plc

## Bronze Sponsors

- Depwise Inc.
- Carl Zeiss AG

## Travel Bursary Support

- MICCAI Society
- National Institutes of Health, USA
- EPSRC-NIHR Medical Image Analysis Network (EP/N026993/1), UK

# Contents – Part III

**Diffusion Tensor Imaging and Functional MRI: Functional MRI**

**Neuroimaging and Brain Segmentation Methods: Neuroimaging**

# Diffusion Tensor Imaging and Functional MRI: Diffusion Tensor Imaging

# Multimodal Fusion of Brain Networks with Longitudinal Couplings

Wen Zhang, Kai Shu, Suhang Wang, Huan Liu, and Yalin Wang[(⊠)]

School of Computing, Informatics and Decision Systems Engineering,
Arizona State University, Tempe, AZ, USA
ylwang@asu.edu

**Abstract.** In recent years, brain network analysis has attracted considerable interests in the field of neuroimaging analysis. It plays a vital role in understanding biologically fundamental mechanisms of human brains. As the upward trend of multi-source in neuroimaging data collection, effective learning from the different types of data sources, e.g. multimodal and longitudinal data, is much in demand. In this paper, we propose a general coupling framework, the multimodal neuroimaging network fusion with longitudinal couplings (**MMLC**), to learn the latent representations of brain networks. Specifically, we jointly factorize multimodal networks, assuming a linear relationship to couple network variance across time. Experimental results on two large datasets demonstrate the effectiveness of the proposed framework. The new approach integrates information from longitudinal, multimodal neuroimaging data and boosts statistical power to predict psychometric evaluation measures.

**Keywords:** Multimodality · Longitudinal · Brain network fusion
Representation

## 1 Introduction

It is widely accepted that brain has one of the most complex networks known to man. One of the modern approaches in neuroscience is considering the brain regional interactions as a graph network, referred to as brain connectome or brain network [7]. By learning network properties, researchers could draw a broad picture about how the brain controls and regulates information through the orderly transfer of neural signals within brain regions [9]. However, effectively learning features from few noisy brain networks is difficult. It is a common belief that the exploitation of auxiliary and complementary information from the multimodal and longitudinal data would be highly beneficial to improve the effectiveness of brain network analysis.

In practice, brain networks can be analyzed from two perspectives, i.e., multimodal and longitudinal. For multimodal data, functional magnetic resonance imaging (fMRI) and diffusion tensor imaging (DTI) are the most widely used

© Springer Nature Switzerland AG 2018
A. F. Frangi et al. (Eds.): MICCAI 2018, LNCS 11072, pp. 3–11, 2018.
https://doi.org/10.1007/978-3-030-00931-1_1

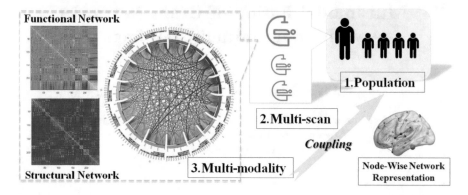

**Fig. 1.** The proposed MMLC framework. Three levels of information couplings are considered: the cross-sectional coupling (Population), the longitudinal coupling (Multiscan) and the multimodal coupling (Multi-modality).

modalities which could yield complementary information of brain networks. Several studies proposed the linear [1] or non-linear [10] multimodal integration approaches for brain network analysis, which significantly improve the detection power of brain structural or mental changes. On the other hand, the longitudinal analysis of brain network portrays progressive changes in brain functional activities or anatomical architectures. For example, as a supplement to baseline analysis, the longitudinal study of age-related changes in the topological organization of structural brain networks effectively explored new patterns of normal aging [15]. It worth noting that the multimodal and longitudinal studies are also inherently related. However, existing works on brain networks usually only consider one perspective of brain networks. It remains a major challenge to achieve multilevel fusion of neuroimages by exploiting these two different types of brain networks.

In this paper, we aim to empower the longitudinal multimodal neuroimaging data fusion with brain network representation learning. We develop a theoretical model (MMLC) which achieves multimodal and longitudinal coupling simultaneously. It bases on two assumptions: (1) both functional and structural networks reflect the regional interactions inside the brain. Thus they should share a similar basis in network topology; (2) the longitudinal changes should relate to the previous as well as the next brain stages. To the best of our knowledge, it is the first work that fuses multimodal and longitudinal brain networks simultaneously. We expect such data fusion model will significantly improve the statistical power of brain network analysis.

## 2   Brian Network Representation Learning

Our brain network fusion framework is presented in Fig. 1, which consists of three coupling components: cross-sectional, multimodal and longitudinal couplings.

## 2.1  Modeling Cross-Sectional Coupling

It is natural to couple subjects in a population to achieve the individual and group properties. In the cross-sectional coupling, we use the consensus matrix to model group properties. We defined a graph $g = (V, X)$ for the individual brain network, where $V$ is the set of all n brain regions and $X$ is the connectivity matrix which records the connectivity strength between pairs of the regions. For each subject, we have two connectivity matrices, $X^f$ and $X^d$, for functional and structural network respectively. For the whole dataset, suppose we have $N$ subjects and each subject was scanned $T$ times, we have a set of functional and structural connectivity matrices, $\{X_{i,j}^f, X_{i,j}^d | i = 1, ..., N, j = 1, ..., T\}$. We conduct network embedding with matrix factorization to map the graph data into a lower dimensional latent space. $X_{i,j} \approx U_{i,j} V_{i,j}^T$ and $V_{i,j}^T \in \mathbb{R}^{N \times P}$ will be the new representation of network in the latent $P$-dimensional space. To look for the consistent patterns among a population of subjects, we adopt the concept of consensus matrix. If subjects are from the same group, it is natural to hypothesize that it has the similar representations in all graph matrix but only differ on new basis matrix $(U_{i,\cdot})$. Hence, $V_{i,\cdot}$ should be the target to measure disagreement of the network patterns. Here, we introduce a new variable $V^* \in \mathbb{R}^{N \times P}$ named the consensus matrix of $V_{i,\cdot}$ and the following loss function to couple the network representations within a group of items, $\sum_{i=1}^N \parallel V_{i,\cdot} - V^* \parallel_F^2$. Note that all $V_{i,\cdot}$ are at the same scale because entries of all the original network matrix $X_{i,\cdot}$ locate within the range of $[-1, 1]$. Even though the general factorization problem might exist multiple solutions due to the uncertainty of basis matrix, the inter-subjects coupling problem described as below is solvable with the added consensus matrix term and assumption that $X_i$ is within the same range.

$$\min_{U_{i,\cdot}; V_{i,\cdot}} \sum_{i=1}^N \parallel X_{i,\cdot} - U_{i,\cdot} V_{i,\cdot}^T \parallel_F^2 + \lambda \parallel V_{i,\cdot} - V^* \parallel_F^2 . \tag{1}$$

## 2.2  Modeling Multimodal Coupling

The structural network builds the anatomical foundations of the functional network. Therefore, there should be the topological similarities between them. Here, we share the basis matrix in Eq. 1 across modality. In this study, we chose a newly proposed brain atlas which has 246 brain regions and reflects both functional and structural connectivity patterns [4]. Ideally, the brain network created from fine-grained atlas contains more details of the regional interaction [16]. After matrix factorization, we get two new network representations for each subject $i$ as $V_{i,\cdot}^f$ and $V_{i,\cdot}^d$ with the corresponding basis matrices $U_{i,\cdot}^f$ and $U_{i,\cdot}^d$. We hypothesize that there are quite significant similarities between those two kinds of networks that can be interpreted with specific presentations in latent space. In other words, given a subject $i$, the functional and structural networks share the same basis matrix $U_{i,\cdot}^f = U_{i,\cdot}^d = U_{i,\cdot}$. Though the similarities of multimodal brain networks might be deciphered as the linear or non-linear relationships, we only focus on the linear relationship in this study. For future study, we could model the non-linear relationship with the high-order of basis matrix. Eventually, together with the inter-subject coupling model, we propose the multi-modality coupling model as below:

$$\min_{U_{i,\cdot};V_{i,\cdot}} \sum_{i=1}^{N} \parallel X_{i,\cdot}^f - U_{i,\cdot}(V_{i,\cdot}^f)^T \parallel_F^2 + \alpha \parallel X_{i,\cdot}^d - U_{i,\cdot}(V_{i,\cdot}^d)^T \parallel_F^2$$
$$+ \lambda_1 \parallel V_{i,\cdot}^f - V^{f*} \parallel_F^2 + \lambda_1 \parallel V_{i,\cdot}^d - V^{d*} \parallel_F^2 . \tag{2}$$

### 2.3   Modeling Longitudinal Coupling

In the longitudinal coupling, we track how the brain evolves over time. We model the smooth variation of brain networks as a Procrustes problem which maps the consensus matrix to a same effect space [8]. First, we improve the model by adding an orthogonality constraint $(V_{\cdot}^{d*})^T V_{\cdot}^{d*} = I$ to the consensus matrix in the structural network. The matrix factorization with orthogonality constraint plays a similar role as the clustering [12] which is consistent with the subgraph organization in human brain networks. After adding the orthogonality constraint in every time point, we further model the relationships between each consecutive pairs of consensus matrix. Suppose we have two consensus matrixes $V_j^{d*}$ and $V_{j+1}^{d*}$ from two consecutive time points $j$ and $j + 1$. Due to the relative stability of structural network within subject, we expect that $V_j^{d*}$ shares a rotation relationship with $V_{j+1}^{d*}$, i.e. $V_j^{d*} = R_{j,j+1}V_{j+1}^{d*}$. $R_{j,j+1} \in \mathbb{R}^{n \times n}$ is a rotation matrix thus $det(R_{j,j+1}) = 1$ and $R_{j+1,j} = R_{j,j+1}^{-1} = R_{j,j+1}^T$. As $(V_j^{d*})^T V_j^{d*} = I$ is satisfied for all $j$, taking the rotation relationship into such orthogonality, we could construct a new symmetric matrix, $\tilde{M}_{j+1,j} = R_{j+1,j}V_{j+1}^{d*}(V_{j+1}^{d*})^T R_{j+1,j}^T$, that $(V_j^{d*})^T \tilde{M}_{j+1,j} V_j^{d*} = I$. Then we merge the consecutive relations from time $j$ to $j + 1$ of $V_j^{d*}$ into a single time variable $j$ as $M_j = (\tilde{M}_{j-1,j} + \tilde{M}_{j+1,j})/2$. The constraint, $(V_j^{d*})^T \tilde{M}_{j+1,j} V_j^{d*} = I$, is called the generalized Stiefel constraint [2] with the mass matrix $M_{i,j}$. In this paper, we applied an optimization algorithm involves the Stiefel manifold based on the Cayley transform for preserving the constraints [6].

### 2.4   The Proposed Model - MMLC

Now let's reformulate the problem for all $N$ subjects and $T$ time points. The multimodal brain network fusion with longitudinal coupling framework is to solve the problem of minimizing the following objective function, $L$, with the corresponding constraint:

$$\min_{\substack{U,V^f,V^d, \\ V^{f*},V^{d*}}} \sum_{i=1}^{N} \sum_{j=1}^{T} \parallel X_{i,j}^f - U_{i,j}(V_{i,j}^f)^T \parallel_F^2 + \alpha \parallel X_{i,j}^d - U_{i,j}(V_{i,j}^d)^T \parallel_F^2$$
$$+ \lambda_1 \parallel V_{i,j}^f - V_j^{f*} \parallel_F^2 + \lambda_1 \parallel V_{i,j}^d - V_j^{d*} \parallel_F^2 + \lambda_2 G \tag{3}$$
$$s.t. \ (V_j^{d*})^T M_{i,j} V_j^{d*} = I$$

where $G = \| U_{i,j} \|_F^2 + \| V_{i,j}^f \|_F^2 + \| V_{i,j}^d \|_F^2 + \| V_j^{f*} \|_F^2 + \| V_j^{d*} \|_F^2$ is the regularization term to prevent overfitting.

## 3    An Optimization Framework for MMLC

The problem proposed in Eq. 3 is a non-convex problem which is difficult to optimize directly. To estimate the optimal, we propose an iterative update procedure together with a Stochastic block coordinate descent algorithm.

**Fixing $V_j^{f*}$ and $V_j^{d*}$, minimize $L$ over $U_{i,j}, V_{i,j}^f$ and $V_{i,j}^d$.** For brevity in this subsection, we use $U, V^f, V^d, V^{f*}$ and $V^{d*}$ to represent $U_{i,j}, V_{i,j}^f, V_{i,j}^d, V_j^{f*}$ and $V_j^{d*}$. First, we fix $V^f$ and $V^d$ to update $U$. For a given subject $i$ and time point $j$, we could take the derivative of $L$ with respect to $U$.

$$\frac{\partial L}{\partial U} = 2(U(V^f)^T V^f - X^f V^f) + 2\alpha(U(V^d)^T V^d - X^d V^d)) + \lambda_2 G'(U). \qquad (4)$$

Here, $G'(U)$ is the derivative of $U$ with respect to $U$. Given a step size $l$, we update $U$ as $U_{new} = U_{pre} - l * \frac{\partial L_1}{\partial U_{pre}}$. Then, we fix $V^d$ and $U$ to update $V^f$. The gradient of $L$ with respect to $V^f$ is:

$$\frac{\partial L}{\partial V^f} = 2(V^f U^T U - X^f U) + 2\lambda_1(V^f - V^{f*}) + \lambda_2 G'(V^f). \qquad (5)$$

Similarly, we update $V^d$ with the same procedure as $V^f$,

$$\frac{\partial L}{\partial V^d} = 2\alpha(V^d(U)^T U - X^d U) + 2\lambda_1(V^d - V^{d*}) + \lambda_2 G'(V^d). \qquad (6)$$

**Fixing $U_{i,j}, V_{i,j}^f$ and $V_{i,j}^d$, Minimize $L$ over $V_j^{f*}$ and $V_j^{d*}$.** For brevity, we use $V_i^f, V_i^d, V^{f*}$ and $V^{d*}$ to represent $V_{i,j}^f, V_{i,j}^d, V_j^{f*}$ and $V_j^{d*}$. We observe that for each time $j$, the framework will generate a group-wise $V_j^{f*}$ and $V_j^{d*}$. After updating all individual $U_i, V_i^f$ and $V_i^d$, we could take the derivative of $L$ with respect to $V^{f*}$.

$$\frac{\partial L}{\partial V^{f*}} = 2\lambda_1 \sum_{i=1}^{N}(V^{f*} - V_i^f) + \lambda_2 G'(V^{f*}). \qquad (7)$$

For $V^{d*}$, an equality constraint $(V^{d*})^T M V^{d*} = I$ will regulate the gradient direction of $L$ with respect to $V^{d*}$, which makes the solution difficult. Instead of directly finding an optimal direction with gradient descent on the surface described by original object function, we construct the descent curves on the constraint-based Stiefel manifold [6]. Specifically, $V^{d*}$ will be divided into two submatrixes $V^{d*} = [V_1^{d*}; V_2^{d*}]$, where $V_1^{d*} \in \mathbb{R}^{s \times p}$ is the free variable to be solved and $V_2^{d*} \in \mathbb{R}^{(n-s) \times p}$ is the fixed variable treated as constants. Then we rearrange the constraint as:

$$\begin{bmatrix} V_1^{d*} \\ V_2^{d*} \end{bmatrix}^T \begin{bmatrix} M_{11} & M_{12} \\ M_{12}^T & M_{22} \end{bmatrix} \begin{bmatrix} V_1^{d*} \\ V_2^{d*} \end{bmatrix} = I. \tag{8}$$

It is easy to conclude that $M_{11}$ is a full rank positive definite matrix. Then a descent curve based on the previous $V^{d*}$ will be constructed and it starts at the point $P_s = V_1^{d*} + M_{11}^{-\frac{1}{2}} M_{12} V_2^{d*}$ which is the initial point for the line search on the generalized Stiefel manifold. Given the descending gradient $-L'(P) = -\frac{\partial L}{\partial V^{d*}} \circ \frac{\partial V^{d*}}{\partial P}$ at point $P$, we further project $-L'(P)$ onto the tangent space of the Stiefel manifold by constructing a skew-symmetric matrix $A = L'(P) P_s^T - P_s L'(P)^T$. This will lead to a curve function $Y(\tau)$ by the Crank-Nicolson-like design as in paper [14].

$$Y(\tau) = (I + \frac{\tau}{2} A M_{11})^{-1} (I - \frac{\tau}{2} A M_{11}) P_s. \tag{9}$$

The above function gives a linear search procedure of updating point $P$ by $P_{new} = Y(\tau)$ for small $\tau$ which results sufficient decrease in $L_2$. Finally, the next feasible $V_{new}^{d*}$ will be given as:

$$V_{new}^{d*}(P) = \begin{bmatrix} P - M_{11}^{-\frac{1}{2}} M_{12} V_2^{d*} \\ V_2^{d*} \end{bmatrix}. \tag{10}$$

## 4   Experiment

### 4.1   Experimental Settings and Baseline Methods

In this paper, we use two datasets to test the effectiveness of the proposed method in predicting anxiety and depression scores, respectively. They are all from the SLIM Repository (Southwest University Longitudinal Imaging Multimodal Brain Data)[1]. Dataset1 contains 105 healthy subjects and the trait anxiety score (TAIs) based on the State Trait Anxiety Inventory (STAI) was set as the predicted variables. Dataset2 includes 77 subjects with the assessment of ATQ (Automatic Thoughts Questionnaire) self-report measures which reflect mental states associated with depression and was the predicted variables in this trial. All subjects in two datasets are right-handed college students with the close age gaps and similar education level. We preprocessed fMRI data by using AFNI software[2] and using FSL[3] for DTI data. Then we construct the functional and structural networks with brain connectivity toolbox[4]. The gap between each pair of consecutive imaging scans is 1.5 years and 3 scans per person are collected. We aim to predict the score of anxiety and depression, which is a regression task.

[1]  http://fcon_1000.projects.nitrc.org/indi/retro/southwestuni_qiu_index.html.
[2]  https://afni.nimh.nih.gov/afni/doc/program_help/.
[3]  https://fsl.fmrib.ox.ac.uk/fsl/fslwiki/FSL.
[4]  https://sites.google.com/site/bctnet/.

To evaluate the regression performance, we adopt the two wildly used evaluation metrics, i.e., root mean square error (RMSE) and mean absolute error (MAE).

We compare the performance of MMLC with representative and state-of-the-art brain network representation learning algorithms as follows: **MFCSMC**: the collective matrix factorization [13] on multiple modalities with the cross-sectional coupling; **EigLC** [5]: a state-of-the-art method which models the longitudinal coupling on a single modality, i.e., structural networks; **sMFLC**: a variant of EigLC by removing the terms concerning functional networks in our model and remaining the matrix factorization (MF) with longitudinal coupling of structural networks; **MFCSLC**: MF with cross-sectional and Longitudinal coupling without sharing the basis matrix $U$. It is a variant of MMLC. Subject $i$ will obtain his new feature set $[V_{i,j}^f, V_{i,j}^d]$ at time point $j$ as the new representation of his original network graphs. We feed the vectorized features and the corresponding scores into a linear regression model for training by using LIBSVM toolbox [3]. In this experiment, each time we randomly pick 70% subjects as the training set and test on the rest. We repeat this process 30 times and the average performance and standard deviation are reported.

## 4.2    Results

We present the regression performance of our proposed model as opposite to other compared methods in Table 1. We set $\alpha = 1, \lambda_1 = 20$ for Dataset 1 and $\alpha = 5, \lambda_1 = 20$ for Dataset 2 according to the cross validation and empirically chose $\lambda_2$ to be 0.1 and dimension of the new feature space $P = 10$. From Table 1, we see that our proposed model outperforms other methods in two datasets. For example, sMFLC, the single modality version of our proposed model with longitudinal coupling have better performance than EigLC significantly. The difference of regression performance between the full coupling and partial coupling of multimodal or longitudinal fusions is small but significant. For example, in Dataset 2, the result of the partial coupling of brain networks in the multimodal fusion (MFCSMC) has the higher value of MAE ($p < 0.0006$, two-sample $t$-test) and RMSE ($p < 0.153$) than MMLC. Besides, MMLC outperforms MFCSLC which contains only longitudinal coupling in two datasets. In all, MMLC has the smallest MAE and RMSE values together with the relatively low variance compared with other testing methods in our experiments.

The proposed framework has two important parameters, i.e., $\alpha$ controlling the contribution of multimodal and $\lambda_1$ controlling cross-sectional coupling. In this section, we evaluate the affects of $\alpha$ and $\lambda_1$ on MMLC. Specifically, we vary $\alpha$ as $\{0.5, 1, 5, 10, 20\}$ and $\lambda_1$ as $\{1, 10, 20, 30, 40\}$. The results in terms of MAE and RMSE are shown in Figs. 2 and 3. From the figure, we observe that: (i) In Dataset 1, both MAE and RMSE value reach the relatively lowest points when $\alpha = 1$. Then, there is a tendency that MAE and RMSE increases along with the increase of $\alpha$. In Dataset 2, the variation of MAE and RMSE is more apparent when $\alpha$ increases to a higher value, e.g., $\alpha > 10$ and meanwhile the prediction performance has a significant drop. (ii) Moreover, when $\lambda_1 = 20$, our framework has the relatively best performance in both datasets. In this

**Table 1.** Regression performance on anxiety and depression datasets

| Methods | Dataset 1 (anxiety) | | Dataset 2 (depression) | |
|---|---|---|---|---|
| | MAE | RMSE | MAE | RMSE |
| EigLC | $3.401 \pm 0.931$ | $5.037 \pm 1.464$ | $3.536 \pm 0.729$ | $5.227 \pm 0.886$ |
| sMFLC | $2.329 \pm 0.135$ | $4.422 \pm 0.770$ | $1.872 \pm 0.556$ | $3.117 \pm 1.405$ |
| MFCSLC | $2.225 \pm 0.185$ | $4.223 \pm 1.120$ | $1.788 \pm 0.092$ | $3.050 \pm 0.504$ |
| MFCSMC | $2.285 \pm 0.151$ | $4.340 \pm 0.918$ | $1.741 \pm 0.080$ | $2.924 \pm 0.392$ |
| MMLC | $\mathbf{2.173 \pm 0.098^*}$ | $\mathbf{4.015 \pm 0.751^*}$ | $\mathbf{1.672 \pm 0.067^*}$ | $\mathbf{2.773 \pm 0.416^*}$ |

\* Significant with $p < 0.05$, two-sample $t$-test of MMLC with the other methods.

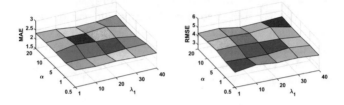

**Fig. 2.** Performance on Dataset 1 (anxiety)

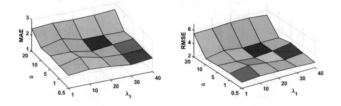

**Fig. 3.** Performance on Dataset 2 (depression)

experiment, we observe the different optimal parameter settings on tasks from two different domains, i.e., anxiety and depression. This is consistent with the previous discovery in brain networks that the distinct roles of functional and structural connectivities shift in different cognition tasks [11].

## 5    Conclusion and Future Works

This paper describes a novel network fusion framework which simultaneously considers multiple levels of information such as relationships between brain functional and structural organizations and longitudinal brain development. We test our proposed framework with two large cohorts and experimental results demonstrate the effectiveness of the generated network representations for prediction of cognitive status. There are several interesting directions that are warranted for further investigation. For example, in the future, we could add the non-linear relationship of multimodal networks to our framework or investigate subnetwork patterns that can be derived from the learned network representations.

**Acknowledgement.** This work was supported by the grants from NIH (R21AG0 49216, RF1AG051710, R01EB025032) and NSF (IIS-1421165, IIS-1217466, 1614576).

# References

1. Abdelnour, F., Voss, H.U., Raj, A.: Network diffusion accurately models the relationship between structural and functional brain connectivity networks. Neuroimage **90**, 335–347 (2014)
2. Absil, P.A., Mahony, R., Sepulchre, R.: Optimization Algorithms on Matrix Manifolds. Princeton University Press (2009)
3. Chang, C.C., Lin, C.J.: LIBSVM: a library for support vector machines. ACM Trans. Intell. Syst. Technol. (TIST) **2**(3), 27 (2011)
4. Fan, L., et al.: The human brainnetome atlas: a new brain atlas based on connectional architecture. Cereb. Cortex **26**(8), 3508–3526 (2016)
5. Jae Hwang, S., et al.: Coupled harmonic bases for longitudinal characterization of brain networks. In: CVPR, pp. 2517–2525 (2016)
6. Jae Hwang, S., et al.: A projection free method for generalized eigenvalue problem with a nonsmooth regularizer. In: ICCV, pp. 1841–1849 (2015)
7. Kong, X., Yu, P.S.: Brain network analysis: a data mining perspective. ACM SIGKDD Explor. Newslett. **15**(2), 30–38 (2014)
8. McIntosh, A.R., Lobaugh, N.J.: Partial least squares analysis of neuroimaging data: applications and advances. Neuroimage **23**, S250–S263 (2004)
9. Mesulam, M.: Brain, mind, and the evolution of connectivity. Brain Cognit. **42**(1), 4–6 (2000)
10. Ng, B., Varoquaux, G., Poline, J.-B., Thirion, B.: A novel sparse graphical approach for multimodal brain connectivity inference. In: Ayache, N., Delingette, H., Golland, P., Mori, K. (eds.) MICCAI 2012. LNCS, vol. 7510, pp. 707–714. Springer, Heidelberg (2012). https://doi.org/10.1007/978-3-642-33415-3_87
11. Park, H.J., Friston, K.: Structural and functional brain networks: from connections to cognition. Science **342**(6158), 1238411 (2013)
12. Pompili, F., Gillis, N., Absil, P.A., Glineur, F.: Two algorithms for orthogonal nonnegative matrix factorization with application to clustering. Neurocomputing **141**, 15–25 (2014)
13. Singh, A.P., Gordon, G.J.: Relational learning via collective matrix factorization. In: SIGKDD, pp. 650–658. ACM (2008)
14. Wen, Z., Yin, W.: A feasible method for optimization with orthogonality constraints. Math. Program. **142**(1–2), 397–434 (2013)
15. Wu, K., Taki, Y., Sato, K., Qi, H., Kawashima, R., Fukuda, H.: A longitudinal study of structural brain network changes with normal aging. Front. Hum. Neurosci. **7**, 113 (2013)
16. Zhang, W., et al.: Functional organization of the fusiform gyrus revealed with connectivity profiles. Hum. Brain Mapp. **37**(8), 3003–3016 (2016)

# Penalized Geodesic Tractography for Mitigating Gyral Bias

Ye Wu[1,2], Yuanjing Feng[1], Dinggang Shen[2], and Pew-Thian Yap[2(✉)]

[1] Institute of Information Processing and Automation,
Zhejiang University of Technology, Hangzhou, China
[2] Department of Radiology and BRIC, University of North Carolina,
Chapel Hill, USA
ptyap@med.unc.edu

**Abstract.** In this paper, we introduce a penalized geodesic tractography (PGT) algorithm for mitigating gyral bias in cortical tractography, which is essential for improving cortical connectomics. Unlike deterministic and probabilistic tractography algorithms that perform one-way tracking, PGT solves a global optimization problem in estimating the pathways connecting multiple regions, instead of local step-by-step orientation tracing. PGT is unconfounded by local false-positive or false-negative fiber orientations and ensures that fiber streamlines that are intended to connect two regions do not terminate prematurely. We show that PGT reduces gyral bias by allowing streamlines to make sharper turns into the cortical gyral matter and results in a significantly more uniform spatial distribution of cortical connections.

## 1 Introduction

Diffusion magnetic resonance imaging (DMRI) [1] allows brain white matter pathways to be reconstructed in-vivo and non-invasively by tracing local fiber orientations, which capture the diffusion patterns of water molecules. In connectomic studies that investigate cortico-cortical connectivity [2,3], fiber tracking often needs to be performed across gray-white matter boundaries in gyral blades of complex cortical convolutions. This process can be challenging due to factors such as partial volume effects and gyral bias [1].

A vast majority of tractography algorithms begin tracking from a seed point and propagate a streamline until some termination criterion is met [4,5]. This is a one-way tracking (OWT) process and can cause unsatisfactory outcome in noisy conditions due to error accumulation, often resulting in prematurely terminated tracts that fall short of reaching the cortex. OWT methods are also

This work was supported by NIH grants (NS093842, EB022880, EB006733, EB009634, AG041721, MH100217, and AA012388) and NSFC grants (61379020, 61703369).

A. F. Frangi et al. (Eds.): MICCAI 2018, LNCS 11072, pp. 12–19, 2018.
https://doi.org/10.1007/978-3-030-00931-1_2

susceptible to gyral bias with streamlines preferentially terminating at gyral crowns rather sulcal banks or fundi. The bias can be observed for a wide range of tractography algorithms (deterministic and probabilistic), diffusion models, diffusion weightings, and even for gradient directions with very high angular resolution [6].

In this work, we introduce a novel algorithm, called penalized geodesic tractography (PGT), for determining the shortest streamlines that connect different brain regions. We show that PGT overcomes the gyral bias problem, giving a significantly more uniform spatial distribution pattern of cortical connections. PGT determines the shortest paths connecting the regions via backtracking of the distance maps computed by solving the eikonal equation on a Finsler manifold with an anisotropic metric induced by the fiber orientation distribution function (FODF). Unlike existing methods that solve for shortest paths based on local costs [7–10], PGT determines a globally minimal path by minimizing a cost function that penalizes both tract curvature and length using a penalized manifold metric. This avoids the error stemming from the step-by-step tracing of the distance map and provides a global solution of the shortest path. Using in-vivo data, we demonstrate that PGT overcomes gyral bias and results in an improved coverage of the cortex.

## 2   Methods

### 2.1   Geodesic Tractography

The aim of geodesic tractography is to determine the shortest paths (i.e., geodesics) connecting a seed region and a target region based on an orientation field. The shortest paths can be determined using a distance map in relation to the seed region. The distance map is calculated by solving an eikonal equation, which is a partial differential equation (PDE) that describes the time of arrival at each point in space as a function of the local speed based on the orientation field.

In this work, the local speed is determined based on the FODF. This is reflected by the Finsler metric $\mathcal{F}$ [11], which defines at each point $x \in \mathbb{R}^3$ a positive norm $\mathcal{F}_x(u_x) \equiv \mathcal{F}(x, u_x) > 0$ whenever orientation $u_x \in \mathbb{S}^2$ is nonzero. The Finsler metric is more flexible than the Riemannian metric and naturally provides the geometric generalization suitable for analysis of more complex fiber configurations, such as crossings [11]. Given a Lipschitz regular path $\gamma(t) \in \mathbb{R}^3$ with boundary conditions $\gamma(0) = x_0$ and $\gamma(T) = x_T$, the length of the path $\ell(\gamma)$ can be measured as

$$\ell(\gamma) = \int_0^T \mathcal{F}(\gamma(t), \dot{\gamma}(t)) \, dt, \tag{1}$$

with the convention $\dot{\gamma}(t) := \frac{d}{dt}\gamma(t)$. Let $U(x)$ be the distance function starting from an arbitrary seed point $x_0 \in \mathbb{R}^3$. $U(x_T)$ is then the minimal value of (1) for a geodesic connecting $x_0$ and $x_T$. The distance function satisfies the eikonal equation

$$\nabla U^T \mathcal{F}^{-1} \nabla U = 1 \quad \text{s.t.} \quad U(x_0) = 0 \tag{2}$$

where $\nabla U$ is the spatial gradient of $U$. Solving for $U$ via (2) can be done effectively using the fast marching algorithm. The shortest path from $x_0$ to $x_T$ can be determined by solving an ordinary differential equation (ODE):

$$\dot{\gamma} \propto \mathcal{F}^{-1} \nabla U. \tag{3}$$

Note that for implementation, the paths are lifted into the configuration space $\mathbb{R}^3 \times \mathbb{S}^2$ of positions and orientations [12].

## 2.2 Penalized Finsler Metric

To ensure a well-behaving distance map that is continuous and smooth, we employ a penalized Finsler metric in the position-orientation space [10, 12]:

$$\mathcal{F}^2_{(x,u_x)}(\dot{x}, \dot{u_x}) := c(x, u_x)^2 (\langle u_x, \dot{x} \rangle^2_+ + \varepsilon^{-2} \langle u_x, \dot{x} \rangle^2_- + \varepsilon^{-2} ||u_x \times \dot{x}||^2 + \zeta^2 ||\dot{u_x}||^2), \tag{4}$$

where $(x, u_x) \in \mathbb{R}^3 \times \mathbb{S}^2$ is a position in the configuration space and $(\dot{x}, \dot{u_x})$ is a tangent vector at this position. Note that $\langle u_x, \dot{u_x} \rangle = 0$ by construction since $u_x \in \mathbb{S}^2$. The parameter $\zeta > 0$ balances the cost of motion in the position and angular spaces, in other words, the amount of curvature penalization. The term $\varepsilon^{-2} ||u_x \times \dot{x}||^2$ penalizes any position motion that is not colinear with the current orientation $u$, and forbids such motion in the limit as $\varepsilon \to 0$. The negative part of their scalar product $\langle u_x, \dot{x} \rangle$ is penalized to forbid reverse motion. We define the local cost function at each position $(x, u_x)$ based on the FODF field $\mathcal{V}$ as

$$c(x, u_x) = \left\{ 1 + \delta \left| \frac{\mathcal{V}_+(x, u_x)}{||\mathcal{V}_+||_\infty} \right| \right\}^{-1}, \tag{5}$$

where $\delta > 0$, $|| \cdot ||_\infty$ the supremum norm and $\mathcal{V}_+(x, u_x) = \max\{0, \mathcal{V}(x, u_x)\}$. Based on the penalized Finsler metric, the distance map $U$ can be computed by solving the eikonal Eq. (2) using Hamilton fast marching with adaptive stencils [13].

## 2.3 Tractography with Optimized Shortest Path

Given the distance map $U$, PGT determines a shortest path connecting a seed position to a region of interest (ROI) with curvature penalization. We define the curvature-penalized cost of a path $\gamma(t)$ as

$$\psi(\gamma) = \int_0^T \mathcal{F}(\gamma(t), \dot{\gamma}(t)) \mathcal{C}(||\ddot{\gamma}(t)||) dt, \tag{6}$$

where $\mathcal{C}(||\ddot{\gamma}(t)||) = \sqrt{1 + (\zeta ||\ddot{\gamma}(t)||)^2}$ is the Reeds-Shepp cost with a typical radius of curvature $\zeta$ [12]. Given the seeds $x_0 \in \mathbb{R}^3$, target region $x_T \in \mathbb{R}^3$, and tangents $u_{x_0}, u_{x_T} \in \mathbb{S}^2$, PGT aims to find the path $\gamma : [0, T] \to \mathcal{W} \subset \mathbb{R}^3 \times \mathbb{S}^2$

**Table 1.** Fiber statistics for the various tractography approaches.

| | Subcortical-cortical | | | Whole brain | | |
|---|---|---|---|---|---|---|
| | Initial | Valid | % | Initial | Valid | % |
| DET | 1M | 703709 | 70.37% | 50M | 36772925 | 73.55% |
| iFOD1 | 1M | 718306 | 71.83% | 50M | 41630557 | 83.26% |
| iFOD2 | 1M | 676357 | 67.64% | 50M | 38935000 | 77.87% |
| PGT | 953956 | 936923 | **98.21%** | 42335648 | 41624624 | **98.32%** |

obeying the boundary conditions and globally minimizing the cost $\psi(\gamma)$ by solving

$$\min_{\gamma} \psi(\gamma) \quad \text{s.t.} \quad \begin{cases} \gamma : [0, T] \to \mathcal{W}, \ \forall t, \ \|\dot{\gamma}(t)\| = 1, \\ \gamma(0) = x_0, \ \gamma(T) = x_T, \ \dot{\gamma}(0) = u_{x_0}, \ \dot{\gamma}(T) = u_{x_T}. \end{cases} \quad (7)$$

The shortest path can be determined (7) with the following equations:

$$\dot{\gamma}(t) = L(t-1)\mathcal{F}_{\gamma(t)}^{-1} dU(\gamma(t)) \text{ with } \begin{cases} \dot{u}_x(t) = L(t-1)c(\gamma(t))^{-1}\nabla_{\mathbb{S}^2}U(\gamma(t)) \\ \dot{x}(t) = L(t-1)c(\gamma(t))^{-1}D_{u_x(t)}\nabla_{\mathbb{R}^3}U(\gamma(t)) \\ \gamma(0) = x_0, \ \gamma(T) = x_T \\ \dot{\gamma}(0) = u_{x_0}, \ \dot{\gamma}(T) = u_{x_T} \end{cases}$$

$$(8)$$

where $L(t)$ is the length of the path at $t$, $D_{u_x}$ is the $3 \times 3$ symmetric positive definite matrix with eigenvalue 1 in the direction $u_x$, and eigenvalue $\varepsilon^2$ in the orthogonal directions. That is, $D_{u_x} := u_x \otimes u_x + \varepsilon^2(3I - u_x \otimes u_x)$ [10]. Solving for the optimal path can be done effectively using the forward Euler integration of the partial differential equations in (8).

## 2.4   Datasets and Processing

A DMRI dataset (Subject ID: 105923) from the Human Connectome Project (HCP) [14] was used to validate our method. The dataset has an isotropic spatial resolution of 1.25 mm, with three $b$-values ($b = 1000, 2000, 3000 \, \text{s/mm}^2$), a total of 270 gradient directions (90 per shell), and 18 non-diffusion-weighted images. Cortical and sub-cortical parcellation was performed on the T1-weighted image using FreeSurfer. The cortical GM map generated by FSL using the T1-weighted MR image was used to define tractography seed voxels. The FODFs were estimated using multi-tissue spherical deconvolution (MTSD) [15]. We employed the WM FODFs to compute the local costs in (5). Qualitative and quantitative comparisons were performed with respect to three algorithms in MRtrix3 [16], i.e., iFOD1 [16], iFOD2 [17], and a SD-based deterministic (DET) [16], with default parameters. 50 million streamlines were generated with each of the three methods. For PGT, 3 seed points were randomly selected for each GM voxel.

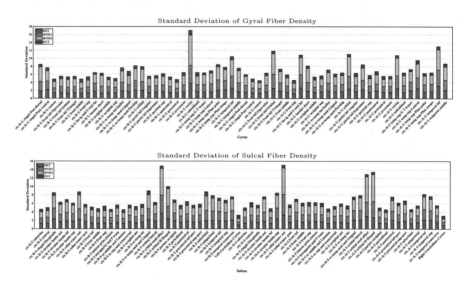

**Fig. 1.** Standard deviations of fiber densities at WM-GM boundaries at gyri and sulci.

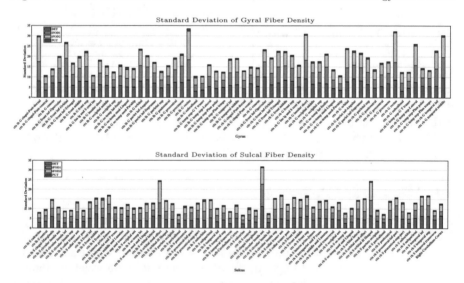

**Fig. 2.** Standard deviations of fiber densities of valid streamlines at WM-GM boundaries at gyri and sulci.

## 3    Results

Table 1 shows the quantitative results for iFOD1, iFOD2, DET, and PGT. Here, a streamline is said to be valid when its endpoint is within a radius of 2 mm from a cortical GM voxel. PGT yields significantly more streamlines that reach the cortical regions with a much smaller rate of prematurely terminated streamlines.

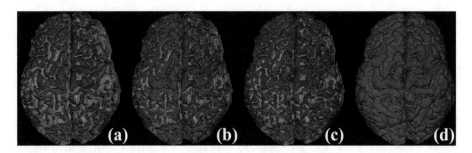

**Fig. 3.** Coverage of streamline endpoints at WM-GM interface: (a) DET; (b) iFOD1; (c) iFOD2; (d) PGT.

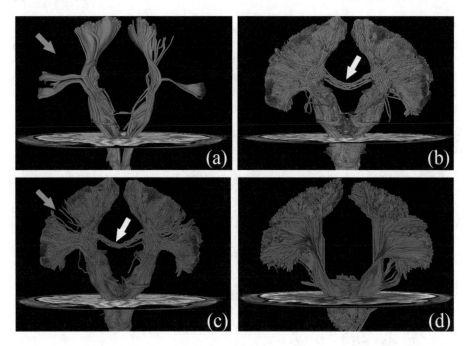

**Fig. 4.** Cortiospinal Tract (CST): (a) DET; (b) iFOD1; (c) iFOD2; (d) PGT.

To evaluate the uniformity of the spatial distribution of the streamline endpoints at the cortex, we computed the standard deviations of the fiber densities (count per volume) across voxels in different cortical regions. The results in Fig. 1 indicate that PGT yields much lower standard deviations than other methods. The same analysis was performed with only the valid streamlines. The results, shown in Fig. 2, indicate that the density variability increases significantly at both gyral and sulcal for DET, iFOD1, and iFOD2.

Figure 3 shows a mapping of the endpoints of the streamlines onto the cortical surface, which confirms that PGT gives a fuller coverage of the cortex. Other methods are generally biased towards the gyral crowns.

We extracted the corticospinal tract (CST) using the white matter query language (WMQL) [18], which is based on FreeSurfer cortical parcellations. In WMQL, the CST streamlines are defined as those connecting the postcentral and precentral gyri and the brain stem. The results, shown in Fig. 4, indicate that PGT yields fiber streamlines that fully cover both postcentral and precentral gyri (orange arrows). PGT also reduces false-positive streamlines (white arrows).

## 4   Conclusion

We have presented penalized geodesic tractography (PGT) for cortical tractography and have demonstrated that PGT can be used to overcome gyral bias. Our method extends the fast marching method by employing a penalized manifold metric that balances the costs in positions and orientations. We have shown with HCP data that PGT improves tract coverage of the cortex and results in a spatially uniform tract density distribution.

## References

1. Johansen-Berg, H., Behrens, T.E.: Diffusion MRI: From Quantitative Measurement to in Vivo Neuroanatomy. Academic Press (2013)
2. Schilling, K., Gao, Y., Janve, V., Stepniewska, I., Landman, B.A., Anderson, A.W.: Confirmation of a gyral bias in diffusion MRI fiber tractography. Hum. Brain Mapp. **39**(3), 1449–1466 (2018)
3. Wedeen, V.J., et al.: The geometric structure of the brain fiber pathways. Science **335**(6076), 1628–1634 (2012)
4. Basser, P.J., Pajevic, S., Pierpaoli, C., Duda, J., Aldroubi, A.: In vivo fiber tractography using DT-MRI data. Magn. Reson. Med. **44**(4), 625–632 (2000)
5. Behrens, T.E., Berg, H.J., Jbabdi, S., Rushworth, M.F., Woolrich, M.W.: Probabilistic diffusion tractography with multiple fibre orientations: what can we gain? NeuroImage **34**(1), 144–155 (2007)
6. Schilling, K., Gao, Y., Janve, V., Stepniewska, I., Landman, B.A., Anderson, A.W.: Confirmation of a gyral bias in diffusion MRI fiber tractography. Hum. Brain Mapp. **39**, 1449–1466 (2017)
7. Jbabdi, S., Bellec, P., Toro, R., Daunizeau, J., Pélégrini-Issac, M., Benali, H.: Accurate anisotropic fast marching for diffusion-based geodesic tractography. J. Biomed. Imaging **2008**, 2 (2008)
8. Fuster, A., Tristan-Vega, A., Haije, T.D., Westin, C.-F., Florack, L.: A novel Riemannian metric for geodesic tractography in DTI. In: Schultz, T., Nedjati-Gilani, G., Venkataraman, A., O'Donnell, L., Panagiotaki, E. (eds.) Computational Diffusion MRI and Brain Connectivity, pp. 97–104. Springer, Cham (2014). https://doi.org/10.1007/978-3-319-02475-2_9
9. Fuster, A., Haije, T.D., Tristán-Vega, A., Plantinga, B., Westin, C.F., Florack, L.: Adjugate diffusion tensors for geodesic tractography in white matter. J. Math. Imaging Vis. **54**(1), 1–14 (2016)

10. Duits, R., Meesters, S.P., Mirebeau, J.M., Portegies, J.M.: Optimal paths for variants of the 2D and 3D Reeds-Shepp car with applications in image analysis. arXiv preprint arXiv:1612.06137 (2016)

11. Florack, L., Fuster, A.: Riemann-Finsler geometry for diffusion weighted magnetic resonance imaging. In: Westin, C.-F., Vilanova, A., Burgeth, B. (eds.) Visualization and Processing of Tensors and Higher Order Descriptors for Multi-Valued Data. MV, pp. 189–208. Springer, Heidelberg (2014). https://doi.org/10.1007/978-3-642-54301-2_8

12. Mirebeau, J.M.: Fast-marching methods for curvature penalized shortest paths. J. Math. Imaging Vis. **60**, 1–32 (2017)

13. Mirebeau, J.M.: Efficient fast marching with Finsler metrics. Numerische Mathematik **126**(3), 515–557 (2014)

14. Van Essen, D.C., Smith, S.M., Barch, D.M., Behrens, T.E., Yacoub, E., Ugurbil, K.: The WU-Minn human connectome project: an overview. NeuroImage **80**, 62–79 (2013)

15. Jeurissen, B., Tournier, J.D., Dhollander, T., Connelly, A., Sijbers, J.: Multi-tissue constrained spherical deconvolution for improved analysis of multi-shell diffusion MRI data. NeuroImage **103**, 411–426 (2014)

16. Tournier, J., Calamante, F., Connelly, A.: Mrtrix: diffusion tractography in crossing fiber regions. Int. J. Imaging Syst. Technol. **22**(1), 53–66 (2012)

17. Tournier, J.D., Calamante, F., Connelly, A.: Improved probabilistic streamlines tractography by 2nd order integration over fibre orientation distributions. In: Annual Meeting of the International Society of Magnetic Resonance in Medicine (ISMRM), p. 1670 (2010)

18. Wassermann, D., et al.: The white matter query language: a novel approach for describing human white matter anatomy. Brain Struct. Funct. **221**(9), 4705–4721 (2016)

# Anchor-Constrained Plausibility (ACP): A Novel Concept for Assessing Tractography and Reducing False-Positives

Peter F. Neher[1(✉)], Bram Stieltjes[2], and Klaus H. Maier-Hein[1,3]

[1] Division of Medical Image Computing, German Cancer Research Center (DKFZ),
Heidelberg, Germany
p.neher@dkfz.de
[2] Radiology and Nuclear Medicine Clinic, University Hospital Basel, Basel,
Switzerland
[3] Section for Automated Image Analysis, Heidelberg University Hospital, Heidelberg,
Germany

**Abstract.** Diffusion tractography suffers from a difficult sensitivity-specificity trade-off. We present an approach that leverages knowledge about anatomically well-known tracts (anchor tracts) in a tractogram to quantitatively assess the remaining tracts (candidate tracts) according to their plausibility in conjunction with this context information. We show that our approach has the potential for greatly reducing the number of false positive tracts in fiber tractography while maintaining high sensitivities using phantom experiments (AUC 0.91). To investigate the applicability of the approach *in vivo*, we analyze 110 subjects of the Human Connectome Project young adult study. We demonstrate how the approach may be used for structured analysis of *in vivo* tractography and show supporting evidence for tracts previously discussed in the literature, while potentially sparking discussions about the role of others.

## 1 Introduction

The problem of false positives in fiber tractography is one of the grand challenges in the research area of diffusion-weighted magnetic resonance imaging (dMRI). Facing fundamental ambiguities especially in bottleneck situations, tractography generates huge numbers of theoretically possible candidate tracts [1,2]. Only a fraction of these candidates is likely to correspond to the true fiber configuration, posing a difficult sensitivity-specificity trade-off. For example for the field of connectomics, which traditionally focuses on the high sensitivity regime, a recent study showed that specificity is crucial and twice as important as sensitivity when performing certain network analyses [3].

Current methods address the issue of false-positive tracts either by focusing exclusively on well-known fiber bundles using prior knowledge [4,5] or by using tract filtering techniques based on the image signal [6,7]. Currently, the link

A. F. Frangi et al. (Eds.): MICCAI 2018, LNCS 11072, pp. 20–27, 2018.
https://doi.org/10.1007/978-3-030-00931-1_3

between these two choices of purely data driven and prior knowledge based approaches is missing.

We propose a novel concept that rigorously exploits prior knowledge about the existence of anatomically known tracts (anchor tracts) to reduce the degrees of freedom of a successive data-driven filtering of the remaining candidate tracts: anchor-constrained plausibility (ACP). This approach is based on the hypothesis that information about the presence or absence of each anchor influences the plausibility of the candidates and thereby reduces the ambiguities in the problem. We demonstrate the potential of this concept to better handle the tractography sensitivity-specificity trade-off in a series of phantom experiments. Since quantitative *in vivo* evaluations of false-positive reduction rates would require a *ground truth* which does not exist, we concentrate on assessing the capabilities of ACP in enabling a structured and objective analysis of tractograms. Therefore we analyzed ACP scores in 110 subjects of the Human Connectome Project (HCP) young adult study and discuss the results in light of existing neuroanatomical knowledge, providing detailed data-driven insights into what we might be missing when focusing only on anatomically known tracts.

## 2   Methods

Essentially, our method scores the candidate tracts by assessing their contribution to the signal, subject to constraints imposed by the anchor tracts. The process consists of three steps:

**Preprocessing:** The input tractogram is filtered using tissue segmentations to discard streamlines that terminate inside the white matter (WM) or that enter the corticospinal fluid (CSF). Based on prior knowledge, anchor tracts are identified and extracted from the filtered tractogram. A variety of open-source tract selection methods such as *TractQuerier* [4], *RecoBundles* [5], *AFQ* [8] or the recently presented *TractSeg* [9] (github.com/MIC-DKFZ/TractSeg) are available for this step. The remaining streamlines are then clustered into individual bundles that represent the candidate tracts. This is achieved in a reproducible way by using cortex parcellations, assigning the streamlines to bundles according to their endpoint labels. Alternatively, e.g. in phantom images, clustering can be employed. Here, we used *QuickBundles* for this purpose [10].

**Residual Calculation:** It is now assessed which parts of the image can be explained by the anchor tracts, using a method similar to *LiFE* [7] but on a fixel [11] instead of a raw signal basis. This is done by fitting a scalar weight for each anchor streamline by minimizing the mean squared error (MSE) between the fiber orientation distribution function (fODF) peak magnitudes calculated from the input image and the corresponding streamline fixels: $\text{MSE}(\mathbf{x}) = \|A\mathbf{x}-\mathbf{b}\|^2/n$, where $\mathbf{x}$ is the streamline weight vector, $A$ is the fixel magnitudes matrix with one column per streamline and one row per fODF peak direction and voxel and $\mathbf{b}$ is the peak magnitudes vector representing the fODF image. The residual vector $\mathbf{r}$ of this sparse system contains the fractions of the fODF peak magnitudes that cannot be explained by the anchor tracts: $\mathbf{r} = \mathbf{b} - A\mathbf{x}$, with all negative elements set to zero to retain only the unexplained parts of $\mathbf{b}$.

**Candidate Scoring:** Now it is analyzed which parts of the residual vector of the previous step can be explained by the candidate tracts. To this end the error $\mathrm{MSE}(\mathbf{y}) = \|B\mathbf{y}-\mathbf{r}\|^2/n$ of a second linear system is minimized, where $B$ represents the candidate streamlines. As in *LiFE*, we define the score of each candidate tract as the root MSE it reduces: $\sqrt{\mathrm{MSE}(\mathbf{y}_i)} - \sqrt{\mathrm{MSE}(\mathbf{y})}$, where $\mathbf{y}$ is the weight vector of all candidate streamlines and $\mathbf{y}_i$ is the modified weight vector with entries corresponding to candidate tract $i$ set to zero. Albeit determined by the weights of the individual candidate streamlines, this score is tract- and not streamline-specific. The procedure follows the intuition that it is plausible to assume a candidate's existence if it can explain parts of the signal that are not explained by any known tract. On the other hand, candidate tracts that are exclusively composed of parts of the anchor tracts – which is a typical cause for false-positives [1] – receive a lower score. The score is interpreted as the candidate's support by the data, given boundary conditions in form of anchor tracts (prior knowledge). It therefore represents a plausibility score for assuming a tract's existence under consideration of the yet unexplained parts of the signal.

## 3   Experiments and Results

### 3.1   In Silico Experiments and Results

We performed three phantom experiments with different degrees of complexity. Experiment 1 is intended as an illustration of the proposed method. The purpose of the other two experiments is to assess the capabilities of ACP in context of the sensitivity-specificity trade-off described in Sect. 1. For each phantom dataset, one test-tractogram was obtained using probabilistic constrained spherical deconvolution (CSD) tractography [12]. Since the ground truth is known in these cases, the anchor tracts could be simply defined using the binary masks of the respective ground truth tracts.

**Experiment 1:** Figure 1 illustrates the principle of the ACP analysis on an example consisting of two crossing fibers simulated with *Fiberfox* [13]. The Invalid Bundle Ratio (IVR) of the original tractogram was 3.8 (the invalid tracts outnumbered the valid tracts by a factor of 3.8). The experiment was performed once with each ground truth tract as anchor. For both configurations, this correctly resulted in the candidates corresponding to the second ground truth tract receiving the highest scores.

**Experiment 2:** For this experiment we employed a simulated replication of the *FiberCup* phantom consisting of seven individual tracts mimicking a coronal slice through the brain [13]. The IVR of the original tractogram was 1.9. The experiment was repeated five times with four out of seven randomly selected anchor tracts in each repetition. The three candidate tracts corresponding to the ground truth tracts were ranked highest in all repetitions (see Fig. 2a).

**Experiment 3:** The main phantom experiment is based on the brain-like phantom simulated with *Fiberfox* used in the ISMRM Tractography Challenge 2015 [1,14]. The IVR of the original tractogram was 7.7. The experiment was repeated fifty times with 50% of the ground truth bundles extracted from the

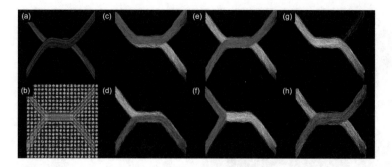

**Fig. 1.** Experiment 1, illustrating the proposed method. (a) and (b) show the ground truth: two crossing fiber tracts and their corresponding simulated ODF representation. (c-g) show exemplary candidate tracts (blue) obtained from the probabilistic tractogram using streamline clustering with the selected anchor tract (white). (h) shows the candidate tract (red) that was correctly ranked highest by the proposed method together with the anchor tract (white).

input tractogram randomly selected as anchor tracts in each repetition. Due to its large extent and dominance, the Corpus Callosum was always included in the set of anchor tracts. For comparison, another fifty repetitions were performed without any anchor tracts. Additional benchmarks were obtained using a volume-based ranking of the candidate tracts, as well as a streamline-weight-based ranking obtained with *LiFE* [7]. The resulting ROC curves are shown in Fig. 2b. The proposed method ($AUC = 0.91$) performed significantly better than the benchmarks without anchor tracts ($AUC = 0.78$, t-test: $p = 1.7^{-25}$) and with volume-based scoring ($AUC = 0.7$, t-test: $p = 1.5^{-29}$). The *LiFE* streamline-weight-based ranking performed similar to random guessing ($AUC = 0.5$).

## 3.2 In Vivo Experiments and Results

*In vivo* experiments were performed on 110 subjects of the HCP young adult study. For each subject we performed probabilistic CSD tractography with and without anatomical constraints (MRtrix) as well as deterministic peak tractography with anatomical constraints (MITK Diffusion). We used multiple tractography methods and joined the results for increased sensitivity and to mitigate tractography biases. From these whole brain tractograms we extracted 63 anchor tracts using overlap and streamline shape criteria, similar to *RecoBundles* [5], on basis of the reference tracts of the same subjects published by Wasserthal et al. [15]. Successively we generated the candidate tracts from the remaining streamlines by grouping them based on their endpoint locations with respect to the *FreeSurfer* Desikan-Killiany atlas cortex parcellations readily available for all HCP subjects [16]. To remove spurious streamlines from these tracts, a simple tract density based filtering was applied. Furthermore, streamlines that connect the same start and end label (loops) as well as very sparse tracts containing less than 50 streamlines were excluded from the subsequent analysis. This process

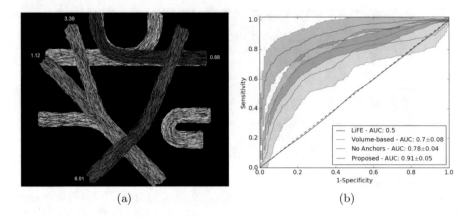

(a)                                              (b)

**Fig. 2.** (a) Experiment 2: Exemplary results on the simulated FiberCup dataset. The four highest ranked candidate tracts (colored) are labeled with their ACP score. The anchor tracts are colored white. The three real tracts received the highest scores. (b) Experiment 3: ROC curves on the ISMRM 2015 Tractography Challenge phantom obtained with the proposed approach (green), the same approach without anchor tracts (red), simple tract-volume-based ranking (gray) and *LiFE* streamline-weight-based ranking (blue).

resulted in an average number of 416 candidate tracts per subject that were included in the subsequent ACP analysis. In the results presented in the remainder of this section, we only included tracts that were detected in at least 90% of all subjects, resulting in 151 reproducible candidate tracts. These candidates are ranked per subject according to their ACP score.

42% of the reproducible candidate tracts consisted of cortical U-fibers, i.e. tracts that connect neighboring gyri. Out of the top-ten ranking candidates, nine are U-fibers. This confirms the intended behaviour of our approach: U-fibers were not included in the set of anchor tract, but are known to exists. This is well reflected by their high ranking.

The top-ten ranking non-U-fiber candidates (see Fig. 3a) include several well known tracts such as the frontal aslant tract (FAT) (left and right hemisphere, parcellation labels 1018–1028 and 2018–2028, see Fig. 3b), tracts connecting the hippocampus and the thalamus (left and right hemisphere, 10–17 and 49–53), which are arguably part of the Fornix but missing from the anchor tracts, as well as connecting the hippocampus and the entorhinal cortex (left hemisphere, 17–1006), which might also include parts of the lower cingulum or the stria terminalis. Furthermore ranked in the top-ten, albeit with a larger ranking variance across subjects compared to the aforementioned tracts, are vertical tracts between the lingual gyrus in the occipital lobe and the superior parietal lobule (left and right hemisphere, labels 1013–1029 and 2013–2029) as well as fibers from the inferior parietal to the inferior temporal (right hemisphere, labels 2008–2009) and fusiform gyrus (right hemisphere, labels 2007–2008). The corresponding contralateral tracts that did not reach the top-ten are still ranked relatively high (top-twenty). The overall highest ranked tract across all subjects consists of fibers connecting the left and right cerebellar cortex (parcellation labels 8 and 47), jumping from one cerebellar hemisphere to the other. In this case, ACP ranking was helpful in identifying a systematic tractography artifact that arises from the strong left-right anisotropy in this region and the limited image resolution. The gap between the two hemispheres, which are tightly pressed against each other, is not adequately resolved and makes the region appear as continuous tissue (see Fig. 3c).

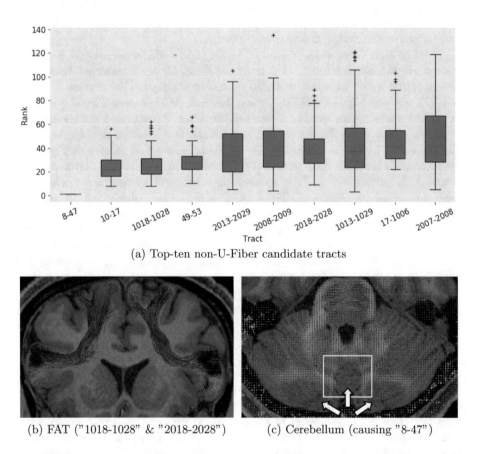

(a) Top-ten non-U-Fiber candidate tracts

(b) FAT ("1018-1028" & "2018-2028")          (c) Cerebellum (causing "8-47")

**Fig. 3.** (a) Top-ten ranked non-U-fiber candidate tracts. The tracts are named according to their parcellation labels described in the text. (b) Coronal view of left and right frontal aslant tract (FAT) of a random subject. (c) Axial view on the cerebellum T1 with overlaid tensor glyphs. The arrows indicate areas with relatively high left-right anisotropy (FA> 0.15). The area in the white box caused systematic false positives in the tractograms (see text).

# 4 Discussion and Conclusion

We proposed a novel concept, anchor-constrained plausibility analysis (ACP), that derives quantitative candidate plausibility scores by jointly assessing tract-based signal contribution levels and prior knowledge in the form of anchor tracts. We evaluate the concept in multiple phantom experiments, showing that this approach has the potential to greatly improve the sensitivity-specificity trade-off in tractography, which is a central issue of current tractography pipelines [1,3]. Our *in vivo* experiments in a cohort of 110 subjects of the HCP project showed that the presented approach yields valuable information for a structured and objective analysis of tractography results.

Even though there is no *ground truth* for the *in vivo* evaluation, the experiments yielded several interesting insights. First, it was reassuring that well-known tracts which were not included as anchor tracts, i.e. the cortical U-fibers or FAT, received high plausibility scores. Second, ACP scoring turned out to be helpful in assessing the quality of the existing anchor tracts and the tractogram in general: parts of the Fornix and smaller connections between the hippocampus and the entorhinal cortex were missing in the reference and consistently popped up as tracts with high ACP scores [17]. Another tract that was systematically scored high turned out to be a systematic artifact of tractography, which we were previously not aware of. Third, ACP scores could play a role in ongoing discussions on brain anatomy. The high ranked vertical association tracts, for example, connecting the lingual gyrus in the occipital lobe and the superior parietal lobule, seem to be associated to the structure identified as vertical occipital fasciculus (VOF) by Yeatman et al. [18]. Other high ranked tracts could not directly be associated with known anatomy, such as connections between inferior parietal and inferior temporal gyrus, which visually reminds of a U-fiber bundle, or between the inferior parietal and fusiform gyrus. Both examples seem to be unrelated to prominent functional connections, but they are consistently important to explain the image data.

In all these considerations, though, it is important to keep in mind that a low score does not necessarily indicate that the respective tract is a false positive but only that its existence is not essential for explaining the measured data. Vice versa, a tract with high score is required to explain the data and it is therefore often plausible to assume its existence. Nevertheless, in some cases, as demonstrated for the cerebellum, additional factors such as the limited image resolution, missing streamlines in the tractogram and prior anatomical knowledge have to be taken into account to assess a tract's overall plausibility.

All methods described in this work are available online in the open-source Medical Imaging Interaction Toolkit (`mitk.org/wiki/DiffusionImaging`), MRtrix (`mrtrix.org`) or Dipy (`nipy.org/dipy`). Future work will investigate the proposed approach in conjunction with connectomics analyses where recent studies have highlighted the disruptive impact of insufficiently specific tractography on global network measures [3], as well as the influence of the proposed filtering on the relationship between the structural and functional connectome.

**Acknowledgements.** This work was supported by the German Research Foundation (DFG) grant numbers MA 6340/10-1 and MA6340/12-1. Data were provided in part by the Human Connectome Project, WU-Minn Consortium (Principal Investigators: David Van Essen and Kamil Ugurbil; 1U54-MH091657) funded by the 16 NIH Institutes and Centers that support the NIH Blueprint for Neuroscience Research; and by the McDonnell Center for Systems Neuroscience at Washington University.

# References

1. Maier-Hein, K.H., et al.: The challenge of mapping the human connectome based on diffusion tractography. Nat. Commun. **8**(1), 1349 (2017)
2. Sinke, M.R.T., et al.: Diffusion MRI-based cortical connectome reconstruction: dependency on tractography procedures and neuroanatomical characteristics. In: Brain Structure and Function (2018)
3. Zalesky, A., et al.: Connectome sensitivity or specificity: which is more important? Neuroimage **142**, 407–420 (2016)
4. Wassermann, D., et al.: The white matter query language: a novel approach for describing human white matter anatomy. Brain Struct. Funct. **221**(9), 4705–4721 (2016)
5. Garyfallidis, E.: Recognition of white matter bundles using local and global streamline-based registration and clustering. Neuroimage **170**, 283–295 (2017)
6. Sherbondy, A.J., et al.: ConTrack: finding the most likely pathways between brain regions using diffusion tractography. J. Vis. **8**(9), 15.1–16 (2008)
7. Pestilli, F., et al.: Evaluation and statistical inference for human connectomes. Nat. Methods **11**(10), 1058–1063 (2014)
8. Yeatman, J.D., et al.: Tract profiles of white matter properties: automating fiber-tract quantification. PLoS ONE **7**(11), e49790 (2012)
9. Wasserthal, J., et al.: TractSeg - fast and accurate white matter tract segmentation. Neuroimage **183**, 239–253 (2018)
10. Garyfallidis, E., et al.: Quickbundles, a method for tractography simplification. Front. Neurosci. **6**, 175 (2012)
11. Raffelt, D.A., et al.: Connectivity-based fixel enhancement: whole-brain statistical analysis of diffusion MRI measures in the presence of crossing fibres. NeuroImage **117**, 40–55 (2015)
12. Tournier, J.D., et al.: MRtrix: diffusion tractography in crossing fiber regions. Int. J. Imaging Syst. Technol. **22**(1), 53–66 (2012)
13. Neher, P.F., et al.: Fiberfox: facilitating the creation of realistic white matter software phantoms. Magn. Reson. Med. **72**(5), 1460–1470 (2014)
14. Maier-Hein, K., et al.: Tractography Challenge ISMRM 2015 Data (2015). type: dataset. https://doi.org/10.5281/zenodo.572345
15. Wasserthal, J., et al.: High quality white matter reference tracts (2017). type: dataset. https://doi.org/10.5281/zenodo.1088278
16. Desikan, R.S., et al.: An automated labeling system for subdividing the human cerebral cortex on MRI scans into gyral based regions of interest. Neuroimage **31**(3), 968–980 (2006)
17. Nieuwenhuys, R., et al.: The Human Central Nervous System: A Synopsis and Atlas. Springer, Heidelberg (1988). https://doi.org/10.1007/978-3-540-34686-9
18. Yeatman, J.D., et al.: The vertical occipital fasciculus: a century of controversy resolved by in vivo measurements. PNAS **111**(48), E5214–E5223 (2014)

# Tract-Specific Group Analysis in Fetal Cohorts Using *in utero* Diffusion Tensor Imaging

Shadab Khan[1]([✉]), Caitlin K. Rollins[1], Cynthia M. Ortinau[2], Onur Afacan[1], Simon K. Warfield[1], and Ali Gholipour[1]

[1] Boston Children's Hospital and Harvard Medical School, 360 Longwood Avenue, Boston, MA, USA
shadab.khan@childrens.harvard.edu
[2] Washington University School of Medicine, St. Louis, MO, USA

**Abstract.** Diffusion tensor imaging (DTI) based group analysis has helped uncover the impact of white matter injuries in a wide range of studies involving subjects from preterm neonates to adults. The application of these methods to fetal cohorts, however, has been hampered by the challenging nature of *in utero* fetal DTI caused by unconstrained fetal motion, limited scan times, and limited signal-to-noise ratio. We present a framework that addresses these issues to systematically evaluate group differences in fetal cohorts. A motion-robust DTI computation approach with a new unbiased DTI template construction method is unified with kernel-regression in age and tensor-specific registration to normalize DTI volumes in an unbiased space. A robust statistical approach is used to map region-specific group differences to the medial representation of the tracts of interest. The proposed approach was applied and showed, for the first time, differences in local white matter fractional anisotropy based on *in utero* DTI of fetuses with congenital heart disease and age-matched healthy controls. This paper suggests the need for fetal-specific pipelines to be used for DTI-based group analysis involving fetal cohorts.

## 1 Introduction

Diffusion Tensor Imaging (DTI) is widely used to image brain microstructure and to compare groups of subjects and patients based on the diffusion properties of the white matter (WM). Unlike *ex utero* imaging where methods such as immobilization devices or sedation can be employed to limit motion, *in utero* fetal DTI is hampered by motion, low SNR, and fewer gradient directions applied to minimize maternal discomfort and the adverse impact of fetal motion.

To account for fetal motion, DTI computation methods that utilize volume-to-volume registration (VVR) and slice-to-volume registration (SVR) have been proposed [5,6,9]. Compared to VVR-based methods, SVR methods have shown

This work was supported by the McKnight Foundation, the Fetal Health Foundation, NIH R01 EB018988, NIH K23NS101120, and Mend A Heart Foundation.

© Springer Nature Switzerland AG 2018
A. F. Frangi et al. (Eds.): MICCAI 2018, LNCS 11072, pp. 28–35, 2018.
https://doi.org/10.1007/978-3-030-00931-1_4

more accurate reconstruction of the microstructure in moving subjects [5,9]. The resulting DTIs, however, would still be noisier than those obtained from motion-free scans since a large number of fetal DWI scans or slices are dropped and the remaining motion-corrected slices are interpolated. This raises challenges and a critical need for robust and unbiased processing of fetal DTI for group analysis. Moreover, in addition to the data being scarce in this domain, there are no publicly available *in utero* fetal DTI templates or atlases. As a result, DTI-based group comparison of fetal cohorts remains largely unreported.

In related work, the Tract-Based Spatial Statistics (TBSS) approach [13] has been used in neonatal studies to evaluate impact of preterm birth on WM development [2], among many other applications. Small region-of-interest (ROI) based comparisons, though frequently reported, are inadequate for generalization over the tracts of interest and are prone to operator-induced placement errors. An improved ROI-based strategy called Tract-Specific Analysis (TSA) [15] that considers entire tract of interest has also been proposed and recently applied to preterm cohorts [11].

In this study, we use an outlier-robust motion-tracking based SVR approach for fetal DTI computation, and tensor-specific registration to construct unbiased kernel-regressed templates for TSA. Kernel-regression in age is necessary given the fast changing shape of the fetal brain and the proposed template construction approach was needed as part of the group analysis pipeline because no existing DTI template construction approaches integrate kernel-regression in age with tensor-specific registration. Our approach accounts for gestational age (GA) and computes an unbiased template; this is followed by GA-adjusted group statistical TSA. The proposed approach was applied to evaluate group differences, for the first time, between *in utero* congenital heart disease (CHD) and healthy control fetuses. The results show differences that extend the previous findings in neonates and adolescents [10,12] suggesting that WM alterations in CHD start *in-utero* as early as 28 weeks GA.

## 2   Methods

Given multiple motion-corrupted, low-SNR diffusion-weighted imaging (DWI) volumes of fetuses in different planes, we aim to study differences present in the DTIs of fetal cohorts. A DTI-based group analysis pipeline including TBSS, ROI-based, and TSA comprises: (1) DTI computation in a standard space, (2) registration of subject DTIs with an existing or study-specific template, and (3) statistical analysis following DTI volume normalization. Our proposed pipeline for fetal brain DTI group analysis comprising these three steps is shown in Fig. 1.

### 2.1   DWI Processing and DTI Computation

Fetal DTIs are computed using a motion and outlier-robust method that computes tensors using motion-corrected diffusion data in a standard space. The slices from $b = 0$ and $b \neq 0$ images are first used to compute composite *B0*

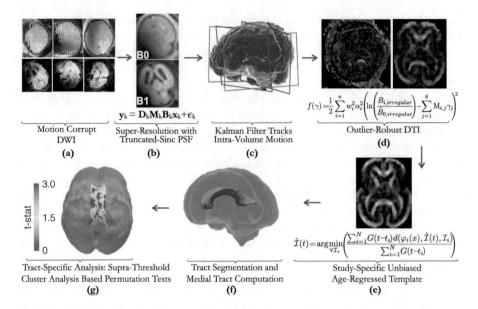

$$\mathbf{y}_k = \mathbf{D}_k \mathbf{M}_k \mathbf{B}_k \mathbf{x}_k + \epsilon_k$$

Motion Corrupt
DWI
**(a)**

Super-Resolution with
Truncated-Sinc PSF
**(b)**

Kalman Filter Tracks
Intra-Volume Motion
**(c)**

$$f(\gamma) = \frac{1}{2}\sum_{i=1}^{n} w_i^2 \alpha_i^2 \left( \ln\left(\frac{B_{i,irregular}}{\tilde{B}_{0,irregular}}\right) - \sum_{j=1}^{6} M_{i,j}\gamma_j \right)^2$$

Outlier-Robust DTI
**(d)**

Tract-Specific Analysis: Supra-Threshold
Cluster Analysis Based Permutation Tests
**(g)**

Tract Segmentation and
Medial Tract Computation
**(f)**

$$\hat{\mathcal{I}}(t) = \underset{\forall \mathcal{I}_i}{\arg\min}\left(\frac{\sum_{i=1}^{N} G(t-t_i)d(\varphi_i(x),\hat{\mathcal{I}}(t),\mathcal{I}_i)}{\sum_{i=1}^{N} G(t-t_i)}\right)$$

Study-Specific Unbiased
Age-Regressed Template
**(e)**

**Fig. 1.** Proposed fetal brain DTI group analysis pipeline: The acquired DWI data (often multi-planar), (a) is used to compute composite super-resolved B0 and B1 images (b). Motion-tracking based slice-to-volume registration (c) is used to map $b = 0$ and $b \neq 0$ slices to standard space [9] where DTI (d) is computed directly from motion-corrected data (shown next to a color FA image obtained directly from the scanner without motion correction). Template construction (e) provides an unbiased spatial frame where group statistical differences (f) are computed and localized (g).

and *B1* images using an SVR-based super-resolution reconstruction approach with a truncated Sinc point spread function (PSF) [7]. A rigid transformation is computed between the composite B0 and B1 images and the T2w image of the fetal brain reconstructed using the SVR approach. Slice-level motion correction is performed for the $b \neq 0$ slices through a robust registration-based state-space motion tracking algorithm [9]. Using the slice to B0, B0 to T2w, and T2w to standard space transformations, each slice from $b = 0$ and $b \neq 0$ image set is rigidly mapped to the standard space. A constrained weighted locally linear least squares method is used to fit single tensor per voxel directly from scattered data, obtained using slice-level motion-corrected DWI, with a Gaussian PSF [9].

## 2.2   Study-Specific Template and Normalization

To spatially normalize DTI volumes to an unbiased space, an age-regressed template is computed by applying a large deformation diffeomorphic metric mapping that optimizes a tensor metric to register DTIs, as described below.

DTI Registration: Registration of two DTIs is defined as a mapping of a manifold $\mathcal{M}$ under a transformation $\varphi : \mathcal{M} \mapsto \mathcal{M}$ that lies in the infinite-dimensional space of diffeomorphism $\mathcal{H}$. A diffeomorphism exists with a smooth inverse and satisfies the Lagrangian ordinary differential equation: $\frac{d}{dt}\varphi(x) = v_t(\varphi(x))$ where $v_t(\varphi(x)) \in R^3$ represents smooth velocity fields that are integrated forward in time to compute the transformations; a corresponding inverse $\varphi^{-1}(x)$ can be computed by integrating velocity fields backwards in time. The image registration cost function $(\xi)$ can thus be defined as follows:

$$\xi = \Psi(\mathcal{I}_f(x) \circ \varphi^{-1}(x), \mathcal{I}_m(x)) + D(\varphi(x)) \tag{1}$$

where $\Psi$ is a matching metric computed between the fixed DTI $\mathcal{I}_f(x)$ and the moving DTI $\mathcal{I}_m(x)$, and $D(\varphi(x))$ is a metric on the transformation that is formulated based on the exact model of deformation used to approximate diffeomorphism and is used in 1 to regularize $\xi$. We choose $\Psi$ to be a tensor similarity metric, i.e. the Euclidean Distance Squared between Deviatoric tensors (DDS).

Template Construction: Since the anatomic variability among fetuses and across age cannot be characterized in Hilbert space, the set of DTIs are considered time-indexed observations in a Riemannian manifold $\mathcal{M}$. A time-regressed template at time $t = \tau$ can be computed as an empirical Fréchet mean, under the distribution of manifold-valued DTIs, using generalized Nadaraya-Watson kernel regression in age [4]. To do this, we use a minimum-distance template computation approach: Given $N$ DTIs denoted as $\mathcal{I}_i(x), i \in [1...N]$, we express the template $\hat{\mathcal{I}}(t)$ computation at time $t$ as the following minimization problem:

$$\hat{\mathcal{I}}(t) = \arg\min_{\forall \mathcal{I}_i} \left( \frac{\sum_{i=1}^{N} G(t - t_i) \, d(\varphi_i(x), \hat{\mathcal{I}}(t), \mathcal{I}_i)}{\sum_{i=1}^{N} G(t - t_i)} \right), \tag{2}$$

where $G(t - t_i) = 1/(\kappa\sqrt{2\pi}) \exp(-(t - t_i)^2/2\kappa^2)$ for $t \le \kappa$ and 0 otherwise, $\kappa$ is a time bandwidth parameter; and $d(\varphi_i(x), \hat{\mathcal{I}}(t), \mathcal{I}_i)$ is a DTI matching metric. This minimization problem is solved using the iterative algorithm described below:

1. Initialize $\hat{\mathcal{I}}(t) : \hat{\mathcal{I}}(t) \leftarrow \exp\left(\sum_{i=1}^{N} w_i \log(\mathcal{I}_i)\right)$ and $w_i = G(t - t_i)/\sum_{i}^{N} G(t - t_i)$.

2. Initialize $\varphi_i(x) \forall \mathcal{I}_i$ using a 3D rigid and affine transform.

3. Estimate $\varphi_i(x)$: time domain is discretized into $T$ steps and $\varphi_i(x)$ is computed by integrating velocity fields $v_i^t(x)$ computed at each time step $t(t = 1...T)$.

4. Compute optimal $\hat{\varphi}_i(x) : \hat{\varphi}_i(x) \leftarrow w_i\hat{\varphi}_i(x)$.

5. Update $\hat{\mathcal{I}}(t) : \hat{\mathcal{I}}(t) \leftarrow \exp\left(\sum_{i=1}^{N} w_i \log(\mathcal{I}_i \circ \varphi_i(x))\right)$.

6. Repeat steps 3 to 5 until $\varphi_i(x)$ and $\hat{\mathcal{I}}(t)$ converge (change less than thresholds).

Spatial Normalization: The DTIs are registered to the computed template using the tensor-specific approach. The rigid and affine transforms are computed using normalized mutual information (NMI) metric and used to initialize the nonrigid transform. All transformations are applied to DTIs in a single step.

## 2.3   Statistical Analysis

We apply TSA for group analysis, which is based on a general linear model and evaluates the distribution of DTI scalars over medial template of interest [16]. TSA was shown to be more robust to errors in registration than other group comparison methods such as TBSS [11]. To perform TSA, a tract of interest is segmented in the template space and its medial template is computed using continuous medial representation [14]. The spatially normalized DTIs are sampled along normals to the medial template and the DTI-scalars are computed. The data from cohorts being compared are projected on the medial template and compared using permutation-based non-parametric suprathreshold cluster analysis that controls the family-wise error rate (FWER). This analysis reveals clusters that are deemed significant at cluster-size threshold $\alpha$ under the null hypothesis. A $t$-statistic threshold is used to construct the null distribution.

# 3   Application and Results

Application: We used our framework to compare a cohort of healthy fetuses (control) to a cohort of fetuses identified with congenital heart disease (CHD). Fetal MRIs were acquired *in utero* at Boston Children's Hospital using an Institutional Review Board approved protocol, with informed consent from pregnant women volunteers. Subjects with at least 2 DWI scans without extreme motion-induced or distortion artifacts were chosen for this study. The MRIs were acquired on 3T Siemens Skyra scanners without maternal sedation or breath-hold. Each scan session included multiple orthogonal T2 HASTE scans using TR/TE = 1.4–2 s/100–120 ms, 0.9–1.1 mm in-plane and 2 mm out-of-plane resolution with 2/4-slice interleaved acquisition, and DWI comprising 2–4 scans in axial or coronal planes with respect to the fetal head with 1 or 2 b = 0 s/mm$^2$ images, and 12-direction b = 500 s/mm$^2$ acquired using TR/TE = 3–4 s/60 ms, 2 mm in-plane and 2–4 mm out-of-plane resolution. Total imaging time was <45 min (3–10 min for DWI).

Processing: We computed DTIs for 8 control and 9 CHD subjects scanned at GAs between 26 week 4 days (26w4d) and 30w0d. A template of control DTIs was computed at $\tau = 28$w3d with 1.2 mm isotropic resolution. The time-bandwidth $\kappa$ was set to 1w6d and $\varphi(x)$ was computed using $T = 6$. Ten iterations of the template construction algorithm were completed to compute $\hat{\mathcal{I}}(t)$ with error minimization in (1) solved at step 3. We used DTITK [17] for tensor-to-tensor registration. The final template was computed by composing linear and non-linear transformations and transforming DTIs in a single step to reduce interpolation error. All DTIs were non-linearly registered to this template and subsequently transformed to the template space in one step.

Tract Medial Representation: In this study, we evaluated the differences in Corpus Callosum (CC) – a large tract that develops early in the 1$^{st}$ trimester. Since the cortical projections in late 2$^{nd}$ trimester are early in their development, we only segmented the anterio-posteriorly oriented region of the CC. Figure 2

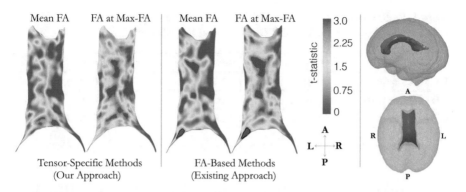

**Fig. 2.** TSA: Significant clusters (green edge) in t-statistic maps computed over the CC medial template for mean-FA and max-FA comparisons in medial templates obtained using tensor-based (proposed) and FA-based (existing) pipelines. These results suggest that the FA near splenium and the body of CC is likely lower in subjects suffering from CHD in comparison to the controls within the clusters shown. FA-based method missed the differences near splenium observed using tensor-specific approach. The segmented CC region (top) and fitted (cmrep) medial template (bottom) is shown in the right.

shows the segmented CC region along with *cmrep-fit* [14] computed medial template.

Statistical Analysis: As described earlier, we used permutation tests for group comparison. A contrast of $[1 \ -1 \ 0]$ was used to compare control against CHD group while adjusting for age as a covariate. Clusters were obtained from the t-statistic image computed for the contrast using $t$-threshold $= 2.62$. Permutation tests were done with number of permutations $N = 10000$, and $\alpha = 0.05$. Clusters computed for true labeling of the data larger than $\alpha N$ percentile of the null distribution were marked significant. Summary statistics of DTI-derived metrics over CC comparing control and CHD groups have been shown in Table 1. Using TSA, projections of group differences in DTI scalars on the CC medial template were computed. We evaluated differences using mean and max-FA projections.

The summary statistics in Table 1 were computed using the control and CHD DTI templates. Consistent with a previous study of neonates [10], we observed that in comparison to control group, CHD group exhibited lower FA and AD but higher MD and RD. This observation suggests that the CC tracts in control group were likely more mature than the CC tracts in CHD group, as it has been shown [3] that brain maturation results in increased FA and reduced MD in CC.

In TSA, we found significant clusters (encircled by green line, Fig. 2) only in FA projections over the medial template along the body of the CC and near the splenium using max-FA projection, and in the splenium using mean projection. For comparison, results obtained by using existing FA-based registration and template construction methods have been shown as well. FA maps from DTIs were computed and used to construct an FA template that was registered to the FA map of the DTI template computed using our method. Linear and non-linear

**Table 1.** Summary statistics of the DTI-derived scalars: significant differences in mean of DTI scalars between control and CHD groups were observed.

|  | Control ($\mu_a \pm \sigma_a$) | CHD ($\mu_b \pm \sigma_b$) | p-value ($H_0 : \mu_a = \mu_b$) | 95% CI ($\mu_a - \mu_b$) |
|---|---|---|---|---|
| FA | $0.25 \pm 0.08$ | $0.21 \pm 0.07$ | <1e-4 | $(0.039, 0.048)$ |
| MD | $1.41 \pm 0.16$ | $1.44 \pm 0.16$ | <1e-4 | $(-0.036, -0.017)$ |
| AD | $1.82 \pm 0.20$ | $1.77 \pm 0.20$ | <1e-4 | $(0.033, 0.058)$ |
| RD | $1.20 \pm 0.17$ | $1.27 \pm 0.17$ | <1e-4 | $(-0.073, -0.052)$ |

transformations from FA registrations were applied in one step to DTIs. ANTS was used for FA-based processing [1] and TSA was applied exactly as described previously. Only 1 large cluster was found significant in comparisons using the FA-based approach in the body of the CC. Compared to the proposed approach, FA-based approach results in larger clusters during permutation tests and leads to larger threshold for significance for true labeling of the data. This results in smaller, more focal differences to be missed; the differences near splenium observed using the proposed pipeline were missed by the FA-based approach even though the t-statistic image shows presence of hyper-intense clusters.

The posterior region of the CC (including splenium) has been reported to develop slower than the anterior region in 2nd trimester of pregnancy [6]. One hypothesis to explain these results is that the altered cardiac physiology of CHD results in impaired oxygenation and nutrient delivery to the developing brain, leading to alterations in the typical trajectory of microstructural development. While similar observations were made in a recent study that reported migrating neurons were delayed in CHD neonates [8], these differences were not studied by *in utero* DTI. We present the first evidence to warrant further studies using DWI to investigate altered brain development in fetuses with CHD and provide a unique, unified framework that addresses many of the challenges of fetal DTI.

## 4     Conclusion

We present comprehensive methods to evaluate tract-specific group differences within the developing fetal brain. DTIs reconstructed using an outlier-robust motion-tracking method were co-registered to construct a study-specific unbiased template incorporating kernel-regression in age. Tract-specific analysis (TSA) was applied to the volume-normalized fetal DTIs and permutation tests were used to evaluate the group differences between CHD and healthy *in utero* fetal cohort while adjusting for age. Significant clusters were found in the body and splenium of corpus callosum. The methods presented in this study enable group-comparison studies using *in utero* acquired DWI of the fetal brain.

# References

1. Avants, B.B., Epstein, C.L., Grossman, M., Gee, J.C.: Symmetric diffeomorphic image registration with cross-correlation: evaluating automated labeling of elderly and neurodegenerative brain. Med. Image Anal. **12**(1), 26–41 (2008)
2. Ball, G., et al.: An optimised tract-based spatial statistics protocol for neonates: applications to prematurity and chronic lung disease. NeuroImage **53**(1), 94–102 (2010)
3. Braga, R.M., et al.: Development of the corticospinal and callosal tracts from extremely premature birth up to 2 years of age. PLoS ONE **10**(5), 1–15 (2015)
4. Davis, B.C., Fletcher, P.T., Bullitt, E., Joshi, S.: Population shape regression from random design data. Int. J. Comput. Vis. **90**, 255–266 (2010)
5. Fogtmann, M., et al.: A unified approach to diffusion direction sensitive slice registration and 3-D DTI reconstruction from moving fetal brain anatomy. IEEE Trans. Med. Imaging **33**(2), 272–289 (2014)
6. Jiang, S., et al.: Diffusion tensor imaging (DTI) of the brain in moving subjects: application to in-utero fetal and ex-utero studies. Mag. Reson. Med. **62**(3), 645–655 (2009)
7. Kainz, B., et al.: Fast volume reconstruction from motion corrupted stacks of 2D slices. IEEE Trans. Med. Imaging **34**(9), 1901–1913 (2015)
8. Kelly, C.J., et al.: Impaired development of the cerebral cortex in infants with congenital heart disease is correlated to reduced cerebral oxygen delivery. Sci. Rep. **7**(1), 15088 (2017)
9. Marami, B., et al.: Temporal slice registration and robust diffusion-tensor reconstruction for improved fetal brain structural connectivity analysis. NeuroImage **156**, 475–488 (2017)
10. Miller, S.P., et al.: Abnormal brain development in newborns with congenital heart disease. New Engl. J. Med. **357**(19), 1928–1938 (2007)
11. Pecheva, D., et al.: A tract-specific approach to assessing white matter in preterm infants. NeuroImage **157**, 675–694 (2017)
12. Rollins, C.K., et al.: White matter microstructure and cognition in adolescents with congenital heart disease. J. Pediatr. **165**(5), 936–944 (2014)
13. Smith, S.M., et al.: Tract-based spatial statistics: voxelwise analysis of multi-subject diffusion data. Neuroimage **31**(4), 1487–1505 (2006)
14. Yushkevich, P.A.: Continuous medial representation of brain structures using the biharmonic PDE. NeuroImage **45**(1), S99–S110 (2009)
15. Yushkevich, P.A., Zhang, H., Simon, T.J., Gee, J.C.: Structure-specific statistical mapping of white matter tracts. NeuroImage **41**(2), 448–461 (2008)
16. Zhang, H., et al.: A tract-specific framework for white matter morphometry combining macroscopic and microscopic tract features. Med. Image Anal. **14**(5), 666–673 (2010)
17. Zhang, H., Yushkevich, P.A., Alexander, D.C., Gee, J.C.: Deformable registration of diffusion tensor MR images with explicit orientation optimization. Med. Image Anal. **10**(5), 764–785 (2006)

# Tract Orientation Mapping
# for Bundle-Specific Tractography

Jakob Wasserthal[1,2], Peter F. Neher[1], and Klaus H. Maier-Hein[1,3(✉)]

[1] Division of Medical Image Computing, German Cancer Research Center (DKFZ),
Heidelberg, Germany
k.maier-hein@dkfz.de
[2] Medical Faculty, Heidelberg University, Heidelberg, Germany
[3] Section for Automated Image Analysis, Heidelberg University Hospital, Heidelberg,
Germany

**Abstract.** While the major white matter tracts are of great interest
to numerous studies in neuroscience and medicine, their manual dissection in larger cohorts from diffusion MRI tractograms is time-consuming,
requires expert knowledge and is hard to reproduce. Tract orientation
mapping (TOM) is a novel concept that facilitates bundle-specific tractography based on a learned mapping from the original fiber orientation distribution function (fODF) peaks to a list of tract orientation
maps (also abbr. TOM). Each TOM represents one of the known tracts
with each voxel containing no more than one orientation vector. TOMs
can act as a prior or even as direct input for tractography. We use an
encoder-decoder fully-convolutional neural network architecture to learn
the required mapping. In comparison to previous concepts for the reconstruction of specific bundles, the presented one avoids various cumber-
some processing steps like whole brain tractography, atlas registration
or clustering. We compare it to four state of the art bundle recognition methods on 20 different bundles in a total of 105 subjects from the
Human Connectome Project. Results are anatomically convincing even
for difficult tracts, while reaching low angular errors, unprecedented run-
times and top accuracy values (Dice). Our code and our data are openly
available.

**Keywords:** Diffusion MRI · Tractography · Deep learning

## 1 Introduction

Diffusion tractography would be much simpler to solve if there existed only one of
the major tracts in the brain. In reality, though, multiple tracts co-exist and overlap, resulting in multiple fiber orientation distribution function (fODF) peaks

**Electronic supplementary material** The online version of this chapter (https://
doi.org/10.1007/978-3-030-00931-1_5) contains supplementary material, which is
available to authorized users.

A. F. Frangi et al. (Eds.): MICCAI 2018, LNCS 11072, pp. 36–44, 2018.
https://doi.org/10.1007/978-3-030-00931-1_5

per voxel and larger bottleneck situations with tracts per voxel outnumbering the peaks per voxel. In consequence, tractography is highly susceptible to false positives [4,5]. The only safe solution around false positives today is the explicit dissection of anatomically well-known tracts. While manual dissection protocols [10] can be considered the current gold standard, a variety of approaches was already developed for automating the process: *Region-of-interest-based* approaches filter streamlines based on their spatial relation to cortical or other anatomically defined regions, which are typically transferred to subject space via atlas registration and segmentation techniques [12,14]. *Clustering-based* approaches group and select streamlines by measuring intra- and inter-subject streamline similarities, referring to existing reference bundles in atlas space [2,6,7].

Concept-wise, many previous approaches have opted for performing a rather blind whole brain tractography and then investing the effort in streamline space, clearing the tractograms from spurious streamlines and grouping the remaining ones. We propose a novel concept, Tract Orientation Mapping (TOM), that approaches the problem before doing tractography by learning tract-specific peak images (tract orientation maps). These can be used as a prior – relating them to Rheault *et al.* [8], who employed registered atlas information as a tract-specific prior – or directly as orientation field for tractography. The larger-scale quantitative evaluation of such approaches is challenging due to the effort required to manually produce high quality reference tracts. To address this, an interactive process in form of auxiliary tools was designed in support of the expert. This helped us achieving high quality semi-automatic reference dissections of 20 bundles in a total of 105 Human Connectome Project (HCP) subjects. On basis of this novel data set, TOM was compared to several state of the art methods and was able to set new standards in terms of quality, quantity and runtime.

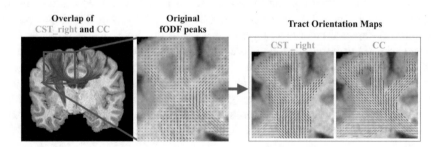

**Fig. 1.** Exemplary depiction of a slice through two of the reference tracts, the original fODF peak image and the corresponding reference TOMs (cf. Fig. 3 for abbr.).

## 2   Methods

Given a set of reference tracts, TOM is based on a learned mapping from the original fODF peak image to a list of tract orientation maps (cf. Fig. 1). Please note that TOM might refer to the general concept (tract orientation mapping) as well as to one of the tract orientation maps itself. Each TOM represents one tract, and each voxel contains one orientation vector representing the local tract orientation, i.e. the local mean streamline orientation of the corresponding reference tract. The mapping is learned via training a fully convolutional neural network (FCNN) with the original fODF peak image as input to regress the different TOMs as output channels. The network is then used to predict estimated TOMs for previously unseen subjects. These can be used as a tract-specific prior or employed directly for tractography (Fig. 2).

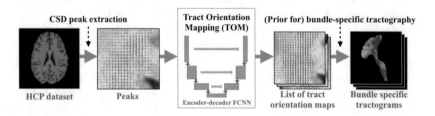

**Fig. 2.** Pipeline overview: constrained spherical deconvolution (CSD) is applied to obtain the three principal fODF directions per voxel. An encoder-decoder FCNN maps the original fODF peak image to bundle-specific peak images, i.e. TOMs, which are then employed as a prior for or as direct input for bundle-specific tractography.

**Model.** We implemented TOM using an encoder-decoder FCNN with long-range skip connections. Our FCNN uses the same number of layers and filters as the U-Net architecture [9], but with a input size of $144 \times 144$ and leaky ReLU activations. It has 9 input channels (the three principal fODF peaks) and 60 output channels (one 3D vector for each of the 20 bundles). The high dimensionality and spatial resolution of our data let us opt for a memory-efficient 2D instead of a 3D architecture. 2D slices were sampled along the y-axis. The decision of using fODF peaks as an input rather than raw image values or parametric representations of the signal, such as spherical harmonics coefficients, was also driven by the desire to reduce memory demand, with the side effect of becoming more independent of the acquisition scheme. fODF peaks were extracted using constrained spherical deconvolution (CSD) and peak extraction in MRtrix [3].

**Training.** Since we are solving a regression task, we employed linear activation functions in the last layer. As loss function we used weighted cosine similarity combined with mean squared error of the $l^2$-norm

$$loss(\hat{y}, y, w) = \frac{1}{N} \sum_{i=0}^{N} \left[ -\frac{1}{w} * \left| \frac{\langle \hat{y}_i, y_i \rangle}{\|\hat{y}_i\|_2 * \|y_i\|_2} \right| + \frac{w * (\|\hat{y}_i\|_2 - \|y_i\|_2)^2}{w_{max}} \right] \quad (1)$$

with $N$ being the number of classes, $y$ the training target, $\hat{y}$ the network output and $w$ a weighting factor which was used to handle the class imbalance between background and bundles and reinforce the training signal of the bundles. We set $w = 10$ at epoch 0 and linearly reduced it to $w = 3$ at epoch 300. For $y = 0$, we set $w = 0$. We used a learning rate of 0.001, an Adamax optimizer, 300 epochs of training and a batch size of 44. The input images were normalized to zero mean and standard deviation one. In the experiments, we applied 5-fold cross validation with splits for training, validation and testing. Hyperparameters were optimized on the validation set, network parameters producing optimal Dice scores were used for testing. Dice scores were calculated by thresholding the $l^2$-norm of the peaks.

**Bundle-Specific Tractography.** There are different flavors of bundle-specific tractography on basis of the estimated TOMs. The maps can be used for direct *TOM tractography*. Here, we applied deterministic MITK Diffusion tractography for this purpose, min. length 50 mm, one seed per voxel. Alternatively, a *TOM prior* can be applied during tractography on the original data. Here, we implemented this by amplifying the bundle-specific peaks using a weighted mean between original and prior peaks, similar to [8]. Last but not least, completely sticking to the original data, a *TOM-based peak selection* might be performed for tractography on the best-matching original peaks. For all three flavors, streamlines are stopped whenever the TOM becomes zero, which is the case outside of the respective bundle.

**Reference Data.** The approach was evaluated on a newly created database of 105 HCP subjects, each with a semi-automatic dissection of 20 prominent white matter tracts [13]. HCP imaging parameters were: 1.25 mm isotropic resolution, 270 gradient directions, three b-values $b = [1000; 2000; 3000] \, s/mm^2$. Starting point per subject was a 10 million streamline tractogram generated using MRtrix CSD and iFOD2 (min. length: 40 mm) [11]. TractQuerier [12] was then used to extract a first approximation of each bundle. Several interactive auxiliary tools were then employed and interactively combined in a workflow implemented in MITK Diffusion to achieve high quality reference dissections of each tract: (a) Manual definition of inclusion and exclusion ROIs, (b) QuickBundles [1] clustering for detection of small or spurious streamline clusters, (c) detection of streamlines that run through low fiber density voxels. Additionally, streamlines that run back and forth inside the target bundle, i.e. making 180° turns within less than 30 mm distance, were also removed. To make the high quality of this dataset accessible, we openly published our reference tracts for all 105 subjects: https://doi.org/10.5281/zenodo.1088277.

**Reference Methods.** We use four state of the art automatic tract delineation methods as baseline, including clustering- and as ROI-based approaches: *RecoBundles* [2] registers the tractogram to a reference subject and uses clustering to detect streamlines that are similar to the reference tracts. It was run with default parameters on 5 different reference subjects, aggregated by taking the mean. Using all 63 training subjects as reference subjects was computationally infeasible due to the long runtime of RecoBundles for 10 million fibers.

*WhiteMatterAnalysis* (WMA) [6,7] clusters streamlines across several subjects and produces a corresponding atlas. Each cluster in the atlas is assigned to a specific anatomical bundle. Registering new subjects to the atlas enables automated bundle delineation. We use the pretrained WMA atlas containing 800 clusters. We manually optimized the nine predefined mappings of clusters to anatomical tracts to better align it with our reference. During this process, we were inherently limited by the finite set of distinct clusters offered by the atlas. We chose not to extend the mapping to the 11 remaining reference tracts, which would require considerable effort given the amount of clusters. Applying WMA to 10 million streamlines requires >100 GB of memory, producing clusters with substantial amounts of false positives. Thus, streamline counts were reduced to 500k, requiring approx. 30 GB of memory and creating cleaner clusters.

*TractQuerier* [12] extracts tracts based on the regions the streamlines have to start in, end in and (not) run through defined in its own query language. We used the same queries as used in our pipeline for extracting the reference data (but without all the subsequent filtering).

*Atlas Registration* was additionally implemented as an in-house reference method. All training subjects were iteratively registered to the same space using FA-based symmetric diffeomorphic registration. The binary reference tract masks were averaged in atlas space and used for removing fibers exiting the masks in unseen registered test subjects.

## 3   Results

**Quantitative.** Figure 3 shows overall mean as well as tract-specific Dice scores for each of the methods. These were calculated on basis of the binary tract masks in comparison to the reference. For the overall means, differences between TOM and all other methods were statistically significant according to Wilcoxon signed-rank tests ($p < 0.001$). The mean angular error for the peaks predicted by TOM in comparison to the reference TOM (i.e. the mean streamline orientation in the reference) was $16.7° \pm 0.5$. This error is smaller than the mean angular error between the reference TOM and the best-matching original fODF peak in each voxel ($18.8° \pm 0.5$).

**Qualitative.** Figure 4 shows qualitative results for the corticospinal tract, the commissure anterior and the inferior occipito-frontal fascicle. We also show the results of RecoBundles which had the best Dice score of all reference methods in our quantitative evaluation. *TOM tractography* shows spatially very coherent and complete bundles, without spurious streamlines. RecoBundles on a deterministic tractogram, however, shows incomplete reconstructions, missing critical parts like the lateral projections of the CST. RecoBundles on a probabilistic tractogram finds the complete bundles (except for the left part of the CA), but introduces false positive and spatially incoherent streamlines.

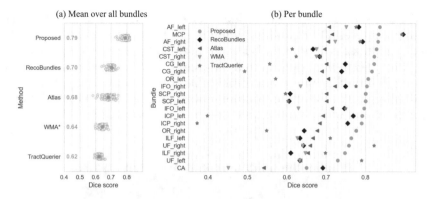

**Fig. 3.** Results of 5-fold cross validation on 105 subjects for the analyzed methods. (AF: arcuate fascicle, CA: commissure anterior, CST: corticospinal tract, CG: cingulum, ICP: inferior cerebellar peduncle, MCP: middle cerebellar peduncle, SCP: superior cerebellar peduncle, ILF: inferior longitudinal fascicle, IFO: inferior occipito-frontal fascicle, OR: optic radiation, UF: uncinate fascicle) *: mean score over the nine tracts provided by WMA analysis, see methods.

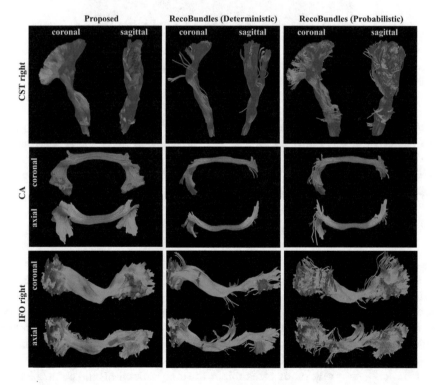

**Fig. 4.** Qualitative comparison on a random subject of the proposed method with RecoBundles, applied to deterministic and probabilistic MITK Diffusion tractography, min. length 50 mm, 1 million streamlines, fODF from MRtrix multi-shell multi-tissue CSD.

We also evaluated how the different flavors of TOM-based bundle-specific tractography affect the reconstructions (Fig. 5). *TOM-based peak selection* failed to reconstruct some difficult parts like the lateral projections of CST. Tractography with a *TOM prior* resulted in a complete CST reconstruction. Similar but more smooth results were obtained using direct *TOM tractography*. This finding is well-aligned with our above-reported observation regarding the angular errors of predicted and original peaks when compared to the reference peaks.

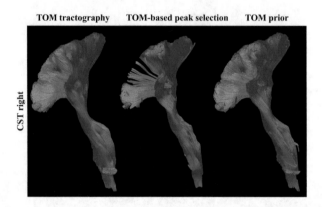

**Fig. 5.** Right CST of a random test subject reconstructed using the different variants of TOM-based bundle-specific tractography.

**Runtime.** For the reconstruction of 20 bundles in one subject, TOM required less than 2 min, making it more than five times faster than the second fastest method Atlas Registration (approx. 11 min). TractQuerier, RecoBundles and WMA required much longer processing times with around 17, 97 and 938 min, respectively (see supplementary materials for more information).

## 4   Discussion and Conclusion

We presented a novel concept for learning-based bundle-specific tractography that employs estimated bundle-specific peak images, i.e. TOMs. The results are highly encouraging when considering quantitative, qualitative and runtime measures. One interesting finding was, in comparison to the reference, the lower angular error of the estimated TOM in comparison to the voxel-wise best-matching fODF peaks. This was reflected in the experiments by improved performances with increasing influence of the TOM. While the fODF represents a probability distribution generating many streamlines during probabilistic tractography, the TOM represents only those streamlines that were selected as a reference.

A limitation of our evaluation was the use of TractQuerier both during the creation of the reference and during validation, inducing a potential positive bias for the method. Furthermore, WMA evaluation was only available for nine out of 20 tracts, which is not necessarily comparable. The margin between WMA and the proposed method remains, though, when restricting TOM to these same nine tracts. Regarding the reference data, although we applied extensive efforts to mitigate the limitations of tractography, our tracts do not represent a real *ground truth*. They are also subject to slight variations in the detailed anatomical definition of tracts, e.g. when it comes to exact start and end regions. Despite these limitations, to the best of our knowledge, the employed data set represents one of the best existing in-vivo approximations of known white matter anatomy in a cohort of that size. The presented approach is based on supervised learning, bearing the inherent limitation of depending on the availability and quality of training data. This is similar to RecoBundles, WMA and Atlas Registration, which also require matching atlas information with defined reference tracts. We have not yet studied the limits of TOM with respect to minimizing the amount of training data. Moreover, we have not yet studied the feasibility of applying HCP-trained TOM to non-HCP datasets, potentially with the help of domain adaptation techniques. While first in-house experiments on schizophrenia patients seem promising, this needs more evaluation and remains a potentially rewarding line of research. The code of our method is available as an easy to use python package: https://github.com/MIC-DKFZ/TractSeg

# References

1. Garyfallidis, E., et al.: QuickBundles, a method for tractography simplification. Front. Neurosci. **6** (2012)
2. Garyfallidis, E., et al.: Recognition of white matter bundles using local and global streamline-based registration and clustering. NeuroImage **170**, 283–295 (2017)
3. Jeurissen, B.: Multi-tissue constrained spherical deconvolution for improved analysis of multi-shell diffusion MRI data. NeuroImage **103**, 411–426 (2014)
4. Knösche, T.R., et al.: Validation of tractography: comparison with manganese tracing. Hum. Brain Mapp. **36**(10), 4116–4134 (2015)
5. Maier-Hein, K.H., et al.: The challenge of mapping the human connectome based on diffusion tractography. Nat. Commun. **8**(1), 1349 (2017)
6. O'Donnell, L.J., et al.: Automatic tractography segmentation using a high-dimensional white matter atlas. IEEE Trans. Med. Imaging **26**(11), 1562–1575 (2007)
7. O'Donnell, L.J., et al.: Automated white matter fiber tract identification in patients with brain tumors. Neuroimage: Clin. **13**, 138–153 (2017)
8. Rheault, F., et al.: Bundle-specific tractography using voxel-wise orientation priors. In: ISMRM (2018)
9. Ronneberger, O., Fischer, P., Brox, T.: U-net: convolutional networks for biomedical image segmentation. In: Navab, N., Hornegger, J., Wells, W.M., Frangi, A.F. (eds.) MICCAI 2015. LNCS, vol. 9351, pp. 234–241. Springer, Cham (2015). https://doi.org/10.1007/978-3-319-24574-4_28
10. Stieltjes, B., Brunner, R.M., Fritzsche, K., Laun, F.: Diffusion Tensor Imaging - Introduction and Atlas. Springer, Heidelberg (2013). https://doi.org/10.1007/978-3-642-20456-2
11. Tournier, J.D., et al.: Improved probabilistic streamlines tractography by 2nd order integration over fibre orientation distributions. In: ISMRM, p. 1670 (2010)

12. Wassermann, D., et al.: The white matter query language: a novel approach for describing human white matter anatomy. Brain Struct. Funct. **221**(9), 4705–4721 (2016)
13. Wasserthal, J., et al.: Tractseg - fast and accurate white matter tract segmentation. arXiv:1805.07103 (2018)
14. Yendiki, A., et al.: Automated probabilistic reconstruction of white-matter pathways in health and disease using an atlas of the underlying anatomy. Front. Neuroinform. **5**(23), 12–23 (2011)

# A Multi-Tissue Global Estimation Framework for Asymmetric Fiber Orientation Distributions

Ye Wu[1,2], Yuanjing Feng[1], Dinggang Shen[2], and Pew-Thian Yap[2(✉)]

[1] Institute of Information Processing and Automation,
Zhejiang University of Technology, Hangzhou, China
[2] Department of Radiology and BRIC, University of North Carolina,
Chapel Hill, USA
ptyap@med.unc.edu

**Abstract.** In connectomics, tractography involves tracing connections across gray-white matter boundaries in gyral blades of complex cortical convolutions. To date, most tractography algorithms exhibit gyral bias with fiber streamlines preferentially terminating at gyral crowns rather than sulcal banks or fundi. In this work, we will demonstrate that a multi-tissue global estimation framework of the asymmetric fiber orientation distribution function (AFODF) will mitigate the effects of gyral bias and will allow fiber streamlines at gyral blades to make sharper turns into the cortical gray matter. This is validated using in-vivo data from the Human Connectome Project (HCP), showing that, in a typical gyral blade with high curvature, the fiber streamlines estimated using AFODFs bend more naturally into the cortex than FODFs. Furthermore, we show that AFODF tractography results in better cortico-cortical connectivity.

## 1 Introduction

Diffusion magnetic resonance imaging (DMRI) [1] is a non-invasive technique that allows reconstruction of white matter (WM) pathways by tracing the directional patterns of water diffusion. In connectomics, this involves tracing across gray-white matter boundaries in gyral blades of complex cortical convolutions [1]. To date, most tractography algorithms exhibit *gyral bias* with streamlines preferentially terminating at gyral crowns rather than sulcal banks or fundi. This bias can be observed even for high angular resolution diffusion imaging (HARDI) data [2,3].

Local fiber orientations, commonly represented via fiber orientation distribution functions (FODFs), are usually assumed to be antipodal symmetric, even though in reality fiber tracts do not necessarily transverse each voxel in a symmetric fashion [4–6], especially in WM with complex configurations such as crossing, bending, bifurcation, kissing, and fanning [3]. In the cortex, a symmetric

This work was supported in part by NIH grants (NS093842, EB022880, EB006733, EB009634, AG041721, MH100217, and AA012388) and NSFC grants (61379020, 61703369).

© Springer Nature Switzerland AG 2018
A. F. Frangi et al. (Eds.): MICCAI 2018, LNCS 11072, pp. 45–52, 2018.
https://doi.org/10.1007/978-3-030-00931-1_6

representation of orientations hinders the estimated streamlines from curving significantly to enter the cortical gray matter (GM). In this work, we demonstrate that asymmetric fiber orientation distribution functions (AFODFs) can potentially mitigate gyral bias in cortical tractography.

To date, asymmetric FODF estimation techniques [4,5] have been mainly formulated in a voxel-wise manner. They are also not designed specifically to address the gyral bias problem. In this work, we introduce a global framework for AFODF estimation, dealing with partial volume effects that are prominent in tissue boundaries. We model the diffusion signal as the convolution of an AFODF with a multi-tissue response function [6]. The multi-tissue model accounts for signal contributions from white matter (WM), gray matter (GM), and cerebrospinal fluid (CSF).

In order to capture the asymmetry of the underlying fiber geometry in a local neighborhood, we extend the multi-tissue model to account for positive orientations (POs) and negative orientations (NOs). The AFODF at each voxel is estimated by enforcing orientation continuity across voxels. To understand the continuity constraint, we consider the scenario where a fiber leaves a voxel in a PO and enters a neighboring voxel in a NO. The constraint minimizes the difference between the PO and the NO. The AODFs are estimated globally across voxels, reducing sensitivity to noise. Unlike other global FODF estimation schemes [7,8], our method is initialization independent and incorporates FODF asymmetry. Using in-vivo data, we demonstrate that the proposed method mitigates the effects of gyral bias and allows streamlines at gyral blades to make sharper turns into cortical GM.

## 2   Method

### 2.1   Asymmetric Fiber Orientation Distribution

The diffusion signal $s_p(g)$ for gradient direction $g \in \mathbb{S}^2$ and location $p \in \mathbb{R}^3$ can be decomposed into $M \in \mathbb{N}$ tissue types, each of which is characterized by an axially symmetric response function (RF) $R_i(g, u)$ [6,9]. The signal contribution of each tissue $i$ in a voxel can be computed as the spherical convolution of its RF $R_i(g, u)$ and the fiber orientation distribution function (FODF) $F_{p,i}(u)$:

$$s_p(g) = \int_{u \in \mathbb{S}^2} \sum_{i=1}^{M} R_i(g, u) F_{p,i}(u) du, \qquad (1)$$

where the FODF is generally modeled using even order spherical harmonics (SHs). Due to the inherent antipodal symmetry of the FODF, i.e., $F_{p,i}(u) = F_{p,i}(-u)$, $u \in \mathbb{S}^2$, sub-voxel fiber fanning and bending cannot be distinguished using symmetric FODFs.

To circumvent the limitation of antipodal symmetry, we augment the multi-tissue constrained spherical deconvolution (CSD) [6] framework by incorporating information from neighbouring voxels $\mathcal{N}_p$, allowing asymmetry in FODF estimation, giving us the AFODF. Based on fiber continuity, a fiber leaving the current voxel $p$ with direction $u$ should enter the next voxel $q \in \mathcal{N}_p$ along its negative direction $-u$ [4]. Based on this observation, the discontinuity of $F_p(\cdot)$ in direction $u$ can be measured by

$$\phi(p, u) = \left| F_p(u) - \frac{1}{K_{p,u}} \sum_{q \in \mathcal{N}_p} W(\langle \hat{v}_{p,q}, u \rangle) F_q(u) \right|, \qquad (2)$$

where $W(\langle \hat{v}_{p,q}, u \rangle)$ is a directional probability distribution function (PDF) that is related to the angular difference between $u$ and $\hat{v}_{p,q} = \frac{q-p}{\|q-p\|}$. $K_{p,u} = \sum_{q \in \mathcal{N}_p} W(\langle \hat{v}_{p,q}, u \rangle)$ is a normalization term. We choose $W(\langle \hat{v}_{p,q}, u \rangle)$ to be a Gaussian PDF with reference direction $\hat{v}_{p,q}$. We represent the positive and negative hemispheres of the WM AFODF using two independent even-order SH bases. The odd-order SH basis typically captures only noise and is therefore not included in the representation [10]. More specifically, the AFODF can be written as

$$F_p(u) = \begin{bmatrix} \mathbf{Y}^+(u \in \mathbb{S}_+^2) & 0 \\ 0 & \mathbf{Y}^-(u \in \mathbb{S}_-^2) \end{bmatrix} \begin{bmatrix} \mathbf{X}_{\text{WM}}^+(p) \\ \mathbf{X}_{\text{WM}}^-(p) \end{bmatrix}, \qquad (3)$$

where $\mathbf{Y}^+(u \in \mathbb{S}_+^2)$ and $\mathbf{Y}^-(u \in \mathbb{S}_-^2)$ are the real symmetric SH bases defined on different hemispheres of $\mathbb{S}^2$. $\mathbf{X}_{\text{WM}}^+(p)$ and $\mathbf{X}_{\text{WM}}^-(p)$ are the corresponding SH coefficients.

## 2.2   Global Estimation Framework

We solve the AFODFs of a set of voxels $\mathbb{P} = \{p_1, p_2, \ldots, p_N\}$ jointly. We group the signal vectors of the $N$ voxels as columns in matrix $\mathbf{S}$ and the SH coefficients of the AODFs of these voxels as columns in matrix $\mathbf{X}$. We assume that three tissue types (WM, GM, and CSF) contribute to the signal. The WM FODF is anisotropic and can be represented using an SH series up to order $l$. The GM and CSF FODFs are isotropic and can be represented using zeroth order SH terms. With a set of directions $\mathbb{U} = \{u_1, u_2, \ldots, u_K\}$, AFODFs of the $N$ voxels are solved simultaneously via

$$\hat{\mathbf{X}} = \arg \min_{\mathbf{X}} \|\mathbf{R}\mathbf{X} - \mathbf{S}\|_F^2 \quad \text{s.t.} \quad \mathbf{A}\mathbf{X} \succeq 0 \quad \text{and} \quad \sum_{p \in \mathbb{P}, u \in \mathbb{U}} \phi(p, u) \leq \epsilon, \qquad (4)$$

where

$$\mathbf{X} = \begin{bmatrix} \mathbf{X}_{\text{WM}}^+(p_1) & \ldots & \mathbf{X}_{\text{WM}}^+(p_N) \\ \mathbf{X}_{\text{WM}}^-(p_1) & \ldots & \mathbf{X}_{\text{WM}}^-(p_N) \\ \mathbf{X}_{\text{GM}}(p_1) & \ldots & \mathbf{X}_{\text{CSF}}(p_N) \\ \mathbf{X}_{\text{CSF}}(p_1) & \ldots & \mathbf{X}_{\text{CSF}}(p_N) \end{bmatrix} \quad \text{and} \quad \mathbf{R} = \begin{bmatrix} \mathbf{R}_{\text{WM}}^+, \mathbf{R}_{\text{WM}}^-, \mathbf{R}_{\text{GM}}, \mathbf{R}_{\text{CSF}} \end{bmatrix}.$$

$$(5)$$

Matrix $\mathbf{R}$ maps the AFODF coefficients to the DW signal by means of spherical convolution. Matrix $\mathbf{A}$ maps the AFODF coefficients to the AFODF amplitudes and the constraint $\mathbf{A}\mathbf{X} \succeq 0$ imposes AFODF non-negativity [6]. Function $\phi(\cdot)$ is as defined in (2), but is computed only for the WM portion of the AFODF.

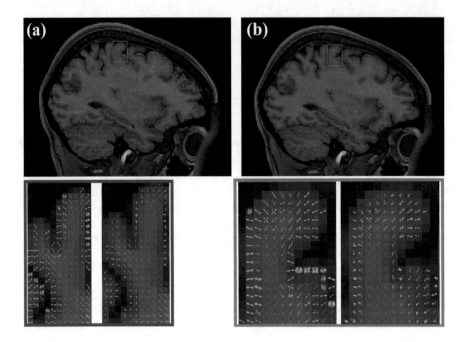

**Fig. 1.** (a) FODFs and (b) AFODFs with close-up views shown for gyral blades.

### 2.3   Optimization

The constrained linear least-squares problem can be cast as a general strictly convex quadratic programming (QP) problem:

$$\hat{\mathbf{X}} = \arg\min_{\mathbf{X}} \left( \text{trace} \left[ \frac{1}{2} \mathbf{X}^\top \mathbf{H} \mathbf{X} + \mathbf{Q}^\top \mathbf{X} \right] \right)$$
$$\text{s.t. } \mathbf{A}\mathbf{X} \succeq 0 \text{ and } \sum_{p \in \mathbb{P}, u \in \mathbb{U}} \phi(p, u) \leq \epsilon, \tag{6}$$

with $\mathbf{H} = \mathbf{R}^\top \mathbf{R}$ and $\mathbf{Q} = -\mathbf{R}^\top \mathbf{S}$. $\hat{\mathbf{X}}$ is obtained via convex optimization using CVX[1] with MOSEK[2].

### 2.4   Materials and Implementation Details

A DMRI dataset (Subject ID: 105923) from the Human Connectome Project (HCP) [11] was used to validate our method. The dataset has an isotropic spatial resolution of 1.25 mm, with 3 $b$-values ($b = 1000, 2000, 3000\,\text{s/mm}^2$), a total of 270 gradient directions (90 per shell), and 18 non-diffusion-weighted images. Cortical and sub-cortical parcellation was performed on the T1-weighted MR

---

[1] http://cvxr.com/.

[2] http://www.mosek.com.

**Table 1.** Valid streamlines.

|         | Deterministic |         |         |        | Probabilistic |         |         |        |
|---------|---------------|---------|---------|--------|---------------|---------|---------|--------|
|         | Seeds         | Initial | Valid   | %      | Seeds         | Initial | Valid   | %      |
| AFODF   | 1020987       | 709353  | **534032** | **75.28%** | 1020987     | 709353  | **539338** | **76.03%** |
| FODF    | 1020987       | 565731  | **169893** | **30.03%** | 1020987     | 565731  | **160897** | **28.44%** |

image using FreeSurfer. The cortical GM map generated by FSL using the T1-weighted MR image was used to define tractography seed voxels in the cortex.

SHs up to order 8, amounting to 45 coefficients, are used to represent the positive/negative hemispheres of the AFODF. The AFODF non-negativity and fiber continuity constraints are enforced on 724 uniformly distributed orientations. The response function for each tissue type is estimated for each shell with the help of the tissue segmentation map generated via FSL using the T1-weighted image.

Whole brain tractography was performed using deterministic and probabilistic algorithms [12] using the peak directions of the AFODFs and FODFs. Tracking was performed with the following parameters: 3 tracts per voxel in the cortical GM, 0.5 voxel step size, 60° maximum turning angle for deterministic tracking and 80° for probabilistic tracking. Tracking was terminated when the fractional anisotropy (FA) value is below 0.1. Streamlines shorter than 10 mm were removed.

# 3    Experimental Results

Figure 1 shows the differences between FODFs and AFODFs. We show the close-up views of gyral blades where fiber pathways exhibit high curvature. It can be observed that the AFODFs reflect the bending and branching characteristics of cortical WM pathways. The asymmetry of the AFODFs becomes more apparent for voxels nearer the cortex.

Figure 2 shows the outcomes of deterministic (columns 1 & 2) and probabilistic (columns 3 & 4) tractography. We also show results after removing invalid fiber streamlines that do not terminate in the cortex (columns 2 & 4). Even though tractography was performed for FODFs and AFODFs using the same number of seeds, AFODFs result in a significantly greater amount of streamlines, even after the removal of invalid streamlines. FODFs, on the other hand, result in a larger fraction of invalid streamlines. Quantitative results in Table 1 confirm this observation, suggesting improvements in tractography across WM-GM boundaries. It can also be observed that AFODFs give a larger amount of streamlines entering the sulcal walls. This increases cortical coverage and reduces gyral bias.

Figure 3 shows the connectivity between 164 regions identified by cortical parcellation [13]. The number of streamlines connecting two regions was recorded and a connectivity matrix was then constructed based on normalized streamlines

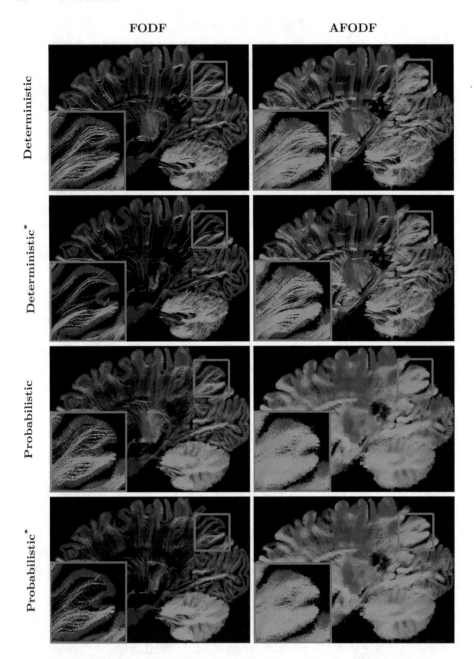

**Fig. 2.** Tractography using FODFs and AFODFs. The asterisks ('⋆') indicate removal of invalid streamlines.

**FODF**                    **AFODF**

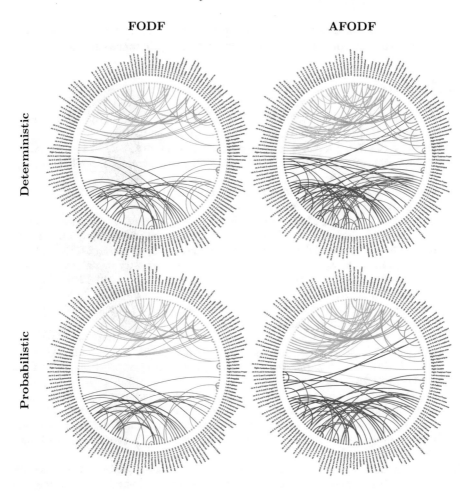

**Fig. 3.** Connectivity between the 164 cortical regions of the Destrieux Atlas [13].

counts. Normalization was performed by pre- and post-multiplication of the connectivity matrix (**C**) with degree matrix (**D**) raised to the power of $-1/2$, i.e., $\mathbf{D}^{-\frac{1}{2}}\mathbf{C}\mathbf{D}^{-\frac{1}{2}}$. All connections with normalized counts greater than 0.1 are shown. Self-connections are removed. The results again confirm that AFODFs yield better cortico-cortical connectivity.

## 4    Conclusion

We have presented a multi-tissue global estimation framework for AFODFs and have demonstrated that AFODFs can be used to mitigate gyral bias in cortical tractography. Our method extends multi-shell multi-tissue CSD by imposing fiber continuity across voxels to resolve orientation asymmetry. We have shown through an HCP dataset that our method resolves realistic sub-voxel fiber configurations and improves deterministic and probabilistic tractography.

# References

1. Johansen-Berg, H., Behrens, T.E.: Diffusion MRI: From Quantitative Measurement to in Vivo Neuroanatomy. Academic Press, Cambridge (2013)
2. Schilling, K., Gao, Y., Janve, V., Stepniewska, I., Landman, B.A., Anderson, A.W.: Confirmation of a gyral bias in diffusion MRI fiber tractography. Hum. Brain Mapp. **39**, 1449–1466 (2018)
3. Maier-Hein, K.H., et al.: The challenge of mapping the human connectome based on diffusion tractography. Nature Commun. **8**, 1349 (2017)
4. Bastiani, M., Cottaar, M., Dikranian, K., Ghosh, A., Zhang, H., Alexander, D.C., Behrens, T.E., Jbabdi, S., Sotiropoulos, S.N.: Improved tractography using asymmetric fibre orientation distributions. NeuroImage **158**, 205–218 (2017)
5. Reisert, M., Kellner, E., Kiselev, V.G.: About the geometry of asymmetric fiber orientation distributions. IEEE Trans. Med. Imaging **31**(6), 1240–1249 (2012)
6. Jeurissen, B., Tournier, J.D., Dhollander, T., Connelly, A., Sijbers, J.: Multi-tissue constrained spherical deconvolution for improved analysis of multi-shell diffusion MRI data. NeuroImage **103**, 411–426 (2014)
7. Schwab, E., Vidal, R., Charon, N.: Spatial-angular sparse coding for HARDI. In: Ourselin, S., Joskowicz, L., Sabuncu, M.R., Unal, G., Wells, W. (eds.) MICCAI 2016. LNCS, vol. 9902, pp. 475–483. Springer, Cham (2016). https://doi.org/10.1007/978-3-319-46726-9_55
8. Ye, C., Zhuo, J., Gullapalli, R.P., Prince, J.L.: Estimation of fiber orientations using neighborhood information. Med. Image Anal. **32**, 243–256 (2016)
9. Yap, P.T., Zhang, Y., Shen, D.: Multi-tissue decomposition of diffusion MRI signals via $\ell_0$ sparse-group estimation. IEEE Trans. Image Process. **25**(9), 4340–4353 (2016)
10. Tournier, J.D., Calamante, F., Connelly, A.: Robust determination of the fibre orientation distribution in diffusion MRI: non-negativity constrained super-resolved spherical deconvolution. NeuroImage **35**(4), 1459–1472 (2007)
11. Van Essen, D.C., et al.: The WU-Minn human connectome project: an overview. NeuroImage **80**, 62–79 (2013)
12. Garyfallidis, E.: Dipy, a library for the analysis of diffusion MRI data. Front. Neuroinformatics **8**, 8 (2014)
13. Destrieux, C., Fischl, B., Dale, A., Halgren, E.: A sulcal depth-based anatomical parcellation of the cerebral cortex. NeuroImage **47**, S151 (2009)

# Diffusion Tensor Imaging and Functional MRI: Diffusion Weighted Imaging

# Better Fiber ODFs from Suboptimal Data with Autoencoder Based Regularization

Kanil Patel[1], Samuel Groeschel[2], and Thomas Schultz[1,3](✉) (iD)

[1] Department of Computer Science, University of Bonn, Bonn, Germany
schultz@cs.uni-bonn.de
[2] Department of Pediatric Neurology and Developmental Medicine and Experimental
Pediatric Neuroimaging, University Children's Hospital Tübingen, Tübingen,
Germany
[3] Bonn-Aachen International Center for Information Technology, Bonn, Germany

**Abstract.** We propose a novel way of estimating fiber orientation distribution functions (fODFs) from diffusion MRI. Our method combines convex optimization with unsupervised learning in a way that preserves the relative benefits of both. In particular, we regularize constrained spherical deconvolution (CSD) with a prior that is derived from an fODF autoencoder, effectively encouraging solutions that are similar to fODFs observed in high-quality training data. Our method improves results on independent test data, especially when only few measurements or relatively weak diffusion weighting (low $b$ values) are available.

## 1   Introduction

Estimating fiber orientation distribution functions (fODFs) is a traditional step in analyzing data from diffusion MRI (dMRI). It is usually performed by spherical deconvolution, which solves an ill-conditioned optimization problem with suitable regularizers and constraints [1,9]. Recently, supervised learning has been explored as an alternative strategy for closely related problems, such as estimating NODDI or diffusional kurtosis [3], microstructure imaging [6], or fiber tractography [7]. Such approaches have been shown to produce robust results, sometimes even from a greatly reduced number of measurements.

This paper aims to harness the power of machine learning also for fODF estimation. A challenge in doing so is the fact that spherical deconvolution involves a response function as a parameter, which is usually estimated individually for each subject [8,10]. In order to ensure that the learned method will generalize to new subjects, we have to design it so that it does not capture specific assumptions about the response function.

We propose a hybrid solution for this problem that combines traditional convex optimization with learning in such a way that the relative strengths of

This work was supported by the DFG under grant SCHU 3040/1-1. We would like to thank Michael Möller (University of Siegen, Germany) for inspiring discussions. Kanil Patel is now with BCAI - Bosch Center for AI.

© Springer Nature Switzerland AG 2018
A. F. Frangi et al. (Eds.): MICCAI 2018, LNCS 11072, pp. 55–62, 2018.
https://doi.org/10.1007/978-3-030-00931-1_7

both approaches are preserved. In particular, we build on an fODF autoencoder, which leverages the ability of unsupervised learning to capture distributions that are too complex to be modeled explicitly: In our case, the distribution of fODFs in high-quality real-world data. We use this autoencoder to regularize optimization-based spherical deconvolution, in which the response function can still be freely exchanged, and traditional constraints can be enforced.

## 2    Background and Related Work

Spherical deconvolution is the standard approach to fODF estimation. It can be formalized as

$$\hat{f} = \text{argmin}_f \|Mf - s\|^2 \tag{1}$$

where vector $s \in \mathbb{R}^k$ contains $k$ diffusion-weighted measurements, solution vector $f$ contains spherical harmonics coefficients describing the fODF, matrix $M = ER$ is the product of a diagonal convolution matrix $R$ and a matrix $E$ that evaluates the spherical harmonics basis functions in the directions of the measurements. Computing $\hat{f}$ via Eq. (1) is ill-conditioned, and it is typically made more robust by imposing non-negativity constraints [1,9].

We have been inspired by recent works which demonstrated that, when dealing with other models in diffusion MRI, replacing optimization-based fitting with supervised regression improves robustness and makes it possible to recover plausible estimates from a significantly reduced set of measurements [3,6]. A straightforward transfer of this idea to Eq. (1) would imply learning a direct mapping from measurements $s$ to model parameters $f$. We decided against this, since it would effectively hard-wire the convolution matrix $R$ into the mapping. Since $R$ is typically adjusted per-subject [8,10], this would impair our ability to transfer the resulting method from one subject to another.

Instead, we pursue an alternative strategy that is based on unsupervised learning. It has been inspired by recent works on image restoration, which is a mathematically analogous problem to Eq. (1), with $s$ taking the role of an image degraded by blur and noise, $f$ a reconstructed image, and $M$ a matrix that includes the characteristics of the blur, similar to how $M$ depends on the response function in our case. Our approach follows [2,5,13] in using autoencoders to define a regularizer for $f$, maintaining full flexibility with respect to $M$. To our knowledge, we are the first to apply this strategy to fODF estimation, and we implement this idea differently than [2,5,13].

## 3    Hybrid fODF Estimation

Our primary goal in regularizing constrained spherical deconvolution is to improve results on suboptimal inputs, such as diffusion MRI data with few gradient directions or low $b$ value. Our proposed method can be written as

$$\tilde{f} = \text{argmin}_f H(f) \quad \text{with} \quad H(f) = \|Mf - s\|^2 + \lambda \|f - AE(\hat{f}; \Theta)\|^2 \tag{2}$$

where $H(f)$ is our novel objective function. We refer to it as a hybrid method, since it combines convex optimization with a learning-based regularizer, rather than replacing optimization with learning as in [3,6,7]. $M$, $f$, $\hat{f}$, and $s$ are defined as in Eq. (1), and $AE(\hat{f}; \Theta)$ denotes application of an autoencoder with parameters $\Theta$ to the preliminary estimate $\hat{f}$ from standard CSD. We will now justify this use of autoencoders and provide details on how they were trained, then explain the computational pipeline of our method and highlight two theoretical advantages of the hybrid approach.

## 3.1   Role and Training of Autoencoders in Our Method

Autoencoders are neural networks that are trained in an unsupervised fashion to reconstruct their own input. They are most frequently used to learn a latent representation of the training data, e.g., for dimensionality reduction. Denoising autoencoders are a variant of this idea, trained to reconstruct a clean output from a noisy input. Their use in Eq. (2) is motivated by the observation in [2] that the difference between the input and output of a denoising autoencoder can be interpreted as the gradient of a regularizer that represents a Gaussian-smoothed version of the data distribution. In other words, when dealing with input that is suboptimal for CSD, our regularizer should guide us to prefer fODFs that agree with those observed in more suitable data. Learning the distribution of fODFs, rather than the mapping from the corresponding measurements, increases our ability to generalize to new subjects and measurement protocols.

All our autoencoders were trained with output fODFs that were estimated with standard CSD from 64 uniformly distributed diffusion-weighted images (DWIs) at $b = 2000$, which is clearly above the minimum number of 45 recommended in [10]. We experimented with three different types of autoencoders:

- A **denoising autoencoder (DAE)** was trained with input fODFs that have been estimated after perturbing the DWIs with additional Rician noise (SNR $= 20$)
- A **subsampling autoencoder (SSAE)** was trained with input fODFs that have been estimated from a subset of 32 DWIs at $b = 2000$
- A **quality transfer autoencoder (QTAE)** was trained with input fODFs that have been estimated from 30 DWIs at $b = 700$, i.e., a measurement setup commonly used for Diffusion Tensor Imaging, but not for CSD

All autoencoders were trained on 122,324 fODFs from the white matter masks of three subjects; a fourth subject was used as a validation dataset. Hyperparameters were tuned individually for each type of autoencoder. Reported results used network architecture $(64 - 32 - 64)$ (i.e., three hidden layers with 64, 32, and 64 neurons, respectively) for the DAE and QTAE, $(128 - 128 - 128 - 128)$ for the SSAE. We obtained better results with sigmoid than with ReLU activation functions, due to the limited size and depth of our networks. Optimization was done using Adadelta [12] with initial learning rate 0.1, decayed by half every 500 epochs, batch size 512, L2 regularization with trade-off parameter 0.01, and dropout with probability 0.01.

### 3.2  Computational Pipeline and Theoretical Guarantees

Our computational pipeline is illustrated in Fig. 1. The autoencoder is trained in a pre-process. Given new measurements, we first solve a standard CSD according to Eq. (1) to obtain a preliminary estimate $\hat{f}$. After passing $\hat{f}$ through the autoencoder, we integrate the regularizer in Eq. (2) into the least squares problem by amending the measurement vector $s$ with $\sqrt{\lambda}\,AE(\hat{f};\Theta)$, and matrix $M$ with $\sqrt{\lambda}\,I$, where the dimension of identity matrix $I$ matches vector $f$.

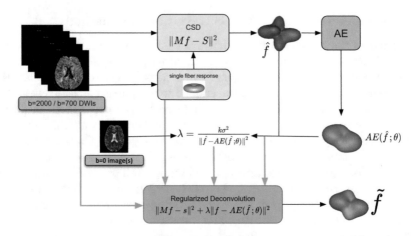

**Fig. 1.** A flow diagram of our method, in which standard constrained spherical deconvolution (CSD) and the output of an fODF autoencoder (AE) feed into a regularized deconvolution that produces the final result $\tilde{f}$.

Unlike pure learning-based methods that directly map measurements to model parameters [3,6,7], our hybrid method allows us to formally guarantee a minimum degree of data fidelity, by maintaining the upper bound

$$\|M\tilde{f} - s\|^2 \le \|M\hat{f} - s\|^2 + k\sigma^2 \tag{3}$$

on the fitting error of the final result $\tilde{f}$. Here, $k$ is the number of diffusion-weighted measurements, and the user's choice of $\sigma$ determines by how much we are willing to increase average residual magnitudes per measurement with respect to the unregularized optimum $\hat{f}$. In our experiments, we set $\sigma$ to 5% of the average $b = 0$ signal intensity within a white matter mask.

The constraint in Eq. (3) is maintained by setting $\lambda = k\sigma^2/\|\hat{f} - AE(\hat{f};\Theta)\|^2$. To see why, assume that a solution $\tilde{f}'$ violates it. Since $\|\tilde{f}' - AE(\hat{f};\Theta)\|^2$ is non-negative, this implies that the value of the objective function from Eq. (2) must be $H(\tilde{f}') > \|M\hat{f} - s\|^2 + k\sigma^2$. However, according to our choice of $\lambda$, $H(\hat{f}) = \|M\hat{f} - s\|^2 + k\sigma^2$, so $H(\tilde{f}') > H(\hat{f})$, and $\tilde{f}'$ cannot be optimal.

Our method is implemented within the framework of a recently proposed variant of CSD, which represents fODFs as fourth-order tensors with a positive definiteness constraint (H-psd) [1]. The problems in Eqs. (1) and (2) are expressed as quadratic cone programs, and solved using the publicly available package CVXOPT [11], as described in more detail in [1]. Note that [1] does not consider regularization based on unsupervised learning, which is our main contribution.

A second theoretical guarantee that is afforded by our hybrid method, but would be non-trivial to integrate into a pure learning-based approach, is preservation of the H-psd constraint. This implies that, while setting $\sigma = 0$ reduces our method to [1], very high values of $\sigma$ do not simply lead to $\tilde{f} \approx AE(\hat{f}; \Theta)$, but rather project $AE(\hat{f}; \Theta)$ onto the set of non-negative fODFs when needed.

# 4 Results and Discussion

## 4.1 Quantitative Results

We quantify the benefit from our method on an independent subject, unrelated to the training or validation data. Within a single session, 64 DWIs at $b = 2000$ and 30 DWIs at $b = 700$ were acquired on a 3T Skyra (Siemens), and corrected for head motion using FSL [4]. From these data, we created four test sets:

1. CSD protocol (64 DWIs at $b = 2000$)
2. A subsample of 32 DWIs from the CSD protocol
3. A subsample of 15 DWIs from the CSD protocol
4. A DTI protocol (30 DWIs at $b = 700$)

**Table 1.** Our regularization resulted in lowest average L2 and angular errors on all test sets, with the most distinct benefits on the suboptimal inputs (2–4).

| Test set | L2 error ($\times 0.1$) | | | | Angular error | | | |
|---|---|---|---|---|---|---|---|---|
| | 1 | 2 | 3 | 4 | 1 | 2 | 3 | 4 |
| Standard CSD | 1.29 | 1.45 | 2.08 | 2.22 | 12.5 | 14.1 | 19.6 | 18.1 |
| DAE | 1.29 | 1.38 | 1.87 | 1.93 | 12.6 | 13.5 | 17.2 | 15.2 |
| SSAE | **1.09** | **1.18** | 1.79 | 1.92 | **11.8** | **12.5** | 17.0 | 15.1 |
| QTAE | 1.17 | 1.27 | **1.74** | **1.85** | 12.3 | 13.0 | **16.4** | **14.2** |

For each set, Table 1 lists results achieved with standard CSD, as well as with CSD regularized by our three different types of autoencoders. Errors are specified with respect to fODFs that have been estimated with standard CSD from an independent measurement of the same subject (64 DWIs at $b = 2000$, on a 3T Prisma) after spatial registration with the Skyra scan, and corresponding rotation of gradient vectors. Error measures are the L2 norm of differences between estimated and reference fODF coefficients, as well as angular differences between up to three strongest fiber directions, weighted by their corresponding volume fractions. All errors are averaged over the 35,470 voxels in the test subject's white matter mask.

For all suboptimal inputs (sets 2–4), all autoencoders resulted in a clear improvement. The best result in each column is marked in boldface. Notably, our regularization allows us to achieve similar angular errors from half the measurements (set 2) as standard CSD on the full set of measurements (set 1). Moreover, we achieve similar angular errors from $b = 700$ data (set 4) as standard CSD does given a similar number of DWIs at $b = 2000$ (set 2). We believe that the fact that the reference dataset has been reconstructed using standard CSD from data similar to set 1 makes it difficult to beat standard CSD on that set. Despite this, the SSAE led to a slight improvement even in this case.

Unsurprisingly, best results were obtained when the protocol for generating training data matched the final use case, i.e., the SSAE worked best on set 2, while the QTAE gave best results on set 4. In all experiments, the SSAE and QTAE dominated the DAE. The fact that all autoencoders improved all challenging use cases 2–4, even ones that did not match their training paradigm, indicates that we succeeded in our attempt to construct a method that, in contrast to pure supervised methods such as [3], does not strictly require an expensive re-training for each new use case.

## 4.2   Practical Benefit from Theoretical Guarantees

We conducted experiments with a simplified variant of our method that directly uses $AE(\hat{f}; \Theta)$ as the final fODF estimate rather than regularizing deconvolution with it. This simplification sacrifices the formal guarantees that were discussed in Sect. 3.2 and thus helps us understand their practical benefit.

**Table 2.** For all autoencoders and test sets, the percentage of voxels for which L2 errors improve over standard CSD is greater when using the autoencoder output to regularize deconvolution, compared to using it directly as an estimate ("CSD+AE").

| Test set | CSD+AE | | | | Regularized deconvolution | | | |
|---|---|---|---|---|---|---|---|---|
| | 1 | 2 | 3 | 4 | 1 | 2 | 3 | 4 |
| DAE | 44.5% | 52.9% | 69.2% | 77.4% | **51.2%** | **60.2%** | **75.9%** | **86.6%** |
| SSAE | 72.7% | 81.4% | 85.7% | 90.0% | **76.9%** | **84.8%** | **88.1%** | **91.5%** |
| QTAE | 56.4% | 64.4% | 83.4% | 92.9% | **66.3%** | **74.6%** | **89.1%** | **93.7%** |

We found that average L2 and angular errors of the two variants were very similar, and none of the two variants dominated the other one. However, we do not only strive for high accuracy on average, but also to achieve an improvement over the baseline for as many individual voxels as possible. As can be seen in Table 2, regularized deconvolution makes a clear contribution towards this goal, since it consistently increases the fraction of voxels for which an improvement over standard CSD is obtained.

This implies that we can often benefit from regularizing CSD with an fODF that has a higher error than the unregularized result, which may seem counter-intuitive. We observed that autoencoders tend to oversmooth the fODFs, sometimes to an extent that increases the L2 error beyond the baseline. In this case, the regularization better balances data fidelity with the level of smoothing.

### 4.3    Iterating Our Method

Since the image restoration works that have inspired our approach [2,5,13] apply the autoencoder repeatedly within an iterative optimization, it was natural to attempt iterating Eq. (2) so that the result $\tilde{f}$ from the previous iteration replaces the original estimate $\hat{f}$ in the next one. However, this increased L2 errors almost always, which we found to be due to the onset of oversmoothing.

### 4.4    Qualitative Results

We also validated our method qualitatively by performing tractography on the resulting fODFs. Figure 2 provides an example from the brainstem region. A tractography on our reference dataset is in (a), while (b) and (c) are based on test set 3 with merely 15 DWIs. As expected from Table 1, our regularizer (c) cannot obtain a perfect reconstruction from this strong subsampling, but it does restore several fibers missing from the unregularized result, marked with ellipses in (b), and it eliminates some spurious fibers, marked with arrows.

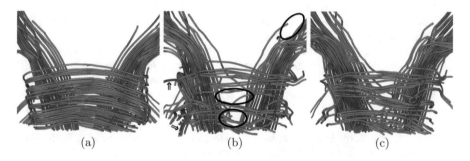

(a)                    (b)                    (c)

**Fig. 2.** Compared to a tractography on our reference dataset (a), our regularizer (c) restores some plausible tracts missing from a standard reconstruction of test set 3 (ellipses in b), and it eliminates some false positives (arrows in b).

## 5    Conclusion

We have presented a novel method for estimating improved fiber ODFs from suboptimal diffusion MRI data, by using fODF autoencoders to learn a suitable regularizer in an unsupervised manner. On independent test data, our regularizer allowed us to achieve similar angular accuracy from only 32 gradient directions as with the standard approach from 64 directions. Similarly, we achieve comparable angular accuracy from 30 directions at $b = 700$ as the standard approach from 32 directions at $b = 2000$. Our current regularizer only uses information from the same voxel; we hope to extend it to also account for spatial neighborhoods.

On a conceptual level, our work addresses more general challenges such as: How can we use machine learning to produce accurate results not just on average, but in as many individual cases as possible? (By imposing suitable constraints.) How can we enforce formal guarantees on the results of a learning based approach? (By combining it with constrained optimization.) How can we eliminate the need for frequent re-training? (By learning something more general, e.g., the natural distribution of fODFs rather than the mapping from subject- or scanner-specific data to fODFs.) In our future work, we plan to continue exploring strategies for these conceptual challenges, within and beyond this specific application.

# References

1. Ankele, M., Lim, L.H., Gröschel, S., Schultz, T.: Versatile, robust, and efficient tractography with constrained higher-order tensor fODFs. Int. J. Comput. Assist. Radiol. Surg. **12**(8), 1257–1270 (2017)
2. Bigdeli, S.A., Zwicker, M., Favaro, P., Jin, M.: Deep mean-shift priors for image restoration. In: Advances in Neural Information Processing Systems (NIPS), pp. 763–772 (2017)
3. Golkov, V., et al.: q-space deep learning: twelve-fold shorter and model-free diffusion MRI scans. IEEE Trans. Med. Imaging **35**(5), 1344–1351 (2016)
4. Jenkinson, M., Beckmann, C.F., Behrens, T.E., Woolrich, M.W., Smith, S.M.: FSL. NeuroImage **62**(2), 782–790 (2012)
5. Meinardt, T., Moeller, M., Hazirbas, C., Cremers, D.: Learning proximal operators: using denoising networks for regularizing inverse imaging problems. In: IEEE International Conference on Computer Vision, pp. 1799–1808 (2017)
6. Nedjati-Gilani, G.L., et al.: Machine learning based compartment models with permeability for white matter microstructure imaging. NeuroImage **150**, 119–135 (2017)
7. Neher, P.F., Côté, M.A., Houde, J.C., Descoteaux, M., Maier-Hein, K.H.: Fiber tractography using machine learning. NeuroImage **158**, 417–429 (2017)
8. Tax, C.M., Jeurissen, B., Vos, S.B., Viergever, M.A., Leemans, A.: Recursive calibration of the fiber response function for spherical deconvolution of diffusion MRI data. NeuroImage **86**, 67–80 (2014)
9. Tournier, J.D., Calamante, F., Connelly, A.: Robust determination of the fibre orientation distribution in diffusion MRI: non-negativity constrained super-resolved spherical deconvolution. NuroImage **35**, 1459–1472 (2007)
10. Tournier, J.D., Calamante, F., Connelly, A.: Determination of the appropriate b value and number of gradient directions for high-angular-resolution diffusion-weighted imaging. NMR Biomed. **26**(12), 1775–1786 (2013)
11. Vandenberghe, L.: The CVXOPT linear and quadratic cone program solvers. Technical report, UCLA Electrical Engineering Department (2010). http://www.seas.ucla.edu/~vandenbe/publications/coneprog.pdf
12. Zeiler, M.D.: ADADELTA: an adaptive learning rate method. CoRR abs/1212.5701 (2012). http://arxiv.org/abs/1212.5701
13. Zhang, K., Zuo, W., Gu, S., Zhang, L.: Learning deep CNN denoiser prior for image restoration. In: IEEE Conference on Computer Vision and Pattern Recognition, pp. 2808–2817 (2017)

# Identification of Gadolinium Contrast Enhanced Regions in MS Lesions Using Brain Tissue Microstructure Information Obtained from Diffusion and T2 Relaxometry MRI

Sudhanya Chatterjee[1]([✉]), Olivier Commowick[1], Onur Afacan[2],
Simon K. Warfield[2], and Christian Barillot[1]

[1] Univ Rennes, CNRS, Inria, Inserm, IRISA UMR 6074, VisAGeS U1228,
Rennes, France
sudhanya.chatterjee@irisa.fr
[2] CRL, Boston Children's Hospital, Harvard Medical School, Boston, MA, USA

**Abstract.** A multiple sclerosis (MS) lesion at an early stage undergoes active blood brain barrier (BBB) breakdown. Identifying MS lesions in a patient which are undergoing active BBB breakdown is of critical importance for MS burden evaluation and treatment planning. However in non-contrast enhanced structural magnetic resonance imaging (MRI) the regions of the lesion undergoing active BBB breakdown cannot be distinguished from the other parts of the lesion. Hence gadolinium (Gd) contrast enhanced T1-weighted MR images are used for this task. However some side effects of Gd injection into patients have been increasingly reported recently. The BBB breakdown is reflected by the condition of tissue microstructure such as increased inflammation, presence of higher extra-cellular matter and debris. We thus propose a framework to predict enhancing regions in MS lesions using tissue microstructure information derived from T2 relaxometry and diffusion MRI (dMRI) multi-compartment models. We show that combination of the dMRI and T2 relaxometry microstructure information can distinguish the Gd enhancing lesion regions from the other regions in MS lesions.

**Keywords:** Diffusion MRI · T2 relaxometry · Microstructure
Brain · Multiple sclerosis

## 1 Introduction

Multiple sclerosis (MS) patients have multiple focal lesions in the brain. In the early stages, the MS lesions undergo active blood brain barrier (BBB) breakdown [1,2]. Lesions in this stage are referred to as enhancing lesions and the clinical significance of its identification is well established. Gadolinium based

© Springer Nature Switzerland AG 2018
A. F. Frangi et al. (Eds.): MICCAI 2018, LNCS 11072, pp. 63–71, 2018.
https://doi.org/10.1007/978-3-030-00931-1_8

contrast agents (GBCA) are popularly used by radiologists to identify enhancing MS lesions. A T1-weighted MRI acquired post GBCA injection is a part of the recommended MRI protocols for diagnosis and follow-up examinations of MS patients [3]. However, the use of GBCAs has been a recent topic of debate, primarily due to reports of Gd deposition in the brain [4,5]. Suggestions insisting on greater debate before GBCA administration has gained traction due to observed MRI signal changes in the brain tissues due to repeated GBCA administrations and other possible health issues [6]. A MS lesions in its enhancing stage has different pathological traits as compared to late or non-enhancing MS lesions [2]. Regions of lesion undergoing active BBB breakdown has a higher water content owing to the undergoing tissue damages such as functional impairment of the morphologically intact endothelial cells [1]. Hence by the virtue of the brain tissue microstructure characteristics of MS lesions, we may identify enhancing regions in the lesion (if any). Although MRI effectively provides in-vivo images of the brain, it is constrained by the limited imaging resolution it can provide. This limitation is primarily attributed to hardware limits and the need for adhering to reasonable scan times for clinical implementations. However, advanced MRI methods, such as diffusion MRI (dMRI) and T2 relaxometry help us obtain estimates on condition of brain tissue microstructure. The multi-compartment models (MCM) [7] in dMRI provide information on the organization of the nerve fibers in the brain. MS lesions disrupt the normal organization of fibers in the brain. The extent of this damage may be assessed from the tissue microstrucutre estimates derived from dMRI MCMs. Myelin is also a critical biomarker in neurodegenerative diseases such as MS [2,8,9]. Demyelination marks the onset of MS [2]. Myelin has a very short T2 relaxation time ($<50$ ms) due to its tightly wrapped structure [10]. Due to higher TEs in dMRI, the myelin information is not present in the dMRI signals. However, myelin information can be obtained by estimating the myelin water fraction (MWF) from T2 relaxometry MRI signal [8,9]. The inflammation in MS lesions can be assessed from T2 relaxometry and dMRI signal. Hence, by combining the tissue microstructure information from these two MRI methods, we can obtain considerable information on brain tissue health. In this work we combine the microstructure information obtained from T2 relaxometry and dMRI to identify Gd enhanced regions in MS lesions. We performed experiments to evaluate whether combining the tissue microstructures is advantageous as compared to using them alone. The observations from this experiment is carried forward to the next stage where we perform enhancing lesion region predictions in a MS patient.

## 2   Materials and Methods

### 2.1   Multi-Compartment T2 Relaxometry Model (MCT2)

Three T2 relaxometry compartments are considered in a voxel based on their T2 relaxations times and are referred to as short-, medium- and high-T2. The short-T2 compartment conveys information on brain tissues with T2 relaxation times shorter than 50 ms. These tissues primarily include myelin and highly myelinated axons [10]. The high-T2 compartment represents the tissues with T2 values greater than 1000ms, comprising primarily of free fluids (CSF in healthy volunteer) and water accumulated in tissues due to pathology (edema regions in MS lesions). The medium-T2 compartment is a mixed pool and conveys information on intracellular matter (such as unmyelinated axons and glia), intra and extracellular fluids [10]. The T2 space is modeled as a weighted mixture of the three compartments, where each compartment is represented by a continuous probability density function (PDF). The signal of a voxel at the $i$-th echo in the MCT2 is modeled as:

$$s\left(t_i\right) = \sum_{j=1}^{3} \alpha_j \int_{0}^{\infty} f_j\left(T_2; \mathbf{p_j}\right) EPG\left(T_2, \triangle TE, i, B_1\right) dT_2 \tag{1}$$

In Eq. (1), $f_j\left(T_2; \mathbf{p}\right)$ is the chosen PDF to represent the $j$-th compartment with parameters $\mathbf{p_j} = \{p_{j_1}, p_{j_2}, \ldots\}$. We used the 2D multislice Carr-Purcell-Meiboom-Gill (CPMG) sequence to acquire T2 relaxometry data. CPMG sequences suffer from the effect of the stimulated echoes due to imperfect refocusing. It is important to address this effect as this leads to errors in T2 estimation [11]. Here we tackle is the problem of stimulated echoes using the iterative technique of Extended Phase Graph (EPG) algorithm [12]. Each compartment's weight is obtained as, $w_j = \alpha_j / \sum_i \alpha_i$. Simultaneous estimation of the weights and parameters of the distributions of such multi-compartment models is non-trivial and not reliable in terms of robustness and accuracy [13]. Hence we choose to fix the PDF parameters. In this work, the $\{f_j(\cdot)\}_{j=1}^{3}$ are chosen as Gaussian PDFs. Their mean and standard deviation are fixed based on the findings from the literature [8–10] and are set as $\mu = \{20, 100, 2000\}$ and $\sigma = \{5, 10, 80\}$ (all values in milliseconds). Estimating the model thus resorts to finding the optimal $\{\alpha_j\}_{j=1}^{3}$ and $B_1$ for the following least squares problem:

$$\left(\hat{\alpha}, \hat{B}_1\right) = \arg\min_{\alpha, B_1} \sum_{i=1}^{m} \left(y_i - \sum_{j=1}^{3} \alpha_j \lambda_j\left(t_i; B_1\right)\right)^2 \tag{2}$$

where $\mathbf{Y} \in \mathbb{R}^m$ is the observed signal; $m$ is the number of echoes; $\alpha \in \mathbb{R}^{+3}$. Although the optimization of $\alpha$ and $B_1$ are linear and non-linear in nature, these variables are linearly separable. $\alpha$ and $B_1$ are computed by non-negative least squares and BOBYQA optimization respectively. The weights of short-T2 ($w_s$), medium-T2 ($w_m$) and high-T2 ($w_h$) compartment for every voxel are used as a feature for each voxel.

## 2.2   Multi-Compartment Diffusion Model (MCDiff)

For diffusion MRI (dMRI), we considered the recently introduced MCM as they provide an intuitive way of describing the different fascicles, cells and free water contributions to each voxel. MCM are defined as a weighted sum of several compartments each describing a fascicle (i.e. a dense set of fibers sharing the same orientation) or isotropic matter (such as free water or water trapped in cell bodies). Similar to T2 relaxometry, each compartment in MCDiff is defined by a PDF. However in MCDiff, the PDFs describe water diffusion probability inside the compartments. A variety of compartment types may be defined [7] based on the assumed white matter microstructure and the acquired dMRI data (the more gradient directions and b-values, the more information may be extracted). Our proposed method is quite independent of this choice and can be applied generically to all parameters that may be extracted from MCM. In the specific study in Sect. 3, we have focused on the following model:

$$p(x) = f_w p_{FW}(x) + \sum_{i=1}^{N=3} a_i p_i(x) \tag{3}$$

where $f_w + \sum_i a_i = 1$ are the weights of the individual compartments, $p_{FW}$ is an isotropic Gaussian PDF specific to free water (i.e. with a variance of $3.0 \times 10^{-3} \, mm^2 \, s^{-1}$), $p_i(x)$ denotes the $i$-th fascicle compartment PDF here defined as a stick model (i.e. a Gaussian PDF with equal secondary eigenvalues, fixed from an outside reference). We have specifically chosen this ball and stick model for our experiments as it can be estimated reliably on our clinical data (see Sect. 2.3). This estimation was performed using the method proposed by Stamm et al. [14], which uses a variable projection on the linear elements of the cost function (the compartment weights) to perform a fast maximum likelihood estimation of the model parameters with Levenberg-Marquardt optimization. We then defined different parameters from this MCM which describe the white matter microstructure inside the voxel. First, each anisotropic compartment $p_i$ is defined as a constrained tensor. Hence we can extract the usual tensor scalar maps for each anisotropic compartment. However, to enable comparison between voxels, we need to average those compartment specific values over all anisotropic compartments. We have thus computed the weighted average of those values (using the weights $a_i$) to get the following scalar maps: fractional anisotropy $(FA_{mc})$, apparent diffusion coefficient $(ADC_{mc})$ and axial diffusivity $(AD_{mc})$. In addition to those maps, the weight of isotropic free water $(f_w)$ is a crucial one that could identify edema or other free water related phenomena and we therefore included it in the parameters as well.

## 2.3   Data

All acquisitions were made on a 3T MRI scanner. The T2 relaxometry data was obtained using a 2D multislice CPMG sequence with the following specifications: first echo time (TE) = 13.8 ms; echo spacing = 13.8 ms; 7 echoes; repetition time (TR) = 4530 ms; $1.33 \times 1.33 \times 3 \, mm^3$ voxel resolution; acquisition time was just less than 7 min. The dMRI acquisition was performed with 30 directions on a single shell of b-value at $1000 \, s/mm^2$, with a $2 \times 2 \times 2 \, mm^3$ voxel resolution, on a $128 \times 128 \times 60$ matrix with TE and TR of 94 ms and 9.3 s respectively. Transverse SE T1-w images $(1 \times 1 \times 3 \, mm^3)$ post Gd contrast agent infusion

(0.1 mmol/kg gadopentetate dimeglumine) were acquired to find Gd enhanced lesions. A T1-w image was also acquired for performing the distortion correction in diffusion images. A 3D MPRAGE image was acquired with inversion time, TR and TE of 900 ms, 1900 ms and 2.98 ms respectively. The voxel resolution was $1 \times 1 \times 1$ mm$^3$ on a $256 \times 256 \times 160$ matrix. Lesions were segmented on T2-w images by radiologist. Hence all images were registered to the T2 image ($1 \times 1 \times 3$ mm$^3$ voxel resolution) linear registration using a block-matching algorithm [15, 16]. Our data set consisted of 10 MS patient datasets demonstrating clinically isolated syndrome (CIS) condition. There were a total of 227 MS lesions in all patients, out of which 28 lesions had gadolinium enhancing regions. The voxels are divided into two groups: *(a)* (E+): voxels appearing on Gd enhanced T1 SE images and *(b)* (L−): lesion voxels which are hyperintense on T2-w images but do not appear on Gd enhanced T1-weighted images. The protocols were approved by the institutional review board, and all participants gave their written consent.

## 2.4    Identifying Enhancing Voxels in Lesions

We performed enhancing voxel identification using the MCT2 and MCDiff estimates. In our database, we had 15012 and 3904 (L−) and (E+) voxels respectively. We adopted a random shuffle and repeat strategy to compensate for the imbalance in the class. 5000 (L−) and 3400 (E+) voxels are randomly selected from the dataset to train the classifier. The remaining (L−) and (E+) voxels are then used to evaluate the classifier performance. This method is repeated 100 times to avoid any bias in sampling the data set for model training. The accuracy statistics are recorded for every repetition. It shall be noted that the model from one repetition is not retained for the next. For a new repetition, the model is trained and validated on a different dataset. We then observe the validation error of the classifier over 100 repitions. We used support vector machine classifier (with radial basis function kernel) in this work [17].

*Experiment-1.* The predictions are performed using three features sets: **(a)** MCT2 derived microstructure information: $\mathcal{F}_R = \{w_s, w_m, w_h,\} \in \mathbb{R}^3$, **(b)** MDiff derived microstructure information: $\mathcal{F}_D = \{f_w, FA_{mc}, ADC_{mc}, AD_{mc}\} \in \mathbb{R}^4$ and **(c)** a features set containing both MCT2 and MCDiff derived microstructure information $(\mathcal{F}_{RD} \in \mathbb{R}^7)$. The aim of this experiment is to observe whether combining the diffusion and T2 relaxometry derived microstructure increases the accuracy of prediction. The observations from this experiment will help us comment on complimentary nature of the feature sets (if any).

*Experiment-2.* In this experiment we illustrate the application of the proposed method on a MS patient. We maintain a MS patient dataset which was never used for training the data set in any of the repetitions. The classifier trained in every repetition is used to predict to which category the voxels in the validation image belong. We perform a majority voting on the 100 predictions to decide the final prediction for each voxel. Subsequently we compute the dice measure on the (E+) and (L−) masks to judge the performance of the classifier. The classifier implementation was performed using the scikit-learn package in Python v2.7.

## 3   Results

*Experiment-1.* We show in Fig. 1 comparison of the prediction performance of the classifier on validation sets over 100 repititions using features sets derived from: (a) MCT2 model only ($\mathcal{F}_R$), (b) MCDiff model only ($\mathcal{F}_D$) and (c) combination of both ($\mathcal{F}_{RD}$). The mean overall accuracy of prediction when using $\mathcal{F}_{RD}$, $\mathcal{F}_R$ and $\mathcal{F}_D$ are 85.57%, 84.24% and 81.73% respectively. From the overall accuracy plot shown in Fig. 1 we observe that combining the microstructure measures from MCT2 and MCDiff model yields better (E+) detection. The true positive rate (TPR) and true negative rate (TNR) plots in Fig. 1 show that MCT2 and MCDiff features are better than the other at detecting non-enhanced and enhanced voxels respectively in MS lesions. However, combining both features ($\mathcal{F}_{RD}$) yields better prediction results. Hence we use $\mathcal{F}_{RD}$ for performing predictions in experiment-2.

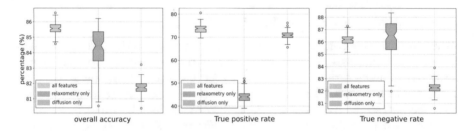

**Fig. 1.** (Left to right): overall accuracy, true positive rate and true negative rate of the predictions of the validation set over 100 iterations.

*Experiment-2.* Results are shown in Fig. 2. This MS patient had 18 MS lesions out of which 3 of them had Gd enhancing regions. The dice score for (E+) and (L−) voxel prediction was 0.64 and 0.86 respectively. The top row in Fig. 2 shows a lesion which had only (L−) voxels and the proposed method successfully predicted that there were no (E+) voxels in the lesion. The second and third row of Fig. 2 shows performance of the method in presence of Gd enhanced voxels in the lesion. Our method identified the (E+) voxels in the lesion. However there were false positives around the Gd enhanced core of the lesions where non-enhancing voxels were identified as belonging to the enhancing region of the lesion.

## 4   Discussion

Our analysis shows that combining tissue microstructure information from multi-compartment T2 relaxometry and diffusion MRI model helps at yielding better prediction accuracy as compared to each feature being used alone. The higher TPR of features derived from MCDiff might be attributed to the fact that during an active BBB breakdown, there is a greater presence of inter and extra cellular

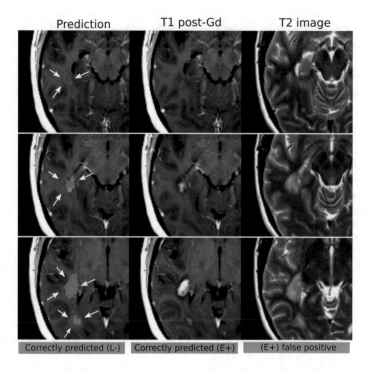

**Fig. 2.** The (E+) prediction results on a test image is shown here. Legend for the segmentation labels are shown below the illustrations.

fluid matters and inflammation as compared to non-enhancing parts of the lesion. The high-T2 water fraction from MCT2 model is capable of identifying higher inflammation. However, the medium-T2 water fraction, as described Sect. 2.1 is heterogeneous in terms of inter and extra-cellular fluids. However, MCDiff models are able to better explain such scenarios. The non-enhancing regions are demyelinated regions with inflammation which can be explained using MCT2 microstructure information.

Our study has certain limitations. The clinical data used in this work did not favour use of state of the art multi-compartment models. Single b-value data with 30 directions limited us to use a realtively simpler MCM model for MCDiff for reliable estimations. Higher number of echoes and shorter echo times for the T2 relaxometry data will facilitate MCT2 models. However, this study illustrates that even with clinical protocols, we have good detection of enhanced lesion regions using only MCT2 and MCDiff features.

## 5   Conclusion

We proposed a method to identify MS lesion voxels in which the brain tissues are undergoing active BBB breakdown using brain tissue microstructure information derived from advanced MRI techniques. The proposed method shows promise and motivation to work on improvement of the model from its current form. In the future work, we plan to have uncertainty measures on the predictions so that we can tackle the issue of false positive detection effectively. We also plan to test it on higher quality data to realize the true potential of the current framework.

# References

1. Guttmann, C., Ahn, S.S., Hsu, L., Kikinis, R., Jolesz, F.A.: The evolution of multiple sclerosis lesions on serial MR. AJNR **16**(7), 1481–1491 (1995)
2. Lassmann, H., Bruck, W., Lucchinetti, C.: Heterogeneity of multiple sclerosis pathogenesis: implications for diagnosis and therapy. Trends Mol. Med. **7**(3), 115–121 (2001)
3. Brownlee, W.J., Hardy, T.A., Fazekas, F., Miller, D.H.: Diagnosis of multiple sclerosis: progress and challenges. Lancet **389**(10076), 1336–1346 (2017)
4. Kanda, T., Ishii, K., Kawaguchi, H., Kitajima, K., Takenaka, D.: High signal intensity in the dentate nucleus and globus pallidus on unenhanced T1-weighted MR images: relationship with increasing cumulative dose of a gadolinium-based contrast material. Radiology **270**(3), 834–841 (2013)
5. Olchowy, C., et al.: The presence of the gadolinium-based contrast agent depositions in the brain and symptoms of gadolinium neurotoxicity-a systematic review. PLoS One **12**(2), e0171704 (2017)
6. Gulani, V., Calamante, F., Shellock, F.G., Kanal, E., Reeder, S.B., et al.: Gadolinium deposition in the brain: summary of evidence and recommendations. Lancet Neurol. **16**(7), 564–570 (2017)
7. Panagiotaki, E., et al.: Compartment models of the diffusion MR signal in brain white matter: a taxonomy and comparison. Neuroimage **59**(3), 2241–2254 (2012)
8. Laule, C., et al.: Magnetic resonance imaging of myelin. Neurotherapeutics **4**(3), 460–484 (2007)
9. MacKay, A.L., Laule, C.: Magnetic resonance of myelin water: an in vivo marker for myelin. Brain Plast. **2**(1), 71–91 (2016)
10. Lancaster, J.L., Andrews, T., Hardies, L.J., Dodd, S., Fox, P.T.: Three-pool model of white matter. J. Magn. Reson. Imaging **17**(1), 1–10 (2003)
11. Crawley, A., Henkelman, R.: Errors in T2 estimation using multislice multiple-echo imaging. MRM **4**(1), 34–47 (1987)
12. Prasloski, T., Mädler, B., Xiang, Q.S., MacKay, A., Jones, C.: Applications of stimulated echo correction to multicomponent T2 analysis. MRM **67**(6), 1803–1814 (2012)
13. Layton, K.J., Morelande, M., Wright, D., Farrell, P.M., Moran, B., Johnston, L.A.: Modelling and estimation of multicomponent T2 distributions. IEEE TMI **32**(8), 1423–1434 (2013)
14. Stamm, A., Commowick, O., Warfield, S.K., Vantini, S.: Comprehensive maximum likelihood estimation of diffusion compartment models towards reliable mapping of brain microstructure. In: Ourselin, S., Joskowicz, L., Sabuncu, M.R., Unal, G., Wells, W. (eds.) MICCAI 2016. LNCS, vol. 9902, pp. 622–630. Springer, Cham (2016). https://doi.org/10.1007/978-3-319-46726-9_72
15. Commowick, O., Wiest-Daesslé, N., Prima, S.: Block-matching strategies for rigid registration of multimodal medical images. In: 9th ISBI, pp. 700–703. IEEE (2012)

16. Ourselin, S., Roche, A., Prima, S., Ayache, N.: Block matching: a general framework to improve robustness of rigid registration of medical images. In: Delp, S.L., DiGoia, A.M., Jaramaz, B. (eds.) MICCAI 2000. LNCS, vol. 1935, pp. 557–566. Springer, Heidelberg (2000). https://doi.org/10.1007/978-3-540-40899-4_57
17. Cortes, C., Vapnik, V.: Support-vector networks. Mach. Learn. **20**(3), 273–297 (1995)

# A Bayes Hilbert Space for Compartment Model Computing in Diffusion MRI

Aymeric Stamm[1,2]([✉]), Olivier Commowick[3], Alessandra Menafoglio[4], and Simon K. Warfield[2]

[1] CADS, Human Technopole, Milan, MI, Italy
aymeric.stamm@htechnopole.it
[2] CRL, Boston Children's Hospital, Harvard Medical School, Boston, MA, USA
[3] Univ. Rennes 1, CNRS, Inria, Inserm, IRISA UMR 6074, VisAGeS U1228, Rennes, France
[4] MOX, Department of Mathematics, Politecnico di Milano, Milan, Italy

**Abstract.** The single diffusion tensor model for mapping the brain white matter microstructure has long been criticized as providing sensitive yet non-specific clinical biomarkers for neurodegenerative diseases because (i) voxels in diffusion images actually contain more than one homogeneous tissue population and (ii) diffusion in a single homogeneous tissue can be non-Gaussian. Analytic models for compartmental diffusion signals have thus naturally emerged but there is surprisingly little for processing such images (estimation, smoothing, registration, atlasing, statistical analysis). We propose to embed these signals into a Bayes Hilbert space that we properly define and motivate. This provides a unified framework for compartment diffusion image computing. Experiments show that (i) interpolation in Bayes space features improved robustness to noise compared to the widely used log-Euclidean space for tensors and (ii) it is possible to trace complex key pathways such as the pyramidal tract using basic deterministic tractography thanks to the combined use of Bayes interpolation and multi-compartment diffusion models.

## 1 Introduction

Diffusion MRI is a unique in-vivo and non-invasive imaging technique that probes the microstructure of tissues by tracking water diffusion [5]. The diffusion process is a probability distribution that describes the random 3-dimensional displacements of water molecules due to thermal agitation. The signal recovered in diffusion MRI is the Fourier transform of the diffusion density evaluated at the magnetic field spatial gradient to which the imaged brain was subjected.

A practical approach to access the diffusion density pertains to resorting to parametric models that lead to an analytic expression of the observed diffusion signals. The first parametric model was introduced two decades ago in [2] and is called *single tensor* (ST) model. The ST model essentially assumes that the diffusion in the voxel is well characterized by a zero-mean Gaussian distribution with

© Springer Nature Switzerland AG 2018
A. F. Frangi et al. (Eds.): MICCAI 2018, LNCS 11072, pp. 72–80, 2018.
https://doi.org/10.1007/978-3-030-00931-1_9

covariance matrix proportional to the so-called diffusion tensor. It has become widely used in clinical practice, which raised the demand for a solid mathematical framework for processing tensor images (estimation, smoothing, registration, atlasing, statistical analysis). As a response, several frameworks have been proposed, including a Riemannian affine-invariant framework [6,9] and the popular log-Euclidean space for tensors, embedding them in a Lie group structure [1].

However, because of the low spatial resolution in diffusion MRI, it has been shown that the ST model is an average of several diffusion processes arising from multiple populations of tissues in the voxel, ultimately leading to an inaccurate description of the microstructure in most parts of the brain white matter. In addition, even within a single homogeneous tissue population, non-Gaussian diffusion has been observed [13]. This has led to the extension of the ST model to mixture models, often called multi-compartment models (MCM), where the voxelwise diffusion signal is modeled as a linear combination of compartmental diffusion signals arising from underlying homogeneous diffusion processes [8]. To the best of our knowledge, the only mathematical framework available for multi-compartment image processing [12] is however limited to multi-tensor images (mixture of single-tensor signals, see [10]) as it relies on log-Euclidean geometry.

In this work, we present an alternative to the log-Euclidean space on tensors called the Bayes Hilbert space [3] for processing compartmental diffusion signals instead of tensors. It provides a unified framework that can accommodate analytic compartment diffusion signals of any type (Gaussian and non-Gaussian). Section 2 gives a general introduction to Bayes Hilbert spaces with motivation and setup in diffusion MRI. Section 3 describes a simulation study and a tractography application on real data to compare Euclidean, log-Euclidean and Bayes interpolation. Results commented in Sect. 4 show that Bayes interpolation features improved robustness to noise and yields better pyramidal tract reconstructions when combined with MCM-based deterministic tractography to account for streamline atlas priors when diffusion models cannot be locally trusted.

## 2 Theory

### 2.1 Bayes Hilbert Spaces and Compartmental Diffusion

The signal in diffusion MRI, hereafter diffusion signal, observed after the application of a diffusion gradient $\mathbf{q}$, where $\|\mathbf{q}\|^2 = b$ is the well known $b$-value, undergoes a signal decay starting from a baseline signal. The form of this decay depends on how water molecules diffuse in tissues in the vicinity of the spatial location where the signal is observed. In general, in a voxel composed of $K$ different tissue structures, the compound diffusion signal reads:

$$S(\mathbf{q}) = \sum_{j=1}^{K} S_j(\mathbf{q}) = \sum_{j=1}^{K} S_{0j} A_j(\mathbf{q}), \quad S_{0j} > 0, \quad A_j \in [0, 1], \tag{1}$$

in which all the information pertaining to the microstructure (set of surrounding tissue structures) is grasped by the signal attenuations $A_j$'s. This is the signal decomposition at the foundation of diffusion compartment imaging [8]. Hence, up to some multiplicative constant, the compartmental diffusion signal $S_j$ can be interpreted as the density of a probability measure $\nu_j$ on the space of diffusion gradients, which is absolutely continuous w.r.t. the Lebesgue measure $\lambda$ such that $\frac{d\nu_j}{d\lambda}$ (Radon-Nykodym derivative w.r.t. the Lebesgue measure $\lambda$) is proportional to $S_j$. Furthermore, this multiplicative constant is related to the $T_2$ relaxation time of tissue $j$ and thus not relevant for microstructure mapping.

To account for this, we propose to embed the probability measures induced by compartmental diffusion signals into the Bayes space $B^2(P)$ of (classes of equivalence of $\mathcal{B}(\mathbb{R}^3)$-finite) measures on the Borel space $(\Omega, \mathcal{B}(\Omega))$, with square-integrable log-densities w.r.t. a reference measure $P$ [3], which reads:

$$B^2(P) = \left\{ \nu : \int_{\mathbb{R}^3} \log^2 \left( \frac{d\nu}{dP} \right) dP < \infty, \nu > 0 \right\}. \tag{2}$$

In this view, if a diffusion signal $S_{j1}$ (or, equivalently a probability measure $\nu_1$ induced by $S_{j1}$) carries a piece of information about the structure of tissue $j$, then a diffusion signal $S_{j2}$ proportional to $S_{j1}$ does not carry additional information about the structure of tissue $j$. Hence, $S_{j1}$ and $S_{j2}$ are regarded as equivalent for microstructure mapping. This key property, known as *scale invariance*, is accounted for in the Bayes space $B^2(P)$ by the induced classes of equivalence. Not every Bayes space $B^2(P)$ however can accommodate compartmental diffusion signals because, depending on the choice of the reference measure $P$, the logarithm of compartmental diffusion signals might not be square-integrable. The choice of the reference measure of the Bayes space is thus critical. Provided that an appropriate reference measure $P$ exists, embedding diffusion signals in $B^2(P)$ provides a unified framework for processing any type of analytic microstructure compartment models [8] at the cost of numerical integrations.

The space $B^2(P)$ is a vector space when endowed with the perturbation and powering operations $(\oplus, \odot)$, defined for $\nu_1, \nu_2 \in B^2(P)$, $A \in \mathcal{B}(\mathbb{R}^3)$, $\alpha \in \mathbb{R}$ as:

$$(\nu_1 \oplus \nu_2)(A) =_{B(P)} \int_A \frac{d\nu_1}{dP} \frac{d\nu_2}{dP} dP, \quad (\alpha \odot \nu_1)(A) =_{B(P)} \int_A \left( \frac{d\nu_1}{dP} \right)^{\alpha} dP,$$

and becomes a separable Hilbert space when equipped with a proper inner product [4]. In the following subsection, we focus on the special case of Gaussian compartmental diffusion for which analytic expressions can be obtained for the perturbation and powering operations as well as for the distance.

## 2.2    Gaussian Compartmental Diffusion

The diffusion signal arising from Gaussian compartmental diffusion reads [2]:

$$S(\mathbf{q}) = S_0 e^{-\mathbf{q}^\top D \mathbf{q}}, \tag{3}$$

where $D$ is the so-called diffusion tensor, i.e. a $3 \times 3$ symmetric positive definite (SPD) matrix. Hence, the probability measure $\nu$ induced by $S$ is a Gaussian distribution with mean $\mathbf{0}$ and covariance matrix $D^{-1}/2$ and we can write, for the purpose of notation, $S \sim \mathcal{N}(\mathbf{0}, D^{-1}/2)$. As a result, Gaussian diffusion signals cannot be embedded in $B^2(\lambda)$ because choosing the Lebesgue measure as reference makes the integral of the squared logarithm of $S$ diverge. However since all moments of the Gaussian distribution are finite, if we define the reference measure $P$ as a Gaussian distribution with mean $\mathbf{0}$ and covariance matrix $D_{\mathrm{ref}}$ SPD, i.e. $P \sim \mathcal{N}(\mathbf{0}, D_{\mathrm{ref}})$, then Gaussian diffusion signals can be embedded in $B^2(P)$. Furthermore, it is possible to obtain analytic expression for the operations $\oplus$ and $\odot$. The perturbation of measure $\nu_1$ by measure $\nu_2$ reads:

$$(\nu_1 \oplus \nu_2)(A) =_B \int_A \frac{d\nu_1}{d\lambda} \frac{d\nu_2}{d\lambda} \frac{d\lambda}{dP} d\lambda =_B \int_A e^{-\mathbf{q}^\top (D_1 + D_2 - D_{\mathrm{ref}})\mathbf{q}} d\lambda(\mathbf{q}).$$

Hence, $(\nu_1 \oplus \nu_2) \sim \mathcal{N}(\mathbf{0}, D_{\mathrm{ref}} + (D_1 - D_{\mathrm{ref}}) + (D_2 - D_{\mathrm{ref}}))$. The result is another Gaussian distribution whose covariance matrix is centered around the one of the reference Gaussian distribution but perturbed by the covariance information in $D_1$ and $D_2$ that are not already present in the reference covariance structure. The multiplication of measure $\nu$ by a scalar $\alpha \in \mathbb{R}$ reads:

$$(\alpha \odot \nu)(A) =_B \int_A \left(\frac{d\nu}{d\lambda}\right)^\alpha \left(\frac{d\lambda}{dP}\right)^{\alpha - 1} d\lambda =_B \int_A e^{-\mathbf{q}^\top (\alpha D + (1-\alpha)D_{\mathrm{ref}})\mathbf{q}} d\lambda(\mathbf{q}).$$

Hence, $(\alpha \odot \nu) \sim \mathcal{N}(\mathbf{0}, \alpha D + (1 - \alpha)D_{\mathrm{ref}})$. The result is another Gaussian distribution whose covariance matrix is a linear combination between the covariances of $\nu$ and $P$. This offers the possibility of giving more or less weight to the information content of a diffusion signal in terms of microstructure. If one think of $\alpha \in [0, 1]$, then $(\alpha \odot \nu)$ means that less weight is put on the information content of $\nu$ (which could come from a poor model estimate, low SNR, etc.) in favor of the information content of the reference $P$. In practice, it shrinks the diffusion tensor of the signal towards the one of the reference.

Note that writing the results of the operations of $\oplus$ and $\odot$ as Gaussian distributions is an abuse of notation. In effect, the space of diffusion signals is a subspace of $B^2(P)$ that is not closed in the Bayes geometry. This is because operations on diffusion signals in $B^2(P)$ might yield non positive "covariance" matrices. However, most applications in which analytic diffusion model computing is required involve weighted average operations, i.e. linear combinations with positive weights whose sum is less than one, which always produce SDP tensors in $B^2(P)$. In effect, it is easy to show that:

$$\bigoplus_{i=1}^N w_i \odot \nu_i \sim \mathcal{N}\left(\mathbf{0}, \sum_{i=1}^N w_i D_i + \left(1 - \sum_{i=1}^N w_i\right) D_{\mathrm{ref}}\right). \tag{4}$$

In the rare events that operations in Bayes space produce non-positive matrices, one can perform an orthogonal projection of the non-positive matrix back into the space of diffusion signals by solving the following minimization problem:

$$\min_{D \in \mathcal{B}(P) \text{ s.t. } D > 0} a d_{\mathcal{B}(P)}^2(M, D) + (1 - a)d_{\mathcal{B}(P)}^2(D_{\mathrm{ref}}, D), \quad a \in [0, 1], \tag{5}$$

where $M$ is the symmetric matrix assumed to have some negative eigenvalues and $d_{\mathcal{B}(P)}$ is the distance on $B^2(P)$ given by:

$$\|\nu_1 \ominus \nu_2\|_{\mathcal{B}(P)}^2 := \text{Tr}\left((D_1 - D_2)D_{\text{ref}}^{-1}(D_1 - D_2)D_{\text{ref}}^{-1}\right) = \|(D_1 - D_2)D_{\text{ref}}^{-1}\|_F^2.$$

## 3    Experimental Setup

### 3.1    Simulations: Robustness to Noise

The goal of the simulated data is to assess the robustness of Bayes space interpolation to MRI-induced noise, compared to more traditional approaches that focus directly on the tensor and embed it either in Euclidean or log-Euclidean space. In Bayes space, as shown in Eq. (4), the tensor associated with the interpolated signal is a convex combination of the tensor interpolated in Euclidean space and the tensor of the reference signal, where the weight $w_{\text{data}}$ associated with the tensor interpolated in Euclidean space shall indicate how much we are willing to trust its information content. For this purpose, we set $w_{\text{data}} := 1 - e^{-\text{SNR}/\beta}$ and include in the comparison three Bayes spaces (Bayes10, Bayes20, Bayes30), using $\beta = 10, 20, 30$ respectively. We generate (i) a set of 31 noise-free diffusion signals according to Eq. (3) with a baseline signal $S_0 = 1$ and 31 diffusion gradients $\mathbf{q}$ uniformly distributed on the hemisphere of radius $\sqrt{b}$, with $b = 1000 \, \text{s/mm}^2$ and (ii) a set of $n = 8$ normalized weights to define spatial weights of the 8 neighbors. Next, for a given SNR and a given method, we obtain a Monte-Carlo estimate of the mean-square error (MSE) of the interpolated tensor by averaging the squared distances between the ground truth tensor and $R = 1000$ replicates of interpolated tensors from noisy neighboring tensors produced as follows:

(a) Add Rician noise to the noise-free signals using $\sigma = S_0/\text{SNR}$;
(b) Get a realistic noisy tensor field by estimating a diffusion tensor in each neighboring voxel via maximum likelihood estimation [11];
(c) Interpolate the tensor in the central voxel from the tensors in the 8 neighboring voxels to which we associate the initially simulated spatial weights;
(d) Compute distance between the interpolated and ground truth tensor.

We used 8 SNR $\in [6.25, 50]$ and three metrics to compare interpolations with ground truth: the angle between principal orientations (*direction* distance), the Euclidean distance between axial diffusivities (*diffusivity* distance) and the Euclidean distance between radial diffusivities (*radius* distance as it is often used as a proxy to measure axon radii). We chose these metrics as they focus on microstructural parameters of direct clinical relevance and are independent from the compared spaces. For interpolation in Bayes space, for each SNR, we set the reference tensor as a noisy version of the ground truth tensor using the same procedure as above using $\text{SNR}\sqrt{N}$. This is because, often, the available data to use as reference comes from atlases generated on similar data but averaged on multiple subjects ($N$). Hence, there is still uncertainty in the reference tensor but its variance is likely to be divided by $N$. In this simulation, we set $N = 20$.

## 3.2 Case Study: Tractography of the Pyramidal Tract

The pyramidal tract (PT) is of primary importance as it handles volontary motion. Neurons initiate in the primary motor cortex (R0), goes successively through the corona radiata (R1), genu and posterior limb of the internal capsule (R2) and cerebral peduncles (R3) to eventually enter the spinal cord (R4). The PT is difficult to reconstruct from tractography due to the large spreading of neurons on the cortex. We aim at showing that interpolation in Bayes space makes deterministic FACT tractography [7] feasible for PT reconstruction. In details, we compare four approaches obtained combining single- or multi-tensor FACT tractography (ST or MT) with log-Euclidean or Bayes interpolation (ST-LogEuclidean, ST-Bayes, MT-LogEuclidean and MT-Bayes).

We scanned a healthy subject on a 3T Siemens Verio magnet for a $T_1$ MPRAGE image at $1\,\mathrm{mm}^3$ resolution and diffusion data at $2\,\mathrm{mm}^3$ resolution using the same gradient table as in the simulations. We used an available PT atlas in MNI space based on 20 healthy subjects that underwent the same acquisition protocol. We estimated ST and MT models from the diffusion data [11] and brought the resulting images into MNI space where FACT tractography was performed. In the case of MT-FACT, at a given point, a single tensor was picked from each neighboring mixture according to the highest orientation similarity w.r.t. arrival direction so that we could proceed as in ST-FACT. We performed tractography of the PT following regions of interest (ROI). We used R0 as seeding mask, stopped a streamline generation when FA fell below 0.1 (for ST-FACT) or when linearly interpolated fraction of free water exceeded 0.8 (for MT-FACT) and filtered the resulting streamlines progressively through R1, R2, R3 and R4. We defined the reference weight at position $\mathbf{x}$ for Bayes interpolation as:

$$w_{\mathrm{ref}}(\mathbf{x}, \mathrm{FA}, \mathrm{SNR}, \mathbf{x}_{\mathrm{ref}}; \beta, \delta) := \left(1 - \mathrm{FA}(1 - e^{-\mathrm{SNR}/\beta})\right) e^{-\|\mathbf{x} - \mathbf{x}_{\mathrm{ref}}\|/\delta},$$

where FA is the fractional anisotropy of the tensor interpolated in Euclidean space, $\mathbf{x}_{\mathrm{ref}}$ is the position of the closest point on the PT atlas and $(\beta, \delta)$ are user-defined parameters that control the decay velocity of SNR and distance weights. In essence, we put more trust in the data when both SNR and orientation coherence between neighbors are high or when the reference is too far.

## 4 Results and Discussion

**Simulations: Robustness to Noise.** Figure 1 shows the MSE between the ground truth tensor in central voxel and the interpolated tensor from noisy neighbors, with increasing amount of noise. For all metrics, interpolation in Bayes space performs uniformly better in terms of robustness to MRI-induced noise for recovering microstructural parameters. The MSE curves for the three Bayes spaces that give increasing importance to the reference measure (from Bayes10 to Bayes30) reveals that even when we trust mostly the data (Bayes10), interpolation in Bayes space is preferable.

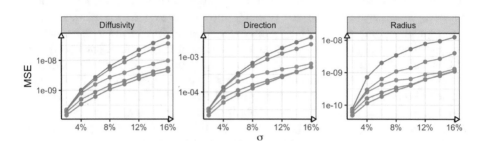

**Fig. 1.** MSE of interpolators as a function of noise. Axial diffusivity recovery (left), principal eigenvector recovery (middle), radial diffusivity recovery (right). Ground truth tensor: orientation $(\sqrt{2}/2, \sqrt{2}/2, 0)^\top$; diffusivities $10^{-3}(1.71, 0.3, 0.1)^\top$ mm$^2$/s.

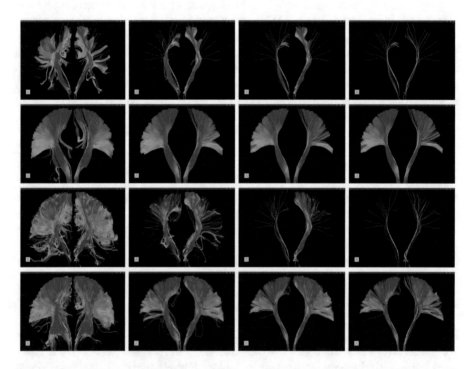

**Fig. 2.** Coronal View of Reconstructed Pyramidal Tracts (with overlaid PT atlas). Columns correspond to increasing number of filtering ROIs: R1 (1st column), R1 + R2 (2nd column), R1 + R2 + R3 (3rd column), R1 + R2 + R3 + R4 (4th column). Rows correspond to the methods (from top to bottom): single-tensor FACT – log-Euclidean, and Bayes, multi-tensor FACT – log-Euclidean, and Bayes.

**Case Study: Tractography of the Pyramidal Tract.** Figure 2 shows PT reconstruction with increasing number of filtering ROIs (from left to right) for all four methods. First, observe that only methods based on Bayes interpolation (rows 2 and 4) successfully manage to reconstruct streamlines that go through all four ROIs and thus are more likely to belong to the PT. Also, in general, deterministic FACT tractography fails to recover PT streamlines with the traditional log-Euclidean interpolation (rows 1 and 3). This is well documented in the literature on ST-FACT. We hypothesize that MT-FACT has a lot more directional information for tracking but, without prior anatomical knowledge, the FACT algorithm (which stepwise follows the most collinear direction) has no mechanism to channel streamlines into following the PT shape. This explanation is supported by the sequential filtering that shows high streamline variability when filtering only by the corona radiata (1st column, 3rd row) and almost no remaining streamlines at the end of the entire filtering process (4th column, 3rd row). The ST-Bayes method seems to mainly follow the shape of atlas streamlines. This is due to an inherent model mis-specification in most parts of the white matter where the ST model provides an insufficient description of the microstructure and therefore presents an articially low FA which uniformly inflates the weight of the reference measure. Conversely, the MT-Bayes version nicely preserves the PT streamlines after complete filtering without heavy influence of the atlas since the MT model provides an accurate description of the microstructure and thus the interpolated tensor has low FA only when neighboring tensors have heterogeneous orientations.

# References

1. Arsigny, V., Fillard, P., Pennec, X., Ayache, N.: Log-Euclidean metrics for fast and simple calculus on diffusion tensors. MRM **56**(2), 411–421 (2006)
2. Basser, P.J., Mattiello, J., LeBihan, D.: Estimation of the effective self-diffusion tensor from the NMR spin echo. J. Magn. Reson. B **103**(3), 247–254 (1994)
3. van den Boogaart, K.G., Egozcue, J.J., Pawlowsky-Glahn, V.: Bayes Hilbert spaces. Aust. N. Z. J. Stat. **56**(2), 171–194 (2014)
4. Egozcue, J.J., Pawlowsky-Glahn, V., Tolosana-Delgado, R., Ortego, M.I., van den Boogaart, K.G.: Bayes spaces: use of improper distributions and exponential families. Rev. Real Acad. Ciencias Exactas, Fis. Nat. Ser. A Mat. **107**(2), 475–486 (2013)
5. Le Bihan, D.: Looking into the functional architecture of the brain with diffusion MRI. Nat. Rev. Neurosci. **4**(6), 469–480 (2003)
6. Lenglet, C., et al.: Statistics on the manifold of multivariate normal distributions: theory and application to diffusion tensor MRI processing. JMIV **25**(3), 423–444 (2006)
7. Mori, S., Crain, B., et al.: Three-dimensional tracking of axonal projections in the brain by magnetic resonance imaging. Ann. Neurol. **45**(2), 265–269 (1999)
8. Panagiotaki, E.: Compartment models of the diffusion MR signal in brain white matter: a taxonomy and comparison. Neuroimage **59**(3), 2241–2254 (2012)
9. Pennec, X., Fillard, P., Ayache, N.: A riemannian framework for tensor computing. Int. J. Comput. Vis. **66**(1), 41–66 (2006)

10. Scherrer, B., Warfield, S.K.: Parametric representation of multiple white matter fascicles from cube and sphere diffusion MRI. PLoS One **7**(11), e48232 (2012)
11. Stamm, A., Commowick, O., Warfield, S.K., Vantini, S.: Comprehensive maximum likelihood estimation of diffusion compartment models towards reliable mapping of brain microstructure. In: Ourselin, S., Joskowicz, L., Sabuncu, M.R., Unal, G., Wells, W. (eds.) MICCAI 2016. LNCS, vol. 9902, pp. 622–630. Springer, Cham (2016). https://doi.org/10.1007/978-3-319-46726-9_72
12. Taquet, M., Scherrer, B., et al.: A mathematical framework for the registration and analysis of multi-fascicle models for population studies of the brain microstructure. IEEE TMI **33**(2), 504–517 (2014)
13. Wang, B., et al.: When Brownian diffusion is not Gaussian. Nat. Mater. **11**(6), 481–485 (2012)

# Detection and Delineation of Acute Cerebral Infarct on DWI Using Weakly Supervised Machine Learning

Stefano Pedemonte[1,2,3](✉), Bernardo Bizzo[1,2,3], Stuart Pomerantz[1,2,3],
Neil Tenenholtz[1], Bradley Wright[1], Mark Walters[1], Sean Doyle[1],
Adam McCarthy[1,4], Renata Rocha De Almeida[1,2,3], Katherine Andriole[1,2,3],
Mark Michalski[1,2,3], and R. Gilberto Gonzalez[2,3]

[1] Center for Clinical Data Science, Boston, MA, USA
stefano.pedemonte@gmail.com
[2] Harvard Medical School, Boston, MA, USA
[3] Massachusetts General Hospital, Boston, MA, USA
[4] Department of Computer Science, University of Oxford, Oxford, UK
https://ccds.io

**Abstract.** Improved outcome in patients with ischemic stroke is achieved through acute diagnosis and early restoration of cerebral flow in appropriate patients. Diffusion-weighted MR imaging (DWI) plays a central role in these efforts by enabling rapid early localization and quantification of ischemic lesions. Automated detection and quantification can potentially accelerate diagnosis, improve treatment safety and efficacy and reduce costs. However, the manual quantification of acute ischemic stroke volumes for algorithm training is time consuming and imprecise. We present YNet as a novel fully-automated deep learning algorithm for detection and volumetric segmentation and quantification of acute cerebral ischemic lesions from DWI. The algorithm is a semi-supervised multi-tasking deep neural network architecture we developed that enables the combination of both weak labels derived from radiology report classification and manually delineated pixel level training data. The model is trained on a very large dataset of 10000 studies, achieves detection sensitivity 0.981, detection specificity 0.980 and segmentation Dice score 0.623 on a heterogeneous test set.

**Keywords:** Weakly supervised deep learning · Stroke detection
Segmentation · Diffusion-weighted MRI

## 1 Introduction

Acute ischemic stroke remains a leading cause of death and disability. However, a number of recent multicenter trials have demonstrated significantly improved neurological outcomes when the latest generation of thrombectomy devices are used to restore cerebral blood flow within 6 h of symptom onset [1,5]. Extension

© Springer Nature Switzerland AG 2018
A. F. Frangi et al. (Eds.): MICCAI 2018, LNCS 11072, pp. 81–88, 2018.
https://doi.org/10.1007/978-3-030-00931-1_10

of the time window for successful treatment out to 24 h has been more recently demonstrated [6]. In such trials, proper patient selection with various neuroimaging modalities, including CT, catheter angiography, and MRI, has been essential for treatment success. With MRI, diffusion-weighted imaging (DWI) sequences provide sensitive detection and localization of even the tiniest ischemic brain lesions within minutes of vessel occlusion. From DWI images, the total amount of ischemic tissue can also be measured. While critical for safe patient selection and for prediction of long term disability, such volume measurements are challenging. Automated methods for DWI lesion detection and measurement have great potential for improving stroke treatment by reducing the time necessary to administer therapy to appropriately selected patients. Though several semi-automated approaches have been proposed in the literature, including adaptive thresholding [8], watershed [12] and fuzzy clustering [13], fully automated solutions are preferable to reduce time-consuming and error-prone steps in the clinical workflow. However, a key bottleneck in building fully-automated solutions is the segmentation models require voxel-level image annotation of each image which is expensive and time-consuming. For example, Chen *et al.* introduced a fully-automated deep learning segmentation algorithm for DWI ischemic lesions based on 2D convolutional neural networks (CNN) which achieved good results of dice score of 0.67 and sensitivity of 0.94, but required training on 741 manually segmented DWI images. Binary classification of diagnostic radiology reports for brain MRI studies as DWI-positive or DWI-negative can be used as a source of weak image-level labels and are far easier to generate than detailed pixel-level annotations. In this work, we introduce a semi-supervised multi-tasking deep learning CNN model trained on explicit classification and segmentation tasks. The model learns to segment and to classify from a heterogeneous set of annotated and weakly labeled images.

## 2    Data

Data for this project was collected from Massachusetts General Hospital and Brigham Women Hospital. All data were collected retrospectively. Ethical approval was granted by the institutional review board. Data included MR studies and the official diagnostic reports.

### 2.1    Weak Labels

From a database containing all radiology reports over a ten year period (2007–2017), parsing of the Impression section text of Brain MRI study reports was performed to create a balanced set of 5000 negative studies and 5000 positive studies for acute ischemia. Parsing methods included keyword and sentence matching and creation of a simple text classifier based on N-grams and Support Vector Machines (SVM). Initial classification results were manually validated and corrected by a trained radiologist (B.B). Studies with keyword matches for terms indicating the presence of post-surgical findings or brain tumor pathology were excluded to reduce the effect of confounding image patterns from non-ischemic disease. The resultant scans were obtained from a variety of scanner manufacturers, magnet strengths, acquisition parameters and patient demographics. Patient age was limited to 20–80 years.

## 2.2 Manual Segmentation

Out of the 5000 positively labeled studies, 500 were manually segmented by two trained radiologists (B.B., R.R.) using the software Osirix version 9.0, by drawing contours of acute ischemic lesions on DWI and ADC series slices. In total, 1423 slices were manually segmented. The average manual segmentation time per volume was 10 min.

## 2.3 Imaging Data

The majority of ischemic lesion segmentation algorithms in the literature are based on DWI [2,12,13]. In this study, in order to mitigate shine-through effect from non-acute lesions, we incorporated both DWI and derived Apparent Diffusion Coefficient (ADC), following clinical practice. DWI and ADC images were acquired using 2D multi-slice MR acquisition modalities with variation in acquisition parameters (number of pixels, number of slices, pixels size, slice thickness, repetition time, echo time) related to different manufacturer platforms. Images were re-sampled in three dimensions on a grid of $256 \times 256 \times 32$ with resolution $1.2\,mm \times 1.2\,mm \times 9\,mm$. To compensate for the large variation in image intensity across studies, a normalization algorithm was applied independently to DWI and ADC image intensities. The algorithm detects the peak corresponding to white matter in the image histogram and scales the image to assign an average pixel value of 1.0 in the white matter.

# 3 Model

We developed YNet, an extension of the UNet architecture [3,11], successfully employed by other authors for semantic segmentation of MR images [9], for our semi-supervised learning. The model operates on multichannel (DWI and ADC) image data at full resolution ($256 \times 256 \times 32 \times 2$).

## 3.1 Semi-Supervised Learning

Figure 1 shows the architecture of the proposed YNet. The network is composed of three branches: encoding, decoding and classification. The encoding branch is composed of a cascade of convolutional blocks, each comprising two convolutional layers followed by max-pooling, batch normalization and activation. The decoding branch is composed of deconvolutional blocks, each comprising an upsampling layer followed by two convolutional layers, batch normalization and activation. The classification branch, connected to the output of the encoder, is

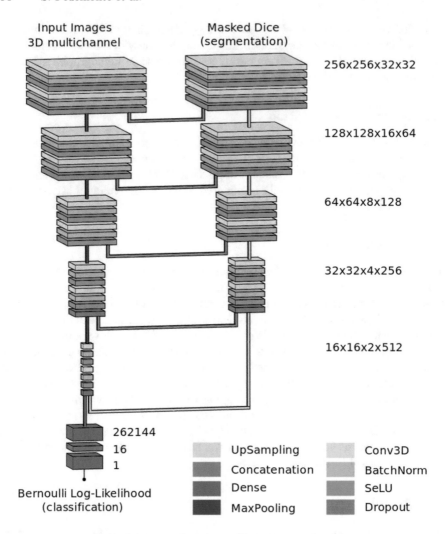

**Fig. 1.** YNet semi-supervised learning architecture.

composed of two fully connected layers. The model loss is the sum of two terms: the negative Bernoulli log-likelihood associated to the output of the classification branch and the negative Dice score which measures the similarity of the output of the decoder branch to the manual segmentation. The Dice score previously used in [2,9] as segmentation score, was modified as follows:

$$Dice(X,Y) = \frac{2\sum_i x_i y_i m + \epsilon}{\sum_i (x_i + y_i)m + \epsilon}, \tag{1}$$

where $x_i$ is the output of the decoder at voxel $i$; $y_i$ is the manual segmentation at voxel $i$; $\epsilon$ is a small number (set to 1.0 in experiments) for numerical stability; and $m \in \{0, 1\}$ is a binary mask variable that enables end-to-end training with mixed image-level and pixel-level annotations. The variable is set to 1 for positive manually segmented and negative cases, to 0 for positive non-segmented cases. This has the effect of disabling gradient propagation in the decoder when the pixels-level labeling is missing. From the perspective of image segmentation, the image-level labels provide semi-supervision. From the perspective of classification, the pixel-level labels act as a supervised focusing mechanism. A similar idea is applied in [10] to the semantic annotation of natural images. We implemented the model in Keras version 2.0 with TensorFlow back-end version 1.3.

## 4    Experiments and Results

The YNet model was compared with two baseline models: a convolutional neural network (CNN) for classification and a 3D UNet for segmentation. The CNN model for classification consisted of $N$ convolutional layers. $L_2$ regularization, dropout, and learning rates were explored through a distributed random search for 20 epochs. Configurations that performed well were then trained until convergence. After exploring a variety of architectures and hyperparameter configurations, the most performant model was selected based on the validation set accuracy. It consisted of six convolutional layers beginning with 36 filters and growing by 12 additional filters per layer. Test set sensitivity was observed to be 0.979 and specificity 0.958.

Experiments were performed using a 2D UNet to optimize the segmentation architecture. In order to select the optimal size of the receptive field, a 2D UNet with four encoding blocks and three decoding blocks was trained at three levels of resolution on all manually segmented slices: $64 \times 64 \times 32$, $128 \times 128 \times 32$ and $256 \times 256 \times 32$. The networks produced Dice coefficients of 0.527, 0.532, 0.561 respectively. The largest receptive field was selected and used in subsequent experiments.

The effect of various pre-processing techniques on segmentation performance was evaluated. Magnetic field inhomogeneity was corrected using the N4 Bias field correction software [14] and all images were rigidly registered to a reference DWI-ADC image pair. Bias field correction was found to degrade the segmentation performance. Image registration was found to slightly accelerate convergence but did not produce any significant improvement in segmentation performance. Bias field correction and image registration were therefore not used in subsequent experiments. In experiments with 2D segmentation, perturbation of the image intensity using random scale and offset was found to improve performance on the validation set. In subsequent experiments, the dataset was augmented 5-folds by multiplying the image intensity by $\mathcal{N}(1, 0.2)$ and by adding $\mathcal{N}(0, 0.2)$.

The final 3D Unet had a receptive field of $256 \times 256 \times 32$ pixels with two input channels (DWI and ADC), five encoder blocks and four decoder blocks. Convolutional layers used $3 \times 3 \times 3$ kernels and MaxPooling layers had pool size of $2 \times 2 \times 2$. Dropout of 0.2 was used in the bottom layers of the encoding and decoding branches for regularization.

The YNet model architecture was identical to the 3D UNet architecture, with the addition of two dense layers for classification. The dense layers had size 262144 × 16 and 16 × 1 respectively. A dropout layer with dropout rate 0.5 was placed between the two dense layers for regularization.

In experiments, it was found that the performance of both the 3D UNet and the 3D YNet improved when pre-training the models in 2D. A 2D model with the same architecture and parameters as the 3D UNet was first trained on the set of all manually segmented 2D slices. The parameters were then transfered to the 3D UNet and YNet models, as depicted in Fig. 2, by padding the 2D kernels with random numbers generated according to the xavier initialization method [4]. During 2D pre-training, batch size was set to 16 and during 3D training it was reduced to 1 due to memory constraints. In order to compensate for the ineffectiveness of the batch normalization modules with batch size 1, we used SeLu activations, recently introduced in [7]. These were found to very effectively accelerate convergence. The models were optimized using standard stochastic gradient descent with learning rate fixed at 0.001 and momentum set to 0.9. The data set was split in 80% training, 10% validation, 10% testing. The models were trained in turn on an NVidia DGX-1 with 8X P100 GPUs (16 Gb RAM per GPU), with whole model 8× replication (except for the CNN, which was trained on 4 GPUs). The models were pre-trained for 20 epochs and trained for 50 epochs. Training time was 4.6 h for the UNet and 86 h for the YNet. Quantitative results are reported in Table 1. Examples of image segmentations produced by the UNet and the YNet are reported in Fig. 3. The YNet determines an increase of Dice score from 0.581 to 0.623 on the test set, and an improvement in quality of the segmentations. The improvement is visible in particular in small lesions and in the reduction of artifacts. The UNet outperformed the CNN in classification, achieving sensitivity 0.981 and specificity 0.980.

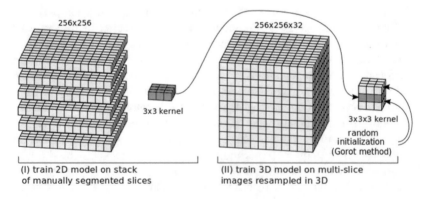

**Fig. 2.** 2D pre-training

**Table 1.** Sensitivity and specificity acute ischemic stroke detection. Dice score acute stroke core segmentation.

|  | CNN | UNet | YNet |
|---|---|---|---|
| Sensitivity | 0.979 | / | 0.981 |
| Specificity | 0.958 | / | 0.980 |
| Dice | / | 0.581 | 0.623 |

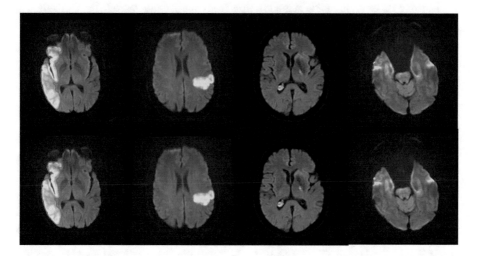

**Fig. 3.** UNet and YNet comparison: the images represent DWI with the overlay of the automated segmentations produced by the neural network. Top: 3D UNet. Bottom: 3D YNet.

## 5    Discussion and Conclusion

In this paper, we have introduced a deep learning architecture to automatically detect and segment acute ischemic stroke lesions on diffusion-weighted MRI to address the challenges of stroke diagnosis and optimal selection of patients for acute stroke therapy. The model is validated on a very large clinical dataset and achieves state-of-the-art performance. To the best of our knowledge, our approach to classify and segment a medical image from both pixel level annotations and weak image level labels and using 3D contextual information is novel. By using end-to-end multi-task learning, the model achieves high classification and segmentation performance, learning to segment from weak labels and to classify from manual annotations. Given that many problems in image interpretation require detection and localization, we believe that the approach explored in this paper can have wide applicability in medical image interpretation, enabling machine learning at large scale with minimum annotation effort. In the future, this method could be extended to account for multiple segmentation and classification labels.

# References

1. Berkhemer, O., et al.: A randomized trial of intraarterial treatment for acute ischemic stroke. N. Engl. J. Med. **372**(1), 11–20 (2015)
2. Chen, L., Bentley, P., Rueckert, D.: Fully automated acute ischemic lesion segmentation in DWI using convolutional neural networks. Neuroimage **15**, 633–643 (2017)
3. Çiçek, Ö., Abdulkadir, A., Lienkamp, S.S., Brox, T., Ronneberger, O.: 3D U-Net: learning dense volumetric segmentation from sparse annotation. In: Ourselin, S., Joskowicz, L., Sabuncu, M.R., Unal, G., Wells, W. (eds.) MICCAI 2016. LNCS, vol. 9901, pp. 424–432. Springer, Cham (2016). https://doi.org/10.1007/978-3-319-46723-8_49
4. Glorot, X., Bengio, Y.: Understanding the difficulty of training deep feedforward neural networks. In: PMLR, vol. 9, pp. 249–256 (2010)
5. Goyal, M., Menon, B., van Zwam, W.: Endovascular thrombectomy after large-vessel ischaemic stroke: a meta-analysis of individual patient data from five randomised trials. Lancet **387**, 1723–1731 (2016)
6. Hacke, W.: A new DAWN for imaging-based selection in the treatment of acute stroke. N. Engl. J. Med. **378**, 81–83 (2018)
7. Klambauer, G., Unterthiner, T., Mayr, A.: Self-normalizing neural networks. In: NIPS, pp. 971–980 (2017)
8. Martel, A.L., Allder, S.J., Delay, G.S., Morgan, P.S., Moody, A.R.: Measurement of infarct volume in stroke patients using adaptive segmentation of diffusion weighted MR images. In: Taylor, C., Colchester, A. (eds.) MICCAI 1999. LNCS, vol. 1679, pp. 22–31. Springer, Heidelberg (1999). https://doi.org/10.1007/10704282_3
9. Milletari, F., Navab, N., Ahmadi, S.: V-Net: fully convolutional neural networks for volumetric medical image segmentation. In: 3DV, pp. 565–571 (2016)
10. Papandreou, G., Chen, L., Murphy, K., Yuille, A.: Weakly- and semi-supervised learning of a DCNN for semantic image segmentation. CoRR 1502.02734 (2015)
11. Ronneberger, O., Fischer, P., Brox, T.: U-Net: convolutional networks for biomedical image segmentation. In: Navab, N., Hornegger, J., Wells, W.M., Frangi, A.F. (eds.) MICCAI 2015. LNCS, vol. 9351, pp. 234–241. Springer, Cham (2015). https://doi.org/10.1007/978-3-319-24574-4_28
12. Subudhi, A., Jena, S., Sabut, S.: Delineation of the ischemic stroke lesion based on watershed and relative fuzzy connectedness in brain MRI. Med. Biol. Eng. Comput. **56**, 1–13 (2017)
13. Tsai, J., et al.: Automated detection and quantification of acute cerebral infarct by fuzzy clustering and histographic characterization on diffusion weighted MR imaging and apparent diffusion coefficient. BioMed. Res. Int. (2014). Article no. 963032
14. Tustison, N.: N4ITK: improved N3 bias correction. IEEE Trans. Med. Imaging **29**(6), 1310–1320 (2010)

# Identification of Species-Preserved Cortical Landmarks

Tuo Zhang[1](✉), Xiao Li[1], Lin Zhao[1], Ying Huang[1], Lei Guo[1], and Tianming Liu[2]

[1] Brain Decoding Research Center, Northwestern Polytechnical University, Xi'an, Shaanxi, China
tuozhang@nwpu.edu.cn
[2] Computer Science Department, The University of Georgia, Athens, GA, USA

**Abstract.** Primate brain evolution has been an intriguing research topic for centuries. Previous comparative studies focused on identification of species-preserved cortical landmarks or axonal pathways via approaches such as registration. However, because of huge cross-species variations, these studies dealt with only a few specific fasciculi and cortices or relied on a predefined brain parcellation shared among species. In this work, we used T1-weighted MRI data and diffusion MRI data to identify novel landmarks on entire cortices based on folding patterns on macaque and human brains and further proposed a pipeline to establish cross-species correspondence for them based on networks derived from streamline fibers. Our experimental results are consistent with the reports in the literature, demonstrating the effectiveness and promise of this framework. The merits of this work lie in not only the identification of a novel, large group of species-preserved cortical landmarks, but also new insights into the relationship between cortical folding patterns and axonal wiring diagrams along the evolution line.

**Keywords:** 3-hinge gyri · Structural connections · Comparative study

## 1 Introduction

Comparative neuroscience and neurology of primate brains have been of an intriguing research topic for decades in that they help to uncover the mechanism mediating the development and evolution of brains [1, 2]. They also provide insight into the nature of both inter-species commonalities for brain regions of basic function and the divergence of brain organization in charge of higher function. Such knowledge provides valuable clues to build animal models to many human neuroscience studies. To this end, effective approaches are needed to facilitate neuroanatomy comparison cross species [3]. However, because the morphology of both the anatomy and axonal pathway are hugely different across species, conventional image registration methods could not apply to cross-species alignment [3, 4]. Therefore, the comparative studies in the literatures either were limited by studying only a few cortical regions and white matter fasciculi, such as those related to vision [4] and language [1], or connectome that is based on brain parcellation scheme shared by species, such as Brodmann areas [5]. In

© Springer Nature Switzerland AG 2018
A. F. Frangi et al. (Eds.): MICCAI 2018, LNCS 11072, pp. 89–97, 2018.
https://doi.org/10.1007/978-3-030-00931-1_11

this respect, identification of a large group of species-preserved anatomical or connective landmarks could be an alternative approach to promoting cross-species comparative studies to a higher level.

Recently, a novel cortical convolution pattern, termed gyral hinge (white dots in Fig. 1(b)), the conjunction of several gyri, was found which possesses unique anatomical and connective profiles [6] in contrast to other cortical patterns. More importantly, correspondences of such gyral hinges were successfully found across human brains with huge variations and even across species. Moreover, corresponding gyral hinges were found to possess consistent white matter fiber morphologies across subjects and species [7]. Along this line, in this paper, we aim to investigate whether these gyral hinges can be used as landmarks for cross-species brain alignment. Using T1-weighted MRI and diffusion MRI (dMRI), we firstly adopted the method in [8] to identify gyral hinges of 3 spokes (3-hinge for short) on the entire cortices of human and macaque individuals. They were then used as nodes to infer dMRI derived individual connective matrix. Finally, a framework was proposed to establish correspondences of these 3-hinges across individuals and species by group-wisely matching the connective matrices. The results demonstrate that the identified corresponding 3-hinges possess consistent connective patterns across subjects and species, and the consistency outperforms those results obtained by conventional image registration methods. The cross-species correspondences were further validated by previous reports. Taken together, it proves promising to use them as landmarks for cross-species alignment, and also suggesting a new relationship between cortical folding and structural connection in species comparative studies.

## 2   Materials and Methods

### 2.1   Overview

Figure 1 illustrates an overview of the analysis framework. In general, the aim of the work is to identify species-preserved 3-hinges in terms of their structural connectivities. The framework was divided to two parts. In the first part (shallow shade in Fig. 1), data processing and analysis were performed on individual space. White matter surfaces were reconstructed from T1-weighted MRI, on which 3-hinges are detected. DMRI derived deterministic streamline fibers were extracted if they connect any two of the 3-hinges, based on which individual connective matrix was constructed. In the second part (dark shade in Fig. 1), 3-hinges were identified which are consistent across subjects within species in terms of their connective patterns. Based on such group-wisely consistent 3-hinges and their connective matrices, the 3-hinges preserved across species were identified if the connective patterns of them are similar between the species.

### 2.2   Data Set Description and Pre-processing

**Human Brain Imaging:** The T1-weighted MRI and dMRI from the Q1 release of WU-Minn Human Connectome Project (HCP) consortium (http://www.humanconnectome.org/) were used in this study. Important imaging parameters are as follows: T1-weighted

MRI: voxel resolution $0.7 \times 0.7 \times 0.7$ mm, TR = 2400 ms, TE = 2.14 ms, flip angle = 8°, image matrix = $260 \times 311 \times 260$. DMRI data: spin-echo EPI sequence; TR = 5520 ms; TE = 89.5 ms; flip angle = 78°; refocusing flip angle = 160°; FOV = $210 \times 180$; matrix = $168 \times 144$; spatial resolution = 1.25 mm× 1.25 mm× 1.25 mm; echo spacing = 0.78 ms. Each gradient table of the dMRI data includes approximately 90 diffusion weighting directions plus 6 $b = 0$ acquisitions. DMRI data consists of 3 shells of $b = 1000, 2000,$ and 3000 s/mm$^2$ with an approximately equal number of acquisitions on each shell within each run. Eighteen randomly selected subjects were used in this study.

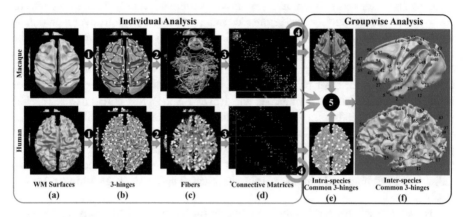

**Fig. 1.** The framework of data processing and analysis. (a)–(d): the data processing steps 1–3 were in individual space and within species. (a) White matter cortical surfaces; (b) white dots represent the locations of 3-hinges and black curves represent gyral crest lines. Surfaces are color-coded by curvatures. Red regions have positive curvatures while blue ones have negative values; (c) streamline fibers that connect 3-hinges; (d) the structural connective matrices. Step 1: 3-hinge identification; step 2: extracting streamline fibers connecting the 3-hinges; step 3: reconstructing individual structural connective matrices based on (c). (e)–(f) intra-species and inter-species group-wise analyses. (e) Identified 3-hinges consistent across subjects within species *via* step 4; (f) Identified 3-hinges that have cross species correspondence *via* step 5.

**Macaque Brain Imaging:** UNC-Wisconsin neurodevelopment rhesus MRI database (T1 weighted MRI and dMRI data) was used in this work (http://www.nitrc.org/projects/uncuw_macdevmri/). 18 different subjects at the age of more than 12 months old were used. The resolution of T1-weighted MRI is $0.27 \times 0.27 \times 0.27$ mm$^3$ and a matrix of $300 \times 350 \times 250$. The basic parameters for diffusion data acquisition were: resolution of $0.65 \times 0.65 \times 1.3$ mm$^3$, a matrix of $256 \times 256 \times 58$, diffusion-weighting gradients applied in 120 directions and $b$ value of 1000 s/mm$^2$, as well as ten images with $b = 0$ s/mm$^2$.

**Preprocessing:** Human data and macaque data share similar pipelines. T1-weighted MRI data was nonlinearly warped to FA map of dMRI data *via* FSL-fnirt [9]. White matter cortical surfaces were reconstructed from the registered T1-weighted MRI *via*

Freesurfer (https://surfer.nmr.mgh.harvard.edu/). For dMRI data, skull-strip and eddy currents were applied firstly, and then the model-free generalized Q-sampling imaging (GQI) method in DSI Studio [10] was used to estimate the density of diffusing water at different orientations. Deterministic fiber tracking algorithm [11] in DSI Studio was used to reconstruct $4 \times 10^4$ fiber tracts for each subject using the default parameters.

## 2.3    Identification of 3-hinges and Reconstruction of Connective Matrix

We adopted the method in [8] to automatically identify the locations of 3-hinges on the entire white matter cortical surfaces ((a), (b) and step 1 in Fig. 1). In brief, the pipeline includes three steps: (1) Given the gyral altitudes (defined in [12]) for each vertex, the white matter cortical surface was segmented into gyral crests (regions over a predefined altitude level) and sulcal basins (regions below the level) *via* the watershed algorithm [13]; (2) A tree marching algorithm was used to construct a tree structure on gyral crests. The roots of the tree are located at the gyral crest centers that have the maximum distance values from crest borders, and all the other gyral crest vertices were recursively connected to them as branches and leaves till the crest borders were reached, following the descending gradients of the distance to the border; (3) We pruned the redundant branches if the path length is shorter than the predefined length threshold to preserve the main trunks as a gyral crest line network (black curves in Fig. 1(b)). The conjunction of the network, the connecting degrees of which are more than 3, are defined as gyral hinges. 3-hinges are of major interest in this work, as 4-hinges are very rarely seen.

To reconstruct structural connective matrix for 3-hinges, dMRI derived streamline fibers were extracted if they pass the neighborhoods (spheres centered at the locations of 3-hinges) of any two of the 3-hinges. The radius of the spheres is 2 mm for macaque and 4 mm for human with regard to their scales. Based on the extracted fibers, a connective matrix was constructed (Fig. 1(c) and (d), step 3 and 4), the nodes of which are the 3-hinges and the connective strength is the number of fibers connecting two 3-hinges.

## 2.4    Identification of Consistent 3-hinge Within Species

In Sect. 2.3, 3-hinge connective matrix was reconstructed on each individual. The numbers and locations of 3-hinges vary on different subjects and have no correspondence. Therefore, we established the correspondence of them within each species, by making the connective matrices consistent across subjects ((d), (e) and step 4 in Fig. 1).

Firstly, surfaces of all subjects were warped to the same spherical space *via* the spherical registration method in Freesurfer, to provide an anatomical constraint for the algorithm that will be detailed later. Then, we randomly selected a subject as a template which has $n$ 3-hinges, and its 3-hinge matrix is denoted by $N$. A global search was performed on the other subjects as testing subjects, respectively. For each of the template's 3-hinge, the 3-hinges on one testing subject (with $m$ 3-hinges and its 3-hinge matrix is denoted by $M$) are in its search scope, or anatomical constraint, when they are within the geodesic distance of $r$ *mm*. Altogether, the testing subject has $n$ groups of candidate 3-hinges. Next, one candidate 3-hinge was selected from each search scope, respectively. No overlap was allowed for the $n$ candidates, and a connective matrix was constructed from them. This is equivalent to extracting an $n \times n$ matrix $N'$ from the original $m \times m$ $M$ matrix. The matrix similarity was defined as $\|N - N'\|_2$. Because of the diagonal symmetry of $N$ and $N'$, only the upper triangular part of them were used. Based on these steps, a similarity value was obtained for each selection of $n$ candidates. The combinations of $n$ candidates with the smallest similarity value gives the identified $n$ 3-hinges out of $m$ original ones on the testing subject that correspond to the $n$ 3-hinges on the template. This algorithm was repeated on all testing subjects, and the final similarity values of all testing subjects were summed up as the score $s$ for this template. It is noted that as $N'$ and $N$ should be of the same size, their dimension was determined by the subject who has the smallest 3-hinge number, $n_s$. Therefore, when this subject was selected as template, the above steps were performed only once. If another subject was selected as template, $n_s$ out of $n$ 3-hinges were repeatedly selected to give the template $n_s \times n_s$ matrix. The smallest similarity value between the selection on the template and the selection on the testing subject was considered as the optimal result. Finally, the template with the smallest score $s$ together with the identified corresponding $n_s$ 3-hinges on all other subjects were obtained. As the computation load was largely determined by repeatedly selecting a $n_s \times n_s$ matrix on either a $n \times n$ template matrix or a $m \times m$ testing subject matrix (310 ± 5 3-hinges for human and 130 ± 2 for macaque), the simulated annealing algorithm was adopted to reduce the computation load [14].

## 2.5    Identification of Species-Preserved 3-hinge

In Sect. 2.4, we obtained $n_s^m$ corresponding 3-hinges for all macaques and $n_s^h$ 3-hinges for all humans. We established the correspondence between the two groups of 3-hinges *via* the similar global search method. Because the template for each species was identified and all subjects within each species are in the same space, warping the human template to the macaque template ensures all subjects are roughly in the same space. Macaque was used as the cross-species template because $n_s^m < n_s^h$. The search scope for macaque 3-hinges on human brains was computed in a group-wise manner. The corresponding $i^{th}$ 3-hinges from all the macaque subjects were used to determine their centroid in the common space, and the geodesic radius $r$ was used with the centroid to give the search scope. For each human subject, the 3-hinges fall in this search scope were selected as candidates, such that each human individual had a candidate set. Only those in the intersection of all individual candidate sets were considered as the effective candidates for the $i^{th}$ 3-hinge on macaque, because the locations of corresponding 3-hinges on different human subjects are varied even in the common space. To conduct the global search, a combination of $n_s^m$ candidates were selected to give 18 individual connective matrices for human. Pairwise matrix similarities between these 18 human matrices and 18 macaque matrices were computed and the mean similarity value was defined as the group-wise cross-species similarity for this candidate combination. Likewise, simulated annealing algorithm was adopted to identify the optimal $n_s^m$ candidates on human with the minimal similarity value.

# 3 Results

## 3.1 Identified Corresponding 3-hinges Within Species

The radius of search scope $r$ is 2 mm for macaque intra-species search scope and 4 mm for human due to their size scales. 128 and 306 3-hinges with correspondence were respectively identified for macaque and human. Their locations, as well as the streamlines fibers connecting them and the connective matrices on four example subjects for each species, are shown in Fig. 2(a)–(c). The inter-individual consistency can be observed with regard to the similar fiber bundle morphologies and connective matrices.

Statistically, for each species, we measured the similarity of the 3-hinge connective matrices for each subject pair. The similarity was defined as the Pearson correlation coefficient of the upper triangular parts of two matrices. The results are shown in Fig. 2 (d). On average, the inter-individual similarity (the diagonal lines in Fig. 2(d) were omitted) is $0.55 \pm 0.11$ for macaque and $0.14 \pm 0.09$ for human. These results were compared to those based on spherical registration method in Freesurfer, by which the 3-hinge correspondence was determined by search for the closest 3-hinge on a registered testing subject to that on the template. Spherical registration method was used for this comparison because of its good performance in cortical pattern alignment within and across species so far [3]. On average, the inter-individual similarity is $0.31 \pm 0.16$ for macaque and $0.09 \pm 0.10$ for human, which are much lower than those by our method.

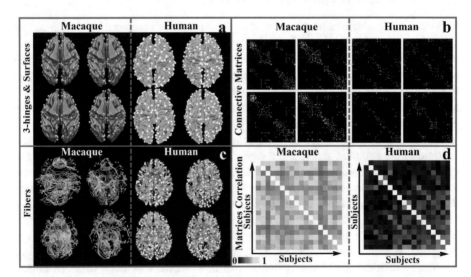

**Fig. 2.** Intra-species analysis results. (a)–(b) four subjects randomly selected for each species are used as examples; (a) identified 3-hinges with correspondence across subjects. Surfaces with curvature map are shown as background; (b) Connective matrices of the 3-hinges in (a). The matrices are binarized (connective strength of the white elements > 0) for ease of visualization. (c) The streamline fibers connecting the 3-hinges in (a). (d) Similarity of connective matrices in (b) of all 18 subjects within each species.

## 3.2    Identified Species-Preserved 3-hinges

As macaque has 128 intra-species consistent 3-hinges, 128 inter-species corresponding 3-hinges were identified in the human brain. The correspondences are visualized in Fig. 3(a). Streamline fibers connecting these 3-hinges on example subjects of both species are in Fig. 3(b). The connective matrices average within species are shown in

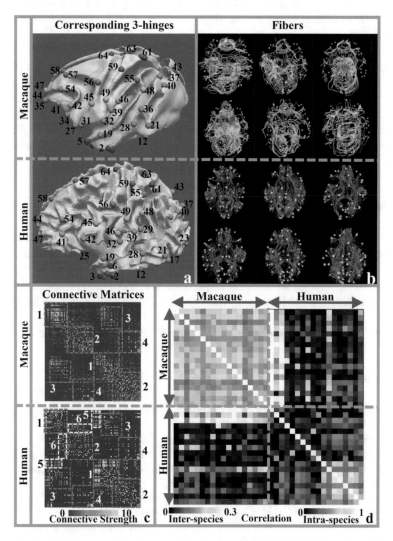

**Fig. 3.** (a) Species-preserved 3-hinges with the corresponding IDs; (b) streamline fibers connecting the 3-hinges in (a) on example subjects; (c) the structural connective matrix averaged over subjects within species. Red boxes highlighted the similarity between two species while white boxes highlight the differences; (d) the inter-individual and cross-species similarity of connective matrices of these 3-hinges in (a). For the ease of visualization, diagonal blocks (intra-species similarity) and off-diagonal-line blocks (inter-species similarity) are at different color scales.

Fig. 3(c) as well. The cross-species similarity is observable with regard to the fiber bundle morphologies in Fig. 3(b) and connective pattern highlighted by red boxes in Fig. 3(c). The inter-individual and inter-species similarity of the connective matrices are shown in Fig. 3(d). On average, the inter-species similarity, averaged over the off-diagonal-line block, is $0.08 \pm 0.06$, and the intra-species similarity for human is $0.23 \pm 0.20$. The intra-species similarity for macaque is the same as the one in Sect. 3.1. These results outperform those based on spherical registration method: $0.05 \pm 0.09$ for the inter-species similarity and $0.09 \pm 0.12$ for the intra-species on human brains.

The identified species-preserved 3-hinges and connective fibers are in line with previous reports [1, 2]. For example, the red boxes highlighted species-preserved blocks #1 and #3 that are the ipsilateral and contralateral connections within temporal lobe and occipital lobe. Blocks #2 and #4 are the ipsilateral and contralateral connections within frontal lobe and parietal lobe. Blocks #5 highlight the connections between temporal lobe and the contralateral temporal pole. Blocks #6 highlight the ipsilateral connections between temporal lobe and inferior frontal lobe/parietal lobe. Blocks #5 and #6 are unique for humans.

## 4   Conclusion

We present a framework to establish cross-species correspondence for cortical 3-hinges in terms of their structural connective patterns, which is in line with previous reports, suggesting that they can be used as landmarks for cross-species alignment. This work could reveal the relation between cortical convolution and structural connection across species, shedding new light to comparative neuroscience studies of primate species.

**Acknowledgements.** T Zhang was supported by NSFC31671005 and NSFC31500798.

## References

1. Rilling, J.K., et al.: The evolution of the arcuate fasciculus revealed with comparative DTI. Nat. Neurosci. **11**(4), 426–428 (2008)
2. de Schotten, M.T., Dell'Acqua, F., Valabregue, R., Catani, M.: Monkey to human comparative anatomy of the frontal lobe association tracts. Cortex **48**(1), 82–96 (2012)
3. Van Essen, D.C.: Surface-based approaches to spatial localization and registration in primate cerebral cortex. Neuroimage **23**, S97–S107 (2004)
4. Orban, G.A., Van Essen, D., Vanduffel, W.: Comparative mapping of higher visual areas in monkeys and humans. Trends Cogn. Sci. **8**(7), 315–324 (2004)
5. Van Essen, D.C., Dierker, D.L.: Surface-based and probabilistic atlases of primate cerebral cortex. Neuron **56**(2), 209–225 (2007)
6. Li, K., et al.: Gyral folding pattern analysis via surface profiling. NeuroImage **52**(4), 1202–1214 (2010)
7. Li, X., et al.: Commonly preserved and species-specific gyral folding patterns across primate brains. Brain Struct. Funct. **222**(5), 2127–2141 (2017)

8. Chen, H., Li, Y., Ge, F., Li, G., Shen, D., Liu, T.: Gyral net: a new representation of cortical folding organization. Med. Image Anal. **42**, 14–25 (2017)
9. Andersson, J.L.R., Jenkinson, M., Smith, S.L.: Non-linear registration, aka spatial normalisation. FMRIB Technical report TR07JA2 (2010)
10. Yeh, F.C., Wedeen, V.J., Tseng, W.Y.I.: Generalized q-sampling imaging. IEEE Trans. Med. Imaging **29**(9), 1626–1635 (2010)
11. Yeh, F.C., Verstynen, T.D., Wang, Y., Fernández-Miranda, J.C., Tseng, W.Y.I.: Deterministic diffusion fiber tracking improved by quantitative anisotropy. PLoS ONE **8**(11), e80713 (2013)
12. Fischl, B., Sereno, M.I., Tootell, R.B., Dale, A.M.: High-resolution intersubject averaging and a coordinate system for the cortical surface. Hum. Brain Mapp. **8**(4), 272–284 (1999)
13. Bertrand, G.: On topological watersheds. J. Math. Imaging Vis. **22**(2–3), 217–230 (2005)
14. Granville, V., Krivánek, M., Rasson, J.P.: Simulated annealing: a proof of convergence. IEEE Trans. Pattern Anal. Mach. Intell. **16**(6), 652–656 (1994)

# Deep Learning with Synthetic Diffusion MRI Data for Free-Water Elimination in Glioblastoma Cases

Miguel Molina-Romero[1,2]([✉]) [iD], Benedikt Wiestler[3] [iD], Pedro A. Gómez[1,2] [iD], Marion I. Menzel[2] [iD], and Bjoern H. Menze[1] [iD]

[1] Computer Science, Technische Universität München, Munich, Germany
miguel.molina@tum.de
[2] GE Healthcare Global Research Organization, Munich, Germany
[3] Department of Neuroradiology, Klinikum rechts der Isar der Technische Universität München, Munich, Germany

**Abstract.** Glioblastoma is the most common and aggressive brain tumor. In clinical practice, diffusion MRI (dMRI) enables tumor infiltration assessment, tumor recurrence prognosis, and identification of white-matter tracks close to the resection volume. However, the vasogenic edema (free-water) surrounding the tumor causes partial volume contamination, which induces a bias in the estimates of the diffusion properties and limits the clinical utility of dMRI.

We introduce a voxel-based deep learning method to map and correct free-water partial volume contamination in dMRI. Our model learns from synthetically generated data a non-parametric forward model that maps free-water partial volume contamination to volume fractions. This is independent of the diffusion protocol and can be used retrospectively. We show its benefits in glioblastoma cases: first, a gain of statistical power; second, quantification of free-water and tissue volume fractions; and third, correction of free-water contaminated diffusion metrics. Free-water elimination yields more relevant information from the available data.

**Keywords:** Glioblastoma · Brain tumor · DTI · Deep learning
Fractional anisotropy · Free-water elimination · Data harmonization

## 1 Introduction

Glioblastomas are the most common primary brain tumor. Ninety percent of these are IDH-wild-type and have a dismal prognosis, with a 5-year survival rate of less than 10%. Diffusion magnetic resonance imaging (dMRI) and the diffusion tensor model (DTI) are used clinically for surgical planing. DTI yields

**Electronic supplementary material** The online version of this chapter (https://doi.org/10.1007/978-3-030-00931-1_12) contains supplementary material, which is available to authorized users.

A. F. Frangi et al. (Eds.): MICCAI 2018, LNCS 11072, pp. 98–106, 2018.
https://doi.org/10.1007/978-3-030-00931-1_12

quantitative estimates of the tissue diffusivity, e.g. mean diffusivity (MD), and fractional anisotropy (FA), an index of tissue microstructure organization. Previous research focused on FA as indicator of tumor grade, tumor cellularity, tumor infiltration and edema assessment [1], and tumor recurrence [2]. However, results are controversial due to reproducibility issues derived from differences in methodology, image acquisition, or post-processing [1].

Data harmonization in dMRI is attracting attention to overcome these problems [3]. One way to reduce uncontrolled variability is to eliminate the free-water signal [4]. Free-water elimination (FWE) uses a two-compartments tissue model composed by tissue (or parenchyma) and free-water [5]. Fitting the diffusion tensor in a two-compartments model is an ill-posed problem that has been solved using spatial regularization [6] or optimized acquisition protocols [7]. Harmonization of image resolution and diffusion directionality can be achieved by image quality transfer (IQT) with non-linear learning algorithms, such as convolutional neural networks [8] or random forest [9]. IQT offers a new dimension in dMRI, enabling learning complex diffusion model on high quality data, to then transfer information captured in signal patterns to low quality data. However, this approach is limited by the availability of rich datasets.

We propose an new method for free-water elimination based on an artificial neural network (ANN), trained with synthetically generated data, that is independent of the number of diffusion shells (b-values) and can be applied retrospectively to any dMRI data. Instead of regularizing the FWE ill-posed inverse problem, we teach a non-parametric forward model to learn the mapping between partial volume contamination and free-water volume fraction from synthetic data. Besides, unlike IQT, our approach works only in the diffusion dimension, enabling an important simplification of the ANN model. We further show the advantages of FWE in glioblastoma cases: (1) a gain of statistical power through data harmonization, (2) complementary information of the tissue microstructure composition, and (3) better assessment of edema, tumor, and tumor infiltrated areas. The source code can be found in https://github.com/mmromero/dry.

## 2   Methods

**Diffusion Signal Modeling:** Following previous work on free-water elimination we modeled the diffusion signal of a single voxel, along the diffusion directions $(b, \boldsymbol{g})$, as the contribution of tissue and free-water compartments:

$$S(TE, b, \boldsymbol{g}) = S_0 \left( \hat{f}_t e^{\frac{-TE}{T_{2_t}}} S_t(b, \boldsymbol{g}) + \hat{f}_{fw} e^{\frac{-TE}{T_{2_{fw}}}} S_{fw}(b, \boldsymbol{g}) \right), \qquad (1)$$

where $b$ and $\boldsymbol{g}$ summarize the gradient effects; $S_0$ is a scaling factor proportional to the proton density; $\hat{f}_t$, $\hat{f}_{fw}$, $T_{2_t}$, and $T_{2_{fw}}$ are the volume fraction and $T_2$ values of tissue and free-water respectively. Since $T_{2_t} < T_{2_{fw}}$, measurements at different echo-times (TE) yield distinct contributions of tissue and free-water.

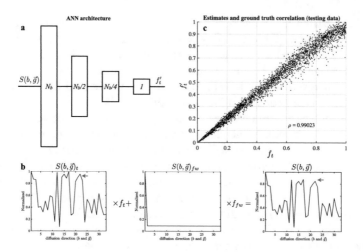

**Fig. 1.** Artificial neural network model. The ANN architecture (a). Training data was synthetically generated following Eq. 2 (b). Free-water partial volume contamination effects were visible (b, red arrows). We ran a correlation analysis for 3000 randomly generated samples (c) reaching a factor of 0.99.

Thus, disentangling the volume fractions from the $T_2$ effects requires measurements at least two different TEs [10]. However, in clinical routine only one TE is acquired simplifying Eq. 1:

$$S(b, \boldsymbol{g}) = S_0 \left( f_t S_t(b, \boldsymbol{g}) + f_{fw} S_{fw}(b, \boldsymbol{g}) \right), \tag{2}$$

where the $T_{2_i}$ and $\hat{f}_i$ effects are integrated in $f_i$, inducing a positive bias towards the new free-water volume fraction ($f_{fw}$) as TE increases. The volume fraction indexes are ratios relative to the signal contribution of each compartment and thus $f_{fw} + f_t = 1$.

**Synthetic Signal Generation:** The diffusion properties of free-water at body temperature are well characterized [5], presenting isotropic behavior and a diffusion coefficient $D_{fw} = 3 \times 10^{-3}$ mm$^2$/s, thus $S_{fw}(b, \boldsymbol{g}) = e^{-bD_{fw}}$. On the other hand, $S_t$ is unknown since it depends on the tissue microstructure organization and the orientation of the diffusion gradients, $\boldsymbol{g}$. Thus, we modeled its behavior with a random variable, $S_t \in \mathbb{R}^{N_b}$, following an uniform distribution, $U(0, 1)$, where $N_b$ is the number of diffusion measures, including non-diffusion-weighted volumes. Furthermore, we also represented the tissue volume fraction of a voxel, $f_t \in [0, 1]$, as a random uniform variable, $U(0, 1)$. Based on the prior knowledge of $D_{fw}$, the models of $S_t$ and $f_t$, and knowing the diffusion protocol, it is possible to generate unlimited synthetic diffusion signals, $S(b, \boldsymbol{g})$, containing free-water partial volume effects (Eq. 2 and Fig. 1b).

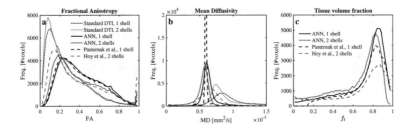

**Fig. 2.** Comparison of FWE with ANN and state of the art methods for two diffusion protocols. The correction effects on FA of ANN for one shell were equivalent to Pasternak's et al. algorithm (a), while ANN MD estimates were less over-regularized (b). Tissue volume fraction estimates were also in agreement although Pasternak's method failed to estimate $f_t < 0.18$ (c). The ANN for two shells and Hoy's et al. algorithm were in good agreement for FA (a) but larger differences were visible for MD and $f_t$ (b and c). See the supplementary material for more information.

**ANN Architecture and Training:** We designed a regression fully connected ANN capable of estimating the tissue volume fraction, $f_t$, directly from the diffusion signal $S(b, g)$ (Eq. 2). The input layer contained as many units as the number of acquired diffusion measures (including non-diffusion-weighted volumes), $N_b$, and a single output unit yielding the estimate of $f_t$. We compared ANN architectures containing from one to five hidden layers, finding an optimum in performance for two hidden layers with $N_b/2$ and $N_b/4$ respectively (Fig. 1a). To train the ANN we used 20000 synthetic signals generated as explained above (70% training, 15% validation, and 15% testing). Convergence for $N_b = 33$ (Fig. 1) was reached after nine epochs and 4.7 s in a consumers laptop (Apple MacBook Pro, Intel Core i5, 8GB RAM; MATLAB, Mathworks, Natwick, MA). The training process depended on the diffusion protocol prescribed. Thus, we trained four networks to match the DWI data used in the experiments below.

## 3 Experiments and Results

**Comparison with State of the Art:** Methods from Pasternak et al. [6] and Hoy et al. [7] are the state of the art for one and two shell acquisitions. For comparison we measured data from a volunteer in a GE 3T MR750w (GE Healthcare, Milwaukee, WI). The protocol comprised first, one diffusion weighted imaging (DWI) acquisition for 30 diffusion direction ($b = 500$ s/mm$^2$) and two $b = 0$ volume ($N_b = 32$); and second a DWI for two shells ($b = 500$ and $1000$ s/mm$^2$) with 30 diffusion directions for each shell and four non-diffusion-weighted volumes ($N_b = 64$). The data was processed with a pipeline including steps for: (1) head motion and eddy current corrections (FSL eddy); (2) denoising based on random matrix theory [11]; and (3) free-water elimination based on ANN, Pasternak's, and Hoy's methods (Fig. 2). The ANN results were comparable to those of the state of the art methods.

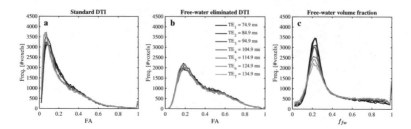

**Fig. 3.** Histogram comparison at several TEs. FA from standard DTI (a) showed larger variability across TEs for FA < 0.4, than the free-water corrected (b). The tissue volume fraction estimates are influenced by $T_2$ decay (Eq. 1) and thus sensitive to TE (c). The free-water elimination step has a TE harmonization effect on the diffusion signal (b), shifting the TE variability into the volume fraction estimate (c).

**Data Harmonization:** The $T_2$ effects described in Eqs. 1 and 2 were investigated using data from a volunteer scanned in the same scanner as before. The protocol consisted of a DWI acquisition for 30 direction ($b = 1000\,\mathrm{s/mm^2}$) and one non-diffusion-weighted volume ($N_b = 31$). This was repeated for seven equispaced TE = 74.9 – 134.9 ms. The data was processed as described before, but only ANN FWE was computed. For comparison two processing lines were created with and without ANN based FWE. Both were fitted with robust DTI (RESTORE) [12] to extract diffusion metrics.

The multi-echo diffusion data showed an increase of free-water (Fig. 3c) and its effects (Fig. 3a, larger low FA peak) with the TE, which agrees with the two-compartments tissue model (Eq. 1) accounting for $T_2$ effects, simplified in Eq. 2. Multi-center studies are often carried out in data acquired with heterogeneous protocols. The prescribed TEs are a source of variability that mostly depend on the gradient strength of the scanner and the image resolution. Data harmonization is important to remove uncontrolled variability and achieve a good statistical power. Free-water elimination plays a double role. First, it accounts for $T_2$ effects in the $f_i$, shifting this variability from the diffusion metrics to the volume fraction estimates (Fig. 3). And second, it removes the "diffusion isotropic noise" from the signal showing the actual tissue anisotropy, eliminating the variability induced by the presence of free-water.

**Glioblastoma Analysis:** Data from 25 patients affected by glioblastoma (IDH wild-type, WHO 2016 classification) were provided by the Department of Neuro-radiology, Klinikum rechts der Isar der Technische Universität München, Munich, Germany. All patients are part of a prospective glioma database, approved by the local ethics committee, and gave written informed consent. They were scanned in a 3 T whole-body scanner (Achieva, Philips Medical Systems, Best, The Netherlands). The protocol included DWI for 32 directions ($b = 800\,\mathrm{s/mm^2}$) and one non-diffusion-weighted volume ($N_b = 33$). Furthermore, $T_2$ turbo spin echo (T2w), $T_2$-FLAIR, and $T_1$ contrast enhanced (CE-T1w) were acquired. The

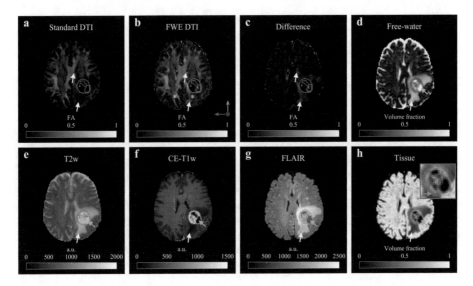

**Fig. 4.** Comparison of metrics and contrasts for one patient. Fractional anisotropy from standard DTI (a) showed a dimmed region corresponding to edema (white arrows). This area was recovered after free-water elimination (b and c). We observed alterations of the white matter integrity compared to the NAWM (b, white asterisks). Free-water (d) and tissue (h) volume fraction maps were estimated by the ANN. Edema regions were well defined and distinguishable from the tumor (red contour). Pools compatible with cytotoxic edema were observable inside the tumor (light-blue arrows). These findings agreed with the observations based on T2w (e), CE-T1w (f), and FLAIR (g). Extended tumor infiltration was derived from the comparison of CE-T1w and tissue volume fraction map (h, green arrow and contour).

DWI data was processed as described before, and T2w, CE-T1w, and FLAIR volumes were registered to the DWI space (Fig. 4).

**Tissue and Free-Water Volume Fraction Estimates:** The free-water and tissue maps computed with ANN (Fig. 4d and h) showed tumor and edema areas that were in agreement with well established methods: T2w, CE-T1w, and FLAIR (Fig. 4e, f, and g). The estimation of tissue and free-water volume fraction maps is an important feature of FWE. They provide complementary information of the underlying tissue organization. Cytotoxic edema and necrosis areas that are not distinguishable can be better identified with the knowledge of the amount of tissue in the voxel (Fig. 4d, e, f, g, and h, light-blue arrows). Furthermore, in clinical routine tumor delineation is based on CE-T1w hyper-intensities (Fig. 4f red contour). When this is compared with tissue volume fraction maps (Fig. 4h zoomed area), we observed an increased tumor region of up to four millimeters that is compatible with tumor infiltration. This agrees with the radiated area after resection (Fig. 4h, green arrow and contour).

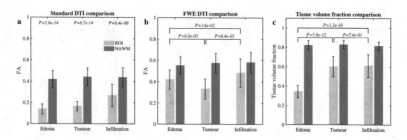

**Fig. 5.** Edema, tumor, and tumor infiltrated regions differentiation. The FA values for standard DTI (a) were statistically different ($\alpha = 0.01$) between NAWM areas. After free-water elimination (b) FA values in edema and tumor infiltrated areas were significantly larger than those for tumor (family-wise error = 0.03), indicating a better organized tissue microstructure. Complementary, the tissue volume fractions in tumor and tumor infiltrated regions were statistically larger than in edema areas, suggesting that the loss in FA in edema was mostly due to free-water infiltration.

**Fractional Anisotropy Recovery:** The comparison between standard and FWE FA maps (Fig. 4a and b) exhibited a recovery of the anisotropic information in the edema region and around the ventricles (Fig. 4c, white arrows). The elimination of the isotropic compartment from the diffusion signal leads to a recovery of the tissue anisotropy captured by the protocol, leading to an enhancement of the FA maps, especially in areas with large partial volume contamination like edema and the border of the ventricles (Fig. 4a, b, and c, white arrows). The correction of the FA maps provides new information of the tissue microstructure integrity hidden by the edema (Fig. 4b asterisks).

**Edema and Infiltration Unmixing:** To assess the impact of FWE in edema, tumor, and tumor infiltrated areas, regions of interests (ROI) were defined by a neuroradiologist for each subject using FLAIR, T2w, CE-T1w, $f_t$, and FWE FA maps (Fig. 4). To minimize the influence of outliers we used the median for each ROI, and ran t-test comparison for FA and tissue volume fraction values across subjects for the three type of ROIs (Fig. 5). For reference, extra ROIs were drawn in normal appearing white mater (NAWM) mostly contralateral.

Comparison of NAWM with tumor and edema areas for the 25 patients showed a statistically significant difference ($\alpha = 0.01$) in FA for standard DTI (Fig. 5a). Tumor infiltration and edema are the driving factors behind the loss in FA. After free-water elimination we compared FWE FA (Fig. 5b) and tissue volume fraction (Fig. 5c) in edema, tumor and tumor infiltrated areas (family-wise error = 0.03). Significantly larger FA was found in edema and tumor infiltrated areas compared to tumor regions. While, tissue volume fraction in edema was statistically lower than in tumor infiltrated and tumor sections.

We hypothesize that the combination of FWE FA and tissue volume fraction yields a better understanding of tissue microstructure integrity (Fig. 5). High FA and low tissue volume fraction might indicate well organized microstructure infiltrated by free-water (vasogenic edema). Low FA and high tissue volume fraction is compatible with the unstructured cellularity found in tumors. Finally, areas with high FA and tissue volume fraction might be caused by tumor infiltration in highly structured white matter bundles.

## Discussion

The presented ANN model combines simplicity with high accuracy. We introduce for the first time an ANN design capable of learning partial volume effect features from synthetic data, and reach comparable results with the state of the art methods but at least 55-fold faster. Besides, our approach, unlike [6] is voxel-based, avoiding blurring artifacts induced by the use of patch based regularization, and can be applied to any diffusion protocol beyond DTI. The robustness of our approach was shown with 25 patients, 2 volunteers, 4 protocols, and 2 scanners from different manufacturers. We attribute these results to the good understanding of the physical signal model mimicked by the ANN.

The addition of free-water elimination in processing pipeline extracts more information from the data, which has a potential benefit for glioblastoma patients in three aspects: diagnosis, surgical planning, and guided radiotherapy. The diagnosis potentially improves with the quantification of the free-water volume fraction map, yielding the severity of the vasogenic edema. The precision of the surgical planning benefits from the corrected FA maps in edema areas providing a better definition of white matter fiber bundles and limiting the resection of healthy tissue. Finally, guided radiotherapy profits from identification of tumor infiltrated areas.

**Acknowledgments.** The authors want to thank Dr. Ofer Pasternak for his support in the comparison of the methods. This work was supported by the TUM Institute of Advanced Study, funded by the German Excellence Initiative, and the European Commission (Grant Agreement Number 605162).

## References

1. Field, A.S.: Diffusion imaging in brain tumors. In: Jones, D.K. (ed.) Diffusion MRI, Chap. 33, pp. 547–563. Oxford University Press, Oxford (2010)
2. Bette, S., et al.: Local fractional anisotropy is reduced in areas with tumor recurrence in glioblastoma. Radiology **283**(2), 499–507 (2017)
3. Eaton-Rosen, Z., et al.: Beyond the resolution limit: diffusion parameter estimation in partial volume. In: Ourselin, S., Joskowicz, L., Sabuncu, M.R., Unal, G., Wells, W. (eds.) MICCAI 2016. LNCS, vol. 9902, pp. 605–612. Springer, Cham (2016). https://doi.org/10.1007/978-3-319-46726-9_70
4. Metzler-Baddeley, C., et al.: How and how not to correct for CSF-contamination in diffusion MRI. Neuroimage **59**(2), 1394–1403 (2012)
5. Pierpaoli, C., Jones, D.K.: Removing CSF contamination in brain DT-MRIs by using a two-compartment tensor model. In: ISMRM, Kyoto, p. 1215 (2004)

6. Pasternak, O., et al.: Free water elimination and mapping from diffusion MRI. Magn. Reson. Med. **730**, 717–730 (2009)
7. Hoy, A.R., et al.: Optimization of a free water elimination two-compartment model for diffusion tensor imaging. Neuroimage **103**, 323–333 (2014)
8. Tanno, R., et al.: Bayesian image quality transfer with CNNs: exploring uncertainty in dMRI super-resolution. In: Descoteaux, M., Maier-Hein, L., Franz, A., Jannin, P., Collins, D.L., Duchesne, S. (eds.) MICCAI 2017. LNCS, vol. 10433, pp. 611–619. Springer, Cham (2017). https://doi.org/10.1007/978-3-319-66182-7_70
9. Alexander, D.C., et al.: Image quality transfer and applications in diffusion MRI. Neuroimage **152**(March), 283–298 (2017)
10. Molina-Romero, M., et al.: A diffusion model-free framework with echo time dependence for free-water elimination and brain tissue microstructure characterization. Magn. Reson. Med. **80**, 2155–2172 (2018)
11. Veraart, J., et al.: Denoising of diffusion MRI using random matrix theory. Neuroimage **142**, 394–406 (2016)
12. Chang, L., et al.: Restore: robust estimation of tensors by outlier rejection. Magn. Reson. Med. **53**(5), 1088–1095 (2005)

# Enhancing Clinical MRI Perfusion Maps with Data-Driven Maps of Complementary Nature for Lesion Outcome Prediction

Adriano Pinto[1,2(✉)], Sérgio Pereira[1,2], Raphael Meier[3], Victor Alves[2],
Roland Wiest[3], Carlos A. Silva[1], and Mauricio Reyes[4]

[1] CMEMS-UMinho Research Unit, University of Minho, Guimarães, Portugal
id6376@alunos.uminho.pt
[2] Centro Algoritmi, University of Minho, Braga, Portugal
[3] Support Center for Advanced Neuroimaging - Institute for Diagnostic and
Interventional Neuroradiology, University Hospital Inselspital and University of Bern,
Bern, Switzerland
[4] Institute for Surgical Technology and Biomechanics, University of Bern,
Bern, Switzerland

**Abstract.** Stroke is the second most common cause of death in developed countries, where rapid clinical intervention can have a major impact on a patient's life. To perform the revascularization procedure, the decision making of physicians considers its risks and benefits based on multimodal MRI and clinical experience. Therefore, automatic prediction of the ischemic stroke lesion outcome has the potential to assist the physician towards a better stroke assessment and information about tissue outcome. Typically, automatic methods consider the information of the standard kinetic models of diffusion and perfusion MRI (e.g. Tmax, TTP, MTT, rCBF, rCBV) to perform lesion outcome prediction. In this work, we propose a deep learning method to fuse this information with an automated data selection of the raw 4D PWI image information, followed by a data-driven deep-learning modeling of the underlying blood flow hemodynamics. We demonstrate the ability of the proposed approach to improve prediction of tissue at risk before therapy, as compared to only using the standard clinical perfusion maps, hence suggesting on the potential benefits of the proposed data-driven raw perfusion data modelling approach.

## 1 Introduction

Stroke ranks second as leading cause of deaths worldwide, with ischemic stroke being the most common type. Ischemic stroke arises from a sudden occlusion of a cerebral artery. Diagnosis and treatment begins with the acquisition of multimodal MRI or CT images, followed by appropriate medical intervention. While several clinical trials have proven the efficacy of mechanical thrombectomy, the

© Springer Nature Switzerland AG 2018
A. F. Frangi et al. (Eds.): MICCAI 2018, LNCS 11072, pp. 107–115, 2018.
https://doi.org/10.1007/978-3-030-00931-1_13

treating physician must carefully evaluate the associated risks and benefits: the volume of ill-perfused tissue potentially salvageable, versus the risk of causing haemorrhage or other complications [9,14]. Hence, predicting the outcome of a stroke lesion (i.e. lesion status at three-month follow-up), and thereby evaluating the effect of a successful or unsuccessful mechanical thrombectomy, has a great potential to guide the decision of the physician.

MRI perfusion and diffusion sequences have been gaining importance in the characterization of ischemic stroke, which provides important information for the revascularization therapy [2,12]. Based on MRI sequences, some machine learning approaches have been recently proposed for ischemic stroke lesion prediction. The majority is based on multivariate linear regression models [5,11], or through more advanced models such as decision trees [7] and CNN-based deep learning architectures [6]. However, up to our knowledge none of the approaches takes into consideration the temporal Perfusion Weighted Imaging data (4D PWI) for stroke lesion prediction. We hypothesize that a data-driven approach to model the raw perfusion imaging data can unveil information complementing the standard clinical perfusion maps derived using kinetic analysis.

In this paper, we propose a novel end-to-end deep learning multi-data branched network that incorporates information from the 4D PWI alongside the standard clinical perfusion and diffusion maps. From the time-stamp acquisitions of the 4D PWI, that characterize the bolus passage, we aim to learn brain blood flow hemodynamics (principal and collateral) to characterize tissue at risk of infarction (penumbra), and the unsalvageable tissue (ischemic core). Since standard perfusion and diffusion maps are generated from kinetic models followed by thresholding (based on clinical knowledge and experience), we hypothesize that there might be loss of relevant information. Hence, we aim to enhance the clinical MRI diffusion and perfusion maps with data-driven maps for stroke lesion outcome prediction. The proposed architecture was evaluated using the publicly-available ISLES 2017 dataset.

The remainder of the paper is organized as follows: Sect. 2 describes the methods of the proposed approach. Section 3 addresses the setup for stroke lesion outcome prediction. Section 4 contains the obtained results and its discussion. Finally, Sect. 5 contains the conclusions.

## 2   Methods

**4D PWI.** At the arrival of the contrast agent to the brain, during the acquisition of the 4D PWI, the healthy tissue will present a drop in the signal intensity value, which then increases as contrast agent starts diluting throughout the system. However, in the presence of ill-perfused tissue the signal intensity values barely changes, since there is no propagation of the contrast agent to the damaged tissue [12]. Figure 1 depicts such signal intensity behaviour in a patient with ischemic stroke. The perfusion blood flow dynamic, captured by the temporal slices of the 4D PWI, is responsible for the generation of the 3D MRI perfusion maps, through the application of kinetic models, deconvolutions in the time space, and clinical thresholding. Therefore, rCBF, rCBV, MTT, TTP, and Tmax perfusion maps can be viewed as surrogate parametric summaries of the raw 4D PWI, encompassing specific blood flow dynamics. From this knowledge emerged the intention to evaluate the encoding of the blood flow hemodynamics directly from 4D PWI, considering altogether complementary information over the diffusion and perfusion maps. Our approach was based on the peak concentration of contrasting agent, which is of extreme importance, since it characterizes the point where the differences of perfusion between healthy tissue and the ill-perfused tissue are higher, allowing a better detection of the penumbra [4]. We developed an automatic approach to detect the time slice where the concentration of contrast agent is higher, which corresponds to a lower signal intensity. We detect automatically the peak of concentration using k-means on the mean signal intensity and standard deviation. The peak is used to define a temporal window to retrieve specific temporal acquisitions regarding the blood dynamics needed to estimate the tissue at risk of infarction. Besides reducing the total number of temporal slices, we also enforce the same spatial-temporal space across patients. Aligning patient data across the peak concentration of contrasting agent yields a common time interval for the retrieval of information. A fixed temporal window size of 26 slices was used, based on the sampling rate of the MRI acquisition.

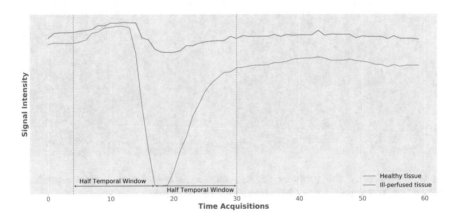

**Fig. 1.** General 4D PWI signal intensity over time acquisitions in a patient with an ischemic stroke lesion. The dashed line corresponds to the temporal slicing performed automatically. The system automatically detects and selects a temporal window of interest, modeling temporal acquisitions characterizing blood dynamics of interest.

**Baseline Architecture.** For stroke outcome lesion prediction, we based our network on the U-Net [10], which has proved to be competitive in many biomedical image segmentation applications. The output of the U-Net is fed to a bi-dimensional GRU layer [3] that processes the information in four directions (superior-inferior, inferior-superior, anterior-posterior and posterior-anterior), to enforce a greater spatial context in stroke lesion outcome prediction. The baseline architecture only considers standard diffusion and perfusion MRI maps (as employed by the state of the art approaches).

**Multi-Data Branched Network.** To merge information from the standard perfusion and diffusion maps and the data-driven 4D PWI data, the proposed architecture fuses two U-Nets as shown in Fig. 2. The top branch U-Net models the raw 4D PWI information, where the temporal information is coded as input channels. The first two layers consist of a feature expansion and feature reduction by 4. Recombining the feature maps allows complex interactions between temporal slices within the temporal window [15]. Since each branch is able to learn different specific features, we then merge the output of each branch, which is fed into a smaller architecture in order to take advantage from complementary information for stroke lesion outcome prediction. The final portion of the network encompasses also a bi-dimensional GRU layer present in both branches of the network.

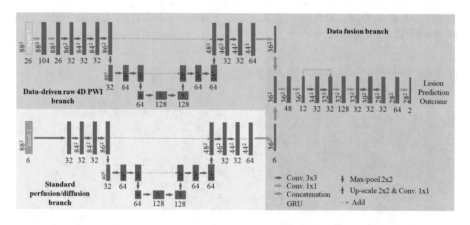

**Fig. 2.** Overview of the proposed architecture for stroke lesion outcome prediction. Blue feature maps result from bi-dimensional convolutions.

## 3   Setup

Our proposal was validated on the publicly available ISLES 2017 database, with a total of 75 cases divided into two datasets: Training dataset (n = 43) and Challenge dataset (n = 32). Each case contains a raw 4D PWI, five 3D MRI perfusion maps (rCBF, rCBV, MTT, TTP, Tmax), one 3D MRI diffusion map (ADC), and the final lesion outcome, which was manually segmented by a clinician on a 90-day follow-up (only available for the training dataset). All MRI maps are already co-registered and skull-stripped [6].

**Table 1.** Methods from ISLES 2017 testing dataset. Each metric contains the mean ± standard deviation.

| | | Dice | Hausdorff Distance | ASSD | Precision | Recall |
|---|---|---|---|---|---|---|
| Challenge | Mok et al.* | 0.32 ± 0.23 | 40.74 ± 27.23 | 8.97 ± 9.52 | 0.34 ± 0.27 | 0.39 ± 0.27 |
| | Kwon et al.* | 0.31 ± 0.23 | 45.26 ± 21.04 | 7.91 ± 7.31 | 0.36 ± 0.27 | 0.45 ± 0.30 |
| | Bertels et al.* | 0.30 ± 0.21 | 33.85 ± 16.82 | 6.81 ± 7.18 | 0.34 ± 0.26 | 0.51 ± 0.32 |
| | Monteiro et al.* | 0.30 ± 0.22 | 46.60 ± 17.50 | 6.31 ± 4.05 | 0.34 ± 0.27 | 0.51 ± 0.30 |
| | Lucas et al.* | 0.29 ± 0.21 | 33.85 ± 16.82 | 6.81 ± 7.18 | 0.34 ± 0.26 | 0.51 ± 0.32 |
| | Choi et al.* | 0.28 ± 0.22 | 43.89 ± 20.70 | 8.88 ± 8.19 | 0.36 ± 0.31 | 0.41 ± 0.31 |
| | Robben et al.* | 0.27 ± 0.22 | 37.84 ± 17.75 | 6.72 ± 4.10 | 0.44 ± 0.32 | 0.39 ± 0.31 |
| | Pisov et al.* | 0.27 ± 0.20 | 49.24 ± 32.15 | 9.49 ± 10.56 | 0.31 ± 0.27 | 0.39 ± 029 |
| | Niu et al.* | 0.26 ± 0.20 | 48.88 ± 11.20 | 6.26 ± 3.02 | 0.28 ± 0.25 | 0.56 ± 0.26 |
| | Sedlar et al.* | 0.20 ± 0.19 | 58.30 ± 20.02 | 11.19 ± 9.10 | 0.23 ± 0.24 | 0.40 ± 0.29 |
| | Rivera et al.* | 0.19 ± 0.16 | 63.58 ± 18.58 | 11.13 ± 7.89 | 0.27 ± 0.25 | 0.21 ± 0.17 |
| | Islam et al.* | 0.19 ± 0.18 | 64.15 ± 28.51 | 14.17 ± 15.80 | 0.29 ± 0.28 | 0.25 ± 0.25 |
| | Chengwei et al.* | 0.18 ± 0.17 | 65.95 ± 25.94 | 9.22 ± 6.99 | 0.37 ± 0.30 | 0.21 ± 0.23 |
| | Yoon et al.* | 0.17 ± 0.16 | 45.23 ± 19.14 | 12.43 ± 11.01 | 0.23 ± 0.27 | 0.36 ± 0.32 |
| | Standard Branch | 0.20 ± 0.19 | 70.04 ± 20.37 | 11.66 ± 7.40 | 0.16 ± 0.20 | 0.61 ± 0.28 |
| | Data-Driven Branch | 0.20 ± 0.18 | 54.59 ± 18.69 | 9.95 ± 5.79 | 0.18 ± 0.21 | 0.61 ± 0.27 |
| | Multi-Data Single Branch | 0.26 ± 0.21 | 46.35 ± 17.59 | 8.37 ± 6.43 | 0.21 ± 0.20 | 0.61 ± 0.28 |
| | Multi-Data Branched | 0.29 ± 0.21 | 41.58 ± 22.04 | 7.69 ± 5.71 | 0.23 ± 0.21 | 0.66 ± 0.29 |

* Static results from [1].

Since MRI acquisitions are from different centers, all maps were resized to the same volume space: $256 \times 256 \times 32$. $T_{max}$ was clipped to $[0, 20s]$, and the ADC was clipped to be within the range $[0, 2600] \times 10^{-6}\text{mm}^2/s$, as values beyond these ranges are known to be biologically meaningless [7]. Afterwards, a linear scaling was performed between $[0, 255]$. Bias field correction was performed to the 4D PWI using the N4ITK method [13] before the resizing and scaling steps.

The training dataset was divided into 36 cases for training and 7 cases for validation. For each case 550 patches of size $88 \times 88$ were randomly extracted. The network was trained with ADAM optimizer (learning rate$= 1e-5$), with a mini-batch of size 4, using as loss function the soft-dice loss [8]. The sum is performed for the $N$ voxels of the patch both in the binary prediction $p_i \in P$ and the ground truth $g_i \in G$. The gradient of the Dice score for the $j-th$ voxel of prediction, was calculated as in Eq. 1.

$$\frac{\delta Dice}{\delta p_j} = \frac{g_j(\sum_i^N p_i^2 + \sum_i^N g_i^2) - 2p_j \sum_i^N p_i g_i}{(\sum_i^N p_i^2 + \sum_i^N g_i^2)^2} \tag{1}$$

The tests were conducted using Keras with Theano, on a Nvidia GeForce GTX-1070 8 GB, where each prediction took around 30 s.

We compare the performance of our proposal with three different studies: Standard Branch, Data-Driven Branch, and Multi-Data Single Branch. The Standard Branch architecture considers diffusion and perfusion maps. The Data-Driven Branch studies the 4D PWI. In the Multi-Data Single Branch we combined the inputs from both branches into a single network. In Table 1 we report results on ISLES 2017 test dataset, which enables us to compare with state of the art methods.

## 4    Results and Discussion

Table 1 contains the results obtained on the ISLES 2017 testing dataset. The Standard Branch and the Data-Driven Branch achieved the same Dice score, but with differences in the distance metrics. The Data-Driven Branch was capable of predicting the lesion outcome with higher robustness, since the Hausdorff and ASSD are lower when compared to the Standard Branch. Nevertheless, both approaches are not capable of reaching state of the art performance. However, when we fuse the information of both models, as proposed in this paper, we observe an improvement on the average Dice score, but also on Hausdorff and ASSD. Being so, both models provide distinct information of value for stroke lesion outcome prediction. Additionally, we also study the performance of a single model with all the inputs combined, referred to as the Multi-Data Single Branch architecture. Such approach reached higher Dice score than the two branches separately, and with lower distance metrics. However, it was not capable of reaching the same performance of our proposal. Therefore, having 2 U-Nets for different input data has benefits on modularity and specificity on how the information is modelled. Using separate deep learning models to directly learn intrinsic biological phenomena allowed a higher robustness and accuracy in lesion outcome location and delineation, which is sustained by the lower distance metric values and higher Dice score.

In addition, we compare our proposal with other approaches. However, such comparison needs to consider the fact that top rank approaches use multiple models (i.e. ensembling). Our proposal was able to achieve a Dice score among the top 5 ranking methods, being within the same Hausdorff range of those methods, with just a single network. From this comparison we highlight the low distance metric values obtained, and also a Dice score in the same level of the 4th ranked method. Figure 3 shows the average Dice score and respective Hausdorff for each method.

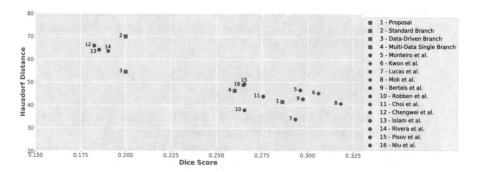

**Fig. 3.** Hausdorff Distance versus Dice score from methods of ISLES 2017 in the testing set. Note that closer to the x axis and further away from the origin is better. Ensemble methods are marked with a cross.

From this analysis we emphasize the benefits of the proposed approach to extract and model information that might not be fully characterized by the standard perfusion and diffusion maps. To assess the complementarity of the data-driven perfusion maps, we computed the normalized mutual information between all the standard perfusion maps and each of the learned feature maps from the data-driven raw 4D PWI branch. As shown in Fig. 4, low association values (less than 0.2) were obtained for all the extracted feature maps, meaning that both branches introduce new and complementary features.

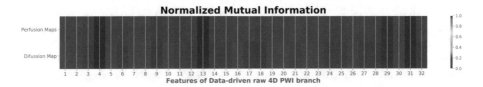

**Fig. 4.** Normalized Mutual Information between the standard perfusion/diffusion maps and the feature maps from the data-driven branch, in training case 36. Values closer to 0 mean low mutual information and closer to 1 represent a high association.

Figure 5 shows the extracted features from the data-driven raw 4D PWI branch. Visually, feature 10 can reflect some descriptions of collateral blood flow, where features 16 and 18 focus on the surrounding lesion area itself in a

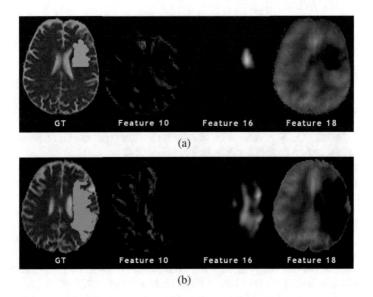

**Fig. 5.** Example of extracted features for training case 33 (a) and case 36 (b), alongside the corresponding ground-truth overlapped over the ADC map.

complementary way. This analysis is particularly important since it can provide complementary information in the decision making process performed by the clinician, providing a better prediction of potentially salvageable tissue.

## 5    Conclusions and Future Work

Parametric perfusion maps can be affected by intrinsic patient physiology [12]. To cope with this effect mathematical models are applied to standardize the behaviour of the contrast bolus passing. Nonetheless, it cannot be independent from patient specific blood flow hemodynamics, which can highly affect the perfusion parametric maps by adding a wide variability in the penumbra delineation [12]. Therefore, in this work, we propose a deep learning architecture, that can process the information from raw 4D PWI data and generate complementary information to the perfusion parametric sequences, which as shown here can increase stroke lesion outcome prediction.

In the future, we intend to perform an interpretability analysis of the data-driven perfusion maps as well as an analysis of the learning patterns of the architecture to ensure the correctness of the predictions with respect to the data being used to drive the lesion outcome predictions.

**Acknowledgments.** Adriano Pinto was supported by a scholarship from the Fundação para a Ciência e Tecnologia (FCT), Portugal (scholarship number PD/BD/113968/2015). This work has been supported by COMPETE: POCI-01-0145-FEDER-007043 and FCT - Fundação para a Ciência e Tecnologia within the Project Scope: UID/CEC/00319/2013.

## References

1. ISLES 2017 Challenge. https://www.smir.ch/ISLES/Start2017. Accessed 08 Feb 2018
2. Barber, P., et al.: Identification of major ischemic change: diffusion-weighted imaging versus computed tomography. Stroke **30**(10), 2059–2065 (1999)
3. Cho, K., et al.: On the properties of neural machine translation: Encoder-decoder approaches. arXiv preprint arXiv:1409.1259 (2014)
4. Hosseini, M.B., Liebeskind, D.S.: The role of neuroimaging in elucidating the pathophysiology of cerebral ischemia. Neuropharmacology **134**, 249–258 (2017)
5. Kemmling, A., et al.: Multivariate dynamic prediction of ischemic infarction and tissue salvage as a function of time and degree of recanalization. J. Cereb. Blood Flow Metab **35**(9), 1397–1405 (2015)
6. Maier, O., et al.: ISLES 2015-a public evaluation benchmark for ischemic stroke lesion segmentation from multispectral MRI. Med. Image Anal. **35**, 250–269 (2017)
7. McKinley, R., et al.: Fully automated stroke tissue estimation using random forest classifiers (faster). J. Cereb. Blood Flow Metab. **37**(8), 2728–2741 (2017)
8. Milletari, F., et al.: V-net: fully convolutional neural networks for volumetric medical image segmentation. In: 2016 Fourth International Conference on 3DV, pp. 565–571. IEEE (2016)

9. World Health Organization, et al.: Global status report on noncommunicable diseases 2014. World Health Organization (2014)
10. Ronneberger, O., Fischer, P., Brox, T.: U-Net: convolutional networks for biomedical image segmentation. In: Navab, N., Hornegger, J., Wells, W.M., Frangi, A.F. (eds.) MICCAI 2015. LNCS, vol. 9351, pp. 234–241. Springer, Cham (2015). https://doi.org/10.1007/978-3-319-24574-4_28
11. Scalzo, F., et al.: Regional prediction of tissue fate in acute ischemic stroke. Ann. Biomed. Eng. **40**(10), 2177–2187 (2012)
12. Song, S., et al.: Temporal similarity perfusion mapping: a standardized and model-free method for detecting perfusion deficits in stroke. PloS one **12**(10), e0185552 (2017)
13. Tustison, N.J., et al.: N4ITK: improved N3 bias correction. IEEE Trans. Med. Imaging **29**(6), 1310–1320 (2010)
14. Wardlaw, J.: Neuroimaging in acute ischaemic stroke: insights into unanswered questions of pathophysiology. J. Intern. Med. **267**(2), 172–190 (2010)
15. Xie, S., et al.: Aggregated residual transformations for deep neural networks. In: 2017 IEEE Conference on CVPR, pp. 5987–5995. IEEE (2017)

# Harmonizing Diffusion MRI Data Across Magnetic Field Strengths

Suheyla Cetin Karayumak$^{(\boxtimes)}$, Marek Kubicki, and Yogesh Rathi

Brigham and Women's Hospital, Harvard Medical School, Boston, MA, USA
{skarayumak,kubicki,yogesh}@bwh.harvard.edu

**Abstract.** Diffusion MRI (dMRI) data is increasingly being acquired on multiple scanners as part of large multi-center neuroimaging studies. However, diffusion imaging is particularly sensitive to scanner-specific differences in coil sensitivity, reconstruction algorithms, acquisition parameters as well as the scanner magnetic field strength, which precludes joint analysis of such multi-site data. Earlier works on dMRI data harmonization were limited to data acquired on different scanners but with the same magnetic field strength (3T). In this work, we explore the possibility of harmonizing dMRI data acquired on scanners with different magnetic field strengths, i.e., 3T and 7T. We propose a linear and several machine learning based non-linear mapping algorithms that use rotation invariant spherical harmonic (RISH) features to map the dMRI data (the raw signal) between scanners without changing the fiber orientations. We extensively validate our algorithms on *in-vivo* data from the Human Connectome Project (HCP) where we used data from 40 subjects with scans done on both 7T and 3T scanners (10 training + 30 test). Using several quantitative metrics such as the root mean squared error (RMSE) in the harmonized dMRI signal and diffusion measures as well as a fiber bundle overlap measure, our preliminary results on 30 test subjects shows that the convolutional neural network (CNN) based algorithm can reliably harmonize the raw dMRI signal across magnetic field strengths. The algorithms proposed are general and can be used for dMRI data harmonization in multi-site studies.

**Keywords:** Harmonization · Diffusion MRI · Machine learning (CNN)

## 1 Introduction

In recent years, several large-scale multi-site neuroimaging studies have been initiated to collect MRI data pertaining to neurodevelopment as well as disease [13,14] to increase statistical power. However, directly pooling dMRI data

---

The authors would like to acknowledge the following grants which supported this work: R01MH102377 (PI: Dr. Marek Kubicki), R01MH097979 (PI: Dr. Yogesh Rathi).

© Springer Nature Switzerland AG 2018
A. F. Frangi et al. (Eds.): MICCAI 2018, LNCS 11072, pp. 116–124, 2018.
https://doi.org/10.1007/978-3-030-00931-1_14

acquired from multiple scanners is fraught with problems due to significant differences in dMRI measures of the same subjects scanned on different scanners [15]. On the other hand, better scanner technologies as well as higher field strength scanners are becoming more popular as they provide better contrast and resolution in diffusion-weighted (DW) imaging [3,11]. For instance, data from a 7T scanner reveals details of tissue properties not visible at 3T [16]. However, data from scanners with different field strengths need to be harmonized to be used jointly.

Recently, several methods have been proposed to harmonize multi-site data, or boost the resolution and quality of dMRI data. Mirzaalian et al. [8] provide a framework for multi-site (3T) harmonization of a single shell (single b-value) dMRI data with similar acquisition parameters (b-values, number of gradients, spatial resolution) and magnetic field strength using rotation invariant spherical harmonics (RISH) features. In [9], the authors use a correction factor for each diffusion tensor derived measure (fractional anisotropy (FA), mean diffusivity (MD)) within a region, while the work in [5] uses a location specific statistical adjustment factor to account for scanner differences. Both these methods perform data harmonization on the final model derived measures (e.g. DTI measures) and not the dMRI signal itself. Consequently, data harmonization has to be done several times independently for each measure, unlike the model-independent method proposed in this work. We also note that, a few works [1,12] have proposed an image quality transfer method which utilizes nonlinear regression to estimate high resolution DTIs or higher order model parameters. These model specific methods however have not been used in the context of data harmonization, but are potential candidates. Consequently, we compare our methods with the work of [1]. Furthermore, the harmonization between multiple field strengths remains unaddressed.

In this work, we harmonize multi-shell (multiple b-values) dMRI data by predicting 7T-like diffusion MRI signal from 3T data by mapping their corresponding RISH features. In particular, we propose to learn an efficient mapping of multi-shell dMRI signal with different spatial resolution and magnetic field strength (3T and 7T) using deep Convolutional Neural Networks (CNN). We investigate and propose two methods: voxel-wise linear mapping, and patch-based non-linear mapping using deep Convolutional Neural Networks (CNN) which are explained in the following sections.

## 2    Method

### 2.1    RISH Features

We represent the dMRI signal $\mathbf{S}$ in a basis of spherical harmonics (SH): $\mathbf{S} \approx \sum_l \sum_m C_{lm} Y_{lm}$, and construct the rotation invariant spherical harmonic (RISH) features which can be appropriately scaled to modify the dMRI signal without changing the principal diffusion directions of the fibers. Thus, our goal is to estimate a voxel-wise linear or a patch-based non-linear mapping of the RISH features between 3T and 7T data from the same set of subjects, which can then be used on test subjects to validate the quality of the mapping.

The following processing was common for both linear and non-linear methods in Sect. 2.2. Due to differences in spatial resolution between 3T and 7T dMRI data, we first upsample each DW volume using a 7th-order B-spline which was shown to perform better than other interpolation schemes [4]. Next, we use a recently proposed unringing method [6] to remove Gibbs ringing artifact from each DW volume. Five RISH feature maps $\mathbf{C}_l^{b,s}(\mathbf{x}; i)$ for each b-value shell with SH orders of $l = \{0, 2, 4, 6, 8\}$ are computed at each voxel location $\mathbf{x} = (x, y, z) \in \mathbb{R}^3$ for each scanner $s$ as follows:

$$\mathbf{C}_l^{b,s}(\mathbf{x}; t) = \sum_{m=1}^{2l+1} C_{lm}(\mathbf{x})^2, \tag{1}$$

where $t$ is the subject number and $b = \{1000, 2000\}$ is the b-value.

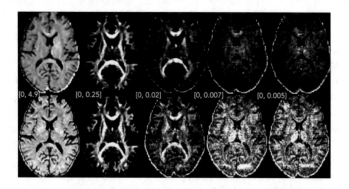

**Fig. 1.** RISH Features of $b = 1000$ shell for SH orders of $l = \{0, 2, 4, 6, 8\}$ are depicted in each sub-figure from left to right for 3T (top row) and 7T scans (bottom row) for HCP Subject ID: 102311 ($[\cdot, \cdot]$: the range of intensities chosen for visualization).

Figure 1 shows the RISH features of the same HCP subject scanned on 3T (top) and 7T (bottom) scanner for $b = 1000$. Each RISH feature captures a different aspect (frequency content) of the diffusion signal. Note the significantly increased energy (contrast) in higher order RISH features in 7T data, that is not quite visible in the 3T data.

## 2.2   Learning the Mapping of RISH Features from 3T to 7T

**Voxel-Wise Linear Mapping:** Using 3T RISH features as input, our goal is to learn the voxel-wise linear mapping of 3T to 7T. To achieve this, first, the RISH features in the training set are used to create multi-modal RISH feature templates (antsMultiVariateTemplateConstruction [2]). Once the template space is constructed separately for each shell, we define the expected value of the voxel-wise RISH features as the sample mean $\mathbb{E}_l^{b,s}(\mathbf{x}') \approx \sum_{t=1}^{N_s} \mathbf{C}_l^{b,s}(\mathbf{x}';t)/N_s$ over the number of training subjects $N_s$, where $s$ is 3T or 7T scanner and $\mathbf{x}'$ is the voxel location in the template space. Next, we compute the voxel-wise linear (scaling only) maps between RISH features of 3T and 7T data in the template space using: $\mathfrak{S}_l(\mathbf{x}') = \sqrt{\frac{\mathbb{E}_l^{b,7T}(\mathbf{x}')}{\mathbb{E}_l^{b,3T}(\mathbf{x}')+\epsilon}}$. We apply this linear map learned from the training data set to new subjects from the test data, by non-rigid transformation of scale maps to the subject space. The 7T-like dMRI signal is estimated by scaling the SH coefficients of the signal at each voxel in the subject space as follows: $\hat{\mathbf{C}}_{lm}(\mathbf{x}) = \hat{\mathfrak{S}}_l(\mathbf{x})\,C_{lm}(\mathbf{x})$, where $\hat{\mathfrak{S}}_l(\mathbf{x})$ is the scale map in the subject space and $\hat{\mathbf{C}}_{lm}(\mathbf{x})$ is the scaled SH coefficients. The final diffusion signal is then computed using:

$$\hat{\mathbf{S}}(\mathbf{x}) = \sum_l \sum_m \hat{\mathbf{C}}_{lm}(\mathbf{x})Y_{lm}. \tag{2}$$

**Patch-Based Non-linear Mapping Using Deep CNN:** Using 3T RISH features as input, our goal is to learn a nonlinear mapping of 3T to 7T as a patch-wise regression problem. Such mapping can be learned using the paired 3T and 7T RISH features of training data. We first align 3T and 7T data as follows: First, we register b0 maps of 3T and 7T data through rigid registration [2]. The estimated transformation is then applied to each DW volume. Next, the gradient vectors are rotated using the rotation matrix estimated through rigid registration. After 3T and 7T DW data are aligned, we compute RISH features as in Eq. 1. To learn the mapping from 3T to 7T, we construct our deep CNN with five convolutional layers. Specifically, we used an $9 \times 9$ RISH feature patch to learn the mapping.

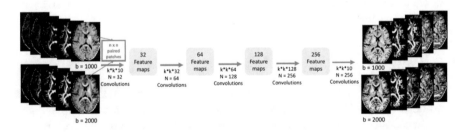

**Fig. 2.** Our proposed deep CNN architecture for learning the mapping from 3T to 7T RISH features for multi-shell data. We used $k = 3$ as the filter size with five layers.

Figure 2 summarizes our deep CNN architecture. In the first layer, the aim is to learn a feature representation of the input 3T RISH feature patches. It includes convolution filters with size of 32 followed by ReLU activation function. In each layer, RISH features are convolved with a $3 \times 3$ kernel with 32, 64, 128,

**Table 1.** Average RMSE (percentage) in FA, MD, GFA and DWI signal in 30 test subjects using different 3T to 7T mapping techniques for each shell separately.

| RMSE: % | Original 3T-7T | Linear | RF-DTI | RF-RISH | CNN-RISH |
|---|---|---|---|---|---|
| $b = 1000$ | | | | | |
| FA | $36.49 \pm 2.22$ | $6.94 \pm 0.019$ | $6.23 \pm 0.023$ | $6.01 \pm 0.011$ | $\mathbf{5.41 \pm 0.010}$ |
| MD | $40.32 \pm 2.67$ | $4.02 \pm 0.012$ | $3.44 \pm 0.021$ | $3.01 \pm 0.008$ | $\mathbf{2.33 \pm 0.013}$ |
| GFA | $38.53 \pm 2.13$ | $7.11 \pm 0.021$ | - | $6.55 \pm 0.015$ | $\mathbf{5.95 \pm 0.011}$ |
| DWI | $39.33 \pm 2.45$ | $7.32 \pm 0.023$ | - | $6.61 \pm 0.012$ | $\mathbf{5.94 \pm 0.010}$ |
| $b = 2000$ | | | | | |
| FA | $34.32 \pm 2.10$ | $5.97 \pm 0.013$ | $5.86 \pm 0.019$ | $5.51 \pm 0.012$ | $\mathbf{4.95 \pm 0.008}$ |
| MD | $35.44 \pm 2.14$ | $3.23 \pm 0.012$ | $2.97 \pm 0.017$ | $2.76 \pm 0.014$ | $\mathbf{2.12 \pm 0.011}$ |
| GFA | $36.66 \pm 2.77$ | $6.53 \pm 0.014$ | - | $6.03 \pm 0.011$ | $\mathbf{5.23 \pm 0.010}$ |
| DWI | $39.43 \pm 2.21$ | $7.08 \pm 0.021$ | - | $6.51 \pm 0.015$ | $\mathbf{5.42 \pm 0.012}$ |

**Table 2.** Overlap (percentage) in various fiber bundles: CST, CB, AF and IOFF traced in the original 3T and 7T, and the harmonized 7T-like data in 30 test subjects.

| Fiber bundle: % | 7T vs 3T | 7T-like vs 7T | 7T-like vs 3T |
|---|---|---|---|
| CST | $93.59 \pm 3.36$ | $92.98 \pm 3.28$ | $93.90 \pm 5.35$ |
| CB | $95.35 \pm 3.03$ | $94.22 \pm 5.50$ | $96.60 \pm 2.26$ |
| AF | $96.53 \pm 1.48$ | $96.80 \pm 1.52$ | $97.61 \pm 1.32$ |
| IOFF | $94.96 \pm 2.42$ | $96.35 \pm 2.21$ | $95.60 \pm 2.81$ |

256 and 256 convolutional filters. After each convolution step, ReLU operation is applied. In training, we used ADAM optimizer with a learning rate $10^{-4}$ and epoch size is selected as 100.

## 3   Results

We used 10 HCP subjects [13] as training subjects with dMRI scans obtained from both 7T and 3T scanners. Another independent set of unseen 30 HCP subjects (with data from both 3T and 7T) were used to evaluate the performance of all the methods. 7T data had the following acquisition parameters: 1.05 mm isotropic spatial resolution, two-shells ($b = 1000$, 2000) with 65 gradient directions on each shell; while 3T data had: 1.25 mm isotropic spatial resolution, three-shells ($b = 1000$, 2000, 3000) with 90 gradient directions on each shell. In this work, we learnt the mapping only for $b = 1000$ and $b = 2000$ shells from 3T to 7T. We compared our methods with another non-linear learning method: Regression Forest (RF) method which was presented by [1] to improve DTI data quality (RF-DTI). Note that, this method was not used in the context of data harmonization, yet we found it relevant to compare our work with it. In this paper, we also introduce RF-RISH (Regression Forest (RF) with RISH features) to provide a fair comparison between our RISH feature based and RF-DTI based method. Using our methods, subject-specific mapping between 3T and 7T was obtained and the final signal was estimated using Eq. 2.

We computed whole brain FA and MD to compare the learning performance between the RISH features based methods and RF-DTI [1]. Root Mean Squared Error (RMSE) on 30 test subjects was computed for DTI specific measures of FA and MD as well non-model specific measure of generalized FA (GFA) and the dMRI signal itself. RMSE was computed between our prediction and the ground truth data that was acquired on 7T from the same set of subjects. Average accuracy and precision values for estimation of FA, MD, GFA and DWI signal are given in Table 1. In Fig. 3(a) top row, we show the estimated FA results using our methods (Linear-RISH, RF-RISH and CNN-RISH) and RF-DTI for $b = 1000$. In Fig. 3(a) bottom row, we show the error maps (RMSE) in FA between the predicted data and the actual scanner acquired 7T data. Figure 3(b) shows error maps in the raw dMRI signal, with most of the error using CNN-RISH confined to the CSF regions of the brain. Even though FA and MD are directly derived from DTI model, RISH features based non-linear methods performed better when compared to RF-DTI. As seen in Table 1 and Fig. 3, our deep CNN-RISH method gives the best performance, with lowest error in several metrics (FA, GFA and dMRI signal error). Thus, our method is tissue model-independent and directly reconstructs the dMRI signal, which can then be used in further analysis.

In order to ensure that our deep CNN method does not change the fiber orientation, we performed whole brain tractography using a multi-tensor unscented Kalman filter (UKF) method [7]. Next, we use the White Matter Query Language (WMQL) [17] to extract specific anatomical white matter bundles from the whole brain tracts. Figure 4 depicts WMQL results for corticospinal tract (CST) and cingulum bundle (CB). After extracting the tracts from the original 3T and 7T, and the harmonized 7T-like data, we used the Bhattacharyya overlap distance $(B)$ to quantify the agreement between the tracts [10]: $B = \frac{1}{3}\left(\int \sqrt{P_h(x)P(x)}dx + \int \sqrt{P_h(y)P(y)}dy + \int \sqrt{P_h(z)P(z)}dz\right)$, where $P(.)$ represents the ground truth probability distribution of the fiber bundle, $P_h(.)$ is the probability distribution of the tracts from the harmonized data and $\mathbf{x} = (x, y, z) \in \mathbb{R}^3$ are the fiber coordinates. $B$ is 1 for a perfect match between two fiber bundles and 0 for no overlap at all. In Table 2, we provide the Bhattacharyya overlap measure for: (i) the original 7T vs the original 3T; (ii) the estimated 7T-like vs the original 7T; (iii) the estimated 7T-like vs the original 3T data. Due to the space limitations, we only show the results for CST, CB, arcuate fasciculus (AF) and the inferior occipito-frontal fascicle (IOFF) tracts. We observed very high overlap of 93–97% for all fiber bundles indicating that fiber orientation is preserved by the harmonization algorithm.

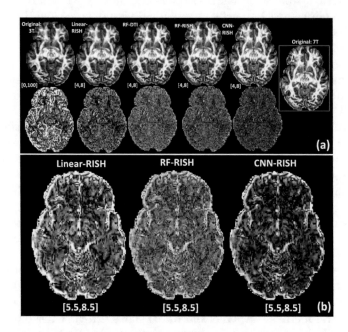

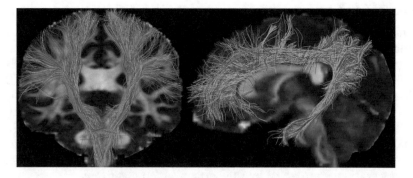

**Fig. 3.** (a) FA comparison between our methods (Linear-RISH, RF-RISH and CNN-RISH) and RF-DTI for $b = 1000$ are shown in the top row. FA map of the original 7T data is depicted in the rightmost figure. RMSE maps in FA for different methods as well as the original 3T data itself (leftmost) are depicted in the bottom row. All results are computed and shown for the same subject; (b) RMSE maps between DW volumes estimated using our methods (Linear-RISH, RF-RISH and CNN-RISH) and the original 7T data for a single subject. ($[\cdot, \cdot]$: the range of intensities chosen for visualization).

**Fig. 4.** Significant (>93%) overlap is seen in CST and CB extracted from the original 3T (blue) and 7T (magenta), and harmonized 7T-like (green) dMRI data.

# 4    Conclusion

In this paper, we proposed a linear and a nonlinear machine learning method to harmonize the raw dMRI data acquired on scanners with very different magnetic field strengths (3T and 7T). We validated our algorithm on 30 test subjects, and demonstrated the efficacy of using this technique to harmonize dMRI data from vastly different scanners in a model-free manner. Even though FA and MD are directly related DTI, both qualitative and quantitative results show that our methods perform better or close (for linear regression) when compared to RF-DTI. The tractography results also prove that our deep CNN method matches the dMRI signal between scanners while preserving the fiber orientations. The proposed method can also be useful to improve the quality and resolution of dMRI data. As a first step, we have demonstrated and validated the utility of this work in harmonizing data from healthy subjects while more validation needs to be done on subjects with gross tissue pathology.

# References

1. Alexander, D.C., et al.: Image quality transfer and applications in diffusion MRI. NeuroImage **152**(Suppl. C), 283–298 (2017)
2. Avants, B.B., et al.: The optimal template effect in hippocampus studies of diseased populations. NeuroImage **49**(3), 2457–2466 (2010)
3. Chilla, G.S., Tan, C.H., Xu, C., Poh, C.L.: Diffusion weighted magnetic resonance imaging and its recent trend-a survey. Quant. Imaging Med. Surg. **5**(3), 407 (2015)
4. Dyrby, T.B.: Interpolation of diffusion weighted imaging datasets. NeuroImage **103**(Suppl. C), 202–213 (2014)
5. Fortin, J.P., et al.: Harmonization of multi-site diffusion tensor imaging data. Neuroimage **161**, 149–170 (2017)
6. Kellner, E., Dhital, B., Kiselev, V., Reisert, M.: Gibbs-ringing artifact removal based on local subvoxel-shifts. Magn. Reson. Med. **76**(5), 1574–1581 (2015)
7. Malcolm, J.G., Shenton, M.E., Rathi, Y.: Filtered multitensor tractography. IEEE Trans. Med. Imaging **29**(9), 1664–1675 (2010)
8. Mirzaalian, H., et al.: Multi-site harmonization of diffusion MRI data in a registration framework. Brain Imaging Behav. **12**, 1–12 (2016)
9. Pohl, K.M., et al.: Harmonizing DTI measurements across scanners to examine the development of white matter microstructure in 803 adolescents of the NCANDA study. Neuroimage **130**, 194–213 (2016)
10. Rathi, Y., Gagoski, B., Setsompop, K., Michailovich, O., Grant, P.E., Westin, C.-F.: Diffusion propagator estimation from sparse measurements in a tractography framework. In: Mori, K., Sakuma, I., Sato, Y., Barillot, C., Navab, N. (eds.) MICCAI 2013. LNCS, vol. 8151, pp. 510–517. Springer, Heidelberg (2013). https://doi.org/10.1007/978-3-642-40760-4_64
11. Sotiropoulos, S.N., et al.: Fusion in diffusion MRI for improved fibre orientation estimation: an application to the 3T and 7T data of the human connectome project. NeuroImage **134**, 396–409 (2016)
12. Tanno, R., et al.: Bayesian image quality transfer with CNNs: exploring uncertainty in dMRI super-resolution. In: Descoteaux, M., Maier-Hein, L., Franz, A., Jannin, P., Collins, D.L., Duchesne, S. (eds.) MICCAI 2017. LNCS, vol. 10433, pp. 611–619. Springer, Cham (2017). https://doi.org/10.1007/978-3-319-66182-7_70

13. Van Essen, D.C., et al.: The WU-Minn human connectome project: an overview. Neuroimage **80**, 62–79 (2013)
14. Volkow, N.D., et al.: The conception of the ABCD study: from substance use to a broad NIH collaboration. Dev. Cogn. Neurosci. **32**, 4–7 (2017). ISSN 1878-9293
15. Vollmar, C., et al.: Identical, but not the same: Intra-site and inter-site reproducibility of fractional anisotropy measures on two 3.0 T scanners. NeuroImage **51**, 1384–1394 (2010)
16. Vu, A.T., et al.: High resolution whole brain diffusion imaging at 7T for the human connectome project. Neuroimage **122**, 318–331 (2015)
17. Wassermann, D., et al.: The white matter query language: a novel approach for describing human white matter anatomy. Brain Struct. Funct. **221**(9), 4705–4721 (2016)

# Diffusion Tensor Imaging and Functional MRI: Functional MRI

# Normative Modeling of Neuroimaging Data Using Scalable Multi-task Gaussian Processes

Seyed Mostafa Kia[1,2(✉)] and Andre Marquand[1,2]

[1] Department of Cognitive Neuroscience, Radboud University Medical Centre, Nijmegen, The Netherlands
[2] Donders Centre for Cognitive Neuroimaging, Donders Institute for Brain, Cognition and Behaviour, Radboud University, Nijmegen, The Netherlands
{s.kia,a.marquand}@donders.ru.nl

**Abstract.** Normative modeling has recently been proposed as an alternative for the case-control approach in modeling heterogeneity within clinical cohorts. Normative modeling is based on single-output Gaussian process regression that provides coherent estimates of uncertainty required by the method but does not consider spatial covariance structure. Here, we introduce a scalable multi-task Gaussian process regression (S-MTGPR) approach to address this problem. To this end, we exploit a combination of a low-rank approximation of the spatial covariance matrix with algebraic properties of Kronecker product in order to reduce the computational complexity of Gaussian process regression in high-dimensional output spaces. On a public fMRI dataset, we show that S-MTGPR: (1) leads to substantial computational improvements that allow us to estimate normative models for high-dimensional fMRI data whilst accounting for spatial structure in data; (2) by modeling both spatial and across-sample variances, it provides higher sensitivity in novelty detection scenarios.

**Keywords:** Gaussian processes · Multi-task learning
Normative modeling · Neuroimaging · fMRI · Clinical neuroscience
Novelty detection

## 1 Introduction

Understanding the underlying biological mechanisms of psychiatric disorders constitutes a significant step toward developing more effective and individualized treatments (*i.e., precision medicine* [11]). Recent advances in neuroimaging and machine learning provide an exceptional opportunity to employ brain-derived biological measures for this purpose. While symptoms and biological

---

**Electronic supplementary material** The online version of this chapter (https://doi.org/10.1007/978-3-030-00931-1_15) contains supplementary material, which is available to authorized users.

A. F. Frangi et al. (Eds.): MICCAI 2018, LNCS 11072, pp. 127–135, 2018.
https://doi.org/10.1007/978-3-030-00931-1_15

underpinnings of mental diseases are known to be highly heterogeneous, data-driven approaches play an important role in stratifying clinical groups into more homogeneous subgroups. Currently, off-the-shelf clustering algorithms are the most predominant approaches for stratifying clinical cohorts. However, the high-dimensionality and complexity of data beside the use of heuristics to find optimal clustering solutions negatively affect the reproducibility and reliability of resulting clusters [10]. Normative modeling [9] offers an alternative approach to model biological variations within clinical cohorts without needing to assume cleanly separable clusters or cohorts. This approach is applicable to most types of neuroimaging data such as structural/functional magnetic resonance imaging (s/fMRI).

Normative modeling employs Gaussian process regression (GPR) [16] to predict neuroimaging data on the basis of clinical and/or behavioral covariates. GPR, and in general Bayesian inference, can be seen as an indispensable part of the normative modeling as it provides coherent estimates of predictive confidence. These measures of predictive uncertainty are important for quantifying centiles of variation in a population [9]. GPR also provides the possibility to accommodate both linear and nonlinear relationships between clinical covariates and neuroimaging data.

The variant of GPR originally employed for normative modeling aims to model only a single output variable. Thus in normative modeling, one should independently train separate GPR models for each unit of measurement (e.g., for each voxel in a mass-univariate fashion). Such a simplification ignores the possibility of modeling and capitalizing on the existing spatial structure in the output space. However, GPR can be extended to perform a joint prediction across multiple outputs in order to account for correlations between variables in neuroimaging data (for example different voxels in fMRI data). Boyle and Frean [6] proposed to employ convolutional processes to express each output as the convolution between a smoothing kernel and a latent function. This idea is later adopted by Bonilla et al. [5] to extend the classical single-task GPR (STGPR) to multi-task GPR (MTGPR) by coupling a set of latent functions with a shared GP prior in order to directly induce correlation between output variables (tasks). They proposed to disentangle the full cross-covariance matrix into the Kronecker product of the sample (in input space) and task (in output space) covariance matrices. This technique provides the possibility to model both across-sample and across-task variations. Despite its effectiveness in modeling structures in data, MTGPR comes with extra computational overheads in time and space, especially when dealing with high-dimensional neuroimaging data. We briefly review recent efforts toward alleviating these computational burdens.

## 1.1   Toward Efficient and Scalable MTGPR

For $N$ samples and $T$ tasks, the time and space complexity of MTGPR are $\mathcal{O}(N^3 T^3)$ and $\mathcal{O}(N^2 T^2)$, respectively. These high computational demands (compared to STGPR with $\mathcal{O}(N^3 T)$ and $\mathcal{O}(N^2 T)$) are mainly due to the need for computing the inverse cross-covariance matrix in learning and inference phases.

In neuroimaging problems that we consider, these can both be relatively high where $N$ in generally in the order of $10^2 - 10^4$ and $T$ is in the order of $10^4 - 10^5$ or even higher. Therefore, improving the computational efficiency of MTGPR is crucial for certain problems, and there have been several approaches proposed for this in the machine learning literature [3, 13]. Here we briefly review two main directions to address the computational tractability issue of MTGPR.

In the first set of approaches, approximation techniques are used to improve estimation efficiency. Bonilla *et al.* [5] made one of the earliest efforts in this direction, in which they proposed to use Nyström approximation on $M$ inducing inputs [13] out of $N$ samples in combination with the probabilistic principal component analysis, in order to approximate reduced $M$-rank and $P$-rank sample and task covariance matrices, respectively. Their approximation reduced the time complexity of hyperparameter learning to $\mathcal{O}(NTM^2P^2)$. Elsewhere, Alvarez and Lawrence [2] proposed to approximate a sparse version of MTGPR, assuming conditional independence between each output variable with all others given the input process. This assumption besides using $M$ out of $N$ input samples as inducing inputs reduces the computational complexity of MTGPR to $\mathcal{O}(N^3T+NTM^2)$ and $\mathcal{O}(N^2T+NTM)$ in time and storage, where for $N = M$ is the same as a set of $T$ independent STGPRs. Alvarez *et al.* in [4] extended their previous work by developing the concept of inducing function rather than inducing input. Their new approach so-called variational inducing kernels achieves time complexity of $\mathcal{O}(NTM^2)$.

The second set of approaches utilize properties of Kronecker product [8] to reduce the time and space complexity in computing the exact (and not approximated) inverse covariance matrix. Stegle *et al.* [15] proposed to use these properties in combination with eigenvalue decomposition of input and task covariance matrices for efficient parameter estimation, and likelihood evaluation/optimization in MTGPR. In this method, the joint covariance matrix is defined as a Kronecker product between the input and task covariance matrices. This approach reduces the time and space complexity of MTGPR to $\mathcal{O}(N^3+T^3)$ and $\mathcal{O}(N^2 + T^2)$, respectively. To account also for structured noise, Rakitsch *et al.* [14] extended this method by using two separate Kronecker products for the signal and noise. Importantly, this provides a significant reduction in computational complexity using all samples (*i.e.*, not just inducing inputs), and is exact in the sense that it does not require any approximation or relaxing assumptions.

**Our Contribution:** In spite of all aforementioned efforts, applications of MTGPR in encoding neuroimaging data from a set of clinically relevant covariates remained very limited, mainly due to the high dimensionality of the output space (*i.e.*, very large $T$). Our main contribution in this text addresses this problem and extends MTGPR to the normative modeling of neuroimaging data. To this end, we use a combination of low-rank approximation of the task covariance matrix with algebraic properties of Kronecker product in order to reduce the computational complexity of MTGPR. Furthermore, on a public fMRI dataset, we show that: (1) our method makes MTGPR possible on very high-dimensional output spaces; (2) it enables us to model both across-space and across-subjects variations, hence provides more sensitivity for the resulting normative model in novelty detection.

## 2   Methods

### 2.1   Notation

Boldface capital letters, $\mathbf{A}$, and capital letters, $A$, are used to denote matrices and scalar numbers. We denote the vertical vector which is resulted from collapsing columns of a matrix $\mathbf{A} \in \mathbb{R}^{N \times T}$ with $vec(\mathbf{A}) \in \mathbb{R}^{NT}$. In the remaining text, we use $\otimes$ and $\odot$ to respectively denote Kronecker and the element-wise matrix products. We denote an identity matrix by $\mathbf{I}$; and the determinant, diagonal elements, and the trace of matrix $\mathbf{A}$ with $|\mathbf{A}|$, $diag(\mathbf{A})$, and $Tr[\mathbf{A}]$, respectively.

### 2.2   Scalable Multi-task Gaussian Process Regression

Let $\mathbf{X} \in \mathbb{R}^{N \times F}$ be the input matrix with $N$ samples and $F$ covariates. Let $\mathbf{Y} \in \mathbb{R}^{N \times T}$ represent a matrix of response variables with $N$ samples and $T$ tasks (here, neuroimaging data with $T$ voxels). The multi-task Kronecker Gaussian process model (MT-Kronprod) [15] is defined as:

$$p(\mathbf{Y} \mid \mathbf{D}, \mathbf{R}, \sigma^2) = \mathcal{N}(\mathbf{Y} \mid \mathbf{0}, \mathbf{D} \otimes \mathbf{R} + \sigma^2 \mathbf{I}) \quad , \tag{1}$$

where $\mathbf{D} \in \mathbb{R}^{T \times T}$ and $\mathbf{R} \in \mathbb{R}^{N \times N}$ are respectively the task and sample covariance matrices (here, modeling correlations across voxels and samples separately). Despite its effectiveness in modeling both samples and tasks variations, the application of MT-Kronprod is limited when dealing with very large output spaces, such as neuroimaging data, mainly due to the high computational complexity of matrix diagonalisation operations in the optimization and inference phases. We propose to address this problem by using a low-rank approximation of $\mathbf{D}$.

Let $\varPhi : \mathbf{Y} \to \mathbf{Z}$ be an orthogonal linear transformation, *e.g.*, principal component analysis (PCA), that transforms $\mathbf{Y}$ to a reduced latent space $\mathbf{Z} \in \mathbb{R}^{N \times P}$, where $P < T$, and we have $\mathbf{Z} = \varPhi(\mathbf{Y}) = \mathbf{YB}$. Here, columns of $\mathbf{B} \in \mathbb{R}^{T \times P}$ represent a set of $P$ orthogonal basis functions. Assuming a zero-mean matrix normal distribution for $\mathbf{Z}$, by factorizing its rows and columns we have:

$$p(\mathbf{Z} \mid \mathbf{C}, \mathbf{R}) = \mathcal{MN}(\mathbf{0}, \mathbf{C} \otimes \mathbf{R}) = \frac{\exp(-\frac{1}{2}Tr[\mathbf{C}^{-1}\mathbf{B}^\top\mathbf{Y}^\top\mathbf{R}^{-1}\mathbf{YB}])}{\sqrt{(2\pi)^{NP} |\mathbf{C}|^P |\mathbf{R}|^N}} \quad , \tag{2}$$

where $\mathbf{C} \in \mathbb{R}^{P \times P}$ and $\mathbf{R} \in \mathbb{R}^{N \times N}$ are column and row covariance matrices of $\mathbf{Z}$. Using the trace invariance property under cyclic permutations, the noise-free multivariate normal distribution of $\mathbf{Y}$ can be approximated from Eq. 2:

$$p(\mathbf{Y} \mid \mathbf{D}, \mathbf{R}) \approx p(\mathbf{Y} \mid \mathbf{C}, \mathbf{B}, \mathbf{R}) = \frac{\exp(-\frac{1}{2}Tr[\mathbf{BC}^{-1}\mathbf{B}^\top\mathbf{Y}^\top\mathbf{R}^{-1}\mathbf{Y}])}{\sqrt{(2\pi)^{NT} |\mathbf{BCB}^\top|^T |\mathbf{R}|^N}} \quad , \tag{3}$$

where $\mathbf{D}$ is approximated by $\mathbf{BCB}^\top$. Our scalable multi-task Gaussian process regression (S-MTGPR) model is then derived by marginalizing over noisy samples:

$$p(\mathbf{Y} \mid \mathbf{D}, \mathbf{R}, \sigma^2) \approx p(\mathbf{Y} \mid \mathbf{C}, \mathbf{B}, \mathbf{R}, \sigma^2) = \mathcal{N}(\mathbf{Y} \mid \mathbf{0}, \mathbf{BCB}^\top \otimes \mathbf{R} + \sigma^2 \mathbf{I}) \quad . \tag{4}$$

**Predictive Distribution:** Following the standard GPR framework [16] and setting $\tilde{\mathbf{D}} = \mathbf{BCB}^\top$, the mean and variance of the predictive distribution of unseen samples, i.e., $p(vec(\mathbf{Y})^* \mid vec(\mathbf{M}^*), \mathbf{V}^*)$, can be computed as follows:

$$vec(\mathbf{M}^*) = (\tilde{\mathbf{D}} \otimes \mathbf{R}^*)(\tilde{\mathbf{D}} \otimes \mathbf{R} + \sigma^2 \mathbf{I})^{-1} vec(\mathbf{Y}), \tag{5a}$$

$$\mathbf{V}^* = (\tilde{\mathbf{D}} \otimes \mathbf{R}^{**}) - (\tilde{\mathbf{D}} \otimes \mathbf{R}^*)(\tilde{\mathbf{D}} \otimes \mathbf{R} + \sigma^2 \mathbf{I})^{-1}(\tilde{\mathbf{D}} \otimes \mathbf{R}^{*\top}), \tag{5b}$$

where $\mathbf{R}^{**} \in \mathbb{R}^{N^* \times N^*}$ is the covariance matrix of $N^*$ test samples , and $\mathbf{R}^* \in \mathbb{R}^{N^* \times N}$ is the cross-covariance matrix between test and training samples.

**Efficient Prediction and Optimization:** For efficient prediction, and fast optimization of the log-likelihood, we extend the approach proposed in [14,15] by exploiting properties of Kronecker product, and eigenvalue decomposition for diagonalizing the covariance matrices. Then the predictive mean and variance can be efficiently computed by:

$$\mathbf{M}^* = \mathbf{R}^* \mathbf{U}_\mathbf{R} \tilde{\mathbf{Y}} \mathbf{U}_\mathbf{C}^\top \mathbf{CB}^\top, \tag{6a}$$

$$\mathbf{V}^* = (\tilde{\mathbf{D}} \otimes \mathbf{R}^{**}) - (\mathbf{BCU}_\mathbf{C} \otimes \mathbf{R}^* \mathbf{U}_\mathbf{R}) \tilde{\mathbf{K}}^{-1} (\mathbf{U}_\mathbf{C}^\top \mathbf{CB}^\top \otimes \mathbf{U}_\mathbf{R}^\top \mathbf{R}^{*\top}), \tag{6b}$$

where $\mathbf{C} = \mathbf{U}_\mathbf{C} \mathbf{S}_\mathbf{C} \mathbf{U}_\mathbf{C}^\top$ and $\mathbf{R} = \mathbf{U}_\mathbf{R} \mathbf{S}_\mathbf{R} \mathbf{U}_\mathbf{R}^\top$ are eigenvalue decomposition of covariance matrices, $\tilde{\mathbf{K}} = \mathbf{S}_\mathbf{C} \otimes \mathbf{S}_\mathbf{R} + \sigma^2 \mathbf{I}$, and $vec(\tilde{\mathbf{Y}}) = diag(\tilde{\mathbf{K}}^{-1}) \odot vec(\mathbf{U}_\mathbf{R}^\top \mathbf{Y} \mathbf{B} \mathbf{U}_\mathbf{C})^1$. Based on our assumption on the orthogonality of components in $\mathbf{B}$, we set $\mathbf{B}^{-1} = \mathbf{B}^\top$ and $\mathbf{B}^\top \mathbf{B} = \mathbf{I}$. Note that in the new parsimonious formulation, heavy time and space complexities of computing the inverse kernel matrix is reduced to computing the inverse of a diagonal matrix, i.e., reciprocals of diagonal elements of $\tilde{\mathbf{K}}$. For the predictive variance, explicit computation of the Kronecker product is still necessary but this can easily be overcome by computing the predictions in mini-batches. For the negative log marginal likelihood of Eq. 4, we have:

$$\mathcal{L} = -\frac{N \times T}{2} \ln(2\pi) - \frac{1}{2} \ln \left| \tilde{\mathbf{K}} \right| - \frac{1}{2} vec(\mathbf{U}_\mathbf{R}^\top \mathbf{Y} \mathbf{B} \mathbf{U}_\mathbf{C})^\top vec(\tilde{\mathbf{Y}}) \quad . \tag{7}$$

The proposed S-MTGPR model has three sets of parameters plus one hyper-parameter: (1) reduced task covariance matrix parameters $\Theta_\mathbf{C}$, (2) input covariance matrix parameters $\Theta_\mathbf{R}$, (3) noise variance $\sigma^2$ that is parametrized on $\Theta_{\sigma^2}$, and (4) $P$ that decides the number of components in $\mathbf{B}$. While the latter should be decided by means of model selection, the first three sets are optimized by maximizing $\mathcal{L}$.

---

[1] See supplementary materials for more descriptive derivations of all equations.

**Computational Complexity:** The time complexity of the proposed method is $\mathcal{O}(N^2T + NT^2 + N^3 + P^3)$. The first two terms are related to the matrix multiplication in computing the squared term in Eq. 7. The last two terms belong to the eigenvalue decomposition of $\mathbf{R}$ and $\mathbf{C}$. The $P^3$ term can be excluded because always $P \leq min(N, T)$. Thus, for $N > T$ and $N < T$ the time complexity is reduced to $\mathcal{O}(N^3)$ and $\mathcal{O}(NT^2)$, respectively. Thus when $N > T$ or $N < T < N^2$, our approach is analytically even faster than the baseline STGPR approach applied independently to each output variable in a mass-univariate fashion. For $N \ll T$, our method is faster than other Kronecker based MTGPRs by a factor of $T/N$. Such improvement not only facilitates the application of MTGPR on neuroimaging data but also it provides the possibility of accounting for the existing spatial structures across different brain regions. In comparison to the related work, the proposed method provides a substantial speed improvement, especially when dealing with a large number of tasks. This is while unlike other approximation approaches, we fully use the potential of all available samples.

## 3 Experiments and Results

### 3.1 Experimental Materials and Setup

In our experiments, we use a public fMRI dataset collected for reconstructing visual stimuli (black and white letters and symbols) from fMRI data [12]. In this dataset, fMRI responses were measured while $10 \times 10$ checkerboard patch images were presented to subjects according to a blocked design. Checkerboard patches constituted random (1320 trials) and geometrically meaningful patterns (720 trials). We use the preprocessed data available in Nilearn package [1] wherein the fMRI data are detrended and masked for the occipital lobe (5438 voxels).[2] Whilst our approach is quite general, we demonstrate S-MTGPR by simulating normative modeling for novelty detection. Therefore, we aim to predict the masked fMRI 3D-volume from the presented visual stimuli in an encoding setting. To this end, we randomly selected 600 random pattern trials, for training the encoding model. The model then learns to represent this reference or normative class such that anomalous or abnormal samples can be detected and characterised. The rest of non-random patterns (720 trials) and random patterns (720 trials) are used for evaluating the encoding model and testing anomaly-detection performance, achieved by fitting a generalised extreme value distribution to the most deviating voxels. In our experiments, we use PCA to transform the fMRI data in the training set from the voxel space to $\mathbf{Z}$, and the resulting $P = 10, 25, 50, 100, 250, 500, 1000$ PCA components are used as basis matrix $\mathbf{B}$ in the optimization and inference.

We benchmark the proposed method against the STGPR (*i.e.*, mass-univariate) and MT-Kronprod models in terms of their runtime, performance of the regression, and quality of resulting normative models. In all models, we use

---

[2] See http://nilearn.github.io/auto_examples/02_decoding/plot_miyawaki_reconstruction.html.

**Table 1.** Three benchmarked methods in our experiments.

| Method | Time complexity | No. parameters | Parameter description |
|---|---|---|---|
| STGPR | $\mathcal{O}(N^3T)$ | 21752 | 1 for linear and 2 for squared exponential kernels, 1 for Gaussian likelihood; multiplied by the number of tasks (5438) |
| MT-Kronprod | $\mathcal{O}(T^3)$ | 9 | 1 for linear, 2 for squared exponential, and 1 for diagonal isotropic kernels; multiplied by 2 (for sample and task covariance functions); plus 1 for Gaussian likelihood |
| S-MTGPR | $\mathcal{O}(NT^2)$ | 10 | Same as MT-Kronprod, plus 1 hyperparameter for the number of PCA bases |

a summation of a linear, a squared exponential, and a diagonal isotropic covariance functions for sample and task covariance matrices in order to accommodate both linear and non-linear relationships. In all cases, we use an isotropic Gaussian likelihood function. This likelihood function has different functionality in the STGPR versus MTGPR settings. In STGPR, it is defined independently for each voxel, thus it handles heteroscedastic, *i.e.*, spatially varying noise. While in MTGPR a single noise parameter is shared for all voxels, hence it merely considers homoscedastic, *i.e.*, spatially stationary, noise. The truncated Newton algorithm is used for optimizing the parameters. Table 1 summarizes the time complexity and the number of parameters of three benchmarked methods in our experiments.

We use the coefficient of determination ($R^2$) to evaluate the explained variance by regression models. In normative modeling, the top 5% values in normative probability maps are used to fit the generalized extreme value distribution (see [9]). To evaluate resulting normative models, we employ area under the curve (AUC) to measure the performance of the model in distinguishing between normal (here random patterns) from abnormal samples (here non-random patterns). All the steps (random sampling, modeling, and evaluation) are repeated 10 times in order to estimate the mean and standard deviation of the runtime, $R^2$, and AUC. All experiments are performed on a system with Intel®Xeon®E5-1620 0 @3.60GHz CPU and 16GB of RAM[3].

## 3.2   Results and Discussion

Figure 1 compares the runtime, $R^2$, and AUC of STGPR and MT-Kronprod, with those of S-MTGPR for different number of bases. As illustrated in Fig. 1(a) S-MTGPR is faster than other approaches where the total runtime of MT-Kronprod (3 days) and STGPR (6 h) can be reduced to 16 min for $P = 25$.

---

[3] The experimental codes are available at https://github.com/smkia/MTNorm.

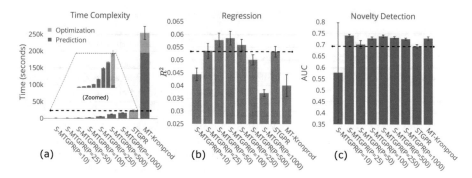

**Fig. 1.** Comparison between S-MTGPR, STGPR, and MT-Kronprod in terms of: (a) optimization and prediction runtime, (b) average regression performance ($R^2$) across all voxels, and (c) AUC in abnormal sample detection using normative modeling.

This difference in runtime is even more pronounced in case of the optimization time where S-MTGPR is at least (for $P = 1000$) 33 and 89 times faster than STGPR and MT-Kronprod, respectively. The multi-task approaches are slower than STGPR in the prediction phase mainly due to the mini-batch implementation of the prediction variance computation (to avoid memory overflow). Figure 1(b) shows this computational efficiency is achieved without penalty to the regression performance; where for certain number of bases the S-MTGPR shows equivalent and even better $R^2$ than STGPR and MT-Kronprod. Furthermore, Fig. 1(c) demonstrates that multi-task learning, by considering spatial structures, generally provides a more accurate normative model of fMRI data in that it more accurately detects samples that were derived from a different distribution to those used to train the model. This fact is well-reflected in higher AUC values for S-MTGPR at $P = 25, 100, 250, 500, 1000$. It is worthwhile to emphasize that these improvements are achieved by reducing the degree-of-freedom of the normative model from 21752 for STGPR to 10 for S-MTGPR (see Table 1).

## 4    Conclusions and Future Work

Assuming a matrix normal distribution on a reduced latent output space, we introduced an efficient and scalable multi-task Gaussian process regression approach to learning complex association between external covariates and high-dimensional neuroimaging data. Our experiments on an fMRI dataset demonstrate the superiority of the proposed approach against other single-task and multi-task alternatives in terms of the computational time complexity. This superiority was achieved without compromising the regression performance, and even with higher sensitivity to abnormal samples in the normative modeling paradigm. Our methodological contribution advances the current practices in the normative modeling from the single-voxel modeling to multi-voxel structural learning. For future work, we will consider enriching the proposed approach by embedding more biologically meaningful basis functions [7], structural modeling of non-stationary noise, and applying our method to clinical cohorts.

# References

1. Abraham, A., et al.: Machine learning for neuroimaging with scikit-learn. Front. Neuroinf. **8**, 14 (2014)
2. Alvarez, M., Lawrence, N.D.: Sparse convolved Gaussian processes for multi-output regression. In: Advances in Neural Information Processing Systems, pp. 57–64 (2009)
3. Álvarez, M.A., Lawrence, N.D.: Computationally efficient convolved multiple output Gaussian processes. J. Mach. Learn. Res. **12**, 1459–1500 (2011)
4. Alvarez, M.A., Luengo, D., Titsias, M.K., Lawrence, N.D.: Efficient multioutput Gaussian processes through variational inducing kernels. In: International Conference on Artificial Intelligence and Statistics, pp. 25–32 (2010)
5. Bonilla, E.V., Chai, K.M., Williams, C.: Multi-task Gaussian process prediction. In: Advances in Neural Information Processing Systems, pp. 153–160 (2008)
6. Boyle, P., Frean, M.: Multiple output Gaussian process regression. Technical report (2005)
7. Huertas, I., et al.: A Bayesian spatial model for neuroimaging data based on biologically informed basis functions. NeuroImage **161**, 134–148 (2017)
8. Loan, C.F.: The ubiquitous Kronecker product. J. Comput. Appl. Math. **123**(1), 85–100 (2000)
9. Marquand, A.F., Rezek, I., Buitelaar, J., Beckmann, C.F.: Understanding heterogeneity in clinical cohorts using normative models: beyond case-control studies. Biol. Psychiatry **80**(7), 552–561 (2016)
10. Marquand, A.F., Wolfers, T., Mennes, M., Buitelaar, J., Beckmann, C.F.: Beyond lumping and splitting: a review of computational approaches for stratifying psychiatric disorders. Biol. Psychiatry **1**(5), 433–447 (2016)
11. Mirnezami, R., Nicholson, J., Darzi, A.: Preparing for precision medicine. New Engl. J. Med. **366**(6), 489–491 (2012). pMID: 22256780
12. Miyawaki, Y., et al.: Visual image reconstruction from human brain activity using a combination of multiscale local image decoders. Neuron **60**(5), 915–929 (2008)
13. Quinonero-Candela, J., Williams, C.K.: Approximation methods for Gaussian process regression. Large-Scale Kernel Mach. 203–224 (2007)
14. Rakitsch, B., Lippert, C., Borgwardt, K., Stegle, O.: It is all in the noise: efficient multi-task Gaussian process inference with structured residuals. In: Advances in Neural Information Processing Systems, pp. 1466–1474 (2013)
15. Stegle, O., Lippert, C., Mooij, J.M., Lawrence, N.D., Borgwardt, K.M.: Efficient inference in matrix-variate Gaussian models with iid observation noise. In: Advances in Neural Information Processing Systems, pp. 630–638 (2011)
16. Williams, C.K., Rasmussen, C.E.: Gaussian processes for regression. In: Advances in Neural Information Processing Systems, pp. 514–520 (1996)

# Multi-layer Large-Scale Functional Connectome Reveals Infant Brain Developmental Patterns

Han Zhang[1], Natalie Stanley[2], Peter J. Mucha[2], Weiyan Yin[3], Weili Lin[1], and Dinggang Shen[1(✉)]

[1] Department of Radiology and Biomedical Research Imaging Center (BRIC), University of North Carolina at Chapel Hill, Chapel Hill, NC 27599, USA
dgshen@med.unc.edu
[2] Carolina Center for Interdisciplinary Applied Mathematics, Department of Mathematics, University of North Carolina at Chapel Hill, Chapel Hill, NC 27599, USA
[3] Department of Biomedical Engineering and Biomedical Research Imaging Center (BRIC), University of North Carolina at Chapel Hill, Chapel Hill, NC 27599, USA

**Abstract.** Understanding human brain functional development in the very early ages is of great importance for charting normative development and detecting early neurodevelopmental disorders, but it is very challenging. We propose a *group-constrained*, *robust* community detection method for better understanding of developing brain functional connectome from neonate to two-year-old. For such a multi-subject, multi-age-group network topology study, we build a *multi-layer* functional network by adding inter-subject edges, and detect modular structure (communities) to explore topological changes of multiple functional systems at different ages and across subjects. This "Multi-Layer Inter-Subject-Constrained Modularity Analysis (MLISMA)" can detect group consistent modules without losing individual information, thus allowing assessment of individual variability in the brain modular topology, a key metric for developmental individualized fingerprinting. We propose a heuristic parameter optimization strategy to wisely determine the necessary parameters that define the modular configuration. Our method is validated to be feasible using longitudinal 0–1–2 year's old infant brain functional MRI data, and reveals novel developmental trajectories of brain functional connectome. This work wassupported by the NIH grants, EB022880, 1U01MH110274, and MH100217.

**Keywords:** Brain network · Connectome · Modularity · Development
Infant

## 1 Introduction

Human infant brain is a rapidly developing complexity both structurally and functionally. While anatomical changes during the first two years of life have been extensively studied [1], the functional developmental changes in this pivotal stage are

© Springer Nature Switzerland AG 2018
A. F. Frangi et al. (Eds.): MICCAI 2018, LNCS 11072, pp. 136–144, 2018.
https://doi.org/10.1007/978-3-030-00931-1_16

still elusive. Understanding how the brain is functionally organized as a large-scale "functional connectome" and its evolution in the very early ages will shed light on the behavioral, cognitive, neurophysiological, neurological, and neuropsychiatric studies in the elder ages and facilitate early detection of developmental disorders [2]. However, the studies on the neonatal and early infancy dynamic maturing processes in the scale of whole-brain networks are still scarce [3–6].

There are three major difficulties. *(1)* Neonate/infant functional Magnetic Resonance Imaging (fMRI) is noisier than the adults' fMRI, which poses a great challenge to robustly model network topological properties. *(2)* Inter-subject variability information is usually lost in traditional averaging-based group-level network analysis, but it is essential for individualized developmental fingerprinting and charting [7]. *(3)* For longitudinal studies of brain development, it is difficult to generate temporally consistent network topological properties using traditional cross-sectional network analysis. In this paper, we propose a new method, namely, Multi-Layer Inter-Subject-Constrained Modularity Analysis (MLISMA), for a ***robust*** network community detection with ***well-preserved*** individual variability dedicated for longitudinal functional connectome development studies. MLISMA probes early brain development along the dimensions of space (brain regions and communities), time (age groups), and subject.

In MLISMA, we build multi-layer networks connecting together the data from all subjects at the same age, instead of the traditional, single-layer group-averaged network. A generalized Louvain (GenLouvain) algorithm [8] is applied to detect community structures or modules. The innovation here is two-fold. First, two key parameters that control inter-subject consistency and modular resolution are ***jointly*** optimized based on multiple empirical metrics, instead of an arbitrary parameter selection. We observed that such a multi-task parameter optimization could eventually lead to temporally consistent parameter settings and brain modularity. Second, we can use MLISMA to ***both*** achieve inter-subject consistency ***and*** probe individual variability that represents the unique brain connectome topology for each subject.

To demonstrate the effectiveness of our method, we applied it to characterize developmental changes in modules (reflecting different brain functional systems) in the neonates' and infants' brains based on a 0–1–2 year's old longitudinal resting-state fMRI (rs-fMRI) dataset. The results suggest a different story that the human brain connectome may develop via conservative rewiring that minimally affects the quantity of the brain functional networks. The individual variability in modular structure may significantly decrease from neonate to 1-year-old and keep stable at 2-year-old. Furthermore, we detect the brain regions with large inter-subject differences in modular participation and their spatiotemporal changes during development. The results provide potential targets for neurodevelopmental monitoring for early abnormality detection.

## 2    Materials and Methods

### 2.1    Multi-layer Network

Each subject's brain functional network can be represented by a functional connectivity (FC) matrix or adjacency matrix with each element representing temporal synchronization of rs-fMRI blood-oxygenation-level-dependent (BOLD) signals from a pair of spatially distant brain regions. Traditional network neuroscience analyses predominantly focus on either each subject's FC matrix or a group-mean FC matrix averaging across all subjects [9], each of which essentially transforms the data into a *single-layer* network analysis that might blur the network topology, be sensitive to individual noise, and can neither detect nor account for individual variability.

By adding edges linking corresponding nodes of different single-layer networks, a multi-layer network can be constructed [10]. Figure 1 shows a schematic illustration of the modularity analysis based on a multi-layer network constituted by 3 subjects, each having a 7-node 2-module FC network. By adding inter-layer edges between each subject pair, we create a multi-layer network that corresponds to a bigger "supra-adjacency matrix". Compared to the single-layer network, a multi-layer representation has many good properties. *(1)* It allows a group-level network analysis while considering every single network's contribution. *(2)* It makes use of inter-subject constraints

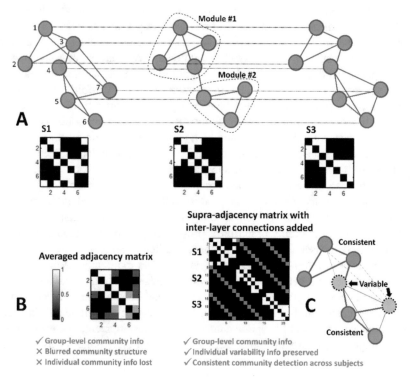

**Fig. 1.** Illustration of the module detection from a supra-adjacency matrix with each subject's adjacency matrix in the diagonal blocks and inter-layer (inter-subject) connections added. (**A**) A simple illustration of 3 subjects' FC networks and their corresponding adjacency matrices with different modular configurations. (**B**) A group averaged adjacency matrix with blurred community structure. (**C**) A multi-layer network (the connectivity strength across different subjects is set to 0.5 in this example), based on which group-consistent modules are detected and the nodes with consistent and variable modular assignment across subjects can be determined.

to achieve robust and consistent modular detection. *(3)* Under the premise of group consistency, each subject's network properties (e.g., modularity) can be individually evaluated while simultaneously contributing to the global population-level analysis, allowing individual variability analysis.

## 2.2    Multi-layer Network-Based Modularity Analysis

Various network metrics can be calculated to describe multi-layer network properties [10]. For brain functional connectome study, one of the most important metrics is network communities, or modules [11]. The brain network modular organization can be derived from a graph partition based on network topology by maximizing the modularity quality function ($Q$) as has been typically used to optimize the concentration of the FC edges within the modules [12]. The modules naturally represent various functional systems, each of which has independent function. For development studies, understanding how modules change spatiotemporally is a key for understanding how different functional systems develop and how functional segregation/integration evolve [13].

This is the first study using multi-layer network-based modularity analysis to reveal brain functional development in the early life (from neonate to 2-year-old). One of the most important questions in developmental neuroscience is how different neonates have different brain network structures and their developmental trajectories that make each subject different from others. Therefore, assessment of the individual variability in the network topology in such a pivotal period of life is essential for both normative charting and early abnormality detection. In this paper, we proposed *MLISMA* (Multi-Layer Inter-Subject-Constrained Modularity Analysis) for this purpose. The flowchart is depicted in Fig. 2, which consists of 4 steps:

**Step-1 (Multi-layer Network Construction):** We construct each subject's FC network based on pair-wise correlation of regional rs-fMRI time series. We then build a multi-layer network for each age group by adding cross-layer edges connecting corresponding brain regions between any pair of subjects of the same age. This exerts an inter-subject consistency constraint to the subsequent module detection.

**Step-2 (MLISMA Module Detection):** GenLouvain [8, 14] was used to group brain regions across all subjects in each age group into group-level modules. Two important parameters that exert a significant effect on the detected modules are the resolution or scaling parameter ($\gamma$) and inter-subject coupling parameter ($\omega$). Previous adult brain connectome studies often used selected, fixed values of $\gamma$ and/or $\omega$ [15], the selections of which risk accidentally ignoring the fundamental, underlying network topology patterns. One recently developed strategy for selecting the parameters was presented in [16]; however, in this work, we sought to more directly select the parameters according to multiple modularity metrics for our data. To this end, we devise *Heuristic Parameter Optimization*, based on the number of modules ($K$) and the individual variability in modular structure (inversely proportional to *NMI*, normalized mutual information of whole-brain modular participation averaged pairwise across all subjects). During this procedure, the stochastic nature of the GenLouvain algorithm was considered. Developmentally-consistent parameters are

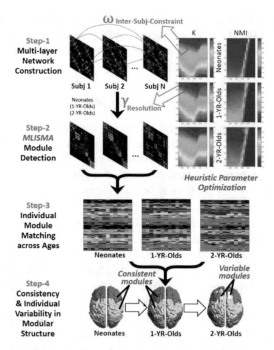

**Fig. 2.** Flowchart of Multi-Layer Inter-Subject-Constrained Modularity Analysis (MLISMA) and its application to infant's brain functional development study. The entire framework consists of 4 steps from the construction of multi-layer networks to the detection of consistent and variable modular belongingness for each brain region. A "Heuristic Parameter Optimization" strategy (detailed in main text) is proposed to determine the two key parameters in such a multi-subject, multi-age-group modular structure analysis. MLISMA introduces a group constraint to individual-level module detection, allowing robust determination of the individual-level modules with group correspondence and consistency toward individual variability analysis.

determined according to the heatmaps in the parameter plane. For each age group, all subjects' modules are generated.

**Step-3 (Module Matching across Ages):** After obtaining the group-constrained individual modules, group-level modular matching is conducted to match corresponding modules across different age groups for the following post-analysis.

**Step-4 (Consistency & Individual Variability Analysis):** Specific brain regions with variable modular assignment across all subjects of the same age can be detected by focusing on individual modular structures as output from GenLouvain. We detect *both* regions with group-consistent modular assignment *and* regions with high individual variability in modular assignment (e.g., by calculating whether more than 50% subjects have the same modular attribute on the same brain region).

### 2.3    Heuristic Parameter Optimization

It is essential to optimize $\gamma$ and $\omega$ *jointly* and *reasonably*. Both parameters will affect each other's optimal setting, thus, they should be estimated jointly. They should lead to reasonable network modular configurations based on the following criteria. We calculate heatmaps of $K$ and $NMI$ by varying $\gamma$ and $\omega$. Since GenLouvain can produce a stochastic result, we repeat the calculation 100 times at each set of parameters to generate averaged heatmaps. The three age groups show astonishingly similar patterns (Fig. 3). We regard extremely small or large modular quantity ($K < 4$ or $>10$) and extremely little individual difference ($NMI > 0.7$) as unreasonable results based on the widely accepted previous findings [3, 4, 11, 13, 15]. According to the leftover area in the $K$ and $NMI$ heatmaps, we determine a zone of parameters for which we are confident about the results and select $\gamma$ and $\omega$ as the center. The result of parameter selection is also assessed with $Q$ heatmap, showing reasonable and consistent across the age groups.

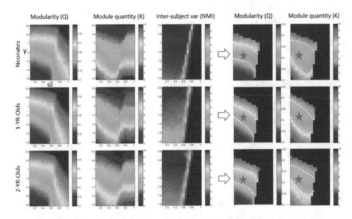

**Fig. 3.** Heuristic Parameter Optimization. Heatmap-based joint optimization of $\gamma$ and $\omega$ based on module quantity ($K$) and inter-subject similarity ($NMI$). The dark blue areas in the right panel indicate unreasonable parameter combinations. The stars are the optimized values of $\gamma$ and $\omega$.

## 2.4 Post-MLISMA Analysis

Different measurements can be employed to qualitatively and quantitatively evaluate developmental brain networks, including *(1)* a summary of the MLISMA result with color-coded modular index modulated by grey-scale-coded individual variability, *(2)* a spatial evolving pattern for each module, *(3)* module size evolution that quantifies how many brain regions are included in different modules at each age, and *(4)* individual variability changes along development (Fig. 4).

# 3    Experiments and Results

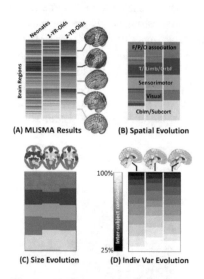

(A) MLISMA Results    (B) Spatial Evolution

(C) Size Evolution    (D) Indiv Var Evolution

**Fig. 4.** MLISMA results show different aspects of evolving brain functional networks. Four different following-up analyses are provided for the MLISMA outputs, see details in the main text.

We used natural sleeping rs-fMRI data from "Multi-visit Advanced Pediatric brain imaging study for characterizing structural and functional development (MAP Study)" with repetitive scans on 13 subjects in the ages of 1-week, 12-, and 24-months [17]. Individual FC network was constructed based on a widely used 268-region atlas [18]. The optimized $\gamma$ and $\omega$ were determined as 1.3 and 0.3, respectively. Modularity $Q$ was found to be slightly increased in the first year of life, but modular quantity $K$ was largely stable ($\sim 5$). Results indicate relatively stable functional integration/segregation notwithstanding the spatial pattern of modules continuously changes. We found prominently reduced individual variability in the modular structure in the age of one year (Fig. 4D). By visiting each brain region for its modular participation and comparing it among different age groups, we identified three major developmental patterns: *(1)* increasing diversity at striatal, frontal, parietal and occipital regions (Fig. 5A), *(2)* decreasing diversity at the frontal-temporal-parietal association areas (Fig. 5B, C), and *(3)* stable module assignment in the primary visual and motor networks (Fig. 5D). The spatial location of the regions with diverse modular participation across subjects generally move from the medial structures to lateral association areas (especially the default mode network).

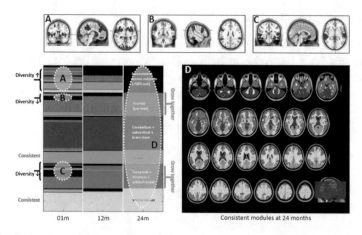

**Fig. 5.** Regions with increasing and decreasing diversity of modular assignment.

Figure 6 shows quantitative results. *(1)* Modular quantity *K* is stable across different ages. *(2)* Modularity *Q* significantly ($p < 0.05$, corrected) increases from neonate to 1-year-old and keeps stable later. *(3)* Individual variability is significantly ($p < 0.05$, corrected) reduced from neonate to 1-year-old and keeps stable thereafter. Interestingly, when assessing the individual variability of modular structure across the brain regions within each lobe rather than the whole brain, we found more developmental details. For example, frontal, temporal and parietal lobes have first increased but then decreased individual similarity, whereas occipital lobe has continuously increased individual similarity in the modular attribute.

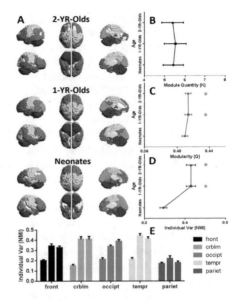

**Fig. 6.** Modular structure development. Asterisks indicate significant changes.

## 4   Discussion

MLISMA avoids previous brute-force group averaging-based module detection and achieves subject-consistent module detection result while preserving and respecting individual variability, which is essential for developmental study. We found a stable developmental pattern in terms of modularity, a new finding compared to a previous report with increasing module quantity. We found novel developmental changes in individual variability of modular attributes. Our study indicates that early brain functional development could be rather stable and inherently consistent for maintaining a balance. The individual variability could reflect unique myelination and pruning processes modulated by environmental/genetic factors [4].

## References

1. Giedd, J.N., Rapoport, J.L.: Structural MRI of pediatric brain development: what have we learned and where are we going? Neuron **67**, 728–734 (2010)
2. Emerson, R.W., et al.: Functional neuroimaging of high-risk 6-month-old infants predicts a diagnosis of autism at 24 months of age. Sci. Transl. Med. **9**(393), eaag2882 (2017)
3. Cao, M., et al.: Toward developmental connectomics of the human brain. Front. Neuroanat. **10**, 25 (2016)
4. Gao, W., et al.: Functional connectivity of the infant human brain: plastic and modifiable. Neuroscientist **23**(2), 169–184 (2016)
5. Zuo, X.N., et al.: Human connectomics across the life span. Trends Cogn. Sci. **21**, 32–45 (2017)

6. Di Martino, A., et al.: Unraveling the miswired connectome: a developmental perspective. Neuron **83**, 1335–1353 (2014)
7. Gao, W., et al.: Intersubject variability of and genetic effects on the brain's functional connectivity during infancy. J. Neurosci. **34**, 11288–11296 (2014)
8. Jeub, L.G.S., et al.: A generalized Louvain method for community detection implemented in MATLAB (2011–2017). http://netwiki.amath.unc.edu/GenLouvain
9. Meunier, D., et al.: Hierarchical modularity in human brain functional networks. Front. Neuroinform. **3**, 37 (2009)
10. Kivela, M., et al.: Multilayer networks. J. Complex Netw. **2**, 203–271 (2014)
11. Bullmore, E., Sporns, O.: Complex brain networks: graph theoretical analysis of structural and functional systems. Nat. Rev. Neurosci. **10**, 186–198 (2009)
12. Newman, M.E.J.: Detecting community structure in networks. Eur. Phys. J. B **38**, 321–330 (2004)
13. Cao, M., et al.: Topological organization of the human brain functional connectome across the lifespan. Dev. Cogn. Neurosci. **7**, 76–93 (2014)
14. Mucha, P.J., et al.: Community structure in time-dependent, multiscale, and multiplex networks. Science **328**, 876–878 (2010)
15. Bassett, D.S., et al.: Robust detection of dynamic community structure in networks. Chaos **23**(1), 013142 (2013)
16. Weir, W.H., et al.: Post-processing partitions to identify domains of modularity optimization. Algorithms **10**, 93 (2017)
17. Zhang, H., Yin, W., Lin, W., Shen, D.: Early brain functional segregation and integration predict later cognitive performance. In: Wu, G., Laurienti, P., Bonilha, L., Munsell, Brent C. (eds.) CNI 2017. LNCS, vol. 10511, pp. 116–124. Springer, Cham (2017). https://doi.org/10.1007/978-3-319-67159-8_14
18. Finn, E.S., et al.: Functional connectome fingerprinting: identifying individuals using patterns of brain connectivity. Nat. Neurosci. **18**(11), 1664–1671 (2015)

# A Riemannian Framework
# for Longitudinal Analysis of Resting-State
# Functional Connectivity

Qingyu Zhao[1(✉)], Dongjin Kwon[1,2], and Kilian M. Pohl[2]

[1] Department of Psychiatry and Behavioral Sciences, Stanford University,
Stanford, USA
qingyuz@stanford.edu
[2] Center of Health Sciences, SRI International, Menlo Park, USA

**Abstract.** Even though the number of longitudinal resting-state-fMRI studies is increasing, accurately characterizing the changes in functional connectivity across visits is a largely unexplored topic. To improve characterization, we design a Riemannian framework that represents the functional connectivity pattern of a subject at a visit as a point on a Riemannian manifold. Geodesic regression across the 'sample' points of a subject on that manifold then defines the longitudinal trajectory of their connectivity pattern. To identify group differences specific to regions of interest (ROI), we map the resulting trajectories of all subjects to a common tangent space via the Lie group action. We account for the uncertainty in choosing the common tangent space by proposing a test procedure based on the theory of latent $p$-values. Unlike existing methods, our proposed approach identifies sex differences across 246 subjects, each of them being characterized by three rs-fMRI scans.

## 1 Introduction

Longitudinal resting-state(rs)-fMRI studies, in which participants are scanned at multiple visits, have been increasingly used for investigating functional connectivity changes and development in human brains [1,2]. However, current methods for rs-fMRI group analysis are mostly designed for cross-sectional studies [3,4]. In this paper, we propose a framework for performing group analysis on longitudinal rs-fMRI data.

Cross-sectional studies often encode the functional connectivity of a subject as an $n \times n$ covariance matrix $C$ of BOLD (blood-oxygen-level dependent) time courses associated with $n$ ROIs. To identify differences in functional connectivity between two groups, a univariate group test is typically applied to each connection (each element in the upper triangle of $C$). One problem in this practice is that the univariate tests neglect the strong statistical dependence among matrix elements: a covariance matrix $C$ is confined by the positive-definite constraint $C = C^{\mathbf{T}}, \mathbf{x}C\mathbf{x}^{\mathbf{T}} > 0$ for all non-zero $\mathbf{x} \in \mathbb{R}^n$. One way to alleviate this problem is to leverage the fact that covariance matrices form a Riemannian manifold [5].

© Springer Nature Switzerland AG 2018
A. F. Frangi et al. (Eds.): MICCAI 2018, LNCS 11072, pp. 145–153, 2018.
https://doi.org/10.1007/978-3-030-00931-1_17

Previous rs-fMRI connectivity studies have reported improved detection sensitivity by performing group analysis directly on that manifold [4]. Based on this observation, we design a Riemannian framework to analyze the change in connectivity patterns captured by longitudinal studies.

In a longitudinal study, each subject is characterized by a series of covariance matrices representing connectivity patterns at multiple visits. Motivated by previous works [6], our longitudinal framework is composed of two parts: (a) fitting a longitudinal trajectory on the covariance matrices of each subject via geodesic regression on the manifold [7] (Sect. 2); and (b) comparing subject-specific trajectories across groups (Sect. 3). The challenge in (b) is that trajectories of different subjects are not directly comparable as they are essentially tangent vectors defined in different tangent spaces. Several methods have been explored to handle this problem, either by directly performing group analysis on the 'tangent bundle space' [6] or by designing mixed-effect models for manifolds [8]. These methods, however, consider the object of interest (in our case a covariance matrix) as a single manifold-valued variable, such that group difference can only be identified with respect to the entire matrix (the whole brain connectivity) instead of each matrix element (connectivity between two ROIs). In order to enable ROI-specific analysis, we map all tangent vectors to a common tangent space of a template point on the manifold. This enables univariate testing to each matrix element across all mapped tangent vectors.

In addition, we define the mapping function based on the Lie group action and briefly discuss its favorable properties to the popular parallel transport mechanism [9, 10]. We define the common tangent space via the identity matrix [9] and the Fréchet mean [4], and we argue that the latter is preferred in the context of connectivity analysis. Finally, our group analysis accounts for the uncertainty in estimating the template via a robust test procedure based on the theory of latent $p$-values. We finally validate our proposed Riemannian framework using both synthetic and real rs-fMRI datasets.

## 2  Computing Subject-Specific Trajectory

Recall that the space of $n \times n$ covariance matrices forms a Riemannian manifold $\mathcal{M}$ [5]. Let $\boldsymbol{A}$ be a point on $\mathcal{M}$, $T_{\boldsymbol{A}}\mathcal{M}$ the tangent space at $\boldsymbol{A}$, and $\boldsymbol{X} \in T_{\boldsymbol{A}}\mathcal{M}$ a tangent vector. There is a unique geodesic curve $\gamma$ (a locally length-minimizing curve on the manifold) with $\gamma(0) = \boldsymbol{A}$ and $\gamma(0)' = \boldsymbol{X}$. The analytical equation of a geodesic is defined by the *exponential map*,

$$\gamma(t) = Exp_{\boldsymbol{A}}(t\boldsymbol{X}) := \boldsymbol{A}^{\frac{1}{2}} expm(\boldsymbol{A}^{-\frac{1}{2}}(t\boldsymbol{X})\boldsymbol{A}^{-\frac{1}{2}})\boldsymbol{A}^{\frac{1}{2}}, \tag{1}$$

where $expm$ is the matrix exponential operator. In other words, the exponential map $Exp_{\boldsymbol{A}}(\boldsymbol{X})$ at an initial point $\boldsymbol{A}$ projects a tangent vector $\boldsymbol{X}$ to a point on the manifold at $\gamma(1)$ along the geodesic $\gamma$ defined by $(\boldsymbol{A}, \boldsymbol{X})$. The inverse mapping of $Exp_{\boldsymbol{A}}(\boldsymbol{X})$ is called the *log map*. It projects a point $\boldsymbol{B} \in \mathcal{M}$ back to a tangent vector at $\boldsymbol{A}$ via $Log_{\boldsymbol{A}}(\boldsymbol{B}) := \boldsymbol{A}^{\frac{1}{2}} logm(\boldsymbol{A}^{-\frac{1}{2}}\boldsymbol{B}\boldsymbol{A}^{-\frac{1}{2}})\boldsymbol{A}^{\frac{1}{2}}$.

Now let us consider $M$ covariance matrices of a subject $\{C^1, \ldots, C^M\}$ measured at $M$ visits. Let $\{t^1, \ldots, t^M\}$ be the time associated with those visits. Without loss of generality, we translate $\{t^i\}$ such that $t^1 = 0$. We then use *geodesic regression* [7] to characterize the change of $\{C^i\}$ over time (the relationship between $t^i$ and $C^i$). Specifically, to find a geodesic curve $(\hat{A}, \hat{X})$ that optimally fits the data (Fig. 1), we minimize the following objective function:

$$(\hat{A}, \hat{X}) = \arg\min_{A, X} \sum_{i=1}^{M} d(Exp_A(t^i X), C^i)^2, \qquad (2)$$

where $d(A, B)$ measures the *geodesic distance* between $A, B \in \mathcal{M}$ via the Riemannian metric $d(A, B) := \|Log_A(B)\|_A := \sqrt{\mathrm{tr}(logm(A^{-\frac{1}{2}} B A^{-\frac{1}{2}}))}$.

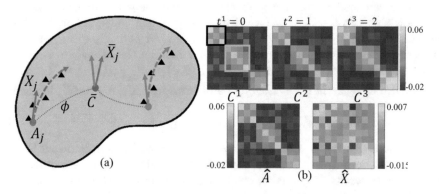

**Fig. 1.** (a) A Riemannian framework for longitudinal connectivity analysis: the $j^{th}$ subject's trajectory (red dashed curve) is fitted via geodesic regression on subject-specific covariance matrices (triangles on the left). All resulting subject-specific tangent vectors are mapped via function $\phi$ to the tangent space of $\bar{C}$ to perform group analysis. (b) Top: 3 covariance matrices $\{C^1, C^2, C^3\}$ of a subject with $t = 0, 1, 2$. Bottom: The optimal geodesic $(\hat{A}, \hat{X})$ derived from geodesic regression.

In particular, the initial point $\hat{A}$ characterizes the connectivity pattern at baseline. The tangent vector $\hat{X}$ characterizes the longitudinal trajectory of that subject's connectivity. Note that the tangent space $T_{\hat{A}}\mathcal{M}$ is the vector space of $n \times n$ symmetric matrices. Therefore, unlike $\hat{A}$, the matrix elements in the upper triangle of $\hat{X}$ are mutually independent for univariate tests.

## 3   Group Analysis for Trajectories

Now let $\{(A_j, X_j)|j = 1, \ldots, N\}$ represent the geodesics of $N$ different subjects. While cross-sectional analysis is interested in analyzing baseline matrices $\{A_j\}$ [4], here we aim to identify group differences in $\{X_j\}$, *i.e.*, the difference across subjects in their longitudinal changes. As mentioned, these tangent vectors are defined in different tangent spaces (e.g. the red and green vector in Fig. 1a), therefore not directly comparable. To handle this problem, we map all tangent vectors $\{X_j | j = 1, \ldots, N\}$ to a common tangent space $T_{\bar{C}}\mathcal{M}$. This step essentially requires the definition of (a) a mapping function $\phi_{A_j \to \bar{C}}(X_j) = \bar{X}_j$ and (b) a template point $\bar{C}$. Finally, we estimate the $p$-values of univariate testing for all $n(n+1)/2$ matrix elements across $\{\bar{X}_j | j = 1, \ldots, N\}$.

**Choosing the Mapping Function $\phi$.** A popular choice of $\phi$ is the *parallel transport* [9,10], which transports tangent vectors on a manifold such that they stay parallel with respect to the affine connection. Despite its appealing geometrical meaning, the major drawback of parallel transport is its path-dependency: transporting a tangent vector along two different curves with the same start and end point generally results in two different 'copies' of the vector. This phenomenon can lead to ambiguity in choosing the template for group analysis. To show this effect, let $\phi_{A \to B}^p(X)$ denote the parallel transport of $X \in T_A \mathcal{M}$ along the geodesic from $A$ to $B$. Let $X_1 \in T_{A_1}\mathcal{M}$ and $X_2 \in T_{A_2}\mathcal{M}$ be two subject-specific tangent vectors. We further assume that they are equivalent when transported to $\bar{C}$, i.e., $\bar{X}_1 = \phi_{A_1 \to \bar{C}}^p(X_1) = \phi_{A_2 \to \bar{C}}^p(X_2) = \bar{X}_2$. Now if we perturb $\bar{C}$ by $\varepsilon$ via $\bar{C}^* = Exp_{\bar{C}}(\varepsilon)$ and transport $X_1, X_2$ to $\bar{C}^*$, in general,

$$\bar{X}_1^* = \phi_{A_1 \to \bar{C}^*}^p(\phi_{\bar{C} \to A_1}^p(\bar{X}_1)) \neq \phi_{A_2 \to \bar{C}^*}^p(\phi_{\bar{C} \to A_2}^p(\bar{X}_2)) = \bar{X}_2^* \qquad (3)$$

as $\phi^p$ is path-dependent. Contradictory to our previous assumption that the two subject-specific tangents are equivalent, Eq. 3 reveals that they are different at $\bar{C}^*$. In other words, (in)equality relationships among $\{X_j\}$ are variant to the template selection, which can lead to serious ambiguity in the comparison of subject-specific trajectories.

To resolve this problem, we exploit the fact that $\mathcal{M}$ is equipped with an affine-invariant Riemannian metric [5]. To be specific, let $GL_n$ denote the Lie group of all $n \times n$ invertible matrices. This group acts on $\mathcal{M}$ via a smooth mapping function $\psi : GL_n \times \mathcal{M} \Rightarrow \mathcal{M}$, $\psi_G(A) := GAG^T = B$, where $G \in GL_n$ and $A, B \in \mathcal{M}$. This group action can be naturally extended to tangent vectors via its derivative map $d\psi_G(X) := GXG^T = Y$, where $X \in T_A \mathcal{M}$ and $Y \in T_B \mathcal{M}$. In other words, $d\psi_G$ achieves the mapping of tangent vectors across different tangent spaces based on the aforementioned smooth group action. With this construction, we propose the following mapping function

$$\phi_{A \to B}^g(X) = d\psi_G(X), \; G = B^{\frac{1}{2}} A^{-\frac{1}{2}}. \qquad (4)$$

Since the group action $\psi$ is transitive [5], so is $\phi^g$, i.e., $\phi_{B \to C}^g(\phi_{A \to B}^g(X)) = \phi_{A \to C}^g(X)$. This property avoids the path-dependent assumption as required in parallel transport, so that (in)equality relationships among $\{\bar{X}_j\}$ are invariant to the choice of the template.

**Choosing the Template $\bar{C}$.** In the context of connectivity analysis, we argue that the Fréchet mean [4] is a more appropriate template compared to the

---

**Algorithm 1.** Latent $p$-value map

---

1: Resample $\{A_j | j = 1, \ldots, N\}$ 500 times to compute the empirical bootstrap distribution of the Fréchet mean $\{\bar{C}^1, \ldots, \bar{C}^{500}\}$.

2: For each bootstrapped $\bar{C}^i$, map $\{X_j | j = 1, \ldots, N\}$ to it to achieve $\{\bar{X}_j^i | j = 1, \ldots, N\}$. Apply univariate tests to each matrix element to get a $p$-value map $P^i$.

3: The final latent $p$-value map $P = \frac{1}{500} \sum_{i=1}^{500} P^i$.

---

identity matrix $I$ [9]. Recall that $X(u,v)$, the $(u,v)^{th}$ element in a subject-specific tangent vector $X$, encodes the longitudinal information about connectivity between the $u^{th}$ and $v^{th}$ ROI. Since $\phi_{A \to \bar{C}}^g$ (or $\phi_{A \to \bar{C}}^p$) is a general linear transformation, $\bar{X}(u,v)$ is a linear combination of $X(u,v)$ and other matrix elements. Consequently, $\bar{X}(u,v)$ no longer precisely relates the two ROIs. This reveals one critical trade-off: only $X$ models true subject-specific ROI information, whereas only $\bar{X}$ is geometrically comparable with other tangent vectors. To alleviate this issue, we realize that when $\bar{C}$ is close to $A$, $\bar{C}^{\frac{1}{2}} A^{-\frac{1}{2}}$ is close to an identity transformation ($\bar{X}(u,v) \approx X(u,v)$). This motivates us to choose $\bar{C}$ as the Fréchet mean, because it is the 'closest' matrix to all $\{A_j\}$ so that true ROI information can be optimally preserved for all subjects.

**A Robust Test Procedure.** In the end, a univariate test is applied to each connection (upper triangular matrix element) across $\{\bar{X}_j\}$. However, with different choices of $\bar{C}$, both the null distribution and observed values of $\{\bar{X}_j(u,v)\}$ vary accordingly to the associated $\phi_{A \to \bar{C}}^g$. Therefore, the uncertainty in estimating the Fréchet mean can lead to unstable $p$-values. To solve this problem, we resort to the theory of latent $p$-values [11]. Recall that the $p$-value is defined as $p = \text{pr}(\bar{\mathcal{X}}_0 > \bar{X} | \mathcal{H}_0)$, i.e., the probability of obtaining a result larger (right-tailed) than the observed $\bar{X}$ under the null hypothesis $\mathcal{H}_0$. Recently, statisticians also interpret $p$-values as random variables [11]. Specifically, the latent $p$-value considers both the strength of the evidence against the null hypothesis, and the uncertainty in the evidence and null distribution. Here we regard both $\bar{\mathcal{X}}_0$ and $\bar{X}$ as unobservable latent variables. As described in [11], the latent $p$-value can be approximated by generating Monte-Carlo realizations of $\bar{\mathcal{X}}_0$. We then let $\bar{\mathcal{X}}_0$ and $\bar{X}$ be dependent on a latent template $\bar{C}$, and we marginalize $\bar{C}$ by $p = \int_{\bar{C}} \text{pr}(\bar{\mathcal{X}}_0(\bar{C}) > \bar{X}(\bar{C}) | \bar{C}, \mathcal{H}_0) \text{pr}(\bar{C})$. Finally, we can approximate this integration by sampling $\bar{C}$ via bootstrapping. Intuitively, instead of regarding $\bar{X}$ and $\bar{\mathcal{X}}_0$ as deterministic variables, we perform multiple test procedures based on templates sampled from $\text{pr}(\bar{C})$ (Algorithm 1).

## 4  Experiments

**Synthetic Data.** In this experiment, we simulated covariance matrices for 2 groups of subjects. Each group has $N = 20$ subjects, and each subject has $M = 3$ covariance matrices. We added longitudinal changes to the 3 matrices of each subject in *Group B*. We validated our framework based on the accuracy in identifying those effects between the two groups.

We started from simulating BOLD time series using SimTB [12]. We randomized simulation parameters for each subject and derived covariance matrices from simulated BOLD signals. To be specific, we defined $n = 10$ ROIs and randomly grouped them into 3 independent networks (Fig. 1b). Only ROIs of the same network activated simultaneously, thus having non-zero covariances. The activation amplitude of each ROI was sampled from the standard Gaussian $\mathcal{N}(0,1)$. The unique activation (functional noise) probability of each ROI was $u = 0.35$. We added Riccan noise (imaging noise) with a Contrast-to-Noise Ratio (CNR) of 1.5. The above simulation parameters were then shared across subjects. To synthesize subject variability, we perturbed the activation amplitudes of each subject with $\mathcal{N}(0, 0.05)$. Then, for each subject in *Group A*, BOLD time courses at 3 time points ($t = 0, 1, 2$) were generated using those simulation parameters. The only difference in simulating *Group B* subjects was that the activation amplitudes of the $3^{rd}$ network successively changed at $\{t^1, t^2, t^3\}$, with a changing rate sampled from $\mathcal{N}(0, 0.25)$. Consequently, only the covariances within that network could change longitudinally. Figure 1b shows one example, where the 9 elements within the red square (the $3^{rd}$ network) are true positives to be identified by group analysis. Unlike the simulation in [4], our strategy relies only on the basic fMRI signal generation mechanism, so that the simulated matrices do not bias any particular covariance modeling technique.

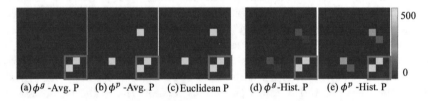

(a) $\phi^g$ -Avg. P      (b) $\phi^p$ -Avg. P      (c) Euclidean P      (d) $\phi^g$-Hist. P      (e) $\phi^p$ -Hist. P

**Fig. 2.** (a)(b)(c) Group differences (yellow) identified by the three methods. (d)(e) Histogram of identified elements counted using the 500 $p$-value maps.

Figure 2a, b, c show the identified group differences (yellow) derived from three longitudinal analysis methods: ($a$) our proposed Riemannian framework with the mapping function $\phi^g$ (Eq. 4); ($b$) Riemannian framework with parallel transport $\phi^p$ (defined in [13]); and ($c$) linear regression on each element across subject-specific covariance matrices $\{C^i\}$ (Euclidean). The final univariate test was the two-sample t-test. The latent $p$-value map (Sect. 3) was used in *Method a* and *b*, whereas *Method c* only required a deterministic $p$-value map as it directly compared the trajectories (linear slopes) of different subjects. Significant matrix elements were identified at $p \leq 0.05$ after correcting for multiple comparison via the Bonferroni procedure. Due to the influence of ROI unique activation and the Riccan noise, none of the three methods identified all true positives. However, only *Method a* yielded no false positive. To show the non-deterministic nature of the univariate tests in *Method a* and *b*, we identified group differences using each of the 500 $p$-value maps and counted the frequency of each element being significant. The spread of the two resulting histograms (Fig. 2d, e) indicates the latent $p$-value map is preferred over any particular deterministic $p$-value map. Moreover, our proposed mapping function $\phi^g$ (Fig. 2d) achieved more consistent group test results compared to the parallel transport (Fig. 2e). Next, to show the importance of using the Fréchet mean as the template, we sampled 1000 $\bar{C}$ via Principal Geodesic Analysis, and generated a deterministic $p$-value map for each sampled $\bar{C}$. We sorted and binned the 1000 test results according to the distance from the sampled $\bar{C}$ to the Fréchet mean. Figure 3a indicates when $\bar{C}$ was near the Fréchet mean, the test accuracy significantly increased. Finally, we tested the three methods' robustness against noise ($u$ and CNR). Under each noise level, we repeated the experiments (including re-grouping ROIs and re-randomizing activation amplitudes) 10 times. Figure 3b shows that the Riemannian framework always outperformed the traditional Euclidean method, and our proposed mapping function $\phi^g$ always achieved the best result.

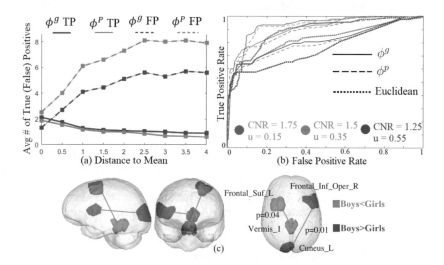

**Fig. 3.** (a) Number of true (false) positives averaged over templates sampled within a certain distance range. (b) ROC of the three methods under three noise levels. (c) The application of our proposed framework to the NCANDA dataset revealed sex effects in two connections. The $p$-values were corrected for multiple comparison.

**The NCANDA Dataset.** We applied our framework to rs-fMRI data of 246 normal adolescents (age 12–21; 117 boys and 129 girls) from the National Consortium on Alcohol and Neurodevelopment in Adolescence (NCANDA) [14]. Each participant in this dataset was scanned three times (baseline, 1-year follow-up and 2-year follow-up). We adopted the same preprocessing procedure as described in [14]. The brain was parcelated into 100 ROIs using [15]. The final univariate test was a general linear model (GLM) accounting for sex, age, site and race. Using *Method a*, we identified sex effects in two connections (Fig. 3c). Specifically, the connectivity between *Frontal_Suf_L* and *Vermis_1* increases faster in girls. The connectivity between *Frontal_Inf_Oper_R* and *Cuneus_L* increases faster in boys. *Method b* and *c* did not yield any significant finding.

## 5   Conclusion

Based on the Riemannian modeling of covariance matrices, we introduced a framework for performing group analysis on longitudinal rs-fMRI. Importantly, our proposed tangent vector mapping function and latent $p$-value strategy aim to best compromise between the geometry of data and clinical interpretation. Both synthetic and real-data experiments indicated the potential of our proposed framework in longitudinal connectivity analysis. Nevertheless, a theoretical combination of Riemannian geometry and consistent statistics still remains an open topic. In addition, we need to further explore the challenging issue of characterizing multi-level longitudinality, *i.e.*, dynamics in the time courses of each single visit vs. development across visits.

**Acknowledgement.** This research was supported in part by NIH grants U24AA021697-06, AA005965, AA013521, AA017168.

## References

1. Odish, O., et al.: Longitudinal resting state fMRI analysis in healthy controls and premanifest Huntington's disease gene carriers: a three-year follow-up study. Hum. Brain Mapp. **36**(1), 110–119 (2015)
2. van der Horn, H., et al.: The default mode network as a biomarker of persistent complaints after mild traumatic brain injury: a longitudinal functional magnetic resonance imaging study. J. Neurotrauma **34**(23), 3262–3269 (2017)
3. Beckmann, C., Mackay, C., Filippini, N., Smith, S.: Group comparison of resting-state FMRI data using multi-subject ICA and dual regression. OHBM (2009)
4. Varoquaux, G., Baronnet, F., Kleinschmidt, A., Fillard, P., Thirion, B.: Detection of brain functional-connectivity difference in post-stroke patients using group-level covariance modeling. In: Jiang, T., Navab, N., Pluim, J.P.W., Viergever, M.A. (eds.) MICCAI 2010. LNCS, vol. 6361, pp. 200–208. Springer, Heidelberg (2010). https://doi.org/10.1007/978-3-642-15705-9_25
5. Pennec, X., Fillard, P., Ayache, N.: A Riemannian framework for tensor computing. IJCV **66**(1), 41–66 (2006)
6. Hong, Y., Singh, N., Kwitt, R., Niethammer, M.: Group testing for longitudinal data. In: Ourselin, S., Alexander, D.C., Westin, C.-F., Cardoso, M.J. (eds.) IPMI 2015. LNCS, vol. 9123, pp. 139–151. Springer, Cham (2015). https://doi.org/10.1007/978-3-319-19992-4_11
7. Fletcher, P.T.: Geodesic regression and the theory of least squares on Riemannian manifolds. IJCV **105**(2), 171–185 (2013)
8. Kim, H.J., Adluru, N., Suri, H., Vemuri, B.C., Johnson, S.C., Singh, V.: Riemannian nonlinear mixed effects models: analyzing longitudinal deformations in neuroimaging. In: CVPR, pp. 172–181 (2017)

9. Ng, B., Dressler, M., Varoquaux, G., Poline, J.B., Greicius, M., Thirion, B.: Transport on Riemannian manifold for functional connectivity-based classification. In: Golland, P., Hata, N., Barillot, C., Hornegger, J., Howe, R. (eds.) MICCAI 2014. LNCS, vol. 8674, pp. 405–412. Springer, Cham (2014). https://doi.org/10.1007/978-3-319-10470-6_51

10. Campbell, K.M., Fletcher, P.T.: Efficient parallel transport in the group of diffeomorphisms via reduction to the lie algebra. In: Cardoso, M.J. (ed.) GRAIL/MFCA/MICGen -2017. LNCS, vol. 10551, pp. 186–198. Springer, Cham (2017). https://doi.org/10.1007/978-3-319-67675-3_17

11. Thompson, E.A., Geyer, C.J.: Fuzzy p-values in latent variable problems. Biometrika **94**(1), 49–60 (2007)

12. Erhardt, E., Allen, E., Wei, Y., Eichele, T., Calhoun, V.: SimTB, a simulation toolbox for fMRI data under a model of spatiotemporal separability. NeuroImage **59**(4), 4160–4167 (2012)

13. Ferreira, R., Xavier, J., Costeira, J.P., Barroso, V.: Newton method for Riemannian centroid computation in naturally reductive homogeneous spaces. In: ICASSP (2006)

14. Müller-Oehring, E., et al.: Influences of age, sex, and moderate alcohol drinking on the intrinsic functional architecture of adolescent brains. Cereb. Cortex **28**(3), 1049–1063 (2018)

15. Craddock, R., James, G., Holtzheimer, P., Hu, X., Mayberg, H.: A whole brain fMRI atlas generated via spatially constrained spectral clustering. Hum. Brain Mapp. **33**(8), 1914–1928 (2012)

# Elastic Registration of Single Subject Task Based fMRI Signals

David S. Lee[✉], Joana Loureiro, Katherine L. Narr, Roger P. Woods, and Shantanu H. Joshi

Ahmanson-Lovelace Brain Mapping Center, Department of Neurology, University of California Los Angeles, Los Angeles, CA, USA
dalee@mednet.ucla.edu

**Abstract.** Single subject task-based fMRI analyses generally suffer from low detection sensitivity with parameter estimates from the general linear model (GLM) lying below the significance threshold especially for similar contrasts or conditions. In this paper, we present a shape-based approach for alignment of condition-specific time course activity for single subject task-based fMRI. Our approach extracts signals for each condition from the entire time course, constructs an unbiased average of those signals, and warps each signal to the mean. As the warping is diffeomorphic, nonlinear and allows large deformations of time series if required, we term this approach as elastic functional registration. On a single subject level, our method significantly detects more clusters and more activated voxels in relevant subcortical regions in healthy controls.

## 1 Introduction

Task based functional magnetic resonance imaging (fMRI) is a powerful approach for examining time varying changes in brain metabolism as a response to a predetermined stimulus [3]. Two main challenges for fMRI analysis are (i) estimation of the shape or pattern of the response (ii) detection of the blood oxygenation level dependent (BOLD) activity, as it only represents a small percentage of the variance of the signal. Non-neuronal contributions such as physiological noise, thermal noise and hardware instabilities, hamper small changes in brain activity from being detected. Low statistical power is therefore mostly caused due to a low sample size or due to a low effect size of the conditions being studied. In order to increase statistical power, research studies typically increase the sample size and perform a group-level analysis to improve both detection and estimation. In this work we focus on the problem of improving statistical power of activations for single subject fMRI analysis by performing functional alignment.

Recently, there have been several interesting approaches [4,5,9,10] that have proposed the synchronization or alignment of fMRI signals primarily for resting state data. Although the end goal in all these methodologies is aligning fMRI time courses to each other, in our work we adopt a different approach. Particularly in the task-based fMRI setting, we would expect the alignment to achieve

© Springer Nature Switzerland AG 2018
A. F. Frangi et al. (Eds.): MICCAI 2018, LNCS 11072, pp. 154–162, 2018.
https://doi.org/10.1007/978-3-030-00931-1_18

increased consensus between parts of the time course that correspond to the same stimulus condition. There are several alternatives for solving this problem. (i) Given a set of time series to be aligned, one could choose any arbitrary time course randomly from this set as a template and align all functions to it. This can introduce a potential bias towards the arbitrary chosen template. (ii) On the other hand, one can achieve all possible registrations of functions and choose a particular time series template that satisfies a certain optimality condition. This issue is similar to the one faced in image/shape registration, where the choice of the template or atlas is critical. Further, the process of alignment itself can be achieved either using a least squares criterion [4,5] or using the dynamic time warping (DTW) approach [10]. The process of least squares alignment is usually non-elastic and thus doesn't account for temporal reparameterizations. The DTW approach does provide an elegant way of recovering temporal misalignment, but does not directly allow for the computation of an intrinsic average, which is important for template estimation on the space of time series functions. Thus one still needs to resort to extrinsic (generally Euclidean) minimization criteria to determine the optimal template.

In our work, we provide a single solution to the above two problems. We use an elastic time series matching approach to construct an unbiased template of this shape pattern and align the condition-specific time course activity to this template. We treat the problem of aligning fMRI data as a shape matching problem [9] and to achieve elastic alignment between time courses to bring the BOLD responses into correspondence. On a single subject level, our method significantly detects more clusters and more activated voxels in relevant subcortical regions in healthy controls. To our knowledge, this is the first work that performs nonlinear elastic temporal registration of task-based fMRI signals.

## 2    Single Subject Temporal Alignment

This section outlines the main approach. Figure 1 shows a schematic of the procedure. The notation and discussion below is described for a fMRI time series acquisition using a blocked design at a single voxel in the brain. Let the conditions for the task and the number of blocks per condition be denoted by $C_i, i = 1, \ldots, k$ and $B_j, j = 1, \ldots, n$ respectively. Generally we assume that the total number of blocks stay fixed across all conditions. Each block is assumed to have a boxcar shape that stays zero throughout the resting period and has a unit magnitude for the duration of the task. Thus for a given task condition $C_i$, the function for block $B_j$ is denoted by $g_{B_j}^{C_i}(t) = U(t - t_{j1}^i) - U(t - t_{j2}^i), \forall t \in [0, T]$, where $T$ is the total acquisition time. Here $t_{j1}^i$ and $t_{j2}^i$ are the stimulus on and off times for condition $C_i$ and block $B_j$. Thus the blocked design stimulus function for the entire duration of the task is given by $g_B^C(t) = \sum_{i=1}^{k} \sum_{j=1}^{n} g_{B_j}^{C_i}(t), \forall t \in [0, T]$. Let the corresponding acquired fMRI time course due to the BOLD hemodynamic response be given by $h(t), \forall t \in [0, T]$. Our goal is to achieve within-condition temporal reparameterization or alignment across the entire fMRI time series

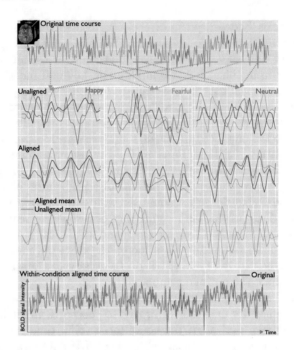

**Fig. 1.** Schematic of the elastic hyperalignment approach. Top row: the original time course color-coded by the fMRI conditions that included *happy, fearful,* and *neutral.* Row 2, 3: the unaligned and the aligned time courses for those conditions. Row 4: the elastically aligned and the unaligned mean respectively. Row 5: within-condition aligned time course.

given by $h$. Therefore for each condition $C_i$, we extract the BOLD response functions as the set $\{f_j^i(t)\}$, where

$$
\begin{aligned}
f_j^i(t) &= 0, && \text{where } 0 \le t \le t_{j1}^i \\
&= g_{B_j}^{C_i}(t)h(t), && \text{where } t_{j1}^i \le t \le t_{j2}^i, i = 1, \ldots, k, j = 1, \ldots, n. \quad (1) \\
&= 0, && \text{where } t \ge t_{j2}^i
\end{aligned}
$$

We now seek an optimal temporal alignment or registration among the set of functions given by $\{f_j^i(t)\}$. This is achieved using the square root velocity parameterization [6,7,12] for the representation of functions. We describe this idea in brief below and refer the reader to [6,7,12] for more details. For the discussion below, we drop the subscripts $i, j$ for convenience. For a given fMRI time course signal $f : I \equiv [0, 1] \to \mathbb{R}$, and its velocity $\dot{f}(t) = \frac{df}{dt}$ and magnitude $|\dot{f}(t)|$, we define its functional representation by the square-root velocity field (SRVF) map $q$ given by $q : [0, 1] \to \mathbb{R}$, $q(t) = \frac{\dot{f}(t)}{\sqrt{|\dot{f}(t)|}}$. For an absolutely continuous $f$, the SRVF transformation ensures that $q$ is square integrable. The set of SRVFs is then given by $\mathbb{L}^2([0, 1], \mathbb{R})$, which is a Hilbert space. To recover the time domain fMRI signal, we use $f(t) = f(0) + \int_0^t q(\tau)|q(\tau)|d\tau$. The SRVF mapping is invertible up to a given $f(0)$. We assume $f(0) = 0$ as the initial condition of the fMRI signal at time $t = 0$. One can impose an analogous unit length constraint on the $q$ function by obtaining $\tilde{q} = \frac{q}{||q||}$. This unit length transformation forces $q$ to lie on a Hilbert sphere denoted by $\mathbb{S}_f$. The space $\mathbb{S}_f$ is defined as $\mathbb{S}_f \equiv \left\{ q \in \mathbb{L}^2 | \int_0^1 (q(s), q(s))_{\mathbb{R}^2} ds = 1, q(s) : [0,1] \to \mathbb{R}^2 \right\}$.

## 2.1   Unbiased Within-Condition Time Series Template Estimation

To match the shape of the responses within conditions, we use the idea of time reparameterization to shift the signals to maximize peaks and troughs of functional activity. The time reparameterization function is represented by diffeomorphic function $\gamma : I \to I$, where $\dot{\gamma} > 0, \forall t \in I$. To change the time domain parameterization, one can compose $f$ with $\gamma$ as

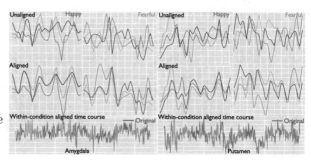

**Fig. 2.** Examples of fMRI hyperalignment for time courses in two voxels from the amygdala and the putamen respectively.

$f \circ \gamma$. In the SRVF domain, this is given by $q \cdot \gamma = \frac{(\dot{f} \circ \gamma) \dot{\gamma}}{\sqrt{|(\dot{f} \circ \gamma) \dot{\gamma}|}} = (q \circ \gamma) \sqrt{\dot{\gamma}}$. To compute an unbiased optimal template that represents an average functional response related to the condition of the task, we exploit the geometry of the space of these time course representations by defining the notion of a metric on the space of $q$ functions. The Riemannian metric on the tangent space of this sphere $T_q(\mathbb{S}_f)$ assumes an elastic form (invariance to temporal reparameterization) and is given by $\langle q_1, q_2 \rangle = \int_t q_1 q_2 dt$. The spherical geometry of the space of functions provides a convenient closed form solution for geodesics between two functions $q_1$ and $q_2$, which are given by $\chi_t(q_1; v) = \cos \left( t \cos^{-1} \langle q_1, q_2 \rangle \right) q_1 + \sin \left( t \cos^{-1} \langle q_1, q_2 \rangle \right) v$, where $t \in [0, 1]$ and the initial tangent vector $v \in T_{q_1}(\mathcal{Q})$ is given by $v = q_2 - \langle q_1, q_2 \rangle q_1$. Then the geodesic distance between the two shapes $q_1$ and $q_2$ on $\mathbb{S}_f$ is given by $d(q_1, q_2) = \int_0^1 \sqrt{\langle \dot{\chi}_t, \dot{\chi}_t \rangle} dt$. Following, the optimal temporal alignment is obtained as $\hat{\gamma} = \arg\min_{\gamma \in \Gamma} d(q_1, q_2 \cdot \gamma)$ by a combination of dynamic programming and gradient descent. This optimal $\hat{\gamma}$ yields the temporal reparameterization (alignment) between two time course signal portions.

Now for the time series, for a given task condition $k$, we find the Karcher mean [8] of the set $\{f_j^k\}$ as solving $\mu_k = \arg\min_q d(q, q_j^k)^2$, which gives us the unbiased within-condition time series template. We align all the time courses for the condition to this mean $\mu$ to achieve temporal alignment of all the partial time courses for that condition. Algorithm 1 outlines the procedure for performing within-subject blocked condition-wise optimal reparameterization for task based fMRI time courses.

---

**Algorithm 1:** Elastic Hyperalignment of Single Subject fMRI

---

**Input**: 4D fMRI time course image.
**Output**: 4D fMRI warped (*within-condition*) time course image.
1 **foreach** *3D voxel* **do**
2 |    Let $h(t), t \in [0, T]$ be the fMRI time course.
3 |    **foreach** *condition $C_i$, $i = 1, \ldots, k$* **do**
4 |    |    Extract the set of blocked time course portions as
$$\{f_j^i(t)\} = g_{B_j}^{C_i}(t) h(t), \qquad \text{where } t_{j1}^i \leq t \leq t_{j2}^i, i = 1, \ldots, k, j = 1, \ldots, n.$$
5 |    |    Map each of the $\{f_j^i\}$ on the Hilbert sphere $\mathbb{S}_f$ as the set $\{q_j^i(t)\}$.
6 |    |    Compute the Karcher mean $\mu_i$ as $\mathrm{argmin}_q \, d(q, q_j^i)^2$ [8] .
7 |    |    Find the temporal alignment to the mean as $\hat{\gamma}_j^i = \mathrm{argmin}_\gamma \, d(\mu, q_j^i \cdot \gamma)$
8 |    |    Construct warped functions as $\{\hat{f}_j^i\} \equiv \{f_j^i(t) \circ \hat{\gamma}_j^i\}, \forall t \in [t_{j1}^i, \leq t_{j2}^i]$.
9 |    |    Construct the reparameterized fMRI time courses $h_j^i(t)$ as

$$h_j^i(t) = 0, \qquad\qquad\qquad\qquad \text{where } 0 \leq t \leq t_{j1}^i \quad (2)$$
$$= h_j^i(t) \quad \text{where } t_{j1}^i \leq t \leq t_{j2}^i, i = 1, \ldots, k, j = 1, \ldots, n.$$
$$= 0, \qquad\qquad\qquad\qquad \text{where } t \geq t_{j2}^i$$

10 |    **end**
11 |    Set $h(t)$ to $\sum_{i=1}^{k} \sum_{j=1}^{n} \hat{h}_j^i(t), t \in [0, T]$.
12 **end**

---

## 3    Results

**Data:** Imaging data was acquired from ten healthy subjects on a Siemens 3T PRISMA scanner. Task fMRI scans were acquired using a multiband (MB) EPI sequence (TE $= 37$ ms, TR $= 800$ ms, voxel size $= 2 \times 2 \times 2$ mm$^3$, FA $= 52°$, MB accl. factor $= 8$) and a T1w structural scan (TE $= 4.6$ ms, TR $= 9.9$ ms, voxel size $= 0.8 \times 0.8 \times 0.8$ mm$^3$, FA $= 2°$) was also acquired. The functional task consisted of a previously validated face-matching paradigm [1], where four different types of images are presented to the subject: neutral, happy, and fearful faces, and objects. In this work, we exclusively considered all the 60 subcortical regions of interest (ROIs) extracted from a widely used automated segmentation protocol for our analysis [2]. After performing standard preprocessing and within-subject function to structure registration steps [11], Algorithm 1 was applied to each individual subject for four contrasts; happy > neutral, fearful > neutral, happy > fearful, and fearful > happy face stimuli. Figure 2 shows examples of fMRI hyperalignment for time courses from the amygdala and the putamen. Single subject GLM analyses were conducted using FEAT [11] and statistical maps were generated for both the original and the warped volumes.

**Single Subject Analysis:** Results show robust differences in the activation maps of all 10 individual subjects before and after alignment. As an example, Fig. 3 shows the activation maps for two different subjects for the original data (in blue) and for the warped data (in orange) for two different contrasts (happy > neutral faces and fearful > neutral faces).

Here, for the happy > neutral contrast, Subject 1 shows activated clusters in the right and left putamen (cluster label a in Fig. 3) for the warped data, but not for the original data. For the same subject, both the original and the warped data show a cluster in the right putamen (b in Fig. 3), but only the warped data shows an activated cluster in the left caudate (c in Fig. 3) for the fearful > neutral contrast. For sub-

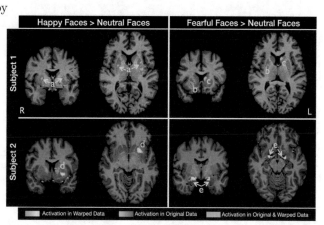

**Fig. 3.** Activated clusters in two different subjects for two different conditions (happy > neutral faces and fearful > neutral faces) for the original data (blue clusters) and for the warped data (orange clusters). The overlapping clusters between the two datasets are shown in green (e.g cluster b). Cluster locations: (a) bilateral putamen; (b) right putamen; (c) left caudate; (d) left caudate; (e) bilateral amygdala.

ject 2, only the warped data revealed clusters in the left putamen for happy > neutral and in the bilateral amygdala for fearful > neutral. The original data did not contain any activated cluster for both contrasts. These results agree with previously published literature implicating the putamen in positive emotional processing, and the role of amygdala in negative facial emotional processing [1] and suggest greater signal to noise for detecting regional differences in BOLD response after applying the warping procedure. Table 1 reports the significant clusters represented in Fig. 3.

**Table 1.** Significantly activated clusters from FSL FEAT for 2 subjects for the Happy > Neutral and Fearful > Neutral contrasts compared for the original and warped data. NA denotes no activated clusters.

| | | Data Type | Cluster Max. | Fig. labels | Voxels | P | Z.MAX | Z Max. coord. (MNI) | Cope.Max | Cope. Max. coord. (MNI) | Cope. Mean |
|---|---|---|---|---|---|---|---|---|---|---|---|
| Subject1 | Happy > Neutral | Original | NA | NA | NA | NA | NA | NA | NA | NA | NA |
| | | Warped | Left-Putamen | a | 141 | 1.66E-14 | 3.58 | (60, 55, 41) | 119 | (60,55,41) | 81.5 |
| | | | Right-Putamen | a | 18 | 0.0289 | 3.04 | (32,63,39) | 87.8 | (33,63,40) | 65.4 |
| | Fearful > Neutral | Original | Right-Putamen | b | 205 | 0.0112 | 3.87 | (33,64,38) | 112 | (33,66,30) | 70.6 |
| | | Warped | Right-Putamen | b | 235 | 3.65E-21 | 4.32 | (33,64,38) | 139 | (33,66,30) | 73.4 |
| | | | Left-Caudate | c | 41 | 2.66E-05 | 3.58 | (49,70,40) | 82.4 | (49,70,40) | 64.7 |
| Subject2 | Happy > Neutral | Original | NA | NA | NA | NA | NA | NA | NA | NA | NA |
| | | Warped | Left-Putamen | d | 92 | 8.15E-10 | 4.36 | (57,67,33) | 76.3 | (57,67,33) | 50 |
| | Fearful > Neutral | Original | NA | NA | NA | NA | NA | NA | NA | NA | NA |
| | | Warped | Right-Amygdala | e | 96 | 3.76E-10 | 4.45 | (34,61,29) | 77.9 | (36,61,29) | 51 |
| | | | Left-Amygdala | e | 58 | 1.01E-06 | 4.86 | (54,60,28) | 92.5 | (54,61,28) | 58.7 |

## Within Group Mean Activations:

To evaluate the group differences between the original data and the warped data we computed cluster thresholded averaged GLM maps across subjects for two contrasts (happy > neutral and fearful > objects) (Fig. 4). From these maps both the original data and the warped data show activated clusters in the cerebellum (cluster a in Fig. 4), in the right amygdala (cluster b in Fig. 4), in the bilateral putamen (cluster c in Fig. 4), and in the bilateral hippocampus (cluster d in Fig. 4). Interestingly only the warped data showed activated clusters in the bilateral caudate (cluster e in Fig. 4) as well as an extended cluster in the right putamen that followed its anatomical boundary.

**Table 2.** Paired t-test for comparison of number of clusters and total activated voxels between original and warped data for 4 contrasts for N = 10 subjects. Bonferroni corrected p-values are denoted by *.

| | test | t-value | df | p-value | 95% confidence int. | mean of diferences |
|---|---|---|---|---|---|---|
| Happy > Neutral | no. clusters | -4.9209 | 9 | 0.0064* | [-6.8606 -2.5394] | -4.7 |
| | no. voxels | -3.8818 | 9 | 0.0296* | [-330.7957 -87.20432] | -209 |
| Fearful > Neutral | no. clusters | -3.3806 | 9 | 0.06495* | [-6.6766 -1.3234] | -4 |
| | no. voxels | -3.8602 | 9 | 0.0304* | [-356.0610 -92.93903] | -224.5 |
| Happy > Objects | no. clusters | -5.262 | 9 | 0.004154* | [-12.2972 -4.9028] | -8.6 |
| | no. voxels | -4.6202 | 9 | 0.0104* | [-1010.1137 -346.0863] | -678.1 |
| Fearful > Objects | no. clusters | -4.5843 | 9 | 0.0104* | [-14.0385 -4.7615] | -9.4 |
| | no. voxels | -3.8551 | 9 | 0.0312* | [-662.9628 -172.6372] | -417.8 |

## Paired Statistics Between Warped and Unwarped Images:

Finally, paired t-tests were computed to evaluate the change in the number of clusters and the number of voxels activated before and after performing elastic alignment for the four different conditions (happy > neutral; fearful > neutral; happy > objects and fearful > objects). There was a significant increase (Bonferroni corrected) in the number of clusters and number of voxels activated after applying warping (Table 2).

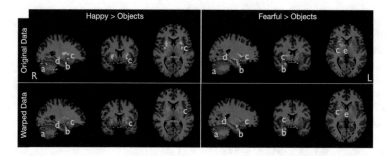

**Fig. 4.** Mean activation maps cluster thresholded for the original data (top row) and for the warped data (bottom row) for two different contrasts (happy faces > objects and fearful faces > objects). Cluster locations: (a) cerebellar cortex; (b) right amygdala; (c) bilateral putamen; (d) hippocampus; (e) right caudate.

## 4    Discussion

We proposed an elastic time series matching approach to construct an unbiased condition-specific template of the BOLD shape patterns. While the warped data shows increased number of clusters and voxels, we want to caution against the possibility of registration and enhancement of noisy activations. This can be potentially validated with repeated randomization tests and test-retest reliability statistics in the future. Further, the Riemannian metric used here promotes both shrinking, stretching and bending. Experiments that control the parameters of this metric will also yield useful data addressing this question. Finally, additional testing of this method on event based fMRI paradigms or simple motor tasks that generate reliable activations will be necessary to understand the behavior and potential utility of this method.

**Acknowledgments.** This research was supported by the NIH/NIAAA award K25AA024192, and the NIH/NIMH award U01MH110008.

## References

1. Chai, X.J., et al.: Functional and structural brain correlates of risk for major depression in children with familial depression. NeuroImage Clin. **8**, 398–407 (2015)
2. Fischl, B., et al.: Automatically parcellating the human cerebral cortex. Cereb. Cortex **14**(1), 11–22 (2004)
3. Glover, G.H.: Overview of functional magnetic resonance imaging. Neurosurg. Clin. **22**(2), 133–139 (2011)
4. Guntupalli, J.S., Hanke, M., Halchenko, Y.O., Connolly, A.C., Ramadge, P.J., Haxby, J.V.: A model of representational spaces in human cortex. Cereb. Cortex **26**(6), 2919–2934 (2016)
5. Joshi, A.A., Chong, M., Li, J., Choi, S., Leahy, R.M.: Are you thinking what i'm thinking? Synchronization of resting fMRI time-series across subjects. NeuroImage **172**, 740–752 (2018)

6. Joshi, S.H., Klassen, E., Srivastava, A., Jermyn, I.: A novel representation for Riemannian analysis of elastic curves in $R^n$. In: IEEE Conference on Computer Vision and Pattern Recognition (CVPR), pp. 1–7. IEEE (2007)
7. Joshi, S.H., Klassen, E., Srivastava, A., Jermyn, I.: Removing shape-preserving transformations in square-root elastic (SRE) framework for shape analysis of curves. In: Yuille, A.L., Zhu, S.-C., Cremers, D., Wang, Y. (eds.) EMMCVPR 2007. LNCS, vol. 4679, pp. 387–398. Springer, Heidelberg (2007). https://doi.org/10.1007/978-3-540-74198-5_30
8. Karcher, H.: Riemannian center of mass and mollifier smoothing. Commun. Pure Appl. Math. **30**, 509–541 (1977)
9. Lee, D.S., Leaver, A.M., Narr, K.L., Woods, R.P., Joshi, S.H.: Measuring brain connectivity via shape analysis of fMRI time courses and spectra. In: Wu, G., Laurienti, P., Bonilha, L., Munsell, B.C. (eds.) CNI 2017. LNCS, vol. 10511, pp. 125–133. Springer, Cham (2017). https://doi.org/10.1007/978-3-319-67159-8_15
10. Meszlényi, R.J., Hermann, P., Buza, K., Gál, V., Vidnyánszky, Z.: Resting state fMRI functional connectivity analysis using dynamic time warping. Front. Neurosci. **11**, 75 (2017)
11. Smith, S.M., et al.: Advances in functional and structural mr image analysis and implementation as FSL. Neuroimage **23**, S208–S219 (2004)
12. Srivastava, A., Wu, W., Kurtek, S., Klassen, E., Marron, J.S.: Registration of functional data using Fisher-Rao metric. arXiv:1103.3817v2 (2011)

# A Generative-Discriminative Basis Learning Framework to Predict Clinical Severity from Resting State Functional MRI Data

Niharika Shimona D'Souza[1(✉)], Mary Beth Nebel[2,3], Nicholas Wymbs[2,3], Stewart Mostofsky[2,3,4], and Archana Venkataraman[1]

[1] Department of Electrical and Computer Engineering, Johns Hopkins University, Baltimore, USA
shimona.niharika.dsouza@jhu.edu
[2] Center for Neurodevelopmental Medicine and Research, Kennedy Krieger Institute, Baltimore, USA
[3] Department of Neurology, Johns Hopkins School of Medicine, Baltimore, USA
[4] Department of Pediatrics, Johns Hopkins School of Medicine, Baltimore, USA

**Abstract.** We propose a matrix factorization technique that decomposes the resting state fMRI (rs-fMRI) correlation matrices for a patient population into a sparse set of representative subnetworks, as modeled by rank one outer products. The subnetworks are combined using patient specific non-negative coefficients; these coefficients are also used to model, and subsequently predict the clinical severity of a given patient via a linear regression. Our generative-discriminative framework is able to exploit the structure of rs-fMRI correlation matrices to capture group level effects, while simultaneously accounting for patient variability. We employ ten fold cross validation to demonstrate the predictive power of our model on a cohort of fifty eight patients diagnosed with Autism Spectrum Disorder. Our method outperforms classical semi-supervised frameworks, which perform dimensionality reduction on the correlation features followed by non-linear regression to predict the clinical scores.

## 1 Introduction

Resting state fMRI (rs-fMRI) allows us to assess brain activity and localize critical functions through steady state patterns of co-activation [1]. Building predictive models at the patient level remains an open challenge due to the high data dimensionality and to the considerable inter-subject variability. Predictive analysis methods usually follow a two step procedure. First, feature selection is applied to the raw correlation values; examples include graph theoretic measures, statistical measures and embedding features obtained from unsupervised learning techniques such as PCA, k-PCA or ICA [2]. As a second step, conventional regression techniques such as Random Forests or Support Vector Regression are applied to the feature space representation to predict the clinical severity. These

© Springer Nature Switzerland AG 2018
A. F. Frangi et al. (Eds.): MICCAI 2018, LNCS 11072, pp. 163–171, 2018.
https://doi.org/10.1007/978-3-030-00931-1_19

strategies adequately capture the group-averaged functional connectivity across the cohort but fail to account for individual variability. Consequently, the generalization power of these techniques is limited.

The recent success of Bayesian [3] and dictionary learning [4] studies on rs-fMRI data is largely based on their ability to simultaneously model the patient and group level information. [4] introduces a basis learning framework for patient subtype classification, which reduces the dimensionality of T1 MR voxel based morphometry data while preserving the anatomical interpretability. [5] introduces a correlation matrix decomposition strategy, where multiple rank one matrix outer products capturing the underlying 'generative' basis are combined using patient specific coefficients. The sparse basis networks identify meaningful co-activation patterns common to all the patients, and the coefficients model the patient variability. Our main contribution lies in exploiting the 'discriminative' nature of rs-fMRI correlation matrices. We estimate the clinical severity of every patient by constructing a regression model which maps the behavioral scores to the functional data space. We jointly optimize for each of the hidden variables in the model, i.e. the basis, coefficients and regression weights. We refine the validation process by quantifying the model generalizability in terms of the regression performance on unseen data, as opposed to the correlation fit measure in [5]. Hence, our framework is less prone to overfitting.

We validate our framework on a population study of Autism Spectrum Disorder (ASD). Patient variability manifests as a spectrum of impairments, typically quantified by a "behavioral score" of clinical severity obtained from an expert assessment. Identifying sub-networks predictive of ASD severity is the key link to understanding the social and behavioral implications of the disorder. Our inclusion of behavioral data into the optimization framework guides the identification of representative networks specific to resting state ASD characterization.

## 2    A Joint Model for Connectomics and Clinical Severity

Let $N$ be the number of patients and $M$ be the number of regions in our brain parcellation. We decompose the patient correlation matrices $\boldsymbol{\Gamma}_n \in \mathcal{R}^{M \times M}$ into a non-negative combination of a $K$ basis subnetworks $\mathbf{b}_k \mathbf{b}_k^T$. The sparse vector $\mathbf{b}_k$ indicates the relative contribution of each brain region to network $k$. The vector $\mathbf{c}_n$ denotes the non-negative contribution of each subnetwork for patient $n$. The coefficients $\mathbf{c}_n$ are subsequently used to model the clinical severity score $y_n$ via the regression weight vector $\mathbf{w} \in \mathcal{R}^K$. We concatenate the subnetworks into a basis matrix $\mathbf{B} \in \mathcal{R}^{M \times K}$, the coefficients into the matrix $\mathbf{C} \in \mathcal{R}^{K \times N}$, and the scores into a vector $\mathbf{y} \in \mathcal{R}^N$. Our combined objective can be written as follows:

$$\mathcal{J}(\mathbf{B}, \mathbf{C}, \mathbf{W}) = \sum_n ||\boldsymbol{\Gamma}_n - \mathbf{B}\mathbf{diag}(\mathbf{c}_n)\mathbf{B}^T||_F^2 + \gamma||\mathbf{y} - \mathbf{C}^T\mathbf{w}||_2^2 \quad s.t. \ \ \mathbf{c}_{nk} \geq 0, \ (1)$$

Here, $\gamma$ is the tradeoff between the behavioral and functional data terms, and $\mathbf{diag}(\mathbf{c}_n)$ is a matrix with the elements of $\mathbf{c}_n$ on its leading diagonal, and off diagonal elements as 0. We impose an $\ell_1$ penalty upon the matrix $\mathbf{B}$ in order to recover a sparse set of subnetworks. Since the objective in Eq. (1) is ill posed, we add quadratic penalty terms on $\mathbf{C}$ and $\mathbf{w}$ which act as regularizers.

$$\lambda_1||\mathbf{B}||_1 + \lambda_2||\mathbf{C}||_2^2 + \lambda_3||\mathbf{w}||_2^2. \tag{2}$$

Equation (2) is added to the overall objective with $\lambda_1$, $\lambda_2$ and $\lambda_3$ being the sparsity, norm penalty on $\mathbf{C}$, and the penalty on $\mathbf{w}$ respectively.

## 2.1   Optimization Strategy

We employ a fixed point alternating minimization strategy to optimize $\mathbf{B}$, $\mathbf{C}$ and $\mathbf{w}$. At every iteration, the optimal solution for one variable is calculated assuming the other variables are held constant. Proximal gradient descent [6] is an effective strategy of optimizing a non-differentiable sparsity penalty such as the one in Eq. (2), when the supporting terms are convex in the variable of interest. However, the expansion of the first Frobenius norm gives rise to non-convex bi-quadratic terms in $\mathbf{B}$, which prevents us from directly computing a proximal solution. Therefore, we introduce $N$ constraints of the form $\mathbf{D}_n = \mathbf{Bdiag}(\mathbf{c}_n)$, which are enforced by an Augmented Lagrangian penalty:

$$\mathcal{J}(\mathbf{B}, \mathbf{C}, \mathbf{w}, \mathbf{D}_n, \mathbf{\Lambda}_n) = \sum_n ||\mathbf{\Gamma}_n - \mathbf{D}_n\mathbf{B}^T||_F^2 + \sum_n \mathrm{Tr}\left[\mathbf{\Lambda}_n^T(\mathbf{D}_n - \mathbf{Bdiag}(\mathbf{c}_n))\right]$$

$$+ \sum_n \frac{1}{2}||\mathbf{D}_n - \mathbf{Bdiag}(\mathbf{c}_n)||_F^2 + \gamma||\mathbf{y} - \mathbf{C}^T\mathbf{w}||_2^2 \quad s.t. \ \ \mathbf{c}_{nk} \geq 0 \tag{3}$$

where, each $\mathbf{\Lambda}_n$ is a matrix of Lagrangians and each of the supporting Frobenius norm terms are regularizers on the Lagrangian constraints. The objective in Eq. (3) is convex in $\mathbf{B}$ and the set $\{\mathbf{D}_n\}$ separately. Our optimization begins by randomly initializing $\mathbf{B}$, $\mathbf{C}$ and $\mathbf{w}$ and setting $\mathbf{D}_n = \mathbf{Bdiag}(\mathbf{c}_n)$ and $\mathbf{\Lambda}_n = \mathbf{0}$. We then iterate through the following four steps until global convergence.

**Step 1 - Optimizing B via Proximal Gradient Descent.** Given the fixed learning rate parameter $t$, the proximal update for $\mathbf{B}$ is:

$$\mathbf{B}^{k+1} = \mathbf{sgn}(\mathbf{X}).^*(\mathbf{max}(|\mathbf{X}| - t, 0)) \ \ s.t. \ \ \mathbf{X} = \mathbf{B}^k - (t/\lambda_1)\frac{\partial \mathcal{J}}{\partial \mathbf{B}} \tag{4}$$

The derivative of $\mathcal{J}$ with respect to $\mathbf{B}$, where $\mathbf{V}_n = \mathbf{diag}(\mathbf{c}_n)$, is computed as:

$$\frac{\partial \mathcal{J}}{\partial \mathbf{B}} = \sum_n \left[2\left[\mathbf{BD}_n^T\mathbf{D}_n - \mathbf{\Gamma}_n\mathbf{D}_n\right] - \mathbf{D}_n\mathbf{V}_n + \mathbf{BV}_n^2 - \mathbf{\Lambda}_n\mathbf{V}_n\right] \tag{5}$$

As seen, the non-smoothness of the $||\mathbf{B}||_1$ penalty is handled by performing iterative shrinkage thresholding applied on a locally smooth quadratic model.

**Step 2 - Optimizing C using Quadratic Programming.** The objective is quadratic in $\mathbf{C}$ when $\mathbf{B}$ and $\mathbf{w}$ are held constant. Furthermore, the $\mathbf{diag}(\mathbf{c}_n)$

term decouples the updates for $\mathbf{c}_n$ across patients. Hence, we use $N$ quadratic programs of the form below to solve for the vectors $\{\mathbf{c}_n\}$:

$$\frac{1}{2}\mathbf{c}_n^T\mathbf{H}_n\mathbf{c}_n + \mathbf{f}_n^T\mathbf{c}_n \quad s.t. \quad \mathbf{A}_n\mathbf{c}_n \geq \mathbf{b}_n \tag{6}$$

The Quadratic Programming parameters for our problem are given by:

$$\mathbf{H}_n = \mathbf{diag}(\mathbf{B}^T\mathbf{B}) + 2\gamma\mathbf{w}\mathbf{w}^T + 2\lambda_2\mathcal{I}_K$$
$$\mathbf{f}_n = -\mathbf{diag}(\mathbf{D}_n^T\mathbf{B}) - \mathbf{diag}(\mathbf{\Lambda}_n^T\mathbf{B}) - 2\gamma y_n\mathbf{w}; \quad \mathbf{A}_n = -\mathcal{I}_K \quad \mathbf{b}_n = \mathbf{0}$$

This strategy helps us find the globally optimal solutions for $\mathbf{c}_n$, as projected onto the $K$ dimensional space of positive real numbers.

***Step 3 - Closed Form Update for*** $\mathbf{w}$. The global minimizer of $\mathbf{w}$ computed at the first order stationary point can be expressed as:

$$\mathbf{w} = (\mathbf{C}\mathbf{C}^T + \frac{\lambda_3}{\gamma}\mathcal{I}_K)^{-1}(\mathbf{C}\mathbf{y}) \tag{7}$$

***Step 4 - Optimizing the Constraint Variables*** $\mathbf{D}_n$ ***and*** $\mathbf{\Lambda}_n$. Each of the primal variables $\{\mathbf{D}_n\}$ has a closed form solution given by:

$$\mathbf{D}_n = (\mathbf{diag}(\mathbf{c}_n)\mathbf{B}^T + 2\mathbf{\Gamma}_n\mathbf{B} - \mathbf{\Lambda}_n)(\mathcal{I}_K + 2\mathbf{B}^T\mathbf{B})^{-1} \tag{8}$$

In contrast, we update the dual variables $\{\mathbf{\Lambda}_n\}$ using gradient ascent:

$$\mathbf{\Lambda}_n^{k+1} = \mathbf{\Lambda}_n^k + \eta_k(\mathbf{D}_n - \mathbf{B}\mathbf{diag}(\mathbf{c}_n)) \tag{9}$$

The updates for $\mathbf{D}_n$ and $\mathbf{\Lambda}_n$ ensure that the proximal constraints are satisfied with increasing certainty at each iteration. The learning rate parameter $\eta_k$ for the gradient ascent step of the augmented Lagrangian is chosen to guarantee sufficient decrease for every iteration of alternating minimization. In practice, we initialize this value to 0.001, and scale it by 0.75 at each iteration.

In all of our derivations, $\mathrm{Tr}(\mathbf{M})$ is the trace operator and gives the sum of the diagonal elements of a matrix $\mathbf{M}$, and $\mathcal{I}_K$ is the $K \times K$ identity matrix.

## 2.2   Predicting Symptom Severity

We use cross validation to evaluate the predictive power of our model. Specifically, we compute the optimal $\{\mathbf{B}^\star, \mathbf{w}^\star\}$ based on the training dataset. We can then estimate the coefficients $\mathbf{c}_{test}$ for a new patient by re-solving the quadratic program in **Step 2** using the previously computed $\{\mathbf{B}^\star, \mathbf{w}^\star\}$. Notice that we must set the data term $\gamma\|\mathbf{C}^T\mathbf{w} - \mathbf{y}\|_2^2$ to 0 in the testing experiments, since the severity $y_{test}$ is unknown. Also, we assume that the constraint $\mathbf{D}_{test} = \mathbf{B}^\star\mathbf{diag}(\mathbf{c}_{test})$ is satisfied exactly for the conditions of the proximal operator to hold. Finally, $y_{test} = \mathbf{c}_{test}^T\mathbf{w}^\star$ is the estimate of the behavioral score for the unseen test patient.

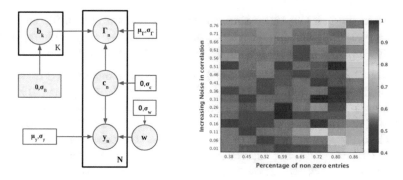

**Fig. 1.** (a) The graphical model from which we generate synthetic data (b) The network recovery performance of our algorithm for varying levels of sparsity and correlation matrix noise variance. For our experiments, we fix the rest of the parameters of the model at $\sigma_{\mathbf{B}} = 0.2$, $\mu_{\mathbf{\Gamma}_n} = \mathbf{B}\mathbf{diag}(\mathbf{c}_n)\mathbf{B}^T$, $\sigma_{\mathbf{c}} = 0.1$, $\mu_{\mathbf{y}} = \mathbf{c}_n^T\mathbf{w}$, $\sigma_{\mathbf{y}} = 0.2$ and $\sigma_{\mathbf{w}} = 0.1$

## 2.3   Baseline Comparison Methods

We compare our algorithm with a standard machine learning pipeline to predict the target severity score. We first perform dimensionality reduction to concentrate the $\frac{M \times (M-1)}{2}$ rs-fMRI correlation pairs into a small number of basis elements. Then, we construct a non-linear regression model to predict clinical severity. We consider two dimensionality reduction/regression combinations:

1 Principal Component Analysis on the correlation coefficients followed by a Random Forest Regression on the projected features
2 Kernel Principal Component Analysis on the correlation coefficients followed by a Random Forest Regression on the embedding features.

## 3   Experimental Results

***Evaluating Robustness on Synthetic Data.*** Our optimization problem in Eq. (1) suggests an underlying graphical model, depicted in Fig. 1(a). Notice that the $\ell_1$ penalty on $\mathbf{B}$ translates to a Laplacian prior with $\sigma_B$ controlling the potentially overlapping level of sparsity. In contrast, the $\ell_2$ penalties translate into Gaussian distributions, with the mean corresponding to the $\ell_2$ argument and the variance related to the regularization parameters. We use this model to sample the correlation matrices $\{\mathbf{\Gamma}_n\}$ and the behavioral scores $\{y_n\}$, and then infer the latent networks generating the data. Figure 1(b) indicates the performance of network recovery from our algorithm. We quantify the peformance in terms of average inner-product similarity between recovered networks and generating networks, both normalized to unit norm. The number of generating and recovery networks is chosen to be 4. Unsurprisingly, increasing the overlap in the sparsity patterns across networks and increasing the noise in the correlation estimates worsens the recovery performance. However, our optimization procedure is robust in the noise regime estimated from our real-world rs-fMRI correlation matrices (0.01–0.2) and for recovered sparsity levels (0.1–0.4). The experiment also helps us identify stable parameter settings for the next section.

***rs-fMRI Dataset and Preprocessing.*** We evaluate our method on a cohort of 58 children with high-functioning ASD (Age: $10.06 \pm 1.26$, IQ: $110 \pm 14.03$). We acquired rs-fMRI scans on a Phillips 3T Achieva scanner using a single-shot, partially parallel gradient-recalled EPI sequence (TR/TE = $2500/30$ ms, flip angle = $70°$, res = $3.05 \times 3.15 \times 3$ mm, 128 or 156 time samples).

Rs-fMRI preprocessing [3] consisted of slice time correction, rigid body realignment, and normalization to the EPI version of the MNI template using SPM. We use a CompCorr strategy to remove the spatially coherent noise from the white matter, ventricles, and six rigid body realignment parameters. We then spatially smoothed the data (6 mm FWHM Gaussian kernel) and band-pass filtered the time series (0.01–0.1 Hz). We use the Automatic Anatomical Labeling (AAL) atlas to define 116 cortical, subcortical and cerebellar regions. Empirically, we observed a consistent noise component having nearly constant contribution from all the brain regions and low predictive power. Consequently, we subtract out the contribution of the first eigenvector from the correlation matrices and used the residuals $\{\mathbf{\Gamma}_n\}$ as inputs for all the methods.

We consider two measures of clinical severity: Autism Diagnostic Observation Schedule (ADOS) total raw score [7], which captures the social and communicative interaction deficits of the patient along with repetitive behaviors (dynamic range: 0–30), and the Social Responsiveness Scale (SRS) total raw score [7] which characterizes social responsiveness (dynamic range: 70–200).

***Predicting ASD Severity.*** We employ a ten fold cross validation strategy for each of the methods, whereby, we train the model on a 90% data split and evaluate the performance on the unseen 10% test data. We perform a grid search to find the optimal parameter setting for each method. Based on these results, we fix the regression tradeoff at $\gamma = 1$, and the three regularization parameters at $\{\lambda_1 = 40, \lambda_2 = 2, \lambda_3 = 1\}$ for SRS, and $\{\lambda_1 = 30, \lambda_2 = 0.2, \lambda_3 = 1\}$ for ADOS, and the learning rate at $t = 0.001$ for proximal gradient descent. The number of components was fixed at 15 for PCA and 10 for k-PCA. For k-PCA, we use an RBF kernel with the coefficient parameter varied between 0.01–10.

As seen from the Fig. 2, the baseline methods have poor validation performance and track the mean value of the held out data (shown by the black line). In comparison, our method not only consistently fits the training set more faithfully, but also generalizes much better beyond the training data. The major shortcoming of the baseline data-driven analysis techniques is in their failure to identify representative patterns of behavior from the correlation features. In contrast, our basis learning technique exploits the underlying structure of the correlation matrices and leverages patient specific information to map the ASD behavioral space, thus improving the prediction performance. As reported in Table 1, our method quantitatively outperforms the baselines approaches, both in terms of the root median square error (rMSE) and the $R^2$ performance.

**Table 1.** Performance evaluation using **root median square error (rMSE)** & $\mathbf{R^2}$ fit, both for testing & training. Lower MSE & higher $R^2$ score indicate better performance.

| Score | Metric | Our method | PCA + RF Reg | k-PCA + RF Reg |
|-------|--------|-----------|--------------|----------------|
| ADOS | rMSE train | **0.088** | 1.07 | 1.017 |
| | $R^2$ train | **0.99** | 0.94 | 0.96 |
| | rMSE test | **2.53** | 2.93 | 2.70 |
| | $R^2$ test | **0.096** | 0.031 | 0.01 |
| SRS | rMSE train | **0.13** | 6.43 | 6.90 |
| | $R^2$ train | **0.99** | 0.95 | 0.97 |
| | rMSE test | **13.26** | 20.51 | 20.30 |
| | $R^2$ test | **0.052** | 0.023 | 0.008 |

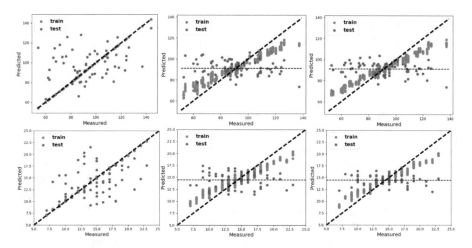

**Fig. 2.** Prediction performance of each method for ADOS (TR) & SRS (BR) **Left:** Our Method ($K = 8$) **Middle:** PCA ($comp = 15$) & RF Regression on the projected data **Right:** k-PCA ($comp = 10$, rbf $C = 0.1$) & RF Regression on the embedding features. Red & Green points correspond to the training & testing performance respectively

***Subnetwork Identification.*** Figure 3 illustrates the basis subnetworks in **B** trained on the ADOS data. The colorbar indicates subnetwork contribution to the AAL regions. Regions storing negative values are anticorrelated with regions storing positive ones. Subnetwork 1 includes competing i.e. anticorrelated contributions from regions of the default mode network (DMN) and somatomotor network (SMN). Abnormal connectivity within the DMN and SMN has been previously reported in ASD [8]. Additionally, subnetwork 5 appears to be comprised of competing contributions from SMN regions and higher order visual processing areas in the occipital and temporal lobes, consistent with behavioral reports of reduced visual-motor integration in ASD. Subnetwork 2 includes competing contributions from prefrontal and subcortical regions (mainly the thalamus,

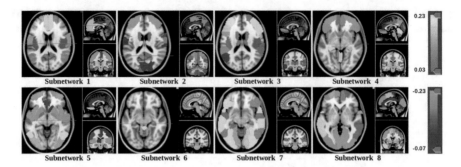

**Fig. 3.** Eight subnetworks identified by our model from ADOS prediction. The blue & green regions are anticorrelated with the red & orange regions for each subnetwork.

amygdala and hippocampus), that may be important for social-emotional regulation in ASD. Finally, subnetwork 3 is comprised of competing contributions from the central executive control network and the insula, which is thought to be critical for switching between self-referential and goal-directed behavior [9].

## 4      Conclusion

Unlike generic machine learning analysis, our matrix decomposition elegantly combines multimodal information from the imaging and behavioral domains. The key to our model is its ability to capture and learn from the structure of correlation matrices. Conventional analysis methods dramatically fall short of unifying the two data viewpoints reliably enough to implicate predictive functional patterns in the brain. Our joint optimization framework robustly identifies brain networks characterizing ASD and provides a key link to quantifying and interpreting the spectrum of manifestation of the disorder across a wide range of population. In the future, we will explore extensions of this model that jointly classify patients versus controls in addition to predicting symptom severity.

**Acknowledgements.** This work was supported by the National Institute of Mental Health (R01 MH085328-09, R01 MH078160-07, K01 MH109766 and R01 MH106564), the National Institute of Neurological Disorders and Stroke (R01 NS048527-08), and the Autism Speaks foundation.

## References

1. Fox, M.D., et al.: Spontaneous fluctuations in brain activity observed with functional magnetic resonance imaging. Nat. Rev. Neurosci. **8**(9), 700 (2007)
2. Murphy, K.P.: Machine Learning: A Probabilistic Perspective. MIT Press, Cambridge (2012)
3. Venkataraman, A., et al.: Bayesian community detection in the space of group-level functional differences. IEEE Trans. Med. Imaging **35**(8), 1866–1882 (2016)

4. Batmanghelich, N.K., et al.: Generative-discriminative basis learning for medical imaging. IEEE Trans. Med. Imaging **31**(1), 51–69 (2012)
5. Eavani, H., et al.: Identifying sparse connectivity patterns in the brain using resting-state fMRI. Neuroimage **105**, 286–299 (2015)
6. Parikh, N., Boyd, S., et al.: Proximal algorithms. Found. Trends® Optim. **1**(3), 127–239 (2014)
7. Payakachat, N., et al.: Autism spectrum disorders: a review of measures for clinical, health services and cost-effectiveness applications. Expert. Rev. Pharmacoeconomics Outcomes Res. **12**(4), 485–503 (2012)
8. Nebel, M.B., et al.: Intrinsic visual-motor synchrony correlates with social deficits in autism. Biol. Psychiatry **79**(8), 633–641 (2016)
9. Sridharan, D., et al.: A critical role for the right fronto-insular cortex in switching between central-executive and default-mode networks. Proc. Natl. Acad. Sci. **105**(34), 12569–12574 (2008)

# 3D Deep Convolutional Neural Network Revealed the Value of Brain Network Overlap in Differentiating Autism Spectrum Disorder from Healthy Controls

Yu Zhao, Fangfei Ge, Shu Zhang, and Tianming Liu$^{(\boxtimes)}$

Department of Computer Science, The University of Georgia,
Athens, GA 30605, USA
tianming.liu@gmail.com

**Abstract.** Spatial distribution patterns of functional brain networks derived from resting state fMRI data have been widely examined in the literature. However, the spatial overlap patterns among those brain networks have been rarely investigated, though spatial overlap is a fundamental principle of functional brain network organization. To bridge this gap, this paper presents an effective 3D convolutional neural network (CNN) framework to derive discriminative and meaningful spatial brain network overlap patterns that can characterize and differentiate Autism Spectrum Disorder (ASD) from healthy controls. Our experimental results demonstrated that the spatial distribution patterns of connectome-scale functional network maps per se have little discrimination power in differentiating ASD from controls via the CNN framework. In contrast, the spatial overlap patterns instead of spatial patterns per se among these connectome-scale networks, learned via the same CNN framework, have remarkable differentiation power in separating ASD from controls. Our work suggested the promise of using CNN deep learning methodologies to discover discriminative and meaningful spatial network overlap patterns and their applications in functional connectomics of brain disorders such as ASD.

**Keywords:** Autism · CNN · Overlap network

## 1 Introduction

Faithful reconstruction and quantitative modeling of brain networks from noisy fMRI data has been of a major neuroscientific research topic for years. Popular brain network reconstruction techniques based on fMRI data include general linear model (GLM) [1] for task-based fMRI (tfMRI), independent component analysis (ICA) [2] for resting state fMRI (rsfMRI), and dictionary learning/sparse representation [3] for both tfMRI and rsfMRI, all of which can reconstruct concurrent network maps from whole brain fMRI data. It has been shown in a variety of literature studies [3] that sparse dictionary learning is superior than other methods in reconstructing spatially overlapping brain networks. For instance, by using the dictionary learning and sparse coding algorithms [3, 4], several hundred of concurrent functional brain networks, characterized by both

© Springer Nature Switzerland AG 2018
A. F. Frangi et al. (Eds.): MICCAI 2018, LNCS 11072, pp. 172–180, 2018.
https://doi.org/10.1007/978-3-030-00931-1_20

spatial maps and associated temporal time series, can be effectively decomposed from either tfMRI or rsfMRI data of an individual brain. However, the spatial overlap patterns among those hundreds of concurrent brain networks have been rarely investigated, though spatial overlap is a fundamental principle of functional brain network organization [5]. It is even a bigger mystery whether those spatial network overlap patterns provide valuable information of brain disorders such as Autism Spectrum Disorder (ASD).

To bridge these technical and knowledge gaps, this paper presents an effective 3D convolutional neural network (CNN) [6, 7] framework to derive discriminative and meaningful spatial brain network overlap patterns that aim to characterize and differentiate ASD from healthy controls. Faced with the challenges of fractionation of the available data in site-specific studies with relatively small sample sizes, in this paper, we utilized multisite (10 sites) rsfMRI datasets from the ABIDE project for the classification problem to improve statistical power, and to accommodate greater variance of ASD and control. In this work, spatial distribution patterns of connectome-scale functional network maps per se were first investigated for classifying ASD patients and controls, and our experimental results demonstrated that they have little discrimination power in differentiating ASD from controls via the CNN framework (less than 50% average accuracy in testing data). In contrast, the spatial overlap patterns among these connectome-scale networks, learned via the same CNN framework, have remarkably better differentiation power in separating ASD from controls (10-fold accuracy: average at 70.5% and peak at 85%) than using functional network spatial patterns per se. Our work also suggested the promise of using CNN deep learning methodologies to discover discriminative and meaningful spatial network overlap patterns and their applications in functional connectomics of brain disorders such as ASD.

## 2   Method

### 2.1   Experimental Data and Preprocessing

Our experimental data were downloaded from the publicly available Autism Brain Imaging Data Exchange (ABIDE: http://fcon_1000.projects.nitrc.org/indi/abide/) with rsfMRI datasets collected from ASD patients and healthy controls. In this study, we investigated ASD patient and healthy control' rsfMRI data from 10 different sites. After manually checking data quality according to preprocessing (e.g. skull removal, registration to standard space) results, 100 ASD patients and 100 healthy controls' data were selected for this experiment. The acquisition parameters vary from different sites: 192–256 mm FOV, 29–47 slices, 1.5–3 s TR, 15–30 ms TE, 60–90° flip angle, (3–3.4) × (3–3.4) × (3–4) mm voxel size. For detailed parameters for each site, please refer to the ABIDE website.

Preprocessing of the rsfMRI data was performed using FSL software tools [8], including skull removal, motion correction (MCFLIRT command in FSL tools was adopted, where 4 mm (voxel level) motion parameter search was conducted to remove micro head-motions [8] and further manual check was performed to remove unsuccessful data), spatial smoothing, temporal pre-whitening, slice time correction, global drift removal, and linear registration to the Montreal Neurological Institute (MNI) standard brain template space, which were implemented by FSL FLIRT and FEAT. Next, the preprocessed rsfMRI data were decomposed into different functional networks using the online dictionary learning algorithm [4].

### 2.2    144 ICNs Generation and Spatial Overlap Feature Selections

As shown in Fig. 3(a), by applying the online dictionary learning and sparse representation techniques in [4], whole brain rsfMRI signals of each individual was decomposed into 200 intrinsic connectivity networks (ICNs). First, whole brain rsfMRI signals were extracted and arranged into a matrix $X \in \Re^{t \times n}$ with n columns representing each voxel and t rows representing each time point of the rsfsMRI scan. By using the online dictionary learning and sparse representation technique [4], each column in $X$ was modeled as a linear combination of atoms of a learned basis dictionary $D$ such that $X = D \times \alpha$, where $D \in \Re^{t \times m}$, $\alpha \in \Re^{m \times n}$, and $m$ is set to 200 (as in [9]) in this paper. Finally, each row in the $\alpha$ matrix was mapped back to the brain volume as a network. Then the 144 ICN templates generated in [9] charactering group-wise consistent ASD/control brain networks were used to search for the correspondence in the 200 decomposed networks – the network having the maximum overlap rate (calculated in Eq. (1)) with the template was taken as its correspondence:

$$\text{overlap rate} = \sum_{k=1}^{|V|} \frac{\min(V_k, W_k)}{(V_k + W_k)/2} \tag{1}$$

where $V_k$ and $W_k$ are the activation intensity of voxel k in network volume maps $V$ and $W$, respectively.

The overlap information among all the 144 ICNs are then gathered, by generating overlap maps of each pair of networks among 144 ICNs using Eq. (2). For 144 ICNs, a total combination number of 10291 overlap maps were generated for each subject. The illustration for overlapped map generation is shown in Fig. 1 by using ICN1 and ICN2 on a randomly selected subject.

$$\text{overlapped  map} = V \cap W \tag{2}$$

where $V$ and $W$ are the binarized ICN volume maps.

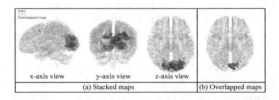

**Fig. 1.** Overlap map illustration using one subject's ICNs. Blue color and yellow color denote ICN 1 and ICN 2 volume map respectively, while red denotes the overlap map.

Among all 10,291 overlap maps, only 150 were selected as representative features for each subject since not all maps are overlapped with each other. In order to select the most overlapped patterns, the 10,291 maps are sorted in a descendant manner using the averaged nonzero voxel numbers in overlapped maps of each group (ASD and control). The first 50 feature maps were selected from maps present in the top of both groups' overlaps (with 1 ASD patient and 1 control subject example shown in Fig. 2(a)); the second 50 feature maps were selected from maps only present in top of control group's overlaps (with 1 ASD patient and 1 control subject example shown in Fig. 2(b)), while the last 50 feature maps were selected from maps only present in the top of ASD group's overlaps (with 1 ASD patient and 1 control subject example shown in Fig. 2 (c)). The 150 overlap maps were then extracted from all the subjects from the two groups as input features for the CNN training and classification.

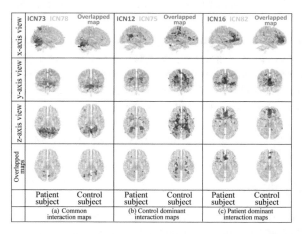

**Fig. 2.** Overlap map feature selections. Blue and yellow maps in each panel are the ICN maps, while red maps are the overlap maps of each pair of ICN maps. (a). Common overlap maps indicate both ASD patient and control groups have these types of overlaps among top 50 patterns. (b). Control-dominant overlap maps (c). ASD-dominant overlap maps.

## 2.3   Computational Framework

After obtaining the 144 intrinsic connectivity networks (ICNs) of each subject (to be described in Sect. 2.3), the spatial overlap patterns of 144 ICNs were used as input features and an fMRI-oriented 3D deep convolutional neural network (CNN) was then designed for the problem of ASD/control subjects classification (Sect. 2.4). The whole framework is briefly illustrated in Fig. 3.

Specifically, sparse representation was first performed on all the preprocessed data to extract concurrent networks at individual level; then the 144 ICNs templates generated in [9] based on the ABIDE datasets were used to search the corresponding ICNs among 200 concurrent networks within each subject. Even though those ICNs were not directly used as input features for final classifications, they will still be used as input features for comparison purpose. Instead, the overlapped maps of each ICN pair were

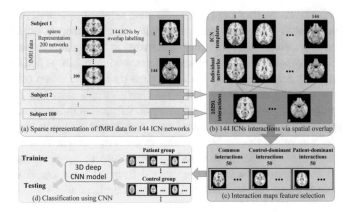

**Fig. 3.** Computational framework for ASD classification.

generated, and 10291 maps in total for each subject were generated, 150 out of which were selected as the input features for the CNN for each subject (Sect. 2.3). Our core idea here is that we hypothesize the networks' overlap patterns represent their functional interactions due to the highly interdigitated and spatially overlapped nature of the functional brain organization [5] and thus the degrees of overlaps carry important functional information of brain networks. Finally, the training and testing phases were done for the proposed deep 3D CNN in a 10-fold cross-validation manner, which will be introduced in Sect. 2.4.

## 2.4   Deep 3D CNN Structure

As mentioned above, the input feature for each subject is a 4D matrix with 150 3D network overlap maps. In order to accommodate the subtle spatial characteristics of the input feature volume and to differentiate them, we adopted and improved an effective 3D CNN framework showing the extraordinary ability in accommodating spatial object pattern representation to extract the hierarchy of useful features from the input volumes. The model was built using Theano (http://deeplearning.net/software/theano/) adopting fully 3D convolutional layer from [10]. The structure of the proposed deep 3D CNN is shown in Fig. 4. This powerful deep 3D multichannel convolutional architecture can well incorporate 3D structure information as hierarchical intrinsic features and then be used for classification.

The neural network weights training was performed by the classic Stochastic Gradient Descent (SGD) with momentum. The objective loss function to be optimized is the multinomial negative log-likelihood with a $\lambda$ (set to 0.001) times the $L_2$ norm of the network weights as regularization term, as shown in Eq. (3).

$$L(\theta) = -\frac{1}{m} \sum\nolimits_{i=1}^{m} \sum\nolimits_{j=1}^{k} 1 \cdot \left\{ y^i = j \right\} \log \left( \theta^T x^i \right)_j + \lambda \| \theta_2 \| \tag{3}$$

where $m$ is the number of samples in one batch (empirically set to 12 to fit memory), and $k$ is the number of the output classes (2 classes for control and patient) and

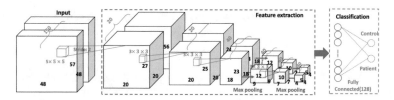

**Fig. 4.** Deep 3D CNN structure for feature extraction and classification. Dimensions of each layer using different colors (green: channel, black: 3D dimension).

$\log\left(\theta^T x^i\right)_j$ is the log-likelihood activation value of the $j_{th}$ output node. The momentum parameter was set to 0.9. In this work, the widely-used dropout technique (with dropout probability 0.2) was adopted for each layer during the training process to reduce the overfitting problem [11]. The convolutional layers were initialized using the similar scheme proposed in [12], and the dense layers were initialized with a Gaussian distribution with $\mu = 0$, $\sigma = 0.01$. Training was performed by utilizing GPU (NVidia Quadro M4000) for 300 epochs. For each training epoch, the total training time is about 2 h. This scale of training time cost makes the proposed 3D CNN framework very suitable for future cognitive and clinical neuroscience applications.

## 3   Results

### 3.1   ASD Patient/Control Classification

Due to the limited sample data, the 3D CNN model may be vulnerable to the overfitting problem. 10-fold cross validation is then operated to make sure the classification accuracy is robust. In total, we have 100 control subjects and 100 ASD subjects, and the 200 subjects were randomly partitioned into 10 equally sized subsamples as testing sets, while the disjoint parts of each testing set are used as the training set. Empirically, the training epochs were set to 300, and learning rate was initially set to 0.01 and then decreased to 0.005 until 66th epoch. It further decreased to 0.001 from 266th epoch to the end. The training accuracy and loss curves for the 10-fold training are shown in Fig. 5 with respect to the number of epochs (300 epochs).

**Fig. 5.** Losses and accuracies during 10-fold training process.

The average, highest and lowest accuracy, sensitivity and specificity are shown in Table 1.

**Table 1.** Accuracy, sensitivity and specificity in ASD/control classification.

| Averaged 10-fold | Accuracy (%) | Sensitivity (%) | Specificity (%) |
|---|---|---|---|
| Using overlapped maps | 70.5 | 74 | 67 |
| Using ICN per se | 45.0 | 51 | 39 |

Among previous research studies on ASD classification using rsfMRI data, accuracy at 60% for ABIDE dataset was reported in [13], and accuracy at $\sim 75\%$ for rsfMRI-based classifiers was reported in [14]. Our method using network overlap information of ICN maps alone from rsfMRI, based on deep 3D CNN classifier, achieved the multi-site robust accuracy of 70.5% on average, and the peak accuracy is 85%. We believe this is a reasonably high accuracy, considering that only network overlap information is used here. In comparison, we also tried using only 144 ICNs as input features from each subject without overlap information using the same CNN structure and training schemes. But using the same CNN structure, the averaged testing classification accuracies were less than 50% (Table 1), which indicates that the 144 ICNs per se are not discriminating between ASD/control groups if spatial ICN overlap information is not incorporated. These results offer a new insight into the abnormal overlaps of functional networks among ASD patient and control groups. In the following section, the analysis on ICN spatial overlap differences between ASD/control groups will be discussed.

## 3.2   Spatial ICN Overlap Differences Between ASD and Controls

The reasonably high accuracy on the 10-fold classification results demonstrated the discriminative features of the ICN's spatial overlap and the ability of spatial pattern description of our proposed deep 3D CNN structure. As previous literature reported [15, 16], regions of the default mode network (DMN) and fusiform gyri are found with decreased functional connectivity among ASD group. It is inspiring that after feature selections by deep 3D CNN described in Fig. 3(c), spatial overlaps involving DMN and fusiform gyrus are found to be dominant only in control group. These features are visualized using randomly selected subjects from both groups in Fig. 6. As we can see, spatial overlaps are substantially stronger in control subjects than ASD subjects, which provides the neuroanatomic substrates for good classification accuracy by CNN.

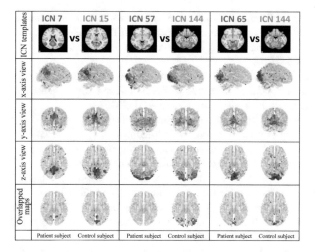

**Fig. 6.** Control-dominant network overlap patterns involving default mode network (ICN 7) and fusiform gyrus (ICN 144).

## 4   Discussion and Conclusion

This work used spatial ICN overlap as features in a deep 3D CNN structure for ASD/control classification and achieved substantially better classification results compared to using only spatial network maps themselves. Our experimental results demonstrated the spatial distribution patterns of connectome-scale functional network maps per se have little discrimination power in differentiating ASD from controls via the CNN framework, while spatial overlap patterns among these connectome-scale networks have remarkable differentiation power. Our work suggested the promise of using CNN deep learning methodologies to discover discriminative and meaningful spatial network overlap patterns and their applications in functional connectomics of brain disorders like ASD in this paper. In the future, additional meaningful features such as the temporal functional interaction patterns among ICNs will be utilized and integrated with the spatial ICN overlap patterns for the purpose of defining more comprehensive, multimodal features to be used to improve the CNN classification accuracy. In addition, improved CNN structures and larger scale of training/testing samples will be explored in the future to further improve the prediction accuracy and to discover potential abnormalities of ICNs' spatial overlaps for diagnosis of brain disorders like ASD.

## References

1. Friston, K.J., et al.: Statistical parametric maps in functional imaging: a general linear approach. Hum. Brain Mapp. **2**, 189–210 (1994)
2. Beckmann, C.F., et al.: Investigations into resting-state connectivity using independent component analysis. Philos. Trans. R. Soc. Lond. B Biol. Sci. **360**, 1001–1013 (2005)

3. Lv, J., et al.: Holistic atlases of functional networks and interactions reveal reciprocal organizational architecture of cortical function. IEEE TBME **62**, 1120–1131 (2015)
4. Mairal, J., et al.: Online learning for matrix factorization and sparse coding. J. Mach. Learn. Res. **11**, 19–60 (2010)
5. Xu, J., et al.: Large-scale functional network overlap is a general property of brain functional organization: reconciling inconsistent fMRI findings from general-linear-model-based analyses. Neurosci. Biobehav. Rev. **71**, 83–100 (2016)
6. Lecun, Y., et al.: Gradient-based learning applied to document recognition RS-SVM reduced-set support vector method. In: Proceedings of IEEE (1998)
7. Krizhevsky, A., et al.: Imagenet classification with deep CNN (2012)
8. Jenkinson, M., et al.: FSL. Neuroimage **62**, 782–790 (2012)
9. Zhao, Y., et al.: Connectome-scale group-wise consistent resting-state network analysis in autism spectrum disorder. Neuroimage: Clin. **12**, 23–33 (2016)
10. Maturana, D., Scherer, S.: VoxNet: a 3D convolutional neural network for real-time object recognition. In: 2015 IEEE/RSJ International Conference on Intelligent Robots and Systems (IROS), pp. 922–928. IEEE (2015)
11. Bell, R.M., Koren, Y.: Lessons from the Netflix prize challenge. ACM SIGKDD Explor. Newsl. **9**, 75 (2007)
12. He, K., et al: Delving deep into rectifiers: surpassing human-level performance on imagenet classification (2015)
13. Nielsen, J.A., et al.: Multisite functional connectivity MRI classification of autism: ABIDE results. Front. Hum. Neurosci. **7**, 599 (2013)
14. Plitt, M., et al.: Functional connectivity classification of autism identifies highly predictive brain features but falls short of biomarker standards. NeuroImage. Clin. **7**, 359–366 (2015)
15. Kleinhans, N.M., et al.: Abnormal functional connectivity in autism spectrum disorders during face processing. Brain **131**, 1000–1012 (2008)
16. Kennedy, D.P., et al.: Failing to deactivate: resting functional abnormalities in autism. Proc. Natl. Acad. Sci. U.S.A. **103**, 8275–8280 (2006)

# Modeling 4D fMRI Data via Spatio-Temporal Convolutional Neural Networks (ST-CNN)

Yu Zhao[1], Xiang Li[2], Wei Zhang[1], Shijie Zhao[3], Milad Makkie[1],
Mo Zhang[4], Quanzheng Li[2,4,5(✉)], and Tianming Liu[1(✉)]

[1] The University of Georgia, Athens, GA 30605, USA
tianming.liu@gmail.com
[2] MGH/BWH Center for Clinical Data Science, Boston, MA 02115, USA
li.quanzheng@mgh.harvard.edu
[3] Northwestern Polytechnical University, Xi'an 710072, Sha'anxi, China
[4] Peking University, Beijing 100080, China
[5] Laboratory for Biomedical Image Analysis, Beijing Institute of Big Data
Research, Beijing 100871, China

**Abstract.** Simultaneous modeling of the spatio-temporal variation patterns of brain functional network from 4D fMRI data has been an important yet challenging problem for the field of cognitive neuroscience and medical image analysis. Inspired by the recent success in applying deep learning for functional brain decoding and encoding, in this work we propose a spatio-temporal convolutional neural network (ST-CNN) to jointly learn the spatial and temporal patterns of targeted network from the training data and perform automatic, pinpointing functional network identification. The proposed ST-CNN is evaluated by the task of identifying the Default Mode Network (DMN) from fMRI data. Results show that while the framework is only trained on one fMRI dataset, it has the sufficient generalizability to identify the DMN from different populations of data as well as different cognitive tasks. Further investigation into the results show that the superior performance of ST-CNN is driven by the jointly-learning scheme, which capture the intrinsic relationship between the spatial and temporal characteristic of DMN and ensures the accurate identification.

**Keywords:** fMRI · Functional brain networks · Deep learning

## 1 Introduction

Recently, analytics of the spatio-temporal variation patterns of functional Magnetic Resonance Imaging fMRI [1] has been substantially advanced through machine learning (e.g. independent component analysis (ICA) [2, 3] or sparse representation [4, 5]) and deep learning methods [6, 7]. As fMRI data are acquired as series of 3D brain volumes during a span of time to capture functional dynamics of the brain, the spatio-

---

Y. Zhao and X. Li—Joint first authors

A. F. Frangi et al. (Eds.): MICCAI 2018, LNCS 11072, pp. 181–189, 2018.
https://doi.org/10.1007/978-3-030-00931-1_21

temporal relationships are intrinsically embedded in the acquired 4D data which need to be characterized and recovered.

In literatures, the spatio-temporal analytics methods can be summarized into two groups: the first group performs the analysis on either spatial or temporal domain based on the corresponding priors, then regress out the variation patterns in the other domain. For example, temporal ICA identifies the temporally independent "signal source" in the 4D fMRI data, then obtains the spatial patterns of those sources through regression. Recently proposed deep learning-based Convolutional Auto-Encoder (CAE) model [8], temporal time series, and spatial maps are regressed later using resulting temporal features. Sparse representation methods, on the other hand, identify the spatially sparse components of the data, while the temporal dynamics of these components are obtained through regression. Works in [9] utilizes Restricted Boltzmann Machine (RBM) for spatial feature analysis ignores the temporal feature.

The second group performs the analysis on spatial and temporal domain simultaneously. For example, [10] applies Hidden Process Model with spatio-temporal "prototypes" to perform the spatio-temporal modeling. Another effective approach to incorporate temporal dynamics (and relationship between time frames) into the network modeling is through Recurrent Neural Network [11]. Inspired by the superior performance and the better interpretability of the simultaneous spatio-temporal modeling, in this work we proposed a deep spatio-temporal convolutional neural network (ST-CNN) to model the 4D fMRI data. The goal of the model is to pinpoint the targeted functional networks (e.g., Default Mode Network DMN) directly from the 4D fMRI data. The framework is based on two simultaneous mappings: the first is the mapping between the input 3D spatial image series and the spatial pattern of the targeted network using a 3D U-Net. The second is the mapping between the regressed temporal pattern of the 3D U-Net output and the temporal dynamics of the targeted network, using a 1D CAE. Summed loss from the two mappings are back-propagated to the two networks in an integrated framework, thus achieving simultaneous modeling of the spatial and temporal domain. Experimental results show that both spatial pattern and temporal dynamics of the DMN can be extracted accurately without hyper-parameter tuning, despite remarkable cortical structural and functional variability in different individuals. Further investigation shows that the framework trained from one fMRI dataset (motor task fMRI) can be effectively applied on other datasets, indicating ST-CNN offers sufficient generalizability for the identification task. With the capability of pin-pointed network identification, ST-CNN can serve as a useful tool for cognitive or clinical neuroscience studies. Further, as the spatio-temporal variation patterns of the data are intrinsically intertwined within an integrated framework, ST-CNN can potentially offer new perspectives for modeling the brain functional architecture.

## 2  Method and Materials

ST-CNN takes 4D fMRI data as input and generates both spatial map and temporal time series of the targeted brain functional network (DMN) as output. Different from CNNs for image classifications (e.g. [12]), ST-CNN consists of a spatial convolution network and a temporal convolution network, as illustrated in Fig. 1(a). The targeted spatial network maps of sparse representation on fMRI data [4] are used to train the spatial network of ST-CNN, while the corresponding temporal dynamics of the spatial networks are used to train the temporal networks.

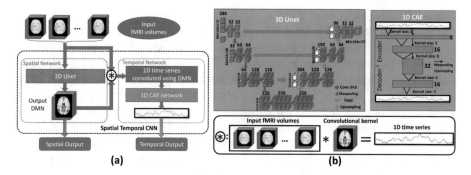

**Fig. 1.** (a). Algorithmic pipeline of ST-CNN; (b). Spatial network structure, temporal network structure, and the combination of the spatial and temporal domain.

## 2.1 Experimental Data and Preprocessing

We use the Human Connectome Project (HCP) Q1 and S900 release datasets [13] for the experiments. Specifically, we use motor task-evoked fMRI (tfMRI) for training the ST-CNN, and test its performance using the motor and emotion tfMRI data from Q1 release and motor task tfMRI data S900 release. The preprocessing pipelines for tfMRI data include skull removal, motion correction, slice time correction, spatial smoothing, global drift removal (high-pass filtering), all implemented by FSL FEAT.

After preprocessing, we apply sparse representation method [4] to decompose tfMRI data into functional networks on both training and testing data sets. The decomposition results consist of both the temporal dynamics (i.e. "dictionary atoms") and spatial patterns (i.e. "sparse weights") of the functional networks. The individual targeted DMN is then manually selected based on the spatial patterns of the resulting networks. The selection process is assisted with sorting the resulting network by their spatial overlap rate with the DMN template (from [14]), measured by Jaccard similarity (i.e. overlap over union). We use the dictionary (1-D time series) of the selected network as ground-truth time series for training the CAE.

## 2.2 ST-CNN Framework

### Spatial Network

The spatial network is inspired from the 2D U-Net [15] for semantic image segmentation. By extending and adapting the 2D classification U-Net to a 3D regression network (Fig. 1(b)), the spatial network takes 4D fMRI data as input, each 3D brain volume along the time frames is assigned to one independent channel. Basically, this 3D U-Net is constructed by a contracting CNN and a expending CNN, where the pooling layers (red arrows in Fig. 1(b)) in the contracting CNN are replaced by upsampling layers (green arrows in Fig. 1(b)). This 3D U-shaped CNN structure contains only convolutional layers without fully connected layers. Loss function for training the spatial network is the mean squared error between the network output which is a 3-D image and the targeted DMN.

**Temporal Network**
The temporal network (Fig. 1(b)) is inspired by the 1-D Convolutional Auto-Encoder (CAE) for fMRI modeling [8]. Both the encoder and decoder of the 1-D CAE have the depth of 3. The encoder starts by taking 1-D signal as input and convolving it with a convolutional kernel size of 3, yielding 8 feature map channels, which are down-sampled using a pooling layer. Then a convolutional layer with kernel size 5 is attached, yielding 16 feature map channels, which are also down-sampled using a pooling layer. The last part of the encoder consists of a convolutional layer with kernel size 8, yielding 32 feature map channels. The decoder takes the output of the encoder as input and symmetrize the encoder as traditional auto-encoder structure. Loss function for training the temporal network is negative Pearson correlation (2) between the temporal CAE output time series with the temporal dynamics of the manually-selected DMN.

$$
\text{Temporal loss} = -\frac{N \sum_1^N xy - \sum_1^N x \sum_1^N y}{\sqrt{\left(N \sum_1^N x^2 - \left(\sum_1^N x\right)^2\right)\left(N \sum_1^N y^2 - \left(\sum_1^N y\right)^2\right)}} \tag{1}
$$

**Combination Joint Operator**
This combination (Fig. 1(b)) procedure connects spatial network and temporal network through a convolution operator. Inputs for the combination are the 4-D fMRI data and 3-D output from the spatial network (i.e. spatial pattern of estimated DMN). The 3-D output will be used as a 3-D convolutional kernel to perform a valid no-padding convolution over each 3-D volume across each time frame of the 4-D fMRI data (3). Since the convolutional kernel size is the same as each 3D brain volume along the 4th (time) dimension, the no-padding convolution will result in a single value at each time frame, thus forming a time series for the estimated DMN. This output time series $ts$ will be used as the input for temporal 1-D CAE, as described above.

$$
ts \in \mathbb{R}^{T \times 1} = \{t_1, t_2, \ldots, t_T | t_i = V_i * DMN \in \mathbb{R}\}, \tag{2}
$$

where $t_i$ is the convolution result at each time frame, $V_i$ is the 3-D fMRI volume at time frame $i$, and DMN is the 3-D spatial network output used as convolution kernel.

## 2.3   Training Process and Model Convergence

Since the temporal network will rely on the DMN spatial map from the spatial network, we split the training process into 3 stages: at the first stage, only spatial network is trained (Fig. 2(a)); at the second stage, temporal network is trained based on the spatial network results (Fig. 2(b)); and finally, the entire ST-CNN is trained for fine-tuning (Fig. 2(c)). As we can see from Fig. 2, the temporal network converges much faster (around 10 times faster) than the spatial network. Thus during the fine-tuning stage, the loss function for ST-CNN is a weighted sum (10:1) of both spatial and temporal loss.

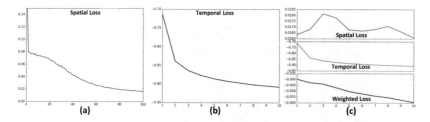

**Fig. 2.** Training losses (y-axis) versus training steps (x-axis). (a). first stage spatial network training loss; (b). second stage temporal network training loss; (c). fine-tuning training loss.

### 2.4 Model Evaluation and Validation

We firstly calculate the spatial overlap rate between the spatial pattern of ST-CNN output and a well-established DMN template to evaluate the performance of spatial network. We then calculate the Pearson correlation of the output time series with ground-truth time series from sparse representation results to evaluate the temporal network. Finally we utilize a supervised dictionary learning method [16] to reconstruct the spatial patterns of the network based on temporal network result to investigate whether the spatio-temporal relationship is correctly captured by the framework.

## 3 Results

We use 52 subjects' motor tfMRI data from HCP Q1 release for training the ST-CNN. We test the same trained network on three datasets: (1) motor tfMRI data from the rest of 13 subjects. (2) motor tfMRI data from 100 randomly -selected subjects in the HCP S900 release. (3) emotion tfMRI data from 67 subjects from HCP Q1 release. Testing results show consistently good performance for DMN identification, demonstrating that trained network is not limited to specific population and specific cognitive tasks.

### 3.1 MOTOR Task Testing Results

The trained ST-CNN is tested on 2 different motor task datasets: 13 subjects from HCP Q1 and 100 subjects from HCP S900, respectively. As shown in Fig. 3, the resulting spatial and temporal patterns are consistent with the ground-truth. Quantitative analyses shown in Table 1 demonstrates that the ST-CNN performs better than sparse representation method, although it is trained from the manually-selected results of sparse representation. The rationale is that the ST-CNN can better adapt to the input data by the co-learned spatial and temporal networks, while sparse representation relies on the simple sparsity prior which can be invalid in certain cases. As shown in Fig. 4, sparse representation cannot identify DMN from certain subjects while ST-CNN can. In HCP Q1 dataset, we have observed 20% (13 out of 65 subjects) of cases where sparse representation fails while ST-CNN succeeds. Considering the fact that DMN is supposed to be consistently presented in the functioning brain regardless of task, this is an intriguing and desired characteristic of the ST-CNN model.

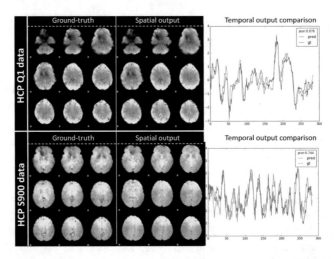

**Fig. 3.** Examples of comparisons between ST-CNN outputs and ground-truth from sparse representation. Here we showed 2 subjects' comparison results from two different datasets (1 HCP Q1 subjects and 1 HCP S900 subjects). Spatial maps are very similar and time series have Pearson correlation coefficient values 0.878 in HCP Q1 data, and 0.744 in HCP S900 data. Red curves are ground-truth. Blue curves are ST-CNN temporal outputs.

### 3.2    EMOTION Task Testing Results

The 67 subjects' emotion task-evoked fMRI data (HCP Q1) were further tested to demonstrate that our trained network based on motor task is not prone to specific cognitive tasks. The ability to extract DMN both spatially and temporally of our framework showed that the intrinsic features of DMN were well captured. As shown in Fig. 5, the spatial maps resemble with the ground-truth sparse representation results and so do the temporal outputs. Quantitative analyses in Table 1 showed that our outputs also had larger spatial overlap with DMN templates than outputs from sparse representation. The temporal outputs were also shown accurate, with an average Pearson correlation coefficient of 0.51.

### 3.3    Spatial Output and Temporal Output Relationship

For further validation, supervised sparse representation [16] is applied on 13 testing subjects' HCP Q1 motor task fMRI data. We set the temporal output of ST-CNN as predefined dictionary atoms to obtain the sparse representation on the data by learning the rest of the dictionaries. The resulting network corresponding to the predefined atom,

**Table 1.** Performance of ST-CNN measured by spatial overlap rate

| Datasets | Spatial overlap with DMN template | | Temporal similarity (Pearson correlation) |
|---|---|---|---|
| | Sparse representation | ST-CNN | |
| HCP Q1 MOTOR (13 subjects) | 0.115 | **0.172** | 0.55 |
| HCP S900 MOTOR (100 subjects) | **0.070** | 0.066 | 0.53 |
| HCP Q1 EMOTION (67 subjects) | 0.095 | **0.168** | 0.51 |

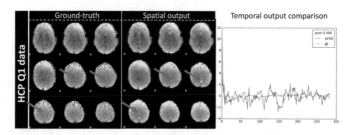

**Fig. 4.** Example of the better DMN identification of ST-CNN than sparse representation (denoted by red arrows). The temporal dynamics of the two networks are also different, where output from sST-CNN (blue) are more reasonable.

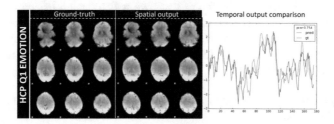

**Fig. 5.** Example of ST-CNN outputs and ground-truth (sparse representation) for EMOTION task. Spatial maps are very similar and time series have Pearson correlation 0.754. Red curve is ground-truth, blue curve is the temporal output by ST-CNN.

which has the fixed temporal dynamics during the learning, are compared with ST-CNN spatial outputs. We found that the temporal output of ST-CNN can lead to an accurate estimation of the DMN spatial patterns as in Fig. 6. The average spatial overlap rate between the supervised results and ST-CNN spatial output is 0.144, suggesting that the spatial output of ST-CNN has close relationship with its temporal output.

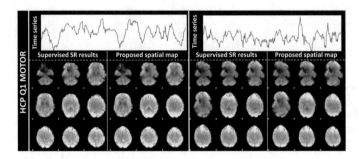

**Fig. 6.** Comparison of the spatial maps between ST-CNN results and supervised sparse representation which takes temporal output of ST-CNN as pre-defined atoms.

## 4    Discussion

In this work, we proposed a novel spatio-temporal CNN model to identify functional networks (DMN as an example) from 4D fMRI data modelling. The effectiveness of ST-CNN is validated by the experimental results on different testing datasets. From an algorithmic perspective, the result shows that ST-CNN embeds the spatial-temporal variation patterns of the 4D fMRI signal into the network, rather than learns the matrix decomposition process by the sparse representation. It is then very important to further refine the framework by training it over DMNs identified by other methods (such as temporal ICA). More importantly, we use DMN as a sample targeted network in the current work, since it should be present in virtually any fMRI data. As detecting the absence/disruption a functional network is as important as identifying it (e.g. for AD/MCI early detection), in the future work we will focus on extending the current framework to pinpoint more functional networks, including task-related networks which should be presented in a limited range of datasets. We will also test ST-CNN on fMRI from abnormal brains for its capability of characterizing the spatio-temporal patterns of the disrupted DMNs.

**Acknowledgments.** Quanzheng Li is supported in part by the National Institutes of Health under Grant R01AG052653.

## References

1. Heeger, D.J., Ress, D.: What does fMRI tell us about neuronal activity? Nat. Rev. Neurosci. **3**, 142–151 (2002)
2. Cole, D.M., Smith, S.M., Beckmann, C.F.: Advances and pitfalls in the analysis and interpretation of resting-state FMRI data. Front. Syst. Neurosci. **4**, 8 (2010)
3. McKeown, M.J., et al.: Independent component analysis of functional MRI: what is signal and what is noise? Curr. Opin. Neurobiol. **13**, 620–629 (2003)
4. Lv, J., et al.: Sparse representation of whole-brain fMRI signals for identification of functional networks. Med. Image Anal. **20**, 112–134 (2015)

5. Zhao, S., et al.: Decoding auditory saliency from brain activity patterns during free listening to naturalistic audio excerpts. Neuroinformatics **16**, 309–324 (2018)
6. Litjens, G., et al.: A survey on deep learning in medical image analysis. Med. Image Anal. **42**, 60–88 (2017)
7. Zhao, Y., et al.: Constructing fine-granularity functional brain network atlases via deep convolutional autoencoder. Med. Image Anal. **42**, 200–211 (2017)
8. Huang, H., et al.: Modeling task fMRI data via deep convolutional autoencoder. IEEE Trans. Med. Imaging. 1 (2017)
9. Hjelm, R.D., et al.: Restricted Boltzmann machines for neuroimaging: an application in identifying intrinsic networks. Neuroimage **96**, 245–260 (2014)
10. Shen, Y., et al.: Spatial–temporal modelling of fMRI data through spatially regularized mixture of hidden process models. Neuroimage **84**, 657–671 (2014)
11. Hjelm, R.D., Plis, S.M., Calhoun, V.: Recurrent neural networks for spatiotemporal dynamics of intrinsic networks from fMRI data. In: NIPS: Brains and Bits (2016)
12. Krizhevsky, A., et al.: Imagenet classification with deep convolutional neural network (2012)
13. Barch, D.M., et al.: Function in the human connectome: task-fMRI and individual differences in behavior. Neuroimage **80**, 169–189 (2013). WU-Minn HCP Consortium
14. Smith, S.M., et al.: Correspondence of the brain's functional architecture during activation and rest. Proc. Natl. Acad. Sci. U.S.A. **106**, 13040–13045 (2009)
15. Ronneberger, O., Fischer, P., Brox, T.: U-Net: convolutional networks for biomedical image segmentation. In: Navab, N., Hornegger, J., Wells, W.M., Frangi, A.F. (eds.) MICCAI 2015. LNCS, vol. 9351, pp. 234–241. Springer, Cham (2015). https://doi.org/10.1007/978-3-319-24574-4_28
16. Zhao, S., et al.: Supervised dictionary learning for inferring concurrent brain networks. IEEE Trans. Med. Imaging **34**, 2036–2045 (2015)

# The Dynamic Measurements of Regional Brain Activity for Resting-State fMRI: d-ALFF, d-fALFF and d-ReHo

Chao Tang, Yuqing Wei, Jiajia Zhao, and Jingxin Nie$^{(\boxtimes)}$

School of Psychology, Center for Studies of Psychological Application,
Institute of Cognitive Neuroscience, South China Normal University,
Guangzhou 510631, China
niejingxin@gmail.com

**Abstract.** The human brain is always in the process of constantly changing. Given that the conventional measurements of regional brain activity are less sensitive to changes over time, three dynamic measurements were proposed to capture the temporal variability of regional brain activity. In this study, dynamic amplitude of low frequency fluctuation (d-ALFF), dynamic fractional amplitude of low frequency fluctuation (d-fALFF) and dynamic regional homogeneity (d-ReHo) were obtained from resting-state functional magnetic resonance imaging (rs-fMRI) of both 238 ADHD and 239 typical developing (TD) subjects. Then, they were applied to detecting the regional activity differences between ADHD and TD group. Compared with the conventional measurements (ALFF, fALFF and ReHo), the dynamic measurements were more sensitive in exploring the differences of regional brain activity between ADHD and TD group. The three new measurements not only enrich the diversity of methods for investigating the dynamic variation of regional brain activity, but also emphasize the significance of detecting the temporal variability of regional brain activity.

**Keywords:** rs-fMRI · Temporal variability · d-ALFF · d-fALFF and d-ReHo

## 1 Introduction

The investigation for dynamic brain activity from resting-state fMRI is mainly based on two patterns: the dynamics of spontaneous brain activities and the dynamics of functional interconnections between spontaneous brain activities [1]. Presently, most of studies for dynamic brain activity are concentrated on exploring the dynamic functional connectivity which mainly involve in the flexible and adjustable configuration between brain networks [2–4]. Several approaches have been proposed for probing time-varying patterns in fMRI imaging data such as sliding window analysis [5], time-frequency analysis [6], point process analysis [7] and temporal graph analysis [8]. Especially, the most common method used for describing brain network dynamics is sliding windows analysis with temporal resolution from seconds to minutes.

Traditionally, voxel-wise methods such as ALFF, fALFF and ReHo have been widely applied to exploring the regional brain activities [9–11]. However, these

© Springer Nature Switzerland AG 2018
A. F. Frangi et al. (Eds.): MICCAI 2018, LNCS 11072, pp. 190–197, 2018.
https://doi.org/10.1007/978-3-030-00931-1_22

measurements don't perfectly capture the dynamic characteristics of regional brain activity over different times.

In view of the limitations of conventional measurements, we proposed three novel dynamic measurements (d-ALFF, d-fALFF and d-ReHo) by combining the sliding window technique with the three conventional measurements (ALFF, fALFF and ReHo) to investigate the temporal variability of voxel-wise brain activity. We hypothesized that our methods were more sensitive than conventional static measurements in detecting the differences of regional brain activity between ADHD and TD group. To verify the effectiveness of our methods, we compared the results obtained by the new measurements with those obtained by the conventional measurements.

## 2    Materials and Methods

### 2.1    Subjects

The resting-state fMRI and structural MRI image data of 238 ADHD and 239 TD were obtained from ADHD-200 [12] and Autism Brain Image Date Exchange (ABIDE) [13] across multiple independent imaging sites. Specially, only the data of typically developing control in the ABIDE were used in this study. For both ADHD and TD subjects, the inclusion criteria included: no history of neurological disease and no diagnosis of either schizophrenia or affective disorder, image at least cover 95% of brain, IQ scores >80. More details of demographic information of these participants were listed in the Table 1.

**Table 1.** Demographic information.

| Sites | ADHD (n = 238) | | | TD (n = 239) | | |
|---|---|---|---|---|---|---|
|  | N | Age (mean) | Gender (M/F) | N | Age (mean) | Gender (M/F) |
| KKI | 17 | 9.64 | 11/6 | 40 | 9.88 | 24/16 |
| NYU | 119 | 10.24 | 88/31 | 114 | 11.86 | 68/46 |
| Peking | 93 | 11.58 | 81/12 | 85 | 11.08 | 50/35 |
| NeuroIMAGE | 9 | 14.67 | 9/0 | | | |
| Total | 238 | 10.89 | 189/49 | 239 | 11.26 | 142/97 |

### 2.2    Data Pre-processing

T1 and resting-state fMRI data were preprocessed with DPARSF [14]. The first ten volumes of rs-fMRI images were discarded for scanner calibration. All images were corrected for within-scan acquisition time differences between slices and were realigned to the middle volume to cut down on inter-scan head motions. Then, these images were registered onto the Montreal Neurological Institute (MNI) standard template and were subsequently resampled to 3 mm isotropic resolution. Spatial smoothing was applied with a Gaussian kernel of 8 mm full-width at half-maximum

(FWHM) to improve the signal-to-noise ratio (SNR). Furthermore, the mean signal of white matter and cerebrospinal fluid were removed as covariates. Linear trend removal coupled with band-pass filtering (0.01–0.1 Hz) were also performed.

## 2.3   Dynamic Measurements

As shown in Fig. 1, our method mainly consists of two parts: segmentation of time windows (Fig. 1A) and calculation of three dynamic measurements (Fig. 1B). In the first part, taking one rs-fMRI image as an example, we firstly divided all BOLD time series of $k$ voxels across the whole brain into $n$ overlapping windows ($W_1$, $W_2$, $W_3$... $W_{n-1}$, $W_n$) each with specific length $l$ and interval $d$ between time windows. In this study, we chose 50 s and 30 s as the length of time window and interval between time windows respectively. In the second part, in order to explore the temporal variability of regional brain activity, we proposed three new measurements dynamic Amplitude of Low Frequency Fluctuation (d-ALFF), dynamic fractional Amplitude of Low Frequency Fluctuation (d-fALFF) and dynamic Regional Homogeneity (d-ReHo) based on the combination of the time windows technique and the conventional measurements (ALFF, fALFF and ReHo).

### d-ALFF Measurement

d-ALFF was calculated based on several different time windows which contained the time series over a specific period of time. Briefly, for a given voxel, the time series was first transformed to the frequency domain using a Fast Fourier Transform (FFT) and the power spectrum was obtained. Subsequently, the square root of the power spectrum was calculated and then averaged over a prespecified frequency band (0.01–0.08 Hz) and the averaged square root was known as ALFF at the given voxel. After calculating ALFF of all voxel in time windows, each participant will get several window-based ALFF maps. Then, we computed the mean and standard deviation of each voxel in all window-based ALFF maps for each subject and further got the corresponding coefficient of variation (CV) which was acquired by dividing the standard deviation by the mean. To better measure the dynamic variation of regional brain activity between different individuals, we used CV as d-ALFF, which represented the temporal variability of absolute energy consumption in low-frequency regional brain activity.

### d-fALFF Measurement

d-fALFF represented the degree of changes over a period of time in a ratio of the power of each frequency at the low-frequency range (0.01–0.08 Hz) to that of the entire frequency range (0–0.25 Hz). Firstly, the linear trend of time series was eliminated. Secondly, the time series for each voxel were transformed to a frequency domain without band-pass filtering. Thirdly, the square root was calculated at each frequency of the power spectrum with FFT. Finally, the sum of amplitude in the 0.01–0.08 Hz frequency range was divided by that in the 0–0.25 Hz frequency range. Similar to d-ALFF, for each subject, the CV value calculated from the mean and standard deviation of multiple windows-based fALFF maps was regarded as the indication of temporal variability of relative energy consumption of regional brain at low frequency band.

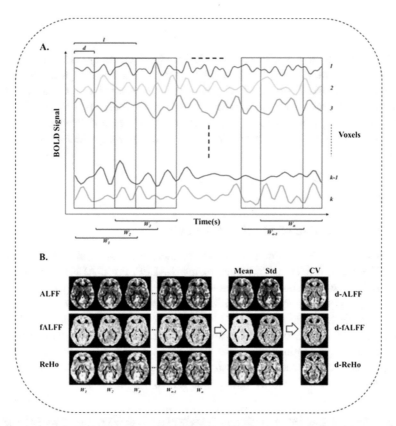

**Fig. 1.** The flow chart of dynamic analysis of regional brain activity. (A) Segmentation of time windows. The colored curve represented the time series across the whole brain and each time series indicated the BOLD signal of each voxel in a participant. The time series were segmented into different time windows and further analysis was constructed on the segmented windows. (B) Calculation of three dynamic measurements. The dynamic measurements (d-ALFF, d-fALFF and d-ReHo) were represented by the coefficient of variation (CV) of the static measurements (ALFF, fALFF and ReHo) calculated from different time windows. The CV was obtained from the mean of static measurements over different time windows divided by corresponding standard deviation (Std).

### d-ReHo Measurement

Similar to ReHo, d-ReHo was also based on the hypothesis that a given voxel's activity was usually correlated to that of its neighbors. Firstly, we got ReHo value of all voxels across different time windows by calculating the Kendall coefficient of concordance (KCC) of time series of a voxel with those of its nearest 26 neighbors. Then, after calculating the mean and standard deviation of each voxel in all time windows, we considered CV value which was equal to the standard deviation divided by the mean as the measurement of temporal variability of regional brain activity, which characterized the dynamic changes of local synchronization of spontaneous fMRI signals within a cluster.

## 3  Statistics Analysis

In order to reduce the effect of noise, spatial smoothing was again carried out with a Gaussian kernel of 8 mm FWHM before further statistics analysis. The dynamic measurements (d-ALFF, d-fALFF and d-ReHo) of regional brain activity between ADHD and TD group were compared by two-sample t-test on each voxel, taking a significant threshold of P < 0.001 (uncorrected), with age and gender as covariates. Voxels with P < 0.001 and cluster size > 540 mm$^3$ were regarded as the significant group differences.

## 4  Results

### 4.1  The Comparison of ALFF and d-ALFF

As shown in the Fig. 2, compared to ALFF, there were more significant group differences between ADHD and TD group using d-ALFF. For d-ALFF, compared to TD group, ADHD group showed increased d-ALFF in the right supramarginal gyrus (SMG), right putamen, right precentral gyrus, left inferior temporal gyrus (ITG), bilateral middle temporal gyrus (MTG) as well as cerebellum, and decreased d-ALFF in the right medial superior frontal gyrus (mSFG), right thalamus, right fusiform, right insula, right posterior cingulate gyrus (PCC), left superior temporal pole, left hippocampus and cerebellum. In contrast, when using ALFF, no significant group difference between ADHD and TD group was found.

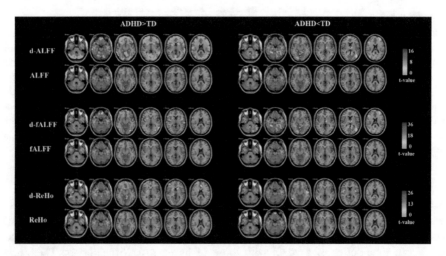

**Fig. 2.** The comparison of static (ALFF, fALFF, ReHo) and dynamic (d-ALFF, d-fALFF, d-ReHo) measurements for regional brain activity between ADHD and TD group. The first and second rows show the group differences between ADHD and TD group using d-ALFF and ALFF respectively. The third and fourth rows show the group differences between ADHD and TD group using d-fALFF and fALFF respectively. The fifth and sixth rows show the group differences between ADHD and TD group using d-ReHo and ReHo respectively.

## 4.2    The Comparison of fALFF and d-fALFF

Clearly, d-fALFF was more sensitive than fALFF in discovering group differences between ADHD and TD group. Using d-fALFF, compared to TD group, ADHD group exhibited obvious increase in the right ITG, bilateral MTG, bilateral putamen, bilateral temporal pole and cerebellum as well as decrease in the left precuneus, bilateral middle frontal gyrus (MFG), bilateral middle frontal gyrus (SFG), right fusiform, right middle occipital gyrus (MOG) and cerebellum. However, using fALFF, there were only a small amount of group differences between ADHD and TD group, such as left lingual gyrus, right angular gyrus, right inferior frontal gyrus (IFG) and cerebellum. More details were shown in the Fig. 2.

## 4.3    The Comparison of ReHo and d-ReHo

As we can see, there were more significant differences between ADHD and TD group using d-ReHo than those of using ReHo (Fig. 2). Compared with TD group, the increased d-ReHo in ADHD group appeared at the left caudate, left orbital frontal gyrus (OFG), left anterior cingulate gyrus (ACC), left angular gyrus, left fusiform, left thalamus, right MOG, left supramarginal gyrus and cerebellum. Besides, the deceased d-ReHo in ADHD group appeared at the left ITG, bilateral MTG, bilateral postcentral gyrus, left insula and right IFG. Nonetheless, no clusters showed increased ReHo in ADHD group and the bilateral IFG, MFG, left superior parietal gyrus (SPG) as well as cerebellum exhibited decreased ReHo in ADHD group compared with TD group.

# 5    Discussion

In this study, three dynamic measurements (d-ALFF, d-fALFF and d-ReHo) were primarily constructed on the combination of the time windows technique and the conventional measurements (ALFF, fALFF and ReHo) which were widely recognized as effective methods to measure regional brain activity in voxel-wise way. For one thing, d-ALFF and d-fALFF described the temporal changes of energy consumption according to the level of oxygen content in spontaneous brain activity. For another, d-ReHo revealed the dynamic variation of functional consistency in local regions by calculating the correlation between time series. Furthermore, we compared the new measurements with the conventional measurements by examining the sensibility of these measurements to distinguish the different regional brain activity between ADHD patients and health people. As predicted, we easily observed more significant differences between ADHD and TD group using the new measurements than those using the conventional measurements. These results were basically consistent with previous anatomical MRI studies [15–17], indicating that the involved regions of abnormal regional brain activities of ADHD patients were awfully wide such as frontal lobe, ACC, caudate and cerebellum.

In summary, the new measurements were suitable to exploring the dynamic changes of regional brain activity between ADHD and TD group. The results showed that the new approaches were more sensitive and stable than the conventional approaches. Although there were a wide range of differences between ADHD and TD group in regional brain activity using new measurements, the effectiveness of results still need further study. For example, the length of each time window and interval between time windows may have a critical influence on the temporal variability of regional brain activity. Besides, although the brain activity in low-frequency band (<0.1 Hz) has been extensively studied [18–20], the changeable and various brain activity is not confined to low-frequency band [21, 22]. Therefore, it is worthwhile to explore the temporal variability of distinct types of brain activity in different frequency ranges. In the aspect of practicability, the three dynamic measurements also need to be retested on more disease studies such as autism and schizophrenia. Altogether, these new measurements not only enrich the diversity of research methods for detecting the regional brain activity, but also contribute to understanding of dynamic measurements.

**Acknowledgements.** This paper was supported by NFSC (National Natural Science Foundation of China) (Grant No. 61403148).

# References

1. Fu, Z., Tu, Y., Di, X., Biswal, B.B., Calhoun, V.D., Zhang, Z.: Associations between functional connectivity dynamics and BOLD dynamics are heterogeneous across brain networks. Front. Hum. Neurosci. **11**, 593 (2017)
2. Zalesky, A., Fornito, A., Cocchi, L., Gollo, L.L., Breakspear, M.: Time-resolved resting-state brain networks. Proc. Natl. Acad. Sci. U.S.A. **111**, 10341–10346 (2014)
3. Hutchison, R.M., et al.: Dynamic functional connectivity: promise, issues, and interpretations. NeuroImage **80**, 360–378 (2013)
4. Di, X., Fu, Z., Chan, S.C., Hung, Y.S., Biswal, B.B., Zhang, Z.G.: Task-related functional connectivity dynamics in a block-designed visual experiment. Front. Hum. Neurosci. **9**, 543 (2015)
5. Hindriks, R., et al.: Can sliding-window correlations reveal dynamic functional connectivity in resting-state fMRI? NeuroImage **127**, 242–256 (2016)
6. Chang, C., Glover, G.H.: Time-frequency dynamics of resting-state brain connectivity measured with fMRI. NeuroImage **50**, 81–98 (2010)
7. Tagliazucchi, E., Siniatchkin, M., Laufs, H., Chialvo, D.R.: The voxel-wise functional connectome can be efficiently derived from co-activations in a sparse spatio-temporal point-process. Front. Neurosci. **10**, 381 (2016)
8. Betzel, R.F., Fukushima, M., He, Y., Zuo, X.N., Sporns, O.: Dynamic fluctuations coincide with periods of high and low modularity in resting-state functional brain networks. NeuroImage **127**, 287–297 (2016)
9. Zang, Y.F., et al.: Altered baseline brain activity in children with ADHD revealed by resting-state functional MRI. Brain Dev. **29**, 83–91 (2007)
10. Zou, Q.H., et al.: An improved approach to detection of amplitude of low-frequency fluctuation (ALFF) for resting-state fMRI: fractional ALFF. J. Neurosci. Methods **172**, 137–141 (2008)
11. Zang, Y., Jiang, T., Lu, Y., He, Y., Tian, L.: Regional homogeneity approach to fMRI data analysis. NeuroImage **22**, 394–400 (2004)
12. Consortium HD: The ADHD-200 Consortium: a model to advance the translational potential of neuroimaging in clinical neuroscience. Front. Syst. Neurosci. **6**, 62 (2012)

13. Di Martino, A., et al.: The autism brain imaging data exchange: towards a large-scale evaluation of the intrinsic brain architecture in autism. Mol. Psychiatry **19**, 659–667 (2014)
14. Yan, C.G., Wang, X.D., Zuo, X.N., Zang, Y.F.: DPABI: data processing & analysis for (resting-state) brain imaging. Neuroinformatics **14**, 339–351 (2016)
15. Mostofsky, S.H., Cooper, K.L., Kates, W.R., Denckla, M.B., Kaufmann, W.E.: Smaller prefrontal and premotor volumes in boys with attention-deficit/hyperactivity disorder. Biol. Psychiatry **52**, 785–794 (2002)
16. Castellanos, F.X., Tannock, R.: Neuroscience of attention-deficit/hyperactivity disorder: the search for endophenotypes. Nat. Rev. Neurosci. **3**, 617–628 (2002)
17. Mackie, S., et al.: Cerebellar development and clinical outcome in attention deficit hyperactivity disorder. Am. J. Psychiatry **164**, 647–655 (2007)
18. Biswal, B., Yetkin, F.Z., Haughton, V.M., Hyde, J.S.: Functional connectivity in the motor cortex of resting human brain using echo-planar MRI. Magn. Reson. Med. **34**, 537–541 (1995)
19. Mantini, D., Perrucci, M.G., Del Gratta, C., Romani, G.L., Corbetta, M.: Electrophysiological signatures of resting state networks in the human brain. Proc. Natl. Acad. Sci. U.S.A. **104**, 13170–13175 (2007)
20. Logothetis, N.K., Pauls, J., Augath, M., Trinath, T., Oeltermann, A.: Neurophysiological investigation of the basis of the fMRI signal. Nature **412**, 150–157 (2001)
21. Trapp, C., Vakamudi, K., Posse, S.: On the detection of high frequency correlations in resting state fMRI. NeuroImage **164**, 202–213 (2018)
22. Buzsaki, G., Draguhn, A.: Neuronal oscillations in cortical networks. Science **304**, 1926–1929 (2004)

# rfDemons: Resting fMRI-Based Cortical Surface Registration Using the BrainSync Transform

Anand A. Joshi[(✉)], Jian Li, Minqi Chong, Haleh Akrami,
and Richard M. Leahy

University of Southern California, Los Angeles, CA, USA
ajoshi@usc.edu

**Abstract.** Cross subject functional studies of cerebral cortex require cortical registration that aligns functional brain regions. While cortical folding patterns are approximate indicators of the underlying cytoarchitecture, coregistration based on these features alone does not accurately align functional regions in cerebral cortex. This paper presents a method for cortical surface registration (rfDemons) based on resting fMRI (rfMRI) data that uses curvature-based anatomical registration as an initialization. In contrast to existing techniques that use connectivity-based features derived from rfMRI, the proposed method uses 'synchronized' resting rfMRI time series directly. The synchronization of rfMRI data is performed using the BrainSync transform which applies an orthogonal transform to the rfMRI time series to temporally align them across subjects. The rfDemons method was applied to rfMRI from the Human Connectome Project and evaluated using task fMRI data to explore the impact of cortical registration performed using resting fMRI data on functional alignment of the cerebral cortex.

## 1 Introduction

Group structural and functional studies of brain imaging data require registration across a population in order to draw inferences at finer scales. For studies involving the cerebral cortex it is often sufficient to perform this registration with respect to a 2D parameterization of the cortical surface. Most cortical surface registration methods are guided either by sulcal and gyral landmarks or curvature maps that reflect cortical folding [1,2]. The resulting registrations are appropriate for quantifying structural characteristics across populations, but there is ample evidence that regions of functional specialization are not accurately aligned across subjects using only anatomical landmarks [3]. Poor alignment can result in reduced statistical power when regions of functional activation do not accurately align. The common practice of spatial smoothing can overcome this problem to some degree, but limits our ability to localize and detect effects at finer scales. Functional regions can be better identified or aligned using a series of functional localizers as is common in fMRI studies of the visual system [4]. But this task-driven approach is

---

This work is supported by the following grants: R01 NS074980, R01 NS089212.

A. F. Frangi et al. (Eds.): MICCAI 2018, LNCS 11072, pp. 198–205, 2018.
https://doi.org/10.1007/978-3-030-00931-1_23

limited by the number of regions that can be mapped in each subject using a discrete set of tasks. A more general approach uses data from subjects watching a movie to drive alignment of the entire cerebral cortex [5]. Another alternative is to use resting fMRI (rfMRI) data. While there is evidence for involvement of a large fraction of cerebral cortex in resting activity, using rfMRI for intersubject alignment presents a challenge because resting time-series cannot be directly compared across subjects as is the case for task fMRI.

A recent series of papers have used the concept of hyperalignment to better align functional data across subjects [6]. The main idea is to use a linear transformation on the data from each subject to maximize the similarity of response profiles in a set of task data. This can be viewed as a method for spatial alignment that does not enforce any topological restrictions on the spatial mapping but instead uses linear combinations of the data from a local neighborhood to produce a spatial inter-subject correspondence. Here we explore an alternative to this approach in which we use a topologically-constrained nonrigid deformation of the cortical surfaces to perform inter-subject registration. An alternative recent method also used rfMRI data for this purpose [7]. In that case, z-score maps derived from ICA analysis of the functional activations were used to perform group-wise cortical surface registration. Spectral features derived from fMRI connectivity matrices have also been used for driving cortical registration [5,8].

An orthogonal transform termed BrainSync performs synchronization of time-series across subjects at homologous locations in the brain [9]. The transform exploits the correlation structure common across subjects to perform the synchronization. When synchronized, rfMRI signals become approximately equal at homologous locations across subjects. The BrainSync transform is lossless and preserves correlation structures. As a result of synchronization we can directly compare time series across subjects. We can then use the aligned time series themselves as a feature to induce functional correspondence through nonrigid registration of the surfaces. This new approach to functional alignment is described here. Starting with the anatomically registered cortical surfaces, each cortical hemispheres is first mapped to a unit square flat map. The rfMRI data are then mapped to these squares and used as features to coregister across subjects using a modified demons algorithm. The distortion in the flat mapping is compensated for using the metric tensor determinant. The method is evaluated using resting and task data from the HCP project database.

## 2   Materials and Methods

As input, we assume structural and rfMRI images for each subject. The structural images are preprocessed to generate cortical surface representations and coregistered to a common atlas. The fMRI data are preprocessed using HCPs minimal processing pipeline [10]. They are then mapped to a common atlas using the mapping computed from the structural images. This preprocessing results in structurally coregistered $V \times T$ data matrices, one per subject, with $V$ vertices in the cortical surface mesh and $T$ time points. We refine this intersubject alignment using the rfMRI by first synchronizing their time-series with Brain-Sync, and then use these time-series as features in the alignment algorithm.

**Flat Mapping and Metric Computation**: We generate a flat map of the cortical surface mesh of the atlas to which the fMRI data have been mapped for each subject. A harmonic map is computed on the unit square for each hemisphere such that the inter-hemispheric fissure that divides the two hemispheres is mapped to the boundary of the square and rest of the surface is mapped to its interior. The fMRI data is resampled onto a $256 \times 256$ regular grid on the square for each hemisphere using linear interpolation, resulting in a $256 \times 256 \times T$ fMRI data representation. For fast computation, the dimensionality of the dataset was reduced using PCA to 20 (chosen based on rank analysis of the data). The SVD of a reference subject's $V \times T$ fMRI data was used to compute a set of 20 temporal basis functions onto which the subject's synchronized fMRI data was projected to produce data of size $256 \times 256 \times 20$. The 3D surface coordinates of the cortical mesh were also resampled to the regular grid and the metric tensor computed using finite differences (Fig. 1).

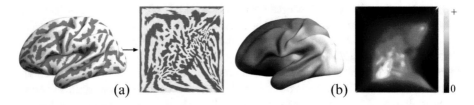

**Fig. 1.** (a) Mean curvature shown on the cortical surface and its flat harmonic map to the unit square; (b) the determinant of the metric tensor induced by the flat harmonic map is shown on the cortex and the flat map.

**rfDemons: rfMRI-Based Cortical Registration**:
We start with Brain-Sync transformed rfMRI data for two individual datasets, with one designated as the 'reference' (or fixed) and the other as the 'subject' (or moving). Each is represented on the square atlas map as a $256 \times 256 \times 20$ data matrix.

Let $F(\boldsymbol{x}(\tau), t)$ denote the rfMRI data at the $t^{\text{th}}$ fMRI time point, at spatial location $\boldsymbol{x}$ in the

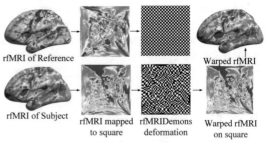

**Fig. 2.** Flowchart of rfDemons algorithm: the cortical surface is mapped to the unit square. The corresponding synchronized rfMRI datasets for subject and reference are then re-sampled on the square. The deformation field that registers the rfMRI data is computed using the metric-compensated symmetrized demons algorithm. The deformation is then applied to the subject rfMRI data to map it back to the cortical surface mesh.

square with $\boldsymbol{x}(\tau)$ a set of deformation fields, parameterized by $\tau$. We define $F(\boldsymbol{x}(\tau), t)$ to be the rfMRI data of the subject when $\tau = 0$ and the subject optimally matched to the reference when $\tau = 1$, so that the subject data are gradually deformed to the reference as $\tau$ goes from 0 to 1. Since the input rfMRI data is synchronized, we can use an optical flow formulation $F(\boldsymbol{x}(\tau), t) = F(\boldsymbol{x}, t)$ [11], i.e. the data are constant with respect to $\tau$, and it is possible to match the data from subject to reference with an appropriate deformation map $\boldsymbol{x}(1)$. Taking the derivative of $F$ with respect to $\tau$, we get $\frac{dF}{d\tau}(\boldsymbol{x}(\tau), t) = \nabla_{\boldsymbol{x}} F(\boldsymbol{x}(\tau), t) \cdot v(\boldsymbol{x})$ where $v(\boldsymbol{x}) = \frac{d\boldsymbol{x}(\tau)}{d\tau}$ is the velocity of point $\boldsymbol{x}$ and $\nabla_{\boldsymbol{x}}$ is the spatial gradient. Our goal is to estimate the velocity $v(\boldsymbol{x})$ at each iteration. The solution of this equation can be obtained in the least squares sense over the cortical surface by minimizing the cost

$$C(v) = \int \int \left( \nabla_{\boldsymbol{x}} F(\boldsymbol{x}(\tau), t) \cdot v(\boldsymbol{x}) - \frac{dF(\boldsymbol{x}(\tau), t)}{d\tau} \right)^2 \det\left( g(\boldsymbol{x}) \right) d\boldsymbol{x} dt \quad (1)$$

over $v$, where $g(\boldsymbol{x})$ indicates the surface metric tensor that encodes the distortion from the surface to the square map of the surface (Fig. 1). Since $g(\boldsymbol{x})$ is slowly varying spatially compared to $F$, we can replace the images $F$ by $F^g = \sqrt{\det\left(g(\boldsymbol{x})\right)} F$. Equation 1 shows that $v$ cannot be uniquely defined, since we only observe the projection of $v$ onto the gradient of $F$. We therefore select the minimum $L_2$ norm solution for $v$. In continuous form using variational calculus we solve the system of equations: $v(\boldsymbol{x}) \int \|\nabla F^g(\boldsymbol{x}(\tau), t)\|^2 dt = \int \frac{dF^g(\boldsymbol{x}(\tau), t)}{d\tau} \nabla F^g(\boldsymbol{x}(\tau), t) dt$ to get

$$v(\boldsymbol{x}) = \int \frac{dF^g(\boldsymbol{x}(\tau), t)}{d\tau} \nabla F^g(\boldsymbol{x}(\tau), t) dt / \int \|\nabla F^g(\boldsymbol{x}(\tau), t)\|^2 dt. \quad (2)$$

Following [11–13], we add the demons force in the denominator for stability when the spatial gradient is small, to obtain:

$$v(\boldsymbol{x}) = \frac{\int \frac{dF^g(\boldsymbol{x}(\tau), t)}{d\tau} \nabla F^g(\boldsymbol{x}(\tau), t) dt}{\int \|\nabla F^g(\boldsymbol{x}(\tau), t)\|^2 dt + \alpha^2 \int \left( \frac{dF^g(\boldsymbol{x}(\tau), t)}{d\tau} \right)^2 dt}.$$

The corresponding discrete version is

$$v^k(\boldsymbol{x}) = \frac{\sum_{n=1}^N (F_T^g(\boldsymbol{x}, n) - F_M^g \circ T^k(\boldsymbol{x})) \cdot \nabla F_T^g(\boldsymbol{x}, n)}{\sum_n \|\nabla F_T^g(\boldsymbol{x})\|^2 + \alpha^2 \sum_n (F_T^g(\boldsymbol{x}) - F_M^g \circ T^k(\boldsymbol{x}))^2}, \quad (3)$$

where $n$ is the fMRI time point and $\circ$ represents the composition operator. $F_T$ is the target image and $F_M$ the moving image, $T^k(\boldsymbol{x})$ is the deformation at the $k^{\text{th}}$ iteration. The hyperparameter $\alpha$ controls for stability and, following [11], it can be shown that the displacement in Eq. 3 is bounded by $1/2\alpha$.

We will consider the symmetric version similar to that proposed in [14]:

$$v^k(\boldsymbol{x}) = \frac{\sum_n (F_T^g(\boldsymbol{x}, n) - F_M^g \circ T^k(\boldsymbol{x})) \cdot \nabla F_T^g(\boldsymbol{x}, n)}{\sum_n ||\nabla F_T^g(\boldsymbol{x}, n)||^2 + \alpha^2 \sum_n (F_T^g(\boldsymbol{x}, n) - F_M^g \circ T^k(\boldsymbol{x}))^2}$$
$$+ \frac{\sum_n (F_T^g(\boldsymbol{x}, n) - F_M^g \circ T^k(\boldsymbol{x}, n)) \cdot \nabla F_M^g(\boldsymbol{x}, n)}{\sum_{n=1}^N ||\nabla F_M^g(\boldsymbol{x}, n)||^2 + \alpha^2 \sum_{n=1}^N (F_T^g(\boldsymbol{x}, n) - F_M^g \circ T^k(\boldsymbol{x}))^2}$$

that leads to an approximately inverse consistent deformation field. The steps of velocity estimation $v(\boldsymbol{x})$, accumulation to the deformation field ($T^{k+1}(\boldsymbol{x}) \leftarrow T^k(\boldsymbol{x}) + v^k(\boldsymbol{x})$), and warping the subject's rfMRI data using this deformation $T(\boldsymbol{x})$, are iterated until the norm of the velocity becomes small (e.g. $10^{-6}$th of a pixel). A flowchart for the rfDemons method is shown in Fig. 2. The execution time for the HCP datasets used below (32K per hemisphere cortical mesh density, 1200 time samples) is 3–4 min on a typical workstation (Pentium V, 16 GB RAM) with a minimum memory requirement of 4 GB. The relatively light computational load of the algorithm is due to the fact that we perform the registration in the flat square space instead of on the sphere (as in spherical demons [8]). Since we compensate for the metric tensor corresponding to the flat map, we minimize regional biases in deformation fields that would otherwise result from metric distortion in the cortical surface maps.

## 3   Validation and Results

**Data**: We used minimally preprocessed (ICA-FIX denoised) resting and task fMRI data from 40 independent subjects (all right handed, age 26–30, 16 male and 24 female), which are publicly available from the Human Connectome Project (HCP) [10, 15]. We used data that was acquired in two independent resting fMRI sessions (with same LR phase encoding) of 15 min each (TR = 720 ms, TE = 33.1 ms, 2 mm × 2 mm × 2 mm voxels) with the subjects asked to relax and fixate on a projected bright cross-hair on a dark background.

**Intra-subject and Inter-subject Consistency**: For the first study, we coregistered two independent sessions of the same subject to a reference and checked for differences in the resulting deformation fields. If individual differences between cortical anatomy and functional specialization, as reflected in the rfMRI signal, are indeed driving the deformation then we would expect to see

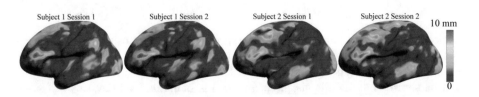

**Fig. 3.** Magnitude of the rfDemons deformation fields for two subjects, two sessions each. Note the cross-session consistency in the deformation fields for each subject.

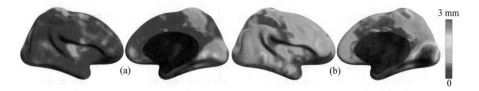

**Fig. 4.** (a) Intra-subject standard deviation maps for rfDemons; (b) Inter-subject standard deviation maps for rfDemons.

similar deformations for two sessions of the same subject. Similarly, we would expect to see differences in these deformation fields for different subjects. In other words, the inter-subject variance should be much larger than the intra-subject variance. To explore this issue, the deformations from the unit square were mapped back to the original surface with metric compensation to represent the true deformations on the cortical surface. Our expectations are confirmed anecdotally in Fig. 3 which shows the magnitudes of these deformations for two sessions each for two subjects. Within subject deformations are very similar, but they differ markedly across subjects. Cross-subject and cross-session differences were further quantified by computing the standard deviations of the deformation difference across sessions for each subject and averaged over subjects (Fig. 4(a)), and also the standard deviation of the deformation differences across subjects, averaged over sessions for all subjects (Fig. 4(b)). Again, the cross-subject difference is much larger than within subject cross-session differences. Somatomotor cortex shows low s.d. of deformation across subjects indicating that anatomical registration also aligns functionally in this region. This is expected since these areas are well defined by the pre and post central sulci. In contrast, the visual areas show much larger inter-subject variability. Again, this is consistent with the known variability of visual functional areas with respect to cortical anatomy, and the reason that functional localizers are frequently used in these areas. The within-subject cross-session maps indicates generally low variability with the exception of areas V1-V4. Performance in this area requires further investigation.

**Task-Data Mapping**: We also used task fMRI data to validate the results of rfDemons registration. We considered z-score maps for the emotion, gambling, language, motor, relational and social tasks in the HCP dataset [15]. In order to validate the cortical alignment obtained from the rfMRI data, we computed the map-

**Table 1.** Correlations between subject and reference z-scores before (structural) and after (rfDemons) functional alignment. We list the median correlation over subjects and interquartile distance together with computed p-values.

| Task | Contrast | Structural | rfDemons | p-value |
|------|----------|-----------|----------|---------|
| Emotion | faces_shapes | 0.32(0.11) | 0.34(0.10) | 5.0E-8 |
| Gambling | punish_reward | 0.03(0.04) | 0.03(0.06) | 0.01 |
| Language | math_story | 0.48(0.09) | 0.54(0.1) | 2.7E-8 |
| Motor | t_avg | 0.27(0.10) | 0.29(0.10) | 1.6E-6 |
| Relational | match_rel | 0.22(0.10) | 0.23(0.13) | 3.2E-7 |
| Social | random_tom | 0.24(0.13) | 0.25(0.14) | 5E-8 |

ping between the subject and reference rfMRI data by rfDemons and applied the resulting deformation map to the z-score maps from the task fMRI data. Our underlying hypothesis is that rfMRI based registration will improve functional alignment of the cortical surface, which will result in turn in better alignment across subjects of the z-score maps from the task data. To perform this comparison, for each task we computed the correlation between the z-score maps for each individual and the reference before and after functional alignment. Median and interquartile values are listed in Table 1.

The before and after alignment correlations were compared using the Wilcoxon ranksum test (1-sided, paired) of the null hypothesis that the paired difference in the correlations has zero median. A significant increase in the correlation of z-score maps of reference and subjects was observed in all cases as shown in Table 1. This result indicates that the resting-fMRI based rfDemons registration is able to improve inter-subject alignment of functional regions as evoked through a series of tasks. Additionally, we averaged the z-score maps across subjects before and after rfDemons alignment. Regions of significant activity lie in the tails of the distribution of these average z-score maps. The ability to more reliably detect this activation should therefore be reflected in a larger number of spatial locations that are outliers in this distribution. We computed the number of vertices on the cortical surface that exceeded a given z-score threshold, computed as a function of that threshold, in Fig. 5 for the language task (math vs. story contrast). As expected, we see a larger fraction of vertices exceeding the higher thresholds after rfDemons alignment.

## 4    Discussion and Conclusion

We have described a novel method (rfDemons) for functional alignment of the cerebral cortex using resting fMRI data. Our studies shows a high degree of within-subject consistency through most of the cortex except in the visual cortex. This latter observation may limit applicability in visual cortex, although the problem could be addressed using data from a combination of resting and visual stimulation. Between subject comparisons indicate a strong spatial dependence on the degree of variability across subjects. Again, this is most pronounced in visual cortex, and is smaller in regions, such as somatomotor cortex,

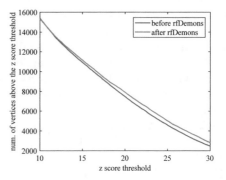

**Fig. 5.** Number of vertices in the cortical surface mesh above given threshold for the averaged z-score maps over subjects, for the language task (math_story) data before and after rfDemons registration.

that are known to be well defined by anatomical landmarks. Through application of the rfDemons registration results to task fMRI data we were able to explore whether functional registration improves the alignment of task-evoked activity. Through correlation studies we see small but significant improvement in the correlation of z-score maps between subjects after rfDemons alignment. This improvement is seen over several different contrasts representing multiple different functional tasks. We also saw an increase in the number of vertices in which the group averaged z-score exceeded a given threshold, indicating the potential for increased sensitivity in detecting task-related activity.

# References

1. Fischl, B.: FreeSurfer. NeuroImage **62**(2), 774–781 (2012)
2. Pantazis, D., et al.: Comparison of landmark-based and automatic methods for cortical surface registration. NeuroImage **49**(3), 2479–2493 (2010)
3. Chong, M., et al.: Individual parcellation of resting fMRI with a group functional connectivity prior. NeuroImage **156**, 87–100 (2017)
4. Sereno, M., Pitzalis, S., Martinez, A.: Mapping of contralateral space in retinotopic coordinates by a parietal cortical area in humans. Science **294**(5545), 1350–1354 (2001)
5. Sabuncu, M.R., Singer, B.D., Conroy, B., Bryan, R.E., Ramadge, P.J., Haxby, J.V.: Function-based intersubject alignment of human cortical anatomy. Cereb. Cortex **20**(1), 130–140 (2010)
6. Guntupalli, J.S., Feilong, M., Haxby, J.V.: A computational model of shared fine-scale structure in the human connectome. PLoS Comput. Biol. **14**(4), e1006120 (2018). https://doi.org/10.1371/journal.pcbi.1006120
7. Zhao, Y., et al.: A novel framework for groupwise registration of FMRI images based on common functional networks. In: Proceedings of ISBI, pp. 485–489. IEEE (2017)
8. Nenning, K.H., Liu, H., Ghosh, S.S., Sabuncu, M.R., Schwartz, E., Langs, G.: Diffeomorphic functional brain surface alignment: functional demons. NeuroImage **156**(Supplement C), 456–465 (2017)
9. Joshi, A.A., Chong, M., Li, J., Choi, S., Leahy, R.M.: Are you thinking what I'm thinking? Synchronization of resting fMRI time-series across subjects. NeuroImage **172**, 740–752 (2018)
10. Glasser, M.F., et al.: The minimal preprocessing pipelines for the Human Connectome Project. NeuroImage **80**, 105–124 (2013)
11. Cachier, P., Pennec, X., Ayache, N.: Fast non rigid matching by gradient descent: study and improvements of the Demons algorithm. Technical report RR-3706, INRIA (1999)
12. Thirion, J.P.: Image matching as a diffusion process: an analogy with Maxwell's demons. Med. Image Anal. **2**(3), 243–260 (1998)
13. Peyrat, J.-M., Delingette, H., Sermesant, M., Pennec, X., Xu, C., Ayache, N.: Registration of 4D time-series of cardiac images with multichannel diffeomorphic demons. In: Metaxas, D., Axel, L., Fichtinger, G., Székely, G. (eds.) MICCAI 2008. LNCS, vol. 5242, pp. 972–979. Springer, Heidelberg (2008). https://doi.org/10.1007/978-3-540-85990-1_117
14. Wang, H., et al.: Validation of an accelerated demons algorithm for deformable image registration in radiation therapy. Phys. Med. Biol. **50**(12), 2887–2905 (2005)
15. Barch, D.M., et al.: Function in the human connectome: task-fMRI and individual differences in behavior. NeuroImage **80**, 169–189 (2013)

# Brain Biomarker Interpretation in ASD Using Deep Learning and fMRI

Xiaoxiao Li[1(✉)], Nicha C. Dvornek[4], Juntang Zhuang[1], Pamela Ventola[5], and James S. Duncan[1,2,3,4]

[1] Biomedical Engineering, Yale University, New Haven, CT, USA
xiaoxiao.li@yale.edu
[2] Electrical Engineering, Yale University, New Haven, CT, USA
[3] Department of Statistics & Data Science, Yale University, New Haven, CT, USA
[4] Radiology and Biomedical Imaging, Yale School of Medicine, New Haven, CT, USA
[5] Child Study Center, Yale School of Medicine, New Haven, CT, USA

**Abstract.** Autism spectrum disorder (ASD) is a complex neurodevelopmental disorder. Finding the biomarkers associated with ASD is extremely helpful to understand the underlying roots of the disorder and can lead to earlier diagnosis and more targeted treatment. Although Deep Neural Networks (DNNs) have been applied in functional magnetic resonance imaging (fMRI) to identify ASD, understanding the data driven computational decision making procedure has not been previously explored. Therefore, in this work, we address the problem of interpreting reliable biomarkers associated with identifying ASD; specifically, we propose a 2-stage method that classifies ASD and control subjects using fMRI images and interprets the saliency features activated by the classifier. First, we trained an accurate DNN classifier. Then, for detecting the biomarkers, different from the DNN visualization works in computer vision, we take advantage of the anatomical structure of brain fMRI and develop a frequency-normalized sampling method to corrupt images. Furthermore, in the ASD vs. control subjects classification scenario, we provide a new approach to detect and characterize important brain features into three categories. The biomarkers we found by the proposed method are robust and consistent with previous findings in the literature. We also validate the detected biomarkers by neurological function decoding and comparing with the DNN activation maps.

## 1 Introduction

Autism spectrum disorder (ASD) affects the structure and function of the brain. To better target the underlying roots of ASD for diagnosis and treatment, efforts to identify reliable biomarkers are growing [1]. Significant progress has been made using functional magnetic resonance imaging (fMRI) to characterize the brain changes that occur in ASD [2].

---

This work was supported by NIH Grant 5R01 NS035193.

A. F. Frangi et al. (Eds.): MICCAI 2018, LNCS 11072, pp. 206–214, 2018.
https://doi.org/10.1007/978-3-030-00931-1_24

Recently, many deep neural networks (DNNs) have been effective at identifying ASD using fMRI [3,4]. However, these methods lack model transparency. Despite promising results, the clinicians typically want to know if the model is trustable and how to interpret the results. Motivated by this, here we focus on developing the interpretation method for deciphering the regions in fMRI brain images that can distinguish ASD vs. control by the deep neural networks.

There are three main approaches for interpreting the important features detected by DNNs. One approach is using gradient ascent methods to generate an image that best represents the class [5]. However, this method cannot handle nonlinear DNNs well. The second approach is to visualize how the network responds to a specific corrupted input image in order to explain a particular classification made by the network [6]. The third one uses the intermediate outputs of the network to visualize the feature patterns [7]. However, all of these existing methods tend to end up with blurred and imprecise saliency maps.

The goal of our work is to identity biomarkers for ASD, defined as important regions of interest (ROIs) in the brain that distinguish autistic and healthy controls. Different from traditional brain biomarker detection methods, by utilizing the high dimensional feature capturing ability of DNNs and brain structure, we propose an innovative 2-stage pipeline to interpret biomarkers. Different from above DNN visualization methods, our main contribution includes a ROI-based image corruption and generating procedure. In addition, we analyze the feature importance using the distribution of DNN predictions and statistical hypothesis testing. We applied the proposed method on multiple datasets and validated our robust findings by decoding neurological function of biomarkers, viewing DNN intermediate outputs and comparing literature reports.

## 2   Method

### 2.1   Two-Stage Pipeline with Deep Neural Network Classifier

We propose a corrupting strategy to find the important regions activated by a well-trained ASD classifier (Fig. 1). The first stage is to train a DNN classifier for classifying ASD vs. control subjects. The DNN we use (2CC3D) has 6 convolutional, 4 max-pooling and 2 fully connected layers, followed by a sigmoid output layer [4] as shown in the middle of Fig. 1. The number of kernels are denoted on each layer in Fig. 1. Dropout and l2 regularization are applied to avoid overfitting. The study in [4] demonstrated that we can achieve higher accuracy using the 2CC3D framework, since it integrates spatial-temporal information of 4D fMRI. Each frame of 3D fMRI is downsampled to $32 \times 32 \times 32$. We use sliding-windows with size $w$ and stride length $stride$ to move along the time dimension of the 4D fMRI sequence and calculate the mean and standard deviation (std) for each voxel's time series within the sliding window. Given $T$ frames in each 4D fMRI sequence, by this method, $\lfloor \frac{T-w}{stride} \rfloor + 1$ 2-channel images (mean and std fMRI images) are generated for each subject. We define the original fMRI sequence as $I(x,y,z,t)$, the mean-channel sequence as $\tilde{I}(x,y,z,t)$ and the std-channel as $\hat{I}(x,y,z,t)$. For any $x,y,z$ in $\{0,1,\cdots,31\}$,

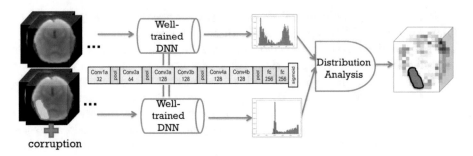

**Fig. 1.** Pipeline for interpreting important features from a DNN

---

**Algorithm 1.** Important Feature Detection For Binary Classification

---

**Input:** $X^0$, a group of images from class 0; $X^1$, a group of images from class 1; $f$, DNN classification model.

1: $\mathbb{P}^0_o \leftarrow f(X^0)$ **and** $\mathbb{P}^1_o \leftarrow f(X^1)$
2: $JSD^o_{+/-} \leftarrow JSD(\mathbb{P}^0_o, \mathbb{P}^1_o)$                                             ▷ by bootstrapping
3: **for** $r$ **in ROIs do**
4:     $\mathbb{P}^0_c \leftarrow f(X^0_{\backslash r})$, $\mathbb{P}^1_c \leftarrow f(X^1_{\backslash r})$                                   ▷ by sampling
5:     $JSD^c_{+/-} \leftarrow JSD(\mathbb{P}^0_c, \mathbb{P}^1_c)$, $Shift^0 \leftarrow \mathbb{P}^0_c - \mathbb{P}^0_o$, $Shift^1 \leftarrow \mathbb{P}^1_c - \mathbb{P}^1_o$
6:     **if** $JSD^c_+ < JSD^c_-$ **or** $median(\mathbb{P}^0_c) > median(\mathbb{P}^1_c)$ **then**     ▷ fool the classifier
7:         **do Wilcoxon(Shift) one tailed test**
8:         **if** $\mathbb{P}^0 \Rightarrow 1$ **and** $\mathbb{P}^1 \Rightarrow 0$ **then**
9:             **r** is an important feature for both classes
10:        **else if** only $\mathbb{P}^0 \Rightarrow 1$ **then**
11:            **r** is an important feature for class 0
12:        **else if** only $\mathbb{P}^1 \Rightarrow 0$ **then**
13:            **r** is an important feature for class 1
14:        **end if**                                                        ▷ $\Rightarrow$ means significant shift
15:    **end if**
16: **end for**

---

$$\tilde{I}(x, y, z, t) = \frac{\sum_{\tau=t+1-w}^{t} I(x, y, z, \tau)}{w} \tag{1}$$

$$\hat{I}(x,y,z,t) = \sqrt{\frac{\sum_{\tau=t+1-w}^{t} [I(x,y,z,\tau) - \tilde{I}(x,y,z,t)]^2}{w - 1}}. \tag{2}$$

The outputs are probabilistic predictions ranging in $[0, 1]$. The second stage is to interpret the output differences after corrupting the image. We corrupt a ROI of the original image and put it in the well-trained DNN classifier to get a new prediction (Sect. 2.2). Based on the prediction difference, we use a statistical method to interpret the importance of the ROI (Sect. 2.3).

## 2.2 Prediction Difference Analysis

We use a heuristic method to estimate the feature (an image ROI) importance by analyzing the probability of the correct class predicted by the corrupted image.

In the DNN classifier case, the probability of the abnormal class $c$ given the original image $\boldsymbol{X}$ is estimated from the predictive score of the DNN model $f$ : $f(\boldsymbol{X}) = p(c|\boldsymbol{X})$. Denote the image corrupted at ROI $\boldsymbol{r}$ as $\boldsymbol{X}_{\backslash \boldsymbol{r}}$. The prediction of the corrupted image is $p(c|\boldsymbol{X}_{\backslash \boldsymbol{r}})$. To calculate $p(c|\boldsymbol{X}_{\backslash \boldsymbol{r}})$, we need to marginalize out the corrupted ROI $\boldsymbol{r}$:

$$p(c|\boldsymbol{X}_{\backslash \boldsymbol{r}}) = \mathbb{E}_{\boldsymbol{x}_r \sim p(\boldsymbol{x}_r|\boldsymbol{X}_{\backslash r})} p(c|\boldsymbol{X}_{\backslash \boldsymbol{r}}, \boldsymbol{x}_r), \tag{3}$$

where $\boldsymbol{x}_r$ is a sample of ROI $r$. Modeling $p(\boldsymbol{x}_r|\boldsymbol{X}_{\backslash r})$ by a generative model can be computationally intensive and may not be feasible. We assumed that an important ROI contains features that cannot be easily sampled from the same ROI of other classes and is predictive for predicting its own class. Hence, we approximated $p(\boldsymbol{x}_r|\boldsymbol{X}_{\backslash r})$ by sampling $\boldsymbol{x}_r$ from each ROI $\boldsymbol{r}$ in the whole sample set. In fMRI study, each brain can be registered to the same atlas, so the same ROI in different images have the same spatial location and number of voxels. Therefore, we can directly sample $\hat{\boldsymbol{x}}_r$s and replace $\boldsymbol{x}_r$ with them. Then we flatten the $\hat{\boldsymbol{x}}_r$ and $\boldsymbol{x}_r$ as vectors $\overrightarrow{\hat{\boldsymbol{x}}_r}$ and $\overrightarrow{\boldsymbol{x}_r}$. From the $K$ sampled $\hat{\boldsymbol{x}}_r^k$s, we calculate the Pearson correlation coefficient $\rho_k = cov(\overrightarrow{\hat{\boldsymbol{x}}_r^k}, \overrightarrow{\boldsymbol{x}_r})/\sigma_{\overrightarrow{\hat{\boldsymbol{x}}_r^k}}\sigma_{\overrightarrow{\boldsymbol{x}_r}}$, where $k \in \{1, 2, \ldots, K\}, \rho \in [-1, 1]$. Because sample size of each class may be biased, we will de-emphasize the samples that can be easily sampled, since $p(c|\boldsymbol{X}_{\backslash \boldsymbol{r}})$ should be irrelevant to the sample set. Therefore, we will do a frequency-normalized transformation. We divide $[-1,1]$ into $N$ equal-length intervals. Each $\rho_k$ will fall in one of the intervals. After $K$ samplings, we calculate $N_i$, the number of sample correlations in interval i, where $i \in \{1, 2, \ldots, N\}$. For the $\rho_k$ located in interval $i$, the frequency-normalized weight is $w_k = \frac{1}{N \cdot N_i}$. Denote $\boldsymbol{X}_k'$ as the image $\boldsymbol{X}$ replacing $\boldsymbol{x}_r$ with $\hat{\boldsymbol{x}}_r^k$. Hence, we approximate $p(c|\boldsymbol{X}_{\backslash \boldsymbol{r}})$ as

$$p(c|\boldsymbol{X}_{\backslash \boldsymbol{r}}) \approx \sum_k w_k p(c|\boldsymbol{X}_k'). \tag{4}$$

## 2.3 Important Feature Interpretation

In the binary classification scenario, we label the reference class as 0 and the experiment class as 1. The original prediction probability of the two classes are denoted as $\mathbb{P}_o^0$ and $\mathbb{P}_o^1$, which are two vectors containing the prediction results $p(c|\boldsymbol{X})$s for each sample in the two classes respectively. Similarly, we have $\mathbb{P}_c^0$ and $\mathbb{P}_c^1$ containing $p(c|\boldsymbol{X}_{\backslash r})$s for the corrupted images. We assume that corrupting an important feature will make the classifier perform worse. One extreme case is that the two distributions shift across each other, which can be approximately measured by $median(\mathbb{P}_c^0) > median(\mathbb{P}_c^1)$. If this is not the case, we use Jensen-Shannon Divergence ($\boldsymbol{JSD}$) to measure the distance of two distributions:

$$JSD(\mathbb{P}_0, \mathbb{P}_1) = \frac{1}{2}KL(\mathbb{P}_0 \parallel \frac{\mathbb{P}_0 + \mathbb{P}_1}{2}) + \frac{1}{2}KL(\mathbb{P}_1 \parallel \frac{\mathbb{P}_0 + \mathbb{P}_1}{2}) \tag{5}$$

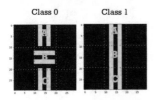

**Fig. 2.** Synthetic images

**Table 1.** Misclassification rate when corrupting patch B

|           | Class 0         | Class 1         |
|-----------|-----------------|-----------------|
| Equal     | $0.10 \pm 0.01$ | $0.91 \pm 0.03$ |
| Normalize | $0.49 \pm 0.02$ | $0.50 \pm 0.01$ |

where $KL(\mathbb{P}_0 \parallel \mathbb{P}_1) = -\sum_i \mathbb{P}_0(i)log(\mathbb{P}_1(i)/\mathbb{P}_0(i))$. Given two distributions $\mathbb{P}_0$ and $\mathbb{P}_1$, we use bootstrap method to calculate the upper bound $JSD_+$ and the lower bound $JSD_-$ with confidence level, $1 - \alpha_{JSD}$. We classify the important ROIs into different categories based on the shift of the prediction distribution before and after corruption. The one-tailed Wilcoxon paired difference test [8] is applied to investigate whether the shift is significant. We use false discovery rate (FDR) controlling procedure to handle testing the large number of ROIs. FDR adjusted q-value is used to compare with the significance level $\alpha_W$. The method to evaluate the feature importance is shown in Algorithm 1.

## 3    Experiments and Results

### 3.1    Experiment 1: Synthetic Data Model

We used simulated experiments to demonstrate that our frequency-normalized resampling algorithm recovers the ground truth patch importance. We simulated two classes of images as shown in Fig. 2, with background = 0 and strips = 1 and Gaussian noise ($\mu = 0, \sigma = 0.01$). They can be gridded into 9 patches. We assumed that patch B of class 0 and 1 are **equally important** to human understanding. However, in our synthetic model, the sample set was biased with 900 images in class 0 and 100 images in class 1. A simple 2-layer convolutional neural network was used as the image classifier, which achieved 100% classification accuracy. Since the shift of corrupted images was obvious, we used misclassification rate to measure whether $p(c|\boldsymbol{X}_{\backslash r})$ was approximated reasonably by equally weighted sampling (which means $w_i = 1/K$) or by our frequency-normalized sampling. In Table 1, our frequency-normalized sampling approach ('Normalize') is superior to the equally weighted one ('Equal') in treating patch B equally in both classes.

### 3.2    Experiment 2: Task-fMRI Experiment

We tested our methods on a group of 82 ASD children and 48 age and IQ-matched healthy controls. Each subject underwent a task fMRI scan (BOLD, TR $= 2000$ ms, TE $= 25$ ms, flip angle $= 60°$, voxel size $3.44 \times 3.44 \times 4$ mm$^3$) acquired on a Siemens MAGNETOM Trio TIM 3T scanner.

For the fMRI scans, subjects performed the "biopoint" task, viewed point light animations of coherent and scrambled biological motion in a block design

[2] (24 s per block). The fMRI data was preprocessed using FSL [9] for (1) motion correction, (2) interleaved slice timing correction, (3) BET brain extraction, (4) spatial smoothing (FWHM = 5 mm), and (5) high-pass temporal filtering. The functional and anatomical data were registered and parcellated by AAL atlas [10] resulting in 116 ROIs. We applied a sliding window ($w = 3$) along the time dimension of the 4D fMRI, generating 144 3D volume pairs (mean and std) for each subject.

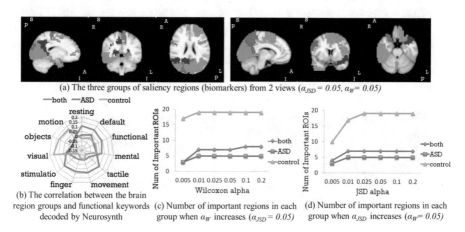

(a) The three groups of saliency regions (biomarkers) from 2 views ($\alpha_{JSD} = 0.05$, $\alpha_w = 0.05$)

(b) The correlation between the brain region groups and functional keywords decoded by Neurosynth

(c) Number of important regions in each group when $\alpha_W$ increases ($\alpha_{JSD} = 0.05$)

(d) Number of important regions in each group when $\alpha_{JSD}$ increases ($\alpha_W = 0.05$)

**Fig. 3.** Important biomarkers detected in biopoint dataset

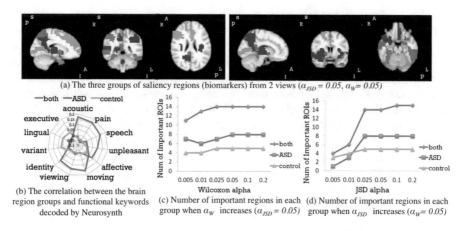

(a) The three groups of saliency regions (biomarkers) from 2 views ($\alpha_{JSD} = 0.05$, $\alpha_w = 0.05$)

(b) The correlation between the brain region groups and functional keywords decoded by Neurosynth

(c) Number of important regions in each group when $\alpha_w$ increases ($\alpha_{JSD} = 0.05$)

(d) Number of important regions in each group when $\alpha_{JSD}$ increases ($\alpha_W = 0.05$)

**Fig. 4.** Important biomarkers detected in ABIDE dataset

We split 85% subjects (around 16k 3D volume pairs) as training set, 7% as validation set for early stopping and 8% as testing set, stratified by class.

The model achieved 87.1% accuracy when evaluated on each 3D pair input of the testing set. Figure 3(a) and (b) give two views of the important ROIs brain map ($\alpha_{JSD} = 0.05$, $\alpha_W = 0.05$). Blue ROIs are associated with identifying both ASD and control. Red ROIs are associated with identifying ASD only and green ROIs are associated with identifying control only. By decoding the neurological functions of the important ROIs with Neurosynth [11], we found (1) regions related to default mode and functional connectivity are significant in classifying both individuals with ASD and controls, which is consistent with prior literature related to executive functioning and problem-solving in ASD [2]; (2) regions associated with finger movement are relevant in classifying individuals with ASD, and (3) visual regions were involved in classifying controls, perhaps because controls may attend to the visual features more closely, whereas ASD subjects tend to count the dots on the video [12].

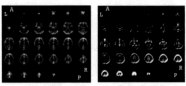 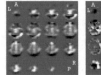

(a)  ASD group outputs from Conv1a          (b) Difference of outputs from Conv2a (Control-ASD)
Left: Biopoint Dataset   Right: ABIDE Dataset          Left: Biopoint Dataset   Right: ABIDE Dataset

**Fig. 5.** Intermediate outputs (activation maps) of DNN

### 3.3    Experiment 3: Resting-State fMRI

We also performed experiments on data from the ABIDE I cohort UM site [9,13].This resulted in 41 ASD subjects and 54 healthy controls. Each subject initially had 293 frames. As in the task-fMRI experiment, we generated 2-channel images. We used the weights of the pre-trained 2CC3D networks in experiment 2 as our initial network weights. We split 33 ASD subjects and 43 controls for training (around 22k 3D volume pairs). 9 subjects were used as validation data for early stopping. The classifier achieved 85.3% accuracy in identifying individual 3D volume on the 10 subjects testing set. The biomarker detection results are shown in Fig. 4: (1) emotion related regions colored in blue are highlighted for both groups; (2) regions colored in red (viewing and moving related) are associated with identifying ASD; and (3) green regions (related to executive and lingual) are associated with identifying control.

### 3.4    Results Analysis

In experiment 2, since the subjects were under visual task, visual patterns were detected. Whereas in experiment 3, subjects were in resting state, so no visual regions were detected. In addition, we found many common ROIs in both experiments: frontal (motivation related), precuneus (execution related), etc. Previous research [2] also indicated these regions are associated with identifying ASD vs. control. Moreover, from the sub-figure (c), (d) of Figs. 3 and 4, the groups of detected important regions are very stable when tuning JSD confidence level (1-$\alpha_{JSD}$) and Wilcoxon testing threshold $\alpha_W$, except when $\alpha_{JSD}$ is very small. This is likely because the original prediction distribution is fat tailed. Furthermore, we validate the results with the activation maps from the 1st and 2nd layers of the DNN. The output of each filter was averaged for 10 controls and for 10 ASD subjects. The 1st convolutional layer captured structural information and distinguished gray vs. white matter (Fig. 5(a)). Its outputs are similar in both control and ASD group. The outputs of the 2nd convolutional layer showed significant differences between groups in Fig. 5(b). Regions darkened and highlighted in Fig. 5(b) correspond to many regions detected by our proposed method.

## 4     Conclusions

We designed a 2-stage (DNN + prediction distribution analysis) pipeline to detect brain region saliency for identifying ASD and control subjects. Our sampling and significance testing scheme along with the accurate DNN classifier ensure reliable biomarker detection results. Our method was designed for interpreting important ROIs for registered images, since the traditional machine learning feature selection methods can not be directly used in interpreting DNNs. Moreover, our proposed method can be directly used to interpret any other machine learning classifiers. Overall, the proposed method provides an efficient and objective way of interpreting the deep learning model applied to neuro-images.

## References

1. Goldani, A.A., et al.: Biomarkers in autism. Front. Psychiatry **5**, 100 (2014)
2. Kaiser, M.D., et al.: Neural signatures of autism. In: PNAS (2010)
3. Iidaka, T.: Resting state functional magnetic resonance imaging and neural network classified autism and control. Cortex **63**, 55–67 (2015)
4. Li, X., et al.: 2-channel convolutional 3D deep neural network (2CC3D) for fMRI analysis: ASD classification and feature learning. In: ISBI (2018)
5. Yosinski, J., et al.: Understanding neural networks through deep visualization. arXiv preprint arXiv:1506.06579 (2015)
6. Zintgraf, L.M., et al.: Visualizing deep neural network decisions: prediction difference analysis. arXiv preprint arXiv:1702.04595 (2017)
7. Zhou, B., et al.: Learning deep features for discriminative localization. In: CVPR. IEEE (2016)
8. Whitley, E., et al.: Statistics review 6: nonparametric methods. Crit. Care **6**, 509 (2002)
9. Smith, S.M., et al.: Advances in functional and structural MR image analysis and implementation as FSL. Neuroimage **23**, S208–S219 (2004)
10. Tzourio-Mazoyer, N., et al.: Automated anatomical labeling of activations in SPM using a macroscopic anatomical parcellation of the MNI MRI single-subject brain. Neuroimage **15**, 273–289 (2002)

11. Yarkoni, T., et al.: Large-scale automated synthesis of human functional neuroimaging data. Nat. Methods **8**, 665 (2011)
12. Ventola, P., et al.: Differentiating between autism spectrum disorders and other developmental disabilities in children who failed a screening instrument for ASD. J. Autism Dev. Disord. **37**, 425–436 (2007)
13. Di Martino, A., et al.: The autism brain imaging data exchange: towards a large-scale evaluation of the intrinsic brain architecture in autism. Mol. Psychiatry **19**, 659 (2014)

# Neural Activation Estimation in Brain Networks During Task and Rest Using BOLD-fMRI

Michael Hütel[1,2(✉)], Andrew Melbourne[1,2], and Sebastien Ourselin[1,2]

[1] Department of Medical Physics and Biomedical Engineering, UCL, London, UK
michael.hutel.13@ucl.ac.uk
[2] School of Biomedical Engineering and Imaging Sciences, KCL, London, UK

**Abstract.** Since the introduction of BOLD (Blood Oxygen Level Dependent) imaging, the hemodynamic response model has remained the standard analysis approach to relate activated brain areas to extrinsic task conditions. Ongoing brain activity unrelated to the task is neglected and considered noise. By contrast, model-free blind source separation techniques such as Independent Component Analysis (ICA) have been used in intrinsic task-free experiments to reveal functional systems usually referred to as "resting-state" networks. However, matrix factorization techniques applied to BOLD imaging do not model the translation of neuronal activity into BOLD fluctuations and depend on arbitrarily chosen regularization measures such as statistical independence or sparsity. We present a novel neurobiologically-driven matrix factorization approach. Our matrix factorization model incorporates the hemodynamic response function that enables the estimation of underlying neural activity in individual brain networks that present during task- and task-free BOLD-fMRI experiments. We validate our model on the recently published Midnight Scanning Club dataset including five hours of task-free and six hours of various task experiments per subject. The resulting temporal and spatial activation patterns obtained from our matrix factorization technique resemble individual task profiles and known functional brain networks, which are either correlated with the task or spontaneously activating unrelated to the task.

## 1 Introduction

The most common experimental approach in BOLD-fMRI is to design task experiments that stimulate individual brain areas causing increased local neuronal activity and changes in local blood flow. A binary function of neural activity comprises onset and duration of the task, and is convolved with a hemodynamic response function (HRF), which relates neural activity to hemodynamic changes. On the basis of this established link between neuronal activity and

**Electronic supplementary material** The online version of this chapter (https://doi.org/10.1007/978-3-030-00931-1_25) contains supplementary material, which is available to authorized users.

A. F. Frangi et al. (Eds.): MICCAI 2018, LNCS 11072, pp. 215–222, 2018.
https://doi.org/10.1007/978-3-030-00931-1_25

BOLD signal, several task experiments have been developed to map out underlying functional systems in humans. By contrast, ongoing spontaneous neural activity from task-free fMRI experiments have been found to organize into so called resting-state networks of intrinsic functional connectivity. In contrast to task-fMRI, model-free approaches like Independent Component Analysis (ICA) or seed-based analysis have been used to study this intrinsic functional connectivity, but these do not model the relation between underlying neuronal activity and spontaneous BOLD signal fluctuations. Recent studies [1] that investigated the link between resting-state and task-fMRI have provided evidence that there is a common functional brain architecture that exhibits increased or decreased neural activity dependent on the intrinsic or extrinsic task demands. We propose a novel generative model embedded within a neural network optimization framework to obtain characteristic brain networks and their corresponding neural activation profiles during a given task. The framework only requires minimal pre-processing and is regularized by the known HRF.

## 2    Materials and Methods

We propose a Convolutional Hemodynamic Autoencoder (CHA) that results in components which we loosely associate with *functional networks*. Each functional network consists of three parts: a spatial map; a neuronal activity time course; and a corresponding BOLD time course. We test our proposed technique on both simulated and real imaging data with minimal preprocessing.

### 2.1    Functional Imaging Data

We use simulated data from[1] and compare our results to the method of Total Activation (TA) [2]. We further evaluate our method on the Midnight Scanning club (MSC) data[2] that comprises 10 healthy adult subjects with six hours of task-free and five hours of task-based BOLD-fMRI experiments including motor tasks, incidental memory and a mixed design task. More details can be found in the respective references [2,3].

### 2.2    Minimal Preprocessing

For the fMRI sequences of each subject, volumes were realigned to the first volume to correct for head motion. The first volume was linearly registered to its corresponding bias-corrected T1-weighted anatomical scan. The intra-subject affine registration and non-linear registration to the Talairach template were combined to map all fMRI volumes with one re-sampling into the Talairach space sampled in a 4 mm isotropic resolution. Time courses of voxels within brain tissue were extracted. Time courses were high-pass filtered (0.01 Hz cutoff) to remove signal drifts from scanner instabilities. Time courses are centered

---

[1] https://miplab.epfl.ch/software/TA/TA.zip.
[2] https://openfmri.org/dataset/ds000224/.

and variance-normalized. The functional 4D volume set per subject is reshaped into a matrix $X \in \mathbb{R}^{t \times v}$ with $t$ time points and $v$ voxels.

## 2.3  Generative Model

Our CHA assumes that BOLD-fMRI data can be modeled as a compressed representation of $c$ components. Each component $i$ consists of a spatial map $h_i \in \mathbb{R}_+^{1 \times v}$ of $v$ voxels, a neural activity time course $N_i \in \mathbb{R}_+^{t \times 1}$ of $t$ time points. The corresponding BOLD time-course $W_i = N_i \circledast \mathcal{H}$ is obtained by convolving $N_i$ with an impulse response function $\mathcal{H}$ (the standard HRF shape model obtained from two Gamma functions). This generative model in matrix notation is given by:

$$X = Wh + b_2 \tag{1}$$

with observed BOLD time course matrix $X \in \mathbb{R}^{t \times v}$ decomposed into BOLD time course matrix $W \in \mathbb{R}^{t \times c}$ and spatial component matrix $h \in \mathbb{R}_+^{c \times v}$. Each BOLD time course in $W$ has a corresponding time course in the neural activity time course matrix $N$. Bias $b_2 \in \mathbb{R}$ creates a negative baseline to compensate for negative time course values after preprocessing described in Sect. 2.2.

## 2.4  Convolutional Hemodynamic Autoencoder

The generative model is embedded into a neural network autoencoder framework. The encoder maps $X$ into the hidden layer $h = f(W^T X + b_1)$ using a rectified linear unit (ReLU) activation function $f : \mathbb{R} \to \mathbb{R}_+$, activation bias $b_1 \in \mathbb{R}^{1 \times t}$ and the transpose of BOLD time course matrix $W$ introduced in Sect. 2.3. The parameters of the autoencoder are neural activity time courses $N$ and biases $b_1$ and $b_2$ resulting in the following cost function:

$$\underset{N, b_1, b_2}{\arg \min} \| X - \hat{X} \|_2^2 \tag{2}$$

minimizing the $l2$ norm between the original data $X$ and its reconstruction $\hat{X}$. Neural activity time courses $N$ are initialized with the absolute values of a random Xavier initialization [4]. Biases $b_1$ and $b_2$ are initialized with zeros. The back-propagation algorithm (chain rule) is applied to derive the corresponding gradient for the cost function. The gradient is optimized with the limited memory Broyden-Fletcher-Goldfarb-Shanno (L-BFGS) optimization scheme.

## 2.5  Hyper-parameter Tuning

The only hyper-parameter to tune is the number of components $c$. We use the following heuristic to find a good solution through model averaging. We run n = 5 random initializations with $c = 1000$ for each of the 10 subjects ($\#subj$) per task[3]. We apply a non-negative matrix factorization (NMF) with non-negative

---

[3] Limited by the available 24 GB GPU memory.

double singular value initialization (NNDSVD) [5] to cluster all combined spatial maps $H \in \mathbb{R}_+^{(c \times n \times \#subj) \times v}$. The NNDSVD speeds up convergence and guarantees deterministic behavior for determining the right number of components $c_{opt}$ in the following NMF cross-validation with Gabriel holdout as described in [6]. The complete processing pipeline is summarized in Fig. 1. Subsequently, subject specific neural activity and BOLD time courses as well as spatial maps are obtained by weighted average informed by the association matrix $W$.

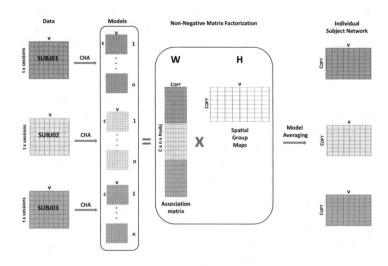

**Fig. 1.** For each task, we concatenate all session data and compute $n$ CHA runs for each subject. We obtain $c \times n \times \#subj$ components. Each component consists of a neural activity time course, BOLD time course and spatial map. We cluster all spatial maps with a NMF and cross-validate to obtain an optimal decomposition with $c_{opt}$ per task. We use the individual weights of the association matrix $W$ to compute a weighted average of neural activity time course, BOLD time course and spatial maps per component.

## 3   Results

### 3.1   Simulation Data

Figure 2 shows the results from simulated data; the simulation data contains four components (top left) and their corresponding neural activity time courses (bottom row). Cross-validation (bottom left) results in a very similar error for four and more components. We obtained neural activity time courses and spatial maps using $c_{opt} = 4$. Our technique is able to find a close estimate of the correct dimensionality of the decomposition and recovers the respective time course and spatial map of each component successfully.

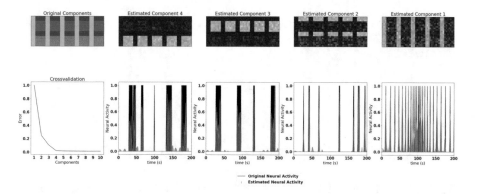

**Fig. 2.** The optimal decomposition into $c_{opt}$ is informed by cross-validation (bottom left). Original spatial maps are recovered by our technique (top row) as well as their corresponding neural activation profiles (bottom row).

## 3.2   Imaging Data

We obtain an optimal group decomposition into $c_{opt}$ components for each task that provides weighted averages of the neural activity time course and spatial map per component in each individual subject as outlined in Fig. 1. For example, the optimal value of $c_{opt}$ for the motor task is 50 components determined by cross-validation with 9-fold Gabriel holdout as depicted in Fig. 3. Six of the 50 components with their corresponding spatial map, neural activation and BOLD time course for subject one are depicted for the motor task in Fig. 4. All first sessions for all three tasks for each subject are included in the Supplementary Material. We found five activated brain networks occurring in all tasks. The Default Mode Network (DMN), Visual Network, Precuneus Network, Salient Network, Right and Left Central Executive Network (RCEN and LCEN). We examined the correlation of the neural activation time course in these five networks with the visual cues in each task for all sessions summarized in box-plots depicted in Fig. 3. The neural activation in the Visual Network follows the visual cues with a small delay as seen in Fig. 4. The Salient Network also exhibits minor correlation to the visual stimuli, while most other networks, such as the DMN, express unrelated neural activity to the task experiment.

## 4   Discussion

We have presented a new technique that simultaneously decomposes the observed BOLD signal into maps of spatial position, neuronal activity time courses, and hemodynamic responses.

This has several advantages over existing methods. In task-based fMRI the general linear model relates the experiment time course to observed BOLD fluctuations, but does not allow for ongoing spontaneous BOLD fluctuations unrelated to the task. Furthermore, it models underlying neural activity as switching

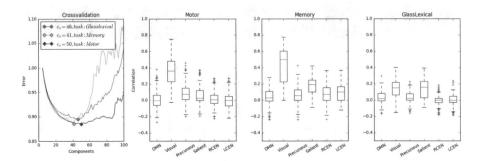

**Fig. 3.** We conducted a 9-fold cross-validation with Gabriel holdout to determine an optimal decomposition for each individual task that fits all subject sessions. The three figures on the right depict the correlation of the neural activation time course of common brain networks to the visual cues in motor, memory and glass-lexical task among all sessions and subjects.

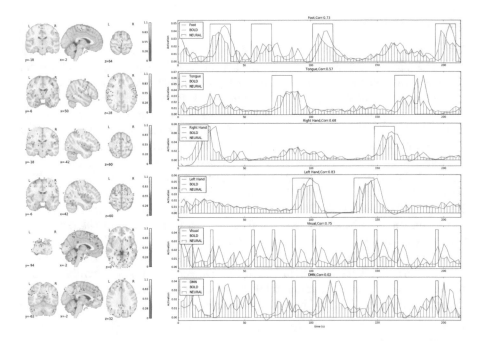

**Fig. 4.** The first four activated brain regions belong to the Sensory-Motor (SMN) network and relate to left and right foot, tongue, left and right hand stimuli, respectively. The fifth and sixth brain network are Visual Network (VN) and Default Mode Network (DMN), respectively. The blue lines represent parts of the block design of the motor task. The red and green line respectively represent neural activity and BOLD signal change in each individual network. We observe that the first five networks follow the respective block stimuli or visual cues given during the task. In contrast, the DMN network exhibits a neural activation profile unrelated to the task but remains strongly detectable.

from off to on instantly. A comprehensive GLM analysis of the several contrasts for the MSC dataset is available[4]. Our generative model estimates the underlying neuronal activity time course and does not know the underlying experimental design and it is therefore truly unsupervised. When compared to resting-state fMRI, matrix decomposition techniques are often used to extract brain networks [7,8]. The disadvantage of these techniques is that they require certain statistical assumptions to hold for the underlying sources of interest. Although effective at removing scanner- or physiological motion-related noise sources, they approximate statistical independence with non-linear functions. The chosen non-linearity has a strong influence on the obtained value distribution of voxel intensities and thus its spatial characteristics in the case of BOLD-fMRI. Additionally, the seminal work by [9] has shown that the commonly used algorithms in ICA, FastICA and InfoMax, tend to produce sparse rather than independent sources in simulated BOLD-fMRI data. This has shifted the focus in the recent years to sparse decomposition techniques [8,9] with weaker model assumptions. However, the these techniques require regularization and therefore hyper-parameter tuning with some sort of cross-validation, which is computationally expensive or intractable depending on the number of parameters to tune. For example, the most similar approach [2] to our proposed CHA generative model requires to tune a spatial and a temporal regularization hyper-parameter. Given that a 5 mm spatial smoothing kernel was applied, the value of the spatial regularization hyper-parameter is weakly motivated and could supposedly not be cross-validated due to the long individual subject processing time (5 h). The spatial smoothing kernel size hyper-parameter varies in studies between 3 to 12 mm. The kernel size thus determines partial-volume effects at both the intra- and inter-individual level. Unique subject network representations are therefore lost or compromised.

Combined, these limitations of common pre- and post-processing techniques potentially make subsequent analysis incomparable. Our proposed generative model delivers smooth spatial network maps with minimal pre-processing and without spatial regularization. Our technique exploits what is known about the neurovascular coupling between neuronal activity and blood perfusion. It does not require artificial spatial or temporal regularization but leverages biological prior information. The decomposition of a BOLD-fMRI scan into spatial network maps, functional time-courses and hemodynamic responses opens the door to sophisticated analyses at both group and single-subject level to examine neural activity in both, spontaneously activating networks and networks engaged in an extrinsic task. The obtained neural activity profile per network provides a means to quantify the activity during a 5 to 10-min scan, opening new analysis routes to clinical diagnosis and drug testing in neurological disorders.

Our technique relies on finding an optimal number of components similar to all matrix factorization techniques. This is challenging because the brain will form networks dynamically and dependent on the task. We find a good

---

[4] https://neurovault.org/collections/2447/.

approximation of individual interacting networks during task and rest by applying sophisticated cross-validation and model averaging strategies.

Our proposed Convolutional Hemodynamic Autoencoder will therefore provide new insights about the underlying cause of BOLD signal change in task and task-free BOLD-fMRI in current and future large fMRI cohort studies.

**Acknowledgements.** We would like to acknowledge the Wellcome Trust (210182/Z/18/Z, 101957/Z/13/Z), EPSRC (NS/A000027/1), MRC (MR/J01107X/1), and the Wolfson Foundation (PR/ylr/18575).

# References

1. Tavor, I., Jones, O.P., Mars, R., Smith, S., Behrens, T., Jbabdi, S.: Task-free MRI predicts individual differences in brain activity during task performance. Science **352**(6282), 216–220 (2016)
2. Karahanoğlu, F.I., Caballero-Gaudes, C., Lazeyras, F., Van De Ville, D.: Total activation: fMRI deconvolution through spatio-temporal regularization. Neuroimage **73**, 121–134 (2013)
3. Gordon, E.M., et al.: Precision functional mapping of individual human brains. Neuron **95**(4), 791–807 (2017)
4. Glorot, X., Bengio, Y.: Understanding the difficulty of training deep feedforward neural networks. In: Proceedings of the Thirteenth International Conference on Artificial Intelligence and Statistics, pp. 249–256 (2010)
5. Boutsidis, C., Gallopoulos, E.: Svd based initialization: a head start for nonnegative matrix factorization. Pattern Recognit. **41**(4), 1350–1362 (2008)
6. Kanagal, B., Sindhwani, V.: Rank selection in low-rank matrix approximations: a study of cross-validation for NMFs. In: Proceedings of the Conference on Advance Neural Information Processing, vol. 1, pp. 10–15 (2010)
7. Beckmann, C.F., Mackay, C.E., Filippini, N., Smith, S.M.: Group comparison of resting-state fMRI data using multi-subject ICA and dual regression. Neuroimage **47**(Suppl. 1), S148 (2009)
8. Abraham, A., Dohmatob, E., Thirion, B., Samaras, D., Varoquaux, G.: Extracting brain regions from rest fMRI with total-variation constrained dictionary learning. In: Mori, K., Sakuma, I., Sato, Y., Barillot, C., Navab, N. (eds.) MICCAI 2013. LNCS, vol. 8150, pp. 607–615. Springer, Heidelberg (2013). https://doi.org/10.1007/978-3-642-40763-5_75
9. Daubechies, I., et al.: Independent component analysis for brain fmri does not select for independence. Proc. Natl. Acad. Sci. **106**(26), 10415–10422 (2009)

# Identification of Multi-scale Hierarchical Brain Functional Networks Using Deep Matrix Factorization

Hongming Li, Xiaofeng Zhu, and Yong Fan$^{(\boxtimes)}$

Center for Biomedical Image Computing and Analytics,
Department of Radiology, Perelman School of Medicine,
University of Pennsylvania, Philadelphia, USA
Yong.Fan@uphs.upenn.edu

**Abstract.** We present a deep semi-nonnegative matrix factorization method for identifying subject-specific functional networks (FNs) at multiple spatial scales with a hierarchical organization from resting state fMRI data. Our method is built upon a deep semi-nonnegative matrix factorization framework to jointly detect the FNs at multiple scales with a hierarchical organization, enhanced by group sparsity regularization that helps identify subject-specific FNs without loss of inter-subject comparability. The proposed method has been validated for predicting subject-specific functional activations based on functional connectivity measures of the hierarchical multi-scale FNs of the same subjects. Experimental results have demonstrated that our method could obtain subject-specific multi-scale hierarchical FNs and their functional connectivity measures across different scales could better predict subject-specific functional activations than those obtained by alternative techniques.

**Keywords:** Brain functional networks · Multi-scale · Hierarchical
Subject-specific · Deep matrix factorization

## 1 Introduction

The human brain can be represented as a multiscale hierarchical network [1, 2]. However, existing functional brain network analysis studies of resting-state functional magnetic resonance imaging (rsfMRI) data typically define network nodes at a specific scale, based on regions of interest (ROIs) obtained from anatomical atlases or functional brain parcellations [3–5]. Recent work has demonstrated important subject-specific variation in the functional neuroanatomy of large-scale brain networks [6], emphasizing the need for tools which can flexibly adapt to individuals' variation while simultaneously maintaining correspondence for group-level analyses.

To capture subject-specific brain network without loss of inter-subject comparability, data-driven brain decomposition methods have been widely adopted to identify spatial intrinsic functional networks (FNs) and estimate functional network connectivity. In order to obtain subject-specific FNs from rsfMRI data of individual subjects while facilitating groupwise inference, independent vector analysis (IVA) and group-information guided ICA (GIGICA) methods have been proposed [7, 8]. More recently,

A. F. Frangi et al. (Eds.): MICCAI 2018, LNCS 11072, pp. 223–231, 2018.
https://doi.org/10.1007/978-3-030-00931-1_26

several methods have been proposed to discover FNs from rsfMRI data with non-independence assumptions [9–11], and directly work on individual subject fMRI data and simultaneously enforce correspondence across FNs of different subjects by assuming that loadings of corresponding FNs of different subjects follow Gaussian [9] or delta-Gaussian [10] distributions. Non-negative matrix decomposition techniques have been adopted to simultaneously compute subject-specific FNs for a group of subjects regularized by group sparsity in order to separate anti-correlated FNs properly so that anti-correlation information between them could be preserved [11, 12]. However, these methods are not equipped to characterize multi-scale hierarchical organization of the brain networks [2]. Although clustering and module detection algorithms could be adopted to detect the hierarchical organization of brain networks, their performance is hinged on the network nodes/FNs used [13, 14].

To address the aforementioned limitations of existing techniques, we develop a novel brain decomposition model based on a collaborative sparse brain decomposition approach [11] and deep matrix factorization techniques [15], aiming to identify subject-specific, multi-scale hierarchal FNs from rsfMRI data. Based on rsfMRI data and task activation maps of unrelated subjects from the HCP dataset [16], we have quantitatively evaluated our method for predicting subject-specific functional activations based on functional connectivity measures of the hierarchical multi-scale FNs of the same subjects. Experimental results have demonstrated that the multi-scale hierarchical subject-specific FNs identified by our method from rsfMRI data could better predict the subject-specific functional activations evoked by different tasks than those identified by alternative techniques.

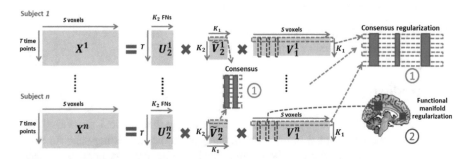

**Fig. 1.** Framework of the deep decomposition model for identifying multi-scale hierarchical subject-specific functional networks (a two-scale decomposition is illustrated).

## 2   Methods

A deep matrix decomposition framework is proposed to identify FNs at multiple scales as schematically illustrated in Fig. 1. Particularly, a deep semi-nonnegative matrix factorization is adopted to jointly detect a hierarchy of FNs from fine to coarse spatial scales in a data-driven way, and a group sparsity regularization term is adopted for FNs of different subjects at each scale to ensure the subject-specific FNs to share similar spatial patterns. Besides enforcing the groupwise correspondence of FNs across subjects, the group sparsity term also encourages FNs at the finest scale to have sparse spatial patterns and FNs at coarse scales to comprise functionally correlated FNs at finer scales. Our decomposition model is further enhanced by a data locality regularization term that makes the decomposition robust to imaging noise and improves spatial smoothness and functional coherences of the subject specific FNs.

## 2.1 Deep Semi-nonnegative Matrix Factorization for Brain Decomposition

Given rsfMRI data $X^i \in R^{T \times S}$ of subject $i$, consisting of $S$ voxels and $T$ time points, we aim to find $K_j$ nonnegative FNs $V_j^i = \left( V_{j:k,s}^i \right) \in R_+^{K_j \times S}$ and their corresponding time courses $U_j^i = \left( U_{j:t,k}^i \right) \in R^{T \times K_j}$ at $j = 1, \ldots, h$ scales, so that $X^i \approx U_1^i V_1^i$; $X^i \approx U_2^i \tilde{V}_2^i V_1^i$, $V_2^i = \tilde{V}_2^i V_1^i$; $\cdots$; $X^i \approx U_h^i \tilde{V}_h^i \cdots V_1^i$, $V_h^i = \tilde{V}_h^i \cdots V_1^i$. The FNs $V_j^i$ at $1 \leq j \leq h$ scales are constrained to have a hierarchical structure and to be non-negative so that each FN does not contain any anti-correlated functional units. A deep semi-nonnegative matrix factorization (DSNMF) framework similar to [15] is adopted to identify the multi-scale hierarchical FNs by optimizing

$$\min_{\{U_j^i, \tilde{V}_j^i\}} \left\| X^i - U_h^i \tilde{V}_h^i \cdots \tilde{V}_1^i \right\|_F^2, s.t. \tilde{V}_j^i \geq 0, V_1^i = \tilde{V}_1^i, \forall 1 \leq j \leq h. \tag{1}$$

The FNs at different scales are represented by $V_h^i = \tilde{V}_h^i \cdots V_1^i, \forall 1 \leq j \leq h$ for subject $i$, and the FNs at 2 consecutive scales are linked hierarchically according to weights determined during the joint decomposition. The decomposition model does not constrain time courses $U_h^i$ to be non-negative so that it can be applied to preprocessed fMRI data with negative values.

## 2.2 DSNMF Based Collaborative Brain Decomposition

Given a group of $n$ subjects, each having fMRI data $X^i \in R^{T \times S}$, $i = 1, \ldots, n$, we identify subject-specific, multiscale hierarchical FNs by optimizing a joint model with integrated data fitting and regularization terms as illustrated in Fig. 1:

$$\min_{\{U_j^i, \tilde{V}_j^i\}} \sum_{i=1}^{n} \left\| X^i - U_h^i \tilde{V}_h^i \cdots \tilde{V}_1^i \right\|_F^2 + \sum_{j}^{h} \lambda_{c,j} R_{c,j} + \lambda_M \sum_{i=1}^{n} R_M^i,$$
$$s.t. \tilde{V}_j^i \geq 0, \left\| \tilde{V}_{j:k_j, \cdot}^i \right\|_\infty = 1, \forall 1 \leq k_j \leq K_j, \forall 1 \leq i \leq n, 1 \leq j \leq h, \tag{2}$$

where $\lambda_{c,j} = \alpha \cdot \frac{n \cdot T}{K_j}$ and $\lambda_M = \beta \cdot \frac{T}{K_1 \cdot n_M}$ are parameters for balancing data fitting and regularization terms, $T$ is the number of time points, $K_j$ is the number of FNs at scale $j$, $n_M$ is the number of spatially neighboring voxels at voxel level, $\alpha$ and $\beta$ are 2 parameters, $R_{c,j}$ and $R_M^i$ are regularization terms. Particularly, $R_{c,j}$ is an inter-subject group sparsity term that enforces FNs of different subjects to have common spatial structures at the same scale $j$. The group sparsity regularization term is defined as

$$R_{c,j} = \sum_{k=1}^{K_j} \left\| \widetilde{V}_{j:k,\cdot}^{1,\ldots,n} \right\|_{2,1} = \sum_{k=1}^{K_j} \frac{\sum_{s=1}^{S_j} \left( \sum_{i=1}^{n} \left( \tilde{V}_{j:k,s}^i \right)^2 \right)^{1/2}}{\left( \sum_{s=1}^{S_j} \sum_{i=1}^{n} \left( \tilde{V}_{j:k,s}^i \right)^2 \right)^{1/2}} \text{ for each row of } \tilde{V}_j^i, \forall 1 \le i \le n,$$

$1 \le j \le h$, where $S_j = K_{j-1}$ when $j > 1$ and $S_1 = S$. The group sparsity regularization (term 1 in Fig. 1) enforces corresponding FNs of different subjects to have non-zero elements at the same spatial locations. Moreover, it encourages FNs to have spatially localized loadings. It is worth noting that the group sparsity term does not force different FNs to be non-overlapping, thus certain functional units may be included in multiple FNs simultaneously at each scale. We also adopt a data locality regularization term (term 2 in Fig. 1), $R_M^i$, to encourage spatial smoothness and functional coherence of the FNs using a graph regularization technique [17] at the finest spatial scale, which is defined as $R_M^i = Tr\left( V_1^i L_M^i \left( V_1^i \right)' \right)$, where $L_M^i = D_M^i - W_M^i$ is a Laplacian matrix for subject $i$, $W_M^i$ is a pairwise affinity matrix to measure spatial closeness or functional similarity between voxels, and $D_M^i$ is its corresponding degree matrix, the affinity between each pair of spatially connected voxels is calculated as $\left( 1 + corr\left( X_{\cdot,a}^i, X_{\cdot,b}^i \right) \right)/2$, where $corr\left( X_{\cdot,a}^i, X_{\cdot,b}^i \right)$ is the Pearson correlation coefficient between their rsfMRI signals, and others are set to zero so that $W_M^i$ has a sparse structure.

We optimize the joint model using an alternative update strategy. When $U_h^i$ and $\tilde{V}_k^i$ ($k \neq j$) are fixed, $\tilde{V}_j^i$ is updated as

$$\tilde{V}_j^i = \tilde{V}_j^i \odot \sqrt{ \frac{ [U_j^{i'} X^i] + \overline{V}_j^{i'} + [U_j^{i'} U_j^i]_- \tilde{V}_j^i \overline{V}_j^{i'} \overline{V}_j^{i'} + \lambda_{c,j} \tilde{V}_j^i \odot G_j^{L21} / (G_j^{L2})^3 }{ [U_j^{i'} X^i]_- \overline{V}_j^{i'} + [U_j^{i'} U_j^i]_+ \tilde{V}_j^i \overline{V}_j^i \overline{V}_j^{i'} + \lambda_{c,j} \tilde{V}_j^i / (G_j \odot G_j^{L2}) } } \text{ if } j > 1$$

$$\text{and } \tilde{V}_1^i = \tilde{V}_1^i \odot \sqrt{ \frac{ [U_1^{i'} X^i]_+ + [U_1^{i'} U_1^i]_- \tilde{V}_1^i + \lambda_M \tilde{V}_1^i W_M^i + \lambda_{c,1} \tilde{V}_1^i \odot G_1^{L21} / (G_1^{L2})^3 }{ [U_1^{i'} X^i]_- + [U_1^{i'} U_1^i]_+ \tilde{V}_1^i + \lambda_M \tilde{V}_1^i D_M^i + \lambda_{c,1} \tilde{V}_1^i / \left( G_1 \odot G_1^{L2} \right) } } \text{ if } j = 1$$

$$G_{j:k,\cdot}^{L21} = repmat\left( \sum_{s=1}^{S} \left( \sum_{i=1}^{n} \left( \tilde{V}_{j:k,s}^i \right)^2 \right)^{1/2}, 1, S \right),$$

$$G_{j:k,\cdot}^{L2} = repmat\left( \left( \sum_{s=1}^{S} \sum_{i=1}^{n} \left( \tilde{V}_{j:k,s}^i \right)^2 \right)^{1/2}, 1, S \right), \qquad (3)$$

$$G_{j:k,s} = \left( \sum_{i=1}^{n} \left( \tilde{V}_{j:k,s}^i \right)^2 \right)^{1/2}, \overline{V}_j^i = \tilde{V}_{j-1}^i \tilde{V}_{j-2}^i \cdots \tilde{V}_1^i,$$

where $\odot$ denotes element-wise multiplication, $repmat(b, r, c)$ denotes matrix obtained by replicating $r$ and $c$ copies of vector $b$ in the row and column dimensions, $[a]_+ = \frac{abs(a)+a}{2}$, $[a]_- = \frac{abs(a)-a}{2}$, $M_{j:k,\cdot}$ denotes the $k$-th row of matrix $M_j$, and $M_{j:k,s}$ denotes the $(k, s)$-th element in the matrix. $\tilde{V}_j^i$ is normalized by the row-wise maximum value along the row dimension after each update iteration.

When $\tilde{V}_k^i$, $1 \leq k \leq j$ are fixed, $U_j^i$ is updated as

$$U_j^i = X_j^i \left( \tilde{V}_j^i \tilde{V}_{j-1}^i \cdots \tilde{V}_1^i \right)^{\dagger}, \tag{4}$$

where $M^{\dagger}$ denotes the Moore-Penrose pseudo-inverse of matrix $M$.

To expedite the convergence of the optimization, a pre-training step at group level is adopted before the joint optimization. In particular, we compute $(U_1, V_1) \leftarrow sparseNMF(X)$ first, where $X$ denotes the temporal concatenated data of $\{X^i, i = 1, .., n\}$, and $(U_2, \tilde{V}_2) \leftarrow sparseNMF(U_1), \cdots, (U_h, \tilde{V}_h) \leftarrow sparseNMF(U_{h-1})$, respectively. $\{ \tilde{V}_1, \tilde{V}_2, \ldots, \tilde{V}_h \}$ are then used to initialize $\{ \tilde{V}_j^i, i = 1, \ldots, n, j = 1, \ldots, h \}$ for the joint optimization. $\{ \tilde{V}_1, \tilde{V}_2, \ldots, \tilde{V}_h \}$ contains a hierarchical structure with overlapping as they are obtained by decomposition in a greedy manner. Once the initialization is done, all FNs at different scales are optimized jointly. We set the parameters $\alpha$ to 1, and $\beta$ to 10 according to [11] in the present study.

## 3    Experimental Results

We evaluated our method based on rsfMRI data and task activation maps of 40 unrelated subjects obtained from the Human Connectome Project (HCP) [16], aiming to evaluate the performance of predicting task-evoked activation responses based on functional connectivity measures of FNs at multiple scales. The FNs derived from rsfMRI data have demonstrated promising performance for predicting task-evoked brain activations [18].

The proposed decomposition model was applied to the minimal-preprocessed, cortical gray-coordinates based rsfMRI data. The number of scales was set to 3, and the number of FNs at the first scale was set to 90, which was estimated by MELODIC automatically [19], the numbers of FNs at the 2nd and 3rd scales were set to 50 and 25 respectively, which were decreased by half approximately along the scales.

We compared the proposed model with two alternative decomposition strategies: multi-scale decomposition performed independently at different scales, and greedy agglomerative hierarchical multi-scale decomposition, with the same setting of scales and the same numbers of FNs. For the independent decomposition, the collaborative sparse decomposition model was adopted to obtain subject-specific FNs independent at 3 scales (with independent initialization). For the agglomerative one, the initialization obtained by greedy decomposition as described Sect. 2.2 was adopted for each scale, but the final decomposition at each scale was obtained using the collaborative sparse model separately.

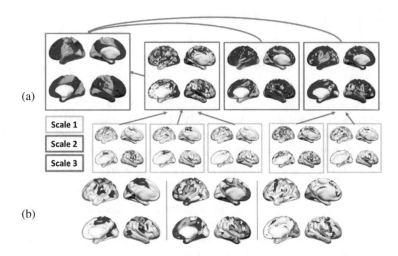

**Fig. 2.** Multi-scale functional networks identified by different methods. (a) A hierarchy of FNs obtained by the proposed method corresponding to sensorimotor networks. FNs at different scales are denoted by bounding boxes in different colors. (b) Sensorimotor regions in separate FNs at the coarsest scale (the $3^{rd}$) identified independently of other scales.

### 3.1 Multi-scale Brain Functional Networks with a Hierarchical Organization

An example hierarchy of FNs (mean of 40 subjects) corresponding to sensorimotor function obtained by the proposed method is illustrated in Fig. 2(a). The FN at the $3^{rd}$ scale (top left) comprised the sensorimotor networks and part of visual networks, and was a weighted composition of FNs at the $2^{nd}$ scale while the FNs at the $2^{nd}$ scale were composed of FNs at the $1^{st}$ scale. This example hierarchy of FNs illustrated that FNs corresponding to sensorimotor function gradually merged from fine to coarse scales in the hierarchy. However, no clear hierarchical organization was observed for FNs independently identified at different scales. Particularly, as shown in Fig. 2(b) sensorimotor regions appeared in separate FNs at the coarsest scale (the $3^{rd}$), instead of forming a single FN. We postulated that the independent decomposition at different scales favored to better data-fitting and therefore was affected by data noise, while the joint decomposition model was more robust to data noise, facilitating accurate identification of FNs with coherent functions, such as the FN comprising sensorimotor regions shown in Fig. 2(a).

### 3.2 Prediction of Task Evoked Activations Based on Multi-scale FN Connectivity

As no ground truth is available for FNs derived from rsfMRI data, we evaluated the multi-scale hierarchical FNs for predicting functional activations evoked by different tasks based on their functional connectivity measures with an assumption that better FNs could provide more discriminative information for predicting the brain activations.

We also compared FNs obtained using different strategies in terms of their prediction performance.

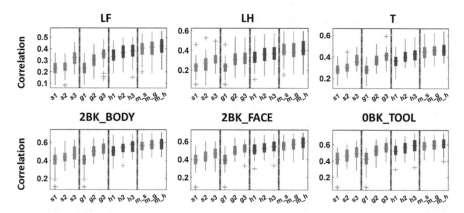

**Fig. 3.** Quantitative comparison of prediction performance for functional activations of different tasks (top: motor task events including left foot, left hand, and tongue; bottom: working memory task events including 2bk_body, 2bk_face, and 0bk_tool). s1 to s3, g1 to g3, and h1 to h3 denote models trained based on FNs at single scale identified independently, in a greedy agglomerative manner, and hierarchically by our method, respectively, while m_s, m_g and m_h denote models trained based on corresponding multi-scale FNs. The models built upon FNs obtained by our method performed significantly better than others (Wilcoxon signed rank test, $p < 0.02$).

Particularly, a whole brain voxelwise functional connectivity (FC) map was obtained for each FN by computing voxelwise Pearson correlation coefficient between the FN's time course and every cortical gray-coordinate's time course of the rsfMRI data. All the FC maps were then transferred to z-score maps using Fisher Z transformation. All the z-score values of FNs on each cortical gray-coordinate were used as features to predict its activation measures under different tasks of the same subject. Similar to [18], the whole cortical surface was divided into 90 parcels according to FNs obtained at the finest scale, and one ordinary least square model was trained for each parcel and every task event. The prediction performance was evaluated using a leave-one-subject-out cross-validation, where one subject's activation was predicted by a model built upon data of the remaining 39 subjects. The prediction was conducted using single-scale FNs (90, 50, 25) or multi-scale FNs (165) obtained using different strategies respectively, and the prediction accuracy was evaluated as the Pearson correlation coefficients between the predicted and real activation maps of 47 task events from 7 tasks categories.

The prediction performance of 6 randomly selected task events is illustrated in Fig. 3. For all task events and FNs identified by different strategies, the prediction models built upon multi-scale FNs outperformed all prediction models built upon any single scale FNs alone, indicating that multi-scale FNs could provide complementary information for the task activation prediction. The prediction models built upon multi-scale hierarchical FNs obtained by our method had significantly better performance than those build upon multi-scale FNs obtained by either the independent decomposition or the greedy agglomerative decomposition (Wilcoxon signed rank test, $p < 0.02$), indicating that the joint optimization of multi-scale hierarchical FNs could benefit from each other and characterize the intrinsic FNs better.

# 4    Conclusions

In this study, we have developed a deep decomposition model to identify multi-scale, hierarchical, subject-specific FNs with group level correspondence across different subjects. Our method is built upon deep semi-nonnegative matrix factorization framework, enhanced by a group sparsity regularization and graph regularization for maintaining inter-subject correspondence and better functional coherence. Experimental results based on rsfMRI data and task activation maps of the same subjects have demonstrated that the multi-scale hierarchical subject-specific FNs could capture informative intrinsic functional networks and improve the prediction performance of task activations evoked by different tasks, compared to FNs identified at different scales independently or in a greedy agglomerative way. In conclusion, our method provides an improved solution for characterizing subject-specific, multi-scale hierarchical organization of the brain functional networks.

**Acknowledgements.** This work was supported in part by National Institutes of Health grants [CA223358, EB022573, DK114786, DA039215, and DA039002].

# References

1. Doucet, G., et al.: Brain activity at rest: a multiscale hierarchical functional organization. J. Neurophysiol. **105**(6), 2753–2763 (2011)
2. Park, H.J., Friston, K.: Structural and functional brain networks: from connections to cognition. Science **342**(6158), 1238411 (2013)
3. Bullmore, E., Sporns, O.: Complex brain networks: graph theoretical analysis of structural and functional systems. Nat. Rev. Neurosci. **10**(3), 186–198 (2009)
4. Honnorat, N., et al.: GraSP: geodesic graph-based segmentation with shape priors for the functional parcellation of the cortex. Neuroimage **106**, 207–221 (2015)
5. Li, H., Fan, Y.: Individualized brain parcellation with integrated functional and morphological information. In: 2016 IEEE 13th International Symposium on Biomedical Imaging (ISBI) (2016)
6. Satterthwaite, T.D., Davatzikos, C.: Towards an individualized delineation of functional neuroanatomy. Neuron **87**(3), 471–473 (2015)
7. Du, Y., Fan, Y.: Group information guided ICA for fMRI data analysis. Neuroimage **69**, 157–197 (2013)
8. Lee, J.H., et al.: Independent vector analysis (IVA): multivariate approach for fMRI group study. Neuroimage **40**(1), 86–109 (2008)

9. Abraham, A., Dohmatob, E., Thirion, B., Samaras, D., Varoquaux, G.: Extracting brain regions from rest fmri with total-variation constrained dictionary learning. In: Mori, K., Sakuma, I., Sato, Y., Barillot, C., Navab, N. (eds.) MICCAI 2013. LNCS, vol. 8150, pp. 607–615. Springer, Heidelberg (2013). https://doi.org/10.1007/978-3-642-40763-5_75

10. Harrison, S.J., et al.: Large-scale probabilistic functional modes from resting state fMRI. Neuroimage **109**, 217–231 (2015)

11. Li, H., Satterthwaite, T.D., Fan, Y.: Large-scale sparse functional networks from resting state fMRI. Neuroimage **156**, 1–13 (2017)

12. Li, H., Satterthwaite, T., Fan, Y.: Identification of subject-specific brain functional networks using a collaborative sparse nonnegative matrix decomposition method. In: 2016 IEEE 13th International Symposium on Biomedical Imaging (ISBI) (2016)

13. Li, H., Fan,Y.: Hierarchical organization of the functional brain identified using floating aggregation of functional signals. In: 2014 IEEE 11th International Symposium on Biomedical Imaging (ISBI) (2014)

14. Li, H., Fan, Y.: Functional brain atlas construction for brain network analysis. In: SPIE Medical Imaging. SPIE (2013)

15. Trigeorgis, G., et al.: A deep matrix factorization method for learning attribute representations. IEEE Trans. Pattern Anal. Mach. Intell. **39**(3), 417–429 (2017)

16. Glasser, M.F., et al.: The minimal preprocessing pipelines for the human connectome project. Neuroimage **80**, 105–124 (2013)

17. Cai, D., et al.: Graph regularized nonnegative matrix factorization for data representation. IEEE Trans. Pattern Anal. Mach. Intell. **33**(8), 1548–1560 (2011)

18. Tavor, I., et al.: Task-free MRI predicts individual differences in brain activity during task performance. Science **352**(6282), 216–220 (2016)

19. Jenkinson, M., et al.: Fsl. Neuroimage **62**(2), 782–790 (2012)

# Identification of Temporal Transition of Functional States Using Recurrent Neural Networks from Functional MRI

Hongming Li and Yong Fan$^{(\boxtimes)}$

Center for Biomedical Image Computing and Analytics, Department of
Radiology, Perelman School of Medicine, University of Pennsylvania,
Philadelphia, USA
Yong.Fan@uphs.upenn.edu

**Abstract.** Dynamic functional connectivity analysis provides valuable information for understanding brain functional activity underlying different cognitive processes. Besides sliding window based approaches, a variety of methods have been developed to automatically split the entire functional MRI scan into segments by detecting change points of functional signals to facilitate better characterization of temporally dynamic functional connectivity patterns. However, these methods are based on certain assumptions for the functional signals, such as Gaussian distribution, which are not necessarily suitable for the fMRI data. In this study, we develop a deep learning based framework for adaptively detecting temporally dynamic functional state transitions in a data-driven way without any explicit modeling assumptions, by leveraging recent advances in recurrent neural networks (RNNs) for sequence modeling. Particularly, we solve this problem in an anomaly detection framework with an assumption that the functional profile of one single time point could be reliably predicted based on its preceding profiles within a stable functional state, while large prediction errors would occur around change points of functional states. We evaluate the proposed method using both task and resting-state fMRI data obtained from the human connectome project and experimental results have demonstrated that the proposed change point detection method could effectively identify change points between different task events and split the resting-state fMRI into segments with distinct functional connectivity patterns.

**Keywords:** Brain fMRI · Functional dynamics · Change point detection
Recurrent neural networks

## 1 Introduction

Brain network analysis based on intrinsic functional connectivity (FC) derived from resting-state functional magnetic resonance imaging (fMRI) data enables us to investigate both static FC, estimated based on the entire fMRI scan, and dynamic FC, varying over the course of a fMRI scan [1, 2].

Existing studies of dynamic FC typically explore temporal dynamics based on network nodes defined by regions of interests (ROIs) based on anatomical atlases or

© Springer Nature Switzerland AG 2018
A. F. Frangi et al. (Eds.): MICCAI 2018, LNCS 11072, pp. 232–239, 2018.
https://doi.org/10.1007/978-3-030-00931-1_27

functional data based brain parcellations, either using sliding-window (SW) based methods [2, 3] or splitting the entire fMRI scan into segments with quasi-static FC patterns [4–6]. In the SW methods, dynamic FC measures are estimated based on data points within multiple time windows, each of them with a fixed width but different starting positions shifted in time by a fixed number of data points. Notably, the SW methods' performance is hinged on the window parameters. Furthermore, it may not be an optimal way to use windows with a fixed width over the entire fMRI scan since the FC states may change at unpredictable intervals [7, 8]. To overcome limitations of the SW methods, a variety of methods have been developed to automatically split the entire fMRI scan into distinct segments, including Dynamic Connectivity Regression (DCR) methods [4], Bayesian inference based methods [6], Vector Autoregressive (VAR) model based methods [9], and statistical test based methods [5]. Different from the SW methods, these methods adaptively detect fMRI signal transitions to split the entire fMRI scan into segments. However, these methods are based on certain assumptions for the fMRI data, such as Gaussian distribution and VAR model, which are not necessarily well suited for fMRI data.

In this study, we develop a deep learning based framework for adaptively detecting dynamic functional state transitions in a data-driven way without any explicit model assumptions, by leveraging recent advances in deep learning based sequence modeling. Deep learning techniques, particularly recurrent neural networks (RNNs) with a long short term memory (LSTM) [10] structure, have achieved remarkable advances in sequence modeling [11], indicating that LSTM-RNNs might be suitable for characterizing fMRI data too. The basic assumption of the proposed deep learning based model is that the functional profile of one single time point could be reliably predicted based on its preceding profiles within a stable functional state, while large prediction errors would occur around change points of functional states. Given the predicted and real functional profiles, the change points are identified as anomaly time points with prediction errors larger than a predefined threshold value. We have applied the proposed method to both resting-state and task fMRI data obtained from the human connectome project (HCP) [12, 13], and experimental results have demonstrated that the proposed method could obtain better detection accuracy compared with state-of-the-art alternative methods on the task fMRI data, and also effectively detect change points that split the resting-state fMRI data into segments with significantly different functional connectivity patterns.

## 2 Methods

To identify temporal functional state transitions from fMRI data, recurrent neural networks (RNNs) with a LSTM structure [10] are trained based on functional profiles from a training cohort, where the functional profiles are extracted using a functional brain decomposition technique [14, 15]. Differences between the predicted functional profiles by the LSTM RNNs and the real ones on a validation cohort are then adopted to determine the optimal threshold for identifying the change points on the testing cohort. The overall framework is schematically illustrated in Fig. 1(a).

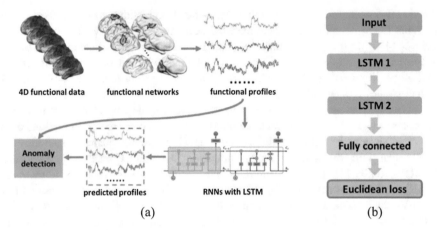

4D functional data    functional networks    functional profiles

Input

LSTM 1

LSTM 2

Fully connected

Anomaly detection

predicted profiles    RNNs with LSTM

Euclidean loss

(a)    (b)

**Fig. 1.** Schematic illustration of our deep learning based change point detection framework. (a) The overall architecture of the proposed model, (b) the RNNs used in the model.

## 2.1    Prediction of Functional Profiles Using LSTM RNNs

Given a group of $n$ subjects, each having a fMRI scan $X^i \in R^{T \times S}$, $i = 1, 2, \dots, n$, consisting of $S$ voxels and $T$ time points, we first obtain $K$ functional networks $V^i \in R_+^{K \times S}$ and its corresponding functional time courses $U^i \in R^{T \times K}$ for each subject using a collaborative sparse brain decomposition method [14, 15] which could identify subject-specific functional networks with group level correspondence for better characterizing the intrinsic functional connectivity at an individual subject level. The functional time courses $U^i$, $i = 1, 2, \dots, n$, are then used as training data to build a LSTM RNNs model for predicting functional profiles.

A LSTM RNNs model $M_{lstm}$ is built to predict the functional profile $U^i(t, \cdot)$ at each time point $t$ using its preceding functional profiles $\{U^i(t_p, \cdot), 1 \le t_p < t\}$ so that

$$U^i(t, \cdot) \approx \tilde{U}^i(t, \cdot) = M_{lstm}\big(\{U^i(t_p, \cdot), 1 \le t_p < t\}\big). \tag{1}$$

Particularly, a LSTM RNNs model with 2 hidden layers is adopted, as shown in Fig. 1(b). Each hidden layer has 256 hidden nodes. A fully connected layer with $K$ output nodes is adopted for predicting the functional profiles. The Euclidean distance between real and predicted functional profiles is used as the objective function to optimize the RNNs model. We implement the model using Tensorflow [16].

## 2.2    Prediction Based Change Point Detection

Given the trained RNNs model $M_{lstm}$, we predict the functional profile for each time point $t(t > 1)$ of every subject $i$, and the prediction error $E^i$ is measured by the deviation from its real functional profiles

$$E^i(t) = \left\| U^i(t, \cdot) - \tilde{U}^i(t, \cdot) \right\|_2. \tag{2}$$

Assuming that the functional profiles could be reliably predicted for each time point based on its preceding functional signals within a quasi-stable functional state, we first detect the anomaly time points as those with relatively large prediction errors

$$A^i(t) = 1 \ if \ E^i(t) > T_v^i, \ and \ A^i(t) = 0 \ \text{otherwise}, \tag{3}$$

where $A^i$ is the vector of length $T$ indicating that the $t$-th time point is one anomaly point if $A^i(t)$ equals to 1, and $T_v^i$ is the threshold value for identifying the predicted anomaly time points, to be determined as

$$T_v^i = mean(E^i) + \lambda * std(E^i), \tag{4}$$

where $mean(x)$ and $std(x)$ denotes the mean and standard deviation of the vector $x$, $\lambda$ is a parameter used to adjust the threshold value.

Due to relatively low signal to noise ratio (SNR) of functional signals from fMRI data, the prediction errors evaluated at individual time points may oscillate a lot even for two consecutive time points. To improve the robustness and specificity of the identified change points, we apply a 1D convolutional operation to $E^i$ as

$$sE^i = conv1D(E^i, w(\sigma)), \tag{5}$$

where $sE^i$ is a smoothed prediction error vector, $w$ is a Gaussian kernel with standard deviation $1/\sigma$, and a larger $\sigma$ corresponding to a narrower kernel. A change point is finally identified as the one with a local maximum $sE^i$ while its $E^i$ value is larger than the threshold $T_v^i$, i.e.,

$$C^i(t) = \begin{cases} 1, \ if \ A^i(t) = 1 \ and \ (sE^i, t) \ is \ a \ Local \ Maximum \\ 0, \ otherwise \end{cases}, \tag{6}$$

where $C^i$ is the vector of length $T$ indicating that the $t$-th time point is one functional change point if $C^i(t)$ equals to 1.

## 3   Experimental Results

We evaluated the proposed method based on both task and resting-state fMRI data of 490 subjects from the HCP [12, 13]. In this study, we focused on two tasks, including motor and working memory tasks. The motor task consisted of 6 events, including 5 movement events, namely left foot (LF), left hand (LH), right foot (RF), right hand (RH), tongue (T), and additionally 1 cue event (CUE) prior to each movement event. The working memory task consisted of 2-back and 0-back task blocks of tool, place, face and body, and a fixation period. The motor task fMRI scan of each subject contained 284 time points, while the working memory fMRI scan contained 405 time points. The resting-state fMRI scan of each subject contained 1200 time points. The fMRI data acquisition and task paradigm were detailed in [12, 13].

We applied the collaborative sparse brain decomposition method [14, 15] to the resting-state fMRI data of 490 subjects and identified 90 subject-specific functional networks (FNs) and their corresponding resting-state time courses. The number of FNs was automatically estimated by MELODIC of FSL [17]. The subject-specific FNs were then used to extract the time courses of task fMRI data for each subject. The proposed change point detection method was then applied to the motor task data, working memory data, and resting-state data respectively. Particularly, we split the whole dataset into training, validation, and testing datasets. The training dataset consisted of data of 400 subjects for training a LSTM-RNNs model for each task, the validation dataset included data of 50 subjects for selecting the optimal $\lambda$ and $\sigma$, and the testing dataset consisted of data of the remaining 40 subjects.

For the task fMRI data, the real change points were defined as the time points when each task event started or ended. The performance of change point detection was quantitatively evaluated using the distance between predicted change points and real ones. For each real change point, the distance to its nearest predicted change point was calculated, and the mean distance across all real change points was used to evaluate the sensitivity of the detection (error_sen). Moreover, the same measure was also calculated between each predicted change point and its nearest real change point to evaluate the specificity of the detection (error_spec). We have compared the proposed method with a Bayesian inference based method [6] in terms of their performance on the task fMRI data. As the Bayesian inference based method could achieve better performance on functional connectivity data with a relative small number of nodes, we picked up the motor and working memory related FNs (13 out of 90, and 24 out of 90 respectively) and applied the two change point detection methods to their functional profiles.

As no ground truth about change points is available on the resting-state fMRI data, two-sample covariance matrix testing [18] was adopted to examine if functional connectivity patterns of two consecutive data segments split by the detected change points were significantly different, and the differences were used as surrogate measures for evaluating the proposed method based on the resting-state fMRI data. The functional profiles of 90 FNs were used for change point detection on the resting-state fMRI data.

### 3.1   Change Point Detection on Task fMRI Data

We first selected the optimal parameters $\lambda$ and $\sigma$ using the validation dataset based on the error_sen and error_spec measures, as shown in Fig. 2(a) for the motor task fMRI data. Figure 2(a, top) demonstrates that the error_sen decreased as $\sigma$ increased, a larger $\sigma$ corresponded a narrower smooth kernel, which led to noisy prediction error vectors and generated more change points. While generating more change points would improve the sensitivity of the detection, its specificity would decrease as shown in Fig. 2(a, bottom). The pattern of prediction errors in term of $\lambda$ had a similar trend as $\sigma$'s. We set $\sigma$ to 6, and $\lambda$ to 0 for the task fMRI data, taking into consideration both error_sen and error_spec, and applied the proposed method to the testing data.

The prediction performance on the motor task fMRI data of two randomly selected testing subjects are illustrated in Fig. 2(b). Most transitions between two consecutive

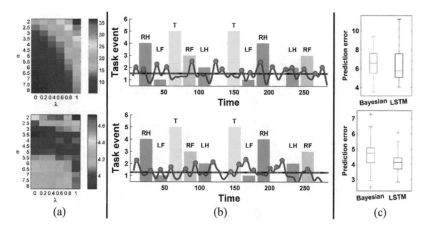

**Fig. 2.** Detection performance on motor task fMRI dataset. (a) Prediction errors on the validation dataset using different parameter settings, top: error_sen, bottom: error_spec, lower is better. (b) Identified change points of two randomly selected testing subjects: the *x*-axis denotes the time points, bar plots with different colors denote different task events ongoing at the located temporal interval, the blue curve is the smoothed prediction error, the dashed black line denotes the threshold used to identify the change points, and red circles denote the identified change points. (c) Prediction errors of 40 testing subjects, top: error_sen, bottom: error_spec.

task events were detected, and the identified change points were largely matched with the starting and ending time points of each task event. The overall prediction performance on the testing dataset is illustrated in Fig. 2(c), our method obtained lower error_sen than the Bayesian method, and the error_spec was significantly lower (Wilcoxon signed rank test, $p < 0.05$).

The prediction performance on the testing dataset of working memory fMRI is illustrated in Fig. 3. The proposed method also obtained better performance on the working memory dataset than the Bayesian inference based method in terms of both detection sensitivity and specificity (Wilcoxon signed rank test, $p < 0.05$).

We also evaluated our method based on the real change points adjusted by a hemodynamic lag of 6 s for the task fMRI data, and our method outperformed the Bayesian inference based method.

## 3.2   Change Point Detection on Resting-State fMRI Data

We finally evaluated the proposed method using the testing dataset of resting-state fMRI data. As no ground truth about change points is available for selecting the optimal parameters $\lambda$ and $\sigma$, we set $\lambda$ to 1 and $\sigma$ to 3, aiming to detect a small number of change points and improve the prediction specificity. The identified change points on the resting-state fMRI data of one randomly selected testing subject are illustrated in Fig. 4(top). The functional connectivity matrices of temporally dynamic segments between consecutive change points, as shown in Fig. 4(bottom), demonstrated that the functional connectivity patterns of consecutive segments were statistically significant

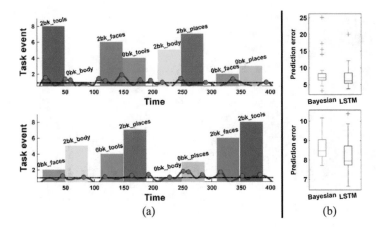

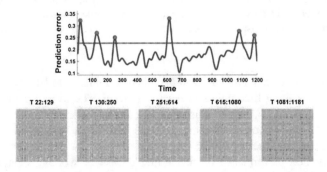

**Fig. 3.** Detection performance on the testing dataset of working memory task fMRI. (a) Identified change points of two randomly selected testing subjects, (b) Prediction error of 40 testing subjects, top: error_sen, bottom: error_spec.

**Fig. 4.** Change point detection on resting-state fMRI data of one randomly selected testing subject. (top) the blue curve is the smoothed prediction error, the dashed black line denotes the threshold used to identify the change points, and red circles denote the identified change points, (bottom) the functional connectivity matrices of temporal segments split by change points.

(two-sample covariance matrix testing, $p < 0.05$), indicating that the change points detected by our method were functionally meaningful.

## 4   Discussion and Conclusions

We propose a LSTM RNNs based change point detection framework for identifying change points of temporal functional state transitions underlying different brain cognitive processes. Different from most of the existing change point detection methods, our learning based prediction model does not rely on any model assumption regarding the underlying functional profiles. The experimental results on the task fMRI data have demonstrated that our method could identify functionally meaningful change points with higher accuracy than a state-of-the-art method. The experimental results on the resting-state fMRI data further demonstrated that our method could effectively capture temporally dynamic functional states with distinct connectivity patterns.

**Acknowledgements.** This work was supported in part by National Institutes of Health grants [CA223358, EB022573, DK114786, DA039215, and DA039002] and a NVIDIA Academic GPU grant.

# References

1. Bullmore, E., Sporns, O.: Complex brain networks: graph theoretical analysis of structural and functional systems. Nat. Rev. Neurosci. **10**(3), 186–198 (2009)
2. Calhoun, V.D., et al.: The chronnectome: time-varying connectivity networks as the next frontier in fMRI data discovery. Neuron **84**(2), 262–274 (2014)
3. Hutchison, R.M., et al.: Dynamic functional connectivity: promise, issues, and interpretations. Neuroimage **80**, 360–378 (2013)
4. Cribben, I., et al.: Dynamic connectivity regression: determining state-related changes in brain connectivity. Neuroimage **61**(4), 907–920 (2012)
5. Jeong, S.O., Pae, C., Park, H.J.: Connectivity-based change point detection for large-size functional networks. Neuroimage **143**, 353–363 (2016)
6. Zhang, J., et al.: Inferring functional interaction and transition patterns via dynamic Bayesian variable partition models. Hum. Brain Mapp. **35**(7), 3314–3331 (2014)
7. Shakil, S., Lee, C.H., Keilholz, S.D.: Evaluation of sliding window correlation performance for characterizing dynamic functional connectivity and brain states. Neuroimage **133**, 111–128 (2016)
8. Hindriks, R., et al.: Can sliding-window correlations reveal dynamic functional connectivity in resting-state fMRI? Neuroimage **127**, 242–256 (2016)
9. Samdin, S.B., et al.: A unified estimation framework for state-related changes in effective brain connectivity. IEEE Trans. Biomed. Eng. **64**(4), 844–858 (2017)
10. Hochreiter, S., Schmidhuber, J.: Long short-term memory. Neural Comput. **9**(8), 1735–1780 (1997)
11. Lipton, Z.C., Berkowitz, J., Elkan, C.: A critical review of recurrent neural networks for sequence learning. arXiv preprint arXiv:1506.00019 (2015)
12. Barch, D.M., et al.: Function in the human connectome: task-fMRI and individual differences in behavior. Neuroimage **80**, 169–189 (2013)
13. Glasser, M.F., et al.: The minimal preprocessing pipelines for the Human Connectome Project. Neuroimage **80**, 105–124 (2013)
14. Li, H., Satterthwaite, T.D., Fan, Y.: Large-scale sparse functional networks from resting state fMRI. Neuroimage **156**, 1–13 (2017)
15. Li, H., Satterthwaite, T., Fan, Y.: Identification of subject-specific brain functional networks using a collaborative sparse nonnegative matrix decomposition method. In: 2016 IEEE 13th International Symposium on Biomedical Imaging (ISBI) (2016)
16. Abadi, M., et al.: TensorFlow: A System for Large-Scale Machine Learning (2016)
17. Jenkinson, M., et al.: FSL. Neuroimage **62**(2), 782–790 (2012)
18. Cai, T., Liu, W., Xia, Y.: Two-sample covariance matrix testing and support recovery in high-dimensional and sparse settings. J. Am. Stat. Assoc. **108**(501), 265–277 (2013)

# Identifying Personalized Autism Related Impairments Using Resting Functional MRI and ADOS Reports

Omar Dekhil[1], Mohamed Ali[1], Ahmed Shalaby[1], Ali Mahmoud[1],
Andy Switala[1], Mohammed Ghazal[1,2], Hassan Hajidiab[2],
Begonya Garcia-Zapirain[5], Adel Elmaghraby[3], Robert Keynton[1],
Gregory Barnes[4], and Ayman El-Baz[1(✉)]

[1] Bioengineering Department, University of Louisville, Louisville, KY, USA
aselba01@louisville.edu
[2] Department of Electrical and Computer Engineering,
Abu Dhabi University, Abu Dhabi, UAE
[3] Computer Engineering and Computer Science Department, Louisville, KY, USA
[4] Department of Neurology, University of Louisville, Louisville, KY, USA
[5] Facultad Ingenieria, Universidad de Deusto, Bilbao, Spain

**Abstract.** In this study, a personalized computer aided diagnosis system for autism spectrum disorder is introduced. The proposed system uses resting state functional MRI data to build local classifiers, global classifier, and correlate the classification findings with ADOS behavioral reports. This system is composed of 3 main phases: (i) Data preprocessing to overcome the motion and timing artifacts and normalize the data to standard MNI152 space, (ii) using a small subset (40 subjects) to extract significant activation components, and (iii) utilize the extracted significant components to build a deep learning based diagnosis system for each component, combine the probabilities for global diagnosis and calculate the correlation with ADOS reports. The deep learning based classification system showed accuracies of more than 80% in the significant components, moreover, the global diagnosis accuracy is 93%. Out of the significant components, 2 components are found to be correlated with neuro-circuits involved in autism related impairments as reported in ADOS reports.

**Keywords:** fMRI · Personalized diagnosis · ADOS · Deep learning

## 1 Introduction

Autism spectrum disorder (ASD) is a neuro-developmental disorder, and it is associated with early-emerging social and communication impairments, in addition to rigid and repetitive patterns of behavior and interests [1,2]. Although

---

O. Dekhil, M. Ali and A. Shalaby—Shared first authorship.

G. Barnes and A. El-Baz—Shared senior authorship.

© Springer Nature Switzerland AG 2018
A. F. Frangi et al. (Eds.): MICCAI 2018, LNCS 11072, pp. 240–248, 2018.
https://doi.org/10.1007/978-3-030-00931-1_28

there are no well established causative factors explaining ASD [3], alterations in functional activity patterns in brain are believed to play important role in explaining and diagnosing ASD [4]. One of the most widely used methods to measure the brain activity is the functional MRI (fMRI) [5,6]. The fMRI is categorized into two main types: resting state fMRI (RfMRI) and task based MRI. In the RfMRI, patients are not asked to do any task, they are only asked to stay awake with open eyes and not to move in the scanner, and Blood Oxygenated Level Dependant (BOLD) signal is recorded overtime [7]. In the task based experiment patients are asked to complete a certain task and the spontaneous low-frequency fluctuations in the BOLD signal in response to this task is recorded [8]. The main purpose of the RfMRI studies is to identify the alternations in the functional connectivity patterns between two groups (i.e., patients VS controls), and it is widely reported in the literature that ASD is associated with such alteration [9,10].

A wide range of studies were concerned about reporting statistical group differences between autistic subjects (ASDs) and typically developed subjects (TDs). In [9], the default mode network activity alternation was examined between 16 ASDs and 15 TDs, where ASDs showed weaker connectivity and this weaker connectivity was positively correlated with ASD main impairments, poorer social skills and increases in restricted and repetitive behaviors and interests, while stronger connectivity in multiple areas were correlated with communication. A more recent study in [10], used a dataset of 84 ASDs (42 males/42 females) and 150 TDs (75 males/75 females). Males with ASD showed patterns of hypo-connectivity, while females with ASD showed hyper-connectivity.

In addition to the importance of reporting group differences between the two groups -ASDs and TDs-, it is also important to try utilizing the RfMRI in autism diagnosis. Although it is a promising direction and initially yielded good results, only few studies were conducted to try diagnosing autism using RfMRI. Two recent examples of these studies are found in [11] and [12]. The main goal in [11] is to identify connectivity networks that are correlated with ASD symptoms severity. In [11], a balanced dataset of 20 ASD subjects and 20 TD subjects was used. Individual salience network maps were created per subject and those maps yielded classification accuracy of 78% between ASDs and TDs, in addition, the salience network was related to the symptoms of restricted and repetitive behaviors with Pearson correlation coefficient, $R^2 = 0.36$ and $P = 0.07$. Another approach for diagnosing using RfMRI was reported in [12], where a longitudinal study used a dataset of 59 subjects at age of 6 months with high familial risk for ASD. The reported functional connectivity at age of 6 months matched the behavioral scores at age of 24 months which demonstrates the power of RfMRI in early ASD diagnosis.

In this study, we are introducing a large scale personalized diagnosis system for ASD. We demonstrate the power of our approach in identifying the symptoms severity by correlating our personalized maps finding with Autism Diagnostic Observation Schedule (ADOS). To build a system with high generalization capability and with no need to run the RfMRI analysis (which is time consuming) for each new unseen subject, we divided the system into two phases, (i) defining areas of statistical significance using small subset of the data and (ii) building a deep learning based system that utilizes those areas in diagnosing unseen subjects and define neuro-circuits of highest correlation with ADOS scores. More details about the approach are discussed in the next sections.

## 2    Materials and Methods

In this study 156 subjects obtained from National Database of Autism Research (NDAR: http://ndar.nih.gov) are used. The dataset is selected to be balanced (78 ASDs and 78 TDs). The used data are obtained from a single study done at George Washington University [13]. All neuroimages were produced by a Siemens Magnetom TrioTim with a 3 T magnet with TR = 2 s, TE = 30 ms and flip angle 90°, to produce images with 3 mm pixel spacing and 4 mm slice spacing. Time to acquire 33 coronal slices spanning the entire brain was 2.01 s. The resting state data were recorded for approximately 6 min. All ASD subjects in the study have ADOS reports. The experiment in this study is divided into 3 main steps explained in details in the next subsections.

### 2.1    Step 1: RfMRI Preprocessing

Prior to RfMRI analysis, there are several preprocessing steps [14]. In this study, the preprocessing was done using SPM12 (http://www.fil.ion.ucl.ac.uk/spm) toolbox. The preprocessing pipeline includes four main steps: (i) image realignment for motion correction, (ii) slice timing correction to overcome the effect of capturing samples at different times, (iii) image normalization to standard MNI152 space resampled every 2 mm, and finally (iv) Gaussian spatial smoothing using filter with full-width half-maximum of 6 mm.

### 2.2    Step 2: RfMRI Analysis and ROIs Extraction

In this step, a balanced subset of 40 subjects (20 ASDs and 20 TDs) are used to obtain the original regions of interest (ROIs). The ROIs were defined as the output components from the Independent Component Analysis (ICA) that showed statistical significance. To extract the ROIs of these 40 subjects, group ICA is used [15]. In the group ICA, time courses of the 40 subjects are concatenated and then decomposed into two matrices (i) group time course multiplied by (ii) group independent spatial components. To extract subject-specific time course and subject-specific spatial components, dual regression algorithm is used.

The intuition behind using ICA is the analogy between the RfMRI analysis and the blind source separation (BSS) problem [16], where it is required to recover set of statistically independent components with minimal error. The BSS problem can be formulated as:

$$x_i(t) = As_i(t) + \mu + \eta_i(t) \tag{1}$$

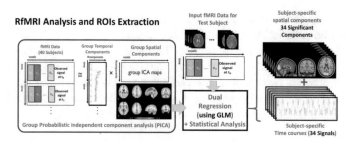

**Fig. 1.** The analysis of 4D-RfMRI data into group spatial components and group time course, then using dual regression to obtain subject specific time courses and subject specific spatial components

Where $x_i$ is the BOLD signal measured over time, $s_i$ is the non-Gaussian source signal, $\eta_i \sim N(0, \sigma^2 \sum_i)$, $A$ is the mixing matrix, $\mu$ is the mean of observations $x_i$ and $i$ is the an index over voxel space To solve this BSS problem it is required to find the unmixing matrix $W$ such that

$$\hat{s}_i = W x_i \tag{2}$$

is a close approximation of the original measured signal. To estimate the unmixing matrix, the it is needed to optimize the rotation matrix $Q$ in the whitened observations space:

$$\hat{s} = Wx = Q\tilde{x} \tag{3}$$

Where:

$$\tilde{x} = (\Lambda_q - \sigma^2 Iq)^{-1/2} U_q^t x \tag{4}$$

are the whitened observations, and $U_q$ and $\Lambda_q$ are the first $q$ eigen-values of $U$ and $\Lambda$, $U$ and $\Lambda$ are the Singular Value decomposition matrices of the observations $X$ and $Q$ is $qxq$ rotation matrix. To solve for the unmixing matrix, and based on the non-Gaussian sources constraint, the algorithm described in [17] is used, where the individual sources are calculated by projecting the the observations $x$ onto the rows of $Q$, thus the $r^{th}$ source is given by the iterative algorithm described in [17]:

$$\hat{q}_r^t \leftarrow \langle (x F'(\hat{s}_r) - F''(\hat{s}_r)) \hat{q}_r \rangle \tag{5}$$

where $\hat{q}_r$ is the $r^{th}$ row of Q and $F$ is general nonquadratic function and $F'$ is the derivative of $F$. To obtain all the rows of Q, this iterative approach is run for q times. For more mathematical details about the solution finding, uniqueness, correctness, and model order, the reader is referred to [15].

After completing the probabilistic ICA analysis of the group concatenated subjects, the output group spatial maps act as a set of spatial regressors and a General Linear Model (GLM) is used to find each individual subject time course. The variance normalized time courses are then fed to another GLM to obtain the subject specific spatial maps [18].

To get the significant ROIs from the output individual spatial maps, a nonparametric permutation test is used with two conditions, (i) ASD > TD and (ii) TD > ASD and 5000 permutations [19]. The output P values are then corrected using Bonferroni correction. Using $\alpha$ level of 0.05 and after the Bonferroni correction, 34 components were found to be significant. The analysis in this study was performed using FSL5.0 package (https://fsl.fmrib.ox.ac.uk/fsl/fslwiki/). The RfMRI analysis and ROIs extraction are illustrated in Fig. 1.

### 2.3    Step 3: Personalized Diagnosis for New Unseen Subjects

With the ROIs having been obtained in the previous step using a subset of 40 subjects, there is no need to repeat the entire pipeline analysis. Our proposed approach is to do the preprocessing steps then feed the new unseen subject to the dual regression to extract both spatial and time components. With this approach it is more feasible and more time efficient. In this study, the time courses and spatial maps of 116 subjects were obtained using the dual regression. These 116 subjects are then used in the personalized diagnosis system.

In the literature, the features used were the individual subjects spatial maps as in [11], but the major drawback of this approach is having number of features much greater than the number of subjects which increases the possibility of having under-determined system and increases the learning difficulty [11]. To overcome this problem in our study we used features derived from the individual temporal components. We used the power spectral density (PSD) of the individual time courses, and the reason behind this selection is the time shift invariance in the PSD. Another advantage in the PSD is the symmetry that allows use of only one half of the signal which increases the amount of features significantly.

As shown in Fig. 2, for each of the 34 ROI, half the signal is fed to a deep learning based classification system which consist of an autoencoder for dimensionality reduction followed by a neural network for classification. To fine tune the model hyper-parameters, i.e., the sparsity proportion, the sparsity regulation, L2 regularization and number autoencoder hidden layers were obtained using a random heuristic search. The searching range for the sparsity propor-

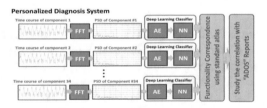

**Fig. 2.** The personalized diagnosis pipeline. The subject specific time course is used to calculate PSD. PSDs are then feed to the deep learning based classification system and the output probabilities are then used for global diagnosis and correlation with ADOS reports

tion is defined between 0.05 and 0.9, for the sparsity reactualization is between 1 and 20, for the L2 regularization is between $10^{-3}$ and $10^{-6}$ and the number of hidden layers is between 10 and 100. The neural network used for classification is a single layer neural network with 10 hidden neurons. The global subject diagnosis is based on a winner-takes-all approach, where the components probabilities are used for voting and the winner class is considered as the global diagnostic decision.

To assess the statistical significance of the classifier, and to ensure its robustness, the significant areas were tested using bootstrapping. The labels of ASD and TD groups were randomly shuffled for 99 times and the classifier was trained and tested using LOSO with this completely uninformative data to ensure random performance in response to randomly shuffled labels.

In order to be able to correlate the significant areas of high accuracies with actual behavioral reports, an atlas defining 34 resting state networks is used. The used atlas is built using 4 local atlases having total of 34 cortical areas and these 4 local atlases are:

1. Parietal cortex atlas [20]: This local atlas, defines both functional connectivity and anatomical connectivity on both humans and macaques. It divides the the parietal cortex into 10 components.
2. Temporoparietal junction (TPJ) [21]: In this local atlas, TPJ is divided into 2 components: (i) anterior TPJ cluster, which showed interaction with ventral prefrontal cortex and anterior insula and (ii) posterior TPJ cluster which showed interaction with posterior cingulate, temporal pole, and anterior medial prefrontal cortex
3. Dorsal frontal cortex [22]: In this local atlas, the human dorsal frontal cortex is parcellated into 10 components They are all between the human inferior frontal sulcus and cingulate cortex
4. Ventral frontal cortex [23]: In this local atlas, the ventral frontal cortex was divided into 11 components, in addition to one more component from ventrolateral frontal pole. The spatial correlation was calculated between the output independent sources and each of the 34 areas defined in the atlas.

## 3   Experimental Results

ASD and TD subgroups were well-matched with respect to gender and age. Out of 78 ASD subjects, 40 were female (51%), while 43 of the 78 TD subjects were female (55%). The gender imbalance was statistically insignificant ($\chi^2 = 0.23$, $p = 0.63$). The mean age of ASD subjects was 13.6 years, while the mean age was 12.8 years for the TD group. The age difference is statistically insignificant ($t = 0.458$, $p = 0.647$). In terms of IQ, the groups were less matched, but the difference in mean score is less than one standard deviation. For ASD group, the mean is 102 and the standard deviation is 21.2, while for the TD group, the mean is 111 and the standard deviation is 15.6.

### 3.1   Significant ROIs Personalized Diagnosis

Out of the 34 components, and after Bonferroni correction, 12 components were found to be statistically significant at $\alpha = 0.01$. These 12 components were then used in building the personalized diagnosis system. Three cross validation techniques are used, 4 folds, 10 folds and leave-one-subject-out. Table 1 shows the

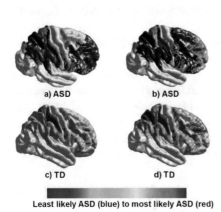

a) ASD          b) ASD

c) TD           d) TD

Least likely ASD (blue) to most likely ASD (red)

**Fig. 3.** The personalized maps of 2 ASDs and 2 TDs. It is obvious that more areas are affected in the ASD cases than in TD cases.

**Table 1.** The Accuracies obtained for the 12 significant components using 4-folds, 10-folds and leave-one-subject-out (LOSO).

| Component | Acc 4-Folds % | Acc 10-Folds % | Acc LOSO % |
|---|---|---|---|
| 1 | 78.5 | 81.1 | 86.3 |
| 4 | 79.3 | 82.4 | 87.0 |
| 8 | 82.7 | 85.6 | 90.2 |
| 11 | 76.2 | 80.4 | 85.06 |
| 13 | 81.1 | 83.7 | 87.6 |
| 14 | 77.1 | 81.8 | 85.7 |
| 17 | 84.3 | 86.1 | 92.8 |
| 18 | 74.2 | 78.6 | 86.3 |
| 22 | 80.5 | 83.7 | 86.3 |
| 23 | 82.9 | 86.5 | 88.9 |
| 25 | 85.8 | 88.1 | 92.8 |
| 34 | 84.1 | 86.5 | 90.2 |

accuracies obtained for the 12 component using the 3 validation techniques, and the most correlated atlas area with each component. Also, Fig. 3 shows sample of personalized diagnosis maps for 2 ASDs and 2 TDs. The overall diagnostic accuracy, sensitivity and specificity after the components fusion are found to be 93%, 91% and 94% respectively.

The result of bootstrap testing done by creating 99 versions of randomly shuffled label yielded a $P$ value of 0.01 for each component diagnosis accuracies, which shows the statistical significance of the classifiers finding.

### 3.2 Correlation Between Personalized Diagnosis and ADOS Reports

To ensure the relevance of these components to ASD, each component output probability is correlated with the Total ADOS score and ADOS severity score. The Pearson correlation coefficient varies between $-0.28$ and $0.27$ for Brodmann area/brain regions involved in neuro-circuits previously reported to be implicated in ASD. Table 2 shows the highly correlated nuero-circuits with ADOS reports and thier anatomical correspondence. To cross validate the relevance of these regions to ASD, each brain region was 347 was correlated with the Total ADOS score and ADOS severity score.

**Table 2.** Mapping between ADOS subscores, Research Domain Criteria (RDoC) neurocircuits, and functional connectivity networks. Anatomical (Brodmann) areas overlapping functional networks are given in parentheses, where Fpl, Cluster 4 is the 4th component of the parietal cortex local atlas and TBJb is the posterior TPJ component.

| Behavioral component | Correlated component | RDoC defined neurocircuit | Anatomical correspondence |
|---|---|---|---|
| Restricted interest/repetitivebehaviors | 4 | Reward learning | Fpl, Cluster 4 (BA10) |
| | 4 | Habit | Fpl, Cluster 4 (BA10) |
| Attention | 4–22 | Ventral attention system | Fpl, Cluster 4 (BA10), TPJb (BA 39–40, 22) |
| Language | 22 | Receptive | TPJb (BA 39–40, 22) |
| Social | 4 | Affiliation and attachment | Fpl, Cluster 4 (BA10) |
| Social | 4–22 | Understanding the mental states of others | Fpl, Cluster 4 (BA10), TPJb (BA 39–40, 22) |
| Executive function | 4 | Working memory | TPJb (BA39–40) |

# 4    Conclusion, and Future Work

This study demonstrate that RfMRI could enhance both local and global diagnostic accuracy of ASD, with increased ability to predict clinical phenotypes, and potential ability to develop better individualized treatments plan. Specific affected networks could be a biomarker for correlation with specific types of behavioral abnormalities. Future research should focus further on using big data technology to integrate multiple datasets from larger populations and multiple modalities (structural MRI, DTI, etc.) to better understand clinically relevant neuro-biological pathways and assess response to personalized treatments in ASD. In addition, integrating information from multiple data sources such as behavioral reports and genetic profiles to get more insight about components of interest observed on each individual subject would be of great importance.

# References

1. Frith, U., et al.: Autism spectrum disorder. Curr. Biol. **15**(19), R786–R790 (2005)
2. Casanova, M.F., et al.: Autism Imaging and Devices. CRC Press, Boca Raton (2017)
3. Vargas, D.L., et al.: Neuroglial activation and neuroinflammation in the brain of patients with autism. Ann. Neurol. **57**(1), 67–81 (2005)

4. Belmonte, M.K., et al.: Autism and abnormal development of brain connectivity. J. Neurosci. **24**(42), 9228–9231 (2004)

5. Huettel, S., et al.: Functional magnetic resonance imaging. Sinauer, Massachusetts, Technical report (2009)

6. El-Baz, A., et al.: Stochastic Modeling for Medical Image Analysis. CRC Press, Boca Raton (2016)

7. Paakki, J.J., et al.: Alterations in regional homogeneity of resting-state brain activity in autism spectrum disorders. Brain Res. **1321**, 169–179 (2010)

8. Kay, K., et al.: GLMdenoise: a fast, automated technique for denoising task-based fMRI data. Front. Neurosci. **7**, 247 (2013)

9. Weng, S.J., et al.: Alterations of resting state functional connectivity in the default network in adolescents with autism spectrum disorders. Brain Res. **1313**, 202–214 (2010)

10. Alaerts, K., et al.: Sex differences in autism: a resting-state fMRI investigation of functional brain connectivity in males and females. Soc. Cogn. Affect. Neurosci. **11**(6), 1002–1016 (2016)

11. Uddin, L.Q., et al.: Salience network-based classification and prediction of symptom severity in children with autism. JAMA Psychiatry **70**(8), 869–879 (2013)

12. Emerson, R.W., et al.: Functional neuroimaging of high-risk 6-month-old infants predicts a diagnosis of autism at 24 months of age. Sci. Transl. Med. **9**(393), eaag2882 (2017)

13. Pelphrey, K.: Multimodal developmental neurogenetics of females with ASD (2014). https://ndar.nih.gov/edit_collection.html?id=2021

14. Dekhil, O., et al.: Using resting state functional MRI to build a personalized autism diagnosis system. In: International Symposium on Biomedical Imaging, ISBI 2018 (2018)

15. Beckmann, C.F., et al.: Probabilistic independent component analysis for functional magnetic resonance imaging. IEEE TMI **23**(2), 137–152 (2004)

16. Zarzoso, V., et al.: Blind separation of independent sources for virtually any source probability density function. IEEE Trans. Sig. Process. **47**(9), 2419–2432 (1999)

17. Minka, T.P.: Automatic choice of dimensionality for PCA. In: Advances in Neural Information Processing Systems (2001)

18. Filippini, N., et al.: Distinct patterns of brain activity in young carriers of the apoe-$\varepsilon$4 allele. Proc. Natl. Acad. Sci. **106**(17), 7209–7214 (2009)

19. Higgins, J.J.: Introduction to modern nonparametric statistics (2003)

20. Mars, R.B., et al.: Diffusion-weighted imaging tractography-based parcellation of the human parietal cortex and comparison with human and macaque resting-state functional connectivity. J. Neurosci. **31**(11), 4087–4100 (2011)

21. Mars, R.B., et al.: Connectivity-based subdivisions of the human right "temporoparietal junction area": evidence for different areas participating in different cortical networks. Cereb. Cortex **22**(8), 1894–1903 (2011)

22. Sallet, J., et al.: The organization of dorsal frontal cortex in humans and macaques. J. Neurosci. **33**(30), 12255–12274 (2013)

23. Neubert, F.X., et al.: Comparison of human ventral frontal cortex areas for cognitive control and language with areas in monkey frontal cortex. Neuron **81**(3), 700–713 (2014)

# Deep Chronnectome Learning via Full Bidirectional Long Short-Term Memory Networks for MCI Diagnosis

Weizheng Yan[1,2,3], Han Zhang[3], Jing Sui[1,2], and Dinggang Shen[3(✉)]

[1] Brainnetome Center and National Laboratory of Pattern Recognition,
Institute of Automation, Chinese Academy of Sciences, Beijing, China
[2] University of Chinese Academy of Sciences, Beijing, China
[3] Department of Radiology and BRIC, University of North Carolina of Chapel Hill,
Chapel Hill, NC, USA
dgshen@med.unc.edu

**Abstract.** Brain functional connectivity (FC) extracted from resting-state fMRI (RS-fMRI) has become a popular approach for disease diagnosis, where discriminating subjects with mild cognitive impairment (MCI) from normal controls (NC) is still one of the most challenging problems. Dynamic functional connectivity (dFC), consisting of time-varying spatiotemporal dynamics, may characterize "chronnectome" diagnostic information for improving MCI classification. However, most of the current dFC studies are based on detecting discrete major "brain status" via spatial clustering, which ignores rich spatiotemporal dynamics contained in such chronnectome. We propose *Deep Chronnectome Learning* for exhaustively mining the comprehensive information, especially the hidden higher-level features, i.e., the dFC time series that may add critical diagnostic power for MCI classification. To this end, we devise a new Fully-connected *bidirectional* Long Short-Term Memory (LSTM) network (Full-BiLSTM) to effectively learn the periodic brain status changes using both past and future information for each brief time segment and then fuse them to form the final output. We have applied our method to a rigorously built large-scale multi-site database (i.e., with 164 data from NCs and 330 from MCIs, which can be further augmented by 25 folds). Our method outperforms other state-of-the-art approaches with an accuracy of 73.6% under solid cross-validations. We also made extensive comparisons among multiple variants of LSTM models. The results suggest high feasibility of our method with promising value also for other brain disorder diagnoses.

## 1 Introduction

Alzheimer's Disease (AD) is an irreversible neurodegenerative disease leading to progressive cognitive and memory deficits. Early diagnosis of its preclinical

---

W. Yan and H. Zhang—Contribute equally to this paper.

© Springer Nature Switzerland AG 2018
A. F. Frangi et al. (Eds.): MICCAI 2018, LNCS 11072, pp. 249–257, 2018.
https://doi.org/10.1007/978-3-030-00931-1_29

stage, mild cognitive impairment (MCI), is of critical value as timely treatment could be the most effective during this stage. Resting-state functional MRI (RS-fMRI) provides an opportunity to assess brain function non-invasively and has been successfully exploited to identify MCI [1]. To capture the time-varying information brain networks, dynamic functional connectivity (dFC) was proposed to characterize the time-resolved connectome, i.e., chronnectome, mostly using sliding-window correlation approach [2,4]. While promising, many current studies have not deeply exploited the rich spatiotemporal information of the chronnectome and utilized it in classification. For example, many studies focused on group comparison by detecting a set of discrete major brain status via clustering time-resolved FC matrices and further calculating their occurrence and dwelling time [4]. Inspired by the new finding that the brain dynamics are hierarchically organized in time (i.e., certain networks are more likely to occur preceding and/or following others [5]), we propose to learn diagnostic features in an end-to-end deep learning framework to better classify MCI.

Recurrent neural networks (RNNs) is a powerful neural sequence learning model for time series analysis. LSTMs are improved RNNs that can effectively solve the "gradient exploding/vanishing" problem by controlling information flow with several gates [6]. It has recently been demonstrated to be able to handle large-scale learning in speech recognition and language translation tasks [7]. However, there is still a significant gap between brain chronnectome modeling and common time series analysis. Directly applying LSTM to dFC-based MCI diagnosis is non-trivial: *(1)* Brain is extraordinary complex whose dynamics could be substantially different from natural language interpretation. *(2)* The background noise is usually more intense in the brain dFC signals, compared to audio/video signals, making it very difficult to capture. *(3)* The brain may continuously use contextual information for guiding higher-level cognitive functions rather than produce an output at the end of the time series with a strict direction. Therefore, a general LSTM could not be suitable for brain chronnectome-based classification. To solve this problem, we propose a new deep learning framework that changes the traditional LSTM in two aspects. *First*, we create Full-LSTM that connects the outputs of all cells to a "fusion" layer to capture a common time-invariant status-switching pattern, based on which the MCI can be diagnosed. *Second*, to excavate the contextual information hidden in the dFC, we further use a bidirectional LSTM (BiLSTM) to access long-range context in both directions [8]. We hereby come out with an end-to-end chronnectome-based classification model, namely *Full-BiLSTM*. The performance of our proposed method has been compared with state-of-the-art methods on ADNI-2 database. As the first "Deep Chronnectome Learning" study, we comprehensively compared the performance of three variants of LSTMs and reported the effect of different hyperparameters. The results support our hypothesis and significantly improved MCI diagnosis.

# 2 Methods

## 2.1 Computing dFC via a Sliding Window Method

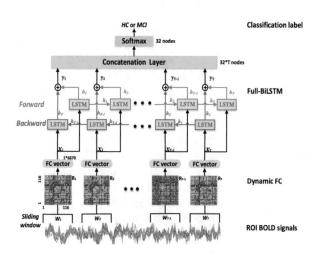

**Fig. 1.** Overview of the Full-BiLSTM for MCI classification.

For each subject, the whole-brain time-varying connectivity matrices are computed based on $M(M = 116)$ ROIs from the automated anatomical labeling (AAL) template using a sliding window approach [3,4]. As shown in Fig. 1, the averaged BOLD time-series $S_i$ in ROI $i$ are first computed. Then, the window $\{W_t\}$ are generated and applied to $S = \{S_i\}$, where $T$ is the total number of sliding windows. Next, for each $W_t$, an FC matrix $R_t$ of size $M * M$ that includes FC strengths between all pairs of $S_{it}$ are calculated. Thus, for each subject, a set of $R_t(t = 1, 2, \ldots, T)$ are obtained, representing the subjects' whole-brain dFC. Due to the symmetry of each $R_t$, all FC strengths in $R_t$ among $M$ ROIs corresponding to a window $t$ are converted to a vector $x_t$ with $M(M - 1)/2$ elements. Therefore, all the dFC time series from the $k_{th}$ subject can be represented by a matrix $X^k = [x_1^k, x_2^k, \ldots, x_t^k]$ with a size of $T * \{M(M - 1)/2\}$ and used as input to Full-BiLSTM classification model.

## 2.2 Fully-Connected Bidirectional LSTM (Full-BiLSTM)

**Long Short-Term Memory (LSTM).** LSTMs incorporates recurrently connected units, each of which receives an input $h_{t-1}$ from its previous unit as well as the current input $x_t$ for the current time point t. Each unit has its memory updating the previous memory $c_{t-1}$ with the current input modulation. The network takes three inputs: $x_t$, $h_{t-1}$, and $c_{t-1}$, and has two outputs: $h_t$ (the output of the current cell state) and $c_t$ (the current cell state). Three gates separately controls input, forget, output. The unit can be expressed as:

$$Input\ Gate: i_t = \sigma(W_{xi}x_t + W_{hi}h_{t-1} + b_i) \tag{1}$$

$$Forget\ Gate: f_t = \sigma(W_{xf}x_t + W_{hf}h_{t-1} + b_f) \tag{2}$$

$$Output\ Gate: o_t = \sigma(W_{xo}x_t + W_{ho}h_{t-1} + b_o) \tag{3}$$

$$Input\ Modulation:\ g_t = \phi(W_{xc}x_t + W_{hc}h_{t-1} + b_c) \tag{4}$$

$$Memory\ Cell\ Update:\ c_t = i_t \odot g_t + f_t \odot c_{t-1} \tag{5}$$

$$Output:\ h_t = o_t \odot \tanh(c_t) \tag{6}$$

Specifically, the input gate $i_t$ controls how much influence the inputs $x_t$ and $h_{t-1}$ exerts to the current memory cell (Eq. 1). The forget gate $f_t$ controls how much influence the previous memory cell $c_{t-1}$ exerts to the current memory cell $c_t$ (Eq. 2). Output gate controls how much influence the current cell $c_t$ has on the hidden state cell $h_t$ (Eq. 3). The memory cell unit $c_t$ is a summation of two components: the previous memory cell unit $c_{t-1}$, which is modulated by $f_t$ and $g_t$ (Eq. 4), and a weighted combination of the current input and the previous hidden state, modulated by the input gate $i_t$ (Eq. 5). Likewise, cell state is filtered with the output gate $o(t)$ for a hidden state updating (Eq. 6), which is the final output from an LSTM cell. With the inputting dFC time series, $W_{x.}$ matrices (containing weights applied to the current input) and $W_{h.}$ matrices (representing weights applied to the previous hidden state) can be learned, $b.$ vectors are biases for each layer, $\sigma$ is sigmoid, $\phi$ is tanh function, and $\odot$ denotes element-wise multiplication.

**Bidirectional LSTM (BiLSTM).** BiLSTM is an effective solution that gets access to both preceding and succeeding information (i.e., context) by involving two separate hidden layers with opposite information flow directions [9]. For a brief description, we denote a process of an LSTM cell as $H$. BiLSTM first computes the forward hidden $\overrightarrow{h}$ and the backward hidden sequence $\overleftarrow{h}$ separately (Eqs. 7–8), and then combines $\overrightarrow{h_t}$ and $\overleftarrow{h_t}$ to generate the final output $y_t$ (Eq. 9). The $W_{x.}$ and $W_{h.}$ matrices in (Eqs. 7–8) are the same as those in (Eqs. 1–4). The $W_{\overrightarrow{h}y}$ (representing weights applied to the forward hidden state) and $W_{\overleftarrow{h}y}$ (representing weights applied to the backward hidden state) are learned with the inputting dFC time series. $b.$ vectors are biases for each layer.

$$Forward\ LSTM:\ \overrightarrow{h}_t = H(W_{x\overrightarrow{h}}x_t + W_{\overrightarrow{hh}}\overrightarrow{h}_{t-1} + b_{\overrightarrow{h}}) \tag{7}$$

$$Backward\ LSTM:\ \overleftarrow{h}_t = H(W_{x\overleftarrow{h}}x_t + W_{\overleftarrow{hh}}\overleftarrow{h}_{t-1} + b_{\overleftarrow{h}}) \tag{8}$$

$$Combined\ Output:\ y_t = H(W_{\overrightarrow{h}y}\overrightarrow{h}_t + W_{\overleftarrow{h}y}\overleftarrow{h}_t + b_y) \tag{9}$$

**Full-BiLSTM.** The traditional BiLSTM classification model usually uses the final state $y_T$ for classification [8]. However, this is insufficient for chronnectome-based diagnosis, because brain may continuously use contextual information to facilitate higher-level cognition and guide status transition, rather than producing a single output at the end of the scanning period. Therefore, the outputs of every repeating cell could be of equally important use and should be concatenated into a dense layer $Y = [y_1, \ldots y_t, \ldots, y_T]$ (see "Concatenation Layer" in Fig. 1).). With this layer, we may abstract a common and time-invariant dynamic transition pattern from all the BiLSTM cells which may represent a constant "trait" information of each subject, instead of the continuously varying brief brain status. While the latter could be of great use in previous status-based studies such as those used *Hidden Markov Chain* for status transition probability modeling in group-level comparison studies [5], it will inevitably lose the precious temporal information which could capture more subtle individual differences for the more challenging disease diagnosis studies. In our framework for MCI diagnosis, the dense layer $Y$ is followed with softmax layer to get the final classification result.

# 3    Experiments and Results

## 3.1    Data Preprocessing

In this study, we use the publicly available Alzheimer's Disease Neuroimaging Initiative dataset (ADNI) to test our method. As shown in Table 1, 143 age- and gender-matched subjects (48 NCs with 164 RS-fMRI scans, and 95 MCIs with 330 RS-fMRI scans) were selected from ADNI-2 database. The goal of ADNI-2 study is to validate the use of various biomarkers including RS-MRI to find the best way to diagnose AD at pre-dementia stage. Each RS-fMRI scan was acquired using 3.0T Philips scanners at different medical centers. All the data were carefully reviewed by the quality control team in Mayo Clinic. ADNI is to date the largest, multi-site, rigorously controlled early AD diagnosis data. The RS-fMRI data were preprocessed following the standard procedure [1].

## 3.2    Dynamic Functional Connectivity Matrix

In this experiment, the window length was 90s (30 volumes) as suggested by previous dFC studies [4]. The window slides in a step of 2 volumes (6s), resulting in 54 segments of BOLD signals. For each subject and each scan, 54 FC matrices were obtained, reflecting the chronnectome. The upper half of the matrix containing 6670 unique dFC links were used and then reshaped into $X^k$ with the size of $54 * 6670$.

## 3.3    Data Augmentation

**Table 1.** Demographic information.

| | NC | MCI |
|---|---|---|
| Number of scans | 164 | 330 |
| Age(mean($\pm$std, yrs)) | 75.4 $\pm$ 6.2 | 72.0 $\pm$ 7.5 |
| Gender(M/F) | 72/92 | 178/152 |

Training deep learning models requires a large number of samples. Fortunately, only part of the dFC time series might be sufficient for discriminating MCIs from NCs because the FC dynamics could happen in a very brief period [5]. This allows us to conduct data augmentation to increase the sample size. Specifically, for each $X^k$, a continuous submatrix of length 30 were cropped as a new sample. By using a sliding window strategy with a stride of 1, the original $X^k$ can be augmented for $54 - 30 + 1 = 25$ times (augmented by a factor of 25). The label of the augmented data from the same subject was kept the same. Of note, all augmented sequences belonging to the same subject were used solely in the training, or validation, or testing phase. In the testing phase, the predicted labels for all the augmented data from the same subject was derived with majority voting to determine the final label for this subject.

### 3.4    Full-BLSTM Parameters and Training Strategy

The Full-BiLSTM model was trained and evaluated using Keras. Data was split into 80% for training and 20% for testing (5-fold cross-validation). 10% of samples from training data were further selected for validation to monitor the training procedure. Training was stopped when the validation loss stopped decreasing for 20 epochs or when the maximum epochs had been executed. The testing data was applied to the trained model to evaluate the performance. The model was trained for minimizing the weighted cross-entropy loss function using stochastic gradient descent (SGD) optimizer. The learning rate (lr) was started from 0.001 and decayed over each update as follow: $lr_t = lr_{t-1}/(1 + decay_{rate} * epochs)$. The $decay_{rate}$ was $10^{-6}$, and the maximum epochs was 200. The batch size was 32. The weights and biases were initialized randomly. To improve the generalization performance of the model and overcome the overfitting problem, we used a dropout method ($dropout = 0.5$) and $l_1 norm$ regularization ($l1 = 0.0005$).

### 3.5    Method Comparison

As dFC is novel in this field, the disease diagnosis works using dFC are quite limited. We compared our approach against various classifiers commonly used. The majority of the dFC studies focus on brain statuses detected by clustering, or the temporal variability of dFC series. Therefore, in the competing methods, we also use these two types of the dFC features for MCI classification. In summary, we compared our method with the classification models using: (1) static FC (sFC); (2) dFC-based brain statuses [4]; and (3) dFC variability [1], as detailed below.

**sFC.** The traditional FC method used in most of the FC studies are based on Pearson's correlation of full-length BOLD signals. After building sFC matrix, an SVM classifier is trained based on the sFC strengths.

**Status-Based.** Group-level chronnectome status is identified by using k-means clustering with all of the dFC matrices in the training data. The occurrence frequency of each status is computed to as features. Then, an SVM classifier is constructed based on the frequency features of all status.

**Variability-Based.** Based on the dFC matrices, the quadratic mean value is computed for each dFC. A total of 6670 features are generated for each subject representing the fluctuation of the signals. The features are further reduced using two-sample t-test. An SVM classifier is constructed based on the dFC variability features.

**Table 2.** Performance of different methods in MCI/NC classification.

| Method | ACC(std)% | SEN(std)% | SPE(std)% | f1(std)% | AUC(std)% |
|---|---|---|---|---|---|
| Static FC + SVM | 61.5(10.0) | 74.0(9.2) | 41.7(14.0) | 70.9(8.2) | 64.2(10.8) |
| dFC-variability | 54.8(12.9) | 54.4(12.3) | 56.8(19.1) | 60.5(12.3) | 49.0(17.0) |
| dFC-status | 61.3(10.0) | 70.8(12.2) | 47.2(13.6) | 69.9(8.6) | 61.9(15.9) |
| *Full-LSTM32* | 71.9(5.9) | 72.3(7.9) | 70.5(15.1) | 76.2(5.3) | 75.9(5.8) |
| *Full-BiLSTM32-Stack* | 69.0(5.0) | 66.7(4.7) | 73.0(9.2) | 73.1(3.5) | 79.2(2.7) |
| *BiLSTM32-Last* | 71.0(10.3) | 76.8(9.6) | 60.9(12.8) | 76.7(8.8) | 75.9(6.0) |
| ***Full-BiLSTM32*** | 73.6(3.7) | 73.9(10.1) | 73.5(7.3) | 77.6(4.4) | 79.8(6.9) |

Notes: Blue-colored methods are the traditional methods; Methods in italic are LSTM-based methods; Our method is in bold italic; Red italic indicates the model without bi-directional LSTM or without Full-LSTM

The performance comparison results are summarized in Table 2 and Fig. 2 showing the ROI curves of all methods. Because of sample imbalance, the area under the ROC curve (AUC) was used as the main metric for comparing the performance of all the methods. Our method achieved 79.8% in AUC and significantly outperformed the traditional sFC and dFC methods. The dFC variability method achieved the lowest result, which could be caused by the severe noise in dFC time series. In contrast, our method could learn the intrinsic brain status transition, thus is more robust to such noise.

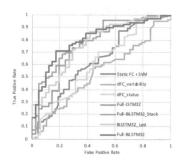

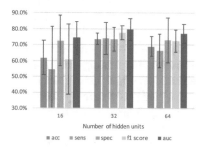

**Fig. 2.** ROC curves of different methods.

**Fig. 3.** Effect of different hidden units

To validate the advantage of Full-BiLSTM, we tested three other LSTM-based architectures. The BiLSTM_Last model uses the output of the last

BiLSTM cell for classification, as used in the traditional sequence processing studies. The Full-LSTM uses the same architecture as our method, but with uni-directional LSTM cells. To investigate whether a deeper BiLSTM layer could increase the performance, the third model is built using stacked Full-BiLSTM (two layers). All these three models use the same parameters as our Full-BiLSTM method. As shown in (Fig. 2), our model still outperformed all these three LSTM-based competing models. Specifically, we observed that (1) BiLSTM outperforms uni-directional LSTM; (2) Full-BiLSTM performs better than BiLSTM_Last; (3) A deeper model does not improve the final performance. In addition, we also compared the performance with and without data augmentation, and found that the accuracy was decreased by 2% without data augmentation. Furthermore, the number of hidden nodes in LSTM may directly affect the learning capacity of an LSTM network. Therefore, we compared the performance of Full-BiLSTM models with a varying number of hidden units, i.e., 16, 32, 64. As shown in Fig. 3, the Full-BiLSTM model with 16 hidden nodes has decreased performance and increased performance variability, compared to the Full-BiLSTM model with 32 hidden nodes. It is likely that 16 hidden units are too limited to store the sequential information of the dFC process. The model with 64 hidden nodes also has suboptimal performance, which could be attributed to overfitting.

The results together indicate that data augmentation and the choice of network structure are crucial for training an effective dFC-based classification model. Most notably, this is the first attempt to use a deep learning framework for individualized disease diagnosis based on dFC. Our results indicate that a sequence model can take advantage of more series information from dFC than the conventional methods. It is also worth noting that our model can be applied to other brain disorder diagnoses.

## 4   Conclusions

In this study, we proposed a new deep learning framework, a Full-BiLSTM model, for brain disease diagnosis using dynamic functional connectivity. To the best of our knowledge, this is the first attempt to propose the "deep chronnetome learning" framework and to prove its feasibility and superiority in a challenging MCI diagnosis task by using time-varying functional information. Comprehensive comparisons among different architectures of the LSTM model were conducted, and the insightful discussions on the influence of the hyperparameters were provided. In summary, the proposed model can not only effectively capture the trait-related brain dynamic changes from the spatiotemporally complex chronnectome, but also can be applied to improve classification of other brain disorders, which shows great promise to be used as a powerful tool to detect potential biomarkers in the community.

**Acknowledgment.** This work was supported in part by NIH grants (AG041721, AG049371, AG042599, and AG053867), NSFC grants (81471367 and 61773380), and the Strategic Priority Research Program of the Chinese Academy of Sciences (XDBS01040102).

# References

1. Chen, X., Zhang, H., Zhang, L., Shen, C., Lee, S., Shen, D.: Extraction of dynamic functional connectivity from brain grey matter and white matter for MCI classification. Hum. Brain Mapp. **38**(10), 5019–5034 (2017)
2. Calhoun, V.D., Miller, R., Pearlson, G., Adali, T.: The chronnectome: time-varying connectivity networks as the next frontier in fMRI data discovery. Neuron **84**(2), 262–274 (2014)
3. Rashid, B., Damaraju, E., Pearlson, G.D., Calhoun, V.D.: Dynamic connectivity states estimated from resting fMRI identify differences among Schizophrenia, bipolar disorder, and healthy control subjects. Front. Hum. Neurosci. **8**, 897 (2014)
4. Allen, E.A., Damaraju, E., Plis, S.M., Erhardt, E.B., Eichele, T., Calhoun, V.D.: Tracking whole-brain connectivity dynamics in the resting state. Cereb. Cortex **24**(3), 663–676 (2014)
5. Vidaurre, D., Smith, S.M., Woolrich, M.W.: Brain network dynamics are hierarchically organized in time. Proc. Natl. Acad. Sci. **114**(48), 12827–12832 (2017)
6. Hochreiter, S., Schmidhuber, J.: Long short-term memory. Neural Comput. **9**(8), 1735–1780 (1997)
7. Sak, H., Senior, A., Beaufays, F.: Long short-term memory recurrent neural network architectures for large scale acoustic modeling. In: INTERSPEECH 2014, pp. 338–342 (2014)
8. Fan, B., Xie, L., Yang, S., Wang, L., Soong, F.A.: A deep bidirectional LSTM approach for video-realistic talking head. Multimed. Tools Appl. **75**(9), 5287–5309 (2016)
9. Graves, A., Schmidhuber, J.: Framewise phoneme classification with bidirectional LSTM networks. IEEE Int. Joint Conf. Neural Netw. **4**, 2047–2052 (2005). https://doi.org/10.1109/IJCNN.2005.1556215

# Structured Deep Generative Model of fMRI Signals for Mental Disorder Diagnosis

Takashi Matsubara$^{(\boxtimes)}$ ⓘ, Tetsuo Tashiro, and Kuniaki Uehara

The Graduate School of System Informatics, Kobe University, 1-1 Rokkodai, Nada, Kobe, Hyogo 657-8601, Japan
matsubara@phoenix.kobe-u.ac.jp
http://www.ai.cs.kobe-u.ac.jp/

**Abstract.** Machine learning-based accurate diagnosis of psychiatric disorders is expected to find their biomarkers and to evaluate the treatments. For this purpose, neuroimaging datasets have required special procedures including feature-selections and dimensional-reductions since they are still composed of a limited number of high-dimensional samples. Recent studies reported a certain success by applying generative models to fMRI data. Generative models can classify small datasets more accurately than discriminative models as long as their assumptions are appropriate. Leveraging our prior knowledge of fMRI signal and the flexibility of deep neural networks, we propose a structured deep generative model, which takes into account fMRI images, disorder, and individual variability. The proposed model estimates the subjects' conditions more accurately than existing diagnostic procedures, general discriminative models, and recently-proposed generative models. Also, it identifies brain regions related to the disorders.

**Keywords:** Deep learning · Generative model
Functional magnetic resonance imaging · Mental-disorder diagnosis
Schizophrenia · Bipolar disorder

## 1 Introduction

With continuously collecting neuroimaging datasets such as functional magnetic resonance imaging (fMRI) [1], many studies have been conducted on machine learning techniques to find specific biomarkers of neurological and psychiatric disorders [2] such as schizophrenia [3,4]. They also provide an opportunity for appropriate treatments and potentially evaluate the effectiveness of the treatments. Since each neuroimaging dataset is still limited in size compared to datasets for other machine-learning tasks, it requires special analysis procedures including hand-crafted features, feature-selections, and dimensional-reductions [3–5].

© Springer Nature Switzerland AG 2018
A. F. Frangi et al. (Eds.): MICCAI 2018, LNCS 11072, pp. 258–266, 2018.
https://doi.org/10.1007/978-3-030-00931-1_30

Recent studies reported a certain success by applying generative models to fMRI images [6–8]. Generative models classify a small-sized dataset better than discriminative models when their assumptions are appropriate [9]. We can leverage our prior knowledge and auxiliary information by constructing the model structure. Suk et al. [6] used hidden Markov models (HMMs) to model the temporal dynamics underlying fMRI signals. Yahata et al. [5] used the sparse canonical correlation analysis (SCCA) to remove features related to known attributes of no interest (e.g., age and sex). Chen et al. [7] employed a linear model composed of a subset shared by all subjects and the remaining adjusted for expressing the functional topography of each subject. These models take into account the individual variability but cannot generalize to an unknown attribute or subject; the generalization is a fundamental problem for diagnosing disorders [10].

On the other hand, *deep neural networks* (DNNs) are attracting attention as flexible machine-learning frameworks (see [11] for a review). DNNs learn high-level features of a given dataset automatically. DNNs have been used as a supervised classifier (a multilayer perceptron; MLP) [4,12] and an unsupervised feature-extractor (an autoencoder; AE) [4,6,12]. Not limited to them, DNNs called *deep neural generative models* (DGMs) build generative models describing relationships between multiple factors in their network structures [13,14]. Tashiro et al. [8] implemented relationships between fMRI images, class label, and scan-wise variability (signals of no interest, such as something in mind) on a DGM and achieved a better diagnostic accuracy than comparative models.

Given the above, we propose a deep generative model dedicatedly structured for fMRI data analysis called *subject-wise DGM* (sw-DGM). The proposed sw-DGM takes into account individual variability (i.e., a subject-wise feature), which is shared by and inferred from all fMRI images obtained from a subject. Thanks to this inference, the proposed sw-DGM generalizes to an unknown subject unlike the study by Chen et al. [7] and potentially deals with unknown attributes unlike the study by Yahata et al. [5].

We evaluate the proposed sw-DGM using resting state fMRI (rs-fMRI) datasets of schizophrenia and bipolar disorders. Our experimental results demonstrate that the proposed sw-DGM provides a more accurate diagnosis than the conventional methods based on the functional connectivity extracted using the Pearson correlation coefficients (PCC) [3,5] and comparative discriminative and generative models; support vector machine (SVM) [15], long short-term memory (LSTM) [16], DGM [8], and AE+HMM [6]. In addition, the proposed sw-DGM identifies brain regions related to the disorders.

## 2    Subject-Wise Deep Neural Generative Model

### 2.1    Subject-Wise Generative Model of FMRI Images

We first propose a structured generative model of a dataset $\mathcal{D} = \{\boldsymbol{x}_i, y_i\}_{i=1}^{N}$ composed of fMRI images $\boldsymbol{x}_i$ and class labels $y_i$ of $N$ subjects indexed by $i$. Each subject $i$ is a control subject ($y_i = 0$) or has the disorder ($y_i = 1$), and provides $T_i$ fMRI images $\boldsymbol{x}_i = \{x_{i,t}\}_{t=1}^{T_i}$. We assume that each subject $i$ has its own

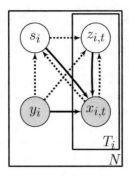

**Fig. 1.** Our proposed generative model composed of fMRI images $x_{i,t}$, a diagnosis $y_i$, a subject-wise feature $s_i$, and scan-wise variabilities $z_{i,t}$.

feature $s_i$ following a prior distribution $p(s)$. The subject-wise feature represents individual variability, which could be brain shape and baseline signal intensity not removed successfully by preprocessing. We also assume that each fMRI image $x_{i,t}$ is associated with the subject's class $y_i$, the subject-wise feature $s_i$, and a latent variable $z_{i,t}$. The latent variable $z_{i,t}$ follows a prior distribution $p(z)$ and represents a scan-wise variability, e.g., brain activity related to something in the subject's mind at that moment, body motion, and so on [8]. Then, we build a generative model $p_\theta$ of fMRI images $x_i$ conditioned by the class label $y_i$ and parameterized by $\theta$. This is depicted in Fig. 1 and expressed as

$$p_\theta(\boldsymbol{x}_i|y_i) = \prod_{t=1}^{T_i} p_\theta(x_{i,t}|y_i) = \prod_{t=1}^{T_i} \int_{s_i} \int_{z_{i,t}} p_\theta(x_{i,t}|z_{i,t}, y_i, s_i) p(z_{i,t}) p(s_i).$$

According to the variational method [13], the model evidence $\log p_\theta(\boldsymbol{x}_i|y_i)$ is bounded using an inference model $q_\phi$ parameterized by $\phi$ as

$$\begin{aligned}
\log p_\theta(\boldsymbol{x}_i|y_i) &\geq \mathbb{E}_{q_\phi(\boldsymbol{z}_i, s_i|\boldsymbol{x}_i, y_i)} \left[ \log \frac{p_\theta(\boldsymbol{x}_i, \boldsymbol{z}_i, s_i|y_i)}{q_\phi(\boldsymbol{z}_i, s_i|\boldsymbol{x}_i, y_i)} \right] \\
&= -D_{KL}(q_\phi(s_i|\boldsymbol{x}_i, y_i)||p(s_i)) \\
&\quad - \sum_{t=1}^{T_i} \mathbb{E}_{q_\phi(s_i|\boldsymbol{x}_i, y_i)} \left[ D_{KL}(q_\phi(z_{i,t}|x_{i,t}, y_i, s_i)||p(z_{i,t})) \right] \\
&\quad + \sum_{t=1}^{T_i} \mathbb{E}_{q_\phi(s_i|\boldsymbol{x}_i, y_i)} \left[ \mathbb{E}_{q_\phi(z_{i,t}|x_{i,t}, y_i, s_i)} \left[ \log p_\theta(x_{i,t}|y_i, z_{i,t}, s_i) \right] \right] \\
&=: \mathcal{L}_g(\boldsymbol{x}_i, y_i),
\end{aligned}$$

$$(1)$$

where $D_{KL}(\cdot||\cdot)$ is the Kullback-Leibler divergence and $\mathcal{L}_g(\boldsymbol{x}_i; y_i)$ is the evidence lower bound (ELBO); the ELBO is the ordinary objective function of the conditional generative model $p_\theta$ and the inference model $q_\phi$ to be maximized.

The ELBO $\mathcal{L}_g(\boldsymbol{x}_i; y)$ is considered to converge to the model evidence $\log p_\theta(\boldsymbol{x}_i|y)$. We estimate the posterior probability $p(y|\boldsymbol{x}_i)$ of the class $y$ of a subject $i$ based on Bayes' rule:

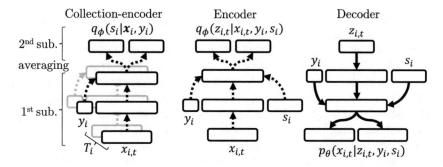

**Fig. 2.** Implementation of the proposed generative model on the deep neural networks.

$$p_\theta(y|\boldsymbol{x}_i) = \frac{p(y)p_\theta(\boldsymbol{x}_i|y)}{\displaystyle\sum_{y'\in\{0,1\}} p(y')p_\theta(\boldsymbol{x}_i|y')} \approx \frac{p(y)\exp\mathcal{L}_g(\boldsymbol{x}_i,y)}{\displaystyle\sum_{y'\in\{0,1\}} p(y')\exp\mathcal{L}_g(\boldsymbol{x}_i,y')} =: \exp\mathcal{L}_d(\boldsymbol{x}_i,y).$$

(2)

We assume the prior probability $p(y)$ of class $y$ to be $p(y=0) = p(y=1) = 0.5$. Hence, if the ELBO $\mathcal{L}_g(\boldsymbol{x}_i, y = 1)$ has a large value, the subject $i$ is more likely to have the disorder.

In addition, the approximation of the log-likelihood of the class label, i.e., $\mathcal{L}_d(\boldsymbol{x}_i, y_i)$, can be an alternative objective function to be maximized, progressing discrimination between the classes [9]. We balanced the two objective functions using the coefficient $\omega \in [0, 1]$ as

$$\mathcal{L}(\boldsymbol{x}_i, y_i) = \omega\mathcal{L}_g(\boldsymbol{x}_i, y_i) + (1 - \omega)\mathcal{L}_d(\boldsymbol{x}_i, y_i).$$

(3)

## 2.2 Implementation on Deep Neural Networks

We implement the generative model $p_\theta$ and inference model $q_\phi$ described above on deep neural networks, and thereby, propose a subject-wise deep generative model (sw-DGM). We assume a preprocessed fMRI signal $x_{i,t}$, a subject-wise feature $s_i$, and a scan-wise variability $z_{i,t}$ as vectors of $n_x$, $n_s$, and $n_z$-dimensions, respectively. The inference model $q_\phi(z_{i,t}|x_{i,t}, y_i, s_i)$ and generative model $p_\theta(x_{i,t}|y_i, s_i, z_{i,t})$ are expressed by multivariate Gaussian distributions with diagonal covariance matrices; their parameters are the outputs of the corresponding DNNs called encoder and decoder (see the right two panels in Fig. 2 and the previous studies [8,13,14] for more detail). The implementation of the inference model $q_\phi(s_i|\boldsymbol{x}_i, y_i)$ requires modification because it accepts a variable-length sequence of fMRI images $\boldsymbol{x}_i = \{x_{i,t}\}_{t=1}^{T_i}$ obtained from a subject $i$. We propose a neural network architecture called *collection-encoder*, which is composed of stacked two sub-networks as depicted in the leftmost panel in Fig. 2. The first sub-network accepts a preprocessed fMRI signal $x_{i,t}$ and the class label $y_i$, and then outputs a hidden activation $h_{i,t}$. The second sub-network accepts the averaged hidden activation $\bar{h}_i = \frac{1}{T_i} \sum_{t=1}^{T_i} [h_{i,t}]$ and outputs the variational posterior $q_\phi(s_i|\boldsymbol{x}_i, y_i)$ of the subject-wise feature $s_i$.

Note that the proposed sw-DGM is not equivalent to other structured DGMs such as Skip Deep Generative Model [14]. They assumed that each sample is generated with more than two latent variables. In contrast, the proposed sw-DGM assumes that the samples $x_{i,t}$ obtained from the same subject $i$ share the subject-wise feature $s_i$ as a latent variable. This assumption potentially gives a good constraint based on a prior knowledge of the fMRI images.

We used three-layered neural networks as the encoder and decoder. We used a two-layered and a single-layered neural networks as the first and the second sub-networks of the collection-encoder, respectively. Each hidden layer of all the DNNs has $u_h$ hidden units followed by the layer normalization [17] and the ReLU activation function [18]. For approximating the expectations in Eq. (1), the subject-wise feature $s_i$ and the scan-wise variability $z_{i,t}$ were sampled from the variational posteriors $q_\phi(s_i|\boldsymbol{x}_i, y_i)$ and $q_\phi(z_{i,t}|x_{i,t}, y_i, s_i)$ once per sample $x_{i,t}$ in the training phase and were substituted with the MAP estimations in the test phase following the previous work [13]. The preprocessed fMRI signals $x_{i,t}$ were augmented using the dropout [19] of a ratio $p$. All the DNNs were jointly trained using the Adam optimization algorithm [20] with parameters $\alpha = 10^{-4}$, $\beta_1 = 0.9$, and $\beta_2 = 0.999$. We selected hyper-parameters from $p \in \{0.0, 0.5\}$, $n_h \in \{50, 100, 200, 400\}$, $n_z = n_s \in \{5, 10, 20, 50, 100\}$ for $n_h > n_z = n_s$, and $\omega \in \{0.0, 0.9, 0.99\}$. We adjusted the imbalance in the classes via oversampling.

## 3    Experiments and Results

### 3.1    Data Acquisition and Comparative Models

We used datasets obtained from the OpenfMRI database. Its accession number is ds000030 (https://openfmri.org/dataset/ds000030/). We performed a preprocessing procedure for rs-fMRI using the SPM12 software package (http://www.fil.ion.ucl.ac.uk/spm/software/spm12/). We discarded the first 10 scans of each subject to ensure magnetization equilibrium. We performed time-slice adjustment, realignment of brain positions via a rigid body rotation, and spatial normalization using the MNI space with a voxel thickness of 2.0 mm. We parcellated each fMRI image into 116 regions of interest (ROIs) using the automated anatomical labeling (AAL) template [21] and averaged intensities of voxels in each ROI region, obtaining a 116-dimensional vector as a preprocessed fMRI signal $x_{i,t}$. As scrubbing, we discarded frames with frame displacements (FD) of more than 1.5 mm or angular rotations of more than $1.5°$ in any direction as well as the following frames. We also discarded subjects who had less than 100 remaining frames and subjects whose fMRI images did not match the MNI template after the spatial normalization. As a result, we obtained 113 control subjects, 44 patients with the schizophrenia, and 45 patients with the bipolar disorder.

As baselines, we evaluated two conventional procedures, which use Pearson's correlation coefficients (PCCs) between the ROIs as the functional connectivities

**Table 1.** Diagnostic accuracies.

| Model | Schizophrenia | | | Bipolar | | |
|---|---|---|---|---|---|---|
| | BACC | SPEC | SEN | BACC | SPEC | SEN |
| PCC+Kendall+PCA+c-means [3] | 0.640 | 0.635 | 0.645 | 0.602 | 0.565 | 0.640 |
| PCC+SCCA+SLR [5] | 0.639 | 0.779 | 0.500 | 0.607 | 0.735 | 0.480 |
| SVM [15] | 0.505 | 0.788 | 0.223 | 0.512 | 0.855 | 0.169 |
| LSTM [16] | 0.661 | 0.854 | 0.467 | 0.571 | 0.802 | 0.340 |
| DGM [8] | 0.722 | 0.920 | 0.524 | 0.619 | 0.650 | 0.587 |
| AE+HMM [6] | 0.618 | 0.554 | 0.682 | 0.616 | 0.490 | 0.742 |
| sw-DGM (proposed) | **0.767** | 0.812 | 0.722 | **0.622** | 0.844 | 0.401 |

(FCs) [3,5]. Following Shen et al. [3], we selected $m$ features in the FCs using the Kendall $\tau$ correlation coefficient, compressed the feature vector into a $d$-dimensional space using the locally linear embedding (LLE) with a parameter of $k$, and clustered them into two classes using the c-means algorithm. This procedure was confirmed to outperform direct classification of the PCCs by the SVM and MLP. Following Yahata et al. [5], we selected $m$ features in the FCs using the SCCA and classified the features using the sparse logistic regression (SLR) with a sparsity determined by automatic relevance determination (ARD). We selected the hyper-parameters from $m \in \{50, 100, 200, 400, 600\}$, $k \in \{5, 8, 10, 12, 15\}$, and $d \in \{2, 5, 10, 20, 50\}$ following the original study [3].

For comparison, we evaluated classifiers; support vector machine (SVM) [15] and long short-term memory (LSTM) [16]. The SVM accepted a single image $x_{i,t}$ and outputted a binary value representing the estimated class using linear kernels. The diagnosis of a subject $i$ is determined by majority voting of $T_i$ estimations, consistent with other comparative models. We selected the hyper-parameter $C$ adjusting the trade-off between classification accuracy and margin maximization from $C \in \{\ldots, 0.1, 0.2, 0.5, 1, 2, 5, 10, \ldots\}$. The LSTM is a recurrently connected neural network, accepting fMRI signals $x_i = \{x_{i,t}\}_{t=1}^{T_i}$ sequentially and outputting the posterior probability $p(y|x_i)$ using the logistic function. The other conditions were the same as those for the proposed sw-DGM.

We also evaluated a simpler DGM proposed in the previous study [8] and hidden Markov model (HMM) with autoencoder (AE) [6]. The DGM modeled relationships between the fMRI signals $x_i$, the class label $t_i$, and the scan-wise variability $z_{i,t}$ using an encoder $q(z_{i,t}|x_{i,t}, y_i)$ and a decoder $p(x_{i,t}|y_i, z_{i,t})$ but does not take into account the subject-wise feature $s_i$ [8]. Following Suk et al. [6], we compressed each fMRI image into a $d$-dimensional space using an AE. Then, we trained a pair of HMMs; $p_\theta(x_{i,t}|y = 1)$ for patients and $p_\theta(x_{i,t}|y = 0)$ for control subjects. Each HMM had Gaussian distributions with full covariance matrices and was trained using Expectation-Maximization (EM) algorithm. We calculated the posterior probability $p(y|x_i)$ using Bayes' rule. We selected the number $n_z$ of units in the bottleneck layer from $n_z \in \{2, 3\}$, the number $n$

**Table 2.** Top 5 contribution weights for diagnosis.

| Schizophrenia | | Bipolar | |
| --- | --- | --- | --- |
| ROI | Weight | ROI | Weight |
| Cerebelum_6_L | 0.0555 | Cingulum_Ant_R | 0.0132 |
| Postcentral_L | 0.0532 | Frontal_Inf_Orb_L | 0.0121 |
| Cingulum_Mid_L | 0.0531 | Cerebelum_7b_R | 0.0116 |
| Lingual_R | 0.0529 | ParaHippocampal_L | 0.0114 |
| Lingual_L | 0.0526 | Temproal_Mid_L | 0.0106 |

of mixture components of the HMM from $n \in \{2, 3, 4, 5, 6, 7\}$, and the hyper-parameters of the AE in the same ranges as the proposed sw-DGM.

### 3.2 Results of Diagnosis and Contribution Weights of ROIs

Since the datasets are imbalanced, we used the following measures; sensitivity $\text{SEN} = \text{TP}/(\text{TP} + \text{FN})$, specificity $\text{SPEC} = \text{TN}/(\text{TN} + \text{FP})$, and balanced accuracy $\text{BACC} = 0.5 \times (\text{SEN} + \text{SPEC})$, where TP, TN, FP, and FN denote true positive, true negative, false positive, and false negative, respectively. We performed 5 trials of 10-fold cross-validation (CV) and summarized the results in Table 1. The proposed sw-DGM achieved the best balanced accuracies among the competitive approaches in the both datasets. Especially, the proposed sw-DGM outperformed or at least performed no worse than the existing DGM [8], implying that the introduction of the subject-wise feature $s_i$ (i.e., individual variability) worked as an appropriate constraint.

As shown in Eq. 2, the diagnosis of a subject $i$ is based on the difference in the conditional log-likelihood $\log p_\theta(\boldsymbol{x}_i|y)$ between the class labels $y = 0$ and $y = 1$. Since each element $x_{i,t,r}$ of an fMRI signal $x_{i,t}$ corresponds to an ROI $r$, we can calculate the ROI-wise average marginal log-likelihoods $\mathbb{E}_{i,t}[\log p_\theta(x_{i,t,r}|y_i)|x_{i,t}]$. An ROI with a large difference in the log-likelihoods between correct and incorrect labels has a large effect on the accurate diagnosis. Hence, we defined

$$W_r = \mathbb{E}_{i,t} \left[ \log p_\theta(x_{i,t,r}|y_j) - \log p_\theta(x_{i,t,r}|1 - y_j)|x_{i,t} \right]$$

as the contribution weight $W_r$ of the ROI $r$ and summarized the ROIs with the top 5 contribution wights in Table 2. Previous studies (e.g., the review paper [22]) have discussed the relationships of some of the listed ROIs to the disorders. The results suggest that the proposed sw-DGM identified the ROIs related to the disorders.

## 4   Conclusion

This study proposed a subject-wise deep generative model (sw-DGM) of fMRI images dedicatedly structured for diagnosing psychiatric disorders. The sw-DGM modeled the joint distribution of rs-fMRI images, class label, individual variability, and scan-wise variability. The individual variability worked as an appropriate constraint, and the sw-DGM achieved a diagnostic accuracy higher than other conventional and comparative approaches. Also, the sw-DGM identified brain regions related to the disorders.

**Acknowledgments.** The authors would like to acknowledge Dr. Ben Seymour, Dr. Kenji Leibnitz, Dr. Hiroaki Mano, and Dr. Ferdinand Peper at CiNet for valuable discussions. This study was supported by the JSPS KAKENHI (16K12487), SEI Group CSR Foundation, and the MIC/SCOPE #172107101.

# References

1. Sejnowski, T.J., et al.: Putting big data to good use in neuroscience. Nat. Neurosci. **17**(11), 1440–1441 (2014)
2. Group, B.D.W.: Biomarkers and surrogate endpoints: preferred definitions and conceptual framework. Clinic. Pharmacol. Ther. **69**(3), 89–95 (2001)
3. Shen, H., et al.: Discriminative analysis of resting-state functional connectivity patterns of schizophrenia using low dimensional embedding of fMRI. NeuroImage **49**(4), 3110–3121 (2010)
4. Castro, E., et al.: Deep independence network analysis of structural brain imaging: application to schizophrenia. IEEE Trans. Med. Imaging **35**(7), 1729–1740 (2016)
5. Yahata, N., et al.: A small number of abnormal brain connections predicts adult autism spectrum disorder. Nat. Commun. **7**(7), 11254 (2016)
6. Suk, H.I., et al.: State-space model with deep learning for functional dynamics estimation in resting-state fMRI. NeuroImage **129**, 292–307 (2016)
7. Chen, P.H., et al.: A Reduced-Dimension fMRI Shared Response Model. In: NIPS. (2015) 460–468
8. Tashiro, T., et al.: Deep neural generative model for fMRI image based diagnosis of mental disorder. In: NOLTA (2017)
9. Lasserre, J., et al.: Principled hybrids of generative and discriminative models. In: CVPR, pp. 87–94 (2006)
10. Abraham, A., et al.: Deriving reproducible biomarkers from multi-site resting-state data: an autism-based example. NeuroImage **147**, 736–745 (2017)
11. Schmidhuber, J.: Deep learning in neural networks: an overview. Neur. Netw. **61**, 85–117 (2015)
12. Liu, S., et al.: Multimodal neuroimaging feature learning for multiclass diagnosis of Alzheimer's disease. IEEE Trans. Biomed. Eng. **62**(4), 1132–1140 (2015)
13. Kingma, D.P., et al.: Semi-supervised learning with deep generative models. In: NIPS, pp. 3581–3589 (2014)
14. Maaløe, L., et al.: Auxiliary deep generative models. In: ICML, vol. 48, pp. 1445–1453 (2015)
15. Pereira, F., et al.: Machine learning classifiers and fMRI: a tutorial overview. NeuroImage **45**, S199–S209 (2009)
16. Dvornek, N.C., et al.: Identifying autism from resting-state fMRI using long short-term memory networks. MLM **I**, 362–370 (2017)
17. Ba, J.L., et al.: Layer normalization, pp. 1–14. arXiv (2016)
18. Nair, V., Hinton, G.E.: Rectified linear units improve restricted Boltzmann machines. In: ICML, pp. 807–814 (2010)

19. Srivastava, N., et al.: Dropout: a simple way to prevent neural networks from overfitting. JMLR **15**, 1929–1958 (2014)
20. Kingma, D.P., Ba, J.: Adam: a method for stochastic optimization. In: ICLR, pp. 1–15 (2015)
21. Tzourio-Mazoyer, N., et al.: Automated anatomical labeling of activations in SPM using a macroscopic anatomical parcellation of the MNI MRI single-subject brain. NeuroImage **15**(1), 273–289 (2002)
22. Andreasen, N.C., Pierson, R.: The role of the cerebellum in schizophrenia. Biol. Psychiatry **64**(2), 81–88 (2008)

# Cardiac Cycle Estimation
# for BOLD-fMRI

Michael Hütel[1,4]([⊠]), Andrew Melbourne[1,4], David L. Thomas[2,3],
and Sebastien Ourselin[1,4]

[1] Department of Medical Physics and Biomedical Engineering, UCL, London, UK
michael.hutel.13@ucl.ac.uk
[2] Leonard Wolfson Experimental Neurology Centre,
UCL Institute of Neurology, London, UK
[3] Department of Brain Repair and Rehabilitation,
UCL Institute of Neurology, London, UK
[4] School of Biomedical Engineering and Imaging Sciences, KCL, London, UK

**Abstract.** Previous studies [1,2] have shown that slow variations in
the cardiac cycle are coupled with signal changes in the blood-oxygen
level dependent (BOLD) contrast. The detection of neurophysiological
hemodynamic changes, driven by neuronal activity, is hampered by such
physiological noise. It is therefore of great importance to model and
remove these physiological artifacts. The cardiac cycle causes pulsatile
arterial blood flow. This pulsation is translated into brain tissue and
fluids bounded by the cranial cavity [3]. We exploit this pulsality effect
in BOLD fMRI volumes to build a reliable cardio surrogate estimate.
We propose a Gaussian Process (GP) heart rate model to build physiological noise regressors for the General Linear Model (GLM) used in
fMRI analysis. The proposed model can also incorporate information
from physiological recordings such as photoplethysmogram or electrocar-
diogram, and is able to learn the temporal interdependence of individual
modalities.

## 1 Introduction

The complex interplay of neurophysiological quantities and neuronal activity
determines cerebral blood flow (CBF). The ratio between oxygenated and deoxygenated hemoglobin of venous cerebral blood indirectly reflects the degree of
neuronal activity in the surrounding tissue. The magnetic resonance (MR) signal is sensitive to blood oxygen level dependent (BOLD) changes due to the
distorting effect of paramagnetic deoxygenated hemoglobin on the homogeneity
of the magnetic field. The functional integrity of the brain depends on precise
control of CBF. Differences in the hemodynamic response between normal and
abnormal cerebrovascular regulation provide important insights for the understanding of brain dysfunction and disease. It is therefore important to track
physiological characteristics of the CBF. The cardiac cycle produces pulsality
artifacts [3] that interfere with the MR signal, impeding the correct estimation

© Springer Nature Switzerland AG 2018
A. F. Frangi et al. (Eds.): MICCAI 2018, LNCS 11072, pp. 267–274, 2018.
https://doi.org/10.1007/978-3-030-00931-1_31

of the hemodynamic response. Pulse oximetry provides a quantitative means to record cardiac cycle during scanning. However, recordings from pulse oximetry sensors can exhibit poor signal quality due to subject motion, are usually not synchronized to scanner triggers, or are regarded as unimportant for fMRI analysis within a clinical setting. Many data-driven methodologies have been proposed when physiological recordings are not available [4]. However, to the best of our knowledge, none of them utilize physiology-derived information. In consequence, we want to directly extract and model a cardiac surrogate from the actual scan. The cardiac surrogate signal can be used as a noise regressor in large clinical cohorts in which physiological monitoring was not available. The cardiac surrogate signal can also be used to supplement physiological recordings within our proposed model. First, we show in highly sampled BOLD-fMRI images with sagittal orientation that areas affected by cardiac and respiratory cycle spatially correlate with areas of high signal variance. We therefore compute an average slice-wise signal with respect to slice ordering and BOLD signal standard deviation. The power spectrum of this cardiac surrogate signal resembles the power spectrum of the signal obtained from pulse oximetry (pulse ox.). We learn the degree of correlation between cardiac surrogate signal and pulse oximetry signal with a Multi-Task Gaussian Process (MTGP) model. Last, we show that this Gaussian Process CARdio Estimation (GPCARE) model produces robust estimates of the cardiac cycle and can also be used to fill in periods of measurement failure.

## 2     A Multi-Task Gaussian Process Model for Cardiac Cycle Estimation

Prior knowledge of the behavior of an examined system can be expressed within a Gaussian Process model. Furthermore, by formulating our problem in a Bayesian framework, we can infer the probabilities of our model parameters in the presence of noise, measurement failure and incompleteness of data, as often occurs within a clinical setting [5,6].

**Gaussian Process Regression Model.** We consider our cardiac cycle estimation as a supervised learning problem. We detect peaks in the cardiac signals, and use the obtained time stamps $\mathbf{x} = \{x_i | i = 1, \ldots, n\}$ as well as the period between adjacent time stamps $\mathbf{y} = \{y_i | i = 1, \ldots, n\}$ as our training data. This data is used to learn a generative model $\mathbf{y} = f(x) + \epsilon$ with latent function $f(x)$ and noise $\epsilon \sim \mathcal{N}(0, \sigma^2)$. For given test time stamps $\mathbf{x}^* = \{x_i^* | i = 1, \ldots, k\}$, i.e. at every slice acquisition time, we want to predict estimates of the unknown heart rate $\mathbf{y}^* = \{y_i^* | i = 1, \ldots, k\}$. We describe the latent function $f(x)$ with probability distributions over functions [7] given by

$$f \sim GP(\mu, \mathbf{\Sigma}), \tag{1}$$

so that the behavior of the function is completely specified by its mean $\mu$ and covariance $\mathbf{\Sigma}$. We encode our prior knowledge about the cardiac signal behavior with a squared-exponential (SE) covariance function $k_{SE}(x, x')$ and constant mean function $\mu(x) = c$, assuming that during rest the heart rate fluctuates slowly around a mean frequency with standard deviation [1] (hyperparameter $\theta_A^2$) and autocorrelation (hyperparameter $\theta_L$), given by

$$k_{SE}(x, x') = \theta_A^2 \exp\left[-\frac{(x - x')^2}{2\theta_L^2}\right]. \tag{2}$$

The individual elements of the covariance matrix $\Sigma(\mathbf{x}, \mathbf{x})$ for a vector $\mathbf{x} \in \mathbb{R}^n$ are given by evaluating the covariance function at $k_{SE}(x, x')$. Prediction on the test time stamps $\mathbf{x}^*$ can be made by averaging over all possible parameter values weighted by their posterior probability. The predictive distribution for $f^* \doteq f(\mathbf{x}^*)$ at $\mathbf{x}^*$ is computed using the posterior distribution

$$p(\mathbf{y}^*|\mathbf{x}^*, \mathbf{x}, \mathbf{y}) \sim \mathcal{N}\left(m(\mathbf{y}^*), var(\mathbf{y}^*)\right) \tag{3}$$

with mean $m(\mathbf{y}^*) = \mu(\mathbf{x}^*) + \Sigma(\mathbf{x}, \mathbf{x}^*)^T \Sigma(\mathbf{x}, \mathbf{x})^{-1}(\mathbf{y} - \mu(\mathbf{x}))$ and variance $var(\mathbf{y}^*) = \Sigma(\mathbf{x}^*, \mathbf{x}^*) - \Sigma(\mathbf{x}, \mathbf{x}^*)^T \Sigma(\mathbf{x}, \mathbf{x})^{-1}\Sigma(\mathbf{x}, \mathbf{x}^*)$. The hyperparameters are optimized by minimizing the negative log marginal likelihood (NLML) defined as

$$-\log p(\mathbf{y}|\mathbf{x}, \theta) = \frac{1}{2}\log|\Sigma| + \frac{1}{2}\mathbf{y}^T\Sigma^{-1}\mathbf{y} + \frac{n}{2}\log 2\pi. \tag{4}$$

Bayesian inference reduces to computing mean and covariance parameters of a multivariate Gaussian posterior distribution.

**Multi-Task Gaussian Process Regression Models.** While one could train several GP models for each individual cardiac modality, one would ideally want to model $m$ modalities simultaneously and exploit the correlation between them. Therefore, we used a Multi-Task Gaussian Process (MTGP) model as proposed in Dürichen et al. [6]. Our training data set extends to $\mathbf{X} = \{x_i^j | i = 1, \ldots, n^j\}$ and $\mathbf{Y} = \{y_i^j | i = 1, \ldots, n^j\}$, in which modality $j$ has $n^j$ number of training data. Furthermore, we require an index $l^j$ as an additional input to identify individual modalities. The covariance function modeling correlation between modalities is given by $k_{CORR}(l, l')$. We assume independence between $k_{CORR}$ and our individual modality covariance function $k_{SE}$ [6] and combine them in

$$k_{MTGP}(x, x', l, l') = k_{CORR}(l, l') \times k_{SE}(x, x'). \tag{5}$$

We rewrite the covariance matrix $\Sigma$ as the Kronecker product of $\Sigma_{CORR}$ and $\Sigma_{SE}$ assuming $n^j = n$ for $j = 1, \ldots, m$ without loss of generality, resulting in

$$\Sigma_{MTGP}(\mathbf{X}, \mathbf{l}, \theta_{CORR}, \theta_{SE}) = \Sigma_{CORR}(\mathbf{l}, \theta_{CORR}) \otimes \Sigma_{SE}(\mathbf{X}, \theta_{SE}). \tag{6}$$

Predictions for test indices $\mathbf{x}^*, \mathbf{l}^*$ are computed using the posterior distribution $p(\mathbf{y}^*|\mathbf{x}^*, \mathbf{l}^*, \mathbf{x}, \mathbf{l}, \mathbf{y})$ similar to an individual GP model [6].

## 3  Experiments

**Cardiac Surrogate Signal.** We acquired five sagittal slices with high temporal resolution using a Multi-Echo BOLD-weighted Echo Planar Imaging (EPI)
sequence with echo times (TE) 8.3, 21.4 and 34 ms and repetition time (TR)
300 ms on a Siemens Trio scanner. Recordings from pulse oximetry and respiration belt were used to build RETROICOR regressors [8]. Physiological respiration and cardiac cycle components can be approximated by their low-order
Fourier series expansion as given by

$$y_\delta = \sum_{m=1}^{M} a_m^c cos(m\phi_c) + b_m^c sin(m\phi_c) + a_m^r cos(m\phi_r) + b_m^r sin(m\phi_r), \quad (7)$$

where $a_m^c, b_m^c$ and $a_m^r, b_m^r$ are the coefficients for cardiac and respiratory function
respectively, and $\phi_c(t)$ and $\phi_r(t)$ are the phases in the respective cardiac and respiratory cycles at time $t$. Figure 1 shows the physiological coefficient estimates
from RETROICOR regressors from one of the acquired slices. Pulsation artifacts due to the cardiac cycle are mostly pronounced in approximate locations
of large vasculature such as orbitofrontal, callosomarginal and pericallosal arter-
ies. Effects from respiration on the MR signal are pronounced in the Superior
Sagittal Sinus and at its crossing with the Transverse Sinus. Figure 1 also shows
the large effect physiology has on the strength of variation in the BOLD time
courses. The short repetition time used to acquire the five sagittal slices is not
applicable for full brain coverage in BOLD fMRI imaging. The cardiac cycle is
thus under-sampled in whole brain fMRI images producing low-frequency alias-
ing artifacts in BOLD time courses. Therefore, there is a great need to obtain
information about heart rate spectrum and variation. We exploit the pulsality
effect the cardiac cycle has on brain tissue and brain fluids, such as CSF, to
build a cardiac surrogate. We have made the following observations in our test
data. First, the power spectrum of this cardiac surrogate resembled the complete
power spectrum of the cardiac cycle during the scan. Second, the cardiac surrogate can be used with or without physiological recordings to compute cardiac
regressors with RETROICOR [8] or nuisance regressors for variations in heart
rate [1,2,9]. And third, correlation between cardiac surrogate signal and actual
physiological recordings can be learned to account for measurement failure as
shown in the following section.

**Cardiac Cycle.** We tested our method on a large fMRI data set that comprises 61 fMRI sessions (cohort comprises healthy adults and adults that were
born preterm, with average age 19 years) acquired with a Philips 3T Achieva
(TR 3000 ms, TE 30 ms, flip angle 80°, voxel size $2.5 \times 2.5 \times 3$ mm$^3$, field of view
(FoV) 240 mm$^2$, 50 oblique transverse slices, slice order descending). From the 61
data sets, 7 fMRI scans were discarded for corrupted physiological recording files
and 3 for strong head movement. Motion correction was performed on all scans.
BOLD time courses were corrected for polynomial trends and z-transformed.
Slices were reordered with respect to their slice acquisition timing. The sagit-
tal FoV was cropped to have an approximate equal contribution of brain tissue

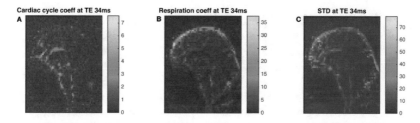

**Fig. 1.** The middle sagittal slice of the acquisition depicts the large influence of physiology on the MR signal. Most variance in the vicinity of large blood vessels can be explained with the RETROICOR cardiac (A) and respiratory (B) regressors. The regions influenced by physiology are also the regions with the greatest standard deviation in BOLD time courses (C).

and cerebral fluids for each slice. The cardiac surrogate signal was obtained by computing the standard-deviation weighted average in each slice resulting in a signal sampled every TR/(number of slices). The high sampling rate captured the entire spectrum of the heart rate. The obtained signal was very noisy due to averaging from slices with different tissue compositions, which resulted in heteroscedasticity. A bandpass filter (0.6 Hz–2.0 Hz) was applied to isolate the cardiac frequency spectrum. The frequency with the greatest power in this band was peak filtered (Gaussian filter) with a bandwidth of 0.5 Hz. The same signal processing was applied to pulse oximetry signals acquired from the Philips physiological monitoring unit. Peak times were extracted from cardiac surrogate and pulse oximetry signal. The individual effects of signal processing are depicted in Fig. 2.1 for scan 19. Individual GPs (covariance function $k_{SE}$) and the MTGP model (covariance function $k_{MTGP}$) were learned on cardio surrogate and pulse ox. peaks depicted in Fig. 2.2. and 2.3, respectively. To simulate missing data, we removed peaks within time interval (84 s, 164 s). The MTGP model coped with missing data using the learned information from the correlation covariance matrix $\Sigma_{CORR}$. The MTGP pulse ox. GP follows the cardiac surrogate GP in the period of missing peaks (Fig. 2.5A) whereas the individual pulse ox. GP goes to the signal mean (Fig. 2.4A). Cardiac pulsation effects are mostly pronounced at the ventricle borders as well as in large vasculature, i.e. the transversal sinus as shown in the statistical parametric mapping of cardiac surrogate GP (individual GP framework) in Fig. 2.6.

We also acquired 3 fMRI sessions with a Siemens Trio 3T (two with TR 2060 ms, multi-echo TEs 8.3, 21.4 and 34 ms, flip angle 90°, voxel size 3 mm³, FoV 192 mm², 36 oblique transverse slices, slice order descending, one with TR 2060 ms, single echo TE 30 ms, flip angle 90°, voxel size 3 mm³, FoV 192 mm², 36 oblique transverse slices, slice order interleaved). Our proposed cardiac surrogate signal was also found in these scans, providing evidence towards applicability to multiple scanner models.

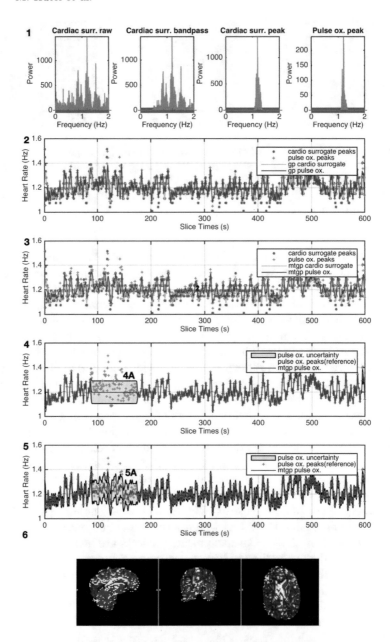

**Fig. 2.** The influence of signal processing on cardiac surrogate as well as final pulse ox. signal for scan 19 of 61 (1). Individual GPs (2) and MTGP model GPs (3) trained on peaks in cardiac surrogate and pulse oximetry signal. Individual GPs (4) and MTGP model GPs (5) simulating measurement failure in interval (84 s, 164 s). Increased uncertainty of the processes in the period of missing measurements (4A and 5A). The statistical parametric mapping (p-value ≤ 0.05) for variance explained by the cardiac surrogate GP of the MTGP model. Coefficients from low in red to high in white (6).

**Individual GPs vs MTGP.** We computed the difference between the frequency with the greatest power in the cardiac surrogate signal and pulse ox. signal. The spectral similarity between both signals was very high in all subjects (median 0.01, $25^{th}$ percentile 0.008, $75^{th}$ percentile 0.078, outliers 4). We computed the difference at slice acquisition times between cardiac surrogate GP and pulse ox. GP within the individual GP framework and in the MTGP framework as depicted in Fig. 3A. We sorted subjects by quality of their pulse ox. power spectrum from left (good) to right (noisy). We computed the frame-wise displacement from the motion realignment parameters of the rigid-body registration for each subject [4]. Some of the subjects that exhibited large head motion also showed a noisy power spectrum of the pulse ox. signal. It was thus assumed that noisy pulse ox. signals resulted mainly from subject motion but also from poor sensor connection or detachment. Nonetheless, some subjects with minor head movement where also found to have noisy pulse ox. signal. Subjects 1–29 had a Gaussian-shaped power spectrum. The pulse ox. recording of subjects 30–51 had a very wide and noisy or bimodal Gaussian-shaped cardiac bandwidth. Large variations in the heart rate are very unusual during rest. The bimodal Gaussian-shaped power spectrum of pulse ox. signals might be due to subjects falling asleep during the scan. The second quartile of differences between GPs in the individual GP framework (Fig. 3A) is below 0.05 Hz for good recordings 1–29 and below 0.1 Hz for most of the noisy recordings 30–51. The MTGP model was able to cope with missing data periods resulting in low errors between MTGP pulse ox. GP (full data) and MTGP pulse ox. GP (missing data) in time interval (84 s, 164 s) as depicted in Fig. 3B.

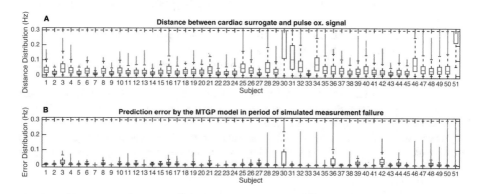

**Fig. 3.** The boxplots show the distribution of differences of cardiac surrogate GP and pulse ox. GP at slice acquisition times in the individual GP model (A) and in the MTGP model (B). The blue frame boxes are the $25^{th}$ percentile to $75^{th}$ percentile. Outliers are indicated in red dots.

# 4   Conclusion

The correlation between surrogate signal and pulse oximetry signal was learned within a MTGP model that produces more robust estimates of the cardiac cycle than individual signals on their own. The cardiac surrogate signal is thus supposed to be used in addition to physiological recordings such as from pulse oximetry. However, there are large clinical fMRI cohorts in which physiological monitoring was not used. For these cohorts, heart rate variation can be extracted from BOLD scans directly using the cardiac surrogate signal. The proposed model presents significant advantages when compared to current data-driven techniques [4] by providing a more realistic joint biophysical noise model of physiologic signals.

**Acknowledgements.** This work was supported by the Wellcome Trust (210182/Z/ 101957/Z/13/Z), EPSRC (NS/A000027/1), MRC (MR/J01107X/1), UCL LWENC (PR/ylr/18575), and the Wolfson Foundation (PR/ylr/18575).

# References

1. Shmueli, K., et al.: Low-frequency fluctuations in the cardiac rate as a source of variance in the resting-state fMRI BOLD signal. NeuroImage **38**, 306–320 (2007)
2. Chang, C., Glover, G.H.: Effects of model-based physiological noise correction on default mode network anti-correlations and correlations. NeuroImage **47**(4), 1448–1459 (2009)
3. Wagshul, M.E., Eide, P.K., Madsen, J.R.: The pulsating brain: a review of experimental and clinical studies of intracranial pulsatility. Fluids Barriers CNS **8**(1), 5 (2011)
4. Murphy, K., Birn, R.M., Bandettini, P.A.: Resting-state fMRI confounds and cleanup. NeuroImage **80**, 349–359 (2013)
5. Stegle, O., Fallert, S.: Gaussian process robust regression for noisy heart rate data. IEEE Trans. Biomed. Eng. **55**(9), 2143–2151 (2008)
6. Dürichen, R., Pimentel, M.A., Clifton, L., Schweikard, A., Clifton, D.A.: Multitask Gaussian processes for multivariate physiological time-series analysis. IEEE Trans. Biomed. Eng. **62**(1), 314–322 (2015)
7. Rasmussen, C.E., Williams, C.K.I.: Gaussian Processes for Machine Learning (2006)
8. Glover, G.H., Li, T.Q., Ress, D.: Image-based method for retrospective correction of physiological motion effects in fMRI: RETROICOR. Magn. Reson. Med. **44**(1), 162–167 (2000)
9. Verstynen, T.D., Deshpande, V.: Using pulse oximetry to account for high and low frequency physiological artifacts in the BOLD signal. NeuroImage **55**(4), 1633–1644 (2011)

# Probabilistic Source Separation
# on Resting-State fMRI and Its Use
# for Early MCI Identification

Eunsong Kang and Heung-Il Suk[✉]

Department of Brain and Cognitive Engineering,
Korea University, Seoul, Republic of Korea
hisuk@korea.ac.kr

**Abstract.** In analyzing rs-fMRI, blind source separation has been studied extensively and various machine-learning techniques have been proposed in the literature. However, to our best knowledge, most of the existing methods do not explicitly separate noise components that naturally corrupt the observed BOLD signals, thus hindering from the understanding of underlying functional mechanisms in a human brain. In this paper, we formulate the problem of latent source separation in a probabilistic manner, where we explicitly separate the observed signals into a true source signal and a noise component. As for the inference of the latent source distribution with respect to an input regional mean signal, we use a stochastic variational Bayesian inference and implement it in a neural network framework. Further, in order for identification of a subject with early mild cognitive impairment (eMCI) rs-fMRI, we also propose to use the relations of the inferred source signals as features, *i.e.*, potential imaging-biomarkers. We presented the validity of the proposed methods by conducting experiments on the publicly available ADNI2 dataset and comparing with the existing methods.

## 1 Introduction

The Alzheimer's disease (AD), the most common cause of dementia, is a neurodegenerative disease and a patient with AD progressively gets worse in cognitive functions over time [1]. In the meantime, mild cognitive impairment (MCI) is characterized as the prodromal stage of dementia. Recent studies have presented that the early diagnosis of MCI can possibly allow to slow its progression to dementia with appropriate and effective clinical treatments, albeit no medicine to cure AD itself available yet. In this regard, the diagnosis of MCI is of paramount importance in the clinic. In order for MCI diagnosis, many studies have investigated the regions related to MCI with neuroimaging data magnetic resonance image (MRI) [9], functional MRI (fMRI) [2] or diffusion tensor image (DTI) [10]. Particularly, researchers have paid more attention to resting-state fMRI (rs-fMRI), which measures blood-oxygenation-level-dependent (BOLD) signals of a subject with no cognitive task involved. From a neurologic perspective, [4]

© Springer Nature Switzerland AG 2018
A. F. Frangi et al. (Eds.): MICCAI 2018, LNCS 11072, pp. 275–283, 2018.
https://doi.org/10.1007/978-3-030-00931-1_32

presented that functional changes in a brain might occur in advance of structural changes, thus it may provide potential imaging-biomarkers for MCI diagnosis.

From a signal-processing and machine-learning, there have been enormous studies in blind source separation (BSS) with rs-fMRI by assuming that the observed rs-fMRI signals are a mixture of unknown sources, which may be relevant to activations of interest [13]. Of various techniques, independent component analysis (ICA) has been the most successfully used in numerous studies. However, it is still compelling for ICA to manually select the important components and distinguish noise from neurophysiological signals. In the meantime, a recent resurgence of deep learning with great success in various applications has been drawing researchers' attention on applying deep learning methods to rs-fMRI analysis. However, due to its unfavorable black-box property, it is limited to use a shallow architecture, *e.g.*, restricted Boltzmann machine (RBM) [7].

While it is commonly accepted that the observed BOLD signals are corrupted by unknown noise sources, to our best knowledge, most of the existing methods in the literature try to estimate the unknown sources without modeling a noise. For instance, ICA decomposes a set of statistically independent signals from the raw observed BOLD signals directly and involves a noise distinguishing step with manual inspection, for which it is required a good neurophysiological knowledge. In the meantime, the RBM-based method [6,11] doesn't explicitly consider the noises at all.

In this work, we propose a novel probabilistic framework of inferring the source distribution for the observed rs-fMRI signals. In particular, we derive a formulation of probabilistic BSS, where the true source signals are explicitly separated from the noisy component, which is represented as an uncertainty or variance term in the distribution. In order to tackle the intractability of the inference problem with respect to the source signal distribution, we apply a stochastic variational Bayesian inference via neural network training with random variable units. Meanwhile, we also propose a novel feature representation from the inferred source signal distribution for early MCI (eMCI) diagnosis with rs-fMRI. We validate the effectiveness of our method by conducting experiments on samples acquired from the ADNI2 cohort.

## 2    Materials and Data Preprocessing

We used rs-fMRI samples acquired from the ANDI2 cohort[1], which consists of 31 eMCI (eMCI) and 30 cognitively normal (CN) subjects. The mean age of the eMCI and CN groups are $73.9 \pm 4.9$ and $73.8 \pm 5.5$, respectively. All subjects were scanned at different imaging centers using 3.0T Philips Achieva scanners with the same scanning protocol and parameters of Repetition Time (TR) = 3,000 ms, Echo Time (TE) = 30ms, flip angle = 80°, acquisition matrix size = 64 × 64, 48 slices, 140 volumes, and a voxel thickness = 3.3 mm.

For magnetization equilibrium, we discarded the first 10 volume images of each subject. The remaining 130 fMRI volume images were processed by applying

---

[1] http://adni.loni.usc.edu.

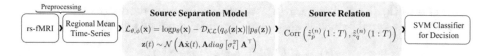

**Fig. 1.** An overview of the proposed framework for eMCI identification with rs-fMRI. We infer the source signals via probabilistic source separation model, and then measure the correlation of sources for classification. For notations, refer to the Sect. 3.

the procedures of slice timing, motion correction, and spatial normalization using SPM8. We considered signals of the gray matter only for further processing. The fMRI brain space was then parcellated into 116 Regions-Of-Interest (ROIs) based on the AAL template. We bandpass-filtered the regional mean time-series with a frequency band of 0.01–0.08 Hz to utilize the low-frequency fluctuation features in rs-fMRI. The representative mean time-series of each ROI was computed by averaging the intensity of all voxels in an ROI, thereby we had a $R(= 116)$ number of regional mean time-series $\mathcal{X}^{(n)} = \left[\mathbf{x}^{(n)}(1), \ldots, \mathbf{x}^{(n)}(T)\right] \in \mathbb{R}^{R \times T}$ for a subject $n$, where $T$ denotes the number of volumes, *i.e.*, $T = 130$ in our case.

## 3  Proposed Method

In this work, we formulate a blind source separation from regional mean time-series of rs-fMRI in a probabilistic manner and infer the distribution of source signals via a neural network framework. Further, we also propose to extract a novel feature representation from the estimated source signals for MCI identification. The overall framework of our proposed method is presented in Fig. 1.

### 3.1  Inferring Latent Source Distribution

With the general assumption that the observed BOLD signal $\mathbf{x}(t) \in \mathbb{R}^R$ is corrupted by noises, we define a regional mean signal of ROIs at time $t$ as follows:

$$\mathbf{x}(t) = \hat{\mathbf{x}}(t) + \boldsymbol{\epsilon}(t) \tag{1}$$

where $\hat{\mathbf{x}}(t)$ is an uncorrupted signal. Without loss of generality, we assume that the noise vector is Gaussian distributed with zero mean and a diagonal covariance as follows:

$$\boldsymbol{\epsilon}(t) \sim \mathcal{N}\left(\mathbf{0}, diag\left[\sigma_i^2\right]\right) \tag{2}$$

where $\sigma_i^2$ denotes an unknown variance of the $i$-th noise variable and $i = \{1, \ldots, R\}$.

A blind source separation is, in general, formulated as follows:

$$\mathbf{z}(t) = \mathbf{A}\mathbf{x}(t) \tag{3}$$

where $\mathbf{z}(t) \in \mathbb{R}^L$ and $\mathbf{A} \in \mathbb{R}^{L \times R}$ denote, respectively, the source signal and the unmixing matrix, and $L$ is the dimension of a source signal. Then by injecting

Eq. (1) into Eq. (3) utilizing Eq. (2), we can analytically derive the distribution of the source signal, which is also Gaussian distributed as follows:

$$\mathbf{z}(t) \sim \mathcal{N}\left(\mathbf{A}\hat{\mathbf{x}}(t), \mathbf{A}diag\left[\sigma_i^2\right]\mathbf{A}^{\top}\right). \tag{4}$$

In Eq. (4), we regard the mean component of $\mathbf{A}\hat{\mathbf{x}}(t)$ as the "*clean*" source signal by filtering out the noisy components into the variance or uncertainty term $\mathbf{A}diag\left[\sigma_i^2\right]\mathbf{A}^{\top}$. It is sufficient to estimate the unmixing matrix $\mathbf{A}$ and the variances $\left\{\sigma_i^2\right\}_{i=1,\ldots,R}$ to fully specify the Gaussian distribution of source signal in Eq. (4). But, rather than estimating the unknown variables of $\mathbf{A}$ and $\left\{\sigma_i^2\right\}_{i=1,\ldots,R}$ and then inferring the source signal distribution, we directly estimate the mean and covariance terms in the Gaussian distribution of the source signal, *i.e.*, $\mathbf{A}\hat{\mathbf{x}}(t)$ and $\mathbf{A}diag\left[\sigma_i^2\right]\mathbf{A}^{\top}$, for the observed signal $\mathbf{x}(t)$.

To estimate the Gaussian distribution of a source signal, we use a stochastic variational Bayesian inference that can be efficiently tackled by means of a variational auto-encoder (VAE) [8] with a two-layer neural network as shown in Fig. 2. Here, we omit the time index $(t)$ without loss of generality for uncluttered. There are two key concepts in building and training our variational auto-encoder; (i) From a generative model's perspective, when we have a joint distribution of $p(\mathbf{x}, \mathbf{z}) = p(\mathbf{x}|\mathbf{z})p(\mathbf{z})$, it is possible to generate an observational signal in the way of first drawing a source sample from the latent source distribution $p(\mathbf{z})$, which corresponds to Eq. (4) in our case, and then drawing an observation sample from the conditional distribution of an observational variable given the value of a source signal $p_\theta(\mathbf{x}|\mathbf{z})$, where $\theta$ denotes the parameters to specify the distribution. In our network, the subnetwork of the hidden layer and the output layer plays this role, thereby regarded as a '*generative network*.' (ii) For the inference of the true latent source distribution $p(\mathbf{z})$, since it is intractable to estimate analytically, we use a variational approximation technique and learn an approximation distribution $q_\phi(\mathbf{z}|\mathbf{x})$ with the subnetwork of the input layer and the hidden layer, which we thus call as an '*inference network*.' Here, $\phi$ denotes the parameters of the inference network.

As for the latent source distribution $p(\mathbf{z})$, we use the definition in Eq. (4) with the independence assumption among the rows of an unmixing matrix $\mathbf{A}$. Thus, we make the mean and covariance terms in Eq. (4) explicit in our network, such that the units in the hidden layer to denote the random variables of a mean vector $\hat{\mathbf{z}}(t) \equiv \mathbf{A}\hat{\mathbf{x}}(t)$ and a covariance matrix $diag\left[\hat{\sigma}_i^2\right] = \mathbf{A}\,diag\left[\sigma_i^2\right]\mathbf{A}^{\top}$, where $i = \{1, \ldots, L\}^2$. Regarding the activation functions in the network, we use an identity function and a softplus function for the hidden units corresponding to the mean and variance, respectively.

The objective function of our network is defined as follows:

$$\mathcal{L}_{\theta,\phi}(\mathbf{x}) = \log p_\theta(\mathbf{x}) - \mathcal{D}_{\mathcal{KL}}(q_\phi(\mathbf{z}|\mathbf{x})||p_\theta(\mathbf{z})) \tag{5}$$

where $p_\theta(\mathbf{x})$ corresponds to an observation likelihood and $\mathcal{D}_{\mathcal{KL}}$ denotes a Kullback-Leibler divergence. The observation likelihood measures how well the

---

[2] Therefore, there are $2L$ hidden units.

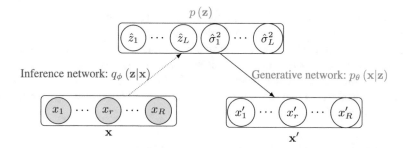

**Fig. 2.** Graphical representation of our stochastic variational inference network that can estimate the distribution of the latent source signals $p(\mathbf{z})$ given an input $\mathbf{x}$, and generate mean regional signals $\mathbf{x}'$. The unshaded circles denote random variables. The black arrows denote full bipartite connections between every pair of units in two rectangles.

generated sample $\mathbf{x}'$ from the generative network, which is conditioned on the sample drawn from the inferred source signal distribution ($\hat{\mathbf{z}}$ and $diag\left[\hat{\sigma}_i^2\right]$) with respect to an input $\mathbf{x}$, is similar to an input $\mathbf{x}$. The network parameters of $\phi$ (inference) and $\theta$ (generative) are trained via stochastic gradient descent and back-propagation with a reparameterization trick [8]. Note that in training our network, a regional mean vector at one time point becomes a single training sample. Thus, for a dataset from $N$ number of subjects with $T$ number of rs-fMRI volumes per subject, we have $N \times T$ number of training samples. After training our network parameters, given a regional mean vector $\mathbf{x}(t)$ at time $t$, the latent source distribution can be inferred by feed-forwarding to the hidden layer.

### 3.2   Source Relations as a Feature Representation

For a regional mean signal $\mathbf{x}(t)$, the source distribution inferred by our network in Fig. 2 can be written as follows:

$$\mathbf{z}(t) \sim \mathcal{N}\left(\hat{\mathbf{z}}(t), diag\left[\hat{\sigma}_i^2\right]\right). \tag{6}$$

Here, it is noteworthy that $\hat{\mathbf{z}}(t)$ corresponds to the clean source signal $\mathbf{A}\hat{\mathbf{x}}(t)$ and the noisy signals are filtered out to the variance component. Then, for a regional mean time-series of a subject $n$, i.e., $\mathcal{X}^{(n)}$, we can obtain the expected "clean" source time-series $\mathcal{S}^{(n)} = \left[\hat{\mathbf{z}}^{(n)}(1), \ldots, \hat{\mathbf{z}}^{(n)}(T)\right] \in \mathbb{R}^{L \times T}$. With this clean source signal, we compute the temporal correlations between source signals defined as follows:

$$\text{Corr}\left(\hat{z}_p^{(n)}(1:T), \hat{z}_q^{(n)}(1:T)\right) = \frac{\text{Cov}\left[\hat{z}_p^{(n)}(1:T), \hat{z}_q^{(n)}(1:T)\right]}{\text{Var}\left[\hat{z}_p^{(n)}(1:T)\right]\text{Var}\left[\hat{z}_q^{(n)}(1:T)\right]} \tag{7}$$

where $\hat{z}_p^{(n)}$ denotes the $p$-th element in the mean vector $\hat{\mathbf{z}}^{(n)}$ and $p, q = \{1, \ldots, L\}$.

When regarding the BOLD signals of each region as low-level information, the source signals can be thought of high-level information. Then it is comparable for the source-wise correlation to the region-wise correlation which is commonly used for functional connectivity estimation. With correlation estimated from the high-level information, it is expected to find more class-discriminative patterns, thereby we propose to use this source correlation as potential imaging-biomarker for eMCI diagnosis.

## 4    Experimental Settings and Results

### 4.1    Comparative Models and Settings

The preprocessed fMRI data was Gaussian normalized for each subject in an ROI-wise manner. As for the number of hidden units, *i.e.*, dimension of source signals, in our network, we empirically set $L$ to $25^3$. To avoid overfitting and to better train of our network, we applied an $\ell_1$ regularization by setting the corresponding hyperparameter to $10^{-3}$.

The inference network and generative network were trained using RMSPropOptimizer with a learning rate of $10^{-4}$, a mini-batch size of 10, and an epoch of 50. The final codes and trained models are open at 'https://github.com/EunsongK/VAE-for-Probabilistic-Source-Separation.'

In order to see the validity and effectiveness of our proposed method, we also compared with three existing models for blind source separation in the literature.

– Independent Component Analysis (ICA) [13]: An ICA is the most popular method for blind source separation with an independence assumption among source signals. Specifically, we used a fastICA by setting the number of sources in $L \in \{20, 25\}$.
– Auto-Encoder (AE) [12]: An AE has the similar structure with our VAE in the sense that it consists of an input layer, a hidden layer, and an output layer. Also, it aims to compress the input data into the hidden units, corresponding to the latent variable in our VAE, and reconstruct the outputs in the same size of inputs. However, unlike our VAE, the units in the network are all deterministic. We trained AE models with the same settings with our VAE.
– Restricted Boltzmann Machine (RBM) [7]: An RBM shares the similarity with our VAE in that it learns the distributions of latent variables and input variables in different layers. Further, as a generative model, it has a power of generating an observed signals. In particular, we used a Gaussian-Bernoulli model with continuous-valued input units and binary-valued hidden units. To compensate for the limitations in representation of binary-valued hidden units[4], we considered a larger number of sources by varying the dimension $L \in \{100, 150, 200, 300\}$. The RBM was trained with a contrastive divergence algorithm [5] with a learning rate of $10^{-4}$, a mini-batch size of 10, and an epoch of 500.

---

[3] We varied the value of $L$ in the space of $\{10, 20, 25, 30, 40\}$.

[4] We used a logistic sigmoid function for an activation of hidden units.

**Table 1.** Performance comparison with the comparative models.

| | AUC | ACC (%) | SEN (%) | SPC (%) | PPV (%) | NPV (%) |
|---|---|---|---|---|---|---|
| Baseline | 0.6241 | 58.66 ± 0.39 | 64.90 ± 1.02 | 52.22 ± 1.11 | 58.40 ± 0.38 | 59.01 ± 0.47 |
| ICA | 0.6847 | 61.52 ± 1.74 | 65.07 ± 1.31 | 57.86 ± 2.40 | 61.49 ± 1.76 | 61.57 ± 1.73 |
| AE | 0.6888 | 63.36 ± 1.53 | 66.86 ± 0.81 | 59.75 ± 2.66 | 63.22 ± 1.74 | 63.54 ± 1.32 |
| RBM | 0.6248 | 59.68 ± 1.74 | 64.82 ± 0.64 | 54.37 ± 3.35 | 59.52 ± 1.77 | 59.88 ± 1.70 |
| Proposed | **0.8335** | **74.45 ± 1.10** | **78.16 ± 1.22** | **71.20 ± 1.45** | **73.72 ± 1.12** | **75.94 ± 1.22** |

## 4.2 Performance Comparison

To validate the superiority of our method, we evaluated the diagnostic performance with 10-fold cross validation, where 9 out 10 folds were used for training and the remaining one fold was used for testing. We repeated this process 10 times to obtain more robust results.

We took the elements of the upper triangle of the $z$-score map, obtained by Fisher's transformation of a source signal correlation matrix, for each subject as features, we have a training set $\left\{ \mathbf{f}^{(n)}, l^{(n)} \right\}_{n=1,\dots,N}$ for classifier training, where $\mathbf{f}^{(n)}$ and $l^{(n)} \in \{-1(\text{CN}), +1(\text{eMCI})\}$ denote, respectively a feature vector and a class label of a subject $n$. For classification, we use a linear Support Vector Machine (SVM) [3]. In SVM training, the hyperparameter $C$ for softmargin was chosen by 5-fold nested cross-validation in the space of $\{10^{-5}, 10^{-4}, \dots, 10^4\}$. For comparison, we also considered a 'Baseline' model that estimated to correlations of the raw regional mean BOLD signals, then it is comparable for the source-wise correlation to the region-wise correlation. For feature extraction, we commonly used a Pearson correlation coefficients of the estimated sources with different models listed in Sect. 4.1. For comparison, we also considered a 'Baseline' model that estimated to correlations of the raw regional mean BOLD signals. For all the competing methods, we transformed the respective correlation matrix into a $z$-score map and used the elements of the upper triangle of the map as input features to a linear SVM[5].

Table 1 shows the performance of the competing methods in different metrics: AUC (Area Under the receiver operator Characteristic Curve), ACC (accuracy), SEN (sensitivity), SPC (specificity), and PPV/NPV (positive/negative predictive value). Note that our proposed method remarkably outperformed all the competing methods with a large margin. Specifically, our method improved the AUC by ∼15% (vs. ICA and AE) and ∼22% (vs. Baseline and RBM). It is also noteworthy that the performance of the source separation methods was consistently superior to the Baseline method, which empirically and indirectly indicates the validity of using high-level source correlation for eMCI diagnosis.

---

[5] We used a package of the 'libSVM-3.21.'

## 5   Conclusion

In this work, we formulated a novel method of representing a blind source distribution inferred from an input signal, where the true source signal and the noise component are separated in the form of mean and covariance terms, respectively. As for the inference of the source distribution with respect to an input regional mean signal, we used a stochastic variational Bayesian inference implemented in a neural network. We also proposed to use the relations of the inferred source signals as features to identify a subject with eMCI. We validated the effectiveness of both the source inference and the source-relation features for eMCI identification by outperforming the competing methods considered in our experiments.

Thanks to the power of separating true source signals and the noise components, it is possible to reconstruct a "clean" or noise-removed input signal from the true source signal only by discarding the noise component. With the noise-removed input signal, it is expected to have better insights and understanding of the underlying mechanisms in brain functions. Thus, it will be our forthcoming research issue.

**Acknowledgement.** This research was supported by the Bio & Medical Technology Development Program of the NRF funded by the Korean government, MSIP(2016941946), and Institute for Information & communications Technology Promotion (IITP) grant funded by the Korea government (MSIT) (No.2017-0-01779, A machine learning and statistical inference framework for explainable artificial intelligence).

## References

1. Alzheimer's Association, et al.: 2017 Alzheimer's disease facts and figures. Alzheimer's Dement. **13**(4), 325–373 (2017)
2. Challis, E., Hurley, P., Serra, L., Bozzali, M., Oliver, S., Cercignani, M.: Gaussian process classification of Alzheimer's disease and mild cognitive impairment from resting-state fMRI. NeuroImage **112**, 232–243 (2015)
3. Chang, C.C., Lin, C.J.: LIVSVM: a library for support vector machines. ACM Trans. Intell. Syst. Technol. (TIST) **2**(3), 27 (2011)
4. Fox, M.D., Raichle, M.E.: Spontaneous fluctuations in brain activity observed with functional magnetic resonance imaging. Nat. Rev. Neurosci. **8**(9), 700 (2007)
5. Hinton, G.E.: Training products of experts by minimizing contrastive divergence. Neural Comput. **14**(8), 1771–1800 (2002)
6. Hjelm, R.D., Calhoun, V.D., Salakhutdinov, R., Allen, E.A., Adali, T., Plis, S.M.: Restricted Boltzmann machines for neuroimaging: an application in identifying intrinsic networks. NeuroImage **96**, 245–260 (2014)
7. Hu, X., et al.: Latent source mining in fMRI via restricted Boltzmann machine. Hum. Brain Mapp. **39**, 2368–2380 (2018)
8. Kingma, D.P., Welling, M.: Auto-encoding variational bayes. In: The International Conference on Learning Representations (ICLR) (2014)
9. Moradi, E., Pepe, A., Gaser, C., Huttunen, H., Tohka, J., Alzheimer's Disease Neuroimaging Initiative.: Machine learning framework for early MRI-based Alzheimer's conversion prediction in MCI subjects. NeuroImage **104**, 398–412 (2015)

10. Neth, B.J., Hughes, T.M., Craft, S.: Metabolic status and family history as determinants of Alzheimer's risk using diffusion tensor imaging. Alzheimer's Dement.: J. Alzheimer's Assoc. **11**(7), P341 (2015)
11. Plis, S.M., et al.: Deep learning for neuroimaging: a validation study. Front. Neurosci. **8**, 229 (2014)
12. Baldi, P.: Autoencoders, unsupervised learning, and deep architectures. In: Guyon, I., Dror, G., Lemaire, V., Taylor, G., Silver, D. (eds.) Proceedings of ICML Workshop on Unsupervised and Transfer Learning. Proceedings of Machine Learning Research, Bellevue, Washington, USA, vol. 27, pp. 37–49. PMLR (2012). http:// proceedings.mlr.press/v27/baldi12a.html
13. de Vos, F., et al.: A comprehensive analysis of resting state fMRI measures to classify individual patients with Alzheimer's disease. NeuroImage **167**, 62–72 (2018)

# Identifying Brain Networks of Multiple Time Scales via Deep Recurrent Neural Network

Yan Cui[1], Shijie Zhao[2], Han Wang[1], Li Xie[1], Yaowu Chen[1],
Junwei Han[2], Lei Guo[2], Fan Zhou[1(✉)], and Tianming Liu[3(✉)]

[1] College of Biomedical Engineering and Instrument Science,
Zhejiang University, Hangzhou, China
fanzhou@mail.bme.zju.edu.cn
[2] School of Automation, Northwestern Polytechnical University, Xi'an, China
[3] Cortical Architecture Imaging and Discovery Lab,
Department of Computer Science and Bioimaging Research Center,
The University of Georgia, Athens, GA, USA
tianming.liu@gmail.com

**Abstract.** For decades, task-based functional magnetic resonance imaging (tfMRI) has been a powerful noninvasive tool to explore the organizational architecture of human brain function. Researchers have developed a variety of brain network analysis methods for tfMRI data, including the general linear model (GLM), independent component analysis (ICA) and sparse representation methods. However, these shallow models are limited in faithful reconstruction and modeling of the hierarchical and temporal structures of brain networks, as demonstrated in more and more studies. Recently, recurrent neural networks (RNNs) exhibit great ability of modeling hierarchical and temporal dependency features in the machine learning field, which might be suitable for tfMRI data modeling. To explore such possible advantages of RNNs for tfMRI data, we propose a novel framework of deep recurrent neural network (DRNN) to model the functional brain networks for tfMRI data. Experimental results on the motor task tfMRI data of Human Connectome Project 900 subjects data release demonstrated that the proposed DRNN can not only faithfully reconstruct functional brain networks, but also identify more meaningful brain networks with multiple time scales which are overlooked by traditional shallow models. In general, this work provides an effective and powerful approach to identifying functional brain networks of multiple time scales from tfMRI data.

**Keywords:** Task fMRI · Brain network · RNN · Deep learning

## 1 Introduction

Exploring the organizational architecture of human brain function has been of great interest in neuroscience community [1]. After decades of active research using non-invasive neuroimaging methods such as functional magnetic resonance imaging

---

Y. Cui and S. Zhao—Co-first authors.

A. F. Frangi et al. (Eds.): MICCAI 2018, LNCS 11072, pp. 284–292, 2018.
https://doi.org/10.1007/978-3-030-00931-1_33

(fMRI), there has been mounting evidence that the brain function is realized by the interaction of multiple concurrent neural process or functional brain networks [2] and these networks are spatially distributed across specific structural substrate of neuroanatomical areas [3]. In these fMRI based studies, researchers developed a variety of brain network reconstruction and modeling techniques, such as the general linear model (GLM) [4], independent component analysis (ICA) [5] and sparse representation/dictionary learning methods [6, 7]. These methods reconstructed many meaningful functional brain networks which are characterized by both spatial maps and corresponding temporal time series from both tfMRI and rsfMRI data sets and greatly advanced our understanding of the regularity and variability of brain functions [4, 5].

However, those existing approaches which are based on *shallow* models are limited in faithful reconstruction and modeling of the *hierarchical and temporal* structures of brain functional networks in tfMRI data [8]. Recently, deep learning methods have attracted much attention in a variety of challenges [9]. The success of deep learning methods lies in the ability of automatically and hierarchically representing the raw data. Inspired by the great success of deep learning methods, more and more researchers applied deep learning methods in functional brain network analysis [10–12]. Although recent works demonstrate the advantages of deep learning methods, information of multiple temporal scales is rarely taken into consideration in these models, although it is known that brain activities have multiple time scales [13].

Recently, recurrent neural networks (RNNs) are gaining more and more attention [14]. Unlike traditional neural networks, RNNs can use their internal memory unit to process arbitrary sequences of inputs and model the sequential and time dependencies on multiple time scales [15]. That is, RNN models make their predictions based on not only the information available at a given time, but also the information that was available in the past. Actually, the brain activity is modulated by long temporal dependencies [16], which quite coincides with the characteristics of RNN models. Therefore, it is quite natural and well justified to adopt RNNs to explore the brain functional networks in tfMRI data. However, it has been rarely explored whether RNNs can be utilized to infer functional brain networks with the whole brain tfMRI data. In order to explore the possible advantages of RNN models, in this study, we proposed a novel, alternative framework of deep recurrent neural network (DRNN) for modeling functional brain networks in tfMRI data. An important characteristic of DRNN framework is that the task stimulus information is sequentially processed through the model and it automatically generates the observed whole brain voxel signals. In this way, the hierarchical and temporal structures of the brain activities are captured and brain networks of multiple time scales (especially time dependency sensitive brain networks) can be identified. We used the motor task tfMRI dataset of HCP 900 subjects data release as a test-bed, and extensive experimental results demonstrated the superiority of the proposed method in identifying functional brain networks of multiple time scales in tfMRI.

## 2    Materials and Methods

### 2.1    Overview

Figure 1 summarizes the proposed deep recurrent neural network (DRNN) model. There are three major steps to model tfMRI functional brain networks using DRNN. First, for each subject, the task design stimulus curves are gathered into a stimulus matrix $X$ ($k$ stimuli with $t$ time series) as the input layer and the whole brain tfMRI signals are aggregated into a big signal matrix $Y$ ($m$ voxels with $t$ time series). Then these task stimulus patterns passed through two hidden layers and each layer is of $n_h$ RNN units, respectively. Next, the response of top hidden layer is connected to the whole brain signal matrix ($m$ voxels' signals with $t$ time series) via a fully connected layer ($[n_h, m]$). Specifically, each hidden node's connection weight vector represents a typical functional brain network, and its corresponding hidden response to specific stimulus patterns represents the temporal activity pattern of the network.

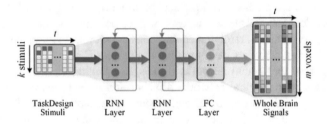

**Fig. 1.**  Overview of the DRNN model.

### 2.2    Data Acquisition and Pre-processing

The Human Connectome Project (HCP) dataset is one of the most systematic and comprehensive neuroimaging data set in current stage which aims to bring data from the major MRI neuroimaging modalities together into a cohesive framework to enable detailed comparisons between brain architecture, connectivity, and function across individual subjects. Importantly, this data set is publicly available which makes it a

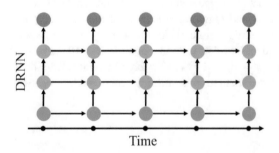

**Fig. 2.**  Illustrative map of DRNN. Blue circle represents input units, green circle represents hidden units, and red circle represents output units.

good test bed for different researchers. In this paper, we adopt motor tfMRI dataset of HCP 900 subjects data release to test our proposed method. The detailed design paradigms of motor task and other tasks are available in [17].

The detailed acquisition parameters of these tfMRI data were set as follows: 220 mm FOV, in-plane FOV: 208 × 180 mm, flip angle = 52, BW = 2290 Hz/Px, 2 × 2×2 mm spatial resolution, 90 × 104 matrix, 72 slices, TR = 0.72 s, TE = 33.1 ms. The preprocessing of the task fMRI data sets includes skull removal, motion correction, slice time correction, spatial smoothing, and global drift removal (high-pass filtering). All these preprocessing steps were implemented in FSL FEAT. All of these individual fMRI datasets are first registered to MNI common space for further study. Besides, the GLM-based activation results are also derived using FSL FEAT for comparison.

## 2.3 Deep Recurrent Neural Network Model

RNNs are feedforward neural networks augmented with edges spanning adjacent time steps where connections between units form a directed cycle. These connections introduce a notation of time and provide memory of past state. In contrast with traditional neural networks which only receive information at the bottom layer and output at the highest layer, RNNs receive input and produce output at each iteration step. However, a common RNN only process information through one layer before going to output, which could not provide hierarchical structure of processing the input information and the temporal hierarchy of input signals is not clear. In order to overcome these limitations, we propose a deep recurrent neural network (DRNN) framework for modeling functional brain networks in tfMRI data. The basic idea of DRNN is stacking RNNs to construct a hierarchical network architecture. Each hidden layer is a recurrent neural network and the hidden state of each layer is the input of next layer. In this way, new information propagates throughout the hierarchy during each network update and temporal context is added in each layer (Fig. 2). As demonstrated in character-based language modelling studies [15], stacking RNNs automatically creates different time scales across different levels and also forms a temporal hierarchical information processing structure.

We define a DRNN with $L$ layers and each layer has $n_i$ hidden units. The input sequence is denoted as $\left(x^{(1)}, x^{(2)}, \ldots, x^{(t)}\right)$ where each data point is a real-valued vector and the target sequence is denoted as $\left(y^{(1)}, y^{(2)}, \ldots, y^{(t)}\right)$ and the hidden state of $i$-th layer is denoted as $h_i^{(t)}$. In order to avoid confusion between the indices of nodes and sequence steps, we use superscripts for time and subscripts for layer index. The output of DRNN model can be modeled as Eq. (1), where $\hat{y}^t$ is the estimated output from the top hidden layer and $V$ is the weight matrix between hidden layer and output, and $\mathbf{b}_i$ is the bias parameters which contain the offset of each node.

$$\hat{y}^t = \sigma\left(Vh_i^t + \mathbf{b}_i\right) \qquad (1)$$

There are different types of RNN architectures and the long short-term memory (LSTM) is among the most popular specialized memory units of RNNs, which is developed for long time series. The first hidden states of an LSTM unit are defined as:

$$\boldsymbol{h}^t = \boldsymbol{o}^t \odot \tanh(\boldsymbol{c}^t) \tag{2}$$

$$\boldsymbol{o}^t = \sigma\left(\boldsymbol{U}_o \boldsymbol{h}^{t-1} + \boldsymbol{W}_o \boldsymbol{x}^t + \boldsymbol{b}_o\right) \tag{3}$$

where $\boldsymbol{c}^t$ is the cell state, $\boldsymbol{o}^t$ are the output gate activities, and $\odot$ denotes elementwise multiplication. Information about the previous time points is stored in the cell state. What information will be retrieved from the cell state is controlled by the output gate. The second-layer and upper hidden states are defined similarly to the first-layer hidden/cell states, except for the input is replaced with the output of first-layer hidden states. The parameters in the DRNN framework is optimized to minimize the mean square error between the whole brain signals and their reconstructions. The TensorFlow [18] system is adopted to implement the models.

### 2.4    Identification of Functional Brain Networks

In the DRNN model, the task design stimulus information is separated in different time points and put into the model step by step in each iteration. In each network update, new information is propagated to the hierarchical structure and temporal context is added in each RNN layer. Each hidden layer in the DRNN is a recurrent neural network and each upper layer receives the hidden state information from previous layer as input. Thus, the output information through the stacking RNNs structure is of different time scales. Finally, the top hidden layer's output is connected to the whole brain signal matrix via a fully connected layer. Specifically, each hidden node's connection weight vector represents a typical functional brain network's spatial distribution and its corresponding hidden response to specific stimulus represents the temporal pattern of the network. In order to compare the derived brain networks with those by other methods, a spatial matching method is adopted to calculate the spatial similarity between the identified networks and the network templates derived from other methods. The spatial similarity is defined as the spatial pattern overlap rate $R$:

$$R(\boldsymbol{S}, \boldsymbol{T}) = \frac{|\boldsymbol{S} \bigcap \boldsymbol{T}|}{|\boldsymbol{T}|} \tag{4}$$

where $\boldsymbol{S}$ and $\boldsymbol{T}$ are cortical spatial maps of a brain network component and the brain network template, respectively.

## 3    Experimental Results

### 3.1    Identified Typical Functional Brain Networks

Figure 3 illustrates a few typical brain networks identified on the motor tfMRI dataset of HCP 900 subjects release using DRNN model. For comparison, we also list the GLM group-wise activation maps on the right column. This figure clearly shows that part of our trained functional networks are quite similar to the corresponding GLM activation maps. In order to quantitatively measure the similarity, we adopt the Eq. (4) to calculate the spatial overlap rate between the identified DRNN networks and the corresponding GLM activation maps, which are listed in the first row of Table 1. In addition, the corresponding temporal patterns are also quite similar to the common HRF response patterns (convolution results of task design paradigm and HRF function). Figure 4 shows that the corresponding temporal response patterns, the task design pattern and the HRF response patterns. It is easy to see that the temporal patterns of DRNN brain networks have high correlations to the HRF responses. Through comparisons, the high spatial overlap rate and close temporal correlation suggest that the proposed DRNN model can identify meaningful and reliable functional networks in an automatic way.

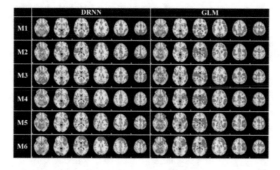

**Fig. 3.** A few identified functional brain networks in the motor task tfMRI dataset of HCP 900 subjects release. The left is the networks identified using DRNN model with LSTM units and the right is the GLM-derived group-wise activation maps. M1–M6 represent different stimuli.

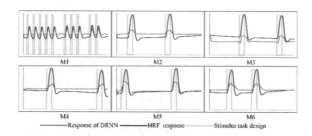

**Fig. 4.** Temporal response patterns corresponding to the identified brain networks in Fig. 3.

**Table 1.** The first row shows the spatial overlap rate between the identified networks by DRNN and the corresponding GLM-derived group-wise activation maps. The second row shows the Pearson correlation between the temporal pattern and the common HRF response patterns.

| M1 | M2 | M3 | M4 | M5 | M6 |
|------|------|------|------|------|------|
| 0.66 | 0.52 | 0.56 | 0.54 | 0.44 | 0.54 |
| 0.93 | 0.94 | 0.90 | 0.95 | 0.94 | 0.85 |

## 3.2    Identified Functional Brain Networks of Multiple Time Scales

During the training stage, the task stimulus information goes through the hierarchical and temporal model iteratively, and the final output naturally reflects the brain network's responses to the original stimulus information crossing multiple time scales. After training stage, we input each stimulus separately and obtain the corresponding temporal patterns for each network. In order to better interpret the identified functional brain networks, we further calculated the correlations between the identified temporal brain activity patterns and the theoretical regressor groups which were adopted in previous literature studies [12]. Essentially, the theoretical regressor groups represent the possible multiple time scale brain responses. Our basic idea is that if a specific temporal pattern is highly correlated with an extended theoretical regressor, the corresponding identified DRNN network should belong to the similar time scale network.

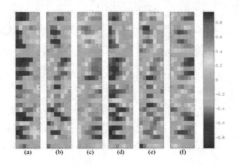

**Fig. 5.** Temporal correlation maps between temporal response patterns of the identified 30 DRNN networks and the extended hypothetical regressor groups in [12]. (a) HRF delay group; (b) derivative form group; (c) integral form group; (d) inversed HRF group; (e) inversed derivative form group; (f) inversed integral form group. In each subfigure, each row represents a DRNN network and 7 columns represent 7 different time delays with an interval of 3 s.

Figure 5 shows the temporal correlation maps between temporal response patterns of the 30 identified DRNN networks using Stimulus M6 and the extended hypothetical regressor groups in [12]. Similarly, we also extended the basic HRF response patterns with multiple delays, derivate, integral and inverse operation. From this figure, we can see that there are a few network temporal patterns highly correlated with the extended hypothetical regressors, and they represent the identified different time scales of brain networks. Figure 6 illustrates a few typical identified different time scales of brain networks and corresponding temporal patterns. From this result, we can see that a variety of time scales of theoretical response networks including multiple delays, multiple inversed HRFs and delays, different derivative and integral operations could be identified. We further checked the spatial patterns of these networks and it is interesting that these networks are similar but not the same. This is reasonable since these networks are evoked by the same stimulus but at different time scales. These multiple time scales of brain networks can be effectively identified with the DRNN framework, which is a major advantage of the proposed model.

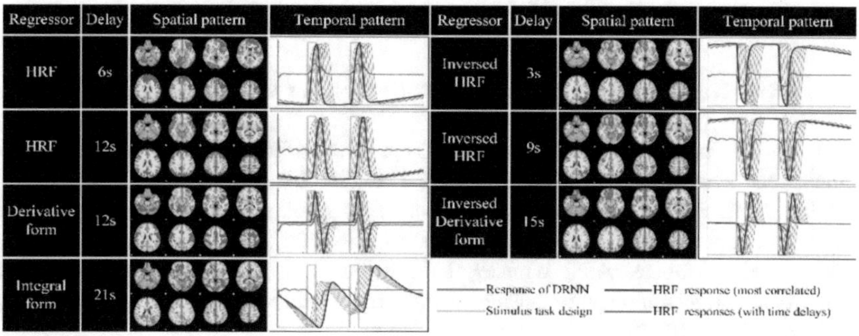

**Fig. 6.** The spatial and temporal patterns of a few identified brain networks of multiple time scales shown in Fig. 5.

The proposed DRNN model was also applied on half of the HCP Q1 release dataset (34 subjects) and obtained similar and consistent results. However, more training data (HCP Q3 release) will improve the reliability and interpretive of the results. L1 and L2 norm regularization were tried during the training stage, but the training loss increased rapidly with either regularization. Therefore, only MSE was taken as the loss function.

## 4 Discussion and Conclusion

In this work, we proposed a novel deep recurrent neural network (DRNN) for modeling functional brain networks in tfMRI data. The DRNN framework naturally combines the common deep neural networks with RNN. Each hidden layer of DRNN is a recurrent neural network and the output of each layer is the input time series of the upper layer. This structure automatically creates different time scales across different levels and thus form a temporal hierarchy. After training with the task stimulus, the whole brain voxel signals are automatically reconstructed with the top hidden layer output. Specifically, the weight vector between the hidden units and the whole brain fMRI signals describes the spatial distribution of this network and the top hidden layer's output under specific stimuli naturally represents the corresponding temporal patterns of the brain network. The hierarchical and temporal information of the brain activities is captured, and different time scales of brain networks can be identified. Extensive experiment results demonstrate the superiority of the proposed DRNN framework.

**Acknowledgements.** This work was supported by National Key R&D Program of China under contract No. 2017YFB1002201, NSF of China 61806167, the Fundamental Research Funds for the Central Universities (2017FZA5021, 3102017zy030) and the China Postdoctoral Science Foundation (2017M613206).

# References

1. Logothetis, N.K.: What we can do and what we cannot do with fMRI. Nature **453**, 869 (2008)
2. Duncan, J.: The multiple-demand (MD) system of the primate brain: mental programs for intelligent behaviour. Trends Cogn. Sci. **14**, 172–179 (2010)
3. Bullmore, E., Sporns, O.: Complex brain networks: graph theoretical analysis of structural and functional systems. Nat. Rev. Neurosci. **10**, 186 (2009)
4. Friston, K.J., Holmes, A.P., Worsley, K.J., Poline, J.P., Frith, C.D., Frackowiak, R.S.: Statistical parametric maps in functional imaging: a general linear approach. Hum. Brain Mapp. **2**, 189–210 (1994)
5. Biswal, B.B., Ulmer, J.L.: Blind source separation of multiple signal sources of fMRI data sets using independent component analysis. J. Comput. Assist. Tomogr. **23**, 265–271 (1999)
6. Lv, J., et al.: Holistic atlases of functional networks and interactions reveal reciprocal organizational architecture of cortical function. IEEE Trans. Biomed. Eng. **62**, 1120–1131 (2015)
7. Zhao, S., et al.: Decoding Auditory Saliency from Brain Activity Patterns during Free Listening to Naturalistic Audio Excerpts. Neuroinformatics **6**, 309–324 (2018)
8. Ferrarini, L., et al.: Hierarchical functional modularity in the resting-state human brain. Hum. Brain Mapp. **30**, 2220–2231 (2009)
9. Russakovsky, O., et al.: Imagenet large scale visual recognition challenge. Int. J. Comput. Vis. **115**, 211–252 (2015)
10. Zhao, Y., et al.: Automatic Recognition of fMRI-derived functional networks using 3D convolutional neural networks. IEEE Trans. Biomed. Eng. **65**, 1975–1984 (2017)
11. Zhao, Y., Ge, F., Liu, T.: Automatic recognition of holistic functional brain networks using iteratively optimized convolutional neural networks (IO-CNN) with weak label initialization. Med. Image Anal. **47**, 111–126 (2018)
12. Huang, H., et al.: Modeling task fMRI data via deep convolutional autoencoder. IEEE Trans. Med. Imaging **37**(7) (2018)
13. Buzsaki, G., Draguhn, A.: Neuronal oscillations in cortical networks. Science **304**, 1926–1929 (2004)
14. Litjens, G., et al.: A survey on deep learning in medical image analysis. Med. Image Anal. **42**, 60–88 (2017)
15. Hermans, M., Schrauwen, B.: Training and analysing deep recurrent neural networks. In: Advances in Neural Information Processing Systems, pp. 190–198 (2013)
16. Güçlü, U., van Gerven, M.A.: Modeling the dynamics of human brain activity with recurrent neural networks. Front. Comput. Neurosci. **11**, 7 (2017)
17. Barch, D.M., et al.: Function in the human connectome: task-fMRI and individual differences in behavior. Neuroimage **80**, 169–189 (2013)
18. Abadi, M., et al.: TensorFlow: a system for large-scale machine learning. In: OSDI, pp. 265–283 (2016)

# A Novel Deep Learning Framework on Brain Functional Networks for Early MCI Diagnosis

Tae-Eui Kam, Han Zhang, and Dinggang Shen[✉]

Department of Radiology and BRIC, University of North Carolina at Chapel Hill,
Chapel Hill, USA
dgshen@med.unc.edu

**Abstract.** Although alternations of brain functional networks (BFNs) derived from resting-state functional magnetic resonance imaging (rs-fMRI) have been considered as promising biomarkers for early Alzheimer's disease (AD) diagnosis, it is still challenging to perform individualized diagnosis, especially at the very early stage of preclinical stage of AD, i.e., early mild cognitive impairment (eMCI). Recently, convolutional neural networks (CNNs) show powerful ability in computer vision and image analysis applications, but there is still a gap for directly applying CNNs to rs-fMRI-based disease diagnosis. In this paper, we propose a novel multiple-BFN-based 3D CNN framework that can *automatically* and *deeply* learn complex, high-level, hierarchical diagnostic features from various independent component analysis-derived BFNs. More importantly, the embedded features of different BFNs could comprehensively support each other towards a more accurate eMCI diagnosis in a unified model. The performance of the proposed method is validated by a large-sample, multisite, rigorously controlled publicly accessible dataset. The proposed framework can also be conveniently and straightforwardly applied to individualized diagnosis of various neurological and psychiatric diseases.

**Keywords:** Diagnosis · Convolutional neural networks · Brain networks
Independent component analysis · Mild cognitive impairment · Deep learning
Resting-state functional MRI

## 1 Introduction

Alzheimer's Disease (AD) is the most common form of dementia with memory and cognitive deficits in elderly people and is to date still irreversible and incurable [1]. Early detection of AD at its preclinical Mild Cognitive Impairment (MCI) stage is of great clinical importance for early intervention. Recently, a few pioneering studies have called for pushing early AD diagnosis to an even earlier stage, i.e., early MCI (eMCI). While promising results based on group-level comparisons were found [2–4]; computer-aided *individualized* eMCI diagnosis is still among the most challenging tasks. Many MCI studies have used resting-state functional Magnetic Resonance Imaging (rs-fMRI)-derived brain functional networks (BFNs) for MCI studies [5, 6]; however, most of them are based on *mass-univariate* analysis, e.g., constructing the default mode network (DMN) and conducting a voxel-wise greedy search for potential

© Springer Nature Switzerland AG 2018
A. F. Frangi et al. (Eds.): MICCAI 2018, LNCS 11072, pp. 293–301, 2018.
https://doi.org/10.1007/978-3-030-00931-1_34

group differences within it [5]. While such traditional BFN-based MCI studies yield some statistically significant brain regions [2–4], they suffer several inherent drawbacks: (1) mass-univariate analysis usually results in loss of statistical power; (2) the voxel-wise analysis may lose rich inter-voxel pattern information in the BFNs; and (3) the severe noise in the rs-fMRI data could overwhelm subtle diagnostic information in eMCI. For example, a study in MCI detected multiple BFNs but only found a small blob in the DMN with statistical differences between MCI and controls [3], which is insufficient to clearly separating the two groups.

MCI could involve complex brain structural and functional changes, which usually involve large-scale functional abnormalities that could encompass *multiple* BFNs, with both local and global changes, as well as with both intra- and inter-BFN association changes. This inspired us to use the modern convolutional neural network (CNN) framework to improve individual diagnosis performance for eMCI because CNNs have been demonstrated to be powerful in automatic learning and capturing hierarchical features by integrating different scales of features with different layers for spatial pattern representation and recognition in computer vision and image analysis applications [7]. Previous works have also demonstrated a promising future of CNN for medical imaging analysis such as disease diagnosis and prognosis [8, 9]. However, directly applying CNNs to rs-fMRI data for disease diagnosis is not straightforward; several key issues need to be solved: (1) As aforementioned, the raw rs-fMRI data is usually noisy; and, more importantly, (2) raw rs-fMRI data is a 4D data with a temporal dimension that does not have correspondence across different subjects, whereas CNN cannot be directly applied to such time-series data. Several studies have proposed a fully connected network to learn patterns of the whole-brain functional connectivity (FC) matrices by treating it as a long FC-feature vector constituting inter-regional rs-fMRI signal correlations [10, 11]. However, they are actually learning an over-simplified BFN without taking voxel-wise-FC details into consideration.

Inspired by independent component analysis (ICA), a widely used data-driven BFN modeling method [12, 13], we propose a novel BFN-based deep learning framework that directly works on the BFNs abstracted by ICA, thus avoiding the technical limitation of directly applying CNN on the noisy raw rs-fMRI time series. Specifically, we use a group-wise ICA algorithm [14] to obtain a set of spatially independent BFNs for each individual, each of which is regarded as a voxel-wise spatial representation of a certain brain functional system mediating specific cognitive function(s) [15]. Instead of directly conducting separate traditional voxel-wise *mass-univariate* comparisons for each BFN (where only superficial features could be used), we train multiple 3D CNNs, each of which learns complex, hierarchical spatial pattern information from a BFN in a layer-by-layer manner; and these CNNs for multiple BFNs are further combined with consequential layers for end-to-end multiple-BFN-based eMCI diagnosis. The innovation of our method is three-fold. (1) It learns diagnostic spatial patterns of each BFN *automatically* and *deeply*, with consideration of both local and global FC features; (2) It takes advantage of different BFNs by letting them support each other toward more accurate disease diagnosis; and (3) It uses widely accepted ICA-derived BFNs and makes the result interpretation relatively intuitive. We demonstrate the feasibility of our method in a challenging eMCI diagnosis problem using a rigorously collected, publicly accessible, multisite Alzheimer's Disease Neuroimaging Initiative 2 (ADNI2)[1] dataset.

## 2    Materials and Preprocessing

We collected raw rs-fMRI data from 49 eMCI subjects (20M/29F, mean age 72.12 ± 7.22 yrs.) and 48 age-, gender-matched Normal Controls (NCs) (20M/28F, mean age 75.50 ± 46.81 yrs.) from the ADNI2 database. The eMCI is the early stage of MCI, whose BFNs are believed to more resemble to the NCs'; thus their classification is more challenging than MCI vs. NC classification. Note that, multiple rs-fMRI scans were longitudinally acquired for most of the subjects. We thus used all the available data as long as they corresponded to stable eMCI or NC status. This allowed us to make use of as much data as possible for more accurate CNN model training. Thus, we used a total of 351 samples (172 and 179 for NCs and eMCIs, respectively), which makes it among one of the largest sample-size studies to date. The data are acquired by 3T Siemens scanners at multiple sites with rigorous quality control (TR = 3,000 ms, TE = 30 ms, voxel size = 3 × 3 × 3.3 mm$^3$, the number of slices = 48, and flip angle = 80°).

We preprocessed the data using a widely-adopted DPARSF toolbox[2] with conventional pipeline as used in [14]. During the preprocessing, the first volumes acquired during the first 10 s. were discarded to ensure magnetization equilibrium and the remaining volumes were head motion corrected, spatially normalized, spatially resampled (voxel resolution = 3 × 3 × 3 mm$^3$), smoothed and band-pass filtered. Nuisance signals including the head motion parameters, mean white matter and cerebrospinal fluid signals are regressed out to further remove noise and artifacts.

## 3    Proposed Method

We propose a novel framework combining Group-Information-Guided ICA (GIG-ICA) [14] and 3D CNNs for BFN-based eMCI diagnosis. GIG-ICA drives subject-specific BFNs by keeping spatial correspondence across subjects under the guidance of the group-level BFN information [14]. Thus, we chose GIG-ICA to generate more robust BFNs for better model training. Of note, other rs-fMRI denoising methods can also be used such as FIX in FSL[3]; while other whole-brain BFN construction methods (e.g., seed-based correlation) can also be as long as robust and reliable BFNs can be extracted [15, 16]. GIG-ICA, as a group-wise ICA method, decomposed 4D rs-fMRI data from all subjects into a set of 3D spatial maps of BFNs for each subject, with which we can construct 3D CNNs to learn sophisticated embedded spatial patterns of BFNs.

---

[1] http://www.adni.loni.usc.edu/.

[2] http://rfmri.org/DPARSF/.

[3] https://fsl.fmrib.ox.ac.uk/fsl/fslwiki/FIX.

Figure 1 illustrates the overall framework of our method. Specifically, from pre-processed rs-fMRI, we first extract 3D subject-specific spatial map of each of the 6 high-level cognitive function-related BFNs that could be affected in eMCIs. For each 3D BFN, we first learn single-BFN-based 3D CNN (SB-CNN), then fuse the multiple SB-CNNs with a multiple-BFN-based 3D CNN (MB-CNN) to make the final decision by merging all the available high-level features from each SB-CNN as a unified framework.

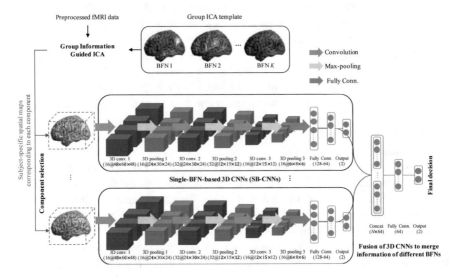

**Fig. 1.** Illustration of the proposed multiple-BFN-based 3D CNN (MB-CNN) framework for early MCI (eMCI) diagnosis. MB-CNN consists of multiple single-BFN-based 3D CNN (SB-CNN) by fusing high-level features of each SB-CNN in a unified framework. Each layer consists of multiple feature maps with (# feature maps@$x \times y \times z$ dimension).

## 3.1    Extraction of Subject-Specific Spatial Maps of BFNs

To extract subject-specific BFNs for inputs of the CNNs, we apply GIG-ICA to the preprocessed rs-fMRI of all subjects. As described above, GIG-ICA uses group information captured by the standard group ICA as a guidance to extract subject-specific BFNs with individually unique BFN information preserved and with inter-subject correspondence ensured. In this work, as the spatial guidance of GIG-ICA, we use the group ICA templates provided by Human Connectome Project (HCP)[4] that consist of 25 components by conducting a standard group ICA with 812 normal subjects.

As the AD pathological attacks in the early stage mainly target on the high-level cognitive function-related BFNs, *e.g.*, the DMN, frontoparietal networks, attention

---

[4] http://www.humanconnectome.org/.

networks, and executive control network; thus, we manually select 6 components which are associated with the higher-level cognitive functions. Figure 2 shows the selected BFNs. We use the GIG-ICA-derived, subject-specific BFN maps of the 6 components for SB-CNN and MB-CNN training.

## 3.2 Learning of Spatial Features of BFNs

We first construct SB-CNN for each BFN, respectively, and then merge those separately trained SB-CNNs as a unified, MB-CNN-based disease diagnosis model (refer to Fig. 1 for more details).

**SB-CNN Construction and Learning.** The subject-specific spatial maps of each BFN are fed to each SB-CNN, respectively, as input. To reduce computational complexity, from each spatial map of $61 \times 73 \times 61$ dimension, we remove the background voxels by cropping the 3D BFN map to only include brain regions of $48 \times 60 \times 48$ dimension. In each SB-CNN, the input is convolved by a series of three convolutional layers with Rectified Linear Unit (ReLU) activation, and each convolutional layer is followed by a max-pooling layer to down-sample the feature maps generated from the previous convolutional layers. The sizes of convolution and max-pooling kernels are set to $3 \times 3 \times 3$ and $2 \times 2 \times 2$, respectively. The last feature maps are fully connected with a series of 2 fully connected layers with 128 and 64 nodes, and an output layer has 2 nodes for binary class labels. The label information is used for back-propagation procedure in model learning. The softmax function is applied to the output units to predict the probability of an input belonging to NC or eMCI group. The model is optimized by the Stochastic Gradient Descent (SGD) algorithm [17].

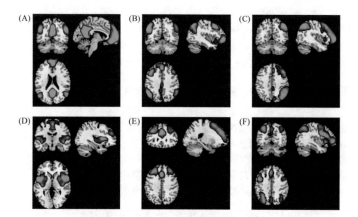

**Fig. 2.** The maps of the selected group-level components of the high-level cognitive function-related BFNs: (A) default model network (DMN), (B, C) frontoparietal networks (FPNs 1&2), (D, E) attention networks (ANs 1&2), and (F) executive control network (ECN), from the group ICA template provided by Human Connectome Project (HCP). We set a threshold of 5 in each map for the visualization purpose.

**MB-CNN Construction and Learning.** The last fully connected layers of each SB-CNN are concatenated, and then an additional fully connected layer with 64 nodes are stacked to merge the high-level features of each SB-CNN together and make them support each other toward better eMCI classification. An output layer with the softmax function is connected to the top of the model for the final decision making. The learned weights of each SB-CNN are adopted as the *initial* weights of MB-CNN, and then all the weights are *refined* together during MB-CNN learning by the SGD algorithm to make the model robust and also make different BFNs affect to each other in a layer-by-layer way.

## 4 Experiments and Results

### 4.1 Experimental Settings

To validate the effectiveness of the proposed MB-CNN method, we compare it with a baseline model, for which we used a traditional voxel-wise mass-univariate analysis to extract features and then fed them in a Support Vector Machine (SVM). For the baseline method, discriminative features from each BFN are extracted by using voxel-wise two-sample $t$-tests on the training set, and then Principal Component Analysis (PCA) is applied to reduce dimension of the features[5]. We also compare the MB-CNN with different SB-CNNs, each of which diagnose eMCI based only on each of the 6 BFNs. For evaluation of the proposed and competing methods, 5-fold cross-validation is adopted. That is, all the subjects are partitioned into 5 subsets, and 4 are used for training and the remaining subset is used for testing. Of note, some subjects have multiple scans in data acquisition (see Sect. 2). For training, all the scans of the subjects included in the training stage are used for better training; but, for the testing subjects, only the *baseline* scan of each subject are used for producing the testing result, to resemble the real application scenario.

### 4.2 Performance Evaluation

For performance evaluation, diagnostic performance is computed by the following quantitative metrics:

- Accuracy (ACC) = $(TP + TN)/(TP + TN + FP + FN)$
- Sensitivity (SEN) = $TP/(TP + FN)$
- Specificity (SPEC) = $TN/(TN + FP)$
- Positive Predictive Value (PPV) = $TP/(TP + FP)$
- Negative Predictive Value (NPV) = $TN/(TN + FN)$

where TP, TN, FP, and FN denote true positive, true negative, false positive, and false negative, respectively.

---

[5] To optimize the SVM hyperparameter, a nested 10-fold cross-validation technique is adopted for the inner cross-validation loop, while the outer loop is 5-fold cross-validation.

In Table 1, we summarize the performance of the proposed and competing methods for eMCI diagnosis. The proposed method shows the best performance for all the performance metrics. Specifically, the proposed MB-CNN shows the best accuracy of 74.23% which significantly improves the performance of the baseline method by more than 9%. Compared to different SB-CNNs only using a single BFN, the proposed method improves the accuracy by ~3–7%. According to the results, the proposed framework shows the effectiveness on extracting spatial features from different BFNs and merging the multiple-BFN features together for more accurate eMCI diagnosis. We also find that the SB-CNNs with DMN and FPN achieved better diagnosis accuracy than those with other BFNs, indicating that these two brain functional systems could be more affected in the early stage of MCI. We also noted that, even using all the 6 BFNs, the traditional method (Baseline) could not compete with any of the CNN-based methods, indicating a promising future of using CNN for disease diagnosis.

**Table 1.** Performance comparison of the proposed and competing methods.

| Method | BFN | ACC (%) | SEN (%) | SPEC (%) | PPV (%) | NPV (%) |
|--------|-----|---------|---------|----------|---------|---------|
| Baseline | All | 64.95 | 69.39 | 60.42 | 64.15 | 65.91 |
| SB-CNN | DMN | 70.10 | 71.43 | 68.75 | 70.00 | 70.21 |
| | FPN1 | 71.13 | 73.47 | 68.75 | 70.59 | 71.74 |
| | FPN2 | 70.10 | 73.47 | 66.67 | 69.23 | 71.11 |
| | AN1 | 68.04 | 71.43 | 64.58 | 67.31 | 68.89 |
| | AN2 | 67.01 | 69.39 | 64.58 | 66.67 | 67.39 |
| | ECN | 67.01 | 67.35 | 66.67 | 67.35 | 66.67 |
| MB-CNN | All | **74.23** | **75.51** | **72.92** | **74.00** | **74.47** |

SB-CNNs are constructed for each BFN, respectively; the baseline and MB-CNN methods use all the six BFNs (DMN: default-mode network; FPN 1&2: two fronto-parietal networks; AN 1&2: two attention networks; ECN: executive control network).

# 5   Discussion and Conclusions

We proposed a novel framework to model spatial patterns of BFNs derived from rs-fMRI by combining GIG-ICA and 3D CNNs in a unified framework for accurate eMCI diagnosis. To our best knowledge, this is the first BFN-based deep learning for the disease diagnosis. Zhao et al., proposed a 3D CNN model for automatic ICA component labeling for each BFN [18]; however, that study used only healthy subjects for a much easier task (component labeling) than the current eMCI diagnosis study. In the proposed method, we first extracted subject-specific spatial maps of high-level cognitive function-related BFNs by using GIG-ICA, and then constructed a 3D CNN for each BFN, respectively, to learn deeply embedded spatial patterns of each BFN. Furthermore, we combined all the CNNs to fuse the deep features of multiple BFNs in a unified framework for joint eMCI diagnosis. Results on a public dataset show the effectiveness of the proposed framework for eMCI diagnosis. In future, we will focus on the time-varying spatial patterns of the BFNs, as the dynamic brain functional connectome could contain more fine-grained information for eMCI diagnosis. The proposed method can also be used for diagnosis of other diseases in the future.

**Acknowledgements.** This work was partially supported by NIH grants (AG041721, AG049371, AG042599, and AG053867).

# References

1. Alzheimer's Association: Alzheimer's disease facts and figures. Alzheimer's Dement. **13**, 325–373 (2017)
2. Celone, K.A., Calhoun, V.D., Dickerson, B.C., et al.: Alterations in memory networks in mild cognitive impairment and Alzheimer's disease: an independent component analysis. J. Neurosci. **4**, 10222–10231 (2006)
3. Binnewijzend, M.A., Schoonheim, M.M., Sanz-Arigita, E., et al.: Resting-state fMRI changes in Alzheimer's disease and mild cognitive impairment. Neurobiol. Aging **33**(9), 2018–2028 (2012)
4. Petrella, J.R., Sheldon, F.C., Prince, S.E., et al.: Default mode network connectivity in stable vs. progressive mild cognitive impairment. Neurology **76**(6), 511–517 (2011)
5. Keiichi, O., Nobuhiro, Y., Kentaro, O., et al.: Can a resting-state functional connectivity index identify patients with Alzheimer's disease and mild cognitive impairment across multiple sites? Brain Connect. **7**(7), 391–400 (2017)
6. Zhao, Q., Lu, H., Metmer, H., et al.: Evaluating functional connectivity of executive control network and frontoparietal network in Alzheimer's disease. Brain Res. **1678**, 262–272 (2018)
7. LeCun, Y., Bengio, Y., Hinton, G.: Deep learning. Nature **521**(7553), 436–444 (2015)
8. Nie, D., Zhang, H., Adeli, E., Liu, L., Shen, D.: 3D deep learning for multi-modal imaging-guided survival time prediction of brain tumor patients. In: Ourselin, S., Joskowicz, L., Sabuncu, M.R., Unal, G., Wells, W. (eds.) MICCAI 2016. LNCS, vol. 9901, pp. 212–220. Springer, Cham (2016). https://doi.org/10.1007/978-3-319-46723-8_25
9. Gao, M., Bagci, U., Lu, L., et al.: Holistic classification of CT attenuation patterns for interstitial lung diseases via deep convolutional neural networks. Comput. Methods Biomech. Biomed. Eng.: Imaging Vis. **6**(1), 1–6 (2018)
10. Hjelm, R.D., Calhoun, V.D., et al.: Restricted Boltzmann machines for neuroimaging: an application in identifying intrinsic networks. NeuroImage **96**, 245–260 (2014)
11. Kam, T.-E., Suk, H.-I., Lee, S.-W.: Multiple functional networks modeling for autism spectrum disorder diagnosis. Hum. Brain Mapp. **38**(11), 5804–5821 (2017)
12. Beckmann, C.F., Deluca, M., et al.: Investigations into resting-state connectivity using independent component analysis. Philos. Trans. R. Soci. B: Biol. Sci. **360**, 1001–1013 (2005)
13. Hafkemeijer, A., Altmann-Schneider, I., Craen, A.J., Slagboom, P.E., Grond, J., Rombouts, S.A.: Associations between age and gray matter volume in anatomical brain networks in middle-aged to older adults. Aging Cell **13**, 1068–1074 (2014)
14. Du, Y., Fan, Y.: Group information guided ICA for fMRI data analysis. NeuroImage **69**, 157–197 (2013)
15. Calhoun, V.D., Adali, T.: Multisubject independent component analysis of fMRI: a decade of intrinsic networks, default mode, and neurodiagnostic discovery. IEEE Rev. Biomed. Eng. **5**, 60–73 (2012)

16. Du, Y., Lin, D., et al.: Comparison of IVA and GIG-ICA in brain functional network estimation using fMRI data. Front. Neurosci. **11**, 267 (2017)
17. Boyd, S., Vandenberghe, L.: Convex Optimization. Cambridge University Press, Cambridge (2004)
18. Zhao, Y., Dong, Q., et al.: Automatic recognition of fMRI-derived functional networks using 3D convolutional neural networks. IEEE Trans. Biomed. Eng. **65**, 1975–1984 (2017)

# A Region-of-Interest-Reweight 3D Convolutional Neural Network for the Analytics of Brain Information Processing

Xiuyan Ni[1(✉)], Zhennan Yan[2], Tingting Wu[3], Jin Fan[3], and Chao Chen[1,4]

[1] The Graduate Center, City University of New York (CUNY), New York, NY, USA
xiuyanni.xn@gmail.com
[2] Siemens Healthineers, Malvern, PA, USA
[3] Department of Psychology, CUNY Queens College, Flushing, NY, USA
[4] Department of Computer Science, CUNY Queens College, Flushing, NY, USA

**Abstract.** We study how human brains activate to process input information and execute necessary cognitive tasks. Understanding the process is crucial in improving our diagnostic and treatment of different neurological disorders. Given functional MRI images recorded when human subjects execute tasks with different levels of information uncertainty, we need to identify the similarity and difference between brain activities at different regions of interest (ROIs), and thus gain insights into the underlying mechanism. To achieve this goal, we propose a new ROI-reweight 3D convolutional neural network (CNN). Our CNN not only learns to classify the task-evoked fMRIs with a high accuracy, but also locates crucial ROIs based on a reweight layer. Our findings reveal several brain regions to be crucial in differentiating brain activity patterns facing tasks of different uncertainty levels.

**Keywords:** Uncertainty representation · Task-evoked fMRI · CNN

## 1 Introduction

Cognitive control is a high-order information processing system in human brain which selects appropriate information, inhibits inappropriate response, and coordinates with actions under guidance of context-specific goals and intention [5,14]. It is implemented by a set of brain regions called the cognitive control network (CCN) (Fig. 1(Left)) and also by its interactions with other domain-specific networks (visual, auditory, somatosensory, motor) and the default mode network (DMN) [2,17,20]. Characterizing how regions of CCN activate, collaborate, and interact with other networks will improve the theoretical understanding of human brains. Furthermore, knowledge in cognitive control will empower us with advanced diagnosis and treatment of various neurological disorders [1,4,6,19].

In this paper, we investigate the uncertainty representation in cognitive control, i.e., how brains process different levels of uncertainty in cognitive tasks.

© Springer Nature Switzerland AG 2018
A. F. Frangi et al. (Eds.): MICCAI 2018, LNCS 11072, pp. 302–310, 2018.
https://doi.org/10.1007/978-3-030-00931-1_35

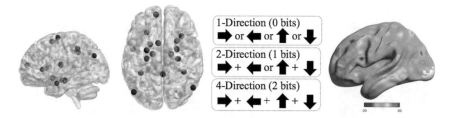

**Fig. 1.** Left and Middle-Left: The cognitive control network (CCN, red) and the default mode network (DMN, blue). Middle-Right: Experiment setting. Human subjects are shown images of arrows and click keys accordingly. The arrows may have one, two, or four possible directions, corresponding to uncertainty levels of zero, one, and two bits measured in Shannon's entropy [18]. More uncertainty is introduced by additional colors and corresponding actions. In total there are six different tasks, with various uncertainty levels. Right: An example of fMRI image mapped from volume to surface.

Early observations, summarized by the Hick-Hyman law [8,10], state that the brain reaction time grows linearly to the uncertainty level, measured in Shannon entropy [18]. In other words, human brains need to work more to execute uncertain tasks. Recently, it has been confirmed that the overall brain activation is linearly correlated to the uncertainty level [5,22]. However, it remains unknown how different regions coordinate and activate to process uncertainty in tasks. To gain region-specific insights, we analyze task-evoked functional MRIs, in particular, fMRI images collected while a human subject is executing choice reaction time (CRT) tasks. These CRT tasks are the same but with different uncertainty levels, corresponding to different numbers of possible choices. See Fig. 1(Middle-Right) and Sect. 3 for more details.

Many learning methods have been leveraged to analyze task-evoked fMRI images. Examples include analysis of variance (ANOVA) [21], general linear model (GLM) [12], and support vector machine (SVM) [3]. Early studies focus on the association of specific regions of interest and the stimuli, and thus miss the information carried by fine scale brain activity patterns. The multi-voxel pattern analysis (MVPA) [7,16] approach was the first to take the whole fMRI image as a multivariable input, and use various classifiers to discover basis patterns crucially related to different stimuli.

In this paper, we use a 3D convolutional neural network (CNN) to capture fine scale activity patterns in the task-evoked fMRI. Recent years have witnessed the success of CNN in computer vision and medical image analysis. In particular, 3D CNN has been used in action recognition [11], object recognition [13], etc. In Neuroimage study, 3D CNN has been used to diagnose Alzheimer's disease (AD) and mild cognitive impairment (MCI) [9], to predict the survival time of brain tumor patients [15], and to reconstruct functional connectivity networks [23]. However, despite its success in achieving the prediction goal, CNN lacks the crucial explainability, namely, the ability to explain the underlying rationale of a prediction. In our problem setting, a standard 3D CNN does not help identify the

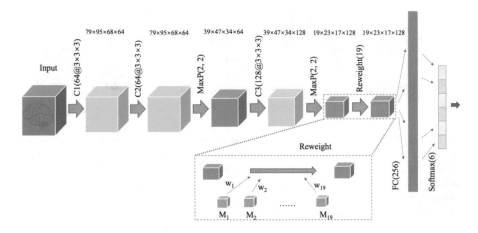

**Fig. 2.** The architecture of the ROI-reweight 3D CNN model.

regions crucial for the differentiation of activity patterns from different cognitive tasks. Unfortunately, this is indeed the primary goal of neuroscientists.

**Our Contributions.** To deliver both prediction power and explainability, we propose a new ROI-reweight CNN. Our method not only classifies fMRIs based CRT tasks, but also learns weights of different ROIs in making the prediction. The idea is to add a novel reweight layer to adjust the high-level representation, i.e., the representation after all convolution and pooling layers. The weight of each element in the high-level representation is determined by nearby ROIs. The ROI reweight layer is detailed in Sect. 2.2. After training, the learned weights of different ROIs measure how important their nearby patterns are in the classification. To the best of our knowledge, our method is the first CNN-based approach to achieve both classification power and explainability in functional MRI study. Our method identifies several regions in both CCN and DMN, i.e., anterior insula (AI), thalamus (TH), posterior cingulate cortex (PCC), etc., as keys to the differentiation of brain activities for different uncertainty levels.

## 2   Method

Our method classifies fMRI images into six different classes, corresponding to six different uncertainty levels. Meanwhile, the method learns weights on ROIs, measuring the significance of each ROI in the classifier. We consider 19 ROIs that are considered significant in cognitive control. These ROIs constitute the cognitive control network and the default mode network (Fig. 1).

We propose a 3D CNN with a reweight layer before the fully connected layers. The reweight layer adjusts the significance of different elements in the high-level representation. But the reweighting tensor is parametrized by 19 weights, associated to 19 ROIs. During the training, these 19 weights are learned. After training, these weights can be used to measure the significance of ROIs in classification. In

Sect. 2.1, we explain the CNN architecture. Detailed explanation of the reweight layer will be given in Sect. 2.2.

## 2.1   3D CNN Architecture

Our network has three convolutional layers, two max-pooling layers, two fully connected layers, and one reweight layer. See Fig. 2 for an overview of the architecture. The input is a single channel fMRI image of size $75 \times 95 \times 68$. The first two convolutional layers, $C1(64@3 \times 3 \times 3)$ and $C2(64@3 \times 3 \times 3)$, both have 64 channels and kernel size $3 \times 3 \times 3$. The third one, $C3(128@3 \times 3 \times 3)$, has 128 channels and kernel size $3 \times 3 \times 3$. The convolution stride is fixed to 1 voxel. The spatial padding of each convolutional layer input is set to 1, such that the spatial size is preserved after convolution. Two max-pooling layers are applied after the second and the third convolutional layers respectively. Max-pooling is performed over a $2 \times 2 \times 2$ voxel window, with stride 2.

The crucial contribution of our paper is a reweight layer following the last max-pooling layer. This layer has the same input and output size; it reweights all input values using a weight tensor controlled by 19 weight parameters, corresponding to 19 brain ROIs. More details will be given in Sect. 2.2. The reweight layer is followed by two fully connected layers, with 256 and 6 channels, respectively. The final layer is the soft-max layer. All hidden layers use the rectified linear unit (ReLU) for non-linearity. The model is trained on cross-entropy loss.

## 2.2   ROI-reweight Layer

The input, $I$, and output, $O$, of the reweight layer have the same size, i.e., $19 \times 23 \times 17 \times 128$. The input, $I$, is the high-level representation produced by the convolutional and max-pooling layers. We can view it as 128-channel signals at $19 \times 23 \times 17$ voxels. The reweight layer uses a $19 \times 23 \times 17$ weight tensor, $\widehat{M}$, to reweight the input. All channels of the $(i, j, k)$-th voxel of $I$ are multiplied by $\widehat{M}(i, j, k)$. Formally, $O(i, j, k, \ell) = I(i, j, k, \ell) \cdot \widehat{M}(i, j, k), \forall \ell = 1, \ldots, 128$.

We parametrize entries of the weight tensor $\widehat{M}$ by weights associated to the ROIs. For generality, we assume $R$ many ROIs. Their weights constitute an $R$ dimensional vector, $\boldsymbol{w} = [w_1, \ldots, w_R]^T$.[1] The relationship between the $\widehat{M}(i, j, k)$ and the ROI weight $w_r$ depends on how much the $r$-th ROI affects the $(i, j, k)$-th entry in the high-level representation. To determine such relationship, we first map the $(i, j, k)$-th entry back to the original image domain by reversing the convolution and max-pooling operation. Each entry $(i, j, k)$ corresponds to a cube in the original image domain. The coordinates of cube center are denoted by $\boldsymbol{p}_{(i,j,k)} \in \mathbb{R}^3$. Let $\boldsymbol{q}_r$ be the center coordinates of the $r$-th ROI. We use 3D radial basis function (RBF) kernel to define the relationship between the two. Formally,

$$\widehat{M}(i, j, k) = \sum_{r=1}^{R} w_r \cdot \exp\left(-\frac{\|\boldsymbol{p}_{(i,j,k)} - \boldsymbol{q}_r\|^2}{2\sigma^2}\right) = \sum_{r=1}^{R} w_r \cdot M_r(i, j, k),$$

---

[1] We use boldface font for vectors, but normal font for matrices and tensors.

$$\widehat{M}(i,j,k) = w_1 \cdot \exp\left(-\frac{d_1^2}{2\sigma^2}\right) + w_2 \cdot \exp\left(-\frac{d_2^2}{2\sigma^2}\right)$$

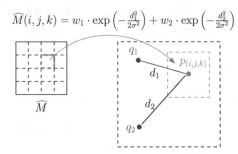

**Fig. 3.** An entry in $\widehat{M}$ is determined by weights of ROIs by mapping it to the original image domain. $d_1$ and $d_2$ are the Euclidean distance between its center and the two ROI centers, $q_1$ and $q_2$.

in which $\sigma$ is tuned on validation set. In other words, the entry $\widehat{M}(i,j,k)$ is the RBF kernel representation of kernels centered at all ROIs and weighted by $\boldsymbol{w}$. See Fig. 3 for an illustration.

Note that the entries of $\widehat{M}$ depend on the weight vector $\boldsymbol{w}$ linearly. Formally, we have $\widehat{M} = \sum_{r=1}^{R} w_r M_r$, in which $M_r$ is a constant tensor of the same size as $\widehat{M}$. It is straightforward to see that if we vectorize the tensor $\widehat{M}$, we have the linear relationship $\widehat{M} = \mathcal{M}\boldsymbol{w}$, in which $\mathcal{M} = [M_1, \ldots, M_R]$ is a matrix with $R$ columns, each corresponds to one vectorized tensor $M_r$. It is easy to see that the partial derivative of the loss, $\mathcal{L}$, w.r.t. the weight vector $\boldsymbol{w}$,

$$\frac{\partial \mathcal{L}}{\partial \boldsymbol{w}} = \frac{\partial \mathcal{L}}{\partial \widehat{M}} \cdot \frac{\partial \widehat{M}}{\partial \boldsymbol{w}} = \frac{\partial \mathcal{L}}{\partial \widehat{M}} \cdot \mathcal{M}^T$$

in which $\partial \widehat{M}/\partial \boldsymbol{w} = \mathcal{M}^T$ is the Jacobian matrix. To train this layer, we just need to update $\boldsymbol{w}$ accordingly at each iteration.

## 3    Experiments and Discussions

We apply our method to task-evoked fMRI images. These images were collected when human subjects executed choice reaction time (CRT) tasks [22]. Each subject performed around 1100 trials. In each trial, the subjects were presented with an arrow and were supposed to respond accordingly. Depending on the possible directions and colors of the arrows, there are six different tasks with different levels of uncertainty. See Fig. 1 for more information. The data was preprocessed using SPM 8. Each gradient-echo planar imaging (EPI) image volume was realigned to the first volume, registered with structural MRI, normalized to the Montréal Neurological Institute (MNI) ICBM152 space, resampled to a voxel size of $2 \times 2 \times 2$ mm, and spatially smoothed. The dimension of each final fMRI image is $79 \times 95 \times 68$. In this study, we focus on 19 ROIs that are considered the most important in cognitive control. These ROIs constitute the two brain networks CCN and DMN (Fig. 1).

**Table 1.** The classification results.

| Classifier | RF | LG | SVM | 3D-CNN | ROI-CNN |
|---|---|---|---|---|---|
| Accuracy(%) | 34.18 | 79.68 | 83.53 | 87.42 | **89.04** |

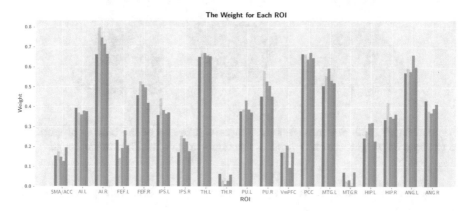

**Fig. 4.** The learned weights of the 19 ROIs. Bars of 5 different colors correspond to 5 different human subjects. SMA/ACC: supplementary motor area extending to anterior cingulate cortex. AI: anterior insular cortex. FEF: frontal eye find. IPS: area around and along the intraparietal sulcus. TH: thalamus. vmPFC: ventral medial prefrontal cortex. PCC: posterior cingulate cortex. MTG: middle temporal gyrus. ANG: angular gyrus. L: ROI located in left hemisphere of the brain. R: ROI located in the right hemisphere.

We train and test our method and other baselines on data from five different human subjects, and report the average accuracy. We randomly reserve 10% of the trials for each subject as the validation set and 10% as the testing set. For each subject individually, all classifiers are trained on the same training set. The best models are selected based on the performances on the validation set.

Random forest (RF), logistic regression (LR), and support vector machine (SVM) are used as baselines. To investigate the effect of the reweight layer, we also apply a 3D CNN without reweight layer (3D-CNN), i.e., our proposed network without the reweight layer, as an ablation. Among all methods, our ROI-reweight CNN (ROI-CNN) achieves the best accuracy. To our surprise, ROI-CNN even outperforms 3D-CNN. We believe the reason is that focusing on the patterns near the 19 crucial ROIs improves the overall performance of the neural network. All results are reported in Table 1.

Our model learns weights on 19 ROIs. We show all learned weights of the five subjects in Fig. 4. We observe high weights on anterior insula (AI), thalamus (TH), posterior cingulate cortex (PCC), etc. This is consistent with the domain knowledge theory in cognitive control [5]. These weights can be used to quantitatively analyze the roles of these ROIs in dealing with uncertain tasks. In Fig. 5,

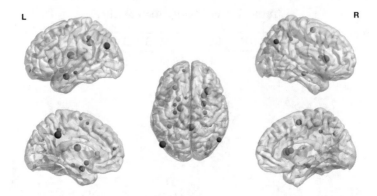

**Fig. 5.** The 19 ROIs drawn in a brain template. The size of the balls is proportional to the learned weight, which represents the importance of an ROI in the tasks.

we visualize these ROIs in their spatial location, with ball radii proportional to their weights.

We observe that the learned weights are highly consistent across different human subjects even though we trained the models separately. Note that due to cross-subject variation, our model trained on one subject usually overfits and cannot perform well on other subjects (the average accuracy is 19.53%, only slightly better than chance level). In other words, despite the fact that the models overfit on individual subjects and cannot generalize to others, they all learn very similar ROI weights. This shows that our model successfully locates the key regions for uncertainty representation. But the activity patterns within each ROI may vary over different human subjects.

## 4    Conclusions and Future Work

In this paper, we propose an ROI-reweight 3D convolutional neural network framework to classify the CRT task-evoked fMRI data, and locate key ROIs. Our framework achieves 89.04% average accuracy in the experiments, and outperforms the existing state-of-the-art linear classifiers and the traditional 3D CNN. In the meantime, it also provides quantitative assessment of the significance of the key ROIs in the brain for uncertainty representation, which could benefit cognitive control study. In the future, we plan to extend our framework to be more robust to cross-subject variation.

**Acknowledgement.** This research is partially supported by the NSF grants IIS 1718802 and CCF 1733866.

# References

1. Castellanos, F.X., Sonuga-Barke, E.J., Milham, M.P., Tannock, R.: Characterizing cognition in ADHD: beyond executive dysfunction. Trends Cogn. Sci. **10**(3), 117–123 (2006)
2. Cole, M.W., Schneider, W.: The cognitive control network: integrated cortical regions with dissociable functions. Neuroimage **37**(1), 343–360 (2007)
3. Cox, D.D., Savoy, R.L.: Functional magnetic resonance imaging (fMRI) "brain reading": detecting and classifying distributed patterns of fMRI activity in human visual cortex. Neuroimage **19**(2), 261–270 (2003)
4. Diamond, A., Barnett, W.S., Thomas, J., Munro, S.: Preschool program improves cognitive control. Sci. (N. Y., NY) **318**(5855), 1387 (2007)
5. Fan, J.: An information theory account of cognitive control. Front. Hum. Neurosci. **8**, 680 (2014)
6. Happé, F.: Autism: cognitive deficit or cognitive style? Trends Cogn. Sci. **3**(6), 216–222 (1999)
7. Haxby, J.V., Gobbini, M.I., Furey, M.L., Ishai, A., Schouten, J.L., Pietrini, P.: Distributed and overlapping representations of faces and objects in ventral temporal cortex. Science **293**(5539), 2425–2430 (2001)
8. Hick, W.E.: On the rate of gain of information. Q. J. Exp. Psychol. **4**(1), 11–26 (1952)
9. Hosseini-Asl, E., Keynton, R., El-Baz, A.: Alzheimer's disease diagnostics by adaptation of 3D convolutional network. In: 2016 IEEE International Conference on Image Processing (ICIP), pp. 126–130. IEEE (2016)
10. Hyman, R.: Stimulus information as a determinant of reaction time. J. Exp. Psychol. **45**(3), 188 (1953)
11. Ji, S., Xu, W., Yang, M., Yu, K.: 3D convolutional neural networks for human action recognition. IEEE Trans. Pattern Anal. Mach. Intell. **35**(1), 221–231 (2013)
12. Josephs, O., Turner, R., Friston, K.: Event-related fMRI. Hum. Brain Mapp. **5**(4), 243–248 (1997)
13. Maturana, D., Scherer, S.: VoxNet: a 3D convolutional neural network for real-time object recognition. In: 2015 IEEE/RSJ International Conference on Intelligent Robots and Systems (IROS), pp. 922–928. IEEE (2015)
14. Miller, E.K.: The prefrontal cortex and cognitive control. Nat. Rev. Neurosci. **1**(1), 59–65 (2000)
15. Nie, D., Zhang, H., Adeli, E., Liu, L., Shen, D.: 3D deep learning for multi-modal imaging-guided survival time prediction of brain tumor patients. In: Ourselin, S., Joskowicz, L., Sabuncu, M.R., Unal, G., Wells, W. (eds.) MICCAI 2016. LNCS, vol. 9901, pp. 212–220. Springer, Cham (2016). https://doi.org/10.1007/978-3-319-46723-8_25
16. Norman, K.A., Polyn, S.M., Detre, G.J., Haxby, J.V.: Beyond mind-reading: multi-voxel pattern analysis of fMRI data. Trends Cogn. Sci. **10**(9), 424–430 (2006)
17. Ridderinkhof, K.R., Ullsperger, M., Crone, E.A., Nieuwenhuis, S.: The role of the medial frontal cortex in cognitive control. Science **306**(5695), 443–447 (2004)
18. Shannon, C.E., Weaver, W.: The Mathematical Theory of Communication. University of Illinois Press, Champaign (1949)
19. Solomon, M., et al.: The neural substrates of cognitive control deficits in autism spectrum disorders. Neuropsychologia **47**(12), 2515–2526 (2009)
20. Sridharan, D., Levitin, D.J., Menon, V.: A critical role for the right fronto-insular cortex in switching between central-executive and default-mode networks. Proc. Natl. Acad. Sci. **105**(34), 12569–12574 (2008)

21. Vuilleumier, P., Armony, J.L., Driver, J., Dolan, R.J.: Effects of attention and emotion on face processing in the human brain: an event-related fMRI study. Neuron **30**(3), 829–841 (2001)
22. Wu, T., et al.: Hick-Hyman law is mediated by the cognitive control network in the brain. Cereb. Cortex **28**, 1–16 (2017)
23. Zhao, Y., et al.: Automatic recognition of fMRI-derived functional networks using 3D convolutional neural networks. IEEE Trans. Biomed. Eng. **65**, 1975–1984 (2017)

# Quantitative Deconvolution of fMRI Data with Multi-echo Sparse Paradigm Free Mapping

César Caballero-Gaudes[1]([⊠]) [iD], Stefano Moia[1] [iD],
Peter A. Bandettini[2,3] [iD], and Javier Gonzalez-Castillo[2] [iD]

[1] Basque Center of Cognition, Brain and Language, 20009 San Sebastian, Spain
c.caballero@bcbl.eu
[2] Section of Functional Imaging Methods, Laboratory of Brain and Cognition,
National Institute of Mental Health, National Institutes of Health,
Bethesda, MD 20892-1148, USA
[3] Functional MRI Core, National Institute of Mental Health,
National Institutes of Health, Bethesda, MD 20892-1148, USA

**Abstract.** This work introduces a novel framework for the deconvolution of the BOLD signal in multi-echo functional MRI (ME-fMRI) data: Multi-echo Sparse Paradigm Free Mapping (ME-SPFM). Building upon a physical model of the ME-fMRI signal and using a sparsity-promoting regularized least squares estimator, this algorithm obtains time-varying maps of the changes in the transverse relaxation rate ($R_2^*$) and the net magnetization ($S_0$) of the signal at the subject level without prior knowledge of the timing of the individual BOLD events. Our results with experimental data demonstrate that the maps of $R_2^*$ changes obtained with ME-SPFM at the times of the stimulus trials closely resemble the maps obtained with standard model-based methods that are aware of the onsets and durations of the experimental events, and considerably improves the accuracy of the deconvolution compared with the SPFM algorithm developed for single-echo fMRI. Furthermore, this method yields estimates of $R_2^*$ changes in physiologically interpretable units ($s^{-1}$), which is a step towards deciphering the dynamic nature of brain activity in a pseudo-quantitative manner in naturalistic paradigms, resting-state or clinical applications with unknown event-timing.

**Keywords:** Multi-echo functional magnetic resonance imaging
Deconvolution · Brain function · Single-trial analysis · Paradigm free mapping

## 1 Introduction

In functional magnetic resonance imaging (fMRI) experiments, data is usually collected with single-echo (SE) sequences that acquire one time series per voxel at a single echo time (*TE*). The *TE* is usually chosen close to the average transverse relaxation time $T_2^*$ of the gray matter region of interest to optimize the sensitivity to the BOLD response and in turn maximize the contrast-to-noise ratio of the signal [1]. However, the $T_2^*$ parameter varies across brain regions. To compensate this variability, a multi-echo (ME) fMRI acquisition can be alternatively deployed where data is acquired at

© Springer Nature Switzerland AG 2018
A. F. Frangi et al. (Eds.): MICCAI 2018, LNCS 11072, pp. 311–319, 2018.
https://doi.org/10.1007/978-3-030-00931-1_36

multiple *TEs* for each time point. The multiple echo signals can then be optimally combined for enhanced BOLD sensitivity [2–4] or input to ME-based denoising techniques [5–8] in order to significantly improve the results obtained with traditional general linear model (GLM) analyses that require a-prior information about all experimental events of interest [9]. Typically, the task-related regressors of the design matrix are defined as the convolution of a hypothesized time series modeling the neuronal activity triggered by the experimental paradigm (e.g. based on the onset, modulation and duration of events) with a hemodynamic response function (HRF).

Recently, however, there has been an increasing interest in methods that detect BOLD events, but do not require any information about the timing, location or cognitive nature of such events. These approaches could provide important insights in cognitive and clinical scenarios with inaccurate or insufficient knowledge of the neuronal activity that underlies the BOLD events (e.g. naturalistic paradigms, resting state). For that, some 'event-detection' methods attempt to deconvolve the neuronal-related signal underlying the BOLD fMRI signal assuming linearity in the BOLD response and only using a given hemodynamic model (e.g. a double gamma variate function). In such cases, the deconvolution becomes an ill-posed inverse problem and existing algorithms mainly vary in the regularized estimator used for the deconvolution. Initial approaches used empirical Bayesian estimators with Gaussian priors [10] or $L_2$-norm regularized least squares estimators such as the ridge regression [11]. More recent algorithms employ sparsity-promoting (i.e. $L_1$-norm based) regularized least squares estimators, such as the least absolute shrinkage and selection operator (LASSO) [12], Fused LASSO [13]; or even structured mixed $L_{2,1}$-norms to perform a spatio-temporal deconvolution [14] or to account for variability in the hemodynamic model [15].

In this work we develop a novel algorithm for the deconvolution of ME-fMRI data: Multi-echo Sparse Paradigm Free Mapping (ME-SPFM). To our knowledge, no algorithm has been previously proposed for deconvolution of ME-fMRI data. Contrary to previous algorithms, which would require the combination of the multiple echoes prior to deconvolution, the proposed approach directly leverages the information available in all echoes. Based on a physical model of the gradient-echo fMRI signal, this algorithm is able to estimate time-varying changes in the transverse relaxation rate ($R_2^* = 1/T_2^*$) and the net magnetization ($S_0$) without prior knowledge of the individual BOLD events. Using ME-fMRI data acquired on 10 subjects during an event-related paradigm [16], we demonstrate that the ME-SPFM algorithm can reliably detect individual BOLD events – namely single-trials at the subject level – with significantly better accuracy than its SE counterpart (i.e. sparse paradigm free mapping, SPFM [12]), and nearly matches the results obtained with a standard trial-based GLM analysis. Besides, ME-SPFM yields voxel-wise quantitative estimates of time-varying changes in $R_2^*$ ($\Delta R_2^*$) in interpretable units ($s^{-1}$), thus enabling a more quantitative study of the brain's time-varying activity with ME-fMRI.

The rest of the paper is organized as follows. In Sect. 2, we describe the signal model of ME-fMRI data and the ME-SPFM algorithm. Details of the ME-fMRI dataset used for evaluation, including MRI data acquisition and preprocessing, and the methods used for evolution are given in Sect. 3. The results of these evaluations are shown in Sect. 4, and conclusions are drawn in Sect. 5.

## 2 Multi-echo Sparse Paradigm Free Mapping

Assuming a mono-exponential decay model in gradient echo fMRI, the signal of a voxel $x$ at time $t$ for an echo time $TE_k$ is given by

$$s(x, t, TE_k) = S_0(x, t)e^{-R_2^*(x,t)TE_k} + n(x, t), \qquad (1)$$

where $S_0(x, t)$ and $R_2^*(x, t)$ are the changes in the net magnetization $S_0$ and the transverse relaxation rate $R_2^*$ of the voxel $x$ at time $t$, respectively, and $n(x, t)$ is a noise term. Hereinafter, the noise term and the voxel index $x$ are neglected to simplify the notation. Let us describe $S_0(t)$ and $R_2^*(t)$ in terms of relative changes with respect to the average values, i.e. $S_0(t) = \overline{S_0} + \Delta S_0(t)$ and $R_2^*(t) = \overline{R_2^*} + \Delta R_2^*(t)$, where $\Delta S_0(t)$ and $\Delta R_2^*(t)$ are typically considerably smaller than $\overline{S_0}$. and $\overline{R_2^*}$. Therefore, it can be shown that signal percentage changes with respect to the mean of the signal (i.e. $\bar{s}(TE_k) = \overline{S_0}e^{-\overline{R_2^*}TE_k}$) can be approximated as [7]

$$y(t, TE_k) \approx \Delta\rho(t) - \Delta R_2^*(t)TE_k. \qquad (2)$$

where $\Delta\rho(t) = \Delta S_0(t)/\overline{S_0}$.

For the purpose of deconvolution of the fMRI signal, let us assume that changes in neuronal-related activity result in changes in $\Delta R_2^*(t)$, which can then be seen as BOLD responses in the voxel time series. Following the linear model usually adopted in fMRI data analyses, changes in $\Delta R_2^*(t)$ can be described as $\Delta R_2^*(t) = h(t) * \Delta a(t)$, where $\Delta a(t)$ represents a signal related to changes in neuronal activity, and $h(t)$ is the hemodynamic response. In practice, the continuous-time MR signal is sampled every repetition time $(TR)$, i.e. $t = nTR$, where $n = 1, \ldots, N$, and $N$ is the number of volumes acquired during the acquisition. Hence, the signal at time point $n$ and echo time $TE_k$ can be approximated as $y_n^k \approx \Delta\rho_n - TE_k(h_n * \Delta a_n)$, where $y_n^k \stackrel{\text{def}}{=} y(nTR, TE_k)$, $\Delta\rho_n \stackrel{\text{def}}{=} \Delta\rho(nTR)$, $h_n \stackrel{\text{def}}{=} h(nTR)$, and $\Delta a_n \stackrel{\text{def}}{=} \Delta a(nTR)$. Gathering all time points as $\mathbf{y}_k = [y_1, \cdots, y_N]^T$, we can write that $\mathbf{y}_k \approx \Delta\rho - TE_k\mathbf{H}\Delta a$, where $\Delta a \in \mathbb{R}^N$ and $\Delta\rho \in \mathbb{R}^N$ denote changes in the neuronal-related signal and the net magnetization, respectively, and $\mathbf{H} \in \mathbb{R}^{N \times N}$ is a Toeplitz matrix whose columns are shifted versions of a hemodynamic response function (HRF) of duration $L$ time points at $TR$ temporal resolution, i.e. $\mathbf{h} = [h_1, \cdots, h_L]$.

If $K$ echoes are acquired at echo times $TE_k, k = 1, \ldots, K$, signal percentage changes of all echoes can be vectorized in a column vector of length $NK$. Considering the neuronal-related signal is identical for all echoes, we can formulate the following ME signal model:

$$\bar{y} \stackrel{\text{def}}{=} \begin{bmatrix} \mathbf{y}_1 \\ \vdots \\ \mathbf{y}_K \end{bmatrix} = \begin{bmatrix} \mathbf{I} \\ \vdots \\ \mathbf{I} \end{bmatrix} \Delta\rho - \begin{bmatrix} TE_1\mathbf{H} \\ \vdots \\ TE_K\mathbf{H} \end{bmatrix} \Delta a \stackrel{\text{def}}{=} \bar{\mathbf{I}}\Delta\rho - \bar{\mathbf{H}}\Delta a, \qquad (3)$$

or simply $\bar{y} = \mathbf{T}x$, where $x^T = \begin{bmatrix} \Delta\rho^T & \Delta a^T \end{bmatrix} \in \mathbb{R}^{2N}$, and $\mathbf{T} = \begin{bmatrix} \bar{\mathbf{I}} & \bar{\mathbf{H}} \end{bmatrix} \in \mathbb{R}^{KN \times 2N}$.

The deconvolution of the changes in the fMRI signal related to neuronal activity, i.e. detection of the BOLD events, involves the estimation of $x$ (i.e. $\Delta a$ and $\Delta \rho$) in Eq. (3). Assuming additive white Gaussian noise, an unbiased estimate of $x$ can be obtained with ordinary least squares. However, in practice, the least squares solution would produce estimates with large variability due to the large collinearity between the columns of $\bar{\mathbf{H}}$. To overcome this issue, in this work we propose to estimate $x$ by means of Basis pursuit denoising [16], equivalent to the LASSO, i.e.

$$\widehat{x} = \arg \min_x \frac{1}{2} \|\bar{y} - \mathbf{T}x\|_2^2 + \lambda \|x\|_1 \tag{4}$$

The $L_1$-norm regularization term encourages sparse estimates with only few non-zero coefficients in $x$, which explain a large variability of the ME-fMRI voxel time series according to the model in Eq. (4), thereby performing both variable selection and regularization to improve prediction accuracy and interpretability of the estimate. We note here that other sparsity-promoting regularized estimators (e.g. Elastic Net or Generalized LASSO) could also be implemented straightforwardly with this formulation.

Instead of setting a fixed regularization parameter $\lambda$, we propose to compute the entire regularization path (e.g. in this work we used the Least-angle regression (LARS) procedure [17] implemented in the *lars* package in R) and select the optimal estimate according to the Bayesian Information Criterion (BIC) as follows:

$$\widehat{x}_{\text{BIC}} = \arg \min_\lambda NK \log (RSS(\lambda)) + \log (NK) \, df (\lambda), \tag{5}$$

where $RSS(\lambda) = \|\bar{y} - \mathbf{T}\hat{x}(\lambda)\|_2^2$ and $df(\lambda)$ are the residual sum of squares and effective degrees of freedom for each estimate as a function of $\lambda$, respectively.

Finally, to compensate for the shrinkage towards zero of the coefficients due to the L1-norm regularization we propose to perform debiasing of the BIC estimate. Debiasing is computed as the least squares estimate on the reduced model corresponding to the subset of non-zero coefficients of the estimate (a.k.a. relaxed LASSO [18]). Specifically, let $\mathcal{A}$ denote the support of $\widehat{x}_{\text{BIC}}$, i.e. $\mathcal{A} = \text{supp}(\widehat{x}_{\text{BIC}}) = \{j, \widehat{x}_{\text{BIC}} \neq 0\}$, the coefficients of the debiased estimate in the support $\mathcal{A}$ are re-computed as $\widehat{x}_{\text{BIC},\mathcal{A}} = (\mathbf{T}_\mathcal{A}^T \mathbf{T}_\mathcal{A})^{-1} \mathbf{T}_\mathcal{A}^T \widehat{y}$, where $\mathbf{T}_\mathcal{A}$ is a matrix including the subset of columns of $\mathbf{T}$ corresponding to the support $\mathcal{A}$, and the coefficients that are not included in $\mathcal{A}$ remain as zero.

# 3  Methods

The ME-SPFM algorithm was evaluated in a ME-fMRI dataset acquired in 10 subjects (5 males, mean $\pm$ SD age $= 25 \pm 3$) with a multi-task rapid event-related paradigm [9]. All participants gave informed consent in compliance with the NIH Combined Neuroscience International Review Board - approved protocol 93-M-1070 in Bethesda, MD.

### 3.1  MRI Data Acquisition and Experimental Paradigm

MRI data was acquired on a General Electric 3 T 750 MRI scanner with a 32-channel receive-only head coil (Waukesha, WI). Functional scans were acquired with a ME GE-EPI sequence (flip angle = 70° for 9 subjects, flip angle = 60° for 1 subject, TEs = 16.3/32.2/48.1 ms, TR = 2 s, 30 axial slices, slice thickness = 4 mm, in-plane resolution = $3 \times 3$ mm$^2$, FOV = 192 mm, acceleration factor 2, number of acquisitions = 220). Six subjects performed two functional runs, and 4 subjects only performed 1 run due to scanning time constraints. MPRAGE and PD images were also acquired for anatomical alignment and visualization purposes (176 axial slices; voxel size = $1 \times 1 \times 1$ mm$^3$).

In each run, the experimental paradigm had 6 trials of each of 5 different tasks (i.e. a total of 30 trials): (1) pressing a button with the left hand, (2) viewing video patterns resembling biological motion, (3) viewing pictures of houses, (4) listening to music, and (5) reading sentences. All trials had a duration of 4 s. Full details are given in [9].

### 3.2  Data Preprocessing

Each ME-fMRI dataset was preprocessed with three pipelines in AFNI [19] including:

1. Second echo dataset (E02, $TE = 32.2$ ms): (1) removal of the initial 10 s. to achieve steady-state magnetization, (2) slice timing correction, (3) volume realignment, registration to anatomical image, and warping to MNI template, (4) spatial normalization to MNI space in a single step, (5) nuisance regression (Legendre polynomials up to 5$^{th}$ order, realignment parameters, their temporal derivatives, and 5 largest principal components of voxels within the lateral ventricles), (6) spatial smoothing (3D Gaussian kernel FWHM = 6 mm), and (7) calculation of signal percentage change. This pipeline was also applied to the first (E01) and third echo (E03) datasets.
2. Optimally combined dataset (OC): Same as E02, but with optimal weighted combination of the three echoes [2] between steps (4) and (5).
3. Multi-echo Independent Analysis dataset (MEICA): Same as E02, but with ME-ICA denoising (v2.5-beta11) between steps (4) and (5).

### 3.3  Data Analysis: ME-SPFM vs. GLM and SPFM

The three preprocessed echo datasets (E01, E02 and E03) were input to the ME-SPFM algorithm described above resulting in voxelwise time-varying estimates of $\Delta R_2^*$ and $\Delta S_0$. The SPM canonical HRF was used as the hemodynamic response function (HRF) to define **H**. ME-SPFM activation maps were created for each trial from the $\Delta R_2^*$ estimates as the average of the volumes when each trial occurs (i.e. 3 TRs). The performance of ME-SPFM was compared with the results of the deconvolution with Sparse Paradigm Free Mapping (SE-SPFM, [12]) and traditional GLM analyses (AFNI *3dREMLfit*) on each of the three types of preprocessed datasets (E02, OC and MEICA).

As for SE-SPFM, datasets were analyzed with *3dPFM* in AFNI using the LASSO algorithm, and also the BIC for selection of the regularization parameter and the SPM canonical HRF [12]. Similar to ME-SPFM, SE-SPFM activation maps were created from the deconvolved coefficients (beta output in *3dPFM*) as the average of the volumes when each trial occurs.

Regarding the GLM analysis, two activation maps were created per task modeling either all task trials together ('task-based') or each trial individually ('trial-based'). The trial-based activation maps were thresholded at uncorrected $p \leq 0.001$ and $p \leq 0.05$. The task-based activation maps were thresholded at FDR-corrected $q \leq 0.05$ to create gold standard maps for computation of sensitivity, specificity and dice coefficients.

## 4    Results

Figure 1 depicts activation maps for four individual events of three different tasks in one representative subject. The $\Delta R_2^*$ maps from ME-SPFM more clearly resemble those from the trial-based GLM analysis in the MEICA dataset. While SE-SPFM maps normally show activations in the same locations as the GLM maps (i.e. high spatial specificity), they exhibit reduced sensitivity.

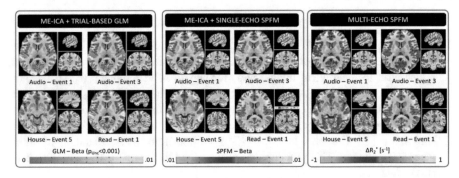

**Fig. 1.** Single-trial activation maps of three different tasks in one representative dataset. The GLM and SE-SPFM maps are obtained from the MEICA dataset. ME-SPFM maps show negative values (blue) as they represent the decrease in $\Delta R_2^*$ resulting in positive BOLD signal changes.

Figure 2 illustrates the preprocessed and fitted signals for trial-based GLM and ME-SPFM analyses in five representative voxels in regions relevant to each task. It can be seen the $\Delta R_2^*$ traces obtained with ME-SPFM are mostly equal to zero in the search of sparsity of the estimate, but negative $\Delta R_2^*$ (deflection of blue trace) are reliably detected at the time of all experimental events (dark-gray bands). Owing to its ability to detect events without knowledge of their timing, ME-SPFM is able to detect additional BOLD events that occur during the acquisition (black arrows), which sometimes coincide with timing of other tasks and could not be revealed by any conventional GLM approach.

Figure 3 shows the average sensitivity, specificity and dice coefficients for the different methods using the task-based GLM activation maps as gold standard. The ME-SPFM algorithm significantly outperforms its SE counterpart, and achieves an average sensitivity and dice coefficients similar to a trial-based GLM analysis with a threshold between $p \leq 0.001$ and $p \leq 0.05$, with a small reduction in specificity.

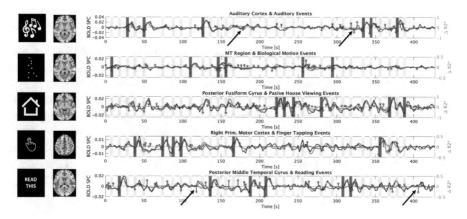

**Fig. 2.** Results for voxels associated with each of the five tasks for a representative dataset. Dashed black traces plot the time series after the ME-ICA preprocessing. Green traces show the fit signals with a trial-based GLM analysis. Blue traces show the estimated $\Delta R_2^*$ time series obtained with ME-SPFM and red traces show the corresponding fit (i.e. after convolution with the HRF). Black arrows mark examples of events with negative $\Delta R_2^*$ missed by the GLM analysis but detected with ME-SPFM. Dark grey bands indicate the timing of events of the task expected to activate the voxel, and light grey bands indicate the timing of events of the other four tasks.

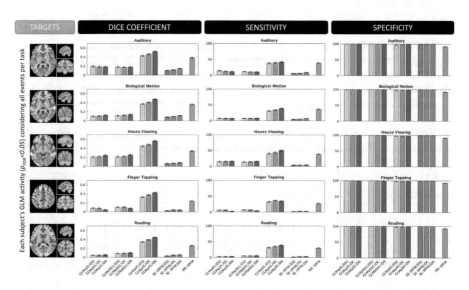

**Fig. 3.** Evaluation using as gold standard the task-based GLM ($q \leq 0.05$) activation maps. Left: Gold standard maps per task for one representative subject. Right: Average dice coefficient, sensitivity, and specificity across datasets (s.e. in error bars) for trial-based GLM at different thresholds and different preprocessing pipelines, as well as for SE-SPFM and ME-SPFM.

# 5  Conclusion

In this work, we developed an algorithm for deconvolution of ME-fMRI data that enables to detect single-trial BOLD responses beyond those caused by events with timing known by the experimenter. Our evaluation in real data showed ME-SPFM obtains significantly better results compared with standard GLM analysis than its SE counterpart. Furthermore, contrary to previous deconvolution algorithms, this approach yields time series of $\Delta R_2^*$ with interpretable units ($s^{-1}$), facilitating a quantitative study of the brain's functional dynamics in unconstrained paradigms. Also, events detected with non-physiologically plausible $\Delta R_2^*$ values can be easily identified and removed for subsequent analyses. The ME-SPFM algorithm will be available in AFNI (*3dMEPFM*).

# References

1. Menon, R.S., Ogawa, S., Tank, D.W., Uğurbil, K.: Tesla gradient recalled echo characteristics of photic stimulation- induced signal changes in the human primary visual cortex. Magn. Reson. Med. **30**(3), 380–386 (1993)
2. Posse, S., et al.: Enhancement of BOLD-contrast sensitivity by single-shot multi-echo functional MR imaging. Magn. Reson. Med. **42**(1), 87–97 (1999)
3. Gowland, P.A., Bowtell, R.W.: Theoretical optimization of multi-echo fMRI data acquisition. Phys. Med. Biol. **52**(7), 1801–1813 (2007)
4. Poser, B.A., Versluis, M.J., Hoogduin, J.M., Norris, D.G.: BOLD contrast sensitivity enhancement and artifact reduction with multiecho EPI: parallel-acquired inhomogeneity-desensitized fMRI. Magn. Reson. Med. **55**(6), 1227–1235 (2006)
5. Bright, M.G., Murphy, K.: Removing motion and physiological artifacts from intrinsic BOLD fluctuations using short echo data. Neuroimage **64**, 526–537 (2013)
6. Kundu, P., Inati, S.J., Evans, J.W., Luh, W., Bandettini, P.A.: Differentiating BOLD and non-BOLD signals in fMRI time series using multi-echo EPI. Neuroimage **60**, 1759–1770 (2012)
7. Kundu, P., Voon, V., Balchandani, P., Lombardo, M.V., Poser, B.A., Bandettini, P.A.: Multi-echo fMRI: a review of applications in fMRI denoising and analysis of BOLD signals. Neuroimage **154**, 59–80 (2017)
8. Caballero-Gaudes, C., Reynolds, R.C.: Methods for cleaning the BOLD fMRI signal. Neuroimage **154**, 128–149 (2017)
9. Gonzalez-Castillo, J., et al.: Evaluation of multi-echo ICA denoising for task based fMRI studies: block designs, rapid event-related designs, and cardiac-gated fMRI. Neuroimage **141**, 452–468 (2016)
10. Gitelman, D.R., Penny, W.D., Ashburner, J., Friston, K.J.: Modeling regional and psychophysiologic interactions in fMRI: the importance of hemodynamic deconvolution. Neuroimage **19**, 200–207 (2003)
11. Caballero-Gaudes, C., Petridou, N., Dryden, I.L., Bai, L., Francis, S.T., Gowland, P.A.: Detection and characterization of single-trial fMRI BOLD responses: paradigm free mapping. Hum. Brain Mapp. **32**(9), 1400–1418 (2011)
12. Caballero-Gaudes, C., Petridou, N., Francis, S.T., Dryden, I.L., Gowland, P.A.: Paradigm free mapping with sparse regression automatically detects single-trial fMRI BOLD responses. Hum. Brain Mapp. **34**(3), 501–518 (2013)

13. Hernandez-Garcia, L., Ulfarsson, M.O.: Neuronal event detection in fMRI time series using iterative deconvolution techniques. Magn. Reson. Imaging **29**(3), 353–364 (2011)
14. Karahanoğlu, F.I., Caballero-Gaudes, C., Lazeyras, F., Van de Ville, D.: Total activation: fMRI deconvolution through spatio-temporal regularization. Neuroimage **73**, 121–134 (2013)
15. Caballero-Gaudes, C., Karahanoğlu, F.I., Lazeyras, F., Van De Ville, D.: Structured sparse deconvolution for paradigm free mapping of functional MRI data. In: Proceedings of 9th IEEE International Symposium on Biomedical Imaging, Barcelona, pp. 322–325 (2012)
16. Chen, S.S., Donoho, D.L., Saunders, M.A.: Atomic decomposition by basis pursuit. SIAM J. Sci. Comput. **20**(1), 33–61 (1998)
17. Efron, B., Hastie, T., Johnstone, I., Tibshirani, R.: Least angle regression. Ann. Stat. **32**(2), 407–499 (2004)
18. Meinshausen, N.: Relaxed lasso. Comput. Stat. Data Anal. **52**(1), 374–393 (2007)
19. Cox, R.W.: AFNI: software for analysis and visualization of functional magnetic resonance neuroimages. Comput. Biomed. Res. **29**(3), 162–173 (1996)

# Brain Decoding from Functional MRI Using Long Short-Term Memory Recurrent Neural Networks

Hongming Li and Yong Fan[(⊠)]

Center for Biomedical Image Computing and Analytics, Department of
Radiology, Perelman School of Medicine, University of Pennsylvania,
Philadelphia, USA
Yong.Fan@uphs.upenn.edu

**Abstract.** Decoding brain functional states underlying different cognitive processes using multivariate pattern recognition techniques has attracted increasing interests in brain imaging studies. Promising performance has been achieved using brain functional connectivity or brain activation signatures for a variety of brain decoding tasks. However, most of existing studies have built decoding models upon features extracted from imaging data at individual time points or temporal windows with a fixed interval, which might not be optimal across different cognitive processes due to varying temporal durations and dependency of different cognitive processes. In this study, we develop a deep learning based framework for brain decoding by leveraging recent advances in sequence modeling using long short-term memory (LSTM) recurrent neural networks (RNNs). Particularly, functional profiles extracted from task functional imaging data based on their corresponding subject-specific intrinsic functional networks are used as features to build brain decoding models, and LSTM RNNs are adopted to learn decoding mappings between functional profiles and brain states. We evaluate the proposed method using task fMRI data from the HCP dataset, and experimental results have demonstrated that the proposed method could effectively distinguish brain states under different task events and obtain higher accuracy than conventional decoding models.

**Keywords:** Brain decoding · Recurrent neural networks
Long short-term memory

## 1 Introduction

Decoding the brain based on functional signatures derived from imaging data using multivariate pattern recognition techniques has become increasingly popular in recent years. With the massive spatiotemporal information provided by the functional brain imaging data, such as functional magnetic resonance imaging (fMRI) data, several strategies have been proposed for the brain decoding [1–7].

Most of the existing fMRI based brain decoding studies focus on identification of functional signatures that are informative for distinguishing different brain states. Particularly, brain activations evoked by task stimuli identified using a general linear model

© Springer Nature Switzerland AG 2018
A. F. Frangi et al. (Eds.): MICCAI 2018, LNCS 11072, pp. 320–328, 2018.
https://doi.org/10.1007/978-3-030-00931-1_37

(GLM) framework are commonly adopted [8]. The procedure of identifying brain activation maps is equivalent to a supervised feature selection procedure, which may improve the sensitivity of the brain decoding. In addition to feature selection using the GLM framework, several studies select regions of interests (ROIs) related to the brain decoding tasks based on *a prior* anatomical/functional knowledge [2]. A two-step strategy [4] that swaps the functional signature identification from spatial domain to temporal domain has recently been proposed to decode fMRI activity in the time domain, aiming to overcome the curse of dimensionality problem caused by spatial functional signatures used for the brain decoding. All these aforementioned methods require knowledge of timing information of task events or types of tasks to carry out the feature selection for the brain decoding, which limits their general application. Other than task-specific functional signatures identified in a supervised manner, several whole-brain functional signatures have been proposed. In particular, whole-brain functional connectivity patterns based on resting-state brain networks identified using independent component analysis (ICA) are adopted for the brain decoding [1]. However, time windows with a properly defined width are required in order to reliably estimate the functional connectivity patterns. Deep belief neural network (DBN) has been adopted to learn a low-dimension representation of 3D fMRI volume for the brain decoding [3], where 3D images are flatten into 1D vectors as features for learning the DBN, losing the spatial structure information of the 3D images. More recently, 3D convolutional neural networks (CNNs) are adopted to learn a latent representation for decoding functional brain task states [5]. Although the CNNs could learn discriminative representations effectively, it is nontrivial to interpret biological meanings of the learned features.

Most of the existing studies perform the brain decoding based on functional signatures computed at individual time points or temporal windows with a fixed length using conventional classification techniques, such as support vector machine (SVM) [9] and logistic regression [2, 4]. These classifiers do not take into consideration the temporal dependency, which is inherently available in the sequential fMRI data and may boost the brain decoding performance. Though functional signatures extracted from time windows [1, 4, 5] may help capture the temporal dependency implicitly, time windows with a fixed width are not necessarily optimal over different brain states since they may change at unpredictable intervals. On the other hand, recurrent neural networks (RNNs) with long short-term memory (LSTM) [10] have achieved remarkable advances in sequence modeling [11], and these techniques might be powerful alternatives for the brain decoding tasks.

In this study, we develop a deep learning based framework for decoding the brain states from task fMRI data, by leveraging recent advances in RNNs. Particularly, we learn mappings between functional signatures and brain states by adopting LSTM RNNs which could capture the temporal dependency adaptively by learning from data. Instead of selecting ROIs or fMRI features using feature selection techniques or *a prior* knowledge of problems under study, we extract functional profiles from task functional imaging data based on subject-specific intrinsic functional networks and the functional profiles are used as features for building LSTM RNNs based brain decoding models. Our method has been evaluated for predicting brain states based on task fMRI data obtained from the human connectome project (HCP) [12], and experimental results have demonstrated that the proposed method could obtain better brain decoding performance than the conventional methods.

## 2    Methods

To decode the brain state from task fMRI data, a prediction model of LSTM RNNs [10] is trained based on functional signatures extracted using a functional brain decomposition technique [13, 14]. The overall framework is illustrated in Fig. 1(a).

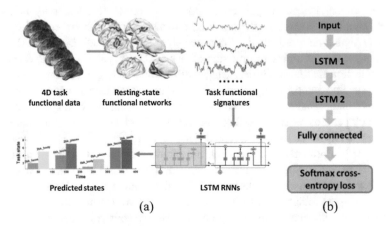

**Fig. 1.** Schematic illustration of the proposed brain decoding framework. (a) The overall architecture of the proposed model, (b) LSTM RNNs used in this study.

### 2.1    Functional Signature Based on Intrinsic Functional Networks

With good correspondence to the task activations [15], intrinsic functional networks (FNs) provided an intuitive and generally applicable means to extract functional signatures for the brain state decoding. Using the FNs, 3D fMRI data could be represented by a low-dimension feature vector, which could alleviate the curse of dimensionality, be general to different brain decoding tasks, and provide better interpretability. Instead of identifying ROIs at a group level [1], we applied a collaborative sparse brain decomposition model [13, 14] to the resting-state fMRI data of all the subjects used for the brain decoding to identify subject-specific FNs.

Given a group of $n$ subjects, each having a resting-state fMRI scan $D^i \in R^{T \times S}$, $i = 1, 2, \ldots, n$, consisting of $S$ voxels and $T$ time points, we first obtain $K$ FNs $V^i \in R_+^{K \times S}$ and its corresponding functional time courses $U^i \in R^{T \times K}$ for each subject using the collaborative sparse brain decomposition model [13, 14], which could identify subject-specific functional networks with inter-subject correspondence and better characterize the intrinsic functional representation at an individual subject level. Based on the subject-specific FNs, the functional signatures $F^i \in R^{T \times K}$ used for the brain decoding are defined as weighted mean time courses of the task fMRI data within individual FNs, and are calculated by

$$F^i = D^i_f \cdot \left(V^i_N\right)', \tag{1}$$

where $D^i_f$ is the task fMRI data of subject $i$ for the brain decoding, $V^i_N$ is the row-wise normalized $V^i$ with its row-wise sum equal to one. Example FNs used in our study are illustrated in Fig. 2.

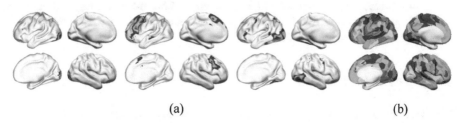

(a)                                                                        (b)

**Fig. 2.** Functional networks used to extract task functional signatures for the brain decoding. (a) Example functional networks, (b) all functional networks encoded in different colors.

### 2.2 Brain Decoding Using LSTM RNNs

Given the functional signatures $F^i$ of a group of $n$ subjects, $i = 1, 2, \ldots, n$, a LSTM RNNs [10] model is built to predict the brain state of each time point based on its functional profile and temporal dependency on its preceding time points. The architecture of the LSTM RNNs used in this study is illustrated in Fig. 1(b), including two hidden LSTM layers and one fully connected layer. Two hidden LSTM layers are used to encode the functional information with temporal dependency for each time point, and the fully connected layer is used to learn a mapping between the learned feature representation and the brain states. The functional representation encoded in each LSTM layer is calculated as

$$
\begin{aligned}
f^l_t &= \sigma\left(W^l_f \cdot \left[h^l_{t-1}, x^l_t\right] + b^l_f\right), \\
i^l_t &= \sigma\left(W^l_i \cdot \left[h^l_{t-1}, x^l_t\right] + b^l_i\right), \\
\tilde{C}^l_t &= tanh\left(W^l_C \cdot \left[h^l_{t-1}, x^l_t\right] + b^l_c\right), \\
C^l_t &= f^l_t * C^l_{t-1} + i^l_t * \tilde{C}^l_t, \\
o^l_t &= \sigma\left(W^l_o \cdot \left[h^l_{t-1}, x^l_t\right] + b^l_o\right), \\
h^l_t &= o^l_t * tanh\left(C^l_t\right),
\end{aligned}
\tag{2}
$$

where $f^l_t$, $i^l_t$, $C^l_t$, $h^l_t$, and $x^l_t$ denote the output of forget gate, input gate, cell state, hidden state, and the input feature vector of the $l$-th LSTM layer ($l = 1, 2$) at the $t$-th time point respectively, and $\sigma$ denotes the sigmoid function. The input features to the first LSTM layer are the functional signatures derived from FNs, and the input to the second LSTM layer is a hidden state vector obtained by the first LSTM layer. A fully connected layer with $S$ output nodes is adopted for predicting the brain state as

$$s_t = softmax(W_s \cdot h_t^2 + b_s), \tag{3}$$

where $S$ is the number of brain states to be decoded, and $h_t^2$ is the hidden state output of the second LSTM layer which encodes the input functional signature at the $t$-th time point and the temporal dependency information encoded in the cell state from its preceding time points.

In this study, each hidden LSTM layer contains 256 hidden nodes, and softmax cross-entropy between real and predicted brain states is used as the objective function to optimize the LSTM RNNs model.

## 3   Experimental Results

We evaluated the proposed method based on task and resting-state fMRI data of 490 subjects from the HCP [12]. In this study, we focused on the working memory task, which consisted of 2-back and 0-back task blocks of tools, places, faces and body, and a fixation period. Each working memory fMRI scan consisted of 405 time points of 3D volumes, and its corresponding resting-state fMRI scan had 1200 time points. The fMRI data acquisition and task paradigm were detailed in [12].

We applied the collaborative sparse brain decomposition model [13, 14] to the resting-state fMRI data of 490 subjects for identifying 90 subject-specific FNs. The number of FNs was estimated by MELODIC [16]. The subject-specific FNs were then used to extract functional signatures of the working memory task fMRI data for each subject, which was a matrix of 405 by 90. The proposed method was then applied to the functional signatures to predict their corresponding brain states. Particularly, we split the whole dataset into training, validation, and testing datasets. The training dataset included data of 400 subjects for training the LSTM RNNs model, the validation dataset included data of 50 subjects for determining the early-stop of the training procedure, and data of the remaining 40 subjects were used as an external testing dataset.

Due to the delay of blood oxygen level dependent (BOLD) response observed in fMRI data, the occurrence of brain response is typically not synchronized with the presentation of stimuli, so the brain state for each time point was adjusted according to the task paradigm and the delay of BOLD signal before training the brain decoding models. Based on an estimated BOLD response delay of 6 s [17], we shifted the task paradigms forward by 8 time points and used them to update the ground truth brain states for training and evaluating the proposed brain state decoding model.

To train a LSTM RNNs model, we have generated training samples by cropping the functional signatures of each subject into clip matrices of 40 by 90, with an overlap of 20 time points between temporally consecutive training clips. We adopted the cropped dataset for training our model for following reasons. Firstly, the task paradigms of most subjects from the HCP dataset shared almost the identical temporal patterns. In other words, the ground truth brain states of most subjects were the same, which may mislead the model training to generate the same output regardless of the functional signatures fed into the LSTM RNNs model if we used their full-length data for training the brain decoding model. In our study, the length of data clips was set to 40 so that each clip contained 2 or 3 different brain states and such randomness could eliminate the aforementioned bias. Secondly, the data clips with temporal overlap also served as data augmentation of the training samples for improving the model training. When evaluating our LSTM RNNs model, we applied the trained model to the full-length functional signatures of the testing subjects to predict brain states of their entire task fMRI scans. We implemented the proposed method using Tensorflow. Particularly, we adopted the ADAM optimizer with a learning rate of 0.001, which was updated every 50,000 training steps with a decay rate of 0.1, and the total number of training steps was set to 200,000. Batch size was set to 32 during the training procedure.

We compared the proposed model with a brain decoding model built using random forests [18], which used the functional signatures at individual time points as features. The random forests classifier was adopted due to its inherent feature selection mechanism and its capability of handling multi-class classification problems. For the random forests based brain decoding model, the number of decision trees and the minimum leaf size of the tree were selected from a set of parameters ({100, 200, 500, 1000} for the number of trees, and {3, 5, 10} for the minimum leaf size) to optimize its brain decoding performance based on the validation dataset.

### 3.1  Brain Decoding on Working Memory Task FMRI Data

The mean normalized confusion matrices of the brain decoding accuracy on the 40 testing subjects obtained by the random forests and the LSTM RNNs models are shown in Fig. 3. The LSTM RNNs model outperformed the random forests model in 5 out of 9 brain states (Wilcoxon signed rank test, $p < 0.002$). The overall accuracy obtained by the LSTM RNNs model was $0.687 \pm 0.371$, while the overall accuracy obtained by the random forests model was $0.628 \pm 0.234$, demonstrating that our method performed significantly better than the random forests based prediction models (Wilcoxon signed rank test, $p < 0.001$). The improved performance indicates that the temporal dependency encoded in the LSTM RNNs model could provide more discriminative information for the brain decoding.

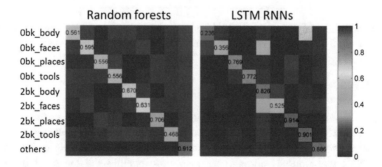

**Fig. 3.** Brain decoding performance of the random forests and LSTM RNNs models on the testing dataset of working memory task fMRI. The colorbar indicates mean decoding accuracy on the 40 testing subjects.

### 3.2 Sensitivity Analysis of the Brain Decoding Model

To understand the LSTM RNNs based decoding model, we have carried out a sensitivity analysis to determine how changes in the functional signatures affect the decoding model based on the 40 testing subjects using a principal component analysis (PCA) based sensitivity analysis method [19]. Particularly, with the trained LSTM RNNs model fixed, functional signatures of 90 FNs were excluded (i.e., their values were set to zero) one by one from the input and changes in the decoding accuracy were recorded. Once all the changes in the brain decoding accuracy with respect to all FNs were obtained for all testing subjects, we obtained a change matrix of $90 \times 40$, encapsulating changes of the brain decoding. We then applied PCA to the change matrix to identify principle components (PCs) that encoded main directions of the prediction changes with respect to changes in the functional signatures of FNs.

The sensitive analysis revealed FNs whose functional signatures were more sensitive than others to the brain decoding on the working memory task fMRI data. Particularly, among top 5 FNs with the largest magnitudes in the first PC as shown in Fig. 4, four of them were corresponding to the working memory evoked activations as demonstrated in [20], indicating that the LSTM RNNs model captured the functional dynamics of the working memory related brain states.

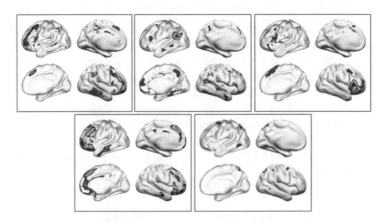

**Fig. 4.** Sensitivity analysis of the brain decoding model on the working memory task fMRI dataset. The top 5 FNs with most sensitive functional signatures are illustrated.

## 4   Conclusions

In this study, we propose a deep learning based model for decoding the brain states underlying different cognitive processes from task fMRI data. Subject-specific intrinsic functional networks are used to extract task related functional signatures, and the LSTM RNNs technique is adopted to adaptively capture the temporal dependency within the functional data as well as the relationship between the learned functional representations and the brain functional states. The experimental results on the working memory task fMRI dataset have demonstrated that the proposed model could obtain improved brain decoding performance compared with a decoding model without considering the temporal dependency.

**Acknowledgements.** This work was supported in part by National Institutes of Health grants [CA223358, EB022573, DK114786, DA039215, and DA039002] and a NVIDIA Academic GPU grant.

# References

1. Shirer, W.R., et al.: Decoding subject-driven cognitive states with whole-brain connectivity patterns. Cereb. Cortex **22**(1), 158–165 (2012)
2. Huth, A.G., et al.: Decoding the semantic content of natural movies from human brain activity. Front. Syst. Neurosci. **10**, 81 (2016)
3. Jang, H., et al.: Task-specific feature extraction and classification of fMRI volumes using a deep neural network initialized with a deep belief network: evaluation using sensorimotor tasks. Neuroimage **145**(Pt B), 314–328 (2017)
4. Loula, J., Varoquaux, G., Thirion, B.: Decoding fMRI activity in the time domain improves classification performance. Neuroimage **180**(Pt A), 203–210 (2018)
5. Wang, X., et al.: Task state decoding and mapping of individual four-dimensional fMRI time series using deep neural network. arXiv preprint arXiv:1801.09858 (2018)
6. Fan, Y., Shen, D., Davatzikos, C.: Detecting cognitive states from fMRI images by machine learning and multivariate classification. In: 2006 Conference on Computer Vision and Pattern Recognition Workshop (CVPRW 2006) (2006)
7. Davatzikos, C., et al.: Classifying spatial patterns of brain activity with machine learning methods: application to lie detection. Neuroimage **28**(3), 663–668 (2005)
8. Mumford, J.A., et al.: Deconvolving BOLD activation in event-related designs for multivoxel pattern classification analyses. Neuroimage **59**(3), 2636–2643 (2012)
9. Shen, G., et al.: Decoding the individual finger movements from single-trial functional magnetic resonance imaging recordings of human brain activity. Eur. J. Neurosci. **39**(12), 2071–2082 (2014)
10. Hochreiter, S., Schmidhuber, J.: Long short-term memory. Neural Comput. **9**(8), 1735–1780 (1997)
11. Lipton, Z.C., Berkowitz, J., Elkan, C.: A critical review of recurrent neural networks for sequence learning. arXiv preprint arXiv:1506.00019 (2015)
12. Glasser, M.F., et al.: The minimal preprocessing pipelines for the Human Connectome Project. Neuroimage **80**, 105–124 (2013)
13. Li, H., Satterthwaite, T.D., Fan, Y.: Large-scale sparse functional networks from resting state fMRI. Neuroimage **156**, 1–13 (2017)
14. Li, H., Satterthwaite, T., Fan, Y.: Identification of subject-specific brain functional networks using a collaborative sparse nonnegative matrix decomposition method. In: 2016 IEEE 13th International Symposium on Biomedical Imaging (ISBI) (2016)
15. Smith, S.M., et al.: Correspondence of the brain's functional architecture during activation and rest. Proc. Natl. Acad. Sci. U.S.A. **106**(31), 13040–13045 (2009)
16. Jenkinson, M., et al.: FSL. Neuroimage **62**(2), 782–790 (2012)
17. Liao, C.H., et al.: Estimating the delay of the fMRI response. Neuroimage **16**(3 Pt 1), 593–606 (2002)
18. Breiman, L.: Random forests. Mach. Learn. **45**(1), 5–32 (2001)

19. Koyamada, S., Koyama, M., Nakae, K., Ishii, S.: Principal sensitivity analysis. In: Cao, T., Lim, E.-P., Zhou, Z.-H., Ho, T.-B., Cheung, D., Motoda, H. (eds.) PAKDD 2015. LNCS (LNAI), vol. 9077, pp. 621–632. Springer, Cham (2015). https://doi.org/10.1007/978-3-319-18038-0_48
20. Barch, D.M., et al.: Function in the human connectome: task-fMRI and individual differences in behavior. Neuroimage **80**, 169–189 (2013)

# Learning Generalizable Recurrent Neural Networks from Small Task-fMRI Datasets

Nicha C. Dvornek[1(✉)], Daniel Yang[3], Pamela Ventola[2],
and James S. Duncan[1,4,5]

[1] Department of Radiology & Biomedical Imaging, Yale School of Medicine,
New Haven, CT, USA
`nicha.dvornek@yale.edu`
[2] Child Study Center, Yale School of Medicine, New Haven, CT, USA
[3] Autism and Neurodevelopmental Disorders Institute,
George Washington University and Children's National Health System,
Washington, DC, USA
[4] Department of Biomedical Engineering, Yale University, New Haven, CT, USA
[5] Department of Electrical Engineering, Yale University, New Haven, CT, USA

**Abstract.** Deep learning has become the new state-of-the-art for many problems in image analysis. However, large datasets are often required for such deep networks to learn effectively. This poses a difficult challenge for many medical image analysis problems in which only a small number of subjects are available, e.g., patients undergoing a new treatment. In this work, we propose a number of approaches for learning generalizable recurrent neural networks from smaller task-fMRI datasets: (1) a resampling method for ROI-based fMRI analysis to create augmented data; (2) inclusion of a small number of non-imaging variables to provide subject-specific initialization of the recurrent neural network; and (3) selection of the most generalizable model from multiple reinitialized training runs using criteria based on only training loss. Using cross-validation to assess model performance, we demonstrate the effectiveness of the proposed methods to train recurrent neural networks from small datasets to predict treatment outcome for children with autism spectrum disorder ($N = 21$) and classify autistic vs. typical control subjects ($N = 40$) from task-fMRI scans.

## 1 Introduction

Deep learning approaches are quickly becoming the machine learning technique of choice for many medical image analysis problems, e.g., image classification, segmentation, and registration [12]. The deep neural networks have a large capacity to learn directly from raw images. However, it is well known that these popular methods can quickly overfit the data, resulting in poor generalization. Thus,

This work was supported by NIH grants R01MH100028 and R01NS035193.

A. F. Frangi et al. (Eds.): MICCAI 2018, LNCS 11072, pp. 329–337, 2018.
https://doi.org/10.1007/978-3-030-00931-1_38

learning useful models generally requires training on very large datasets and using proper model validation techniques.

The large data requirement poses a challenge in analyzing many medical imaging datasets, in which only a small number of subjects may be available. For example, it may not be feasible to gather large amounts of data when studying a specific disease population or a new experimental treatment, or it may be difficult to obtain expert manual annotations of large datasets for training. Recent trends toward open science and data sharing have made larger medical imaging datasets more widely available, e.g., the ABIDE dataset for autism [2]; however, creating large datasets for every disease and medical imaging problem is clearly not possible. While some medical image analysis problems can handle smaller datasets by using patch-based approaches (e.g., in image segmentation [12]) to augment the amount of data, such methods are not as suitable for analyzing neurological data from functional magnetic resonance imaging (fMRI).

In this paper, we propose new strategies that facilitate learning more generalizable neural network models from small fMRI datasets. We first adopt a recurrent neural network with long short-term memory (LSTM) to generate predictions from a whole-brain parcellation of fMRI data. We then use resampling approaches to generate multiple summary time-series for each region in the parcellation, augmenting the original dataset. Next, we utilize available non-imaging variables to provide subject-specific initialization of the LSTM network. Finally, we describe a criteria for selecting the most generalizable model from many training instances on the same data using only training loss, allowing all available data to be used for model training. We apply the proposed strategies and compare them to other approaches to learn from task-fMRI for two small data examples: (1) a regression problem of predicting treatment outcome from 21 children with autism spectrum disorder (ASD), and (2) a classification problem of identifying autistic children vs. typical controls from a dataset of 40 subjects.

## 2    Methods

### 2.1    Base LSTM Architecture for fMRI

LSTMs and related architectures are designed to learn long-term dependencies in time-series data [7]. They have recently been applied to fMRI for modeling brain dynamics [6] and for classification [3]. In addition to the time dependent nature of the model, LSTMs are a nice neural network model specifically for small fMRI datasets, since an "unrolled" LSTM with $T$ timesteps can be thought of as a deep network with $T$ layers that share the same parameters across all the layers. This likely considerably reduces the number of model parameters that need to be learned compared to other standard deep neural network architectures.

Standard fMRI whole-brain analysis involves first summarizing the data in a number of regions of interest (ROIs) according to some brain parcellation. While deep networks are able to learn from raw image inputs, the ROI approach is beneficial in our case of dealing with smaller fMRI datasets, as fMRI data is very noisy and the ROI representation greatly reduces the input data dimension.

Our base LSTM architecture is based on the model in [3], with added regularization and slight changes for regression vs. binary classification. The summary time-series from the ROIs are used as input to an LSTM. For regression, we pass the output from the LSTM at the last timestep to a dense layer to produce the predicted value (Fig. 1(a), blue path); thus, the entire task-fMRI sequence is analyzed before providing a final prediction. For classification, we more closely follow the network in [3]; the LSTM output from every timestep is passed to a shared dense layer with a single node, followed by mean pooling across time and a sigmoid activation function to produce the classification probability (Fig. 1(b), blue path). During training, we include dropout of the LSTM weights [5] and add dropout (with probability 0.5) between the LSTM output and dense layer.

## 2.2  Data Augmentation by Resampling

Standard data augmentation techniques for neural networks to learn from image data include using random croppings and random rotations of the images. However, our LSTM network is designed to use the time-series from the brain ROIs as inputs, and such augmentation techniques are not appropriate for fMRI. We could perform random cropping along the time dimension, but LSTMs learn best from long sequences. Another generic approach is to inject random noise into the inputs [16]; however, it is unclear how to choose the best noise model and associated parameters, and while such approaches may slow down overfitting, it may not be representative of the variation in the fMRI data.

We instead propose a resampling approach to augment the data. Traditional ROI analysis extracts the mean time-series calculated from all voxels in the ROI. To inject variation to the ROI time-series, we propose sampling only a subset of the ROI voxels or sampling all voxels with replacement (bootstrapping) and using the average of the sampled data to summarize the time-series for the ROI.

## 2.3  LSTM Initialization with Non-imaging Variables

An LSTM cell contains two state vectors, the hidden state (i.e., output) $h_t$ and the cell state $c_t$. The state of an LSTM at time $t$ depends on the current input $x_t$ and the cell state from the previous timestep $h_{t-1}$ and $c_{t-1}$:

$$g_t = \sigma\left(W_g x_t + U_g h_{t-1} + b_g\right), \text{with } g \in \{i, f, o\} \tag{1}$$

$$\tilde{c}_t = \tanh\left(W_c x_t + U_c h_{t-1} + b_c\right) \tag{2}$$

$$c_t = i_t * \tilde{c}_t + f_t * c_{t-1} \tag{3}$$

$$h_t = o_t * \tanh\left(c_t\right) \tag{4}$$

where $i$, $f$, and $o$ represent input, forget, and output gates, $\tilde{c}_t$ is the current estimated cell state, and $W$, $U$, and $b$ are the LSTM model parameters.

Unless otherwise specified, the initial state of the LSTM is set to zeros, $h_0 = c_0 = \mathbf{0}$. Simple non-imaging subject information (e.g., age) is often available. We propose initializing the LSTM by feeding such non-imaging information into 2 dense layers, whose outputs are the initial hidden and cell states (Fig. 1, green path). Such initialization approaches have been proposed in other domains, e.g., an LSTM model to generate an image caption was initialized on image features extracted via a convolutional neural network [10]. In our small data setting, conditioning the LSTM on subject-specific parameters helps to incorporate non-imaging variation across subjects with just a small increase in model parameters, unlike other multi-modal fusion techniques for neural networks [4,13].

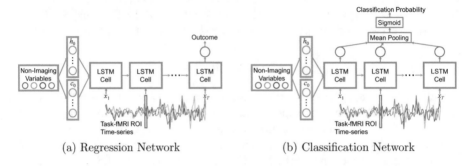

(a) Regression Network          (b) Classification Network

**Fig. 1.** LSTM networks with initialization using non-imaging variables.

## 2.4   Model Selection from Training Loss

Neural network training (in the non transfer-learning case) is generally performed using random initialization of model weights. With large amounts of data, several training runs can be performed with different initializations, and a validation dataset can be set aside to assist with choosing the best trained model. However, with small datasets, we would prefer to use all available data for training. Furthermore, splitting off a small validation set is likely not representative of the test data and thus is not appropriate for model selection.

We propose choosing the best model from several initializations based on the recorded training loss curve. Rather than choosing the model with the lowest loss, which is likely to be the most overfit to the small dataset, we choose the model that fits slowest to the data. We quantify this criteria with the following:

$$\hat{M} = \arg\max_M \left[ median\left(\triangle L_{M,s}\right) \frac{1}{L_M\left(0\right) \times s} \right], \tag{5}$$

where $L_M$ is the training loss curve for model $M$, $L_M\left(0\right)$ is the loss after epoch 0, $\triangle L_{M,s}$ are the first differences of the loss curve from epoch 0 to stopping epoch $s$, and $s$ is the first epoch such that $L_M\left(s\right) < L_M(0)/e$. Thus, the criteria is looking for the model that learns slowest, measured by the median of the first differences over the epochs up to epoch $s$, weighted by the initial loss and the number of epochs to reduce the loss to $1/e$ of the initial loss (borrowing the idea of relaxation time). We only look at the first differences up to $1/e$ since we are more interested in how fast the model fits the data in earlier epochs. Furthermore, the training curve will likely have a very long flat tail due to overfitting to the training set, making it difficult to measure differences in convergence. We scale our criteria by the initial loss, since given two curves with the same rate of convergence but with different initial losses, we would rather choose the model with the higher initial loss, signifying a worse initial fit and overall slower learning of the model. Finally, we scale by the number of epochs for the signal to decay ("relaxation time"), since given two curves with similar convergence measured by the other two metrics, we want the model that takes longer to minimize the loss.

## 3    Experiments

### 3.1    Data and Preprocessing

Data was acquired from 21 children with ASD (ages $6.05 \pm 1.24$ years) and 19 typically-developing controls (TC) (ages $6.42 \pm 1.29$ years). Each subject underwent a T1-weighted MP-RAGE structural MRI ($1 \times 1 \times 1$ mm$^3$ voxel size) and BOLD T2*-weighted fMRI sequence ($3.44 \times 3.44 \times 4.00$ mm$^3$ voxel size) acquired during a biological motion perception task [9]. The fMRI paradigm involved viewing point light animations of coherent and scrambled biological motion in a block design ($\sim$24 s per block, $\sim$5 min scan). Non-imaging information collected included age, sex, IQ, and score on the Social Responsiveness Scale (SRS), 2nd edition [1]. SRS measures severity of social impairment in autism; lower scores signify better social ability. ASD subjects then underwent 16 weeks of Pivotal Response Treatment [11], a behavioral therapy for ASD. SRS score was measured again at the end of treatment.

Images were preprocessed in FSL [8] using the pipeline by Pruim et al. [14], which included motion correction, interleaved slice timing correction, brain extraction, 4D mean intensity normalization, spatial smoothing (5 mm FWHM), data denoising via ICA-AROMA [14], nuisance regression using white matter and cerebrospinal fluid, and high-pass temporal filtering (100 s). Functional MRI were aligned to the standard MNI brain with the aide of the structural MRI. The AAL atlas [15] was applied, resulting in 90 cerebral ROIs from which summary time-series (156 timepoints) were extracted and used as input to the LSTM. Since fMRI absolute signal varies greatly across the brain, each summary time-series was standardized (subtracted mean, divided by standard deviation). The data for each non-imaging variable were normalized to range $[-1, 1]$.

### 3.2    Regression Example: Prediction of Treatment Outcome

We investigated the effectiveness of the proposed learning strategies on the following regression problem: to predict the treatment outcome (i.e., percent change in SRS) for the 21 children with ASD from baseline information. Leave-one-out cross-validation (train on 20, test 1) was used to assess model performance. Mean squared error (MSE), standard deviation (SD) of the squared error, and Pearson's correlation coefficient ($r$) between predicted and true treatment outcome were computed from cross-validation folds. Paired one-tailed t-tests were used to compare the squared errors, and p-values for $r$ provided evidence for non-zero correlation, with a significance level of 0.05. Neural networks were implemented and trained in Keras using the MSE loss function, adadelta optimizer, 8 hidden LSTM units, a maximum of 100 epochs with early stopping (patience of 5 epochs monitoring training loss), and a batch size of 32 unless otherwise specified.

We first directly trained the LSTM network on the 21 fMRI datasets. We varied the batch size (2, 5, 10, 20) to try to improve learning. The best result is shown in Table 1a ("Original"); however, errors between the best result and other batch sizes were not significantly different, and correlations were insignificant.

**Table 1.** Results for predicting treatment outcome.

(a) Data augmentation approaches.

| Dataset | MSE (SD) | $r$ | $p_r$ |
|---|---|---|---|
| Original | 0.097 (0.160) | 0.35 | 0.1204 |
| Repeat | 0.035 (0.049) | 0.45 | 0.0415 |
| Low Noise | 0.034 (0.037)* | 0.47 | 0.0324 |
| High Noise | 0.029 (0.029)* | 0.59 | 0.0050 |
| Sample 10 | 0.029 (0.034)* | 0.58 | 0.0058 |
| Sample 50 | 0.034 (0.036)* | 0.47 | 0.0313 |
| Sample 250 | 0.030 (0.026)* | 0.55 | 0.0101 |
| Bootstrap | 0.031 (0.041)* | 0.53 | 0.0129 |

*Significantly better than original dataset.

(b) Non-imaging data and model selection.

| Dataset | MSE (SD) | $r$ | $p_r$ |
|---|---|---|---|
| Bootstrap (BS) | 0.031 (0.041) | 0.53 | 0.0129 |
| BS + Non-Imaging | 0.020 (0.025) | 0.73 | 0.0002 |
| BS + Top Fusion | 0.035 (0.037) | 0.46 | 0.0339 |
| BS + Model Bag | 0.032 (0.037) | 0.51 | 0.0175 |
| BS + Model Select | 0.028 (0.032) † | 0.60 | 0.0044 |
| BS + Non-Imaging + Model Bag | 0.024 (0.029) † | 0.66 | 0.0011 |
| BS + Non-Imaging + Model Select | 0.018 (0.025)^† | 0.77 | <0.0001 |

^Significantly better than bootstrap dataset.
†Significantly better than one individual model.

We then compared the following data augmentation techniques: (1) repeating the data ("Repeat"), (2) standard noise injection by adding zero-mean Gaussian noise with SD equal to SD of the time-series divided by 10 ("Low Noise") or 2 ("High Noise"), and (3) the proposed resampling approach of randomly sampling 10, 50, or 250 voxels without replacement or bootstrap sampling all voxels from each ROI to compute the summary time-series. We repeated the augmentation approaches 50 times per subject, resulting in 1050 samples. Results are shown in Table 1a. Simply repeating the data resulted in significant correlation, although MSE did not significantly improve. All other augmentation approaches produced significant correlation as well as significantly reduced the MSE. While the high noise augmentation nominally resulted in the highest correlation, there were no significant differences between any noise and sampling methods.

Since errors were not significantly different and the bootstrap sampling does not require any parameter selection, we tested the remaining learning strategies on only the bootstrap-augmented dataset (Table 1b). Initializing the LSTM with non-imaging data dramatically improved the correlation and reduced the MSE by 35% (just missing significance with $p = 0.0572$), at the cost of only a 1% increase in number of parameters. Applying a standard multimodal fusion approach to combine the final fMRI score and non-imaging data in a dense layer, also increasing the number of parameters by 1%, results in worse performance ("Top Fusion"), demonstrating the benefit of our LSTM initialization method.

We tested our model selection approach by assessing the training curves from 2 separate runs, and compared this to averaging the predictions from the 2 runs (bagging). We applied these approaches to the bootstrap dataset and the bootstrap with non-imaging model. Model bagging did not produce significantly lower errors compared to the individual models for the bootstrap dataset. Our model selection approach resulted in significantly lower MSE compared to at least one of the individual models. Furthermore, applying all three of our proposed learning strategies resulted in significantly more accurate predictions compared to data augmentation alone, with the highest correlation with the true outcomes. The effect of adding each proposed learning strategy is illustrated in Fig. 2.

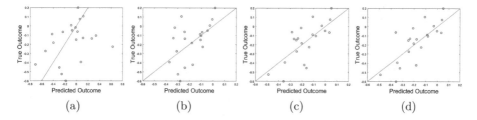

(a)          (b)          (c)          (d)

**Fig. 2.** Plots of true vs. predicted treatment outcome after applying each proposed learning strategy. Perfect predictions would fall on red reference line. (a) Original data. (b) Data augmentation with bootstrap resampling. (c) Bootstrap resampling and LSTM initialization with non-imaging data. (d) Bootstrap resampling, inclusion of non-imaging data, and model selection based on training loss criteria in (5).

### 3.3   Classification Example: Autism vs. Typical Control

We tested the proposed learning strategies on training the LSTM network to classify the 21 ASD and 19 TC subjects (52.5% ASD subjects). Ten-fold cross-validation (train on 36, test 4) was repeated 10 times, and performance of different methods were measured using mean and standard deviation of the cross-validation accuracy, true positive rate (TPR), and true negative rate (TNR). Paired one-tailed t-tests were used to compare cross-validation performance between different methods, with a significance level of 0.05. Training was run with similar Keras setup as above, with maximum number of epochs reduced to 20.

Results quantifying the effects of the proposed learning strategies are shown in Table 2. Learning from the original, non-augmented sample results in chance accuracy. Applying bootstrap resampling (50 resamples, resulting in 2000 total samples) significantly improves the accuracy and TPR. Using non-imaging variables to set the initial LSTM state further improved all performance measures, with significant differences compared to using the original dataset. Finally, additionally including model selection produced the best performing model.

**Table 2.** Results for classifying ASD vs. TC subjects.

| Dataset | Mean (SD) Accuracy (%) | Mean (SD) TPR (%) | Mean (SD) TNR (%) |
|---|---|---|---|
| Original | 51.8 (3.3) | 56.1 (13.3) | 55.1 (12.4) |
| Bootstrap | 64.5 (5.1)* | 70.7 (7.3)* | 60.9 (11.1) |
| Bootstrap + Non-imaging | 67.5 (6.7)* | 72.2 (9.2)* | 64.6 (6.3)* |
| Bootstrap + Non-imaging + Model select | 69.8 (5.5)*$^{\wedge\dagger}$ | 75.1 (8.4)* | 65.5 (6.8)* |

*Significantly better than original dataset. $^{\wedge}$Significantly better than bootstrap dataset. $^{\dagger}$Significantly better than at least one individual model.

## 4    Conclusions

In this work, we presented strategies for training LSTMs on small datasets and demonstrated their effectiveness in learning better generalized models. Our methods for facilitating learning included a data augmentation approach specific to ROI-based analysis, incorporation of subject-specific variations by initializing the LSTM based on each subject's non-imaging parameters, and model selection based on training loss criteria alone to maximize the amount of data available for training. Regression and classification learning from 2 small task-fMRI datasets showed that while naïve training of the LSTM was unable to learn useful models, combining the proposed learning strategies resulted in the successful training of more generalizable LSTMs.

## References

1. Constantino, J.N., Gruber, C.P.: The Social Responsiveness Scale (SRS-2), 2nd edn. Western Psychological Services, Torrance (2012)
2. Di Martino, A., et al.: The autism brain imaging data exchange: towards a large-scale evaluation of the intrinsic brain architecture in autism. Mol. Psychiatry **19**, 659–667 (2014)
3. Dvornek, N.C., Ventola, P., Pelphrey, K.A., Duncan, J.S.: Identifying autism from resting-state fMRI using long short-term memory networks. In: Wang, Q., Shi, Y., Suk, H.-I., Suzuki, K. (eds.) MLMI 2017. LNCS, vol. 10541, pp. 362–370. Springer, Cham (2017). https://doi.org/10.1007/978-3-319-67389-9_42
4. Dvornek, N.C., Ventola, P., Duncan, J.S.: Combining phenotypic and resting-state fMRI data for autism classification with recurrent neural networks. In: ISBI (2018)
5. Gal, Y., Ghahramani, Z.: A theoretically grounded application of dropout in recurrent neural networks. In: NIPS (2016)
6. Güçlü, U., van Gerven, M.A.J.: Modeling the dynamics of human brain activity with recurrent neural networks. Front. Comput. Neurosci. **11**, 7 (2017)
7. Hochreiter, S., Schmidhuber, J.: Long short-term memory. Neural Comput. **9**(8), 1735–1780 (1997)
8. Jenkinson, M., Beckmann, C.F., Behrens, T.E., Woolrich, M.W., Smith, S.M.: FSL. NeuroImage **62**, 782–790 (2012)

9. Kaiser, M., et al.: Neural signatures of autism. Proc. Natl. Acad. Sci. U.S.A. **107**(49), 21223–21228 (2010)
10. Karpathy, A., Fei-Fei, L.: Deep visual-semantic alignments for generating image descriptions. IEEE Trans. Pattern Anal. Mach. Intell. **39**, 664–676 (2017)
11. Koegel, R., Koegel, L.: Pivotal Response Treatments for Autism: Communication, Social, and Academic Development. Brookes Publishing Company, Baltimore (2006)
12. Litjens, G., et al.: A survey on deep learning in medical image analysis. Med. Image Anal. **42**, 60–68 (2017)
13. Ngiam, J., Khosla, A., Kim, M., Nam, J., Lee, H., Ng, A.Y.: Multimodal deep learning. In: The 28th International Conference on Machine Learning (2011)
14. Pruim, R.H., Mennes, M., van Rooij, D., Ller, A., Buitelaar, J.K., Beckmann, C.F.: ICA-AROMA: a robust ICA-based strategy for removing motion artifacts from fMRI data. NeuroImage **112**, 267–277 (2015)
15. Tzourio-Mazoyera, N., et al.: Automated anatomical labeling of activations in SPM using a macroscopic anatomical parcellation of the MNI MRI single-subject brain. NeuroImage **15**, 273–289 (2002)
16. Zur, R.M., Jiang, Y., Pesce, L.L., Drukker, K.: Noise injection for training artificial neural networks: a comparison with weight decay and early stopping. Med. Phys. **36**, 4810–4818 (2009)

# Diffusion Tensor Imaging and Functional MRI: Human Connectome

# Fast Mapping of the Eloquent Cortex by Learning L2 Penalties

Nico Hoffmann[1(✉)], Uwe Petersohn[2], Gabriele Schackert[3], Edmund Koch[4], Stefan Gumhold[1], and Matthias Kirsch[3]

[1] Computer Graphics and Visualisation, Technische Universität Dresden, 01062 Dresden, Germany
nico.hoffmann@tu-dresden.de
[2] Applied Knowledge Representation and Reasoning, Technische Universität Dresden, 01062 Dresden, Germany
[3] Department of Neurosurgery, University Hospital Carl Gustav Carus, 01307 Dresden, Germany
[4] Clinical Sensoring and Monitoring, Technische Universität Dresden, 01307 Dresden, Germany

**Abstract.** The resection of brain tumors beneath eloquent areas of the human brain requires precise delineation of eloquent areas for maximum removal of tumor mass while minimizing the risk for postoperative functional deficits. Non-invasive mapping of eloquent areas can be carried out by intraoperative thermal imaging since neural activity alters the cortical temperature distribution. These characteristic changes in cortical temperature can be modeled by a response function. A prominent choice for this response function is the haemodynamic response function. However, the signal is typically superimposed by various effects such as motion artifacts, physiological effects, sensor drifts as well as autoregulation which have to be compensated.

In this paper, we contribute a regularized semiparametric regression framework that recognizes the response function while it compensates for arbitrary background signals. We achieve this by learning a tailored L2 penalty that basically regularizes the estimated background signal such that it doesn't comprise the characteristics of the response function. The evaluation of this approach is carried out by augmented semisynthetic resting-state- as well as intraoperative thermal imaging data.

## 1 Introduction

Intraoperative thermal imaging of the exposed cerebral cortex can be used to analyze the surface temperature as a surrogate measure of regional cerebral blood flow (rCBF) [1]. The main contributor to the detected temperature variations is changes in rCBF [2]. Additionally, neurovascular coupling links transient neural activity to a corresponding change in rCBF [3], which can be detected by thermal imaging [4]. However, various effects such as heat flow, convection, and surrounded cooling tissue interfere with the measured radiation arising from

© Springer Nature Switzerland AG 2018
A. F. Frangi et al. (Eds.): MICCAI 2018, LNCS 11072, pp. 341–348, 2018.
https://doi.org/10.1007/978-3-030-00931-1_39

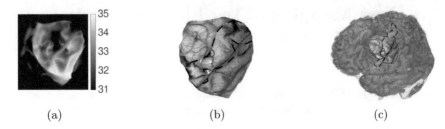

(a)                              (b)                              (c)

**Fig. 1. (a)** Shows a thermal image of the exposed human cortex. The dashed region in **(b)** and **(c)** marks the primary somatosensory cortex. The recognition of small changes in cortical temperature originating from neurovascular coupling enables the mapping of eloquent areas of the brain. For this reason, we induced neural activity by contralateral somatosensory evoked potentials. Areas with statistically significant changes in cortical temperature are highlighted in the RGB image **(b)**.

superficial tissue and thus considerably decrease the signal-to-noise ratio [1]. Sensor drifts and other hysteresis effects of the employed camera system also alter the signal. In medical imaging, physiological effects such as autoregulatory activity, motion artifacts at heart rate or respiration have to be considered as well.

In this paper, we propose a flexible regression framework that recognizes the deterministic signal change induced by focal neural activity while it compensates for arbitrary signal alterations induced by the camera and physiological effects. For this purpose, we introduce a novel machine learning scheme for L2 regularization which extends the semiparametric regression framework [5] such that it is able to approximate arbitrary background signal while the characteristics of the foreground component are being preserved.

## 2    Modeling Superimposed Signals in Terms of Semiparametric Regression

The response function depends on the observed physiological effect. In functional MRI, the haemodynamic response function or a boxcar function are commonly used [6]. In functional thermal imaging, we expect the heat dissipation of neural activity to be bell-shaped. This behavior is approximated by a Gaussian function with $\mu$ being the temporal delay between the onset of the stimulus and the peak in regional cerebral blood flow while $\sigma$ denotes its thermal extent:

$$\gamma(t_i; \mu, \sigma) = exp((-(t_i - \mu)/2\sigma)^2) \qquad (1)$$

This non-linear function is discretized at $m$ time points $[t_1 \cdots t_i \cdots t_m]$

$$\Gamma(\mu, \sigma) = [\gamma(t_1; \mu, \sigma) \cdots \gamma(t_m; \mu, \sigma)]^T \qquad (2)$$

given $t_i \in \mathbb{R}$. In case of $k + 1$ repeated stimuli with uniform period $\Delta_e$ we have

$$X_\Gamma \in \mathbb{R}^m = \sum_{i=0}^{k} [\Gamma(k\Delta_e + \mu, \sigma)] \qquad (3)$$

**Demixing Parametric and Non-parametric Components.** Sparse representations are a popular approach for representing signals or images by orthonormal bases or learned dictionaries [7]. L1 regularization allows to find sparse, compact representations of data, which has been successfully applied to demixing problems [8]. Unfortunately, this regularization method leads to non-differentiable problems meaning that we cannot obtain closed form solutions. Fortunately, the semiparametric regression framework allows us to model data in terms of non-parametric component as well as a priori known deterministic response function (parametric component). The former describe complex temporal autocorrelation patterns that are a priori unknown. An increase in the degrees of freedom of the non-parametric component enhances the compensation of these effects yet might also result in overfitting or absorbing energy of the response function. A length $m$ signal $y(t)$ can be formulated by the semiparametric regression framework. This integrates the response function $\gamma$, a non-parametric $\kappa$-th order B-spline component $B \in \mathbb{R}^{m \times m_S}$, $m_S$ knots and a certain L2 penalty $\mathbb{P} \in \mathbb{R}^{m_S \times m_S}$ as of

$$y(t) = \alpha\gamma(t, \mu, \sigma) + \sum_{j=1}^{m_S} B_{j,\kappa}(t)\beta_j \text{ subject to } ||\mathbb{P}\beta||_2^2 < C \tag{4}$$

with $\beta = [\beta_1 \dots \beta_{m_S}]$. Now, we stack our discretized response function $X_\Gamma$ and the B-spline basis into a single matrix $G = [X_\Gamma \ B]$. The least-squares estimator $\hat{b} = [\hat{\alpha} \ \hat{\beta}]$ can now be obtained by solving

$$\hat{b} = \underset{b}{\text{argmin}} \ ||Gb - y||_2^2 + \lambda||\mathbb{P}\beta||_2^2 \text{ subject to } \lambda \geq 0 \tag{5}$$

Equation 5 is minimized with respect to $b$ given the block diagonal matrix $S = blockdiag(0, 1_{m_S})$. Setting its derivate equal to zero yields the penalized normal equations

$$\hat{b}(\lambda) = (G^T G + \lambda S^T \mathbb{P}^T \mathbb{P}S)^{-1} G^T y \tag{6}$$

Typically, $\lambda$ is optimized by a grid search based on Akaike's Information Criterion (see [5] for details). The presence of our response function can be evaluated by testing the hypothesis $\mathbb{H}_0 : \hat{\alpha} = 0$. This test can be done using the z-statistics $Z = \hat{\alpha}/s.e.(\hat{\alpha})$ and provides us confidence regarding the presence of $\hat{\alpha}$. The standard error can be estimated by $s.e.(\alpha) = \sigma^2 Cov(\alpha)$ under homoscedasticity- and no correlation assumption of the model's residual [5]. The multiple comparison problem is approached by the Bonferroni correction $p_\alpha^{bonf} = p_\alpha/k$ with $k$ being the number of pixels.

Regularization imposes constraints on the local behavior of the model's coefficients. Classically, Tikhonov regularization yields a shrinkage penalty through $\mathbb{P} = I$ given the identity matrix $I \in \mathbb{R}^{m_S \times m_S}$ that penalizes large values. Another common choice is $\mathbb{P} = \Delta_p$ [5] which imposes a roughness penalty by the degree $p$ difference operator $\Delta_p$ resulting in smooth fits. In contrast to those strategies, we want to learn a penalty term $\mathbb{P}$ such that the non-parametric component approximates any signal while it preserves the characteristics of our response function. For this purpose, we will learn a penalty matrix $\mathbb{P} \in \mathbb{R}^{ms \times ms}$ that penalizes those correlation patterns in the B-spline coefficients that relate to our response function.

## 3   Learning L2 Penalties

Suppose we only have limited knowledge about our data meaning that we are just able to specify the response function while the background signal can be any highly variate signal with short- and longterm autocorrelation. Let's assume that we are given an arbitrary resting-state signal $y \in \mathbb{R}^m$. Let further be $\tau \in \mathbb{R}^m$ one instance of our discretized response function such as the discretized thermal response to neural activity as seen in Eq. 1 or the haemodynamic response function. This allows us to sample instances $y^* = y + \tau$ by altering certain characteristics of $\tau$ such as its amplitude or periodicity. Our learning task can now be formulated in terms of the energy function $E$ and the Frobenius norm $||X||_F^2 = tr(X^T X)$ as of

$$\underset{P}{argmin}\; E = \sum_{i=1}^{n} ||\mathbb{P}\mathbb{B}y_i^* - \mathbb{B}\tau_i||_2^2 + \lambda ||\mathbb{P}||_F^2 \;\; \text{subject to}\;\; \lambda \geq 0 \qquad (7)$$

with B-spline fit $\mathbb{B} = (B^T B)^{-1} B^T$ and $n$ samples. The intuition behind the energy function $E$ is, that we can rewrite $(\mathbb{P}\mathbb{B}y_i^*)$ as $(\mathbb{P}\mathbb{B}y_i + \mathbb{P}\mathbb{B}\tau_i)$ which leads to the minimization of $||(\mathbb{P}\mathbb{B}y_i + \mathbb{P}\mathbb{B}\tau_i) - \mathbb{B}\tau_i||_2^2$. This means that $E$ is minimal with respect to $\mathbb{P}$ if we find a matrix $\mathbb{P}$ that filters the energy content of our background signal $y_i$ while it preserves the contribution of the deterministic component $\tau_i$. Thus, our $\mathbb{P}$ has to fulfill

$$\mathbb{P}\mathbb{B}y_i \rightarrow 0 \qquad (8)$$
$$\mathbb{P}\mathbb{B}\tau_i \rightarrow \mathbb{B}\tau_i \qquad (9)$$

in order to minimize $E$ with respect to all $1 \leq i \leq n$. By stacking $\tau_i, y_i, y_i^*$ into $T = [\tau_1, \tau_2, \ldots, \tau_n]^T$, $Y = [y_1, y_2, \ldots, y_n]^T$, $Y^* = [y_1^*, y_2^*, \ldots, y_n^*]^T$ and setting $\delta E / \delta \mathbb{P} = 0$, the least-squares estimate of our penalty matrix becomes

$$\hat{\mathbb{P}} = (\mathbb{B}TY^T \mathbb{B}^T)(\mathbb{B}Y^* Y^{*^T} \mathbb{B}^T + \lambda I)^{-1} \qquad (10)$$

Summarizing, the learned penalty matrix $\hat{\mathbb{P}}$ preserves the energy content of our response function while it filters the contribution of all other components. Thus, plugging $\hat{\mathbb{P}}$ into Eq. 5 constrains the non-parametric B-spline component to solely preserve the characteristics of the response function.

## 4   Experimental Results and Discussion

All intraoperative procedures were allowed by the human ethics committee and all patients provided informed consent. The proposed penalty learning scheme is

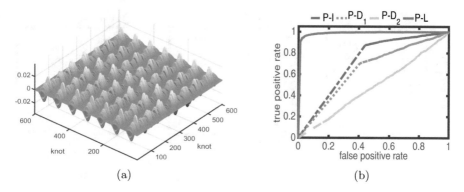

(a)                                                        (b)

**Fig. 2. (a)** The regular pattern in the learned penalty matrix relates to the autocorrelation structure of the response function. **(b)** This penalty matrix (denoted by $P - L$) significantly improves the accuracy compared to classical regularization strategies such as Tikhonov regularization $(P - I)$ or roughness regularization $(P - D_1, P - D_2)$.

evaluated on semisynthetic- as well as intraoperative data. Semisynthetic data was generated by augmenting thermal data at resting state by certain response functions, such as the Gaussian response function (Eq. 1) with randomly sampled $\alpha \in [0.04; 0.08]$ and constant $\mu = 0$ and $\sigma = 10$. The range of $\alpha$ was chosen according to [4] and results in a signal-to-noise ratio of $1.3\overline{3}$ to $2.6\overline{6}$. The dataset consists of short as well as long-term autocorrelation effects that are due to sensor noise, sensor drift, tissue cooling as well as periodic physiological signals. The analysis is quantified in terms of accuracy $ACC = 100 * (TP + TN)/(P + N)$ given the number of true positives $TP$, false positives $FP$, true negatives $TN$, positive samples $P$ and negative samples $N$. The true positive rate $(TPR)$ is defined as $TPR = TP/P$ and false positive rate $(FPR)$ as $FPR = FP/N$. We generated balanced synthetic data meaning that $P = N$. Finally, the root mean square error of the estimate $\hat{y} = G\hat{b}$ is defined as $||y - \hat{y}||_2^2$.

## 4.1   Analysis of the Learning Scheme

Assumptions (8), (9) were validated experimentally given a 3rd order B-Spline basis $B \in \mathbb{R}^{1024 \times 603}$ at 600 knots. $y_i$ denotes a randomly chosen resting-state signal sampled at 2.5 Hz while $\tau_i$ represents our discretized Gaussian response function $(1 \leq i \leq 12,000)$. The learned penalty $\hat{\mathbb{P}}$ compressed $y_i$ by 91.01% (assumption 8) while $\tau_i$ was solely reduced by 9.41% (assumption 9). Therefore, most of the content of the raw signal $y_i$ was filtered while the response function $\tau_i$ was preserved by our linear penalty $\mathbb{P}$. The incorporated and preserved autocorrelation behavior can be seen in Fig. 2a. $\mathbb{P}$ also retains the amplitude of our response function when applied to $y^*$. The blue dots of Fig. 3c resemble $max(\mathbb{P}y_i^*)$. The first 8,000 time-series of resting-state data were superimposed by the response function scaled to several $\alpha$ (dashed line) and the remaining

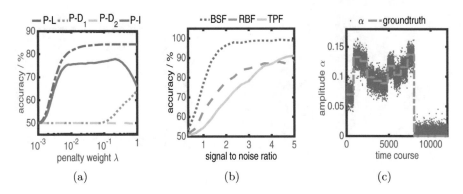

**Fig. 3. (a)** The accuracy should change monotonically depending on $\lambda$ in case of a tailored penalty. **(b)** The B-Spline basis results in best accuracy due to its numerical stability. **(c)** The amplitude $\alpha$ was recovered (blue dots) well by the regression model using the learned penalty (blue dots). In case of no signal ($\alpha = 0$) the estimated signal amplitude is below the noise level of the camera (30 mK).

time-series contain no response function. Therefore, the learned penalty operator succeeded in preserving $\alpha$ while arbitrary background noise was filtered.

**Regularization.** Traditional penalization strategies perform suboptimal in (semi-) parametric regression since those penalties yield a robust B-spline estimate however they don't ensure that the B-spline doesn't absorb energy of the parametric component such as our response function. In general, we found a significant increase in accuracy by using the learned L2 penalty matrix (see Figs. 2b, 3a and Table 1).

**Basis Functions.** Common choices for spline basis functions are truncated polynomials (TPF), radial basis functions (RBF) and B-splines (BSF) [5]. Our results indicate, that the choice of the basis function is crucial to the overall performance. We found that B-splines are more numerical stable (see [9] for details) compared to TPF. The performance of RBF and TPF decreased at some point since we were facing numerical instabilities. This issue arises from the inner term of Eq. 5 which becomes close to singular and therefore yields inaccurate estimates. However, B-splines and TPF result in similar performance in case of signals with mixed amplitudes $\alpha$ (Table 1). BSF also allowed us to use a larger number of knots resulting in a better approximation of the background signal and improved accuracy (see Fig. 3b). Note that the datasets of Fig. 3b) and Table 1 differ. We restricted the analysis to B-splines, TPF and RBF however the approach should be applicable to arbitrary undercomplete basis functions.

**Response Function.** The evaluation comprises of Gaussian function (Sect. 2), the haemodynamic response function (HRF) [6] of SPM12 with parameters (6,

**Table 1.** The accuracy [%] depends on the choice of basis function as well as the desired response function.

| | $P-L$ | $P-I$ | $P-D_1$ | $P-D_2$ | | BSF | RBF | TPF |
|---|---|---|---|---|---|---|---|---|
| Accuracy [%] | **89.2** | 75.2 | 66.4 | 49.8 | Sine | 94.9% | 91.69% | 94.62% |
| TPR [%] | **78.5** | 80.8 | 60.5 | 4.2 | Boxcar | 88.8% | 87.63% | 92.93% |
| FPR [%] | **0.2** | 30.4 | 27.6 | 4.5 | HRF | 96.7% | 89.63% | 96.55% |
| RMSE | **1.558** | 2.869 | 3.632 | 3.519 | Gaussian | 95.75% | 93.33% | 95.65% |

16, 1, 1, 6, 0, 32), a boxcar function [10] as well as a cropped sine function with $f(x) = sin(x)$ and $x \in [0, \pi]$. A penalty matrix was learned for each of these functions. We found superior performance independent of the basis function (Table 1) given a specific configuration of each function. However, the performance decreases when the training data comprises of very heterogeneous configurations or mixtures of response functions. In this case, a more generic penalty matrix is learned which contradicts the idea of a tailored penalty. Therefore, the performance can be maximized by learning a penalty for each configuration.

## 4.2 Intraoperative Data

The cases suffered from a primary brain tumor or metastasis beneath the primary somatosensory cortex. We recorded all intraoperative datasets during the contralateral electrical stimulation of the patient's median nerve. This stimulation evokes neural activity on the somatosensory cortex of the hand areal which we also verified by intraoperative phase reversal. The protocol alternates between repeated 30 s stimulation and 30 s resting periods and had been repeated for seven times. We were able to recognize neural activity using the proposed regression framework using a BSF with 600 knots, Gaussian response function and the learned penalty matrix (see Table 2). In each case, large clusters of neural activity were detected on the postcentral gyrus at $p_\alpha^{bonf} = 0.05/k$ with $k = 307,200$ (as seen in Fig. 1, dashed area). The large heterogeneity of the eloquent areas ($4348 \pm 5347$ pixels) suggests that more medical research is needed in order to unveil patient-specific factors that contribute to this observation. The average runtime was $56.25 \pm 1.1$ s.

**Table 2.** Analysis of intraoperative thermal imaging data.

| Case | Sex/age | Pathology | Location | Preoperative clinical condition | Activation size | Z-statistics |
|---|---|---|---|---|---|---|
| 1 | m/58 | Met. Adeno-ca. | precentral l. | hemiparesis r. (4/5) | 3059 px | $8.5 \pm 2.5$ |
| 2 | f/33 | Met. Mamma-ca. | parietal l. | facioplegia | 14900 px | $15.2 \pm 6.7$ |
| 3 | m/79 | Met. Melanoma | frontoparietal r. | hemiparesis l. arm (3/5), l. leg (4/5) | 2168 px | $-6.3 \pm 0.7$ |
| 4 | m/69 | Glioblastoma IV | parietooccipital l. | symptomatic seizures | 1050 px | $-6.4 \pm 0.9$ |
| 5 | m/72 | Astrocytoma III | parietal l. | hemiparesis brachiofacial | 564 px | $-5.9 \pm 0.5$ |

## 5   Conclusion

We addressed the intraoperative mapping of the eloquent cortex by demixing imaging data into a known response function and arbitrary background signals using the semiparametric regression framework. The background signals were approximated by a non-parametric basis function while the response to evoked neural activity was modeled by a parametric function. The accuracy was significantly improved by restraining the non-parametric component from resembling the characteristics of the response function. We achieve this improvement by learning a penalty that can be seen as a tailored filter operator that preserves the characteristics of the response function. Hereby, we are able to design a generic regression framework that compensates arbitrary background signals and doesn't require any data preprocessing steps. This statistically sound framework is able to recognize heavily superimposed signals even in case of short- and long-term autocorrelation effects. We successfully applied the proposed method to intraoperative thermal imaging datasets in order to visualize focal neural activity. The active areas were further validated using the gold standard in neurosurgery: electrophysiology (phase reversal). Finally, we demonstrated that the proposed L2 penalty learning scheme yields good results regardless of the choice of basis- and response function.

**Acknowledgements.** This work was supported by the European Social Fund (project no. 100270108) and the Saxonian Ministry of Science and Art. The authors would also like to thank the medical staff that supported this research project.

## References

1. Lahiri, B., Bagavathiappan, S., Jayakumar, T., Philip, J.: Medical applications of infrared thermography: a review. Infrared Phys. Technol. **55**(4), 221–235 (2012)
2. Papo, D.: Brain temperature: what it means and what it can do for (cognitive) neuroscientists. Neurons and Cognition (2013)
3. Huneau, C., Benali, H., Chabriat, H.: Investigating human neurovascular coupling using functional neuroimaging: a critical review of dynamic models. Front. Neurosci. **9**, 467 (2015)
4. Gorbach, A.M., et al.: Intraoperative infrared functional imaging of human brain. Ann. Neurol. **54**(3), 297–309 (2003)
5. Ruppert, D., Wand, M.P., Carroll, R.J.: Semiparametric Regression. Cambridge University Press, Cambridge (2003)
6. Friston, K.J., Ashburner, J.T., Kiebel, S., Nichols, T.E., Penny, W.D.: Statistical Parametric Mapping: The Analysis of Functional Brain Images. Academic Press, London (2007)
7. Aharon, M., Elad, M., Bruckstein, A.: $rmK$-SVD: an algorithm for designing overcomplete dictionaries for sparse representation. IEEE Trans. Signal Process. **54**(11), 4311–4322 (2006)
8. Greer, J.B.: Sparse demixing of hyperspectral images. Trans. Image Process. **21**(1), 219–228 (2012)
9. Eilers, P.H.C., Marx, B.D.: Splines, knots, and penalties. WIRES Comput. Stat. **2**(6), 1–26 (2010)
10. Ashby, F.G., Waldschmidt, J.G.: Fitting computational models to fMRI data. Behav. Res. Methods **40**(3), 713–721 (2008)

# Combining Multiple Connectomes via Canonical Correlation Analysis Improves Predictive Models

Siyuan Gao[1]([✉]), Abigail S. Greene[2], R. Todd Constable[3],
and Dustin Scheinost[3]

[1] Department of Biomedical Engineering, Yale University, New Haven, CT, USA
siyuan.gao@yale.edu
[2] Interdepartmental Neuroscience Program, Yale University, New Haven, CT, USA
[3] Department of Radiology and Biomedical Imaging, Yale University, New Haven,
CT, USA

**Abstract.** Generating models from functional connectivity data that predict behavioral measures holds great clinical potential. While the majority of the literature has focused on using only connectivity data from a single source, there is ample evidence that different cognitive conditions amplify individual differences in functional connectivity in a distinct, complementary manner. In this work, we introduce a computational model, labeled multidimensional Connectome-based Predictive Modeling (mCPM), that combines connectivity matrices collected from different task conditions in order to improve behavioral prediction by using complementary information found in different cognitive tasks. We apply our algorithm to data from the Human Connectome Project and UCLA Consortium for Neuropsychiatric Phenomics (CNP) LA5c Study. Using data from multiple tasks, mCPM generated models that better predicted IQ than models generated from any single task. Our results suggest that prediction of behavior can be improved by including multiple task conditions in computational models, that different tasks provide complementary information for prediction, and that mCPM provides a principled method for modeling such data.

## 1 Introduction

Advanced functional magnetic resonance imaging (fMRI) techniques, particularly functional connectivity (FC) analyses, are revealing robust individual differences in patterns of neural activity that predict continuous behavioral measures and clinical symptoms [1,2]. While FC is usually calculated from data acquired during rest, task conditions can perturb specific cognitive circuits in ways that can better reveal individual differences and improve behavioral prediction [3,4]. However, depending on the behavior under study, different tasks may emphasize different features. Thus, methods that incorporate FC information from a spectrum of cognitive tasks into a single predictive model may represent

© Springer Nature Switzerland AG 2018
A. F. Frangi et al. (Eds.): MICCAI 2018, LNCS 11072, pp. 349–356, 2018.
https://doi.org/10.1007/978-3-030-00931-1_40

the best performing and most generalizable methods for prediction of behavior from FC data. In order to combine different task FC matrices into a single predictive model in a principled way, we propose a novel algorithm based on canonical correlation analysis (CCA) and the previously validated Connectome-based Predictive Modeling (CPM) method [5], labeled multidimensional CPM (mCPM). To evaluate mCPM, we use data from two large open-source datasets (the Human Connectome Project (HCP) [6], and the UCLA Consortium for Neuropsychiatric Phenomics (CNP) LA5c Study [7]) with fMRI data collected from multiple task conditions. This paper is organized as follows: Sect. 2 summarizes related work and introduces our mCPM algorithm. Section 3 presents our evaluation of mCPM to predict IQ from connectivity matrices using the HCP and CNP datasets. Finally, Sect. 4 offers some concluding remarks.

## 2   Multidimensional Connectome-Based Predictive Modeling (mCPM)

**Overview:** mCPM is an extension of CPM to handle FC data from multiple sources. mCPM uses CCA to find complementary information from multiple connectivity matrices in order to improve prediction of behavioral measures. Here, we focus on connectivity data derived from multiple fMRI tasks. However, mCPM is agnostic to the type of connectivity data and can easily incorporate structural connectivity data from DTI or FC data from other modalities like EEG.

**Related Work:** Development and application of binary or categorical classification, such as predicting patient groupings, from FC data is a mature area of research with many proposed approaches. A non-exhaustive list includes applications to Alzheimer's disease [8], attention deficit hyperactivity disorder [9], schizophrenia [10], depression [11], and autism [12]. In contrast, prediction of dimensional outcomes from FC data is an emerging field and is considerably more challenging than binary classification of disease state. Prediction of continuous variables requires correct modeling over the whole range of the behavioral measure, whereas classification of binary groups largely requires correct grouping of participants near the margin. Additionally, associations between FC and behavior in healthy participants have substantially lower effect sizes than differences due to disease. Previous approaches to prediction of dimensional outcomes from FC data include support vector regression [1], elastic nets [2], pooled edge strength and linear models [5,13], and partial least squares regression [14]. However, these approaches only use FC data from a single condition (typically resting-state) rather than FC data from multiple conditions.

**Connectome-Based Predictive Modeling (CPM):** CPM [5] is a validated method for extracting and pooling the most relevant features from connectivity data in order to construct linear models to predict behavioral measures. Briefly,

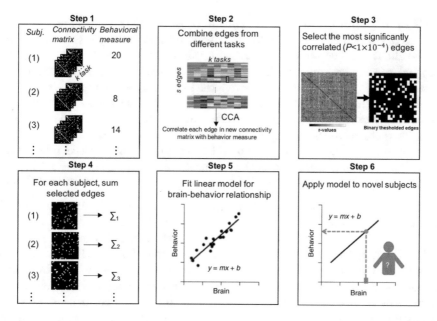

**Fig. 1.** Schematic of mCPM. Inputs to mCPM are connectivity matrices from multiple sources and behavioral measures. Step 1: participants are divided into training and testing sets. Step 2: across all participants in the training set, the same edge from different tasks are combined via CCA and correlated with behavior. Step 3: significant edges are selected for further analysis. Step 4: for each participant, the significant edges are then pooled into a summary value of FC. Step 5: a linear regression model is built between the summary value of FC and the behavioral measure. Step 6: summary values of FC are calculated for each participant in the testing set. This value is then fed into the linear regression model to generate the predicted behavioral measure for the current test participant. Figure modified with permission from [5].

edges of connectivity matrices that are significantly correlated with the behavior of interest are selected. The selected features are then pooled (*e.g.* averaged) and linear regression is used to predict the behavior in novel participants.

**Canonical Correlation Analysis (CCA):** For two sets of observation matrices $\mathbf{X}$ and $\mathbf{Y}$, assuming that the variables are correlated, CCA seeks linear combinations of the columns of these two matrices that maximize their correlation. In other words, we want to find vectors $\mathbf{a}$ and $\mathbf{b}$ such that the random variables $\mathbf{Xa}$ and $\mathbf{Yb}$ maximize the correlation. Assuming that $\mathbf{X}$ and $\mathbf{Y}$ are normalized such that each column of either matrix has mean zero and unit variance, the correlation to be maximized can be expressed by the following equation:

$$\rho = \frac{(\mathbf{Xa})^T(\mathbf{Yb})}{\sqrt{[(\mathbf{Xa})^T(\mathbf{Xa})][(\mathbf{Yb})^T(\mathbf{Yb})]}} \tag{1}$$

**mCPM Pipeline:** The mCPM pipeline consists of six steps (Fig. 1). In the first step, participants are divided into training and testing sets using 10-fold cross-validation. The second step is the combination of task FC edges. For the $(i, j)$ edge, we have a matrix $\mathbf{E}_{i,j} \in \mathbf{R}^{(n-1)*t}$. Rows of the matrix denote each participant's $(i, j)$ edge's different strengths under $t$ different tasks. Using CCA, we can find the canonical coefficients $\mathbf{w}_{i,j} \in \mathbf{R}^t$ for each edge. As each edge matrix $\mathbf{E}_{i,j}$ corresponds to the observation matrix $\mathbf{X}$ in Eq. 1, these coefficients $\mathbf{w}_{i,j}$ corresponds to the vector $\mathbf{a}$, and the observation matrix $\mathbf{Y}$ will store the behavioral measures. We then combine FC matrices from all tasks into a total connectivity matrix using different canonical correlations. For $n - 1$ subjects' combined $(i, j)$ edge, $\mathbf{E}_{total_{i,j}} = \sum_t \mathbf{E}_{i,j}^t w_{i,j}^t$, where the $t$-th column of $\mathbf{E}_{i,j}$, $\mathbf{E}_{i,j}^t$ contains $(i, j)$ edge from $n - 1$ subjects under the $t$-th task and each edge is demeaned across different participants and within the single task. For the third step, we assign the significantly correlated edges to the "correlated network" (CN). The significance of the correlation is found from the CCA. Here, we assume that CCA always maximizes the positive correlation between combined edge strength and the behavioral measure as the sign of the canonical coefficients can trivially be changed to maximize, instead of minimize, the correlation. So edges are therefore always positively correlated with the behavioral measures. Various significance thresholds can be used. In the fourth step, we calculate "network strength" $\mathbf{s}^{CN}$ by pooling (*i.e.* summing) the strength of all CN edges in each participant's total connectivity matrix, yielding a summary value $s_k^{CN}$ for the $k$-th participant:

$$\mathbf{s}^{CN} = \sum_{i,j} m_{i,j}^+ \sum_t \mathbf{E}_{i,j}^t w_{i,j}^t \tag{2}$$

where $\mathbf{s}^{CN}$ is the vector of summary values, $\mathbf{m}^+$ is the binary matrix indexing the edges $(i, j)$ that survived thresholding for the CN. In the fifth step, we use linear regression $(y_{behav_k} = \beta_0 * s_k^{CN} + \beta_1)$ to model the association between "network strength" and the behavioral measure in $n - 1$ participants. In the sixth step, the "network strength" is calculated for the excluded participant, and is submitted to the corresponding regression model to generate a behavioral measure estimate for that participant. This process is repeated iteratively, with different participants in the training and testing datasets.

## 3    Experiments and Results

**Methods:** *Datasets:* We applied mCPM to FC data from HCP and CNP datasets. We restricted our analyses to those subjects who participated in all fMRI runs in each dataset, whose mean frame-to-frame displacement was less than 0.1 mm and whose maximum frame-to-frame displacement was less than 0.15 mm and for whom IQ measures were available. 515 subjects from the HCP dataset and 169 subjects from the CNP dataset were retained for analysis. For initial testing, we used IQ, measured using an abbreviated Raven's progressive matrices form in the HCP and the Matrix Reasoning scale from the Wechsler

**Table 1.** Comparison of prediction performance of CPM and SVR using FC matrices from a single task, and mCPM and SVR using FC data from all available tasks. CPM and mCPM are shown at three different edge selection thresholds. Reported values are the "correlation\MSE" between predicted and observed IQ. mCPM (bolded) produced the best performance as measured by both correlation and MSE.

| | Task | Prediction method | | | |
|---|---|---|---|---|---|
| | | (m)CPM $p < 1\mathrm{E}{-}2$ | (m)CPM $p < 1\mathrm{E}{-}3$ | (m)CPM $p < 5\mathrm{E}{-}4$ | SVR |
| HCP | Gambling | 0.35\17.78 | 0.33\17.90 | 0.32\18.08 | 0.35\18.05 |
| | Language | 0.29\18.59 | 0.29\18.43 | 0.30\18.34 | 0.25\19.40 |
| | Motor | 0.30\18.74 | 0.30\18.59 | 0.29\18.73 | 0.21\20.40 |
| | Relational | 0.23\20.29 | 0.23\20.24 | 0.22\20.45 | 0.26\19.70 |
| | Social | 0.29\18.83 | 0.26\19.46 | 0.25\19.60 | 0.33\18.02 |
| | WM | 0.34\18.07 | 0.32\18.46 | 0.31\18.63 | 0.32\18.70 |
| | Emotion | 0.27\19.50 | 0.28\19.20 | 0.28\19.04 | 0.28\19.33 |
| | All tasks | **0.39\17.15** | **0.39\17.19** | **0.39\17.22** | 0.36\18.00 |
| CNP | BART | 0.26\105.6 | 0.32\100.3 | 0.39\93.09 | 0.23\103.3 |
| | PAMENC | 0.37\95.07 | 0.35\97.51 | 0.35\98.90 | 0.36\93.17 |
| | PAMRET | 0.30\100.3 | 0.26\105.1 | 0.23\108.8 | 0.28\99.04 |
| | SCAP | 0.30\98.80 | 0.22\109.6 | 0.20\113.0 | 0.03\116.2 |
| | SST | 0.22\107.2 | 0.22\108.3 | 0.17\115.2 | 0.36\93.30 |
| | SWITCH | 0.22\106.6 | 0.16\117.0 | 0.06\131.9 | 0.23\101.9 |
| | All tasks | **0.49\82.04** | **0.42\88.31** | **0.40\89.95** | 0.39\91.45 |

Adult Intelligence Scale in the CNP datasets, as the behavior of interest for prediction. With a single behavioral measure, mCPM simplifies to multi-linear regression. The seven tasks from the HCP were the gambling, language, motor, relational, social, WM, and emotion tasks. The six tasks from the CNP were the balloon analogue risk (BART), paired memory encoding (PAMENC), paired memory retrieval (PAMRET), spatial working memory capacity (SCAP), stop-signal (SST), and task-switching (SWITCH) tasks. *fMRI processing:* fMRI data were processed with standard methods and parcellated into 268 nodes using a whole-brain, functional atlas defined previously in a separate sample [13]. Next, the mean timecourses of each node pair were correlated and Fisher transformed, generating six or seven $268 \times 268$ connectivity matrices per subject for the CNP and HCP datasets, respectively. Task FC was calculated based on the "raw" task timecourses, with no regression of task-evoked activity. These matrices were used to generate cross-validated predictive models of IQ. *Prediction methods:* We compared mCPM with two competing methods: CPM and support vector regression (SVR) [1]. CPM and SVR were used on each task independently. In addition, all task connectivity matrices for a single subject were vectorized, concatenated, and used as inputs to a SVR model to test for increased prediction performance asso-

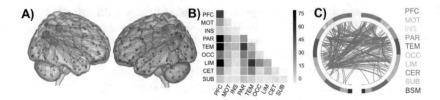

**Fig. 2.** Visualization of predictive edges selected by mCPM (feature threshold of $p <$ 1E−2) in the HCP dataset. **(A)** Brain surfaces: Spheres represent nodes, where the size of the sphere indicates the number of selected edges connected to that node. **(B)** Matrix plots: rows and columns represent canonical networks. The cells represent the total number of selected edges connecting the nodes between the two networks. **(C)** Circle Plots: Nodes are arranged in two half circles reflecting anatomy from anterior to posterior. The nodes are color coded according to the lobes and labeled as: prefrontal (PFC), motor (MOT), insula (INS), parietal (PAR), temporal (TEM), occipital (OCC), limbic (LIM), cerebellum (CER), subcortical (SUB), and brain stem (BSM).

ciated with using all available task data. Because the feature-selection thresholds used in CPM and mCPM are inevitably arbitrary, we tested three different thresholds: ($p <$ 1E−2, $p <$ 1E−3, & $p <$ 5E−4). Except as otherwise noted, all reported results were generated using a feature-selection threshold of $p <$ 1E−2. Model performance was quantified as the Pearson correlation between predicted and true IQ ($r$) and mean squared error (MSE).

**mCPM Improves Prediction Accuracy:** All models from either CPM, SVR, or mCPM (except for the SCAP task using SVR) produced significantly better predictions than chance ($p <$ 0.05, permutation test with 5,000 iterations). Using data from all the tasks, mCPM and SVR generated models that achieved superior predictions compared to models generated by CPM and SVR based on a single task (Table 1). For both HCP and CNP, predictions using the mCPM model with feature threshold $p <$ 1E−2 were significantly ($p <$ 0.05) better than the best-performing CPM model from a single task (*i.e.* the gambling task) and the SVR model using all available task data when assessed with Steiger's Z-test. mCPM is not significantly influenced by different, arbitrary thresholds for edge selection (Table 1). For visualization, we selected the edges that appeared in every cross-validation iteration of mCPM from HCP. In line with previous work [13], predictive edges are mainly located in frontal-parietal networks (Fig. 2).

**Tasks Contribute Differentially to Prediction:** To better understand the influence of each task on the mCPM results, we calculated the average score of each task involved in the prediction. First, we calculated the weighted sums of the selected edges, where weights $\mathbf{w}_{i,j}$ are determined by CCA. Then the average score was defined as the mean of those weighted sums across different participants:

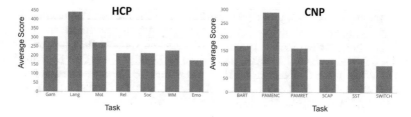

**Fig. 3.** Average score of different tasks. The language and PAMENC tasks show the largest score, highlighting their importance in prediction. The abbreviations can be found in methods.

$$as_t = \frac{\sum\limits_{i,j} \mathbf{E}^t_{i,j} * \mathbf{w}^t_{i,j}}{n} \tag{3}$$

As shown in Fig. 3, all tasks provide positive contributions to our prediction and larger selected edge strengths correspond to larger predicted values of IQ. For the HCP, the largest portion of the summary statistic $\mathbf{s}^{CN}$ was composed of edge strength from the language task. Interestingly, the language task was near the middle of individual task prediction performances. Together, these suggest that the language task, while not the individually most predictive task, contains a larger amount of unique variance than the other tasks, which can be leveraged when the tasks are combined via mCPM. For the CNP, the largest portion of $\mathbf{s}^{CN}$ was composed of edge strength from the PAMENC task. As certain tasks may contain redundant information for prediction, we adopted forward feature selection for both the HCP and UCLA datasets to select the optimal combination of tasks. The optimal set of tasks for HCP was the gambling, motor, relational, social, WM and emotion tasks, with a resulting correlation between observed and predicted IQ of $r = 0.40$. For the UCLA dataset, the optimal set was all the available 6 tasks with a correlation of $r = 0.49$.

## 4    Discussion and Conclusion

We proposed mCPM to combine connectivity matrices from multiple sources using CCA. Using task fMRI data from two large open-source datasets to predict IQ, we demonstrate that mCPM produces more accurate predictive models than either using a single source or conventional approaches (*e.g.* SVR) using multiple sources. We further verified that our algorithm is robust under different feature selection thresholds. These results support the existence of complementary information in different tasks, highlight an opportunity to use multiple task FC matrices to more comprehensively characterize individual differences, and suggest the ability of mCPM to utilize this information in maximizing predictive power. Future work will include using structural connectivity data from diffusion tensor imaging in mCPM, incorporating bagging procedures to optimally select the best combinations of data sources, and, ultimately, predicting clinical measures in patient populations.

**Acknowledgements.** Data were provided in part by the Human Connectome Project, WU-Minn Consortium (Principal Investigators: David Van Essen and Kamil Ugurbil; 1U54 MH091657) funded by the 16 NIH Institutes and Centers that support the NIH Blueprint for Neuroscience Research; and by the McDonnell Center for Systems Neuroscience at Washington University and the Consortium for Neuropsychiatric Phenomics (UL1 DE019580, RL1 MH083268, RL1 MH083269, RL1 DA024853, RL1 MH083270, RL1L M009833, PL1 MH083271, and PL1 NS062410).

# References

1. Dosenbach, N.U.F., et al.: Prediction of individual brain maturity using fMRI. Science **329**(5997), 1358–1361 (2010)
2. Smith, S.M., et al.: A positive-negative mode of population covariation links brain connectivity, demographics and behavior. Nat. Neurosci. **18**(11), 1565–1567 (2015)
3. Finn, E.S., Scheinost, D., Finn, D.M., Shen, X., Papademetris, X., Constable, R.T.: Can brain state be manipulated to emphasize individual differences in functional connectivity? NeuroImage **160**, 140–151 (2017)
4. Vanderwal, T., Eilbott, J., Finn, E.S., Craddock, R.C., Turnbull, A., Castellanos, F.X.: Individual differences in functional connectivity during naturalistic viewing conditions. NeuroImage **157**, 521–530 (2017)
5. Shen, X., et al.: Using connectome-based predictive modeling to predict individual behavior from brain connectivity. Nat. Protoc. **12**(3), 506–518 (2017)
6. Van Essen, D.C., et al.: The WU-Minn human connectome project: an overview. Neuroimage **80**, 62–79 (2013)
7. Poldrack, R.A., et al.: A phenome-wide examination of neural and cognitive function. Sci. Data **3**, 160110 (2016)
8. Chen, G., et al.: Classification of alzheimer disease, mild cognitive impairment, and normal cognitive status with large-scale network analysis based on resting-state functional MR imaging. Radiology **259**(1), 213–221 (2011)
9. Brown, M., et al.: ADHD-200 global competition: diagnosing ADHD using personal characteristic data can outperform resting state fMRI measurements. Front. Syst. Neurosci. **6**, 69 (2012)
10. Arbabshirani, M., Kiehl, K., Pearlson, G., Calhoun, V.: Classification of schizophrenia patients based on resting-state functional network connectivity. Front. Neurosci. **7**, 133 (2013)
11. Zeng, L.-L., et al.: Identifying major depression using whole-brain functional connectivity: a multivariate pattern analysis. Brain **135**(5), 1498–1507 (2012)
12. Plitt, M., Barnes, K.A., Martin, A.: Functional connectivity classification of autism identifies highly predictive brain features but falls short of biomarker standards. NeuroImage: Clin. **7**, 359–366 (2015)
13. Finn, E.S., et al.: Functional connectome fingerprinting: identifying individuals using patterns of brain connectivity. Nat. Neurosci. **18**(11), 1664–1671 (2015)
14. Kosuke Yoshida, Y., et al.: Prediction of clinical depression scores and detection of changes in whole-brain using resting-state functional MRI data with partial least squares regression. PLoS ONE **12**(7), e0179638 (2017)

# Exploring Fiber Skeletons via Joint Representation of Functional Networks and Structural Connectivity

Shu Zhang[1(✉)], Tianming Liu[1], and Dajiang Zhu[2]

[1] Cortical Architecture Imaging and Discovery Laboratory,
University of Georgia, Athens, GA, USA
shuzhang1989@gmail.com
[2] The University of Texas at Arlington, Arlington, TX 76019, USA

**Abstract.** Studying human brain connectome has been an important, yet challenging problem due to the intrinsic complexity of the brain function and structure. Many studies have been done to map the brain connectome, like Human Connectome Project (HCP). However, multi-modality (DTI and fMRI) brain connectome analysis is still under-studied. One challenge is the lack of a framework to efficiently link different modalities together. In this paper, we integrate two research efforts including sparse dictionary learning derived functional networks and structural connectivity into a joint representation of brain connectome. This joint representation then guided the identification of the main skeletons of whole-brain fiber connections, which contributes to a better understanding of brain architecture of structural connectome and its local pathways. We applied our framework on the HCP multimodal DTI/fMRI data and successfully constructed the main skeleton of whole-brain fiber connections. We identified 14 local fiber skeletons that are functionally and structurally consistent across individual brains.

**Keywords:** Structural connectivity · Functional networks · Joint representation
Connectome

## 1 Introduction

Understanding the brain connectome has been significantly important in cognitive and clinical neuroscience [1–3]. It is fundamentally critical for researchers to understand the organizational architecture of human brain from both structural and functional perspective. With advanced neuroimaging techniques such as MRI, we are able to measure and quantify brain structure/function in vivo. When mapping brain connectivity, functional MRI (fMRI) and Diffusion Tensor Imaging (DTI) are two modalities commonly used. Based on fMRI and DTI datasets, many studies have been done to investigate the brain connectome using either functional interactions, e.g., correlations [4], partial correlations [5] and regression [6], or the strength of white matter connections [7]. On the other hand, numerous reports have indicated that the structural connectivity patterns "connectional fingerprint" of brain areas can largely determine what functions they perform [8]. However, multi-modality (DTI and fMRI) brain

© Springer Nature Switzerland AG 2018
A. F. Frangi et al. (Eds.): MICCAI 2018, LNCS 11072, pp. 357–366, 2018.
https://doi.org/10.1007/978-3-030-00931-1_41

connectome analysis is still under-studied, in our view. The challenge is the lack of an efficient framework that can integrate the knowledge from two different modalities together.

Here, our proposed computational framework integrates two lines of research efforts including sparse dictionary learning derived functional networks and DTI derived fiber based structural connectivity into a joint representation of brain connectome. In this way, functional connectivity and structural connectivity can be studied and analyzed simultaneously. As illustrated in Fig. 1, we applied our framework on the Human Connectome Project (HCP) multimodal DTI/fMRI datasets to derive the main skeletons of the whiter matter pathways that are most active when performing different brain functions. We identified 14 major local fiber patterns from the main skeletons which have both functional and structural consistency across multiple individuals. The derived white matter skeleton and its local fiber patterns provide a new way to study brain connectome via multimodalities of MRI and shed novel insights on integrating brain structural and functional information.

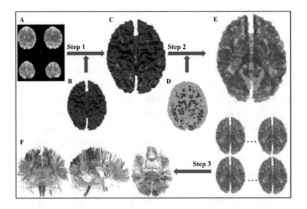

**Fig. 1.** The proposed framework of joint representation of functional networks (based on fMRI data) and structural connectivity (based on DTI data). (A) Functional networks from fMRI. (B) The cortical surface within DTI space. (C) The result of mapping functional networks onto the DTI cortical surface. (D) The whole brain fibers. (E) The fiber bundles connecting to functional activation areas in C. (F) An example of main skeletons of brain connections. Step1: registering the functional networks to DTI space; Step 2: screening fibers which connect the activation areas on the cortical surface; Step 3: using the joint representation profiles from step 1 and 2 for statistical analysis and constructing the main skeletons of the brain connections.

## 2   Materials and Methods

### 2.1   Data Acquisition and Preprocessing

For this study we used the data from HCP Q1 release [1] that includes seven task-fMRI datasets of 68 participants. The tasks include working memory, gambling, motor, language, social cognition, relational processing and emotion processing. For task-fMRI, the acquisition parameters are as follows: 72 slices, TR = 0.72 s, TE = 33.1 ms and 2.0 mm isotropic voxels. The acquisition parameters were as follows: $2 \times 2 \times 2$ mm spatial resolution, 0.72 s temporal resolution and 1,200 time points. For DTI data, spatial resolution = 1.25 mm × 1.25 mm × 1.25 mm. More details of data acquisition and preprocessing may be found in [9].

## 2.2 Representation of Functional Networks

A brain functional network can be defined as the brain regions that functionally "linked" [10]. It has been proven that the dictionary learning and sparse coding approaches are able to successfully identify task-related and resting state brain functional networks even when they have overlaps in both spatial and/or temporal domain [11, 12]. Based on dictionary learning, the whole-brain fMRI signals can be represented as a linear combination of a relatively small number of dictionary signals. The major steps are illustrated in Fig. 2. Firstly, the whole-brain normalized signals are arranged into a matrix $X$ (Fig. 2A) with n columns (n voxels) and each column contains a single fMRI signal with length of t (t time points). Then X is decomposed into two parts: dictionary matrix $D$ (Fig. 2B) and a sparse coefficient matrix $\alpha$ (Fig. 2C). The empirical cost function is summarized in (1), and its aiming of sparse representation using $D$, $\ell(x_i, D)$ is defined in (2), where $\lambda$ is a regularization parameter to trade off the regression residual and sparsity level.

$$f_n(D) \triangleq \frac{1}{n} \sum_{i=1}^{n} \ell(x_i, D) \tag{1}$$

$$\ell(x_i, D) \triangleq \min_{\alpha^i \in R^m} \frac{1}{2} \|x_i - D\alpha_i\|_2^2 + \lambda \|\alpha_i\|_1 \tag{2}$$

Each element of $\alpha$ indicates the extent when the corresponding dictionary atom is involved in representing the actual fMRI signals. As a result, each row of $\alpha$ can be mapped back to the brain volume space as a functional brain network pattern (Fig. 2C). Because 400 was proven to be an appropriate number of dictionary size [11] for HCP Q1 dataset, in this work we also set 400 for all task fMRI data. Thus, for each HCP subject, 2,800 functional networks will be obtained from seven tasks.

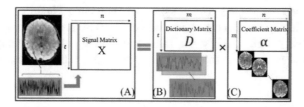

**Fig. 2.** Pipeline of using dictionary learning to derive 2,800 functional networks for each subject.

## 2.3    Representation of Structural Connectivity

In this section, we explore DTI derived fibers connecting to the activated areas of each functional network (step 2 in Fig. 1). Note that the fibers are under the DTI space, so we need to register the individual fMRI data to its own DTI space. Here we adopted a widely used linear registration tool – FLIRT from FSL [13]. White matter surface can be obtained through the DTI tissue segmentation and DTI cortical surface reconstruction algorithms [14]. Then we mapped the voxel from the registered fMRI data to its nearest vertex on the cortical surface and thus each surface vertex can be linked to the corresponding functional intensity in the decomposed coefficient matrix in Sect. 2.2. At last, for each cortical surface labeled with functional intensity values, we will examine the whole brain fibers and extract every fiber if both of its ending locations connected to activated regions on cortical surface (Fig. 1E). A threshold T = 0.5 is used to judge if a vertex on the cortical surface is active or not. Similar to the threshold in task activation detection [11], T is set empirically in this work. In this way, we could extract the fiber bundles which include all the connections from the activation area of different functional networks. A vector $N_i$ can be used to represent the fiber connection of the corresponding functional network $i$:

$$N_i^j = [f_1, f_2, f_3 \cdots f_{n-1}, f_n] \tag{3}$$

where $i$ represents the $i$-th functional network, $j$ represents the subject index, $f$ represents a fiber which is from the whole brain fibers and $n$ is the total number of the fibers of subject $j$. The value $f$ will be set to 1 if this fiber has a connection to the $i$-th functional network and 0 otherwise.

## 2.4    Joint Representation of Functional Networks and Structural Connectivity to Identify Main Skeletons of the Brain Connections

In this section, we introduce a novel joint representation approach to integrate the functional and structural connectivity together to explore the main skeleton of fiber connectomes. In the Sect. 2.3, we can obtain the registered functional networks and the related fiber connections. Here, each fiber connection pattern we achieved was from a single functional network. However, the human brain is widely considered to include a collection of specialized functional networks that flexibly interact when different brain functions are performed [15]. Thus, instead of studying a single connection pattern derived from single functional network, we need a way to discover the fiber connectome in a global vision. In this work, instead of working on the overlap of the functional networks, we focus on the overlaps of fibers. A matrix $Y$ is generated for each subject:

$$Y \in \mathbb{R}^{m*n} \tag{4}$$

$n$ represents the total number of fibers, $m$ is the total number of functional networks (2,800 in this work) for each subject and $N_i$ defined in Sect. 2.3 is one row from $Y$. Each row of $Y$ represents the fiber connections for a single functional network and each column represents the functional networks connecting to the corresponding fiber. Then, we are able to conduct the statistics of the elements in each column of the matrix $Y$, thus a histogram vector $H$ can be computed:

$$H = [h_1, h_2, h_3 \ldots h_{n-1}, h_n], h_i = \sum\nolimits_{j=1}^{2800} y_{j,i} \tag{5}$$

where $h_i$ is the total number of functional networks that fiber $i$ participated, and the more networks $i$ participated, the more activated intensity $i$ is. After we have the fiber connectome matrix $Y$ and its histogram vector $H$, then we can rank those fibers from most activated fibers to the least activated fibers. Thus, we could identify which fibers tend to be more activated in the functional networks and use them to generate main skeletons of the brain connectomes. An example is shown in the Fig. 1F. It contains 5000 most activated fibers across the whole brain. In order to examine the consistency of the skeleton we obtained, we applied our approach on HCP Q1 release data.

## 2.5    Local Connectome Analysis Based on the Main Skeletons of the Brain Connectomes

The skeletons of brain connectomes describe the main connections across the major brain regions. More importantly, they represent the most commonly used fibers and their connection pathways in multiple functional networks. In order to better understand the main skeletons we obtained, we perform further analysis to investigate the local brain areas and connections that the skeletons connected to. To analyze the main skeletons, here we only focus on the fibers from the main skeletons obtained from Sect. 2.4 and examine the relationship between those fibers and functional networks. The main skeleton fiber connection matrix is defined as $Y_s$:

$$Y_s \in R^{m*n'} \tag{6}$$

where $n'$ is the number of fibers from main skeletons. We extracted each row of $Y_s$ and studied the corresponding functional networks and fiber connections as well. Through a simple $k$-means clustering algorithm, typical local fiber pattern will be identified from the equation:

$$c^i := arg\,min_j \|x^i - \mu_j\|^2, \mu_j := \frac{\sum_{i=1}^m 1\{c^i = j\}x^i}{\sum_{i=1}^m 1\{c^i = j\}} \tag{7}$$

where $c$ is the cluster of $i$, $\mu_j$ is the center of cluster $j$, $x$ is the sample data.

# 3    Experimental Results

## 3.1    The Main Skeletons of the Fiber Connections of Human Brain

According to Sects. 2.2, 2.3 and 2.4, we obtained the main skeleton of the fiber connections from one subject at three different connectome levels, which are shown in

Fig. 3. Three different connectome levels have 500 fibers (Fig. 3B), 5,000 fibers (Fig. 3C) and 10,000 fibers (Fig. 3D), respectively. Although the number of the extracted fibers is largely different, we find that the connectome pathway is relative robust. For example, the fiber connections in the frontal lobe are obvious and consistent across those three levels. We named these connectome pathways as the skeletons of the fiber connections of human brain. We want to emphasize that, in the paper, we used the main skeletons with level of 5,000 fibers as the standard and further analyses are also based on this level. The reason we choose level of 5000 is that it has the clearness and robustness of the fiber connectome pathways. In details, level of 500 occupies only 0.25% from whole brain fibers, thus this number is too small to clearly and completely represent the connectome pathway. Level of 10,000 holds about 5% fibers from whole brain, but among those 10,000 fibers, some fibers are not very active. Thus the sparsity of the connection matirx $Y$ is only about 0.0035, which is too small from our experience. In contrast, level of 5,000 accounts for nearly 2.5% fibers and the sparsity of the connection matrix is about 0.008, thus, level of 5000 is chosen.

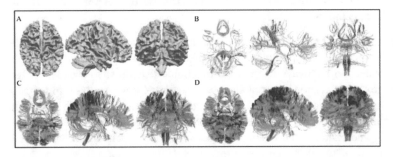

**Fig. 3.** The main skeletons of the fiber connections of an individual case. (A) The cortical surface of the brain. (B) Main skeletons of the fiber connections on level of 500. (C) Main skeletons of the fiber connections on level of 5,000. (D) Main skeletons of the fiber connections on level of 10,000.

## 3.2   The Consistency of the Main Skeletons of the Fiber Connections Across Different Subjects

In order to check the robustness of the main skeletons of the fiber connections we obtained, we adopted our framework on HCP Q1 release dataset. The main skeletons are obtained and we show 10 of them as examples in Fig. 4 to illustrate their consistency across different subjects.

From the Fig. 4, we can see that the main skeleton of the fiber connections is clear and consistent across the subject. Compared with whole brain fibers (as shown in Fig. 1D), these 5,000 most activated fibers describe clear connectome pathways for the fiber connectomes. Those connections represent the most dominant connection patterns under task performances and they connected significant brain regions. This result is interesting because the functional networks and whole brain fibers are from each individual and the way to obtain main skeletons is totally possessed individually.

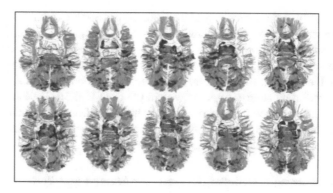

**Fig. 4.** The main skeletons of the fiber connections for 10 subjects. Each main skeleton is shown separately. The sequence of sbj1 to sbj10 is from left to right, and top to bottom.

Impressively, the pattern of the skeleton is quite similar across these subjects. Further analysis for the consistency will be provided in the Sects. 3.3 and 3.4.

### 3.3 Explore Major Local Pattern for the Fiber Bundles from the Main Skeleton of the Fiber Connections

Using the approaches from the Sect. 2.5, we can obtain fiber connections for each functional network at the level of 5,000 fibers from $Y_s$, and we aim to investigate how the main skeletons participated in the functional networks. Thus, for each subject, Eq. 7 will be applied on the corresponding main skeleton fiber connection matrix $Y_s$. After examining the consistency of each fiber bundle cluster across the subjects, 14 major local patterns are identified from the main skeletons and are shown in the Fig. 5. These local patterns are reasonably consistent across subjects, and they compose the main skeletons.

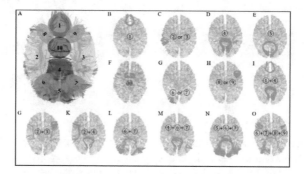

**Fig. 5.** The main connection patterns of the skeleton. (A) The main skeletons of the fiber connections. 10 local regions are highlighted with different colored circles. (B–O) 14 major local patterns are shown to present the contribution of local regions to the main skeletons. The numbers of their local regions from A are also provided in the figure.

From those 14 major patterns, 8 of them are the connections within the single brain region. Another 6 patterns are the combination of those 8 unique connections, meaning that some functional networks have much stronger activation levels and they may include two or more functional networks. This is an evidence for the hierarchical theory of functional networks. Another interesting finding is about the fiber connectome between left and right hemispheres. As we can see from Fig. 5B, D, E and F, the main fiber connections between left and right hemispheres are from corpus callosum. Apart from corpus callosum, there are many fiber bundles connecting left and right hemispheres, however, they do not belong to the main skeleton.

### 3.4    Corresponding Functional Networks for Major Local Patterns

It is interesting to know whether the corresponding functional networks of those local pattern fiber connections are consistent. It is worth noting that local pattern fibers are from the main skeleton fiber connection matrix $Y_s$. However functional networks are corresponding to the whole brain fiber connection matrix $Y$, $Y \gg Y_s$. So it is not necessary that the corresponding functional networks of same local pattern fibers must be consistent. To examine functional consistency of local patterns, we retrieved the fiber connections from $Y_s$ and their corresponding functional networks. We used local pattern from Fig. 5E as an example and illustrated them in Fig. 6. From Fig. 6, the activation areas are consistent, and they are located in the occipital lobe. That is, for those local patterns, their corresponding function networks are also consistent. In addition to the local pattern in Fig. 5E, other major local patterns have similar characteristics. Thus, we have the conclusion that the main skeletons of the fiber connections we obtained have not only structural consistency but also consistent functional networks.

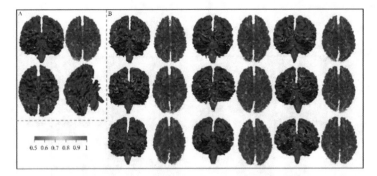

**Fig. 6.** The functional networks for local pattern Fig. 5E across the subjects. (A) An overview of functional network and its fiber connections (green lines) from sbj1. (B) Another 9 examples. The sequence of sbj2 to sbj9 is from left to right, top to bottom. Color bar is shown on the bottom left.

# 4 Conclusion

In this paper, we proposed a novel framework for joint representation of structural connectivity and functional networks to explore the main fiber skeletons of the brain. The major advantage of our framework is that it enables learning connections from multimodality (both fMRI and DTI) to investigate the most activated fibers and then derive the main skeletons of fiber connections. The analysis of our framework on HCP multimodal DTI/fMRI data suggested that main skeletons of the fiber connections can be robustly identified. In addition, through studying the main skeletons of the fiber connections, typical local patterns can be discovered and studied. Those local patterns will help to not only present both functional and structural consistency across different subjects, but also provide a new insight to understand the mechanism of the fiber connectome of the brain.

# References

1. Van Essen, D.C., Smith, S.M., Barch, D.M., et al.: The WU-Minn human connectome project: an overview. Neuroimage **80**, 62–79 (2013)
2. Sporns, O., Tononi, G., Kötter, R.: The human connectome: a structural description of the human brain. PLoS Comput. Biol. **1**(4), e42 (2005)
3. Wang, J., Zuo, X., et al.: Disrupted functional brain connectome in individuals at risk for Alzheimer's disease. Biol. Psychiat. **73**(5), 472–481 (2013)
4. Finn, E.S., Shen, X., Scheinost, D., et al.: Functional connectome fingerprinting: identifying individuals using patterns of brain connectivity. Nat. Neurosci. **18**(11), 1664 (2015)
5. Zhang, J., Wang, J., Wu, Q., et al.: Disrupted brain connectivity networks in drug-naive, first-episode major depressive disorder. Biol. Psychiat. **70**(4), 334–342 (2011)
6. Zhu, D., Li, X., Jiang, X., Chen, H., Shen, D., Liu, T.: Exploring high-order functional interactions via structurally-weighted LASSO models. In: Gee, J.C., Joshi, S., Pohl, K.M., Wells, W.M., Zöllei, L. (eds.) IPMI 2013. LNCS, vol. 7917, pp. 13–24. Springer, Heidelberg (2013). https://doi.org/10.1007/978-3-642-38868-2_2
7. Duffau, H.: Stimulation mapping of white matter tracts to study brain functional connectivity. Nat. Rev. Neurol. **11**(5), 255 (2015)
8. Passingham, R.E., Stephan, K.E., Kötter, R.: The anatomical basis of functional localization in the cortex. Nat. Rev. Neurosci. **3**(8), 606 (2002)
9. Woolrich, M.W., Ripley, B.D., Brady, J.M., Smith, S.M.: Temporal autocorrelation in univariate linear modelling of FMRI data. NeuroImage **14**(6), 1370–1386 (2001)
10. Sporns, O., Chialvo, D.R., Kaiser, M., Hilgetag, C.C.: Organization, development and function of complex brain networks. Trends Cogn. Sci. **8**(9), 418–425 (2004)
11. Lv, J., Jiang, X., et al.: Holistic atlases of functional networks and interactions reveal reciprocal organizational architecture of cortical function. IEEE TBME **62**(4), 1120–1131 (2015)
12. Zhang, S., et al.: Sparse representation of higher-order functional interaction patterns in task-based FMRI data. In: Mori, K., Sakuma, I., Sato, Y., Barillot, C., Navab, N. (eds.) MICCAI 2013. LNCS, vol. 8151, pp. 626–634. Springer, Heidelberg (2013). https://doi.org/10.1007/978-3-642-40760-4_78
13. Jenkinson, M., Smith, S.: A global optimisation method for robust affine registration of brain images. Med. Image Anal. **5**(2), 143–156 (2001)

14. Liu, T., Nie, J., Tarokh, A., Guo, L., Wong, S.T.: Reconstruction of central cortical surface from brain MRI images: method and application. NeuroImage **40**(3), 991–1002 (2008)
15. Fair, D.A., et al.: Functional brain networks develop from a "local to distributed" organization. PLoS Comput. Biol. **5**(5), e1000381 (2009)

# Phase Angle Spatial Embedding (PhASE)

## A Kernel Method for Studying the Topology of the Human Functional Connectome

Zachery Morrissey[1], Liang Zhan[2], Hyekyoung Lee[5], Johnson Keiriz[1], Angus Forbes[3], Olusola Ajilore[1], Alex Leow[1(✉)], and Moo Chung[4]

[1] University of Illinois at Chicago, Chicago, IL 60612, USA
aleow@psych.uic.edu
[2] University of Pittsburgh, Pittsburgh, PA 15260, USA
[3] University of California Santa Cruz, Santa Cruz, CA 95064, USA
[4] University of Wisconsin-Madison, Madison, WI 53706, USA
[5] Seoul National University, Seoul, Republic of Korea

**Abstract.** Modern resting-state functional magnetic resonance imaging (rs-fMRI) provides a wealth of information about the inherent functional connectivity of the human brain. However, understanding the role of negative correlations and the nonlinear topology of rs-fMRI remains a challenge. To address these challenges, we propose a novel graph embedding technique, **phase angle spatial embedding (PhASE)**, to study the "intrinsic geometry" of the functional connectome. PhASE both incorporates negative correlations as well as reformulates the connectome modularity problem as a kernel two-sample test, using a kernel method that induces a *maximum mean discrepancy* (MMD) in a reproducing kernel Hilbert space (RKHS). By solving a graph partition that maximizes this MMD, PhASE identifies the most functionally distinct brain modules. As a test case, we analyzed a public rs-fMRI dataset to compare male and female connectomes using PhASE and minimum spanning tree inferential statistics. These results show statistically significant differences between male and female resting-state brain networks, demonstrating PhASE to be a robust tool for connectome analysis.

## 1 Introduction

Human connectomics has progressed rapidly in the last decade in part due to advances in the application of graph theory and complex network analysis to brain imaging data [2,11]. For example, graph theoretical analysis of resting-state (i.e. task-negative) fMRI has been increasingly used to study the inherent functional connectivity of the brain in both healthy subjects and those with psychiatric or neurodegenerative conditions [8]. Thus, better understanding resting-state connectomes can elucidate how brain networks are organized differently.

In functional connectome analysis, graph theory is used to organize the pairwise correlations for all brain regions of interest (ROIs) into an adjacency matrix with signed edges [2]. However, there are some limitations with this approach.

© Springer Nature Switzerland AG 2018
A. F. Frangi et al. (Eds.): MICCAI 2018, LNCS 11072, pp. 367–374, 2018.
https://doi.org/10.1007/978-3-030-00931-1_42

One issue complicating traditional rs-fMRI graph analysis is how to properly model negative correlations between ROIs [9,14]. While applying thresholds or similar approaches to focus on positive correlations alone can simplify downstream analyses, this potentially removes biologically-relevant connectivity information [11]. Another issue is that while graph metrics can reveal important information regarding global/nodal characteristics of connectivity data, topological features are largely inaccessible to this analysis [13]. Thus, these limitations highlight the need for a way to represent functional connectomes in their native space.

The main contribution of this paper is the development of a kernel method approach for studying the topology of the functional connectome. To this end, we proposed **phase angle spatial embedding (PhASE)**, by modeling the functional coupling between two ROIs as a phase angle (from 0 to $\pi/2$, corresponding to full co-activation vs. anti-activation, respectively). With PhASE, topological features of resting-state connectivity can be intuitively understood by encoding each ROI using all phase angles that relate this ROI to the rest of the brain. This offers a mathematically robust method for working with positive *and* negative edges, as the full range of correlative information is preserved. Further, we equipped phase with a kernel, allowing us to conceptualize connectome modularity as a kernel two-sample problem, via the maximization of the corresponding maximum mean discrepancy (MMD) metric [4,5] in a reproducing kernel Hilbert space (RKHS). Last, we outlined a statistical inference procedure to compare the connectome topology between two groups (e.g., male vs. female) via the corresponding minimum spanning trees (MSTs) induced by the manifold geodesic distances in PhASE [7]. Thus, the functional connectome can be transformed into a new space representative of its "intrinsic geometry", and can be used for state-of-the-art topological data analysis techniques [3].

## 2    Methods

*Rationale of PhASE.* The core procedure of PhASE is to encode the probability of two brain regions ($i$ and $j$) exhibiting a coupling relationship of co- ($p_{ij}^+$) vs. anti- ($p_{ij}^-$) activation using a phase angle between 0 and $\pi/2$ (Fig. 1). One (simple) way of defining probability can be done using individual correlation matrices. In a dataset with $Z$ subjects, if we treat the correlation between brain regions $i$ and $j$ as an edge ($e_{ij}$), then we can define the probability of there being a negative edge between $i$ and $j$ ($p_{ij}^-$) as the average occurrence of observing a negative correlation between $i$ and $j$ across all $Z$ subjects, i.e. $p_{ij}^- = \sum_{z=1}^{Z} \mathbb{I}\left[e_{ij}^z < 0\right]/Z$, where $e_{ij}^z$ is the edge for the $z$-the subject. Computing this probability for all ROIs then forms a *negative probability matrix* (NPM). By taking advantage of the simple relationship that $p_{ij}^- + p_{ij}^+ = 1$, the phase angle between and $i$ and $j$ can be defined as

$$\theta_{ij} = \arctan\left(\frac{p_{ij}^-}{p_{ij}^+}\right)^{1/2}. \tag{1}$$

Furthermore, by fixing $i$ and considering all possible $j$, the PhASE procedure encodes the $i$-th ROI in an $N$-dimensional Euclidean space (where $N$ is the
number of ROIs), such that the phase angle vector $\boldsymbol{\theta}_i = [\theta_{i1}, \theta_{i2}, \ldots, \theta_{iN}]^\top \in$
$[0, \pi/2]^N$. (A phase angle of 0 represents fully co-activating ROIs and $\pi/2$ fully
anti-activating ROIs). Given a PhASE matrix, state-of-the-art nonlinear dimensionality reduction techniques or topological analysis via the geodesic distance
(in the $N$-dimensional PhASE embedding space) induced minimum spanning
tree (MST) can then be used to recover the intrinsic geometry of the functional
connectome.

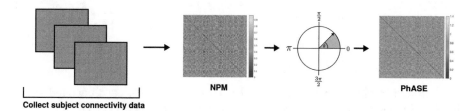

**Fig. 1.** Computing PhASE from rs-fMRI data. Connectivity data is acquired from
multiple subjects, and the frequency of edge negativity is computed for each ROI to
form a negative probability matrix (NPM). Then, the phase angle ($\theta$) is computed for
each ROI, which encodes each ROI's "phase" with every other ROI in the network.

*PhASE-Induced Modularity: PhASE as a Kernel Two-Sample Test.* We further
show that PhASE reformulates the connectome modularity problem as a kernel
two-sample problem. In its simplest form, the connectome modularity problem
seeks to decompose a graph into $M$ modules, or "communities", and is often
computed by maximizing the Q modularity metric using Louvain's method [10,
11]. Here, we illustrate how PhASE tackles this problem from a machine-learning
perspective using the case $M = 2$ as an example of finding two communities.

Following the formulation detailed in Gretton et al. [4], consider the random
variables $x$ and $y$ defined on a metric space $\mathcal{X}$ equipped with the metric $d$,
with the corresponding Borel probabilities $p$ and $q$ (i.e. $x \sim p$ and $y \sim q$).
Given observations $X := \{x_1, \ldots, x_m\}$ and $Y := \{y_1, \ldots, y_n\}$ drawn from the
probability distributions $p$ and $q$, $p = q$ if and only if $\mathbf{E}_x(f(x)) = \mathbf{E}_y(f(y)) \; \forall \, f \in$
$C(\mathcal{X})$, where $C(\mathcal{X})$ is the space of bounded continuous functions on $\mathcal{X}$.

Given this setup, and a class of functions $\mathcal{F}$ such that $f : \mathcal{X} \to \mathbb{R}$, the
*maximum mean discrepancy* (MMD) between $p$ and $q$ with respect to $\mathcal{F}$ is defined
as

$$\text{MMD}[\mathcal{F}, p, q] := \sup_{f \in \mathcal{F}} \left( \mathbf{E}_x[f(x)] - \mathbf{E}_y[f(y)] \right). \tag{2}$$

In PhASE modularity, using the same definitions for $x, y, p, q, X,$ and $Y$, each
ROI is assigned to two possible classes of distribution ($p$ and $q$) such that the

MMD is maximized. Given observations $X$ and $Y$, an empirical estimate of the MMD can therefore be defined as

$$\text{MMD}[\mathcal{F}, X, Y] := \sup_{f \in \mathcal{F}} \left( \frac{1}{m} \sum_{i=1}^{m} f(x_i) - \frac{1}{n} \sum_{i=1}^{n} f(y_i) \right). \tag{3}$$

However, with RKHS it is possible to evaluate the squared MMD using kernel values alone such that

$$\text{MMD}^2[\mathcal{F}, X, Y] = \frac{1}{m(m-1)} \sum_{i=1}^{m} \sum_{j \neq i}^{m} k(x_i, x_j) + \frac{1}{n(n-1)} \sum_{i=1}^{n} \sum_{j \neq i}^{n} k(y_i, y_j) \tag{4}$$

$$- \frac{2}{mn} \sum_{i=1}^{m} \sum_{j=1}^{n} k(x_i, y_j).$$

A standard choice of kernel defined between two ROIs, each encoded using its PhASE phase angle vector $\boldsymbol{\theta}_i$, is the Radial Basis Function (RBF)

$$k(i, j) = \exp \left\{ -\sigma \sum_{\ell=1}^{N} |\theta_{i\ell} - \theta_{j\ell}|^2 \right\}. \tag{5}$$

Another kernel choice, which may be more "natural" for phase angle vectors is

$$k(i, j) = \frac{1}{N} \sum_{\ell=1}^{N} \cos \left( \theta_{i\ell} - \theta_{j\ell} \right), \tag{6}$$

where $N$ is the number of ROIs. Thus, this kernel represents each ROI's phase angle vector similarity in relation to all other ROIs in the network. While these two choices appear quite different, a quick check on their Taylor expansions make it clear that they have very similar leading terms.

This kernel-based approach for representing functional brain connectivity and modularity in a native way can be used in conjunction with topological data analysis on minimum spanning trees, as described in the following section. As a result, these techniques can be complemented together to offer a robust framework for studying differences across group networks.

*Inference on Minimum Spanning Trees.* Most statistical inference methods on MST rely on existing univariate statistical test procedures on scalar graph theory features such as the average path length [12]. Since the probability distribution of such features are often not known, resampling techniques such as the permutation test are frequently used. The permutation test is known as the exact test procedure in statistics. However, it is not exact in practice and only an approximate method due to the computational bottleneck of generating every possible permutation, which can be astronomically large. Thus, the statistical significance is computed using the small subset of all possible permutations, which gives approximate $p$-values. Here, we present a new method for the permutation test, where every possible permutation is enumerated combinatorially.

Let $M_1$ and $M_2$ be the minimum spanning trees (MST) corresponding to $N \times N$ PhASE-derived adjacency matrices $C_1$ and $C_2$. We are interested in testing hypotheses

$$H_0 : M_1 = M_2 \text{ vs. } H_1 : M_1 \neq M_2. \tag{7}$$

The statistic for testing $H_0$ is constructed using Kruskal's algorithm [6]. Kruskal's algorithm is a greedy algorithm with runtime $O(N \log N)$. The algorithm starts with an edge with the smallest weight. Then add an edge with the next smallest weight. This sequential process continues while avoiding a loop and generates a spanning tree with the smallest total edge weights. Thus, the edge weights in MST correspond to the order in which the edges are added in the construction of MST. For a graph with $N$ nodes, MST has $N - 1$ edges.

Let $w_1^1 < w_2^1 < \cdots < w_{N-1}^1$ and $w_1^2 < w_2^2 < \cdots < w_{N-1}^2$ be the ordered edge weights of two MSTs obtained from Kruskal's algorithm. Let $f$ be a monotone function that maps the edge weights to integers such that $\phi(w_j^1) = \psi(w_j^2) = j$. $\phi, \psi$ can be the number of edges in the subtree generated in the $j$-th iteration in Kruskal's algorithm. Under the null hypothesis (7), the monotone functions $\phi(w_j^1)$ and $\psi(w_j^2)$ are interchangeable, thus satisfying the condition for the permutation test. The pseudo-metric $D(M_1, M_2) = \sup_t |\phi(t) - \psi(t)|$ is then used as a test statistic. Under $H_0$, $D(M_1, M_2) = 0$. The larger the value of $D$ is, it is more likely to reject $H_0$. From Theorem 2 in [3], it can be shown that the probability distribution of $D$ is given by

$$P(D \geq d) = 1 - \frac{A_{N-1,N-1}}{\binom{2N-2}{N-1}}, \tag{8}$$

where $A_{u,v}$ satisfies $A_{u,v} = A_{u-1,v} + A_{u,v-1}$ with the boundary condition $A_{0,N} = A_{N,0} = 1$ within band $|u - v| < d$. (8) is used to compute $p$-values in the study.

*Resting-State fMRI Data Acquisition.* For testing PhASE, a publicly available rs-fMRI dataset from the F1000 Project [1] was used. For details regarding data acquisition and processing, see http://fcon_1000.projects.nitrc.org/. The resulting data is a $177 \times 177$ ROI correlation matrix for each subject ($Z = 986$, males $= 426$, females $= 560$). Then, the negative probability matrix was formed using the simple definition of probability described in (2).

## 3   Results and Discussion

As a test case for our phase angle spatial embedding (PhASE) method, PhASE was applied to resting-state functional connectivity data ($177 \times 177$ ROI) [1] to investigate differences between male and female resting-state brain networks. First, we investigated network modularity using MST and anatomical representations. To compare to standard modularity techniques, Louvain's Q modularity method [11] ($\gamma = 1$) was used, generating three communities (Fig. 2A). After applying the PhASE transformation and computing the MMD kernel matrix via (6), two communities were generated (Fig. 2B). Notably, the PhASE modularity boundaries share similar boundaries with those achieved by Q, both in MST and anatomical space. This demonstrates that PhASE can create reproducible results as those achieved by standard methods.

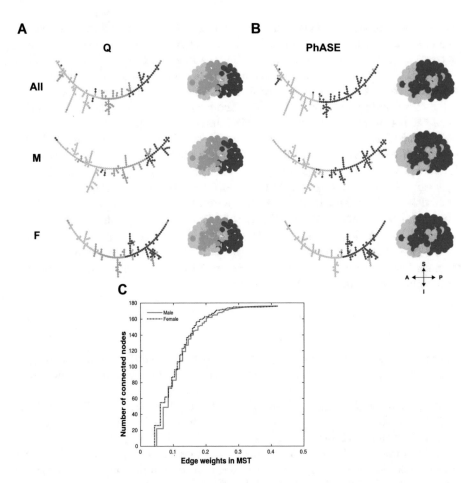

**Fig. 2.** (A–B) Minimum spanning trees (MSTs) and anatomical representations (MNI coordinates) for all, male (M), and female (F) groups, with associated modularity color using Q ($\gamma = 1$) and PhASE. (C) Exact topological inference comparing male and female brain networks. The distance between networks was $D_{\max}$(male, female) = 33, occurring at edge weights 0.0598 and 0.0686, and was statistically significant ($p < 0.001$).

While MSTs can produce informative visual differences between group networks, we sought to quantitatively analyze whether there are statistically significant differences between male and female resting-state brain networks using MSTs. As described above, the null hypothesis (7) that the male and female graphs are equivalent was tested using a nonparametric inference method that does not need to rely on the permutation test. Using the male and female PhASE matrices, MST edge weights were simply the phase angle between ROIS (1). Exact topological inference (8) was then used to compute the distance between male and female MSTs, and was statistically significant ($D_{\max}$(male, female) = 33, $p < 0.001$). Thus, the null hypothesis can be rejected, suggesting that there are intrinsic differences in resting-state connectivity between sexes. These differences can be seen in Fig. 2C, wherein there is a consistent separation between male and female number of connected nodes as a function of MST edge weights. Although the curves are similar in shape, the results indicate that the female MSTs appear to exhibit a higher number of connected nodes at similar edge weights compared to males, suggesting that female subjects' resting-state backbone networks are more "connected" than male subjects—a salient network finding that merits further investigation, and importantly here demonstrates the capability of PhASE for topological analysis.

## 4   Conclusions

Representing and computing the modularity of functional connectomes with negative edges is an ongoing challenge in rs-fMRI analysis. To that end, we presented a kernel-based approach for analyzing functional connectivity using a novel graph embedding method, phase angle spatial embedding (PhASE), that respects negative correlations and reformulates the modularity problem as a two-sample problem. By doing so, it is possible to create a more natural representation of functional connectivity that encodes connective phase information in its native space and can be used with advanced topological analyses. As an initial case, sex differences were examined using PhASE with MST inferential statistical methods, revealing statistically significant differences between male and female rs-fMRI networks. Taken together, these results demonstrate the versatility of PhASE as a connectome analysis framework that can yield greater insight into functional connectivity while preserving the intrinsic geometry of the human connectome.

**Acknowledgments.** This manuscript was supported by NIH T32 AG057468 to ZM, NIH AG056782 to LZ and AL, and NIH EB022856 to MC.

## References

1. Biswal, B.B., et al.: Toward discovery science of human brain function. Proc. Natl. Acad. Sci. **107**(10), 4734–4739 (2010)
2. Bullmore, E., Sporns, O.: Complex brain networks: graph theoretical analysis of structural and functional systems. Nat. Rev. Neurosci. **10**(3), 186–198 (2009)

3. Chung, M.K., Villalta-Gil, V., Lee, H., Rathouz, P.J., Lahey, B.B., Zald, D.H.: Exact topological inference for paired brain networks *via* persistent homology. In: Niethammer, M., et al. (eds.) IPMI 2017. LNCS, vol. 10265, pp. 299–310. Springer, Cham (2017). https://doi.org/10.1007/978-3-319-59050-9_24

4. Gretton, A., Borgwardt, K.M., Rasch, M.J., Schölkopf, B., Smola, A.: A kernel two-sample test. J. Mach. Learn. Res. **13**(Mar), 723–773 (2012)

5. Gretton, A., et al.: Optimal kernel choice for large-scale two-sample tests. In: Advances in Neural Information Processing Systems, pp. 1205–1213 (2012)

6. Lee, H., Kang, H., Chung, M.K., Kim, B.-N., Lee, D.S.: Persistent brain network homology from the perspective of dendrogram. IEEE Trans. Med. Imaging **31**(12), 2267–2277 (2012)

7. Lee, H., Chung, M.K., Kang, H., Kim, B.-N., Lee, D.S.: Computing the shape of brain networks using graph filtration and Gromov-Hausdorff metric. In: Fichtinger, G., Martel, A., Peters, T. (eds.) MICCAI 2011. LNCS, vol. 6892, pp. 302–309. Springer, Heidelberg (2011). https://doi.org/10.1007/978-3-642-23629-7_37

8. Lv, H.: Resting-state functional MRI: everything that nonexperts have always wanted to know. Am. J. Neuroradiol. **39**(8), 1390–1399 (2018). https://doi.org/10.3174/ajnr.A5527. ISSN 0195-6108

9. Murphy, K., Fox, M.D.: Towards a consensus regarding global signal regression for resting state functional connectivity MRI. NeuroImage **154**, 169–173 (2017)

10. Newman, M.E.: Modularity and community structure in networks. Proc. Natl. Acad. Sci. **103**(23), 8577–8582 (2006)

11. Rubinov, M., Sporns, O.: Complex network measures of brain connectivity: uses and interpretations. NeuroImage **52**(3), 1059–1069 (2010)

12. Stam, C., Tewarie, P., Van Dellen, E., van Straaten, E., Hillebrand, A., Van Mieghem, P.: The trees and the forest: characterization of complex brain networks with minimum spanning trees. Int. J. Psychophysiol. **92**(3), 129–138 (2014)

13. Ye, A.Q., et al.: The intrinsic geometry of the human brain connectome. Brain Inf. **2**(4), 197–210 (2015)

14. Zhan, L.: The significance of negative correlations in brain connectivity. J. Comput. Neurol. **525**(15), 3251–3265 (2017)

# Edema-Informed Anatomically Constrained Particle Filter Tractography

Samuel Deslauriers-Gauthier[1(✉)], Drew Parker[3], François Rheault[2],
Rachid Deriche[1], Steven Brem[3], Maxime Descoteaux[2], and Ragini Verma[3]

[1] Inria Sophia Antipolis, Université Côte d'Azur, Sophia Antipolis, France
samuel.deslauriers-gauthier@inria.fr
[2] Université de Sherbrooke, Sherbrooke, Canada
[3] University of Pennsylvania, Philadelphia, USA

**Abstract.** In this work, we propose an edema-informed anatomically
constrained tractography paradigm that enables reconstructing larger
spatial extent of white matter bundles as well as increased cortical coverage in the presence of edema. These improvements will help surgeons
maximize the extent of the resection while minimizing the risk of cognitive deficits. The new paradigm is based on a segmentation of the brain
into gray matter, white matter, corticospinal fluid, edema and tumor
regions which utilizes a tumor growth model. Using this segmentation,
a valid tracking domain is generated and, in combination with anatomically constrained particle filter tractography, allows streamlines to cross
the edema region and reach the cortex. Using subjects with brain tumors,
we show that our edema-informed anatomically constrained tractography paradigm increases the cortico-cortical connections that cross edema-
contaminated regions when compared to traditional fractional anisotropy
thresholded tracking.

## 1 Introduction

Diffusion magnetic resonance imaging (MRI) allows to non-invasively reconstruct
white matter tracts. In the context of pre-operative surgical planning, tractography is used to identify and locate fiber bundles in the vicinity of the lesion.
Equipped with this knowledge, the surgeon attempts to maximize tumor removal
while minimizing damage to functional white matter networks [1]. While the
development efforts on every step of the tractography pipeline are extensive,
most algorithms assume healthy brain and are tested on healthy white matter models. Diffusion tensor imaging tractography has been investigated in the
context of neurosurgery [5,6], but is subject to the well known limitation of the
diffusion tensor. There is thus very little data on the performance of modern tractography algorithms in the presence of edematous or tumoral tissue. One notable
exception is the use of multi-tensor models in the context of unscented Kalman
filter (UKF) tractography, which has been investigated on brain tumor patients

---

M. Descoteaux and R. Verma—Co-senior authors. They have contributed equally.

© Springer Nature Switzerland AG 2018
A. F. Frangi et al. (Eds.): MICCAI 2018, LNCS 11072, pp. 375–382, 2018.
https://doi.org/10.1007/978-3-030-00931-1_43

**Fig. 1.** Segmentation of the edema obtained from GLISTR (first) and the tracking mask obtained by thresholding the FA at 0.2 (second). The edema lowers the FA and isolates the temporal lobe. The include map (third) and exclude map (fourth) show that in EI-PFT the edema can be traversed as it is in black in both masks. The tumor is visible in the exclude map indicating that streamlines cannot end in this region.

[2–4]. The authors show that two-tensor UKF increases the volume of the arcuate fasciculus and corticospinal tract and their coverage of the functional MRI activation sites, especially when used with whole brain seeding. These results highlight the need for further investigation into the use of modern tractography strategies for pre-operative surgical planning.

Tractography algorithms reconstruct streamlines in the white matter of the brain by following the local diffusion directions obtained through diffusion MRI. While several criteria can be used to stop the tracking, the most common is to place a threshold on the fractional anisotropy (FA) map [7]. The rational is that the FA is high in regions with coherent fiber directions and therefore acts as a proxy for the white matter. However, it is well known that many phenomena can lower the measured FA, such as fiber crossings, free water contamination, and pathologies like edema in the presence of tumor [9]. In these scenarios, the tracking may be prematurely terminated yielding a streamline that does not reach the cortex. Furthermore, because the FA does not encode anatomical information, algorithms that rely on it will produce physically impossible streamlines, such as those that start or end in the ventricles. Anatomically constrained tractography (ACT) [8] alleviates this problem by defining a valid tracking domain based on the anatomy of the subject defined by the tissue segmentation. In combination with particle filter tractography, which backtracks and attempts to find alternate routes when it prematurely hits a region outside of the valid tracking domain, ACT was shown to reduce the number of false positive streamlines [10].

Because edema and tumor change the local tissue properties of the brain, tractography based on FA thresholding may not provide an accurate representation of fiber near the lesion. In this paper, we provide a paradigm that integrates anatomical information of both healthy and unhealthy tissues with particle filter tracking to obtain tractography through edema. This has been demonstrated on 4 tumor subjects.

## 2    Methods

### 2.1    Edema-Informed Anatomically Constrained Particle Filter Tractography

Tracking is rendered edema-informed by defining a valid tracking domain which integrates the knowledge of the various tissue types: white matter, gray matter,

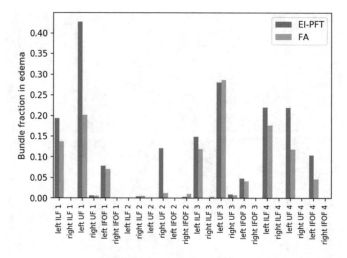

**Fig. 2.** Comparison of the bundle fraction in edema obtained using EI-PFT or FA threshold tractography. Larger values indicates that a greater proportion of the bundle volume intersects with the edema.

corticospinal fluid, edema, and tumor segmentation. More specifically, the tracking domain is defined by 3 probabilistic maps. The first map, referred to as the include map, determines valid start and end points for streamlines. That is, the value of a voxel of the include map gives the probability that a streamline will terminate within this voxel and be included in the final tractogram. An example of an include map is illustrated in the third image of Fig. 1 where white regions indicate valid streamline start or end points. The second map, referred to as the exclude map, determines the probability that a streamline will terminate within a region and be excluded from the final tractogram. An example of an exclude map is illustrated in the fourth image of Fig. 1 where white regions indicates invalid streamlines start or end points. Note that for any given voxel, the sum of the two maps must be between zero and one. Together, the include and exclude maps define a third implicit map which ensure the probabilities sum to one. This third map contains the regions that streamlines are free to cross, but where they cannot terminate. This separation of the tracking domain into 3 maps was used in [10], but their definition of the maps was based solely on the segmentation of healthy tissues, whereas we include edema and tumor information. Specifically, we define the include map to be the gray matter segmentation and the exclude map as $P_{ex} = P_{tumor} + P_{csf} - P_{tumor}P_{csf}$ where $P_{tumor}$ and $P_{csf}$ are the tumor and CSF segmentations, respectively. Note that the edema segmentation map is not used explicitly in the computation of the tracking domain, but it nonetheless affects it implicitly. Indeed, just like the white matter, the edema is considered as a region where the streamlines can traverse but not end. These changes to the tracking domain have two important effects on the final tractogram. First, streamlines entering the tumor will now be excluded, as the

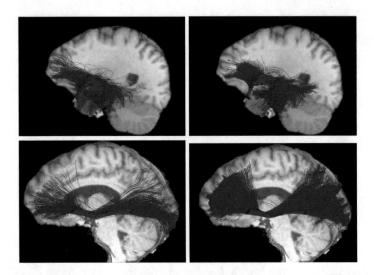

**Fig. 3.** Comparison of the UF of subject 1 (top) and the IFOF of subject 7 (bottom) reconstructed using EI-PFT (left) and FA thresholded tractography (right). For subject 1, the streamlines recovered by EI-PFT penetrate further into the temporal lobe and wrap the tumor. This is consistent with the maps of Fig. 1 where the temporal lobe was isolated by the lowered FA of the edema. For subject 4, the streamlines projecting to the inferior portion of the occipital lobe have a greater intersection with the edema when reconstructed with IE-PFT.

tumor is part of the exclude map. Second, streamlines displaced by the tumor are more likely to be identified because the particle filter will backtrack and attempt to find a way around the tumor in the edema. We refer to this method as edema-informed anatomically constrained particle filter tractography (EI-PFT). It should be noted that common segmentation tools, e.g. FAST (FSL), assume a healthy brain and will therefore misclassify edematous or tumoral tissue as white matter, gray matter, or corticospinal fluid. These misclassifications are of critical importance in our paradigm because they directly influence the tracking domain and thus the final tractogram. For example, considering edematous tissue as gray matter will cause streamlines to prematurely terminate inside the lesion and yield an erroneous connection. Likewise, misclassifying a tumor may allow streamlines to traverse the tumor instead of being displaced by it. To produce the final segmentation, we therefore opted for the fusion of two segmentations. First, the white matter, gray matter, and corticospinal fluid segmentations are based on the partial volumes obtained with FAST (FSL) as it is very reliable in healthy regions away from the lesion. Second, the edema and tumor regions are based on the glioma image segmentation and registration (GLISTR) [11] tool which provides accurate tumor and edema segmentation based on tumor growth modeling. To ensure the maps of the 5 labels sum to one, the corrected maps were computed as $P_* = \bar{P}_* - P_{edema}\bar{P}_* - P_{tumor}\bar{P}_*$ where $P_{edema}$ and $P_{tumor}$ are

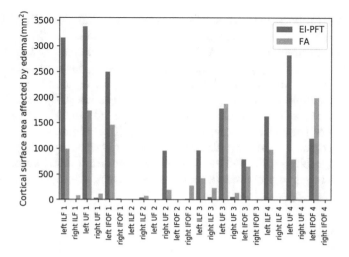

**Fig. 4.** Comparison of the cortical surface area where streamlines intersecting the edema terminate recovered by EI-PFT and FA thresholding tractography. Larger values indicate that more streamlines traverse the edema and reach the cortex.

GLISTR outputs and $\bar{P}_*$ is the white matter, gray matter, or corticospinal fluid map obtained from FAST.

## 2.2   Data Acquisition and Preprocessing

For 4 subjects having a tumor near the temporal stem, multi-shell multi-band DWI [TR/TE = 5,216/100 ms, resolution = 2 mm isotropic, 64 diffusion directions at b = 2,000 s/mm$^2$, 30 diffusion directions at b = 800 s/mm$^2$, 15 diffusion directions at b = 300 s/mm$^2$, and 7 b = 0 images] were acquired on an Siemens 3T TrioTim scanner using a 32-channel head coil. In a separate session on a Siemens 3T TrioTim, T1 MPRAGE, T1 contrast-enhanced $(0.977 \times 0.977 \times 1\,\text{mm})$, T2 and FLAIR images were acquired using a 12-channel head coil. The diffusion MRI volumes were corrected for distortion correction and motion correction using eddy (FSL). Structural images were bias corrected with N3 (ITK), smoothed with susan (FSL) and then coregistered to T1 with flirt (FSL), and then all were brought to FA and b0 space with ANTs. The fiber orientation distribution functions were then computed using constrained spherical deconvolution using a spherical harmonic order of 8. FSL FAST was used to segment the T1 image into white matter, gray matter, and corticospinal fluid partial volume maps. GLISTR was used on the T1, T2, T1 contract-enhanced, and FLAIR to segment the lesion into edema and tumor regions. The output of FAST and GLISTR were used to define the include and exclude maps of the anatomically constrained tractography, as described in the previous section. An example of the include and exclude maps obtained for subject 1 are illustrated in Fig. 1. Probabilistic particle filter tractography was then used to generate the streamlines by using 1 seed per voxel over the white matter mask. Finally, the streamlines were automatically segmented into bundles using recobundles [12], which makes use of streamline geometry instead of anatomical landmarks. This streamlines segmentation strategy is critical as brain tumor subjects may have severe deformations which prevent typical landmarks from being identified. Because our subjects all have a tumor near the temporal stem, we report the results for the inferior longitudinal fasciculus (ILF), uncinate fasciculus (UF), and the inferior fronto-occipital fasciculus (IFOF). These 3 bundles all funnel near the temporal stem and were the most affected by the lesion. To evaluate if EI-PFT is able to traverse edema regions, we also report result obtained by using FA thresholded tractography pipeline. The same fiber orientation distribution functions were used as input to probabilistic tractography where the tracking was stopped when the FA dropped below 0.2.

## 3   Results

To quantify our results, we computed the volume of the recovered bundles that intersect the edema mask. To obtain normalized values, we divide this volume by the total volume of the bundle, yielding a bundle fraction in edema. Since the volume of a bundle can be affected by shorter streamlines, we also compute the cortical surface area overlapping with the streamlines that traverse through the edema and reach the cortical surface. This surface area was estimated by selecting the streamlines that intersect the edema and by counting the number of start and end voxels of each streamline that intersect with the cortical surface. For both the bundle fraction in edema and the cortical surface area of streamlines intersecting the edema, we expect larger values to be obtained by tractography pipelines are able to penetrate edematous regions.

Figure 2 illustrates the bundle fraction in edema for each of our bundles of interest. Out of 9 bundles with a non-negligible bundle fraction in edema (above 5%), 8 had a larger fraction using EI-PFT than FA thresholding tractography. These results are notable because the complete paradigm includes an automatic streamline bundling step that removes streamlines that are not detected in bundles. In other words, EI-PFT explores the edematous region more than FA thresholding while still producing streamlines that are classified as part of a bundle. This is further illustrated in Fig. 3, where the uncinate fasciculus of subject 1 is illustrated. It is clear that both EI-PFT and FA thresholding tractography produce 'bundle like' outputs, but EI-PFT has a greater intersection with the edema. Figure 4 compares the cortical surface area of streamlines intersecting the edema obtained using EI-PFT and FA thresholding tractography. Out of 10 bundles with a non-negligible area (above $500\,\text{mm}^2$), 8 have a larger area when obtained using EI-PFT. This highlights thats EI-PFT explores the edematous tissue while producing cortico-cortical streamlines. Examples of these surfaces are presented in Fig. 5. The images were generated by coloring the cortical endpoints of the streamlines by their overlap with the edema mask. It can be observed that EI-PFT yields broader and more intense cortical surface area indicating regions where streamlines overlapping the edema terminate.

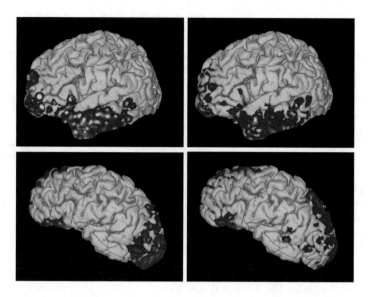

**Fig. 5.** The cortical surface area of streamlines intersecting the area obtained using EI-PFT (left) and FA thresholded tractography (right) for the UF of subject 1 (top) and the IFOF of subject 4 (bottom). Warmer color indicates that more streamlines that intersect the edema terminate in that region of the cortex. EI-PFT has more intense and broader areas in the temporal and frontal poles, indicating areas that may still be functionally active even if they are affected by the lesion.

## 4   Conclusion

The surgical treatment of any intrinsic brain pathology necessitates careful planning to develop a safe trajectory from the cortical surface, through sulci and white matter to reach the tumor boundary while minimizing injury to the white matter tracts. Our evaluation demonstrates that EI-PFT is able to better achieve this in comparison to the FA thresholded tractography that is available in the surgical planning tools, potentially leading to safer resections. Additionally, larger tract volume in the edema, which our paradigm is able to achieve due to tracking through edema, is a precursor to increased margins of resection and hence better patient outcome.

**Acknowledgements.** This work was supported by NIH grant R01NS096606 (PI: Verma, Brem) and has received funding from the European Research Council (ERC) under the European Union's Horizon 2020 research and innovation program (ERC Advanced Grant agreement No 694665: CoBCoM).

# References

1. McGirt, M.J., et al.: Independent association of extent of resection with survival in patients with malignant brain astrocytoma. J. Neurosurg. **110**, 156–162 (2009)
2. Chen, Z., et al.: Reconstruction of the arcuate fasciculus for surgical planning in the setting of peritumoral edema using two-tensor unscented Kalman filter tractography. NeuroImage Clin. **7**, 815–822 (2015)
3. Chen, Z., et al.: Corticospinal tract modeling for neurosurgical planning by tracking through regions of peritumoral edema and crossing fibers using two-tensor unscented Kalman filter tractography. Int. J. CARS **11**, 1475–1486 (2016)
4. Liao, R., et al.: Performance of unscented Kalman filter tractography in edema: analysis of the two-tensor model. NeuroImage Clin. **15**, 819–831 (2017)
5. Mandelli, M., Berger, M.S., Bucci, M., Berman, J., Amirbekian, B., Henry, R.G.: Quantifying accuracy and precision of diffusion MR tractography of the corticospinal tract in brain tumors. J. Neurosurg. **121**, 349–358 (2014)
6. Pujol, S., et al.: The DTI Challenge: towards standardized evaluation of diffusion tensor imaging tractography for neurosurgery. J. Neuroimaging **25**, 875–882 (2015)
7. Jeurissen, B., Descoteaux, M., Mori, S., Leemans, A.: Diffusion MRI fiber tractography of the brain. NMR Biomed. (2017)
8. Smith, R.E., Tournier, J.D., Calamante, F., Connelly, A.: Anatomically-constrained tractography: improved diffusion MRI streamlines tractography through effective use of anatomical information. Neuroimage **62**, 1924–1938 (2012)
9. Duffau, H.: The dangers of magnetic resonance imaging diffusion tensor tractography in brain surgery. World Neurosurg. **81**, 56–58 (2014)
10. Girard, G., Whittingstall, K., Deriche, R., Descoteaux, M.: Towards quantitative connectivity analysis: reducing tractography biases. NeuroImage **98C**, 266–278 (2014)
11. Gooya, A., et al.: GLISTR: glioma image segmentation and registration. IEEE Trans. Med. Imaging **31**, 1941–1954 (2012)
12. Garyfallidis, E., et al.: Recognition of white matter bundles using local and global streamline-based registration and clustering. NeuroImage **170**, 283–295 (2017)

# Thalamic Nuclei Segmentation Using Tractography, Population-Specific Priors and Local Fibre Orientation

Carla Semedo[1]([✉]), M. Jorge Cardoso[1,2], Sjoerd B. Vos[1,3,4],
Carole H. Sudre[1,2,5], Martina Bocchetta[5], Annemie Ribbens[6], Dirk Smeets[6],
Jonathan D. Rohrer[5], and Sebastien Ourselin[2]

[1] Centre for Medical Image Computing (CMIC), University College London,
London, UK
carla.semedo.14@ucl.ac.uk
[2] School of Biomedical Engineering and Imaging Sciences,
King's College London, London, UK
[3] MRI Unit, Epilepsy Society, Chalfont St Peter, UK
[4] Wellcome EPSRC Centre for Interventional and Surgical Sciences (WEISS),
University College London, London, UK
[5] Dementia Research Centre, Department of Neurodegenerative Disease,
Institute of Neurology, University College London, London, UK
[6] icometrix, Leuven, Belgium

**Abstract.** The thalamus is a deep grey matter structure that plays an important role in propagating nerve impulses between subcortical regions and the cerebral cortex. It is composed of distinct nuclei that have unique long-range connectivity. Accurate thalamic nuclei segmentation provides insights about structural connectivity and the neurodegeneration mechanisms occurring in distinct brain disorders, for instance Alzheimer's disease and Frontotemporal dementia (FTD). In this work, we propose a novel thalamic nuclei segmentation approach that relies on tractography, thalamic nuclei priors and local fibre orientation. Validation was performed in a cohort of healthy controls and FTD patients against other thalamus connectivity-based parcellation methods. Results showed that the proposed strategy led to anatomical plausible thalamic nuclei segmentations and was able to detect connectivity differences between controls and FTD patients.

## 1 Introduction

The thalamus is the main gateway of information to the cerebral cortex for: (i) all sensory information, with exception of the olfactory system; (ii) anatomical loops of motor systems, between the cerebellum or basal ganglia and cerebral cortex and (iii) projections from limbic structures to the brain cortex. Additionally, the human thalamus has distinct nuclei that differ in terms of subcortical and cortical neural pathways. These two aspects have led to the development of segmentation techniques to parcellate the thalamus into different nuclei.

© Springer Nature Switzerland AG 2018
A. F. Frangi et al. (Eds.): MICCAI 2018, LNCS 11072, pp. 383–391, 2018.
https://doi.org/10.1007/978-3-030-00931-1_44

Current imaging modalities such as Computed Tomography (CT) and Magnetic Resonance (MR) enable us to identify the thalamus, but they do not provide enough contrast to differentiate between thalamic subnuclei. Conversely, diffusion MRI, a technique which assesses water diffusion within biological tissues, has emerged as a way of exploring the unique white matter (WM) pathways within each thalamic nuclei and their connections with cortical regions. The majority of thalamus parcellation techniques reported to date rely on diffusion MR data to divide the thalamus into its distinct nuclei, by considering either the local diffusion patterns and/or structural connectivity.

A commonly used thalamus tractography-based procedure was developed by Behrens et al. [1], available as part of FSL[1]. Diffusion data analysis and tractography are done in a combination of standard and native spaces through single atlas propagation, which can affect the biological plausibility of connectivity findings. Furthemore, validation was done in a small cohort of normal controls, but its performance has not been tested when dealing with highly pathological data.

A recent strategy [2] addressed these aspects by performing Diffusion Weighted (DW) data processing and probabilistic tractography in the subject's native space, as well as by integrating population-specific thalamic nuclei priors. This method was able to detect connectivity differences between the thalamus and both the frontal and temporal lobes, which is consistent with previous FTD studies. These results could not be replicated with the FSL pipeline [3].

In this work, we propose to improve on the above mentioned native space based strategy by integrating local fibre orientation information to enhance thalamus parcellations.

All the previously mentioned procedures were evaluated on a population of healthy controls and FTD patients. Results showed that the proposed strategy led to robust thalamic nuclei segmentations and was also sensitive to connectivity changes between controls and FTD patients.

## 2   Methods

### 2.1   Data

A group of 55 individuals – participants of a multicentre study – 23 healthy controls (mean age 43.0 years, age range [26.5, 70.6], 16 female) and 32 FTD patients (mean age 50.9 years, age range [20.5, 77.0], 19 female) were considered in this work. Diffusion data consisted of two repeated scans, acquired along 64 non-collinear isotropically distributed directions (b-value of $1000\,s/mm^2$) and four additional b-0 volumes on 1.5T and 3T Philips, Siemens and GE scanners, with $2.5\,mm^3$ isotropic voxel size. Volumetric T1-weighted MPRAGE images were also obtained with an isotropic voxel size of $1.1\,mm^3$.

---

[1] http://fsl.fmrib.ox.ac.uk/.

## 2.2   Pre-processing

T1 images were corrected for bias field using N4 algorithm [4]. DW images were also corrected for artifacts. First, the DW data was motion- and eddy-current corrected by affinely registering them to an average b-0 image, generated through a groupwise registration of the b-0 volumes in each subjects data [5]. Secondly, susceptibility artefacts were addressed through phase unwrapping followed by non-linear registration along the phase encoding direction of the distorted DW images to the T1-weighted image [6].

## 2.3   FSL Pipeline

In brief, Behrens et al. work [1] comprised five main steps. First, the thalamus and six cortical regions of interest (ROIs) – prefrontal, motor, sensory, parietal, occipital and temporal cortices – per hemisphere were extracted from an atlas in MNI152 space, available as part of SPM12, which follows the Neuromorpho-metrics Inc. protocol and includes 135 neuroanatomical labels. Then, all these ROIs were registered from MNI space to each subject's diffusion space by com-posing the intra-subject affine registration between the mean $b = 0$ volume and T1-weighted data (using the FLIRT algorithm [7]) with a non-linear registration from the T1-weighted image into the atlas in MNI152 space (using the FNIRT algorithm [8]). Third, fibre orientations within the dMRI data were inferred using the Ball-and-Stick model. Fibre architecture between the thalamus and the six cortical ROIs was reconstructed through probabilistic tractography and for each hemisphere separately, using the thalamus as seed region and any of the six cortical ROIs as target region. Here, the connectivity between any two points of interest was defined as the proportion of samples that pass between them. Finally, each thalamic voxel was labelled as the cortical region with the highest connection probability.

## 2.4   Thalamus Parcellation Using Tractography, Population-Specific Priors and Local Fibre Orientation

### 2.4.1   Previous Work (PW)

The basis of the strategy proposed in this study was an existing thalamic nuclei segmentation procedure [2], that succinctly consists in:

1. *Data analysis in subject's space:* Each subject T1-weighted image was par-cellated into 143 regions [9]. Thalamus and six cortical regions (prefrontal, motor, sensory, parietal, occipital and temporal cortices) were selected per hemisphere. During DW artefacts correction, the average b-0 volume was registered into the T1-weighted image as one of the steps to correct for sus-ceptibility distortions. The inverse of this transformation was computed and used to map the thalamic and cortical masks from T1 to DW space. Fibre ori-entations distribution (FOD) were inferred directly from the measured DW signal by using a multi-fibre algorithm constrained spherical deconvolution

method [10]. Probabilistic streamline tracking was then performed using the iFOD2 algorithm [11] and repeated for every seed point (thalamus) and target region (any of the six cortical regions), enabling the estimation of the number of streamlines that may exist between them. Finally, a segmentation of the thalamic nuclei was obtained by estimating the probability that each thalamic voxel was connected to a target cortical ROI.

2. *Population-specific thalamic nuclei priors:* Thalamic nuclei priors were derived through an iterative strategy. First, a population-specific space was built through groupwise registration of T1 and FA images of all individuals [5]. Secondly, the connectivity maps of each subject were propagated to the common space and averaged to get estimates of the group thalamic nuclei location. The obtained priors were then mapped back to each individual's space and used to compute the new connection probability between every thalamic voxel and each cortical region. The steps above were repeated until convergence, and then a thalamus parcellation was obtained.

### 2.4.2   Proposed Strategy (PS)

The former segmentation procedure was extended to not only consider tractography and thalamic nuclei priors, but also to incorporate fibre orientation information from each subnuclei. This iterative approach was based on a mixture model of von Mises-Fisher (vMF) and optimised using Expectation-Maximisation (EM). In more detail, it considered:

- **Tractography:** performed as in Sect. 2.4.1, enables the estimation of six connectivity maps, that describe the connectivity probability of each thalamic voxel to a particular cortical region (prefrontal, motor, sensory, parietal, occipital and temporal). This was represented in the PS as $\tau$;
- **Thalamic nuclei priors:** population-specific and derived following the same steps as in Sect. 2.4.1. They provide *a priori* information about the expected anatomical connectivity for each thalamic voxel and a specific cortical region, characterised in PS by $\pi$;
- **Fibre orientation:** the FOD from each voxel, reconstructed as in Sect. 2.4.1, was then segmented into its three major fibre orientations using a peak-finding procedure that relies on multiple starting direction points and Newton optimisation [12]. The principal fibre orientation in each voxel was then selected and described in terms of vMF functions.

The diffusion directional data $V = \{v_1, \ldots, v_N\}$ was modelled by a mixture of $c$ von Mises-Fishers distributions and its probability density function described as:

$$f(v|\theta) = \prod_{i=1}^{N} \sum_{c=1}^{C} \alpha_{ic} f_c(v_i|\theta) \tag{1}$$

The term $\alpha_c$ represents the probability of a certain sample $v_i$ belonging to nucleus $c$ based on information derived from tractography $\tau$ and thalamic nuclei priors $\alpha_c = \tau_{ic}.\pi_{ic}$. On the other hand, $f_c(v|\theta_c)$ is a vMF distribution that models the principal fibre orientation in each subnucleus $c$:

$$f_c(v|\theta) = f_c(v|\mu_c, k_c) = C(k_c)e^{k_c \mu_c^T v} \tag{2}$$

where $\mu_c$ is the mean direction, $k_c$ the concentration parameter, $C(k_c) = \frac{k^{1/2}}{(2\pi)^{3/2} I_{1/2}(k_c)}$ the normalisation constant and $I_{1/2}$ the modified Bessel function of the first kind and order $1/2$.

The model parameters $\theta = \{\mu_c, k_c\}_{c=1,...,C}$ were learned through a maximum likelihood formulation $\hat{\theta} = \arg\max_\theta L(\theta) = \arg\max_\theta \log f(v|\theta)$ and optimised in a iterative way using Expectation-Maximisation (EM) with the following stages:

- E-step: update the membership weights of every element $v_i$ in each cluster $c$ given the parameter set $\theta$ at iteration $t$ $(\theta^t)$

$$p(c|v_i, \theta^t) = \frac{\alpha_c f_c(v_i|\theta^t)}{\sum_{c'=1}^{C} \alpha_{c'} f'_c(v_i|\theta^t)} \tag{3}$$

- M-step: find the parameters $\theta^{t+1}$ that maximise the expectation of the log likelihood $L(\theta)$

$$r_c = \sum_{i=1}^{N} v_i.p(c|v_i, \theta^t) \tag{4}$$

$$\bar{r}_c = \frac{||r_c||}{\sum_{i=1}^{N} p(c|v_i, \theta^t)} \tag{5}$$

$$\mu_c^{t+1} = \frac{r_c}{||r_c||} \tag{6}$$

$$k_c^{t+1} \approx \frac{3\bar{r}_c - \bar{r}_c^3}{1 - \bar{r}_c^2} \tag{7}$$

The steps above are repeated until convergence. The algorithm returns both $\theta = \{\mu_c, k_c\}_{c=1,...,C}$ which are the parameters of $c$ vMF distributions that model the directional data $V$, as well the soft-clustering $p(c|v_i, \theta)$ for all $c$ and $v_i$ samples.

## 3    Results

Thalamus connectivity-based segmentations were derived for the three described strategies (FSL, PW and PS) and individuals – controls and FTD patients – as depicted in Fig. 1. The volumes of each connectivity-defined region (CDR) were normalised to the total intracranial volume (TIV) and subsequently compared between the two groups of subjects, in separate for each hemisphere and method, using the Mann-Whitney test. Then, multiple-comparison correction with false discovery rate (FDR) strategy was performed (Table 1).

The PW and PS strategies led to smoother parcellations and more consistent clusters between hemispheres, both in controls and patients, in contrast to FSL (Fig. 1). Significant volume differences between controls and patients were observed on the prefrontal CDR in both hemispheres by all the strategies, as shown in Table 1. Additionally, PS and PW were also sensitive to connectivity differences in the temporal CDR, which was not detected with FSL.

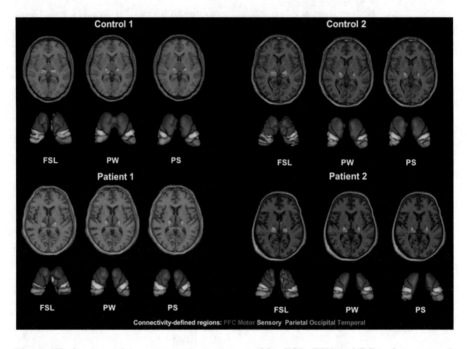

**Fig. 1.** Thalamic nuclei segmentations generated by FSL, PW and PS in four example individuals displayed in MNI152 space.

## 4   Discussion and Conclusion

Previously, Behrens et al. [13] developed a probabilistic tractography algorithm and use it to parcellated the thalamus based on its anatomical connectivity towards distinct cortical regions. The resulting segmentations were in agreement with previous known anatomy, with anterior thalamic regions preferentially connected to prefrontal, motor and sensory cortices, whereas posterior clusters had mainly WM projections towards parietal, occipital and temporal regions [14].

This aspect was also evident on the thalamus parcellations generated by PW and PS approaches, both on normal controls and FTD patients, which did not occur with the segmentations derived with FSL, as depicted in Fig. 1.

**Table 1.** Corrected FDR p-values obtained by comparing the normalised CDRs volumes between controls and FTD patients groups. Significant volumes differences were detected on the prefrontal CDR in both hemispheres with all strategies, whereas for the temporal CDR this was just observed with PW and PS approaches (one-tailed significance test and $P < 0.05$).

| Method | Prefrontal | Motor | Sensory | Parietal | Occipital | Temporal |
|--------|-----------|-------|---------|----------|-----------|----------|
| *Left hemisphere* | | | | | | |
| FSL | 0.039* | 0.992 | 0.840 | 0.651 | 0.992 | 0.110 |
| PW | 0.014* | 0.406 | 0.705 | 0.853 | 0.705 | 0.014* |
| PS | 0.014* | 0.496 | 0.555 | 0.799 | 0.250 | 0.016* |
| *Right hemisphere* | | | | | | |
| FSL | 0.039* | 0.947 | 0.625 | 0.625 | 0.947 | 0.794 |
| PW | 0.012* | 0.904 | 0.794 | 0.687 | 0.236 | 0.019* |
| PS | 0.012* | 0.947 | 0.936 | 0.625 | 0.035 | 0.019* |

The connectivity differences detected by PS and PW were in concordance with previous FTD studies that have reported volume differences both on prefrontal and temporal cortices [15,16], while FSL was only able to pick on the prefrontal region (Table 1).

PW and PS strategies rely both on tractography and thalamic nuclei priors, but additionally PS accounts for the local fibre orientation. Similar results were obtained with these approaches both in terms of the generated segmentations (Fig. 1) and volume comparisons with identical p-values (Table 1). The incorporation of fibre orientation by PS may not seem an obvious advantage, but it complements the global connectivity information obtained with tractography, and hence guarantees a better subdivision of the thalamus. This is particularly relevant in the posterior thalamic region as it is a fibre-crossing area, so the anatomical connectivity reconstruction is more challenging.

Further validation of these results may involve a bigger cohort, as well the performance of reproducibility tests. Future work will enhance the proposed strategy by including other relevant modalities and geometrical constraints in order to generate better thalamic nuclei segmentations and priors, and extend the technique to the study of other neurological pathologies.

**Acknowledgements.** This research was funded by the EPSRC UCL Centre for Doctoral Training in Medical Imaging (EP/L016478/1), NIHR BRC UCLH/UCL High Impact Initiative and Icometrix industrial partner. The Dementia Research Centre is supported by Alzheimer's Research UK, Brain Research Trust, and The Wolfson Foundation. This work was funded by the NIHR Queen Square Dementia Biomedical Research Unit, the NIHR UCL/H Biomedical Research Centre and the Leonard Wolfson Experimental Neurology Centre (LWENC) Clinical Research Facility as well as an Alzheimer's Society grant (AS-PG-16-007). JDR is supported by an MRC Clinician Scientist Fellowship (MR/M008525/1) and has received funding from the NIHR Rare

Disease Translational Research Collaboration (BRC149/NS/MH). This work was supported by the MRC UK GENFI grant (MR/M023664/1) and by The Bluefield Project. CHS receives support from the Alzheimers Society (AS-JF-17-011).

# References

1. Behrens, T.E.J., Berg, H.J., Jbabdi, S., Rushworth, M.F.S., et al.: Probabilistic diffusion tractography with multiple fibre orientations: what can we gain? NeuroImage **34**(1), 144–155 (2007)
2. Semedo, C., Cardoso, M.J., Vos, S.B., et al.: Improved tractography-based segmentation of the human thalamus. In: Proceedings of the International Society for Magnetic Resonance in Medicine (2017)
3. Semedo, C., Cardoso, M.J., Vos, S.B., et al.: Thalamic nuclei segmentation on dementia using tractography and population-specific priors. In: Proceedings of the International Society for Magnetic Resonance in Medicine (2018)
4. Tustison, N.J., Avants, B.B., Cook, P.A.: N4ITK: improved N3 bias correction. IEEE Trans. Med. Imaging **29**(6), 1310–1320 (2010)
5. Modat, M., Ridgway, G.R., Taylor, Z.A., et al.: Fast free-form deformation using graphics processing units. Comput. Methods Programs Biomed. **98**(3), 278–284 (2010)
6. Daga, P., Pendse, T., Modat, M., White, M., et al.: Susceptibility artefact correction using dynamic graph cuts: application to neurosurgery. Med. Image Anal. **18**(7), 1132–1142 (2014)
7. Jenkinson, M., Bannister, P., Brady, M., Smith, S.: Improved optimization for the robust and accurate linear registration and motion correction of brain images. NeuroImage **17**(2), 825–841 (2002)
8. Klein, A., Andersson, J., Ardekani, B.A., Ashburner, J., et al.: Evaluation of 14 nonlinear deformation algorithms applied to human brain MRI registration. NeuroImage **46**(3), 786–802 (2009)
9. Cardoso, M.J., Modat, M., Wolz, R., et al.: Geodesic information flows: spatially-variant graphs and their application to segmentation and fusion. IEEE Trans. Med. Imaging **34**(9), 1976–1988 (2015)
10. Tournier, J.D., Calamante, F., Connelly, A.: Robust determination of the fibre orientation distribution in diffusion MRI: non-negativity constrained super-resolved spherical deconvolution. NeuroImage **35**(4), 1459–1472 (2007)
11. Tournier, J.D., Calamante, F., Connelly, A.: Improved probabilistic streamlines tractography by 2nd order integration over fibre orientation distributions. In: Proceedings of the International Society for Magnetic Resonance in Medicine, vol. 18, p. 1670 (2010)
12. Jeurissen, B., Leemans, A., Tournier, J.D., Jones, D.K., Sijbers, J.: Investigating the prevalence of complex fiber configurations in white matter tissue with diffusion magnetic resonance imaging. Hum. Brain Mapp. **34**(11), 2747–2766 (2013)
13. Behrens, T.E.J., Johansen-Berg, H., Woolrich, M.W., et al.: Non-invasive mapping of connections between human thalamus and cortex using diffusion imaging. Nat. Neurosci. **6**(7), 750–757 (2003)
14. Catani, M., De Schotten, M.T.: Atlas of Human Brain Connections. Oxford University Press, Oxford (2012)

15. Rohrer, J.D., Nicholas, J.M., Cash, D.M., et al.: Presymptomatic cognitive and neuroanatomical changes in genetic frontotemporal dementia in the Genetic Frontotemporal dementia Initiative (GENFI) study: a cross-sectional analysis. Lancet Neurol. **14**(3), 253–262 (2015)
16. Cardenas, V.A., Boxer, A.L., Chao, L.L.: Deformation-based morphometry reveals brain atrophy in frontotemporal dementia. Arch. Neurol. **64**(6), 873–877 (2007)

# On Quantifying Local Geometric Structures of Fiber Tracts

Jian Cheng[1,4,5(✉)], Tao Liu[1], Feng Shi[2], Ruiliang Bai[3], Jicong Zhang[1], Haogang Zhu[1], Dacheng Tao[4], and Peter J. Basser[5]

[1] Beijing Advanced Innovation Center for Big Data-Based Precision Medicine, Beihang University, Beijing, China
jian_cheng@buaa.edu.cn
[2] Shanghai United Imaging Intelligence Co., Ltd., Shanghai, China
[3] Interdisciplinary Institute of Neuroscience and Technology, Zhejiang University, Hangzhou, China
[4] UBTECH Sydney AI Centre, SIT, FEIT, University of Sydney, Sydney, Australia
[5] SQITS, NIBIB, NICHD, National Institutes of Health, Bethesda, USA

**Abstract.** In diffusion MRI, fiber tracts, represented by densely distributed 3D curves, can be estimated from diffusion weighted images using tractography. The spatial geometric structure of white matter fiber tracts is known to be complex in human brain, but it carries intrinsic information of human brain. In this paper, inspired by studies of liquid crystals, we propose tract-based director field analysis (tDFA) with total six rotationally invariant scalar indices to quantify local geometric structures of fiber tracts. The contributions of tDFA include: (1) We propose orientational order (OO) and orientational dispersion (OD) indices to quantify the degree of alignment and dispersion of fiber tracts; (2) We define the local orthogonal frame for a set of unoriented curves, which is proved to be a generalization of the Frenet frame defined for a single oriented curve; (3) With the local orthogonal frame, we propose splay, bend, and twist indices to quantify three types of orientational distortion of local fiber tracts, and a total distortion index to describe distortions of all three types. The proposed tDFA for fiber tracts is a generalization of the voxel-based DFA (vDFA) which was recently proposed for a spherical function field (i.e., an ODF field). To our knowledge, this is the first work to *quantify* orientational distortion (splay, bend, twist, and total distortion) of fiber tracts. Experiments show that the proposed scalar indices are useful descriptors of local geometric structures to visualize and analyze fiber tracts.

## 1 Introduction

Diffusion MRI (dMRI) provides a unique window to non-invasively reveal anatomical connections (i.e., fiber tracts) and white matter tissue properties in human brain [3]. In dMRI, a typical processing pipeline before statistical analysis is: (1) fit a diffusion model (e.g., diffusion tensor imaging (DTI) [2]) to

© Springer Nature Switzerland AG 2018
A. F. Frangi et al. (Eds.): MICCAI 2018, LNCS 11072, pp. 392–400, 2018.
https://doi.org/10.1007/978-3-030-00931-1_45

measured diffusion signals in each voxel; (2) calculate various voxel-wise scalar indices from the model parameters; (3) calculate local fiber directions in each voxel based on the model parameters, and then perform tractography. Scalar indices from diffusion models in step 2 are to quantify different tissue properties at different levels. For example, at the voxel level, fractional anisotropy (FA), generalized FA, return-to-origin probability, mean-squared displacement, and orientation dispersion index are indices for tissue properties inside a single voxel in DTI, Q-ball imaging [11], spherical polar Fourier imaging [6], NODDI [13], etc. At the local neighborhood level, gradient norm [8], sheet probability index [10], and orientational distortion [5] are indices for tissue properties within a local neighborhood.

Fiber tracts estimated by tractography are curves densely distributed in the 3D space. The spatial geometric structure of fiber tracts is known to be very complex in human brain [3]. In a local spatial region, typical geometric structures of fiber tracts include splay (aka diverge, converge, or fanning), bend, twist, crossing, kissing, etc. Grid and sheet structures were also reported in some areas in human brain [10,12]. See Fig. 1 for a demonstration of splay, bend, and twist of fiber tracts. Compared with various voxel-wise indices, only a few tract-based indices have been proposed. Savadjiev et al. [9] proposed the total dispersion index for each point in a tract. Curvature and torsion based on the Frenet frame along the curve are two famous features of a single tract [1,4]. However, the Frenet frame and its features are not designed for a set of curves, thus they cannot quantify local geometric structures of fiber tracts shown in Fig. 1.

Cheng et al. [5] proposed a framework called director field analysis (DFA) to quantify orientational order, dispersion of spherical functions, and orientational distortion (splay, bend, twist, and total distortion) of a spherical function field at the local neighborhood level. Since the DFA in [5] was developed for a spherical function field, we call it as voxel based DFA (vDFA). vDFA does not work for fiber tracts. In this paper, inspired by studies of liquid crystals [7], we propose tract-based director field analysis (tDFA) which generalizes vDFA in [5] to fiber tracts. tDFA is not proposed to replace vDFA, but adopts methods and concepts in vDFA to fiber tracts. In tDFA, we define total 6 scalar indices at each point in fiber tracts, where orientational order and dispersion quantify the alignment and dispersion of fiber tracts, and splay, bend, twist and total orientational distortion quantify local orientational distortion of fiber tracts. tDFA is applied directly on fiber tracts *after* tractography. To our knowledge, this is the first work to quantify orientational distortion of fiber tracts (i.e., splay, bend and twist).

## 2    Method

### 2.1    Directors, Oriented Curves and Unoriented Curves

A *director*, borrowed from studies of liquid crystals [7], is defined as a vector $v$ that is equivalent to its negative $-v$ [5]. A director $v$ can be represented as a dyadic tensor $vv^T$ without sign ambiguity. A local fiber direction in a voxel (i.e., a peak of the ODF in that voxel) is a director by definition. "Sign ambiguity"

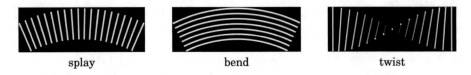

splay                    bend                    twist

**Fig. 1.** Demonstration of three types of distortions of fiber tracts.

also happens for fiber tracts. A fiber tract is mathematically a differentiable curve in $\mathbb{R}^3$ with two endpoints. A differentiable curve can be parameterized as an *oriented curve* $C : [0,1] \mapsto \mathbb{R}^3$, where $C(t)$ is differentiable and $C(0)$ and $C(1)$ are two endpoints. If we set $\gamma(t) = C(1-t)$ (i.e., inverse the orientation of the curve), then $\gamma : [0,1] \mapsto \mathbb{R}^3$ is a new parameterization with the opposite orientation, i.e., $\gamma(0) = C(1)$ and $\gamma(1) = C(0)$. Note that these two oriented curves actually represent the same *unoriented curve* in geometry, however, for a point $C(t)$ (i.e., $\gamma(1-t)$) in the curve, the two tangent vectors in these two parameterization have opposite directions, i.e., $T_{C(t)} = -T_{\gamma(1-t)}$. Thus, tangent vectors of fiber tracts are actually directors.

The sign ambiguity of fiber tracts is not a problem for some cases. For example, the curvature and torsion do not change under the above two parameterization. However, it is indeed a crucial problem if we would like to compare two tangent vectors. Thus, we will use tools of directors in [5] for calculation to avoid this problem. The difference of two directors $\boldsymbol{v}_1$ and $\boldsymbol{v}_2$ in the director representation, denoted as $\mathrm{Diff}_d$, is $\mathrm{Diff}_d(\boldsymbol{v}_1, \boldsymbol{v}_2) = s_1\boldsymbol{v}_1 - s_2\boldsymbol{v}_2$, where $s_i = \pm 1$ such that $s_1 s_2 \boldsymbol{v}_1^T \boldsymbol{v}_2 \geq 0$ [5].

## 2.2  Orientational Order and Dispersion of Fiber Tracts

Considering a set of fiber tracts $\{C_i\}$ which are curves densely distributed in $\mathbb{R}^3$, for each point $\boldsymbol{x} \in C_i$, inspired by liquid crystals [5,7], we define orientational order index (OO) as

$$
\mathrm{OO} = \sum_{\boldsymbol{y} \in \Omega(\boldsymbol{x})} w(\boldsymbol{y}, \boldsymbol{x}) \frac{3(\mathbf{u}_1(\boldsymbol{y})^T \mathbf{u}_1(\boldsymbol{x}))^2 - 1}{2}, \tag{1}
$$

where $\boldsymbol{y}$ is a point in a curve $C_j$ in a spatial neighborhood $\Omega(\boldsymbol{x})$ of $\boldsymbol{x}$, $\mathbf{u}_1(\boldsymbol{x})$ and $\mathbf{u}_1(\boldsymbol{y})$ are unit norm tangent vectors of curves at $\boldsymbol{x}$ and $\boldsymbol{y}$, $w(\boldsymbol{y}, \boldsymbol{x})$ is a spatial weighting function (e.g., uniform or Gaussian weighting) and $\sum_{\boldsymbol{y} \in \Omega(\boldsymbol{x})} w(\boldsymbol{y}, \boldsymbol{x}) = 1$. See Fig. 2. We have $\mathrm{OO} \in [-0.5, 1]$. Then, we define the orientational dispersion index (OD) as $\mathrm{OD} = 1 - \mathrm{OO}$. Thus, $\mathrm{OD} \in [0, 1.5]$. If $\mathbf{u}_1(\boldsymbol{y})$ is parallel to $\mathbf{u}_1(\boldsymbol{x})$, $\forall \boldsymbol{y} \in \Omega(\boldsymbol{x})$, i.e., the least dispersion case, then $\mathrm{OO} = 1$ and $\mathrm{OD} = 0$ at $\boldsymbol{x}$. If $\mathbf{u}_1(\boldsymbol{y})$ is orthogonal to $\mathbf{u}_1(\boldsymbol{x})$, $\forall \boldsymbol{y} \in \Omega(\boldsymbol{x})$, i.e., the most dispersion case, then $\mathrm{OO} = -0.5$ and $\mathrm{OD} = 1.5$ at $\boldsymbol{x}$. Note that the sign ambiguity of tangent vectors $\mathbf{u}_1(\boldsymbol{y})$ and $\mathbf{u}_1(\boldsymbol{x})$ dose not change the values of OO and OD. In practice, we normally set $\Omega_{\boldsymbol{x}}$ as a ball centered at $\boldsymbol{x}$ with a radius $r = 4\,\mathrm{mm}$, considering typical isotropic voxel size of a DW image is $2\,\mathrm{mm}$. Note that the total dispersion in [9] is defined as the mean of differences between tangent vectors in a circle and $\mathbf{u}_1(\boldsymbol{x})$, while our definition of OD is inspired by the order parameter in liquid crystals [5, 7].

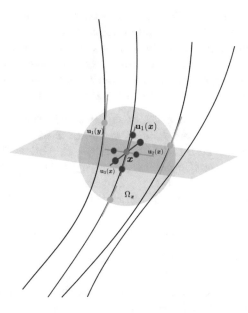

**Fig. 2.** Demonstration of tract-based DFA at a point $\boldsymbol{x}$, denoted as the red point. We use tubes, instead of traditional arrows, to denote tangent vectors because of the sign ambiguity of the unoriented curves and tangent vectors. $\Omega_x$ is a local neighborhood of $\boldsymbol{x}$, denoted as the grey sphere. The yellow points are within $\Omega_x$, and their tangent vectors are used to calculate OO and OD, and construct the second and third directors in the local orthogonal frame. The pink plane is the orthogonal plane of red director $\mathbf{u}_1(\boldsymbol{x})$. The red, green, and blue tubes are the 3 directors $\{\mathbf{u}_1(\boldsymbol{x}), \mathbf{u}_2(\boldsymbol{x}), \mathbf{u}_3(\boldsymbol{x})\}$ in the local orthogonal frame. The 6 purple points $\{\boldsymbol{x} \pm k\mathbf{u}_i\}$ alone these 3 directors in the local orthogonal frame are used to calculate the distortion indices.

## 2.3   Local Orthogonal Frame for a Set of Unoriented Curves

Curvature and torsion are typical features of a single curve based on the Frenet frame [1]. However, the Frenet frame has several limitations: (1) It requires a direction (i.e., orientation) of a curve; (2) It is defined only for a single curve, not for a set of curves; (3) For a straight line, the Frenet frame is not well defined because the curvature of straight lines is 0.

We propose a local orthogonal frame at each point $\boldsymbol{x}$ of fiber tracts. The local orthogonal frame has three directors. The first director, denoted as $\mathbf{u}_1(\boldsymbol{x})$, is the tangent vector at $\boldsymbol{x}$, and the other two directors are in the orthogonal plane of the first director. We project all tangent vectors of fiber tracts in a local neighborhood $\Omega_x$ of $\boldsymbol{x}$ onto the orthogonal plane, and define the covariance matrix of the projected directors as

$$\mathbf{Q}_x = \sum_{y \in \Omega_x} w(\boldsymbol{y}, \boldsymbol{x}) \mathbf{u}_{1,\perp}(\boldsymbol{y}) \mathbf{u}_{1,\perp}^T(\boldsymbol{y}), \qquad \text{where } \mathbf{u}_{1,\perp}(\boldsymbol{y}) = \mathbf{u}_1(\boldsymbol{y}) - (\mathbf{u}_1^T(\boldsymbol{y})\mathbf{u}_1(\boldsymbol{x}))\mathbf{u}_1(\boldsymbol{x}).$$

$$(2)$$

Then we set the second director $\mathbf{u}_2(\boldsymbol{x})$ as the eigenvector associated with the largest eigenvalue of $\mathbf{Q}_x$, i.e., the principal component in PCA. In this way, $\mathbf{u}_2(\boldsymbol{x})$ indicates the direction of the largest change of $\mathbf{u}_1(\boldsymbol{x})$ in the orthogonal plane. Then, the third director is the cross product of the first two directors. See Fig. 2. We prove in Proposition 1 that when $\Omega_x$ tends infinitesimally small, the local orthogonal frame will converge to the Frenet frame.

**Proposition 1.** *Let $\{C_i\}$ be a set of differentiable curves. Let $\boldsymbol{x} \in C_1$ be a point in $C_1$, and $\mathbf{u}_1(\boldsymbol{x})$ is the tangent vector at $\boldsymbol{x}$ in $C_1$. Assume $\boldsymbol{x}$ is not a crossing point. Given a local ball $\Omega_x = \{\boldsymbol{y} \mid \|\boldsymbol{y} - \boldsymbol{x}\| \leq r\}$ with a radius $r$, let the constructed local orthogonal frame be $\{\mathbf{u}_1(\boldsymbol{x}), \mathbf{u}_2(\boldsymbol{x}), \mathbf{u}_3(\boldsymbol{x})\}$. For $\boldsymbol{x} \in C_1$, let the Frenet frame be $\{\mathbf{u}_1(\boldsymbol{x}), \boldsymbol{v}_2(\boldsymbol{x}), \boldsymbol{v}_3(\boldsymbol{x})\}$. The vectors in two frames all have unit norm. Then we have $\lim_{r \to 0} (\mathbf{u}_2(\boldsymbol{x})^T \boldsymbol{v}_2(\boldsymbol{x}))^2 = 1$, $\lim_{r \to 0} (\mathbf{u}_3(\boldsymbol{x})^T \boldsymbol{v}_3(\boldsymbol{x}))^2 = 1$.*

### 2.4   Orientational Distortions (Splay, Bend, and Twist)

Based on distortion analysis of liquid crystals [7], there are 3 types of orientational distortions as showed in Fig. 1. (1) splay: bending occurs perpendicular to the director; (2) bend: bending is parallel to the director; (3) twist: neighboring directors are rotated with respect to one another, rather than aligned. Inspired by liquid crystals [5,7], after we obtain the local orthogonal frame for each point in tracts, we define at each point three scalar indices to describe the three types of local distortions, and a total distortion index:

$$\text{Splay index: } s = \sqrt{(\mathbf{u}_2^T \tfrac{\partial \mathbf{u}_1}{\partial \mathbf{u}_2})^2 + (\mathbf{u}_3^T \tfrac{\partial \mathbf{u}_1}{\partial \mathbf{u}_3})^2}, \tag{3}$$

$$\text{Bend index: } b = \sqrt{(\mathbf{u}_2^T \tfrac{\partial \mathbf{u}_1}{\partial \mathbf{u}_1})^2 + (\mathbf{u}_3^T \tfrac{\partial \mathbf{u}_1}{\partial \mathbf{u}_1})^2}, \tag{4}$$

$$\text{Twist index: } t = \sqrt{(\mathbf{u}_2^T \tfrac{\partial \mathbf{u}_1}{\partial \mathbf{u}_3})^2 + (\mathbf{u}_3^T \tfrac{\partial \mathbf{u}_1}{\partial \mathbf{u}_2})^2}, \tag{5}$$

$$\text{Total distortion index: } \quad d = \sqrt{s^2 + b^2 + t^2}, \tag{6}$$

where $\frac{\partial \mathbf{u}_1}{\partial \mathbf{u}_i}$, $i = 1, 2, 3$, is the directional derivative of $\mathbf{u}_1(\boldsymbol{x})$ along $\mathbf{u}_i(\boldsymbol{x})$, i.e.,

$$\frac{\partial \mathbf{u}_1}{\partial \mathbf{u}_i} \approx \frac{\text{Diff}_d(\mathbf{u}_1(\boldsymbol{x} + k\mathbf{u}_i), \mathbf{u}_1(\boldsymbol{x} - k\mathbf{u}_i))}{2k}, \tag{7}$$

where $k$ is small, and $\text{Diff}_d$ means the difference of two directors in the director representation [5]. See Sect. 2.1. Since the director field is not continuous in practice, we use an interpolation method to estimate $\mathbf{u}_1(\boldsymbol{x}+k\mathbf{u}_i)$ and $\mathbf{u}_1(\boldsymbol{x}-k\mathbf{u}_i)$ from local neighborhoods of $\boldsymbol{x} + k\mathbf{u}_i$ and $\boldsymbol{x} - k\mathbf{u}_i$. Let $\boldsymbol{z} = \boldsymbol{x} \pm k\mathbf{u}_i$, then we estimate $\mathbf{u}_1(\boldsymbol{z})$ from its neighborhood tangent directors of tracts:

$$\mathbf{u}_1(\boldsymbol{z}) = \arg\min_{\mathbf{u},\lambda} \sum_{\boldsymbol{y} \in I(\boldsymbol{z})} \|\lambda \mathbf{u}\mathbf{u}^T - w_s(\boldsymbol{y}, \boldsymbol{z})w_b(\boldsymbol{y}, \boldsymbol{x})\mathbf{u}_1(\boldsymbol{y})\mathbf{u}_1^T(\boldsymbol{y})\|^2, \tag{8}$$

where $I(z)$ is a neighborhood of $z$, $w_s(y, z) = \frac{1}{Z\|y-z\|^2}$, $w_b(y, x)$ is the bundle probability weight, and $Z$ is the normalization factor such that $\sum_{y \in I(z)} w_s(y, z) w_b(y, z) = 1$. we use the inverse distance weight so that if $z$ is a point in a tract, then $w_s = 1$ and $\mathbf{u}_1(z)$ is just the tangent vector of the tract. It can be proved that the interpolated $\mathbf{u}_1(z)$ is the actually the principal eigenvector of $\sum_{y \in I(z)} w_s(y, z) w_b(y, z) \mathbf{u}_1(y) \mathbf{u}_1^T(y)$. We set $I(z)$ (i.e., $I(x \pm k\mathbf{u}_i)$) as a ball with a radius of $2k$, and $k$ is normally set as $1\,mm$, half of the typical voxel size of $2\,mm$. If one would like to calculate indices using all points $y$ from all fiber bundles, then $w_b(y, x) = 1$ as a constant. If one would like to calculate indices using points $y$ only from the same fiber bundle as $x$, then we may set $w_b(y, x)$ as a pre-determined probability of fiber clustering, or we could simply set $w_b(y, x) = 1$ only if the angle between $\mathbf{u}_1(y)$ and $\mathbf{u}_1(x)$ is smaller than a threshold $\theta_0$ (e.g., $45°$), and $w_b(y, x) = 0$, otherwise. If there is no specific mention, we set $w_b(y, x) = 1$ only if the angle is smaller than $45°$.

## 3    Experiments

**Synthetic Data Experiments.** We generated three synthetic fiber tracts (splay, bend, and twist data) which demonstrate these three types of distortions, and then calculated the proposed scalar indices for each point in the fiber tracts. These scalar indices were used to color the fiber tracts. See Fig. 3. We omit OO, because $OO = 1 - OD$. The first three rows in Fig. 3 show that the proposed splay, bend, and twist indices completely separate these 3 datasets, where only one among these 3 values is non-zero in each dataset. We also combined fiber tracts in splay and bend data, then calculated these indices using interpolation with points from all fiber bundles and with points only from the same fiber bundle (angle threshold $\theta_0 = 45°$). The last two rows in the figure show that in a crossing area, it is better to use fiber tracts from the same bundle to calculate bundle specific distortion indices. Note that increased OD happens in crossing areas, compared with single bundle areas, and calculation of OD does not require an interpolation.

**Real Data Experiments.** We performed tractography on a publicly available dataset with a single subject from DIPY (dipy.org). The data has a single shell with $b = 2000\,s/mm^2$ and 150 directions on the shell. The corpus callosum (CC) was extracted based on diffusion ODF and deterministic tracking. Then, all six indices were calculated from the fiber tracts of CC. Figure 4 shows the scalar indices as colors of tracts. In the right subfigure, splay, bend, and twist values are set as red, green and blue color channels, respectively. The bending areas of CC are mainly in green, which means that the bend index is higher than slay and twist indices in those bending areas. We can also see red (high splay values) in fanning areas of CC, and blue (high twist values) when tracts are twisting in the 3D space.

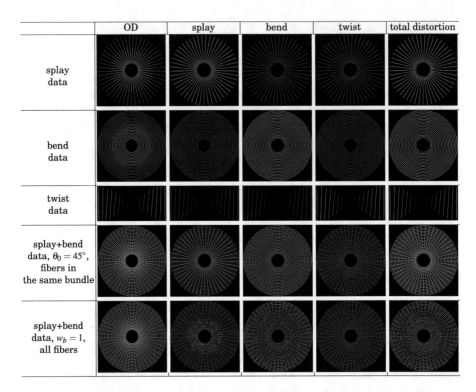

|  | OD | splay | bend | twist | total distortion |
|---|---|---|---|---|---|
| splay data | | | | | |
| bend data | | | | | |
| twist data | | | | | |
| splay+bend data, $\theta_0 = 45°$, fibers in the same bundle | | | | | |
| splay+bend data, $w_b = 1$, all fibers | | | | | |

**Fig. 3.** Synthetic datasets of fiber tracts. Each row is a set of fiber tracts. The OD, splay, bend, twist and total distortion indices are used to color the fiber tracts. Low/high values are in blue/red color. The last two rows are crossing of bend and splay data with two interpolation strategies.

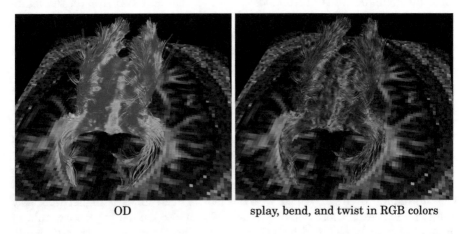

OD                    splay, bend, and twist in RGB colors

**Fig. 4.** The corpus callosum of a real subject. Left: tracts are colored by OD values in points, where low/high values are in blue/red colors. Right: tracts are colored by splay, bend, and twist values in points, where splay, bend, and twist values are set as red, green and blue color channels, respectively.

# 4    Conclusion

We propose a unified mathematical framework called tract-based director field analysis (tDFA) with six scalar indices to quantify local geometric structure of fiber tracts. OD and OO are useful to quantify the degree of alignment and dispersion of fiber tracts; The distortion indices (i.e., splay, bend, twist and total distortion) demonstrate good sensitivity in both synthetic datasets and a real dataset on CC. To our knowledge, this is the first work to *quantify* orientational distortion (splay, bend, and twist) of fiber tracts. The proposed indices are rotationally invariant, because they are calculated from intrinsic quantities (i.e., local orthogonal frames) of tracts.

**Acknowledgement.** This work was supported by Australian Research Council Projects FL-170100117, DP-180103424, LP-150100671, Intramural Research Program of NICHD (ZIA-HD000266), the National Key R&D Program of China (2016YFF0201002), and the Beijing Municipal Science and Technology Commission (BJMSJY-160153).

# References

1. Basser, P.J.: New histological and physiological stains derived from diffusion-tensor MR images. Ann. N. Y. Acad. Sci. **820**, 123–138 (1997)
2. Basser, P.J., Mattiello, J., LeBihan, D.: MR diffusion tensor spectroscropy and imaging. Biophys. J. **66**, 259–267 (1994)
3. Basser, P.J., Pajevic, S., Pierpaoli, C., Duda, J., Aldroubi, A.: In vivo fiber tractography using DT-MRI data. Magn. Reson. Med. **44**, 625–632 (2000)
4. Batchelor, P.G., Calamante, F., Tournier, J.-D., Atkinson, D., Hill, D.L.G., Connelly, A.: Quantification of the shape of fiber tracts. Magn. Reson. Med. **55**(4), 894–903 (2006)
5. Cheng, J., Basser, P.J.: Director Field Analysis (DFA): exploring local white matter geometric structure in diffusion MRI. Med. Image Anal. **43**, 112–128 (2018)
6. Cheng, J., Ghosh, A., Jiang, T., Deriche, R.: Model-free and analytical EAP reconstruction via spherical polar fourier diffusion MRI. In: Jiang, T., Navab, N., Pluim, J.P.W., Viergever, M.A. (eds.) MICCAI 2010. LNCS, vol. 6361, pp. 590–597. Springer, Heidelberg (2010). https://doi.org/10.1007/978-3-642-15705-9_72
7. Collings, P.J., Hird, M.: Introduction to Liquid Crystals: Chemistry and Physics. CRC Press, Boca Raton (1997)
8. Pajevic, S., Aldroubi, A., Basser, P.J.: A continuous tensor field approximation of discrete DT-MRI data for extracting microstructural and architectural features of tissue. J. Magn. Reson. **154**, 85–100 (2002)
9. Savadjiev, P., et al.: Fusion of white and gray matter geometry: a framework for investigating brain development. Med. Image Anal. **18**(8), 1349–1360 (2014)
10. Tax, C.M.W., et al.: Sheet Probability Index (SPI): characterizing the geometrical organization of the white matter with diffusion MRI. NeuroImage **142**, 260–279 (2016)
11. Tuch, D.S.: Q-ball imaging. Magn. Reson. Med. Off. J. Int. Soc. Magn. Reson. Med. **52**, 1358–1372 (2004)

12. Wedeen, V.J., et al.: The geometric structure of the brain fiber pathways. Science **335**(6076), 1628–1634 (2012)
13. Zhang, H., Schneider, T., Wheeler-Kingshott, C.A., Alexander, D.C.: NODDI: practical in vivo neurite orientation dispersion and density imaging of the human brain. Neuroimage **61**(4), 1000–1016 (2012)

# Neuroimaging and Brain Segmentation Methods: Neuroimaging

# Modeling Longitudinal Voxelwise Feature Change in Normal Aging with Spatial-Anatomical Regularization

Zhuo Sun[1,2], Wei Xu[3], Shuhao Wang[3], Junhai Xu[4,5], and Yuchuan Qiao[5(✉)]

[1] Thorough images, Beijing, China
zhuo.sun@thorough.ai
[2] Division of Image Processing, Department of Radiology,
Leiden University Medical Center, Leiden, The Netherlands
[3] Institute for Interdisciplinary Information Sciences, Tsinghua University, Beijing,
China
[4] School of Computer Science and Technology, Tianjin University, Tianjin, China
[5] USC Institute for Neuroimaging and Informatics, Los Angeles, USA
Yuchuan.Qiao@loni.usc.edu

**Abstract.** Image voxel/vertex-wise feature in the brain is widely used for automatic classification or significant region detection of various dementia syndromes. In these studies, the non-imaging variables, such as age, will affect the results, but may be uninterested to the clinical applications. Imaging data can be considered as a combination of the confound variable (e.g. age) and the variable of clinical interest (e.g. AD diagnosis). However, non-imaging confound variable is not well dealt in each voxel. In this paper, we proposed a spatial-anatomical regularized parametric function fitting approach that explicitly modeling the relationship between the voxelwise feature and the confound variable. By adding the spatial-anatomical regularization, our model not only obtains a better voxelwise feature estimation, but also generates a more interpretable parameter map to help understand the effect of confound variable on imaging features. Besides the commonly used linear model, we also develop a spatial-anatomical regularized voxelwise general logistic model to investigate deeper of the aging process in gray matter and white matter density map.

**Keywords:** Voxelwise feature · Spatial-anatomical regularization
General logistic · Longitudinal model · Normal aging

## 1 Introduction

Neuroimaging data (e.g. MRI) is widely used to predict clinical or psychometric interest variable (MMSE sore or AD diagnosis) [1,2]. It is believed that brain pathology or psychology change will lead to brain morphology [3] or functional [3] change in the neuroimaging data. However, these changes can be considered as combinations of effect of the clinical interested variable (e.g. Alzheimer's

© Springer Nature Switzerland AG 2018
A. F. Frangi et al. (Eds.): MICCAI 2018, LNCS 11072, pp. 403–410, 2018.
https://doi.org/10.1007/978-3-030-00931-1_46

Disease) and other factors (e.g. age). Such variables are commonly referred as "Confound" variables, which are highly correlated with the neuroimaging data but independent from the clinical interested variable. Such confound variables may decrease the power of the prediction or significant region detection [4].

To remove the effect of the confound variables, many approaches have been proposed to learn a better predictor. The most common approach is to match the sample subject with respect to the confound variable. However, such approach will largely decrease the number of variable in imaging data and limit the usage of the learned predictor. Another approach is to combine the confound variables and the features from image to jointly learn a predictor [5,6]. However, in many methods, especially for methods using voxelwise feature, the number of imaging feature is much larger than the size of confound variables. This unbalance will likely decrease the correction effect of the confound variables. Lastly, the influence of the confound variables can be filtered out by independently fitting a linear model of the confound variables using the normal control subject and remove them from the image features [7]. After filtering out the image features related to the confound variables, the predictor can focus on the features related to the clinical interested variable and will be not sensitive to the length of features.

Inspired by the success of the filtering out approach [7], we proposed a spatial-anatomical regularized voxelwise model fitting approach to model the effect of the confound variables to the imaging feature. Besides the better estimation of confound variables related feature, the proposed method improves the interpretability of the learned voxelwise model. Another contribution of the proposed method is the portability to different models. Compared to the voxelwise independent linear model used in [7], our approach obtains more influence of the confound variables. In our application, we show the relationship between the voxelwise gray/white density and the age in normal subjects.

## 2    Methods

In this section, we first define the notations before explaining the details. The $F_i(x)$ represents the voxelwise feature in $i$-th sample subject at voxel $x$, $i \in \{1, 2, \cdots, N\}$. The $\theta_k(x)$ represents the value of the $k$-th parameter map at voxel $x$, $k \in \{1, 2, \cdots, K\}$. The $K$ is the number of parameters for each voxel, and it varies for different models, e.g. $K = 2$ is for the linear model while $K = 4$ is for the general logistic model. The $t_i$ represents the age of a sample subject $i$ when the scan was taken.

### 2.1    Age Correction in Voxel Level

For each voxel $x$, we assume that the feature $F_i(x)$ is the combination of the clinical interested variable and the confound variables. In our application, we consider $F_i(x) = D_i(x) + A(x, t_i)$, where $D_i(x)$ is the voxelwise feature related to the disease progress and $A(x, t_i)$ the voxelwise feature related to the normal aging progress. For a normal subject, we can assume that there is no disease related progress and $F_i(x) = 0 + A(x, t_i)$. Therefore, the age correction problem is simplified to model the normal aging part $A(x, t_i)$ by a parametric function, with parameter $\boldsymbol{\theta}$.

Due to the large variations of the brain changes, it is reasonable to assume that different brain regions have different parameters. A voxelwise generative model $f(t_i; \boldsymbol{\theta}(x))$ is built to estimate the normal aging effect and the estimation error over all normal sample subjects observation $F_i(x)$ is minimized. The optimal parameter of a linear model is estimated using:

$$
\begin{aligned}
\widehat{\boldsymbol{\theta}}(x) &= \underset{\boldsymbol{\theta}(x)}{\operatorname{argmin}} \sum_{i=1}^{N} \|A(x, t_i) - f(t_i; \boldsymbol{\theta}(x))\|_2^2 \\
&= \underset{\boldsymbol{\theta}(x)}{\operatorname{argmin}} \sum_{i=1}^{N} \|A(x, t_i) - \boldsymbol{\theta}_1(x) t_i - \boldsymbol{\theta}_2(x))\|_2^2.
\end{aligned}
\tag{1}
$$

The linear function is too weak to model the complex brain changes of normal aging, so a general logistic function with 4 parameters is introduced to model the normal aging related feature. In the general logistic function:

$$
f(t; \{M, N, T_0, \alpha\}) = M + \frac{N}{1 + e^{-\alpha(t - T_0)}},
\tag{2}
$$

the parameter $T_0$ is the time point where the curve change has the highest speed and it is often considered as the key time point to indicate the starting time. The parameter $\alpha$ represents the growth rate of the curve. A large $\alpha$ indicates that the age related change appears in a very short time interval. The $M$ and $N$ limit the range of the response of $f(t)$. Therefore, even for a new sample with age $t$ far from the training set, the general logistic is opt to generate a reasonable response, while the linear model may generate an unreasonable higher or lower response. For the general logistic function, the optimal parameter $\boldsymbol{\theta}(x)$ is estimated using:

$$
\begin{aligned}
\widehat{\boldsymbol{\theta}}(x) &= \underset{\boldsymbol{\theta}(x)}{\operatorname{argmin}} \sum_{i=1}^{N} \|A(x, t_i) - f(t_i; \boldsymbol{\theta}(x))\|_2^2 \\
&= \underset{\boldsymbol{\theta}(x)}{\operatorname{argmin}} \sum_{i=1}^{N} \|A(x, t_i) - \boldsymbol{\theta}_1(x) - \frac{\boldsymbol{\theta}_2(x)}{1 + e^{-\boldsymbol{\theta}_4(x)(t - \boldsymbol{\theta}_3(x))}}\|_2^2.
\end{aligned}
\tag{3}
$$

## 2.2 Spatial-Anatomical Regularization

The voxelwise estimation of the optimal parameter commonly ignores the spatial and anatomical relation among voxels, however, these relations are very important in many medical image analysis applications. For example, the parameters from the voxelwise function fitting can be used to segment the myocardial scar identification [8] and the integration of the spatial-anatomical relations among voxels can improve the classification performance [9]. A spatial-anatomical regularization (SAR) penalty term was integrated in our proposed method which encourages the smoothness of the parameters from spatial-anatomical neighbor voxels.

Given an anatomical segmentation $S$, the spatial-anatomical neighbor voxel relation $R(p, q)$ between point pair $\{p, q\}$ is defined as:

$$R(p, q) = \begin{cases} 1 & \text{if } \|p - q\|_2^2 \leq 3 \quad \text{and} \quad S(p) = S(q) \\ 0 & \text{otherwise,} \end{cases} \tag{4}$$

and the SAR penalty is computed as:

$$SAR = \sum_{k=1}^{K} \sum_{p} \sum_{q} (R(p, q)(\boldsymbol{\theta}_k(p) - \boldsymbol{\theta}_k(q)))^2. \tag{5}$$

The SAR penalty can be simplified as a quadratic term:

$$SAR(\boldsymbol{\theta}) = \sum_{k=1}^{K} (L\boldsymbol{\theta}_i)'(L\boldsymbol{\theta}_i), \tag{6}$$

where $L$ is a sparse matrix. Each row of $L$ encodes one pair of spatial-anatomical neighbor point pair. For the $c$-th spatial-anatomical neighbor point pair $\{p, q\}$ with $R(p, q) = 1$, we set $L[c, p] = 1$ and $L[c, 1] = -1$.

Here we introduce a spatial-anatomical regularization instead of a spatial only regularization for two reasons: (1) voxel belong to different anatomical region may have different aging effect; (2) with SAR, we can divide all voxelwise features into several groups and learn the parameters for each group independently. This way can decrease the GPU memory cost and accelerate the computation by parallel computing.

## 2.3   Optimization and Implementation

With the new defined SAR penalty, the optimal voxelwise parameters are computed using:

$$\widehat{\boldsymbol{\theta}}(:) = \operatorname*{argmin}_{\boldsymbol{\theta}(:)} \sum_{x} \sum_{i=1}^{N} \|A(x, t_i) - f(t_i; \boldsymbol{\theta}(x))\|_2^2 + \lambda \sum_{k=1}^{K} \boldsymbol{\theta}_k'(L'L)\boldsymbol{\theta}_k, \tag{7}$$

where $\lambda$ is a non-negative hyper-parameter to control to contribution of the voxelwise data fitting term and the SAR penalty. In our application, $A(x, t_i)$ is equal to the input voxelwise feature map $F_i(x)$. With the voxelwise parameter $\boldsymbol{\theta}(x)$, we can predict the voxelwise feature for a given age $t$, as $Y(x, t) = f(t; \boldsymbol{\theta}(x))$.

In this paper, the proposed method is implemented using Python and the GPU-based MxNet. In this implementation, we use the autograd functionality to compute the gradient with respect to the voxelwise parameters. It is easy to extend to other models for new applications. The cost function (Eq. 7) is optimized by ADAM optimizer [10], with 1000 iteration. The code will be public available for research usage.

**Table 1.** Demographic characteristics of the studied normal subject dataset (from ADNI). MMSE means Mini-Mental State Examination.

| Group | Number | Age | Gender | Baseline MMSE |
|-------|--------|-----|--------|---------------|
| Training | 430 | $77.1 \pm 6.7$ [58.3 − 95.3] | 240M/190F | $28.9 \pm 1.3$ [24 − 30] |
| Test | 179 | $79.0 \pm 4.7$ [63.1 − 92.8] | 86M/93F | $29.1 \pm 1.3$ [23 − 30] |

## 3   Dataset

In this paper, we use the T1-weighted MR scans of the normal subjects from public available ADNI dataset (http://adni.loni.usc.edu/), up to Nov 27th, 2017. All selected normal subjects must have a CDR score equal or less than 0. The selected scans are divided into a training and a test group according to the ADNI study phase. The scans belong to ADNI 2 and ADNI 3 are used as training set, while the subjects in ADNI 1 and not in the first two year are treated as the test set. Overall, 609 scans were selected, including 430 scans for the training set and 179 scans for the test set. The demographic characteristics of the selected scans are summarized in Table 1.

All images are B1 corrected and preprocessed by a SPM-based Computational Anatomy Toolbox (CAT12 http://www.neuro.uni-jena.de/cat/) with the default parameter in the cross-sectional pipeline to generate gray matter and white matter density map. We use the neuromorphometrics atlas defined at the CAT12 template space as our segmentation $S$, which contains 140 ROIs.

## 4   Experiment and Results

In the experiment, we first trained the voxelwise linear model with ($\lambda > 0$) and without ($\lambda = 0$) SAR penalty. For each method, the voxelwise parameters were learned using the training set on gray matter density feature map, and then the voxelwise gray matter map was predicted for the test set. For each ROI, the mean square error (MSE) between the observed voxelwise gray matter density feature $F$ and the estimated one $Y$ is computed. The difference of two methods is shown in Fig. 1. It is clear to see that, for each ROI on the test set, the linear model with SAR has smaller MSE than the linear model without SAR. The slope parameter ($\theta_1$) of the linear model learned with and without SAR overlaid on the T2 images, is shown in Fig. 2. As we can see, the slope parameter learned from linear model without SAR is very noisy and with no interpretability.

After verifying the effect of SAR penalty, the optimal parameter maps using general logistic model with SAR was learned on the gray density feature map. The barplot of the differences of MSE between two models in each ROI on the test set gray matter density feature is shown in Fig. 3. In the barplot, we found that there is only 1 ROI that linear model with SAR has smaller MSE compared to the general logistic model with SAR. Overall, the general logistic model with SAR beat the linear model on the rest 139 ROIs.

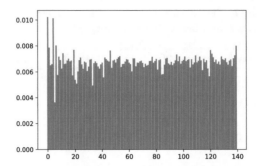

**Fig. 1.** Bar plot of the difference of mean square error (MSE) for each ROI on the test set using linear model without and with SAR. The positive bar indicates that the linear model without SAR has larger MSE than the linear model with SAR.

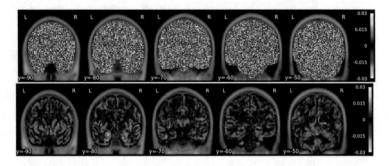

**Fig. 2.** Learned voxelwise slope parameter overlay on T2 images. Top: using linear model without SAR; bottom: using linear model with SAR.

Beside the gray matter density map, we also learned the optimal parameter map using the general logistic with SAR on the white matter density map. When comparing the parameter maps from gray matter and white matter density, we found some interesting points. From Fig. 4, the parameter map ($T_0(\boldsymbol{\theta}_3)$ in the general logistic model), we found that the Caudate and Pallidum regions have smaller $T_0$ in gray matter, which indicates the gray matter changes first occurred in these regions in normal aging process. The growth rate parameter $\alpha(\boldsymbol{\theta}_4)$ maps of gray matter and white matter were plotted in Fig. 5. In both maps, the positive value indicates the voxelwise feature increase while the negative value indicates the opposite effect. We can see that both gray matter and white matter features decreased with time in most brain regions. However, the voxelwise gray matter density tended to increase in regions around ventricle, which may be due to the expansion of ventricle. In both gray matter and white matter, large negative values were observed in the hippocampus region or around, which may indicate a fast degeneration of gray and white matter in these regions during aging process.

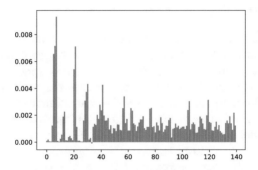

**Fig. 3.** Bar plot of the difference of mean square error (MSE) of each ROI on the test set using linear model with SAR and general logistic model with SAR. The positive bar indicates the linear model with SAR has larger MSE than the general logistic model with SAR.

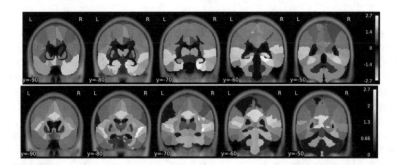

**Fig. 4.** The general logistic model with SAR learned voxelwise starting point parameter $T_0$ $(\boldsymbol{\theta}_3)$ relative to the mean age (77.1) overlaid on image. Top: using gray matter density feature; bottom: using white matter density map.

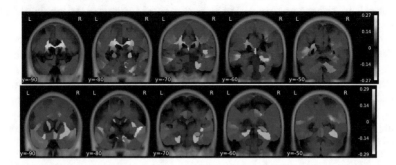

**Fig. 5.** The general logistic model with SAR learned voxelwise growth rate parameter $\alpha(\boldsymbol{\theta}_4)$ overlaid on image. Top: using gray matter density feature; bottom: using white matter density map.

## 5  Conclusion

In this paper, we present a new method to model the aging related changes in voxelwise feature of the brain. It uses the voxelwise 4 parameters general logistic model and a spatial-anatomical regularization penalty term. The cost function is optimized using MxNet, which makes the method easy to extend to other applications. In the given voxelwise feature maps derived from ADNI, our method obtains a better voxelwise feature estimation. More important, our method offers a better interpretability of the learned parameter maps, which may help the clinician to learn the normal aging process from big data of normal population. Currently, only the voxelwise feature derived from T1 weighted MR image was used. It is of great interest to use our approach with multiple types of voxelwise features from different modalities and with starting points from different regions to indicate the aging or disease progression.

## References

1. Sabuncu, M.R., Konukoglu, E.: Clinical prediction from structural brain MRI scans: a large-scale empirical study. Neuroinformatics **13**(1), 31–46 (2015)
2. Stonnington, C.M., Chu, C., Klppel, S., Jack, C.R., Ashburner, J., Frackowiak, R.S.: Predicting clinical scores from magnetic resonance scans in Alzheimer's disease. NeuroImage **51**(4), 1405–1413 (2010)
3. Frisoni, G.B., Fox, N.C., et al.: The clinical use of structural MRI in Alzheimer disease. Nat. Rev. Neurol. **6**(2), 67 (2010)
4. Rao, A., Monteiro, J.M., Mourao-Miranda, J.: Predictive modelling using neuroimaging data in the presence of confounds. NeuroImage **150**, 23–49 (2017)
5. Gould, I.C., Shepherd, A.M., et al.: Multivariate neuroanatomical classification of cognitive subtypes in schizophrenia: a support vector machine learning approach. NeuroImage Clin. **6**, 229–236 (2014)
6. Rao, A., Monteiro, J.M., et al.: A comparison of strategies for incorporating nuisance variables into predictive neuroimaging models. In: 2015 International Workshop on Pattern Recognition in NeuroImaging (PRNI), pp. 61–64. IEEE (2015)
7. Dukart, J., Schroeter, M.L., Mueller, K., et al.: Age correction in dementia-matching to a healthy brain. PLoS One **6**(7), e22193 (2011)
8. Tao, Q., Lamb, H.J., Zeppenfeld, K., van der Geest, R.J.: Myocardial scar identification based on analysis of look-locker and 3D late gadolinium enhanced MRI. Int. J. Cardiovasc. Imaging **30**(5), 925–934 (2014)
9. Cuingnet, R., Glaunès, J.A., Chupin, M., Benali, H., Colliot, O.: Spatial and anatomical regularization of SVM: a general framework for neuroimaging data. IEEE Trans. Pattern Anal. Mach. Intell. **35**(3), 682–696 (2013)
10. Kingma, D.P., Ba, J.: Adam: a method for stochastic optimization. arXiv preprint arXiv:1412.6980 (2014)

# Volume-Based Analysis of 6-Month-Old Infant Brain MRI for Autism Biomarker Identification and Early Diagnosis

Li Wang[1]($\boxtimes$), Gang Li[1], Feng Shi[2], Xiaohuan Cao[1], Chunfeng Lian[1],
Dong Nie[1], Mingxia Liu[1], Han Zhang[1], Guannan Li[1],
Zhengwang Wu[1], Weili Lin[1], and Dinggang Shen[1]($\boxtimes$)

[1] Department of Radiology and BRIC,
University of North Carolina at Chapel Hill, Chapel Hill, USA
{li_wang, dinggang_shen}@med.unc.edu
[2] Shanghai United Imaging Intelligence Co., Ltd., Shanghai, China

**Abstract.** Autism spectrum disorder (ASD) is mainly diagnosed by the observation of core behavioral symptoms. Due to the absence of early biomarkers to detect infants either *with* or *at-risk of* ASD during the first postnatal year of life, diagnosis must rely on behavioral observations long after birth. As a result, the window of opportunity for effective intervention may have passed when the disorder is detected. Therefore, it is clinically urgent to identify imaging-based biomarkers for early diagnosis and intervention. In this paper, *for the first time*, we proposed a volume-based analysis of infant subjects with risk of ASD at very early age, i.e., as early as at 6 months of age. A critical part of volume-based analysis is to accurately segment 6-month-old infant brain MRI scans into different regions of interest, e.g., white matter, gray matter, and cerebrospinal fluid. This is actually very challenging since the tissue contrast at 6-month-old is extremely low, caused by inherent ongoing myelination and maturation. To address this challenge, we propose an anatomy-guided, densely-connected network for accurate tissue segmentation. Based on tissue segmentations, we further perform brain parcellation and statistical analysis to identify those significantly different regions between autistic and normal subjects. Experimental results on National Database for Autism Research (NDAR) show the advantages of our proposed method in terms of both segmentation accuracy and diagnosis accuracy over state-of-the-art results.

**Keywords:** Autism · Infant · Biomarker · Diagnosis · Segmentation

## 1 Introduction

Autism spectrum disorder (ASD) refers to a group of complex neurodevelopmental disorders characterized by repetitive and characteristic patterns of behavior and difficulties with social communication and interaction. The latest analysis from the Centers

---

This work was supported in part by MH109773, MH088520, MH108914, MH117943, MH116225, and MH107815.

A. F. Frangi et al. (Eds.): MICCAI 2018, LNCS 11072, pp. 411–419, 2018.
https://doi.org/10.1007/978-3-030-00931-1_47

for Disease Control and Prevention estimated 1 in 68 children diagnosed with ASD in the U.S.. ASD is mainly diagnosed by the observation of core behavioral symptoms [1]. For many neurodevelopmental disorders, brain dysfunction may precede abnormal behavior by months or even years. However, according to Interagency Autism Coordinating Committee strategic plan, due to the absence of early biomarkers to detect people either with or at-risk of ASD, diagnosis must rely on behavioral observations long after birth. As a result, ASD is not typically diagnosed until around 3–4 years of age in the U.S. [2]. Consequently, intervention efforts may miss a critical developmental window. Thus, it is extremely important to detect ASD earlier in life for better intervention.

Magnetic resonance imaging (MRI) based characterization of ASD has been explored as a complement to the current behavior-based diagnosis [3]. MRI-based brain volumetric studies on children and young adults with ASD found abnormalities in the hippocampus [4], precentral gyrus [5], and anterior cingulate gyrus [6]. However, most of the autistic subjects involved in previous studies are 2+ years old. In fact, the first year of postnatal development represents the most drastic postnatal developing phase, with a rapid tissue growth and milestones of a wide range of cognitive and motor functions [7]. This early period is critical in neurodevelopmental disorders such as ASD [8]. In this work, *for the first time*, we perform a volume-based analysis on infants at 6 months of age with risk of autism. To perform volume-based analysis and further identify possible imaging biomarkers, accurate segmentation of the infant brain into different types of tissue, e.g., white matter (WM), gray matter (GM), and cerebrospinal fluid (CSF), is the most critical step. It will allow reliable quantification of volumetric tissue abnormalities for ASD [9]. However, accurate segmentation of infant brain MRI is challenging, especially for those at a very early age, such as at 6 months old [10], due to the lowest tissue contrast caused by the largely immature myelination [10]. For instance, the 1st column of Fig. 1 shows representative examples of T1-weighted (T1w) and T2-weighted (T2w) MRI scanned at 6 months of age. It can be observed that intensities in WM and GM are within similar range (especially around the cortex), even they were acquired with an imaging protocol dedicated for maximizing tissue contrast [10]. Such isointensity poses a significant challenge for accurate tissue segmentation. To the best of our knowledge, few studies focused on tissue segmentation of 6-month-old infants at-risk of ASD.

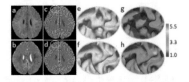

**Fig. 1.** (a) and (b) show T1w and T2w 6-month-old infant brain images with extremely low tissue contrast caused by the inherent ongoing myelination and maturation. (c) and (d) are the segmentation results by [11] and the proposed work with anatomical guidance, respectively, with their corresponding inner surfaces shown in (e) and (f), and cortical thickness (in mm) shown in (g) and (h). Without anatomical guidance, some WM is missing by previous work [11] as indicated by the yellow ellipse (c), which results in abnormal cortical thickness (g).

Recently, deep convolutional neural networks (CNNs) have demonstrated outstanding performances in a wide range of computer vision and image analysis applications. For example, fully convolutional networks (FCNs) [12, 13], as a natural extension of traditional CNNs, are now a common choice for semantic image segmentation in computer vision. FCNs train end-to-end segmentation models by directly optimizing intermediate feature layers. To compensate for the resolution loss induced by pooling layers, U-Net [12] introduces skip connections between their down-sampling and up-sampling paths, thus helping the up-sampling paths recover fine-grained information from the down-sampling layers. To date, many network architectures further incorporate the residual connection [14] or dense connection [15] to CNNs [11]. Although these previous networks can automatically learn effective feature hierarchies in a data-driven manner, most of them ignore the *prior* anatomical knowledge during the segmentation. A typical example can be seen in Fig. 1(c), in which some WM is missing, leading to abnormal cortical thickness in (g). Particularly, in the task of 6-month-old infant brain tissue segmentation, there are two kinds of critical *prior* knowledge, i.e.,

*(1) tissue contrast between CSF and GM is higher than that between GM and WM;*
*(2) cortical thickness is within a certain range.*

Capitalizing on these kinds of *prior* knowledge, we propose an Anatomy-guided and Densely-connected U-Net (ADU-Net) method for accurate segmentation of 6-month-old infant brain MRI. Specifically, we first train an initial ADU-Net to segment CSF and GM+WM, considering that CSF is relatively easier to be distinguished. Then, based on CSF segmentation and the second kind of *prior* knowledge, we estimate the outer cortical surface (i.e., CSF and GM boundary), and use it as guidance to train a cascaded ADU-Net for estimation of the inner cortical surface (i.e., GM and WM boundary). Based on the segmentation results, we parcellate infant brain into 83 ROIs [16] for computing volumetric measures of 6-month-old autistic infants and healthy controls, and finally perform early diagnosis.

## 2   Method

**Dataset and Preprocessing.** The T1w and T2w MR images of 18 de-identified infants were gathered from National Database for Autism Research (NDAR). All images were acquired at around 6 months of age on a Siemens 3T scanner. All scans were acquired while the infants were naturally sleeping and fitted with ear protection, with their heads secured in a vacuum-fixation device. T1w MR images were acquired with 160 sagittal slices using parameters: TR/TE = 2400/3.16 ms and voxel resolution = 111 mm$^3$. T2w MR images were obtained with 160 sagittal slices using parameters: TR/TE = 3200/499 ms and voxel resolution = $1 \times 1 \times 1$ mm$^3$. Note that the imaging protocol has been optimized to maximize tissue contrast [10]. For image preprocessing, T2-weighted images were linearly aligned onto their corresponding T1w MR images. Then, in-house tools were used to perform skull stripping, intensity inhomogeneity correction, and histogram matching for each MR modality.

Accurate manual segmentation is of importance for learning-based segmentation methods. However, due to the low contrast and huge number of voxels in 3D brain image, manual segmentation is extremely time-consuming. Hence, to generate reliable manual segmentations, we first take advantage of longitudinal follow-up 24-month-old scan of the same subject, with high tissue contrast, to generate an initial segmentation for 6-month-old scans by using a publicly available software iBEAT (http://www.nitrc. org/projects/ibeat/). This is based on the fact that, at term birth, the major sulci and gyri in the brain are already presented, and are generally preserved but only fine-tuned during early postnatal brain development [17]. Therefore, we can utilize the late-time-point longitudinal images (e.g., 24-month-old), which can be segmented with high accuracy by using existing segmentation tools, e.g., FreeSurfer [18], to guide the segmentation of early-time-point (e.g., 6-month-old) infant images. Based on the segmentation results generated by iBEAT, manual editing was further performed by an experienced neuroradiologist. For each subject, around 200,000 voxels (24% of total brain volume) were re-labeled. In this way, the potential bias from the automatic segmentations can be largely minimized and also the quality of manual segmentation can be ensured.

### 2.1    Anatomy-Guided Densely-Connected U-Net (ADU-Net)

To derive anatomic guidance from the outer surface, we need to first classify brain images into two classes, i.e., CSF and WM+GM. Inspired by the recent success of densely-connected networks [15] and U-Net [12] in medical image segmentation, we propose an anatomy-guided densely-connected U-Net architecture (shorted as ADU-Net) for the segmentation of 6-month-old infant brain images. The proposed network architecture is shown in Fig. 2, which includes a down-sampling path and an up-sampling path, going through seven dense blocks. Each dense block consists of three BN-ReLU-Conv-Dropout operations, in which each Conv includes 16 kernels and the dropout rate is 0.1. In the down-sampling path, between any two contiguous dense blocks, a transition down block (i.e., Conv-BN-ReLU followed by a max pooling layer) is included to reduce the feature map resolution and increase the receptive field. While in the up-sampling path, a transition up block, consisting of a transposed convolu-

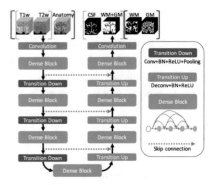

**Fig. 2.** Diagram of our architecture for segmentation. Input 1: T1w and T2w images for (CSF, WM+GM) segmentation to construct anatomy guidance; Input 2: T1w and T2w images and anatomy guidance for (WM, GM) segmentation.

tion, is included between any two contiguous dense blocks. It up-samples the feature maps from the preceding dense block. The up-sampled feature maps are then concatenated with the same level feature maps in the down-sampling path, and then input to the subsequent dense block. The final layer in the network is a convolution layer, followed by a softmax non-linearity to provide the per-class probability at each voxel. For all the convolutional layers, the kernel size is $3 \times 3 \times 3$ with stride size 1 and 0-padding.

**Network Implementation.** We randomly extract $32 \times 32 \times 32$ 3D patches from training images. The loss function is cross-entropy. The kernels are initialized by Xavier, and the bias are initially set to 0. We use SGD optimization strategy. The learning rate is 0.005 and multiplies by 0.1 after each epoch. Training and testing are performed on a NVIDIA Titan X GPU. Basically, training a ADU-Net takes around 96 h and in application stage, segmenting a 3D image requires 70–80 s.

**Anatomical Guidance Generation.** T1w and T2w MR images of training subjects and their corresponding manual segmentations are employed to train the network. To generate the anatomical guidance, as mentioned, we first train an initial ADU-Net to classify the brain images into two classes (i.e., CSF and WM + GM). Figure 3 shows the final estimated CSF and WM + GM segmentation maps for a testing image (in Fig. 1). Based on the segmentation results, it is straightforward to construct a signed distance function (i.e., a level set function) *with respect to* the boundary of GM/CSF, as shown in Fig. 3(c). Basically, the function value at each voxel is the shortest distance to its nearest point on the boundary of GM/CSF, taking positive value for voxels inside of WM+GM, and negative value for voxels outside of WM+GM. Therefore, the zero level set corresponds to the outer surface, as shown in Fig. 3(d).

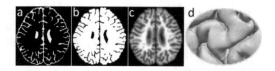

**Fig. 3.** (a) and (b) show the estimated CSF and WM + GM segmentations for a testing image in Fig. 1. (c) illustrates the signed distance function with respect to the outer surface shown in (d).

**Anatomy Guided Tissue Segmentation.** We further classify WM+GM into WM and GM separately by training a cascaded ADU-Net with the same architecture as in Fig. 2. It is worth noting that, besides using T1w and T2w MR images, the level set function (Fig. 3(c)) is also input to the network as an anatomical guidance for tissue segmentation. By recalling the inner cortical surface of the testing image in Fig. 1, we can see that the cortical thickness is incorrect due to the missing WM areas. With anatomical guidance from the outer surface in Fig. 3(d), more WM voxels are expected to keep cortical thickness in a reasonable range. From Fig. 1(d), it can be observed that the WM in the yellow ellipse is recovered, and thus the missing inner gyrus is also recovered (Fig. 1(f)) by the anatomical guidance, resulting in a reasonable cortical thickness in Fig. 1(h). Similarly, the topological errors, i.e., holes or handles, causing abnormal cortical thickness, can also be corrected.

## 2.2    ROI-Based Volumetric Measures as Imaging Biomarkers

To parcellate 6-month-old infant brain image into different regions of interest (ROIs), we employ a multi-atlas strategy. In particular, a total of 33 two-year-old subjects were employed as individual atlases (www.brain-development.org) [16]. Each atlas consists of a T1w MR image and a label image of 83 ROIs. We first employ FreeSurfer [18] to segment each T1w MR image into WM, GM, and CSF. Then, we register all atlases into each 6-month-old infant image space based on their segmentations using ANTs [19]. Finally, we employ majority voting to parcellate each 6-month-old infant brain into 83 ROIs.

We use leave-one-out cross-validation to evaluate the diagnosis performance. Basically, one subject is used for testing and the rest subjects are used for training. For the training data, we perform unpaired $t$-test on each of ROI measures between autistic and normal infant subjects, and find ROI measures with statistically significant differences as potential autism-related biomarkers. With these identified biomarkers, we perform early diagnosis of at-risk infants in the testing data. Following [20], we adopt random forest as the classifier for disease diagnosis. Instead of using just a single ratio of extra-axial fluid to the total cerebral volume as in [10], we employ the volumes of CSF, GM and WM in each identified autism-related ROI as features to construct the diagnosis model. This process is repeated until all subjects are used for testing.

## 3    Experimental Results and Conclusions

**Segmentation Results on 18 6-Month-Old Infant Subjects from NDAR.** We first make comparisons with state-of-the-art methods on 18 6-month-old infant subjects from NDAR. Among these competing methods, SegNet [21] has achieved promising results in natural images segmentation; U-Net [12] has achieved the best performance on ISBI 2012 EM challenge dataset, and Bui et al.'s method [22] has won the 1st prize in iSeg-2017. For all the methods, we perform the same 6-fold cross-validation. In each fold, we use 12, 3 and 3 subjects as training, validation, and testing datasets, respectively. The results on a testing subject by different methods are shown in Fig. 4. The estimated inner surface and cortical thickness by the proposed method are much more consistent with the ground truth from manual segmentation. We further quantitatively evaluate the results by using the Dice ratio and Modified (95th percentile) Hausdorff Distance (MHD). As shown in Table 1, our proposed method achieves a significantly better performance in terms of Dice ratio on WM and GM, and MHD on WM. We then apply our trained model on other 59 infants (29 sex- and age- balanced autistic/30 normal subjects) and further perform ROI-based volumetric measurement and diagnosis as detailed below.

**Diagnosis/Prediction on 6-Month-Old Infant Subjects from NDAR.** Based on the brain parcellation results, we then perform unpaired $t$-test on each ROI between 29 autistic and 30 normal subjects, who were later diagnosed at 24 months old. Among these two groups, we find statistically significant differences in many ROIs ($p$-value < 0.005), e.g., right precentral gyrus, left hippocampus, cingulate gyrus, cuneus, and lateral occipital cortex. Most of the autism-related regions identified by our method were also confirmed with previous reports on other children and young adults. For example, it was found in [23] that ASD is associated with decreased structural connectivity and functional connectivity in the lateral occipital cortex. This disruption may impair the integration of visual communication cues in ASD individuals, thereby impacting their social communications. In [4], it was found that children (7.5–12.5 years of age) with autism had larger left hippocampus than the controls. Although the ages of studied subjects are different, these results encourage us to explore the individualized diagnostic value of such potential biomarker regions. The last column of Table 1 provides the classification accuracy (ACC) and the area under the ROC curve (AUC) using leave-one-out cross-validation. We make a quick comparison with [20], in which they achieved 0.87 ACC and 0.96 AUC on *19* autistic and *19* normal infants. Besides, we also perform ROI-based volumetric measurements based on the segmentations achieved by other methods in Table 1. We found that much less significant regions were identified based on their segmentations. The corresponding ACC and AUC by other methods are also provided in the last column of Table 1. It is clear that our method achieves the highest diagnosis/prediction accuracy, which also indicates a high segmentation accuracy achieved by the proposed ADU-Net.

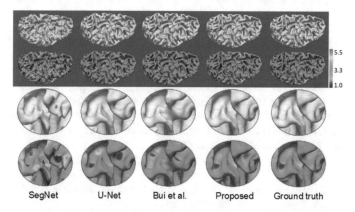

**Fig. 4.** Comparison with state-of-the-art methods on 18 6-month-old infant subjects from NDAR. The first and second rows show the inner surface and corresponding cortical thickness, respectively, with the zoomed views shown in the third and fourth rows. From left to right: results by SegNet [21], U-Net [12], Bui et al.'s method [22], proposed method, and ground truth from manual segmentation. Color bar indicates the thickness in mm.

**Table 1.** Segmentation accuracy on 18 6-month-old infant images, and diagnosis accuracy on other 59 infant subjects (29 autistics and 30 normal controls) from NDAR. The bold values indicate that our proposed method is significantly better than others with $p$-value < 0.005.

| Method | Dice ratio (%) | | | MHD (mm) | | | Accuracy & AUC |
|---|---|---|---|---|---|---|---|
| | CSF | GM | WM | CSF | GM | WM | |
| SegNet [21] | 89.5 ± 1.41 | 75.5 ± 0.96 | 80.6 ± 1.23 | 8.81 ± 1.11 | 7.80 ± 1.49 | 9.11 ± 1.10 | 0.67, 0.43 |
| U-Net [12] | 92.7 ± 0.89 | 89.0 ± 0.56 | 91.9 ± 0.43 | 6.68 ± 0.80 | 6.62 ± 0.74 | 6.79 ± 0.61 | 0.78, 0.53 |
| Bui et al. [22] | 92.3 ± 0.85 | 88.5 ± 0.57 | 91.6 ± 0.44 | 6.59 ± 0.86 | 6.97 ± 1.22 | 8.08 ± 2.09 | 0.87, 0.51 |
| Proposed | 95.8 ± 0.58 | **92.3 ± 0.54** | **93.3 ± 0.43** | 5.59 ± 0.72 | 5.84 ± 0.88 | **5.05 ± 1.10** | **0.90, 0.98** |

To conclude, for the first time, this paper presents a novel volume-based analysis on 6-month-old infant subjects at-risk of autism. We first proposed an anatomy-guided and densely-connected convolutional neural network for accurate tissue segmentation. Based on accurate segmentations, we then perform brain parcellation and statistical analysis to identify significantly different regions between autistic and normal infant subjects for final classification. Comparisons with the state-of-the-art methods demonstrate the advantages of our proposed method in terms of both segmentation accuracy and diagnosis accuracy. Our future work will include improvement of diagnosis accuracy by using surface-based features, e.g., cortical thickness and surface area, and further validation on more infant subjects.

# References

1. http://www.nimh.nih.gov/health/topics/autism-spectrum-disorders-asd/index.shtml
2. Damiano, C.R., et al.: Future directions for research in autism spectrum disorders. J. Clin. Child Adolesc. Psychol. **43**, 828–843 (2014)
3. Yahata, N., et al.: A small number of abnormal brain connections predicts adult autism spectrum disorder. Nat. Commun. **7**, 11254 (2016)
4. Schumann, C.M., et al.: The amygdala is enlarged in children but not adolescents with autism; the hippocampus is enlarged at all ages. J. Neurosci. **24**, 6392–6401 (2004)
5. Greimel, E., et al.: Changes in grey matter development in autism spectrum disorder. Brain Struct. Funct. **218**, 929–942 (2013)
6. Thakkar, K.N., et al.: Response monitoring, repetitive behaviour and anterior cingulate abnormalities in autism spectrum disorders (ASD). Brain **131**, 2464–2478 (2008)
7. Li, G., et al.: Computational neuroanatomy of baby brains: A review. Neuroimage (2018)
8. Knickmeyer, R.C., et al.: A structural MRI study of human brain development from birth to 2 years. J. Neurosci. **28**, 12176–12182 (2008)
9. Anagnostou, E., Taylor, M.J.: Review of neuroimaging in autism spectrum disorders: what have we learned and where we go from here. Mol. Autism **2**, 4 (2011)
10. Hazlett, H.C., et al.: Brain volume findings in six month old infants at high familial risk for autism. Am. J. Psychiatry **169**, 601–608 (2012)
11. Jégou, S., et al.: The one hundred layers tiramisu: fully convolutional densenets for semantic segmentation. arXiv:1611.09326 (2016)
12. Ronneberger, O., Fischer, P., Brox, T.: U-Net: convolutional networks for biomedical image segmentation. In: Navab, N., Hornegger, J., Wells, William M., Frangi, Alejandro F. (eds.) MICCAI 2015. LNCS, vol. 9351, pp. 234–241. Springer, Cham (2015). https://doi.org/10. 1007/978-3-319-24574-4_28
13. Long, J., et al.: Fully convolutional networks for semantic segmentation. In: CVPR, pp. 3431–3440 (2015)
14. He, K., et al.: Deep residual learning for image recognition. In: CVPR, pp. 770–778 (2016)
15. Huang, G., et al.: Densely connected convolutional networks. In: CVPR, pp. 2261–2269 (2017)
16. Gousias, I.S., et al.: Automatic segmentation of brain MRIs of 2-year-olds into 83 regions of interest. Neuroimage **40**, 672–684 (2008)
17. Chi, J., et al.: Gyral development of the human brain. Ann. Neurol. **1**, 86–93 (1977)
18. Fischl, B.: FreeSurfer. Neuroimage **62**, 774–781 (2012)

19. Avants, B.B., et al.: Symmetric diffeomorphic image registration with cross-correlation: evaluating automated labeling of elderly and neurodegenerative brain. Med. Image Anal. **12**, 26–41 (2008)
20. Mostapha, M., Casanova, M.F., Gimel'farb, G., El-Baz, A.: Towards non-invasive image-based early diagnosis of autism. In: Navab, N., Hornegger, J., Wells, W.M., Frangi, A.F. (eds.) MICCAI 2015. LNCS, vol. 9350, pp. 160–168. Springer, Cham (2015). https://doi.org/10.1007/978-3-319-24571-3_20
21. Badrinarayanan, V., et al.: SegNet: a deep convolutional encoder-decoder architecture for image segmentation. arXiv:1511.00561 (2015)
22. Bui, T.D., et al.: 3D densely convolutional networks for volumetric segmentation. arXiv: 1709.03199 (2017)
23. Jung, M., et al.: Decreased structural connectivity and resting-state brain activity in the lateral occipital cortex is associated with social communication deficits in boys with autism spectrum disorder. Neuroimage (2017)

# A Tetrahedron-Based Heat Flux Signature for Cortical Thickness Morphometry Analysis

Yonghui Fan[1], Gang Wang[1,2], Natasha Lepore[3], and Yalin Wang[1(✉)]

[1] School of Computing, Informatics, and Decision Systems Engineering,
Arizona State University, Tempe, AZ, USA
ylwang@asu.edu

[2] School of Information and Electrical Engineering, Ludong University, Yantai, China

[3] CIBORG Lab, Department of Radiology, Children's Hospital Los Angeles,
Los Angeles, CA, USA

**Abstract.** Cortical thickness analysis of brain magnetic resonance images is an important technique in neuroimaging research. There are two main computational paradigms, namely voxel-based and surface-based methods. Recently, a tetrahedron-based volumetric morphometry (TBVM) approach involving proper discretization methods was proposed. The multi-scale and physics-based geometric features generated through such methods may yield stronger statistical power. However, several challenges, such as the lack of well-defined thickness statistics and the difficulty in filling tetrahedrons into the thin and curvy cortex structure, impede the broad application of TBVM. In this paper, we present a universal cortical thickness morphometry analysis approach called *tetrahedron-based Heat Flux Signature* (tHFS) to address these challenges. We define the tetrahedron-based weak form heat equation and Laplace-Beltrami eigen decomposition and give an explicit FEM-based discretization formulation to compute the tHFS. We further show a tHFS metric space with which cortical morphometric distances can be directly visualized. Additionally, we optimize the cortical tetrahedral mesh generation pipeline and fill dense high-quality tetrahedra in the grey matters without sacrificing data integrity. Compared with existing cortical thickness analysis approaches, our experimental results of distinguishing among Alzheimer's disease (AD), cognitively normal (CN) and mild cognitive impairment (MCI) subjects shows that tHFS yields a more accurate representation of cortical thickness morphometry. The tHFS metric experiment provides a more vivid visualization of tHFS's power in separating different clinical groups.

**Keywords:** Cortical thickness · Tetrahedron-based morphometry

## 1 Introduction

Human cortical thickness analysis performed *in vivo* via magnetic resonance imaging (MRI) emerged quickly as a significant neuroimaging biomarker due to

© Springer Nature Switzerland AG 2018
A. F. Frangi et al. (Eds.): MICCAI 2018, LNCS 11072, pp. 420–428, 2018.
https://doi.org/10.1007/978-3-030-00931-1_48

its close association with neurodegenerative and psychiatric diseases [3,6,14]. Cortical thickness morphometry analysis consists in the accurate measurement of the quantative traits of cortical thickness and in the design of reliable morphological signatures from MRI in an operative way [8,9,11].

Conventional voxel-based morphometry (VBM) methods [1,2,5] have been surpassed in recent years by surface-based morphometry (SBM) approaches [6,9] because of the newly emerged shape modeling methods. Tetrahedron-based volumetric morphometry (TBVM) approach is a representative new method for cortical analyses [17,18]. It deals with the measurement of brain structure and changes on tetrahedral mesh. When measuring the variations of bioinformation in the normal direction, SBM implements a localized node-to-node registration with predefined constraints in the tangent space. The Euclidean distance between corresponding nodes is commonly used as a metric, as shown in Fig. 1(a). Compared with SBM, TBVM approach allows for more degrees of freedom, including both tangent space and normal space, hence making it an intuitive choice to model cortical thickness. However, the performance improvement and further application development of TBVM approaches are impeded by several existing challenges: (1) the lack of a well-explained explicit morphometry signature formulation and of an accurate discretization formulation defined in the tetrahedron domain; (2) the difficulty in automatically generating high-quality tetrahedral meshes for the cortex.

In this paper, we propose a universal thickness morphometry analysis approach called *tetrahedral-based Heat Flux Signature* (tHFS) to address these challenges and increase statistical power for cortical thickness estimation and disease severity classification. We firstly define the weak form heat equation and the Laplace-Beltrami eigen decomposition on tetrahedral meshes. Then, two different discretization methods based on FEM are derived for computing pial-white point pairs and the LBO spectrum which are used for defining the tHFS. The general computational framework containing the major steps in the method is illustrated in Fig. 2. We further derive a tHFS distance as a metric for measuring disease severity in the form of a numerical distance between patients from the same or different disease groups. One important usage of tHFS distance is to visualize the disease severity by mapping the mutual distances to a 2D plane.

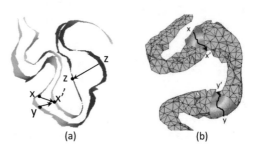

**Fig. 1.** Different modeling methods in mophometry analysis. (a) Surface-based mophometry approach. (b) Tetrahedron-based mophometry approach.

Our experiments focus on distinguishing different brain atrophy levels in Alzheimer's disease. We selected 368 MRI images of initial-visit patients from Alzheimer's Disease Neuroimaging Initiative (ADNI-2) [13] dataset. Experiments

include cerebral cortex tetrahedral mesh generation, clinical group classification
and a tHFS distance demonstration.

## 2   Methods

### 2.1   Cortical Tetrahedral Mesh Generation

Surfaces generated by FreeSurfer are usually error free, but genus-preserving
down-sampling and the process of combining pial and white surfaces always create self-intersections. The folded gyri and sulci also make it harder to generate
multi-layer tetrahedral meshes. Repeated smoothing is a commonly used method
to remove geometric errors [17], but may cause data integrity loss. Our solution
here is a minimally invasive surgery over intersection regions. A small neighborhood of the error nodes is selected and moved along the inward normal direction
by a small step size. This process is repeated until the intersection is removed.
Local smoothing is applied to the selected nodes. Hence, all modified nodes are
marked to differentiate them from the unaltered nodes. This protocol sacrifices
a small number of vertices to maintain the integrity of the whole mesh. The
tetrahedral mesh is then generated by Tetgen [16].

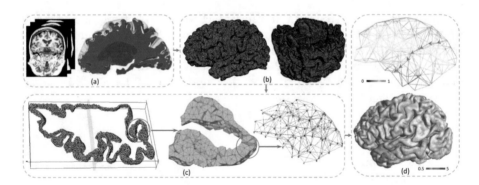

**Fig. 2.** General computation framework. (a) Left: MRI images. Right: pial and white
surfaces. (b) Left: cortical tetrahedral mesh. Right: cut plane of tetrahedral mesh. (c)
Left: one slice cut from the red band part of (b). Middle: zooming in the part of the
entorhinal cortex marked by the red circle on the left. Right: a further zoomed-in
picture and its detailed inner structure. (d) Top: temperature distribution and heat
flux trajectory. Bottom: the cortical thickness map overlaid in color scale.

### 2.2   Tetrahedron-Based Heat Equation

The heat flux of vertex $m$ in direction $s$ per unit time is defined by the
weighted temperature difference between $m$ and the neighboring vertex along
$s$: $f_s^m = -k(\frac{\partial T}{\partial s})$. On tetrahedral meshes, the discretized heat flux is defined by

the weighted Newton's law of cooling: $-k(T_s - T_m)$. Here $k$ is heat conductivity. The minus sign refers to the inversed direction of the temperature gradient. Therefore, the heat flux computation is converted to heat equation problem.

Assuming that the temperature distribution on boundary surfaces is prescribed and static, we apply Galerkin's method to convert the continuous heat equation problem to a discrete weak form problem with Dirichlet boundary conditions [4,15]. We define the discrete harmonic energy matrix $S$ as [18]:

$$
S = \begin{cases} \frac{1}{12}\sum_{v_j \subseteq N(v_i)} \sum_{T_l \subseteq N(v_i, v_j)} L^{(i,j)} \cot\theta_l^{(i,j)}, & if \quad v_j \subseteq N(v_i) \\ 0, & otherwise \end{cases} \tag{1}
$$

where $N(v_i)$ is the set of the neighboring vertices of vertex $v_i$; $N(v_i, v_k)$ is the set of the neighboring tetrahedrons of edge$(v_i, v_k)$; $L^{(i,j)}$ is the length of the opposite edge to edge$(v_i, v_j)$ in tetrahedron $T_l$; and $\theta_l^{i,j}$ is the dihedral angle of edge$(v_i, v_j)$ in tetrahedron $T_l$. In Fig. 3(a), the neighboring vertices of 1 are represented as green points. Neighboring edges of 1 are in yellow. Two tetrahedrons shared by $(1, 2)$ are in green. Figure 3(b) shows the dihedral angle and opposite edge of $(1, 2)$.

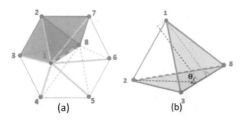

(a)     (b)

**Fig. 3.** Numerical discretization. (a) Neighboring relationship of vertex 1. (b) Dihedral angle $\theta$ and opposite edge of edge(1, 2).

Extracting two matrices from $S$: (1) $W_{ii}$: a square matrix indicates inner nodes $i$; (2)$W_{ib}$: a matrix with inner node $i$ and boundary node $b$. Under the Dirichlet boundary condition, the interior temperature distribution $T$ is: $L_{tet}T = T_{\partial\Omega}$, $L_{tet} = diag(\sum_i W_{ii}) + diag(\sum_i W_{ib}) - W_{ii}$. $T_{\partial\Omega}$ is the temperature on the boundary. $L_{tet}$ is the tetrahedron-based Laplace-Beltrami operator (LBO).

A heat flow trajectory is defined as the path from one point on the pial surface to one point on white surface in the direction of the temperature's gradient descent. It is illustrated as the top figure of Fig. 2(d). Our method is to search for the biggest temperature flux among all neighboring vertices and then move to that vertex as a new starting point. We repeat this process until the new target point falls on the white surface. All the passing edges are the heat trajectory. Here, our definition of cortical thickness is the sum of all the passing edges. Starting vertex $x$ and destination vertex $y$ form a pial-white vertices pair$(x, y)$, which is used in defining the tHFS.

## 2.3   Tetrahedron-Based Laplace-Beltrami Eigen Decomposition

The Eigencompanion Method is employed to define the tHFS. The eigen pairs are computed by solving the tetrahedron-based Laplace-Beltrami eigen problem over the tetrahedral mesh: $L_{tet}\phi = \lambda\phi$. Here, the tetrahedron-based LBO is defined under Neumann boundary conditions: $L_{tet} = B^{-1}(D - S)$, where $S$ is defined in Eq. 1 and $D$ is a diagonal matrix defined as the sum of each row in $S$. The heat capacity matrix $B$ is:

$$
B_{i,j} = \begin{cases} \sum_{T_l \subseteq N(v_i)} \frac{|V_l|}{10} + \sum_{k \subseteq N(v_i)} \sum_{T_l \subseteq N(v_i, v_k)} \frac{|V_l|}{20}, & if \quad i = j \\ \sum_{T_l \subseteq N(v_i, v_j)} \frac{|V_l|}{20}, & if \quad v_j \subseteq N(v_i) \\ 0, & otherwise \end{cases} \tag{2}
$$

where $N(v_i)$ is the set of neighboring tetrahedrons of vertex $i$; $N(v_i, v_k)$ is the set of neighboring tetrahedrons of edge(i, k); $V_l$ is tetrahedron $l$'s volume. Solving the eigen problem of $L_{tet}$, we get eigenvalue and eigenvector pairs $(\lambda_n, \phi_n)$.

## 2.4   Heat Flux Signature

With the pial-white vertices pairs $(x, y)$ calculated in Sect. 2.2 and the eigen pairs $(\lambda_n, \phi_n)$, the tHFS is defined as:

$$
tHFS(x, y|t_0 + \delta t) = \sum_{n}^{N} \frac{\phi_n(x)' \phi_n(y)}{e^{\lambda_n(t_0 + \delta t)}} \tag{3}
$$

where $t_0$ is an initial constant, usually 0 or 1 and $\delta t$ is step size defined by weighted maximum eigenvalue: $\delta t = \frac{2}{0.8 \times \lambda_{max}}$.

tHFS is a matrix with the number of vertices as the horizontal dimension and the number of steps as the vertical dimension. If considering the variable $\delta t$ is a time interval and $t_0$ is the initial time, HFS is a time series matrix. The $i^{th}$ row stands for the heat flux changes along $(x, y)^i$. Stochastic Coordinate Coding (SCC) [12] is then used to organize the features for binary classification.

## 2.5   tHFS Distance

Theoretically, an effective disease severity quantitation method is also convertible to a metric which provides not only a numerical way in describing the relationships of groups and subjects but also a visualization tool via Multi-dimensional Scaling (MDS). Here we introduce a metric derived from tHFS, called tHFS distance. The tHFS distance aims at measuring static tetrahedron-based cortical thickness morphometry relationships. As a static metric, setting $\delta t = 0$ and $t = 1$, the tHFS distance between two samples $M$ and $M'$ is:

$$
d_n(M, \tilde{M}) \approx \inf_{\gamma \in \Gamma(n)} \max \left\{ \frac{1}{Volume(M)} \sum_{i=1}^{N} d(v_i, \tilde{M}) A_i, \frac{1}{Volume(\tilde{M})} \sum_{i=1}^{N} d(M, \tilde{v}_i) \tilde{A}_i \right\} \tag{4}
$$

where $N$ and $N'$ are numbers of vertices in temperature pairs; The eigen spaces of different meshes may be matched by adjusting signs of eigenvectors [10], so we defined $\gamma$ as the sign of eigenvectors; $\Gamma(n)$ is the set of signs. $A_i$ and $\tilde{A}_i$ are the weighted volume sums of surrounding tetrahedrons: $A = \frac{1}{4}\sum_{v_i \in T_l} volume(T_l)$. The vertex to mesh distances $d(x, \tilde{M})$ and $d(M, y)$ are defined as:

$$d(x, \tilde{M}) = \min_{y \in \tilde{M}} \left\| tHFS_M^{\Phi(n)}(x) - tHFS_{\tilde{M}}^{\tilde{\Phi}_\gamma(n)}(y) \right\|_2, \forall x \in M \tag{5}$$

$$d(M, y) = \min_{x \in M} \left\| tHFS_M^{\Phi(n)}(x) - tHFS_{\tilde{M}}^{\tilde{\Phi}_\gamma(n)}(y) \right\|_2, \forall y \in \tilde{M} \tag{6}$$

where $\tilde{\Phi}_\gamma(n)$ is an othogonal basis adjusted by $\Gamma(n)$. Equation 4 provides an optimized $\Gamma(n)$ so that two orthogonal basses are closest. We proved that tHFS distance has the properties of a metric.

## 3    Experimental Results

In experiments, we focus on human brain atrophy severity analysis of Alzheimer's disease patients. Our data contains 94 Alzheimer's disease (AD) patients, 137 normal controls (CN) and 137 mild cognitive impairment (MCI) patients from the Alzheimer's Disease Neuroimaging Initiative phase 2 (ADNI-2) baseline initial-visit dataset. All subjects are scanned using a 3 T scanner. The acquired MRI images are then pre-processed by FreeSurfer[1] to segment pial and white matter surfaces. The pial and white surfaces are downsampled to 120000 faces for the mesh generation. As a universal tool, tHFS can be used to analysis other abnormalities in diseases such as schizophrenia.

### 3.1    Cortical Tetrahedral Mesh Generation

Figures 1(b) and 2(b, c) illustrate the generated tetrahedral mesh. 96.88% subjects yielded high quality cortical tetrahedral mesh after the first iteration, and 100% after the second iteration. Each mesh contains 5.6–5.8 million tetrahedrons and more than 1.5 million vertices.

### 3.2    Cortex Thickness

Although cortical thickness measurement is not our main purpose, we illustrate one thickness map result at the bottom of Fig. 1(d). The color is rendered on the pial surface by scaling with the thickness.

### 3.3    Disease Severity Classification

In experiment settings, 10-fold cross validation is used for avoiding data selection bias. AdaBoost [7] is the classifier. 10-steps tHFS with 5 and 10 pairs of smallest eigenfunctions (removing the first smallest one) are used when comparing with the lumped volumetric method. 5 pairs are used to compare with FreeSurfer thickness method.

---

[1] https://surfer.nmr.mgh.harvard.edu/.

**Table 1.** Classification results comparing with other methods.

| Group | | 5: Lumped | 5: tHFS | 10: Lumped | 10: tHFS | FreeSurfer thickness |
|---|---|---|---|---|---|---|
| AD-CN | ACC | 0.950 | 0.974 | 0.957 | **0.978** | 0.762 |
| | SEN | 0.953 | 0.963 | 0.960 | **0.978** | 0.770 |
| | SPE | 0.940 | **0.989** | 0.948 | 0.968 | 0.785 |
| AD-MCI | ACC | 0.926 | 0.954 | 0.941 | **0.964** | 0.674 |
| | SEN | 0.939 | 0.960 | 0.953 | **0.967** | 0.737 |
| | SPE | 0.905 | 0.894 | 0.922 | **0.960** | 0.776 |
| CN-MCI | ACC | 0.922 | **0.966** | 0.951 | 0.966 | 0.534 |
| | SEN | 0.932 | 0.972 | 0.960 | **0.973** | 0.682 |
| | SPE | 0.905 | **0.959** | 0.940 | **0.948** | 0.716 |

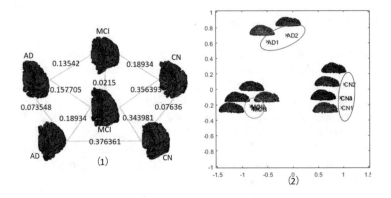

**Fig. 4.** tHFS distances. (1) A mutual distance example. (2) MDS map.

For comparisons, we select the surface-based thickness morphometry method used in FreeSurfer [6] and a lumped tetrahedron-based method [17]. In [17], the lumped discrete volumetric LBO is defined as: $L_p = D^{-1}S$, where $D = diag(d_1, \ldots, d_n)$. $d_n$ is the weighted volume sum of all the tetrahedrons sharing vertex $i$: $d_n = \sum_{T_l \in N(i)} Volume(T_l)/4$.

The classification results are shown in Table 1. Three performance measures, Accuracy (AC), Sensitivity (SEN) and Specificity (SPE), were evaluated. Considering that the number of eigen pairs may affect the results, here we demonstrate the results of 5 eigen pairs and 10 eigen pairs. The accuracy performance converges at about 10 pairs, so the results of 10 pairs are taken as the general selection. As a thickness morphometry signature, tHFS boosts statistical power to distinguish disease severity compared with pure thickness information. This is due to the more precise discretization and better mesh quality.

### 3.4   tHFS Distance

In this experiment, we randomly selected 2 AD patients, 4 CN patients and 4 MCI patients to calculate the mutual tHFS distances. The result is a distance matrix. In Fig. 4(1) we demonstrate a mutual distance example of 2 AD, 2 CN and 2 MCI. We calculate the MDS of the mutual distance matrix with its covariance matrix. The spectrum of the covariance matrices shows the feasibility to find a good configuration of distances using largest two eigenvalues, as the others converge to zero quickly. The MDS map is shown in Fig. 4. Clearly we see that three groups

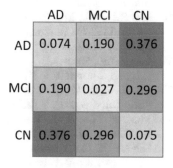

**Fig. 5.** Average tHFS distance.

are centered in different regions, while subjects within the same group have much smaller distances than subjects across different groups. The average group distance is defined by averaging the distances of subjects belonging to the same or different groups. The average distance chart is illustrated in Fig. 5. Both the MDS map and average group distance chart indicate that tHFS is powerful in distinguishing different disease severity groups.

Here we have developed an accurate, robust TBVM approach for cortical thickness analysis. Besides validating it with other disease datasets, we will continue refining it to advance computational anatomy research.

**Acknowledgements.** The research is supported in part by NIH (R21AG049216, RF1AG051710, R01EB025032), NSF (IIS-1421165) and NSFC (61772253).

## References

1. Ashburner, J., Friston, K.J.: Voxel-based morphometrythe methods. Neuroimage. **11**(6), 805–821 (2000)
2. Baron, J., et al.: In vivo mapping of gray matter loss with voxel-based morphometry in mild alzheimer's disease. Neuroimage **14**(2), 298–309 (2001)
3. Das, S.R., Avants, B.B., Grossman, M., Gee, J.C.: Registration based cortical thickness measurement. Neuroimage **45**(3), 867–879 (2009)
4. Delkhosh, M., Delkhosh, M., Jamali, M.: Greens function and its applications. J. Basic. Appl. Sci. Res. **2**(9), 8865–76 (2012)
5. Draganski, B., Gaser, C., Busch, V., Schuierer, G., Bogdahn, U., May, A.: Neuroplasticity: changes in grey matter induced by training. Nature **427**(6972), 311 (2004)
6. Fischl, B., Dale, A.M.: Measuring the thickness of the human cerebral cortex from magnetic resonance images. Proc. Natl. Acad. Sci. **97**(20), 11050–11055 (2000)
7. Freund, Y., Schapire, R., Abe, N.: A short introduction to boosting. J. Japan. Soc. Artif. Intell. **14**(771–780), 1612 (1999)
8. Hutton, C., Draganski, B., Ashburner, J., Weiskopf, N.: A comparison between voxel-based cortical thickness and voxel-based morphometry in normal aging. Neuroimage **48**(2), 371–380 (2009)

9. Jones, S.E., Buchbinder, B.R., Aharon, I.: Three-dimensional mapping of cortical thickness using Laplace's equation. Hum. Brain Mapp. **11**(1), 12–32 (2000)
10. Lai, R., et al.: Metric-induced optimal embedding for intrinsic 3D shape analysis. In: 2010 IEEE Conference on Computer Vision and Pattern Recognition (CVPR), pp. 2871–2878. IEEE (2010)
11. Lerch, J.P., Evans, A.C.: Cortical thickness analysis examined through power analysis and a population simulation. Neuroimage **24**(1), 163–173 (2005)
12. Liu, J., Ji, S., Ye, J., et al.: SLEP: sparse learning with efficient projections. Arizona State Univ. **6**(491), 7 (2009)
13. Mueller, S.G., et al.: The Alzheimer's disease neuroimaging initiative. Neuroimaging Clin. **15**(4), 869–877 (2005)
14. Schwarz, C.G.: A large-scale comparison of cortical thickness and volume methods for measuring Alzheimer's disease severity. NeuroImage Clin. **11**, 802–812 (2016)
15. Shi, Y., Chan, C.H.: Multilevel green's function interpolation method for analysis of 3-D frequency selective structures using volume/surface integral equation. JOSA A **27**(2), 308–318 (2010)
16. Si, H.: TetGen, a Delaunay-based quality tetrahedral mesh generator. ACM Trans. Math. Softw. (TOMS) **41**(2), 11 (2015)
17. Wang, G., Wang, Y.: Towards a holistic cortical thickness descriptor: heat kernel-based grey matter morphology signatures. Neuroimage **147**, 360–380 (2017)
18. Wang, Y., Gu, X., Yau, S.T., et al.: Volumetric harmonic map. Commun. Inf. Syst. **3**(3), 191–202 (2003)

# Graph of Brain Structures Grading for Early Detection of Alzheimer's Disease

Kilian Hett[1,2(✉)], Vinh-Thong Ta[1,2,3], José V. Manjón[4], Pierrick Coupé[1,2], and the Alzheimer's Disease Neuroimaging Initiative

[1] Univ. Bordeaux, LaBRI, UMR 5800, PICTURA, 33400 Talence, France
kilian.hett@labri.fr
[2] CNRS, LaBRI, UMR 5800, PICTURA, 33400 Talence, France
[3] Bordeaux INP, LaBRI, UMR 5800, PICTURA, 33600 Pessac, France
[4] Universitat Politècnia de València, ITACA, 46022 Valencia, Spain

**Abstract.** Alzheimer's disease is the most common dementia leading to an irreversible neurodegenerative process. To date, subject revealed advanced brain structural alterations when the diagnosis is established. Therefore, an earlier diagnosis of this dementia is crucial although it is a challenging task. Recently, many studies have proposed biomarkers to perform early detection of Alzheimer's disease. Some of them have proposed methods based on inter-subject similarity while other approaches have investigated framework using intra-subject variability. In this work, we propose a novel framework combining both approaches within an efficient graph of brain structures grading. Subsequently, we demonstrate the competitive performance of the proposed method compared to state-of-the-art methods.

**Keywords:** Patch-based grading · Intra-subject variability
Inter-subject similarity · Alzheimer's disease classification
Mild cognitive impairment

## 1 Introduction

Alzheimer's disease (AD) is the most common dementia leading to a neurodegenerative process causing mental dysfunctions. According to the world health organization, the number of patients having AD will double in 20 years. Neuroimaging studies performed on AD subjects revealed that brain structural alterations are advanced when diagnosis is established. Indeed, the clinical symptoms of AD is

Data used in preparation of this article were obtained from the Alzheimer's Disease Neuroimaging Initiative (ADNI) database (adni.loni.usc.edu). As such, the investigators within the ADNI contributed to the design and implementation of ADNI and/or provided data but did not participate in analysis or writing of this report. A complete listing of ADNI investigators can be found at: http://adni.loni.usc.edu/wp-content/uploads/how_to_apply/ADNI_Acknowledgement_List.pdf.

A. F. Frangi et al. (Eds.): MICCAI 2018, LNCS 11072, pp. 429–436, 2018.
https://doi.org/10.1007/978-3-030-00931-1_49

preceded by brain changes that stress the need to develop new biomarkers to detect the first stages of the disease. The development of such biomarkers can make easier the design of clinical trials and therefore accelerate the development of new therapies.

Over the past decades, the improvement of magnetic resonance imaging (MRI) has led to the development of new imaging biomarkers [2]. Many works developed biomarkers based on inter-subject similarities to detect anatomical alterations by using group-based comparison (*e.g.*, patients vs. normal controls). Some of them are based on regions of interest (ROI) to capture brain structural alterations at a large scale of analysis. The alterations of specific structures such as the cerebral cortex and hippocampus (HIPP) are usually captured with volume, shape, or cortical thickness (CT) measurements [17]. Other approaches proposed to study the inter-subject similarity between individuals from the same group at a voxel scale. Such methods commonly use voxel-based morphometry (VBM). VBM-based studies showed that the medial temporal lobe (MTL) is a key area to detect the first manifestations of AD [17]. Recently, more advanced methods have been designed to improve computer-aided diagnosis [2]. Among them, patch-based grading (PBG) framework [3] proposed to better analyze inter-subject similarities. PBG demonstrated state-of-the-art results for AD diagnosis and prognosis [3,5,13].

Beside inter-subject similarity approaches, other methods proposed to capture the correlation of brain structures alterations within subjects. Indeed, although similarity-based biomarkers provide helpful tools to detect the first signs of AD, the structural alterations leading to cognitive decline are not homogeneous within a given subject. Such biomarkers assumed that the structural changes caused by the disease may not occur at isolated areas but in several inter-related regions. Therefore, intra-subject variability features provide relevant information. Some methods proposed to capture the relationship of spread cortical atrophy with a network-based framework [16]. Other approaches estimate inter-regional correlation of brain tissues volumes [18]. Recently, convolutional neural network (CNN) have been used to capture relationship between anatomical structures volumes [11]. Finally, some works showed that patch-based strategy can be used to model intra-subject brain alteration [7,12].

The main contribution of this work is the development of a novel representation based on a graph of brain structures grading (GBSG) combining inter-subject pattern similarity and intra-subject variability features to better capture AD signature. First, inter-subject similarities are captured using patch-based grading framework applied over the entire brain. Second, intra-subject variabilities are modeled by a graph representation. In our experiments, we compare the performance of intra-subject variability features (*i.e.*, the edges of our graph) with inter-subject pattern similarity features (*i.e.*, the vertices). Moreover, we demonstrate the capability of intra-subject variability features to early detect AD and show that the combination of both features improves AD prognosis. Finally, we present competitive results of our new method compared to state-of-the-art approaches.

**Table 1.** Description of the ADNI dataset used in this work.

| Characteristic/group | CN | sMCI | pMCI | AD |
|---|---|---|---|---|
| Number of subjects | 228 | 100 | 164 | 191 |
| Ages (years) | $75.8 \pm 5.0$ | $75.3 \pm 7.2$ | $74.2 \pm 6.64$ | $75.26 \pm 7.4$ |
| Sex (M/F) | 117/109 | 150/73 | 101/64 | 98/88 |
| MMSE | $29.05 \pm 0.9$ | $27.1 \pm 2.5$ | $26.3 \pm 2.0$ | $22.8 \pm 2.9$ |

## 2   Materials and Methods

### 2.1   Dataset

Data used in this work were obtained from the Alzheimer's Disease Neuroimaging Initiative (ADNI) dataset[1]. We use all the baseline T1-weighted (T1w) MRI of the ADNI1 phase. This dataset includes AD patients, subjects with mild cognitive impairment (MCI) and cognitive normal (CN) subjects (see Table 1). MCI is a presymptomatic phase of AD composed of subjects who have abnormal memory dysfunctions. In our experiments we consider two groups of MCI. The first group is composed of patients having stable MCI (sMCI) and the second one is composed with patients having MCI symptoms at the baseline and converted to AD into the following 36 months. This group is named progressive MCI (pMCI).

### 2.2   Preprocessing

The data are preprocessed using the following steps: (1) denoising using a spatially adaptive non-local means filter [8], (2) inhomogeneity correction using N4 method [14], (3) low-dimensional non-linear registration to MNI152 space using ANTS software [1], (4) intensity standardization, (5) segmentation using a non-local label fusion [4] and (6) systematic error corrections [15]. The patch-based multi-template segmentation was performed using 35 images manually labeled by Neuromorphometrics, Inc.[2] using the brain-COLOR labeling protocol composed of 134 structures.

### 2.3   Computation of Patch-Based Grading Biomarkers

To capture alterations caused by AD, we use the recently developed patch-based grading framework [3]. PBG framework provides at each voxel a grade between $-1$ and $1$ related to the alteration severity. The grading value $g$ at $x_i$ is defined as:

$$g_{x_i} = \frac{\sum_{t_j \in K_i} w(P_{x_i}, P_{t_j}) p_t}{\sum_{t_j \in K_i} w(P_{x_i}, P_{t_j})}, \tag{1}$$

---

[1] http://adni.loni.ucla.edu.
[2] http://Neuromorphometrics.com.

where $P_{x_i}$ and $P_{t_j}$ represent the patches surrounding the voxel $i$ of the test subject image $x$ and the voxel $j$ of the template image $t$, respectively. The template $t$ comes from a training library composed of CN subjects and AD patients. $p_t$ is the pathological status set to $-1$ for patches extracted from AD patients and to 1 for those extracted from CN subjects. $K_i$ is a set of the most similar $P_{t_j}$ patches to $P_{x_i}$ found in the training library. The anatomical similarity between the test subject $x$ and the training library is estimated by a weight function $w(P_{x_i}, P_{t_j}) = \exp(-||P_{x_i} - P_{t_j}||_2^2/(h^2 + \epsilon))$, where $h = \min_{t_j} ||P_{x_i} - P_{t_j}||_2^2$ and $\epsilon \to 0$.

## 2.4  Graph Construction

In our GBSG method, the grading process is carried out over the entire brain. Afterwards, the corresponding segmentation is used to fuse grading values and to built our graph (see Fig. 1). We define an undirected graph $G = (V, E, \Gamma, \omega)$, where $V = \{v_1, \ldots, v_N\}$ is the set of vertices for the $N$ considered brain structures and $E = V \times V$ is the set of edges. In our work, the vertices are the mean of the grading values for a given structure while the edges are based on grading distributions distances between two structures (see Fig. 1).

To this end, the probability distributions of PBG values (see Eq. 1) are estimated with a histogram $H_v$ for each structure $v$. The number of bins is computed with the Sturge's rule [19]. For each vertex we assign a function $\Gamma : V \to \mathbb{R}$ defined as $\Gamma(v) = \mu_{H_v}$, where $\mu_{H_v}$ is the mean of $H_v$. For each edge we assign a weight given by the function $\omega : E \to \mathbb{R}$ defined as $\omega(v_i, v_j) = \exp(-d(H_{v_i}, H_{v_j})^2/\sigma^2)$ where $d$ is the Wasserstein distance with $L_1$ norm [10] that showed best performance during our experiments.

Graph representation of structure grading provides high-dimensional features (see Fig. 1). In this work we used the Elastic Net regression (EN) method that provides a sparse representation of the most discriminative edges and vertices, and thus enables to reduce the feature dimensionality by capturing the key structures and the key relationships between the different brain structures (see Fig. 2). Thus, after normalization, a concatenation of the two feature vectors is given as input of EN feature selection method.

## 2.5  Details of Implementation

PBG was computed with a patch-match method [4]. We used the parameters proposed in [5] for patch size and size of $K_i$. This results in a whole brain grading in about 10 seconds. Age effect is corrected using linear regression estimated on CN population. The EN method is computed with SLEP package [6]. Two classifiers were used to validate our method - a support vector machine (SVM)[3]. with a linear kernel and a random forest (RF)[4]. The linear SVM has a soft margin parameter, which was optimized in a range of $2^i$, with $i = \{-10, 9, \ldots, 10\}$.

---

[3] http://www.csie.ntu.edu.tw/~cjlin/libsvm.

[4] http://code.google.com/p/randomforest-matlab.

All features were normalized using z-score. In our experiments, we performed sMCI versus pMCI classification. The EN features selection and the classifiers were trained with CN and AD (see Fig. 2). As shown in [13], the use of CN and AD to train the feature selection method and the classifier enables to better discriminate sMCI and pMCI subjects. Moreover, it also enables to get the results without cross-validation step and more importantly to limit bias and overfitting problem. Thus, only one run was performed for the SVM and 30 runs was performed to capture the inner variability of RF. The mean accuracy (ACC), sensibility (SEN), and specificity (SPE) over these 30 iterations are provided as results (see Table 2).

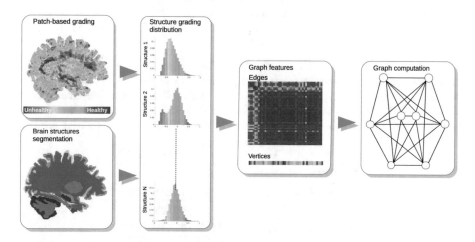

**Fig. 1.** Illustration of the graph construction method. From left to right, for each segmented structure an estimation of the density probability of PBG values are computed. Then, histograms are used to built our graph of brain structure grading. Thus, 134 histograms representing each segmented brain structures are estimated. Edges are the distances between structure grading distribution while vertices are the mean grading value for a given structure (see text for more details).

## 3  Results and Discussions

To investigate the results of our new GBSG method combining inter-subject pattern similarity features (*i.e.*, vertices) and intra-subject variability features (*i.e.*, the edges) several experiments were performed (see Table 2). The original hippocampal grading (HIPP PBG) is used as baseline [3].

First, we estimated the classification performances obtained by each feature separately using SVM. Compared to HIPP PBG, vertices showed an improvement of the specificity while the accuracy and the sensibility did not change. Therefore, additional structures selected by EN did not improve results compared to use HIPP only. On the other hand, the edges feature improved the accuracy

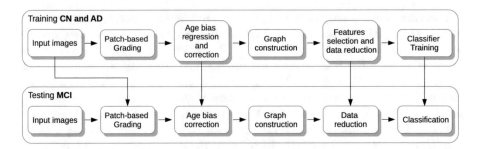

**Fig. 2.** Scheme of the proposed GBSG pipeline. PBG is computed using CN and AD training groups. CN group is also used to correct the bias related to age. This estimation is applied then to AD and MCI subjects. Afterwards, the graph is constructed, the feature selection is trained on CN and AD and is applied to CN, AD and MCI. Finally the classifier is trained with CN and AD.

**Table 2.** Classification of sMCI versus pMCI. Results obtained by inter-subject similarity features (*i.e.*, vertices), intra-subject variability features (*i.e.*, edges) and a combination of both. The original PBG applied on HIPP [3] is used as baseline. Results show that GBSG edge features improve the accuracy and the sensibility as compared to HIPP PBG and GBSG vertices features. Finally, GSBG provides the best results.

| Methods | Classifier | ACC | SEN | SPE |
|---|---|---|---|---|
| HIPP PBG | SVM | 71.5% | 72.5% | 70.0% |
| GBSG vertices | SVM | 71.9% | 71.95% | **72.0%** |
| | RF | 70.1% | 69.6% | 71.1% |
| GBSG edges | SVM | 74.6% | 81.7% | 63.0% |
| | RF | 73.8% | 81.3% | 61.6% |
| GBSG | SVM | 75.8% | 82.3% | 65.0% |
| | RF | **76.5%** | **81.7%** | 68.0% |

and the sensibility but was less specific compared to HIPP PBG and vertices. These results indicate that relevant information is encoded within GBSG edges.

Second, we evaluated the performance of combining vertex and edge features. GBSG provided the best results in terms of accuracy and sensibility. Moreover it improved the specificity compared to the intra-subject variability features. Finally, we compared SVM and RF classifiers to study the stability of our framework. The results obtained with both classifier showed the same tendency. The RF provided the best results with 76.5% of accuracy.

These results obtained with two different classification methods demonstrate the complementarity of inter-subject similarity and intra-subject variability features. Indeed, both information - level of structure degradations and global pattern of key structure modifications - are relevant.

Afterwards, we compared our GBSG method using RF classifier with state-of-the-art methods on similar ADNI1 datasets. First, we included methods

**Table 3.** Comparison of the proposed method with state-of-the-art approaches. These results show the competitive performance of our new GBSG method that obtains the best accuracy on both definitions of sMCI/pMCI populations.

| Method | sMCI/pMCI | Area | Feature | ACC | SEN | SPE |
|---|---|---|---|---|---|---|
| PBG [3] | 238/167 | HIPP | Inter | 71.0% | 70.0% | 71.0% |
| VBM [9] | 100/164 | Brain | Inter | 74.7% | 88.8% | 51.59% |
| aPBG [13] | 129/171 | Brain | Inter | 75.0% | - | - |
| ICT [16] | 111/89 | Cortex | Intra | 75.0% | 63.5% | 84.4% |
| MIL [12] | 238/167 | MTL | Intra | 72.0% | 69.0% | 74.0% |
| CNN [11] | 226/167 | GM | Intra | 74.8% | 70.9% | 78.8% |
| GBSG | 129/171 | Brain | Inter + Intra | **75.2%** | 80.0% | 68.7% |
| | 100/164 | Brain | Inter + Intra | **76.5%** | 81.7% | 68.0% |

modeling inter-subject variability based on PBG within HIPP [3], VBM [9] and an advanced PBG (aPBG) estimated over the entire brain [13]. Second, we included methods capturing intra-subject variability based on last deep learning framework [11], multiple instance learning (MIL) [12] and integrative network of cortical thickness abnormality (ICT) [16]. We applied our GBSG on two definitions of sMCI/pMCI populations as defined in [9] and [13] to perform a fair comparison. Results of this comparison are presented in Table 3. This comparison shows that best methods based on intra-subject or inter-subject obtained similar accuracy around 75% while our GBSG combining both reached 76.5% of accuracy. Compared to VBM on the same dataset [9], our GBSG improved accuracy by 1.8 percent point. However, compared to aPBG [13] GBSG provided similar results on the same dataset. Finally, compared to the CNN-based method proposed in [11] our method obtained competitive performances. These results highlight the efficiency of combining intra-subject and inter-subject features.

## 4   Conclusions

In this paper, we proposed a novel framework based on a promising graph of brain structures grading. Our new method combines inter-subject pattern similarities and intra-subject variabilities to better detect AD alterations. The pattern similarity is estimated with a patch-based grading strategy, while the intra-subject variability between structures grading is based on graph modeling. Our experiments showed the complementarity of both information. Finally, we demonstrated that our method obtains competitive performance compared to the most advanced methods.

**Acknowledgement.** This study has been carried out with financial support from the French State, managed by the French National Research Agency (ANR) in the frame of the Investments for the future Program IdEx Bordeaux (HL-MRI ANR-10-IDEX-03-02), Cluster of excellence CPU and TRAIL (HR-DTI ANR-10-LABX-57).

# References

1. Avants, B.B., et al.: A reproducible evaluation of ANTs similarity metric performance in brain image registration. Neuroimage **54**(3), 2033–2044 (2011)
2. Bron, E.E., et al.: Standardized evaluation of algorithms for computer-aided diagnosis of dementia based on structural MRI: the CADDementia challenge. NeuroImage **111**, 562–579 (2015)
3. Coupé, P., et al.: Scoring by nonlocal image patch estimator for early detection of Alzheimer's disease. NeuroImage Clin. **1**(1), 141–152 (2012)
4. Giraud, R., et al.: An optimized patchmatch for multi-scale and multi-feature label fusion. NeuroImage **124**, 770–782 (2016)
5. Hett, K., Ta, V.-T., Manjón, J.V., Coupé, P.: Adaptive fusion of texture-based grading: application to Alzheimer's disease detection. In: Wu, G., Munsell, B.C., Zhan, Y., Bai, W., Sanroma, G., Coupé, P. (eds.) Patch-MI 2017. LNCS, vol. 10530, pp. 82–89. Springer, Cham (2017). https://doi.org/10.1007/978-3-319-67434-6_10
6. Liu, J., et al.: SLEP: sparse learning with efficient projections. Arizona State Univ. **6**(491), 7 (2009)
7. Liu, M., et al.: Hierarchical fusion of features and classifier decisions for Alzheimer's disease diagnosis. Hum. Brain Mapp. **35**(4), 1305–1319 (2014)
8. Manjón, J.V., et al.: Adaptive non-local means denoising of MR images with spatially varying noise levels. J. Magn. Reson. Imaging **31**(1), 192–203 (2010)
9. Moradi, E., et al.: Machine learning framework for early MRI-based Alzheimer's conversion prediction in MCI subjects. Neuroimage **104**, 398–412 (2015)
10. Rubner, Y., et al.: The earth mover's distance as a metric for image retrieval. Int. J. Comput. Vis. **40**(2), 99–121 (2000)
11. Suk, H.I., et al.: Deep ensemble learning of sparse regression models for brain disease diagnosis. Med. Image Anal. **37**, 101–113 (2017)
12. Tong, T., et al.: Multiple instance learning for classification of dementia in brain MRI. Med. Image Anal. **18**(5), 808–818 (2014)
13. Tong, T., et al.: A novel grading biomarker for the prediction of conversion from mild cognitive impairment to Alzheimer's disease. IEEE Trans. Biomed. Eng. **64**(1), 155–165 (2017)
14. Tustison, N.J., et al.: N4ITK: improved N3 bias correction. IEEE Trans. Med. Imaging **29**(6), 1310–1320 (2010)
15. Wang, H., et al.: A learning-based wrapper method to correct systematic errors in automatic image segmentation: consistently improved performance in hippocampus, cortex and brain segmentation. NeuroImage **55**(3), 968–985 (2011)
16. Wee, C.Y., et al.: Prediction of Alzheimer's disease and mild cognitive impairment using cortical morphological patterns. Hum. Brain Mapp. **34**(12), 3411–3425 (2013)
17. Wolz, R., et al.: Multi-method analysis of MRI images in early diagnostics of Alzheimer's disease. PloS One **6**(10), e25446 (2011)
18. Zhou, L., et al.: Hierarchical anatomical brain networks for MCI prediction: revisiting volumetric measures. PloS One **6**(7), e21935 (2011)
19. Sturges, H.A.: The choice of a class interval. J. Am. Stat. Assoc. **21**(153), 65–66 (1926)

# Joint Prediction and Classification of Brain Image Evolution Trajectories from Baseline Brain Image with Application to Early Dementia

Can Gafuroğlu, Islem Rekik[✉],
and for the Alzheimer's Disease Neuroimaging Initiative

BASIRA Lab, CVIP Group, School of Science and Engineering, Computing,
University of Dundee, Dundee, UK
irekik@dundee.ac.uk
http://www.basira-lab.com

**Abstract.** Despite the large body of existing neuroimaging-based studies on brain dementia, in particular mild cognitive impairment (MCI), *modeling and predicting* the early dynamics of dementia onset and development in healthy brains is somewhat overlooked in the literature. The majority of computer-aided diagnosis tools developed for classifying healthy and demented brains mainly rely on either using single time-point or longitudinal neuroimaging data. Longitudinal brain imaging data offer a larger time window to better capture subtle brain changes in early MCI development, and its utilization has been shown to improve classification and prediction results. However, typical longitudinal studies are challenged by a limited number of acquisition timepoints and the absence of inter-subject matching between timepoints. To address this limitation, we propose a novel framework that learns how to *predict* the developmental trajectory of a brain image from a single acquisition timepoint (i.e., baseline), while *classifying* the predicted trajectory as 'healthy' or 'demented'. To do so, we first rigidly align all training images, then extract 'landmark patches' from training images. Next, to predict the patch-wise trajectory evolution from baseline patch, we propose two novel strategies. The first strategy learns in *a supervised manner* to select a few training atlas patches that best boost the classification accuracy of the target testing patch. The second strategy learns in *an unsupervised manner* to select the set of most similar training atlas patches to the target testing patch using multi-kernel patch manifold learning. Finally, we train a linear classifier for each predicted patch trajectory. To identify the final label of the target subject, we use majority voting to aggregate the labels assigned by our model to all landmark patches' trajectories. Our image prediction model boosted the classification performance by 14% point without further leveraging any enhancing methods such as feature selection.

© Springer Nature Switzerland AG 2018
A. F. Frangi et al. (Eds.): MICCAI 2018, LNCS 11072, pp. 437–445, 2018.
https://doi.org/10.1007/978-3-030-00931-1_50

# 1    Introduction

The early detection of Alzheimer's Disease (AD) allows patients to undergo preventive treatment to minimize the impact of the disease on their lives. In particular, distinguishing early Mild Cognitive Impairment (eMCI), which is the earliest form of AD, from normal controls (NC) is a challenging task, due to the subtlety of the anatomical differences between NC and eMCI subjects. However, this classification problem has great significance due to AD being the most common cause of dementia [1,2]. Presently, AD is diagnosed using neuroimaging along with measures of cognitive performance [2]. This process places a burden on the experts as they must examine structural Magnetic Resonance Imaging (MRI). The atrophy patterns in early MCI patients in particular are subtle and these patients show no signs of cognitive impairment aside from minor memory concerns. Applying machine learning to automate the diagnosis of eMCI can alleviate the burden on these experts and provide consistent interpretations of the MRI data. As such, there have been a vast number of studies that aim to apply machine learning methods on neuroimaging data to make predictions and diagnoses regarding dementia, some of which use data from a single time point [3–5] and some of which utilize data from multiple time points [6–9].

[6] used similarity maps between aligned baseline and follow-up images to predict progression from MCI to AD by classifying stable MCI patients against MCI converters. [7] used incomplete longitudinal data from MCI subjects in conjunction with sparse learning algorithms and found that the longitudinal data improves their classification accuracy for identifying converters. [8] proposed a temporally structured support vector machine (SVM) classifier, which is designed specifically to work with longitudinal MRI data, with no limit on the number of follow-ups that can be used and achieved state-of-the-art performance in classifying MCI converters. [9] used data from multiple modalities measured at multiple time points in order to predict progression from MCI to AD using an SVM classifier. This work nicely modeled the relationship between data acquired at different time points, where the estimations of derived features between time points was introduced in the learning model, such as cortical thinning speed. By incorporating multiple time points into their framework, all of these works leveraged additional relevant information to improve classification accuracy. However, these methods are limited by the requirement of more than a single acquisition timepoint; hence, early MCI diagnosis or MCI conversion to AD prediction from baseline data is impeded. As preventive treatment is more likely to succeed the earlier the disease is detected, requiring subjects to wait for multiple measurements at different time points may hinder their treatment and recovery processes.

On the other hand, several works focused on using a single acquisition timepoint for dementia diagnosis and classification, which avoids the limitation of requiring patients to wait for multiple scans. For instance, [3] used SVMs to classify AD against NC from MRI images. Similarly, [4] used SVM classifiers with data acquired using varying scanning equipments. [5] compared ten previously explored methods of classification on a single dataset with three different

AD-related classification problems: classifying AD against NC, MCI against NC, as well as classifying MCI converters against stable MCI patients. However, these works do not integrate longitudinal data into their frameworks, thereby forego-ing the potential improvement in classification accuracy that could be achieved with information from additional timepoints.

To solve this issue, we propose to diagnose a patient in an early stage from their baseline image; however, more reliably by leveraging longitudinal infor-mation *predicted* at later timepoints. Specifically, we unprecedentedly propose a joint image evolution trajectory prediction and classification framework from a single acquisition timepoint. Our framework comprises four steps. First, we detect key landmarks in the target anatomical region of interest (ROI) across all training subjects at baseline $t_1$. Second, for each baseline patch centered at a specific landmark, we learn how to predict its evolution trajectory at follow-up timepoints. To do so, we propose two different novel strategies. The first strat-egy learns in a supervised way how to select baseline 'atlas patches' that are expected to yield the smallest patch prediction error at a follow-up timepoint $t_2$ for an input testing patch. The second strategy learns in an unsupervised way how to nest both training and testing patches in a high-dimensional mani-fold using multiple-kernel learning. Next, we select the closest baseline training patches (or atlas patches) on the manifold to the testing patch. Following the selection of the best training atlas patches at baseline $t_1$, we simply average their corresponding atlas patches at $t_2$ to predict the testing patch at $t_2$. Ultimately, we train an ensemble classifier to predict the label for each learned patch-wise evolution trajectory, which are then aggregated using majority voting to classify the target subject.

## 2    Joint Prediction and Classification of Brain Image Evolution Trajectories from Baseline

In this section, we present the supervised and unsupervised learning strategies proposed for jointly predicting and labeling brain image evolution trajectories from a single acquisition timepoint. We denote matrices by boldface capital letters, e.g., $\mathbf{X}$, and vectors are denoted by bold lowercase letters, e.g., $\mathbf{x}$. Figure 1 illustrates the pipeline for each of the proposed strategies based on a supervised or unsupervised selection of atlas patches to predict the evolution trajectory of a baseline testing patch. In the training stage, both strategies share a fundamental step, which consists of detecting key landmarks across all training images to learn a landmark-wise prediction and classification model.

**Proposed Landmark Detection Method.** Since brain disorders affect the morphology of a particular anatomical brain region, we expect that the regions at the boundary (or edge) of the target ROIs capture the most discriminative information. Hence, for each training subject, we apply a Sobel filter to the train-ing label map of the target ROI to detect its edge. Next, we average all training edges to generate an 'edge density' at each voxel. Ultimately, by thresholding

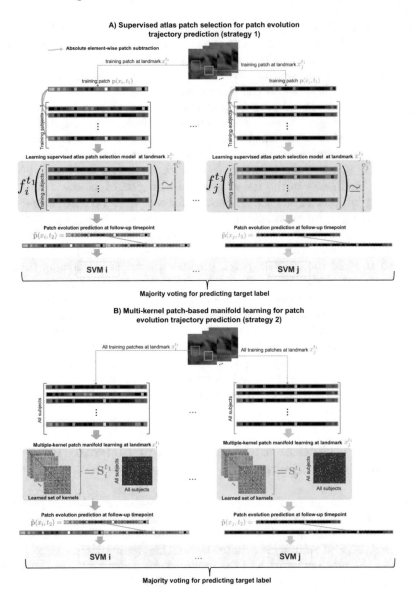

**Fig. 1.** *Pipeline of the proposed joint prediction and classification framework of brain image evolution trajectories from baseline image using two strategies (A) and (B). (A) Supervised atlas patch selection strategy.* A regression function $f_i^{t_1}$ is learned at each landmark $x_i$ to map absolute distance vectors between two patches to an error prediction vector. This allows to *learn* how to select the best atlas patches at $t_1$ for our patch evolution trajectory prediction at $t_2$. *(B) Unsupervised atlas patch selection strategy.* Patch-based multi-kernel manifold learning is used to select the most similar atlas patches to the target testing patch at $t_1$. The predicted follow-up patches at $t_2$ are then concatenated with the baseline patch at $t_1$ and classified by majority voting using an ensemble of SVM classifiers in both approaches.

the edge density map, we generate the key landmarks that represent the centers of our training and testing patches for learning each of our supervised and unsupervised image prediction models.

**Supervised Atlas Patch Selection for Patch Evolution Trajectory Prediction (Strategy 1).** Inspired by the work of [10] on atlas selection learning for brain image segmentation, we propose a patch-based atlas selection strategy that learns in a supervised manner how to minimize the prediction error for a testing patch evolution trajectory. For each target landmark $x_i$, we learn how to select the best training atlas patches at baseline timepoint $t_1$ that minimize the prediction error for a target testing patch at a follow-up timepoint $t_k$, $k \geq 2$. Our learning model assumes that two similar intensity patches at baseline will evolve similarly at follow-up timepoints. Specifically, given $n-1$ training subjects, we first build an intensity disparity matrix of size $(n-2)^2$ at $t_1$, where each row denotes an element-wise absolute difference between two training patches $\mathbf{p}_{i,s}^{t_1}$ and $\mathbf{p}_{i,s'}^{t_1}$. Next, we build a corresponding error vector $\mathbf{e}_i^{t_1}$ that quantifies the prediction error of using $\mathbf{p}_{i,s}^{t_1}$ for subject $s$ at $t_1$ as an atlas patch to predict $\tilde{\mathbf{p}}_{i,s'}^{t_2}$ for subject $s'$ at $t_2$. Specifically, we define $\tilde{\mathbf{p}}_{i,s'}^{t_2} = \alpha \times p_{i,s'}^{t_2}$, where $\alpha = \mathbf{p}_{i,s'}^{t_1}/\mathbf{p}_{i,s}^{t_1}$. If $\mathbf{p}_{i,s}^{t_1}$ has a zero-element, then the corresponding element in vector $\alpha$ is set to a high value. As for the prediction error, we define it as the average absolute difference between the ground truth patch $\mathbf{p}_{i,s'}^{t_2}$ and the predicted patch $\tilde{\mathbf{p}}_{i,s'}^{t_2}$ for the 'testing' subject $s'$. Next, in the *training stage*, we learn a support vector regressor function $f_i^{t_1}$ that maps the intensity disparity matrix at each landmark $x_i$ onto the corresponding prediction error vector (Fig. 1A). In the *testing stage*, we compute the pairwise distance between each training patch $\{\mathbf{p}_{i,s}^{t_1}\}_{s=1}^{n-1}$ and the testing patch $\mathbf{p}_{i,tst}^{t_1}$. Then, by testing the learned regression function for each pairwise absolute distance $f_i^{t_1}(|\mathbf{p}_{i,s}^{t_1} - \mathbf{p}_{i,tst}^{t_1}|)$, we predict the error of using subject $s$ to predict $\mathbf{p}_{i,tst}^{t_2}$. Ultimately, we select the top $K$ atlas patches at $t_1$ with the lowest prediction errors, then average their corresponding patches at $t_2$ to output $\tilde{\mathbf{p}}_{i,tst}^{t_2}$.

**Multi-kernel Patch-Based Manifold Learning for Patch Evolution Trajectory Prediction (Strategy 2).** In this strategy, we use the multi-kernel manifold learning (MKML) method proposed in [11], to identify the baseline atlas patches whose follow-up images best represent the testing baseline patch by learning patch-to-patch similarities (Fig. 1B). In this section, we briefly present the MKML framework introduced in [11] and how we extended it to our aim. MKML learns a distance metric in an unsupervised manner by combining multiple Gaussian kernels, and uses it to output a learned pairwise similarity matrix of size $n \times n$ from an input matrix of size $n \times d$ where $n$ is the number of subjects and $d$ is the size of their vectorized patches. For each landmark $x_i$, to predict $\tilde{\mathbf{p}}_{i,tst}^{t_2}$ for a testing subject, we first learn a baseline similarity matrix $\mathbf{S}_i^{t_1}$ for all training and testing samples. This allows to learn an intensity patch manifold where all patch vectors $\{\mathbf{p}_{i,1}^{t_1}, \ldots, \mathbf{p}_{i,n}^{t_1}\}$ are nested. Instead of using one predefined distance metric which may fail to capture the nonlinear relationship in the patch data, we use multiple Gaussian kernels with learned weights to better explore in depth the similarity patterns among patches centered at a fixed landmark across a set of subjects. In other words, adopting multiple kernels allows to better fit the true underlying statistical distribution of the input matrix of intensity patch features. Additionally, constraints are imposed on kernel weights to avoid a single kernel selection [11]. The Gaussian kernel is expressed as follows: $\mathbf{K}(\mathbf{p}_{i,s}^{t_1}, \mathbf{p}_{i,s'}^{t_1}) = \frac{1}{\epsilon_{s,s'}\sqrt{2\pi}} e^{\left(-\frac{|\mathbf{p}_{i,s}^{t_1}-\mathbf{p}_{i,s'}^{t_1}|^2}{2\epsilon_{s,s'}^2}\right)}$ , where $\epsilon_{s,s'}$ is defined as: $\epsilon_{s,s'} = \sigma(\mu_s + \mu_{s'})/2$, where $\sigma$ is a tuning parameter and $\mu_s = \frac{\sum_{l\in KNN(\mathbf{p}_{i,s}^{t_1})}|\mathbf{p}_{i,s}^{t_1}-\mathbf{p}_{i,s'}^{t_1}|}{k}$, where $KNN(\mathbf{p}_{i,s}^{t_1})$ represents the top $k$ neighboring subjects of subject $s$. The computed kernels are then averaged to further learn the similarity matrix $\mathbf{S}_i^{t_1}$ at landmark $x_i$ and baseline timepoint $t_1$ through an optimization framework formulated as follows:

$$\min_{\mathbf{S}_i^{t_1},\mathbf{L},\mathbf{w}} \sum_{k,j} -w_l\mathbf{K}_l(\mathbf{h}^k,\mathbf{h}^j)\mathbf{S}_i^{t_1}(k,j) + \beta\|\mathbf{S}_i^{t_1}\|_F^2 + \gamma\mathbf{tr}(\mathbf{L}^T(\mathbf{I}_n - \mathbf{S}_i^{t_1})\mathbf{L}) + \rho\sum_l w_l logw_l$$

Subject to: $\sum_l w_l = 1$, $w_l \geq 0$, $\mathbf{L}^T\mathbf{L} = \mathbf{I}_c$, $\sum_j \mathbf{S}_i^{t_1}(k,j) = 1$, and $\mathbf{S}_i^{t_1}(k,j) \geq 0$ for all $(k,j)$, where:

- $\sum_{k,j} -w_l\mathbf{K}_l(\mathbf{h}^k,\mathbf{h}^j)\mathbf{S}_i^{t_1}(k,j)$ refers to the relation between the similarity and the kernel distance with weights $w_l$ between two subject-specific patches. The learned similarity should be small if the distance between a pair of patches is large.
- $\beta\|\mathbf{S}_i^{t_1}\|_F^2$ denotes a regularization term that avoids over-fitting the model to the data.
- $\gamma\mathbf{tr}(\mathbf{L}^T(\mathbf{I}_n - \mathbf{S}_i^{t_1})\mathbf{L})$: $\mathbf{L}$ is the latent matrix of size $n \times c$ where $n$ is the number of subjects and $c$ is the number of clusters. The matrix $(\mathbf{I}_n - \mathbf{S}_i^{t_1})$ denotes the graph Laplacian.
- $\rho\sum_l w_l logw_l$ imposes constraints on the kernel weights to avoid selection of a single kernel.

An alternating convex optimization is adopted where each variable is optimized while fixing the other variables until convergence [11]. Finally, based on the landmark-specific learned matrix $\mathbf{S}_i^{t_1}$, we select the top $K$ training patches (or $K$ atlas patches) that are most similar to the target testing patch at baseline. Finally, we predict $\tilde{\mathbf{p}}_{i,tst}^{t_2}$ as a weighted average of corresponding $K$ atlas patches at follow-up $t_2$.

**Predicted Trajectories' Labeling Using an Ensemble Classifier.** The last shared step in both proposed strategies is to *label* patch evolution trajectory for a testing subejct at each landmark $x_i$ as 'healthy' or 'disordered'. To do so, for each landmark $x_i$, we train a support vector machine (SVM) using the concatenation of baseline training patches $\{\mathbf{p}_{i,s}^{t_1}\}_{s=1}^{n-1}$ and their corresponding predicted patches $\{\tilde{\mathbf{p}}_{i,s}^{t_2}\}_{s=1}^{n-1}$ using strategies 1 or 2. The left out testing subject is then classified using majority voting on predicted labels outputted by all SVMs (i.e., across all landmarks).

## 3 Results and Discussion

**Data and Model Parameters.** We used data from 30 NC (Normal Control) and 30 eMCI subjects acquired from the Alzheimer's Disease Neuroimaging Initiative (ADNI) dataset (adni.loni.usc.edu). All training baseline and follow-up MR images along with their AAL-atlas based label maps are rigidly aligned to a common space. For evaluation, we tested our methods on landmark patches extracted from the left hippocampus and the left lateral ventricle. We selected these preliminary ROIs due to: (1) their prevalence in the dementia literature [5,6], and (2) the fact that atrophy rates were found to be faster in the left hemisphere compared to the right hemisphere [12]. We fixed the patch size to $7 \times 7 \times 7$ across all methods. The number of neighbours (the best atlas patches) used for predicting the follow-up trajectory is set to 2 for both strategies. For MKML parameters, we set the number of clusters to $c = 2$, and $m = 2$ kernels.

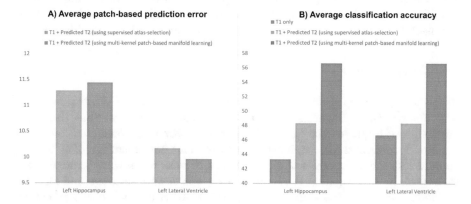

**Fig. 2.** (A) Prediction accuracy for each ROI using proposed strategies for predicting the follow-up image evolution trajectory from baseline. (B) Classification accuracy by our proposed methods.

**Evaluation.** We used leave-one-out cross-validation to evaluate the classification accuracy of the proposed methods. Figure 2A displays the average patch prediction error computed using the mean absolute error (MAE) between ground-truth and predicted patches at the follow-up timepoint. We note that for both of the selected ROIs, both methods led to similar results, where unsupervised prediction slightly outperformed supervised prediction in the hippocampus as opposed to the ventricle. As for the eMCI/NC classification results (Fig. 2B), integrating predicted follow-up timepoints significantly increased by ~5% point (pp) (resp. ~2pp) when using supervised prediction method and by ~14pp (resp. ~10pp) when using unsupervised prediction strategy for the hippocampus (resp. the ventricle). Clearly, multi-kernel patch-based manifold learning for patch evolution prediction consistently produced the best classification results. We would like to highlight that the main contribution of this work is to highlight the great potential of predicting follow-up data from baseline data in boosting classification and disorder diagnosis. Both strategies proposed in Fig. 1 can be boosted by using enhancing methods such as feature selection or deep-learning approaches.

It should be noted that our supervised atlas selection strategy relies on the assumption that the multiplication of an intensity patch of subject $s$ at baseline by a weighting vector $\alpha$ can produce the intensity patch of a different subject $s'$. Through propagating this rule to follow-up timepoints, we can learn how to predict patch evolution trajectory. However, there is no theoretical proof that this assumption holds aside from the fact that the follow-up data predicted in this way improved our classification accuracy when compared to classifying using the baseline images only. We also note that our framework leverages information from only brain MR images, however brain dementia also atrophies the cortical surface [13]. Hence, based on the seminal shape evolution learning models for predicting infant cortical development from a single timepoint [14] and inspired from the joint shape-image regression model proposed in [15], we aim to build a unified model which simultaneously predicts cortical shape and brain image evolution trajectories for a more accurate early diagnosis from baseline data.

## 4    Conclusion

In this work, we proposed a novel joint image prediction and classification framework for the early diagnosis of MCI. Our initial results show that there is a possibility of improvement in classification accuracy when predicted longitudinal data is leveraged for classification even without additional use of enhancing approaches such as feature selection. Our future work will involve improving the framework by investigating the impact of various feature selection and learning methods, including deep learning architectures. Additionally, we wish to extend our framework by investigating whether timepoint-to-timepoint similarity maps such as those proposed in [6] can further improve our classification accuracy when the follow-up images are not measured, but *predicted* instead.

## References

1. Apostolova, L.G., Thompson, P.M.: Mapping progressive brain structural changes in early Alzheimer's disease and mild cognitive impairment. Neuropsychologia 46(6), 1597–1612 (2008)
2. Frisoni, G.B., Fox, N.C., Jack, C.R., Scheltens, P., Thompson, P.M.: The clinical use of structural MRI in Alzheimer disease. Nat. Rev. Neurol. 6(2), 67–77 (2010)
3. Magnin, B., et al.: Support vector machine-based classification of Alzheimer's disease from whole-brain anatomical MRI. Neuroradiology 51(2), 73–83 (2009)
4. Kloppel, S., et al.: Automatic classification of MR scans in Alzheimer's disease. Brain 131(Pt 3), 681–689 (2008)
5. Cuingnet, R., et al.: Automatic classification of patients with Alzheimer's disease from structural MRI: a comparison of ten methods using the ADNI database. NeuroImage 56(2), 766–781 (2011)

6. Sanroma, G., et al.: Early prediction of Alzheimer's disease with non-local patch-based longitudinal descriptors. In: Wu, G., Munsell, B.C., Zhan, Y., Bai, W., Sanroma, G., Coupé, P. (eds.) Patch-MI 2017. LNCS, vol. 10530, pp. 74–81. Springer, Cham (2017). https://doi.org/10.1007/978-3-319-67434-6_9

7. Thung, K.H., Wee, C.Y., Yap, P.T., Shen, D.: Identification of progressive mild cognitive impairment patients using incomplete longitudinal MRI scans. Brain Struct. Funct. **221**(8), 3979–3995 (2016)

8. Zhu, Y., Zhu, X., Kim, M., Shen, D., Wu, G.: Early diagnosis of Alzheimer's disease by joint feature selection and classification on temporally structured support vector machine. In: Ourselin, S., Joskowicz, L., Sabuncu, M.R., Unal, G., Wells, W. (eds.) MICCAI 2016. LNCS, vol. 9900, pp. 264–272. Springer, Cham (2016). https://doi.org/10.1007/978-3-319-46720-7_31

9. Zhang, D., Shen, D., Initiative, A.D.N.: Predicting future clinical changes of MCI patients using longitudinal and multimodal biomarkers. PLOS One **7**(3), 1–15 (2012)

10. Sanroma, G., Wu, G., Gao, Y., Shen, D.: Learning-based atlas selection for multiple-atlas segmentation. In: Proceedings of the IEEE Conference on Computer Vision and Pattern Recognition, pp. 3111–3117 (2014)

11. Wang, B., Zhu, J., Pierson, E., Ramazzotti, D., Batzoglou, S.: Visualization and analysis of single-cell RNA-seq data by kernel-based similarity learning. Nat. Methods **14**, 414 (2017)

12. Thompson, P.M., et al.: Dynamics of gray matter loss in Alzheimer's disease. J. Neurosci. **23**(3), 994–1005 (2003)

13. Möller, C., et al.: Alzheimer disease and behavioral variant frontotemporal dementia: automatic classification based on cortical atrophy for single-subject diagnosis. Radiology **279**(3), 838–848 (2015)

14. Rekik, I., Li, G., Yap, P.T., Chen, G., Lin, W., Shen, D.: Joint prediction of longitudinal development of cortical surfaces and white matter fibers from neonatal MRI. NeuroImage **152**, 411–424 (2017)

15. Fishbaugh, J., Prastawa, M., Gerig, G., Durrleman, S.: Geodesic regression of image and shape data for improved modeling of 4D trajectories. Biomedical Imaging (ISBI), 2014 IEEE 11th International Symposium on (2014) 385–388

# Temporal Correlation Structure Learning for MCI Conversion Prediction

Xiaoqian Wang[1], Weidong Cai[2], Dinggang Shen[3], and Heng Huang[1(✉)]

[1] Department of Electrical and Computer Engineering, University of Pittsburgh, Pittsburgh, USA
henghuanghh@gmail.com
[2] School of Information Technologies, University of Sydney, Sydney, Australia
[3] Department of Radiology and BRIC, University of North Carolina at Chapel Hill, Chapel Hill, USA

**Abstract.** In Alzheimer's research, Mild Cognitive Impairment (MCI) is an important intermediate stage between normal aging and Alzheimer's. How to distinguish MCI samples that finally convert to AD from those do not is an essential problem in the prevention and diagnosis of Alzheimer's. Traditional methods use various classification models to distinguish MCI converters from non-converters, while the performance is usually limited by the small number of available data. Moreover, previous methods only use the data at baseline time for training but ignore the longitudinal information at other time points along the disease progression. To tackle with these problems, we propose a novel deep learning framework that uncovers the temporal correlation structure between adjacent time points in the disease progression. We also construct a generative framework to learn the inherent data distribution so as to produce more reliable data to strengthen the training process. Extensive experiments on the ADNI cohort validate the superiority of our model.

**Keywords:** Deep learning · Temporal correlation structure
MCI conversion prediction · Alzheimer's disease

## 1 Introduction

Alzheimer's disease (AD) is a complex chronic progressive neurodegenerative disease that gradually affects human memory, judgment, and behavior. As an important intermediate stage between normal aging and AD, MCI possesses an increased risk of transiting to AD. That being the case, how to recognize the MCI samples with high potential of switching to AD prior to dementia becomes an essential problem in Alzheimer's prophylaxis and early treatment.

H. Huang—This work was partially supported by U.S. NIH R01 AG049371, NSF IIS 1302675, IIS 1344152, DBI 1356628, IIS 1619308, IIS 1633753.

A. F. Frangi et al. (Eds.): MICCAI 2018, LNCS 11072, pp. 446–454, 2018.
https://doi.org/10.1007/978-3-030-00931-1_51

Neuroimaging provides an effective tool to characterize the structure and functionality of nervous system, thus has greatly contributed to Alzheimer's study [22]. Extensive work has been proposed to predict MCI conversion using neuroimaging data [6,15]. Previous methods usually formulate MCI conversion prediction as a binary classification (distinguishing MCI converters from non-converters) [15] or multi-class classification problem (when considering other classes such as AD or health control (HC)) [6], where the methods take the neuroimaging data at baseline time as the input and classify if the MCI samples will convert to AD in years.

Despite the prosperity and progress achieved in MCI conversion prediction, there are still several problems existing in previous methods. (1) Although we expect the model to be capable of forecasting the MCI conversion years before the change of disease status, the training process should not be limited to just baseline data. In the longitudinal study of AD, usually the data at several time points along the disease progression is available, such as baseline, month 6, month 12, *etc.* However, previous methods only consider the baseline data in the training process, thus ignore the temporal correlation structure among other time points.

(2) The labeling process for Alzheimer's is time-consuming and expensive, so the MCI conversion prediction suffers greatly from limited training data.

To deal with these problems, we propose a novel model for MCI conversion prediction. Firstly, we study the temporal correlation structure among the longitudinal data in Alzheimer's progression. Since AD is a chronically progressive disorder and the neuroimaging features are correlated [11], it can be helpful to analyze the temporal correlation between neuroimaging data in the disease progression as in other nervous system diseases [3]. We construct a regression model to discover such temporal correlation structure between adjacent time points. Our model incorporates the data at all time points along the disease progression and uncovers the variation trend that benefits MCI conversion prediction.

Secondly, we construct a classification model to predict the disease status at each time point. Different from previous classification models that use the baseline data to forecast the progression trend in two or three years, our classification model focuses on adjacent time points. Compared with previous models that require a highly distinguishable conversion pattern appears several years before dementia, our model predicts the progression trend for consecutive time points, thus is more accurate and reliable.

Thirdly, we construct a generative model based on generative adversarial network (GAN) to produce more auxiliary data to improve the training of regression and classification model. GAN model is proposed in [5], which uses the adversarial mechanism to learn the inherent data distribution and generate realistic data. We use the generative model to learn the joint distribution of neuroimaging data at consecutive time points, such that more reliable training data can be obtained to improve the prediction of MCI conversion.

## 2   Temporal Correlation Structure Learning Model

### 2.1   Problem Definition

In MCI conversion prediction, for a certain sample and a time point $t$, we use $\mathbf{x}_t \in \mathbb{R}^p$ to denote the neuroimaging data at time $t$ while $\mathbf{x}_{t+1} \in \mathbb{R}^p$ for the next time point, where $p$ is the number of imaging markers. $\mathbf{y}_t \in \mathbb{R}$ is the label showing the disease status at time $t$ and $t + 1$. Here we define three different classes for $\mathbf{y}_t$: $\mathbf{y}_t = 1$ means the sample is AD at both time $t$ and $t + 1$; $\mathbf{y}_t = 2$ shows MCI at time $t$ while AD at time $t + 1$; while $\mathbf{y}_t = 3$ indicates that the sample is MCI at both time $t$ and $t+1$. In the prediction, given the baseline data of an MCI sample, the goal is to predict whether the MCI sample will finally convert to AD or not.

### 2.2   Revisit GAN Model

GAN model is proposed in [5], which plays an adversarial game between the generator $G$ and discriminator $D$. The generator $G$ takes a random variable $\mathbf{z}$ as the input and outputs the generated data to approximates the inherent data distribution. The discriminator $D$ is proposed to distinguish the data $\mathbf{x}$ from the real distribution and the data produced from the generator. Whereas the generator $G$ is optimized to generate data as realistic as possible to fool the discriminator. The objective function of the GAN model has the following form.

$$\min_{G} \max_{D} \; \mathbb{E}_{\mathbf{x} \sim p(\mathbf{x})} \big[ \log(D(\mathbf{x})) \big] + \mathbb{E}_{\mathbf{z} \sim p(\mathbf{z})} \big[ \log(1 - D(G(\mathbf{z}))) \big],$$

where $p(\mathbf{z})$ denotes the distribution of the random variable and $p(\mathbf{x})$ represents the distribution of real data. The min-max game played between $G$ and $D$ improves the learning of both the generator and discriminator, such that the model can learn the inherent data distribution and generate realistic data.

### 2.3   Illustration of Our Model

Inspired by [2], we propose to approximate the joint distribution of neuroimaging data at consecutive time points and data label $([\mathbf{x}_t, \mathbf{x}_{t+1}], \mathbf{y}_t) \sim p(\mathbf{x}, \mathbf{y})$ by considering the following:

$$\min_{G_t, G_{t+1}} \max_{D} \; \mathbb{E}_{([\mathbf{x}_t, \mathbf{x}_{t+1}], \mathbf{y}_t) \sim p(\mathbf{x}, \mathbf{y})}[\log(D([\mathbf{x}_t, \mathbf{x}_{t+1}], \mathbf{y}_t))]$$
$$+ \mathbb{E}_{\mathbf{z} \sim p(\mathbf{z}), \mathbf{y} \sim p(\mathbf{y})}[\log(1 - D([G_t(\mathbf{z}, \mathbf{y}), G_{t+1}(\mathbf{z}, \mathbf{y})], \mathbf{y}))],$$

where the generators take a random variable $\mathbf{z}$ and a pseudo label $\mathbf{y}$ as the input and output a data pair $([G_t(\mathbf{z}, \mathbf{y}), G_{t+1}(\mathbf{z}, \mathbf{y})], \mathbf{y})$ that is as realistic as possible. Still, the discriminator is optimized to distinguish real from fake data. The construction of such generative model approximates the inherent joint distribution of neuroimaging data at adjacent time points and label, which generates more reliable samples for the training process.

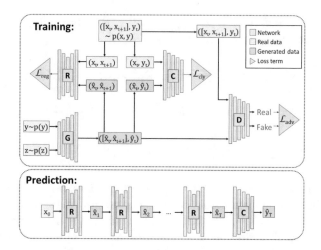

**Fig. 1.** Illustration of our Temporal-GAN model. $\mathbf{x}_t$ and $\mathbf{x}_{t+1}$ are the neuroimaging data at two adjacent time points and $\mathbf{y}_t$ is the label ($\mathbf{y}_t = 1$ if both $\mathbf{x}_t$ and $\mathbf{x}_{t+1}$ are at AD status; $\mathbf{y}_t = 2$ if $\mathbf{x}_t$ is MCI while $\mathbf{x}_{t+1}$ is AD; $\mathbf{y}_t = 3$ if both $\mathbf{x}_t$ and $\mathbf{x}_{t+1}$ are MCI.). The regression network $R$ predicts $\mathbf{x}_{t+1}$ from $\mathbf{x}_t$ so as to uncover the temporal correlation between adjacent time points. The classification network $C$ predicts the label $\mathbf{y}_t$ from $\mathbf{x}_t$. We also construct a generative model with generator $G$ and discriminator $D$ to approximate the joint distribution underlying data pair $([\mathbf{x}_t, \mathbf{x}_{t+1}], \mathbf{y}_t)$ to generate more reliable data for training $R$ and $C$. In the prediction process, the neuroimaging data $\mathbf{x}_0$ at baseline time for MCI samples is given, and we use $R$ and $C$ to predict whether the MCI sample will convert to AD at time $T$.

To uncover the temporal correlation structure among the neuroimaging data between consecutive time points, we construct a regression network $R$ to predict $\mathbf{x}_{t+1}$ from $\mathbf{x}_t$, such that progression trend among neuroimaging data along the disease progression can be learned. The network $R$ takes data from both real distribution and the generators as the input and optimize the following:

$$
\min_{R} \ \mathbb{E}_{([\mathbf{x}_t, \mathbf{x}_{t+1}], \mathbf{y}_t) \sim p(\mathbf{x}, \mathbf{y})}[\|\mathbf{x}_{t+1} - R(\mathbf{x}_t)\|_1]
$$
$$
+ \ \lambda_{reg} \mathbb{E}_{\mathbf{z} \sim p(\mathbf{z}), \mathbf{y} \sim p(\mathbf{y})}[\|G_{t+1}(\mathbf{z}, \mathbf{y}) - R(G_t(\mathbf{z}, \mathbf{y}))\|_1], \tag{1}
$$

where the hyper-parameter $\lambda_{reg}$ balances the importance of real and generated data. We consider $\ell_1$-norm loss to make the model $R$ more robust to outliers.

In addition, we construct a classification structure $C$ to predict the label $\mathbf{y}_t$ given data $\mathbf{x}_t$. The optimization of $C$ is based on the following:

$$
\min_{C} \ -\mathbb{E}_{([\mathbf{x}_t, \mathbf{x}_{t+1}], \mathbf{y}_t) \sim p(\mathbf{x}, \mathbf{y})}[\mathbf{y}_t \log(C(\mathbf{x}_t))]
$$
$$
- \ \lambda_{cly} \mathbb{E}_{\mathbf{z} \sim p(\mathbf{z}), \mathbf{y} \sim p(\mathbf{y})}[\mathbf{y} \log(C(G_t(\mathbf{z})))], \tag{2}
$$

where $\lambda_{cly}$ is a hyper-parameter to balance the role of real and generated data.

Given a set of real data $\{([\mathbf{x}_t^i, \mathbf{x}_{t+1}^i], \mathbf{y}_t^i)\}_{i=1}^n$, the above three loss terms can be approximated by the following empirical loss:

$$\mathcal{L}_{adv} = \frac{1}{n} \sum_{i=1}^n \log(D([\mathbf{x}_t^i, \mathbf{x}_{t+1}^i], \mathbf{y}_t^i)) + \sum_{j=1}^{n_z} \log(D([G_t(\mathbf{z}^j, \mathbf{y}^j), G_{t+1}(\mathbf{z}^j, \mathbf{y}^j)], \mathbf{y}^j)),$$

$$\mathcal{L}_{reg} = \frac{1}{n} \sum_{i=1}^n \|\mathbf{x}_{t+1}^i - R(\mathbf{x}_t^i)\|_1 + \lambda_{reg} \sum_{j=1}^{n_z} \|G_{t+1}(\mathbf{z}^j, \mathbf{y}^j) - R(G_t(\mathbf{z}^j, \mathbf{y}^j))\|_1,$$

$$\mathcal{L}_{cly} = -\frac{1}{n} \sum_{i=1}^n \mathbf{y}_t^i \log(C(\mathbf{x}_t^i)) - \lambda_{cly} \sum_{j=1}^{n_z} \mathbf{y}^j \log(C(G_t(\mathbf{z}^j, \mathbf{y}^j))).$$

For a clear illustration, we plot a figure in Fig. 1 to show the structure of our Temporal-GAN model (temporal correlation structure learning for MCI conversion prediction with GAN). The implement details of the networks can be found in the experimental setting section. The optimization of our model is based on a variant of mini-batch stochastic gradient descent method.

## 3   Experimental Results

### 3.1   Experimental Setting

To evaluate our Temporal-GAN model, we compare with the following methods: **SVM-Linear** (support vector machine with linear kernel), which has been widely applied in MCI conversion prediction [6,15]; **SVM-RBF** (SVM with RBF kernel), as employed in [10,21]; and **SVM-Polynomial** (SVM with polynomial kernel) as used in [10]. Also, to validate the improvement by learning the temporal correlation structure, we compare with the **Neural Network** with exactly the same structure in our classification network (network $C$ in Fig. 1) that only uses baseline data. Besides, we compare with the case where we do not use the GAN model to generate more auxiliary samples, *i.e.*, only using network $C$ and $R$ in Fig. 1, which we call **Temporal-Deep**.

The classification accuracy is used as the evaluation metric. We divide the data into three sets: training data for training the models, validation data for tuning hyper-parameters, and testing data for reporting the results. We tune the hyper-parameter $C$ of SVM-linear, SVM-RBF and SVM-Polynomial methods in the range of $\{10^{-3}, 10^{-2}, \ldots, 10^3\}$. We compare the methods when using different portion of testing samples and report the average performance in five repetitions of random data division.

In our Temporal-GAN model, we use the fully connected neural network structure for all the networks $G$, $D$, $R$ and $C$, where each hidden layer contains 100 hidden units. The implementation detail is as follows: the number of hidden layers in structure $G$, $D$, $R$ and $C$ is $3, 1, 3, 2$ respectively. We use leaky rectified linear unit (LReLU) [12] with leakiness ratio 0.2 as the activation function of all layers except the last layer and consider weight normalization [14] for layer normalization. Also, we utilize the dropout mechanism in the regression structure $R$ with the dropout rate of 0.1. The weight parameters of all layers are initialized using the Xavier approach [4]. We use the ADAM algorithm [9] to update the weight parameters with the hyper-parameters of ADAM algorithm set as default. Both values of $\lambda_{reg}$ in Eq. (1) and $\lambda_{cly}$ in Eq. (2) are set as 0.01.

## 3.2 Data Description

All data were downloaded from the ADNI database (adni.loni.usc.edu). Each MRI T1-weighted image was first anterior commissure (AC) posterior commissure (PC) corrected using MIPAV2, intensity inhomogeneity corrected using the N3 algorithm [17], skull stripped [20] with manual editing, and cerebellum-removed [19]. We then used FAST [23] in the FSL package3 to segment the image into gray matter (GM), white matter (WM), and cerebrospinal fluid (CSF), and used HAMMER [16] to register the images to a common space. GM volumes obtained from 93 ROIs defined in [8], normalized by the total intracranial volume, were extracted as features. Out of the 93 ROIs, 24 disease-related ROIs were involved in the MCI prediction [18]. This experiment includes data from six different time points: baseline (BL), month 6 (M6), month 12 (M12), month 18 (M18), month 24 (M24) and month 36 (M36). All 216 samples with no missing MRI features at BL and M36 time are used by all the comparing methods, where there are 101 MCI converters (MCI at BL time while AD at M36) as well as 115 non-converters (MCI at both BL and M36). Since our Temporal-GAN model can use data at time points other than BL and M36, we include a total of 1419 data pairs with no missing neuroimaging measurement for training the classification, regression and generative model in our Temporal-GAN model. All neuroimaging features in the data are normalized to zero mean and unit variance.

## 3.3 MCI Conversion Prediction

We summarize the MCI conversion classification results in Table 1. The goal of the experiment is to accurately distinguish converter subjects from non-converters among the MCI samples at baseline time. From the comparison we notice that Temporal-GAN outperforms all other methods under all settings, which confirms the effectiveness of our model. Compared with SVM-Linear, SVM-RBF, SVM-Polynomial and Neural Network, the Temporal-GAN and Temporal-Deep model illustrates apparent superiority, which validates that the temporal correlation structure learned in our model substantially improves the prediction of MCI conversion. The training process of our model takes advantage of all the available data along the progression of the disease, which provides more beneficial information for the prediction of MCI conversion. By comparing Temporal-GAN and Temporal-Deep, we can notice that Temporal-GAN always performs better than Temporal-Deep, which indicates that the generative structure in Temporal-GAN could provide reliable auxiliary samples to strengthen the training of regression $R$ and classification $C$ model, thus improves the prediction of MCI conversion.

**Table 1.** MCI conversion prediction with different portion of testing data.

| Methods | 10% | 20% | 50% |
|---|---|---|---|
| SVM-Linear | 0.6273 ± 0.0668 | 0.6558 ± 0.0648 | 0.6093 ± 0.0484 |
| SVM-RBF | 0.5818 ± 0.0881 | 0.5767 ± 0.0770 | 0.5852 ± 0.0381 |
| SVM-Polynomial | 0.6545 ± 0.0617 | 0.5953 ± 0.1253 | 0.5611 ± 0.0429 |
| Neural Network | 0.3727 ± 0.0340 | 0.4233 ± 0.0426 | 0.4685 ± 0.0324 |
| Temporal-Deep | 0.7455 ± 0.0464 | 0.7209 ± 0.0441 | 0.6741 ± 0.0451 |
| Temporal-GAN | **0.7818 ± 0.0445** | **0.7488 ± 0.0342** | **0.7000 ± 0.0570** |

### 3.4   Visualization of the Imaging Markers

We use feature weight visualization in Fig. 2 to validate if our Temporal-GAN can detect disease-related features when using all 93 ROIs in the MCI conversion prediction. We adopt the Layer-wise Relevance Propagation (LRP) [1] method to calculate the importance of neuroimaging features in the testing data. We can notice that our Temporal-GAN model selects several important features from all 93 ROIs. For example, our method identifies fornix as a significant feature in distinguishing MCI non-converters. The fornix is an integral white matter bundle that locates inside the medial diencephalon. [13] reveals the vital role of white matter in Alzheimer's, such that the degradation of fornix indicates essential predictive power in MCI conversion. Moreover, cingulate region has been found by our model to be related with MCI converters. Previous study [7] finds significantly decreased Regional cerebral blood flow (rCBF) measurement in the left posterior cingulate cortex in MCI converters, which serves as an important signal in forecasting the MCI conversion. The replication of these findings proves the validity of our model.

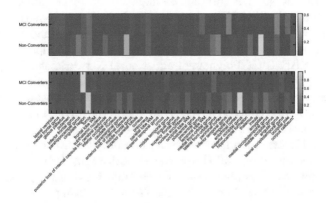

**Fig. 2.** Visualization figure showing the feature weights from our Temporal-GAN model. The upper figure shows features on the left hemisphere while the lower corresponds to the right hemisphere.

# 4   Conclusion

In this paper, we proposed a novel Temporal-GAN model for MCI conversion prediction. Our model considered the data at all time points along the disease progression and uncovered the temporal correlation structure among the neuroimaging data at adjacent time points. We also constructed a generative model to produce more reliable data to strengthen the training process. Our model illustrated superiority in the experiments on the ADNI data.

# References

1. Bach, S., Binder, A., Montavon, G., Klauschen, F., Müller, K.R., Samek, W.: On pixel-wise explanations for non-linear classifier decisions by layer-wise relevance propagation. PloS One **10**(7), e0130140 (2015)
2. Chongxuan, L., Xu, T., Zhu, J., Zhang, B.: Triple generative adversarial nets. In: Advances in Neural Information Processing Systems, pp. 4091–4101 (2017)
3. Fiorini, S., Verri, A., Barla, A., Tacchino, A., Brichetto, G.: Temporal prediction of multiple sclerosis evolution from patient-centered outcomes. In: Machine Learning for Healthcare Conference, pp. 112–125 (2017)
4. Glorot, X., Bengio, Y.: Understanding the difficulty of training deep feedforward neural networks. In: Proceedings of the Thirteenth International Conference on Artificial Intelligence and Statistics, pp. 249–256 (2010)
5. Goodfellow, I., et al.: Generative adversarial nets. In: Advances in Neural Information Processing Systems, pp. 2672–2680 (2014)
6. Hu, K., Wang, Y., Chen, K., Hou, L., Zhang, X.: Multi-scale features extraction from baseline structure MRI for MCI patient classification and AD early diagnosis. Neurocomputing **175**, 132–145 (2016)
7. Huang, C., Wahlund, L.O., Svensson, L., Winblad, B., Julin, P.: Cingulate cortex hypoperfusion predicts Alzheimer's disease in mild cognitive impairment. BMC Neurol. **2**(1), 9 (2002)
8. Kabani, N.J.: 3D anatomical atlas of the human brain. Neuroimage **7**, P-0717 (1998)
9. Kingma, D., Ba, J.: Adam: a method for stochastic optimization. arXiv preprint arXiv:1412.6980 (2014)
10. Lemos, L.: Discriminating Alzheimer's disease from mild cognitive impairment using neuropsychological data. Age (M ± SD) **70**(8.4), 73 (2012)
11. Liu, S., et al.: Multimodal neuroimaging feature learning for multiclass diagnosis of Alzheimer's disease. IEEE Trans. Biomed. Eng. **62**(4), 1132–1140 (2015)
12. Maas, A.L., Hannun, A.Y., Ng, A.Y.: Rectifier nonlinearities improve neural network acoustic models. In: International Conference on Machine Learning (ICML), vol. 30 (2013)
13. Nowrangi, M.A., Rosenberg, P.B.: The fornix in mild cognitive impairment and alzheimers disease. Front. Aging Neurosci. **7**, 1 (2015)
14. Salimans, T., Kingma, D.P.: Weight normalization: a simple reparameterization to accelerate training of deep neural networks. In: Advances in Neural Information Processing Systems (NIPS), pp. 901–909 (2016)
15. Schmitter, D.: An evaluation of volume-based morphometry for prediction of mild cognitive impairment and Alzheimer's disease. NeuroImage: Clin. **7**, 7–17 (2015)

16. Shen, D., Davatzikos, C.: Hammer: hierarchical attribute matching mechanism for elastic registration. IEEE Trans. Med. Imaging **21**(11), 1421–1439 (2002)
17. Sled, J.G., Zijdenbos, A.P., Evans, A.C.: A nonparametric method for automatic correction of intensity nonuniformity in MRI data. IEEE Trans. Med. Imaging **17**(1), 87–97 (1998)
18. Wang, H., et al.: Identifying quantitative trait loci via group-sparse multitask regression and feature selection: an imaging genetics study of the ADNI cohort. Bioinformatics **28**(2), 229–237 (2011)
19. Wang, Y., et al.: Knowledge-guided robust MRI brain extraction for diverse large-scale neuroimaging studies on humans and non-human primates. PloS One **9**(1), e77810 (2014)
20. Wang, Y., Nie, J., Yap, P.-T., Shi, F., Guo, L., Shen, D.: Robust deformable-surface-based skull-stripping for large-scale studies. In: Fichtinger, G., Martel, A., Peters, T. (eds.) MICCAI 2011. LNCS, vol. 6893, pp. 635–642. Springer, Heidelberg (2011). https://doi.org/10.1007/978-3-642-23626-6_78
21. Wei, R., Li, C., Fogelson, N., Li, L.: Prediction of conversion from mild cognitive impairment to Alzheimer's disease using MRI and structural network features. Front. Aging Neurosci. **8**, 76 (2016)
22. Weiner, M.W., et al.: The Alzheimer's disease neuroimaging initiative: a review of papers published since its inception. Alzheimer's Dement. **9**(5), e111–e194 (2013)
23. Zhang, Y., Brady, M., Smith, S.: Segmentation of brain MR images through a hidden Markov random field model and the expectation-maximization algorithm. IEEE Trans. Med. Imaging **20**(1), 45–57 (2001)

# Synthesizing Missing PET from MRI with Cycle-consistent Generative Adversarial Networks for Alzheimer's Disease Diagnosis

Yongsheng Pan[1,2], Mingxia Liu[2], Chunfeng Lian[2], Tao Zhou[2], Yong Xia[1(✉)], and Dinggang Shen[2(✉)]

[1] School of Computer Science and Engineering,
Northwestern Polytechnical University, Xi'an 710072, China
yxia@nwpu.edu.cn
[2] Department of Radiology and BRIC, University of North Carolina at Chapel Hill,
Chapel Hill, NC 27599, USA
dgshen@med.unc.edu

**Abstract.** Multi-modal neuroimages (e.g., MRI and PET) have been widely used for diagnosis of brain diseases such as Alzheimer's disease (AD) by providing complementary information. However, in practice, it is unavoidable to have missing data, i.e., missing PET data for many subjects in the ADNI dataset. A straightforward strategy to tackle this challenge is to simply discard subjects with missing PET, but this will significantly reduce the number of training subjects for learning reliable diagnostic models. On the other hand, since different modalities (i.e., MRI and PET) were acquired from the same subject, there often exist underlying relevance between different modalities. Accordingly, we propose a two-stage deep learning framework for AD diagnosis using both MRI and PET data. Specifically, in the *first* stage, we impute missing PET data based on their corresponding MRI data by using 3D Cycle-consistent Generative Adversarial Networks (3D-cGAN) to capture their underlying relationship. In the *second* stage, with the complete MRI and PET (i.e., after imputation for the case of missing PET), we develop a deep multi-instance neural network for AD diagnosis and also mild cognitive impairment (MCI) conversion prediction. Experimental results on subjects from ADNI demonstrate that our synthesized PET images with 3D-cGAN are reasonable, and also our two-stage deep learning method outperforms the state-of-the-art methods in AD diagnosis.

## 1 Introduction

Structural magnetic resonance imaging (MRI) and positron emission tomography (PET) have been widely used for diagnosis of Alzheimer's disease (AD) as well as prediction of mild cognitive impairment (MCI) conversion to AD. Recent studies have shown that MRI and PET contain complementary information for improving the performance of AD diagnosis [1,2].

© Springer Nature Switzerland AG 2018
A. F. Frangi et al. (Eds.): MICCAI 2018, LNCS 11072, pp. 455–463, 2018.
https://doi.org/10.1007/978-3-030-00931-1_52

**Fig. 1.** Proposed two-stage deep learning framework for brain disease classification with MRI and possibly incomplete PET data. Stage (1): MRI-based PET image synthesis via 3D-cGAN; Stage (2): Landmark-based multi-modal multi-instance learning (LM³IL).

A common challenge in multi-modal studies is the *missing data problem* [3, 4]. For example, in clinical practice, subjects who are willing to be scanned by MRI may reject PET scans, due to high cost of PET scanning or other issues such as concern of radioactive exposure. In the baseline Alzheimer's Disease Neuroimaging Initiative (ADNI-1) database, only approximately half of subjects have PET scans, although all 821 subjects have MRI data. Previous studies usually tackle this challenge by simply discarding subjects without PET data [5]. However, such simple strategy will significantly reduce the number of training subjects for learning the reliable models, thus degrading the diagnosis performance.

To fully utilize all available data, a more reasonable strategy is to impute the missing PET data, rather than simply discarding subjects with missing PET data. Although many data imputing methods have been proposed in the literature [3], most of them focus on imputing missing feature values that are defined by experts for representing PET. Note that, if these hand-crafted features are not discriminative for AD diagnosis (i.e., identifying AD patients from healthy controls (HCs)), the effect of imputing these missing features will be very limited in promoting the learning performance. Therefore, in this work, we focus on imputing missing PET images, rather than hand-crafted PET features.

Recently, the cycle-consistent generative adversarial network (cGAN) [6] has been successfully applied to learning the bi-directional mappings between relevant image domains. Since MR and PET images scanned from the same subjects have underlying relevance, we resort to cGAN to learn bi-directional mappings between MRI and PET, through which missing PET scan can be then synthesized based on its corresponding MRI scan.

Specifically, we propose a two-stage deep learning framework to employ all available MRI and PET for AD diagnosis, with the schematic illustration shown in Fig. 1. In the *first* stage, we impute the missing PET images by learning bi-directional mappings between MRI and PET via 3D-cGAN. In the *second* stage, based on the complete MRI and PET (i.e., after imputation), we develop a landmark-based multi-modal multi-instance learning method (LM³IL) for AD diagnosis, by learning MRI and PET features automatically in a data-driven manner. To the best of our knowledge, this is one of the first attempt to impute 3D PET images using deep learning with cycle-consistent loss in the domain of computer-aided brain disease diagnosis.

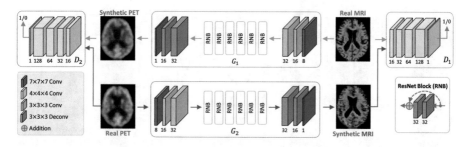

**Fig. 2.** Illustration of our proposed MRI-based PET image synthesis method, by using 3D-cGAN for learning the mappings between MRI and PET.

## 2 Method

**Problem Formulation:** Assume $\left\{\mathbf{X}_M^i, \mathbf{X}_P^i\right\}_{i=1}^N$ is a set consisting of $N$ subjects, where $\mathbf{X}_M^i \in \mathcal{X}_\mathcal{M}$ and $\mathbf{X}_P^i \in \mathcal{X}_\mathcal{P}$ are, respectively, the MRI and PET data for the $i^{\text{th}}$ subject. A multi-modal diagnosis model can then be formulated as $\hat{\mathbf{y}}^i = F\left(\mathbf{X}_M^i, \mathbf{X}_P^i\right)$, where $\hat{\mathbf{y}}^i$ is the predicted label (e.g., AD/HC) for the $i^{\text{th}}$ subject. However, when the $i^{\text{th}}$ subject does not have PET data (i.e., $\mathbf{X}_P^i$ is missing), the model $F\left(\mathbf{X}_M^i, -\right)$ cannot be executed. An intuitive way to address this issue is to use data imputation, e.g., to predict a virtual $\hat{\mathbf{X}}_P^i$ using $\mathbf{X}_M^i$, considering their underlying relevance. Letting $\hat{\mathbf{X}}_P^i = G\left(\mathbf{X}_M^i\right)$ denoting data imputation with the mapping function $G$, the diagnosis model can then be formulated as

$$\hat{\mathbf{y}}^i = F(\mathbf{X}_M^i, \mathbf{X}_P^i) \approx F\left(\mathbf{X}_M^i, G\left(\mathbf{X}_M^i\right)\right). \tag{1}$$

Therefore, there are two *sequential* tasks in the multi-modal diagnosis framework based on incomplete data, i.e., (1) learning a reliable mapping function $G$ for missing data imputation, and (2) learning a classification model $F$ to effectively combine complementary information from multi-modal data for AD diagnosis. To deal with the above two tasks sequentially, we propose a two-stage deep learning framework, consisting of two cascaded networks (i.e., 3D-cGAN and LM³IL as shown in Fig. 1), with the details given below.

**Stage 1: 3D Cycle-consistence Generative Adversarial Network (3D-cGAN).** The first stage aims to synthesize missing PET by learning a mapping function $G: \mathcal{X}_\mathcal{M} \rightarrow \mathcal{X}_\mathcal{P}$. We require $G$ to be a one-to-one mapping, i.e., there should exist a reversed function $G^{-1}: \mathcal{X}_\mathcal{P} \rightarrow \mathcal{X}_\mathcal{M}$ to keep the mapping consistent.

To this end, we propose a 3D-cGAN model, which is an extension of the existing 2D-cGAN [6]. The architecture of our 3D-cGAN model is illustrated in Fig. 2, which includes two generators, i.e., $G_1: \mathcal{X}_\mathcal{M} \rightarrow \mathcal{X}_\mathcal{P}$ and $G_2: \mathcal{X}_\mathcal{P} \rightarrow \mathcal{X}_\mathcal{M}$ ($G_2 = G_1^{-1}$), and also two adversarial discriminators, i.e., $D_1$ and $D_2$. Specifically, each generator (e.g., $G_1$) consists of three sequential (i.e., encoding, transferring and decoding) parts. The encoding part is constructed by three convolutional (Conv) layers (with 8, 16, and 32 channels, respectively) for extracting the knowledge of images in the original domain (e.g., $\mathcal{X}_\mathcal{M}$). The transferring part

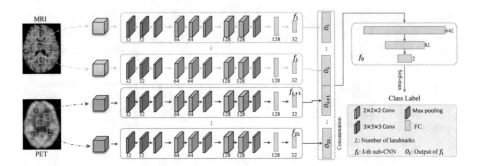

**Fig. 3.** Architecture of our proposed landmark-based multi-modal multi-instance learning for AD diagnosis, including $2L$ patch-level feature extractor (i.e., $\{f_l\}_{l=1}^{2L}$) and a subject-level classifier (i.e., $f_0$).

contains 6 residual network blocks [7] for transferring the knowledge from the original domain (e.g., $\mathcal{X}_\mathcal{M}$) to the target domain (e.g., $\mathcal{X}_\mathcal{P}$). Finally, the decoding part contains 2 deconvolutional (Deconv) layers (with 8 and 16 channels, respectively) and 1 Conv layer (with one channel) for constructing the images in the target domain (e.g., $\mathcal{X}_\mathcal{P}$). Besides, each discriminator (e.g., $D_2$) contains 5 Conv layers, with 16, 32, 64, 128, and 1 channel(s), respectively. It inputs a pair of real image (e.g., $\mathbf{X}_P^i$) and synthetic image (e.g., $G_1(\mathbf{X}_M^i)$), and then outputs a binary indicator to tell us whether the real and its corresponding synthetic images are distinguishable (output = 0) or not (output = 1). To train our 3D-cGAN model with respect to $G_1$, $G_2$, $D_1$, and $D_2$, a hybrid loss function is defined as:

$$\mathcal{L}(G_1, G_2, D_1, D_2) = \mathcal{L}_g(G_1, D_2) + \mathcal{L}_g(G_2, D_1) + \lambda\mathcal{L}_c(G_1, G_2), \qquad (2)$$

where

$$\mathcal{L}_g(G_1, D_2) = \mathbb{E}_{x \in \mathcal{X}_\mathcal{P}} \log(D_2(x)) + \mathbb{E}_{x \in \mathcal{X}_\mathcal{M}} \log(1 - D_2(G_1(x))), \qquad (3)$$

$$\mathcal{L}_c(G_1, G_2) = \mathbb{E}_{x \in \mathcal{X}_\mathcal{M}} \|G_2(G_1(x)) - x\|_1 + \mathbb{E}_{x \in \mathcal{X}_\mathcal{P}} \|G_1(G_2(x)) - x\|_1, \qquad (4)$$

are the adversarial loss and cycle consistency loss [6], respectively. The former ensures the synthetic PET images be similar to the real images, while the latter keeps each synthetic PET be consistent with the corresponding real MRI. Parameter $\lambda$ controls the importance of the consistency.

In our experiments, we empirically set $\lambda = 10$, and then trained $D_1$, $D_2$, $G_1$, and $G_2$ alternatively by minimizing $-\mathcal{L}_g(G_2, D_1)$, $-\mathcal{L}_g(G_1, D_2)$, $\mathcal{L}_g(G_1, D_2) + \lambda\mathcal{L}_c(G_1, G_2)$ and $\mathcal{L}_g(G_2, D_1) + \lambda\mathcal{L}_c(G_1, G_2)$, iteratively. The Adam solver [8] was used with a batch size of 1. The learning rate for the first 100 epochs was kept as $2 \times 10^{-3}$, and was then linearly decayed to 0 during the next 100 epochs.

**Stage 2: Landmark-based Multi-modal Multi-Instance Learning (LM$^3$IL) Network.** In the second stage, we propose the LM$^3$IL model to learn and fuse discriminative features from both MRI and PET for AD diagnosis.

Specifically, we extract $L$ patches (with size of $24 \times 24 \times 24$) centered at $L$ pre-defined disease-related landmarks [9] from each modality. Therefore, for the $i^{\text{th}}$ subject, we have $2L$ patches denoted as $\left\{\mathbf{P}_l^i\right\}_{l=1}^{2L}$, in which the first $L$ patches are extracted from $\mathbf{X}_M^i$, while the next $L$ patches are extracted from $\mathbf{X}_P^i$ or $G_1(\mathbf{X}_M^i)$ when $\mathbf{X}_P^i$ is missing.

By using $\left\{\mathbf{P}_l^i\right\}_{l=1}^{2L}$ as the inputs, the architecture of our LM$^3$IL model is illustrated in Fig. 3, which consists of $2L$ patch-level feature extractors (i.e., $\{f_l\}_{l=1}^{2L}$) and a subject-level classifier (i.e., $f_0$). All $\{f_l\}_{l=1}^{2L}$ have the *same structure* but *different parameters*. Specifically, each of them consists of 6 Conv layers and 2 fully-connected (FC) layers, with the rectified linear unit (ReLU) used as the activation function. The outputs of the 2$^{\text{nd}}$, 4$^{\text{th}}$ and 6$^{\text{th}}$ layers are down-sampled by the max-pooling operations. The size of the Conv kernels is $3 \times 3 \times 3$ in the first two Conv layers, and $2 \times 2 \times 2$ in the remaining four Conv layers. The number of channels is 32 for the 1$^{\text{st}}$, 2$^{\text{nd}}$ and 8$^{\text{th}}$ Conv layers, 64 for 3$^{\text{rd}}$ and 4$^{\text{th}}$ Conv layers, and 128 for the 5$^{\text{th}}$, 6$^{\text{th}}$ and 7$^{\text{th}}$ layers. Each patch $\mathbf{P}_l^i$ ($l \in \{1, \ldots, 2L\}$) is first processed by the corresponding sub-network $f_l$ to produce a patch-level feature vector $\mathbf{O}_l$ (i.e., the outputs of the last FC layer) with 32 elements. After that, feature vectors from all landmark locations in both MRI and PET are concatenated, which are then fed into the subsequent subject-level classifier $f_0$. The subject-level classifier $f_0$ consists of 3 FC layers and a soft-max layer, where the first two layers (with the size of $64L$ and $8L$, respectively) aim to learn a subject-level feature representation to effectively integrate complementary information from different patch locations and also different modalities, based on which the last FC layer (followed by the soft-max operation) outputs the diagnosis label (e.g., AD/HC). For the $i^{\text{th}}$ subject, the whole diagnosis procedure in our LM$^3$IL method can be summarized as:

$$\hat{\mathbf{y}}^i = F(\mathbf{X}_M^i, \mathbf{X}_P^i) = f_o\left(f_1(\mathbf{P}_1^i), \ldots, f_{2L}(\mathbf{P}_{2L}^i)\right). \tag{5}$$

In our experiments, the proposed LM$^3$IL model was trained with *log* loss using the stochastic gradient descent (SGD) algorithm [10], with a momentum coefficient of 0.9 and a learning rate of $10^{-2}$.

## 3    Experiments

**Materials and Image Pre-processing.** We evaluate the proposed method on two subsets of ADNI database [11], including ADNI-1 and ADNI-2. Subjects were divided into four categories: (1) AD, (2) HC, (3) progressive MCI (pMCI) that would progress to MCI within 36 months after baseline time, and (4) static MCI (sMCI) that would not progress to MCI. There are 821 subjects in ADNI-1, including 199 AD, 229 HC, 167 pMCI and 226 sMCI subjects. Also, ADNI-2 contains 636 subjects, including 159 AD, 200 HC, 38 pMCI and 239 sMCI subjects. While all subjects in ADNI-1 and ADNI-2 have baseline MRI data, only 395 subjects in ADNI-1 and 254 subjects in ADNI-2 have PET images.

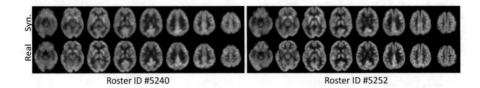

Roster ID #5240                    Roster ID #5252

**Fig. 4.** Illustration of synthetic (Syn.) PET generated by our method for two typical subjects (Roster IDs: 5240, 5252), as well as their corresponding real images.

All MR images were pre-processed via four steps: (1) anterior commissure (AC)-posterior commissure (PC) alignment, (2) skull stripping, (3) intensity correction, (4) cerebellum removal, and (5) linear alignment to a template MRI. Each PET image was also aligned to its corresponding MRI via linear registration. Hence, there is spatial correspondence between MRI and PET for each subject.

**Experimental Settings.** We performed two groups of experiments in this work. In the *first* group, we aim to evaluate the quality of the synthetic images generated by 3D-cGAN. Specifically, we train the 3D-cGAN model using subjects with complete MRI and PET scans in ADNI-1, and test this image synthesis model on the complete subjects (with both MRI and PET) in ADNI-2. The averaged peak signal-to-noise ratio (PSNR) is used to measure the image quality of those synthetic PET and MR images generated by our method.

In the *second* group, we evaluate the proposed LM$^3$IL method on both tasks of AD classification (AD vs. HC) and MCI conversion prediction (pMCI vs. sMCI) using both real multi-modal images and our synthetic PET images. Six metrics are used for performance evaluation, including accuracy (ACC), sensitivity (SEN), specificity (SPE), F1-Score (F1S), the area under receiver operating characteristic (AUC) and Matthews correlation coefficient (MCC) [12]. Subjects from ADNI-1 are used as the training data, while those from ADNI-2 are treated as independent test data. In LM$^3$IL, 30 landmarks are detected for each MRI via a landmark detection algorithm [9], and these landmarks in each MRI are further located in its corresponding PET image. For each subject, we extract 30 image patches ($24 \times 24 \times 24$) centered at 30 landmarks from image of each modality (i.e., MRI and PET) as the input of LM$^3$IL.

Our LM$^3$IL method is compared with five approaches: (1) gray matter (GM) volume within 90 regions-of-interest (denoted as **ROI**) [5], (2) voxel-wise GM density (denoted as **VGD**) [13], (3) landmark-based local energy patterns (**LLEP**) [9], (4) landmark-based deep single-instance learning (**LDSIL**) [14], and (5) landmark-based deep multi-instance learning (**LDMIL**) [14] that can be regarded as a single-modal variant of our LM$^3$IL method using only MRI. To test the effect of our generated PET images, we further compare LM$^3$IL with its variant (denoted as **LM$^3$IL-C**) that use only subjects with complete MRI and PET data. We share the same landmarks and the same size of image patches in LLEP, LDSIL, LDMIL, LM$^3$IL-C and LM$^3$IL. Note that four variants of our

**Table 1.** Performance of seven different methods in both tasks of AD classification (AD vs. HC classification) and MCI conversion prediction (pMCI vs. sMCI classification).

| Method | AD vs. HC classification | | | | | | pMCI vs. sMCI classification | | | | | |
|---|---|---|---|---|---|---|---|---|---|---|---|---|
| | ACC (%) | SEN (%) | SPE (%) | F1S (%) | MCC (%) | AUC (%) | ACC (%) | SEN (%) | SPE (%) | F1S (%) | MCC (%) | AUC (%) |
| ROI | 79.17 | 78.62 | 79.60 | 76.92 | 58.00 | 86.73 | 66.06 | 47.37 | 69.04 | 27.69 | 11.98 | 63.77 |
| VGD | 80.50 | 77.35 | 83.00 | 77.84 | 60.44 | 87.62 | 64.26 | 36.84 | 68.62 | 22.05 | 04.02 | 59.29 |
| LLEP | 82.22 | 77.36 | 86.07 | 79.35 | 63.83 | 88.11 | 68.59 | 39.47 | 73.22 | 25.64 | 09.67 | 63.63 |
| LDSIL | 90.56 | 87.42 | 93.03 | 89.10 | 80.82 | 95.74 | 70.04 | 36.84 | 75.31 | 25.23 | 09.49 | 64.48 |
| LDMIL | 91.09 | 88.05 | 93.50 | 89.74 | 81.91 | 95.86 | 76.90 | 42.11 | 82.43 | 33.33 | 20.74 | **77.64** |
| $LM^3IL$-C | 87.50 | 84.85 | 89.36 | 84.85 | 74.21 | 93.08 | 76.92 | 44.44 | 81.16 | 30.77 | 19.81 | 68.59 |
| $LM^3IL$ | **92.50** | **89.94** | **94.53** | **91.37** | **84.78** | **95.89** | **79.06** | **55.26** | **82.85** | **40.86** | **30.13** | 75.84 |

methods (i.e., LDSIL, LDMIL, $LM^3IL$-C and $LM^3IL$) automatically learn features of MRI/PET via deep network, while the remaining methods (ROI, VGD and LLEP) rely on support vector machines with default parameters.

**Performance of Image Synthesis Model.** To evaluate the quality of the synthetic images generated by 3D-cGAN, we first train the 3D-cGAN model using complete subjects (i.e., containing both PET and MRI) in ADNI-1, and test this image synthesis model on the complete subjects in ADNI-2. Two typical subjects with real and synthetic PET scans are shown in Fig. 4. From Fig. 4, we can observe that our synthetic PET look very similar to their corresponding real images. Also, the mean and standard deviation of PSNR values of synthetic PET images in ADNI-2 are $24.49 \pm 3.46$. These results imply that our trained 3D-cGAN model is reasonable, and the synthetic PET scans have acceptable image quality (in terms of PSNR).

**Results of Disease Classification.** We further evaluate the effectiveness of our two-stage deep learning method in both tasks of AD classification and MCI conversion prediction. The experimental results achieved by seven different methods are reported in Table 1. From Table 1, we can see that the overall performance of our $LM^3IL$ method is superior to six competing methods regarding six evaluation metrics. *Particularly,* our method achieves a significantly improved sensitivity value (i.e., nearly 8% higher than the second best sensitivity achieved by ROI) in pMCI vs. sMCI classification. Since higher sensitivity values indicate higher confidence in disease diagnosis, these results imply that our method is reliable in predicting the progression of MCI patients, which is potentially very useful in practice. *Besides,* as can be seen from Table 1, four methods (i.e., LDSIL, LDMIL, $LM^3IL$-C and $LM^3IL$) using deep-learning-based features of MRI and PET usually outperform the remaining three approaches (i.e., ROI, VGD and LLEP) that use hand-crafted features in both classification tasks. This suggests that integrating feature extraction of MRI and PET and classifier model training into a unified framework (as we do in this work) can boost the performance of AD diagnosis. *Furthermore,* we can see that our $LM^3IL$ method using both MRI and PET generally yields better results than its two single-modal variants (i.e., LDSIL and LDMIL) that use only MRI data. The underlying reason could be that our method can employ the complementary information contained in MRI and PET data. *On the other hand,* our $LM^3IL$ consistently outperforms $LM^3IL$-C that utilize only subjects with complete MRI and PET data. These results clearly demonstrate that the synthetic PET images generated by our 3D-cGAN model are useful in promoting brain disease classification performance.

## 4  Conclusion

In this paper, we have presented a two-stage deep learning framework for AD diagnosis, using incomplete multi-modal imaging data (i.e., MRI and PET). Specifically, in the first stage, to address the issue of missing PET data, we proposed a 3D-cGAN model for imputing those missing PET data based on their corresponding MRI data, considering the relationship between images (i.e., PET and MRI) scanned for the same subject. In the second stage, we developed a landmark-based multi-modal multi-instance neural network for brain disease classification, by using subjects with complete MRI and PET (i.e., both real and synthetic PET). The experimental results demonstrate that the synthetic PET images produced by our method are reasonable, and our proposed two-stage deep learning framework outperforms conventional multi-modal methods for AD classification. Currently, only the synthetic PET images are used for learning the classification models. Using these synthetic MRI data could further augment the training samples for improvement, which will be our future work.

**Acknowledgment.** This research was supported in part by the National Natural Science Foundation of China under Grants 61771397 and 61471297, in part by Innovation Foundation for Doctor Dissertation of NPU under Grants CX201835, and in part by NIH grants EB008374, AG041721, AG042599, EB022880. Data collection and sharing for this project was funded by the Alzheimer's Disease Neuroimaging Initiative (ADNI) and DOD ADNI.

## References

1. Calhoun, V.D., Sui, J.: Multimodal fusion of brain imaging data: a key to finding the missing link(s) in complex mental illness. Biol. Psychiatry: Cogn. Neurosci. Neuroimaging **1**(3), 230–244 (2016)
2. Liu, M., Gao, Y., Yap, P.T., Shen, D.: Multi-hypergraph learning for incomplete multi-modality data. IEEE J. Biomed. Health Inform. **22**(4), 1197–1208 (2017)
3. Parker, R.: Missing Data Problems in Machine Learning. VDM Verlag, Saarbrücken (2010)
4. Liu, M., Zhang, J., Yap, P.T., Shen, D.: View-aligned hypergraph learning for Alzheimer's disease diagnosis with incomplete multi-modality data. Med. Image Anal. **36**, 123–134 (2017)
5. Zhang, D., Shen, D.: Multi-modal multi-task learning for joint prediction of multiple regression and classification variables in Alzheimer's disease. NeuroImage **59**(2), 895–907 (2012)
6. Zhu, J.Y., Park, T., Isola, P., Efros, A.A.: Unpaired image-to-image translation using cycle-consistent adversarial networks. arXiv preprint arXiv:1703.10593 (2017)

7. He, K., Zhang, X., Ren, S., Sun, J.: Deep residual learning for image recognition. In: IEEE Conference on Computer Vision and Pattern Recognition, pp. 770–778 (2016)
8. Kingma, D.P., Ba, J.: Adam: a method for stochastic optimization. arXiv preprint arXiv:1412.6980 (2014)
9. Zhang, J., Gao, Y., Gao, Y.: Detecting anatomical landmarks for fast Alzheimer's disease diagnosis. IEEE Trans. Med. Imaging **35**(12), 2524–2533 (2016)
10. Boyd, S., Vandenberghe, L.: Convex Optimization. Cambridge University Press, Cambridge (2004)
11. Jack, C., Bernstein, M., Fox, N.: The Alzheimer's disease neuroimaging initiative (ADNI): MRI methods. J. Magn. Reson. Imaging **27**(4), 685–691 (2008)
12. Matthews, B.: Comparison of the predicted and observed secondary structure of T4 phage lysozyme. Biochim. Biophys. Acta (BBA) - Protein Struct. **405**(2), 442–451 (1975)
13. Ashburner, J., Friston, K.J.: Voxel-based morphometry - the methods. NeuroImage **11**(6), 805–821 (2000)
14. Liu, M., Zhang, J., Adeli, E., Shen, D.: Landmark-based deep multi-instance learning for brain disease diagnosis. Med. Image Anal. **43**, 157–168 (2018)

# Exploratory Population Analysis with Unbalanced Optimal Transport

Samuel Gerber[1(✉)], Marc Niethammer[2], Martin Styner[2], and Stephen Aylward[1]

[1] Kitware Inc., Carborro, NC 27510, USA
samuel.gerber@kitware.com
[2] University of North Carolina, Chapel Hill, NC 27504, USA

**Abstract.** The plethora of data from neuroimaging studies provide a rich opportunity to discover effects and generate hypotheses through exploratory data analysis. Brain pathologies often manifest in changes in shape along with deterioration and alteration of brain matter, i.e., changes in mass. We propose a morphometry approach using unbalanced optimal transport that detects and localizes changes in mass and separates them from changes due to the location of mass. The approach generates images of mass allocation and mass transport cost for each subject in the population. Voxelwise correlations with clinical variables highlight regions of mass allocation or mass transfer related to the variables. We demonstrate the method on the white and gray matter segmentations from the OASIS brain MRI data set. The separation of white and gray matter ensures that optimal transport does not transfer mass between different tissues types and separates gray and white matter related changes. The OASIS data set includes subjects ranging from healthy to mild and moderate dementia, and the results corroborate known pathology changes related to dementia that are not discovered with traditional voxel-based morphometry. The transport-based morphometry increases the explanatory power of regression on clinical variables compared to traditional voxel-based morphometry, indicating that transport cost and mass allocation images capture a larger portion of pathology induced changes.

## 1 Introduction

Neurological disease and disorder manifest in subtle and varied changes in brain anatomy that can be non-local in nature and effect amounts of white and gray matter as well as relative positioning and shapes of local brain anatomy. To detect and quantify these changes are primary goals of morphometry based population analysis. We propose a morphometry approach based on unbalanced optimal transport, termed UTM, that yields a voxelwise comparison but can detect global and regional deterioration of matter. The UTM formulation explicitly separates changes in matter volume from changes in matter location. This separation of effects leads to stronger correlations and more readily interpretable visualizations.

A. F. Frangi et al. (Eds.): MICCAI 2018, LNCS 11072, pp. 464–472, 2018.
https://doi.org/10.1007/978-3-030-00931-1_53

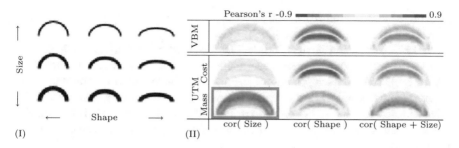

**Fig. 1.** (I) Illustration of unbalanced optimal transport on a toy data set of 100 half-ellipses with different shapes (ratio of major to minor radius) and sizes (black pixel count). (II) Spatial correlations of size, shape and size + shape to VBM (pixel intensity) and UTM (mass allocation and transport costs). Positive and negative correlation indicate an increase and decrease of mass, cost, or intensity, respectively. Only UTM is able to detect size variation (highlighted box) as strongly positively correlated to mass allocation. Both UTM and VMB detect shape variation, with stronger correlations in transport cost than mass allocation in UTM. UTM identifies both change in shape and mass for the variable depending on shape plus size.

Optimal transport, as the name implies, solves the problem of transporting mass from a probability measure $\mu$ to a probability measure $\nu$, such that the cost of moving mass from the source $\mu$ to the target $\nu$ is minimized. Unbalanced optimal transport [3,11] extends optimal transport to measures that do not need to have equal mass by adding a mechanism to add mass to the optimization problem. The solution of the unbalanced optimal transport yields a transport plan, or coupling, that measures local mass allocation and movement between source and target locations. We decompose these transport plans to measure for each subject, at each voxel, mass allocations and costs of mass transfer. These two measures explicitly separate changes due to differences in amount of matter from differences due to changes in location of matter related to relative position and shape of anatomies.

Voxel-based morphometry (VBM) [1] yields spatially localized changes in brain anatomy but has difficulty in discovering regionally or globally occurring changes. Figure 1 demonstrate the capability of UTM to detect changes in size on a toy example of one hundred half-ellipses with different ellipticity and thickness. Correlating the mass allocation and mass transfer images from UTM with either shape or size of the half-ellipses shows that the proposed method is capable of correctly attributing changes to either variation in shape or size. Traditional VBM results in weak correlation with changes in size and cannot attribute the source of the changes to ellipse shape or size.

Deformation-based morphometry (DBM) [2] addresses the issue of detecting global effects by comparing the parameters of non-linear spatial normalizations to a template. DBM results, typically shown as modes of variation, are more difficult to visualize and interpret. UTM combines the strengths of VBM and DBM methods. While the results are based on a voxelwise analysis and are easily interpretable, the quantities compared stem from a global optimization problem which can detect global and regional effects.

A key observation is that VBM, TBM and DBM are driven by local image gradients, and voxelwise measures ultimately lead to very similar results [8, Chap. 6]. UTM solves a global optimization problem based on the distribution of mass, and the allocation of mass is not driven by image gradients. In UTM allocation of mass can be diffuse over a large region without the need for a smooth spatial normalization to distribute the image gradient driven warp over a larger region.

VBM has been shown to be sensitive to the particulars of the spatial normalization [4,6]. UTM still depends on a rigid spatial normalization to bring the subjects into a common coordinate system were voxelwise comparisons are feasible. However, the optimal transport is insensitive to small miss-alignments in the spatial normalization step; small shifts in mass locations incur only small transport costs. After the spatial normalization UTM does not depend on a template, the mass allocation and transport cost images are based on an averaging of quantities derived from pairwise transport plans, which alleviates bias due to registrations to a single template.

The key contributions of this work are:

1. A voxel-wise morphometry approach capable of detecting regional and global changes.
2. A formulation of unbalanced optimal transport to detect and disentangle mass changes.
3. A demonstration of the importance of incorporating mass imbalances to detect and localize neurodegenerative anatomical changes (Sect. 4) from changes in mass location (Sect. 3).
4. An application of optimal transport to gray and white matter masks individually to avoid mass exchange between tissues and incommensurable measurements across subjects in MRI intensities (Sect. 4).

## 2   Related Work

Recent advances in the computation of optimal transport plans [5,9] paved the way for a flurry of applications in machine learning and spurred interest in applications to medical image analysis. Gramfort et al. [10] use optimal transport for improved averaging of neuroimaging data, Feydy et al. [7] use unbalanced optimal transport as a similarity measure for diffeomorphic registration, and Kundu et al. [12] formulate a DBM approach, TrBM, which replaces non-linear warps by optimal transport plans.

UTM integrates optimal transport to morphometric analysis in a different way from TrBM. TrBM yields a parametrization of the transport plans akin to DBM that captures global changes. UTM uses the transport plan to create voxelwise measures and adds mass imbalances into the transport plans, which prove to be an important indicator in neurodegenerative diseases. The TrBM is applied to graylevel MRI intensities, we choose to use the gray and white matter masks instead of MRI intensity values to avoid conversion of white to gray matter in the optimal transport optimization and to avoid difficulties in normalizing MRI intensities across subjects.

# 3    Unbalanced Optimal Transport Morphometry

Morphometry with unbalanced optimal transport follows the standard voxelwise morphometry pipeline but performs voxelwise statistical analysis on the mass allocation and transport cost images derived from the solution of unbalanced optimal transport between the subjects. Section 3.1 describes the unbalanced optimal transport problem and Sect. 3.2 the construction of the mass allocation and transport cost maps for each subject.

## 3.1    Unbalanced Optimal Transport

For two probability measures $\mu$ and $\nu$ on probability spaces $\mathbf{X}$ and $\mathbf{Y}$ respectively, a coupling of $\mu$ and $\nu$ is a measure $\pi$ on $\mathbf{X} \times \mathbf{Y}$ such that the marginals of $\pi$ are $\mu$ and $\nu$. The coupling $\pi$ defines a *transport plan* that captures how much mass $\pi(x, y)$ is transported from any $x \in \mathbf{X}$ to any $y \in \mathbf{Y}$. To define optimal transport and optimal couplings, we need a cost function $\mathsf{c}(x, y)$ on $\mathbf{X} \times \mathbf{Y}$ representing the work or cost needed to move a unit of mass from $x$ to $y$. An optimal coupling $\pi^*$ minimizes this cost over all choices of couplings $\mathcal{C}(\mu, \nu)$ between $\mu$ and $\nu$:

$$\pi^* = \operatorname*{argmin}_{\pi \in \mathcal{C}(\mu, \nu)} \int_{\mathbf{X}} \int_{\mathbf{Y}} \mathsf{c}(x, y) d\pi(x, y). \tag{1}$$

For discrete distributions $\mu = \sum_1^n w(x_i)\delta(x_i)$ and $\nu = \sum_1^m v(y_i)\delta(y_i)$ with $\sum w(x_i) = \sum v(y_i) = 1$ the optimal transport problem can be solved by linear programming. To extend the formulation to deal with arbitrary positive measures $\mu$ and $\nu$ with mass imbalance $\Delta = |\sum_1^n v(y_i) - \sum_1^m w(x_i)|$, the linear program is modified to allow for the creation of mass by adding a new source location $z^s$ and target location $z^t$ and constraints to restrict the amount of mass added to be at most $w(x_i)$ and $v(x_i)$ at any target and source location, respectively. With these modifications the linear program reads:

$$\min_{\pi} \sum_{\substack{i=1,\dots,n \\ j=1,\dots,m}} \mathsf{c}(x_i, y_j)\pi(x_i, y_j) \text{ s.t.} \begin{cases} \sum_j \pi(x_i, y_j) + \pi(x_i, z^t) = \mu(\{x_i\}) = w(x_i) \\ \sum_i \pi(x_i, y_j) + \pi(z^s, y_j) = \nu(\{y_j\}) = v(y_j) \\ \sum_j \pi(x_i, z^t) + \sum_j \pi(z^s, y_j) = \Delta \\ \pi(x_i, y_j) \geq 0, \pi(z^s, y_j) \geq 0, \pi(x_i, z^t) \geq 0 \end{cases}$$

$$\tag{2}$$

The modifications result in a standard optimal transport problem that can, with only minor modifications, be solved by fast approximation algorithms for large data sets such as the Sinkhorn approach [5] or multiscale strategies [9].

An arbitrary cost $\mathsf{c}(x^s, y_j)$ and $\mathsf{c}(x_i, y^t)$ can be assigned to allocate mass, and restriction to only allocate $\Delta$ amount of mass can be relaxed or removed, striking a trade-off between the cost of creating mass and the cost of moving mass. For our application, we choose the allocation of mass at zero cost but only allow for the allocation of exactly $\Delta$ mass. This forces the transfer of all jointly available mass between source and target location and distribute the mass $\Delta$ optimally to reduce the cost of movement.

Solving the unbalanced optimal transport problem is convex and yields a global minimum without any parameter tuning, the only choice is the selection of a cost c.

### 3.2   Construction of Mass Allocation and Transport Cost Images

For an image $X^k$ denote by $X^k(x_i)$ the associated non-negative value at voxel location $x_i$, which defines a measure $\mu^k = \sum_1^n X^k(x_i)\delta(x_i)$. To construct the voxelwise mass allocation and transport cost images we solve for optimal transport plans $\pi_{k,l}^*$ between all images $X^k$ and $X^l$. The variables $z^s$ and $z^t$ in Eq. 2 capture the amount of mass allocated when moving mass from $X^k$ to $X^l$. Denote by $z_{k,l}^s$ and $z_{k,l}^t$ the mass allocation variable associated with the optimal transport plan $\pi_{k,l}^*$. For subject $X_k$ the mass allocation image $M^k$ is constructed by $M^k(x_i) = \sum_l \left( \pi_{k,l}^*(z_{k,l}^s, x_i) - \pi_{k,l}^*(x_i, z_{k,l}^t) \right)$ and the transport cost image by $C^k(x_i) = \sum_l \sum_j \pi_{k,l}^*(x_i, x_j)\mathsf{c}(x_i, y_j)$.

The images $M^k$ and $C^k$ are smoothed with a small Gaussian to increase correlations between neighboring pixels. The smoothed images $M^k$ and $C^k$ replace the smoothed intensity or Jacobian determinant images in the statistical analysis of a VBM or TBM pipeline.

## 4   Application to OASIS Brain MRI

We apply UTM to the OASIS brain data set [13]. Code to replicate the results is located at https://github.com/KitwareMedical/UTM. The OASIS database consists of T1 weighted MRI of 416 subjects aged 18 to 96. One hundred of the subjects over the age of 60 are diagnosed with mild to moderate dementia. The images in the OASIS data set are already skull-stripped, gain-field corrected and registered to the Talaraich atlas space [14] with a 12-parameter affine transform. To focus on dementia related effects we restrict the analysis to 137 patients from age 60 to 80. This set of patients contains 66 healthy patents and 71 patients with a diagnosis of very mild to moderate dementia as established by a clinical dementia rating (CDR, $0 =$ normal, $0.5 =$ very mild, $1 =$ Mild, $2 =$ Moderate). In addition to CDR the data set contains information about Age and a mini mental state examination score (MMSE, count of correct answers with a perfect score of 30).

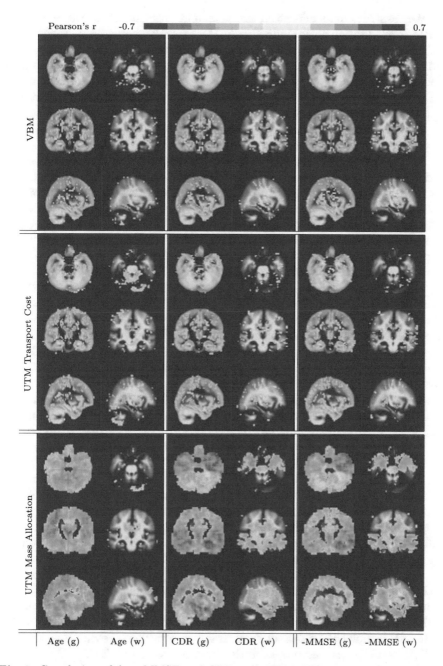

**Fig. 2.** Correlation of Age, MMSE and CDR with VBM, UTM mass imbalances and UTM transport costs on (g) gray and (w) white matter segmentations. Correlations are only shown at locations permutation tested p-value less than 0.05 (1000 permutations). The background image are average white and gray matter segmentations.

We construct mass allocation and transport cost images for white and gray matter segmentation masks individually. However, the method can without modification be applied to any scalar image or with the definition of a corresponding cost function even to vector or tensor valued images. For a sensible application the only requirement is that the measurement at voxels are commensurable across subjects. For the cost c we use the squared euclidean distance between the voxel locations to introduce a preference of many small mass transports over single larger mass transfers. To reduce computation costs, we subsample the images to $44 \times 52 \times 44$, which yields computations times of approximately one second using a multiscale optimal transport solver [9]. Both VBM intensities of the white and gray matter masks and the UTM mass allocation and transport cost images are smoothed with a Gaussian with 3 mm standard deviation before computing correlations or regression models. The multiscale optimal transport solver introduces an additional smoothing of the mass allocation instead of focusing allocations at the boundary regions. This yields visually more pleasing results but does not change the overall results drastically. This effect could also be achieved by limiting the amount of mass allocated at any given location.

Figure 2 shows voxelwise correlations to Age, CDR and negative MMSE and compares to a traditional intensity VBM analysis. The most striking difference is that the loss in overall gray matter is immediately visible in the mass allocation part of UTM. The overall loss related to Age is differently distributed from the loss related to CDR and MMSE. CDR and MMSE have more pronounced gray matter deterioration in the cerebellum and in the temporal lobe, known to be involved with cognitive decline. A second striking difference is the loss of white matter around the ventricles in the area of the hippocampus that is significant in CDR and MMSE but not with respect to Age. The observed associations are consistent with findings reported in the literature on Alzheimer's disease and mild cognitive impairment. The VBM analysis is not capable to discover the global and regionally varying diminishing of gray and white matter.

We compare the explanatory power of UTM to VBM with elastic net regression [15] of transport cost and transport images and intensity images on clinical parameters. We set the elastic net regression penalty trade-off to $\alpha = 0.1$, a large ridge penalty and small sparsity penalty trade-off. Table 1 reports cross-validated root mean square error (RMSE) using regularizations based on the minimal RMSE and the strongest regularization that results in an RMSE within one standard deviation of the minimal RMSE, as advocated by [15]. The minimal regularization appears to overfit with a much smaller $R^2$ but only a minor reduction in RMSE. Regression on UTM images outperforms VBM on all clinical variables. For the weaker regularized models UTM results in a larger reduction in RMSE combined with smaller reduction in $R^2$. These results indicate that UTM captures more neurodegenerative information than VBM.

**Table 1.** Elastic net regression on UTM transport cost and mass allocation images versus regression on VBM intensity images. Reported are 10-fold cross-validated root mean square error (RMSE) and $R^2$ with the amount of regularization either selected by the minimal RMSE (min) or the most parsimonious, strongest regularized, model within one standard deviation of the minimal RMSE (1se).

| Model | RMSE 1se | $R^2$1se | RMSE min | $R^2$min |
|---|---|---|---|---|
| VBM, Age | 4.89 | 0.24 | 4.81 | 0.95 |
| UTM, Age | **4.51** | **0.39** | 4.29 | 0.72 |
| VBM, MMSE | 3.80 | 0.21 | 3.61 | 0.97 |
| UTM, MMSE | **3.61** | **0.25** | 3.27 | 0.54 |
| VBM, CDR | 0.36 | 0.21 | 0.33 | 0.69 |
| UTM, CDR | **0.32** | **0.40** | 0.30 | 0.72 |

## 5   Conclusion

The paper demonstrates that UTM captures changes related to size not detected by traditional VBM and can attribute effects to mass transfer and mass allocation. UTM captures regional and global changes while retaining the benefits of a voxelwise visualization and analysis of results. The results demonstrate the signficance of incorporating mass imbalance into the morphometry framework.

**Acknowledgments.** This work was funded, in part, by NIH grants R01EB021391, R01HD055741, U54HD079124, R42NS086295, R44NS081792, R44CA165621, and R01EB021396 and by NSF grant ECCS-1711776.

## References

1. Ashburner, J., Friston, K.J.: Voxel-based morphometry-the methods. Neuroimage **11**(6), 805–821 (2000)
2. Ashburner, J., Hutton, C., Frackowiak, R., Johnsrude, I., Price, C., Friston, K.: Identifying global anatomical differences: deformation-based morphometry. Hum. Brain Mapp. **6**(5–6), 348–357 (1998)
3. Benamou, J.-D.: Numerical resolution of an "unbalanced" mass transport problem. ESAIM: Math. Model. Numer. Anal. **37**(5), 851–868 (2003)
4. Bookstein, F.L.: "Voxel-based morphometry" should not be used with imperfectly registered images. Neuroimage **14**(6), 1454–1462 (2001)
5. Cuturi, M.: Sinkhorn distances: lightspeed computation of optimal transport. In: Advances in Neural Information Processing Systems, pp. 2292–2300 (2013)
6. Davatzikos, C.: Why voxel-based morphometric analysis should be used with great caution when characterizing group differences. Neuroimage **23**(1), 17–20 (2004)
7. Feydy, J., Charlier, B., Vialard, F.-X., Peyré, G.: Optimal transport for diffeomorphic registration. In: Descoteaux, M., Maier-Hein, L., Franz, A., Jannin, P., Collins, D.L., Duchesne, S. (eds.) MICCAI 2017. LNCS, vol. 10433, pp. 291–299. Springer, Cham (2017). https://doi.org/10.1007/978-3-319-66182-7_34

8. Frackowiak, R.S.: Human Brain Function. Academic Press, London (2004)
9. Gerber, S., Maggioni, M.: Multiscale strategies for computing optimal transport. J. Mach. Learn. Res. **18**(72), 1–32 (2017)
10. Gramfort, A., Peyré, G., Cuturi, M.: Fast optimal transport averaging of neuroimaging data. In: Ourselin, S., Alexander, D.C., Westin, C.-F., Cardoso, M.J. (eds.) IPMI 2015. LNCS, vol. 9123, pp. 261–272. Springer, Cham (2015). https://doi.org/10.1007/978-3-319-19992-4_20
11. Guittet, K.: Extended Kantorovich norms: a tool for optimization. Ph.D. thesis, INRIA (2002)
12. Kundu, S., Kolouri, S., Erickson, K.I., Kramer, A.F., McAuley, E., Rohde, G.K.: Discovery and visualization of structural biomarkers from MRI using transport-based morphometry. NeuroImage **167**, 256–275 (2018)
13. Marcus, D.S., Fotenos, A.F., Csernansky, J.G., Morris, J.C., Buckner, R.L.: Open access series of imaging studies: longitudinal mri data in nondemented and demented older adults. J. Cogn. Neurosci. **22**(12), 2677–2684 (2010)
14. Talairach, J., Tournoux, P.: Co-planar Stereotaxic Atlas of the Human Brain: 3-Dimensional Proportional System: An Approach to Cerebral Imaging. Thieme, New York (1988)
15. Zou, H., Hastie, T.: Regularization and variable selection via the elastic net. J. R. Stat. Soc., Ser. B **67**, 301–320 (2005)

# Multi-modal Synthesis of ASL-MRI Features with KPLS Regression on Heterogeneous Data

Toni Lassila[1]($^\boxtimes$), Helena M. Faria[1], Ali Sarrami-Foroushani[1],
Francesca Meneghello[2], Annalena Venneri[3], and Alejandro F. Frangi[1]

[1] Centre for Computational Imaging and Simulation Technologies in Biomedicine (CISTIB), Department of Electronic and Electrical Engineering, University of Sheffield, Sheffield, UK
`t.lassila@sheffield.ac.uk`
[2] IRCCS Fondazione Ospedale San Camillo, Lido Venice, Italy
[3] Department of Neuroscience, University of Sheffield, Sheffield, UK

**Abstract.** Machine learning classifiers are frequently trained on heterogeneous multi-modal imaging data, where some patients have missing modalities. We address the problem of synthesising arterial spin labelling magnetic resonance imaging (ASL-MRI) - derived cerebral blood flow (CBF) - features in a heterogeneous data set. We synthesise ASL-MRI features using T1-weighted structural MRI (sMRI) and carotid ultrasound flow features. To deal with heterogeneous data, we extend the kernel partial least squares regression (kPLSR) - method to the case where both input and output data have partial coverage. The utility of the synthetic CBF features is tested on a binary classification problem of mild cognitive impairment patients vs. controls. Classifiers based on sMRI and synthetic ASL-MRI features are combined using a maximum probability rule, achieving a balanced accuracy of 92% (sensitivity 100 %, specificity 80 %) in a separate validation set. Comparison is made against support vector machine-classifiers from literature.

## 1 Introduction

Arterial spin-labelling magnetic resonance imaging (ASL-MRI) is a non-invasive blood flow imaging modality that can improve the diagnosis of Alzheimer's disease (AD) by providing estimates of cerebral blood flow (CBF) and identifying regions of chronic cerebral hypoperfusion in individuals. However, ASL-MRI is still not part of clinical routine; many research databases used to train dementia classifiers, such as ADNI (Alzheimer's Disease Neuroimaging Initiative), do not include ASL-MRI as part of their data collection protocol.

Image synthesis refers to the simulation of missing image modalities with machine learning algorithms using image modalities that are available. Synthetic image modalities can in some cases provide additional predictive value beyond the original data used for synthesis [7]. We use kernel partial least squares regression (kPLSR) for synthesising ASL-MRI-based CBF maps using structural MRI

© Springer Nature Switzerland AG 2018
A. F. Frangi et al. (Eds.): MICCAI 2018, LNCS 11072, pp. 473–481, 2018.
https://doi.org/10.1007/978-3-030-00931-1_54

(sMRI) features and carotid ultrasound flow measurements as regressors. Using partial volumes of cortical and sub-cortical regions as features allows the relation between cerebral volume loss and reduction in CBF in dementia patients to be learned by the model. The synthetic CBF maps are used to generate CBF maps of patients for whom no ASL-MRI images are available (*CBF imputation*).

The challenge of building models using multi-modal data is that multiple cohorts may be required to achieve enough coverage and some cohorts will have some modalities missing. We refer to this as the *heterogeneous data problem*. To address it, we modify the NIPALS algorithm for training the kPLSR model to work on heterogeneous data, where some features are missing in part of the input data $X$, and some of the output data $Y$ are also missing. The synthetic CBF maps are utilised as classification features in discriminating mild cognitive impairment-patients (MCIs) from cognitively healthy controls (CHCs).

The MCI vs. CHC-problem is less studied than the AD vs. CHC-problem (see e.g. the review [1]) because sMRI-derived partial volume-features are less informative in the prodromal stage of AD. We apply a simultaneous feature selection and classification strategy based on: (i) use of regional CBF values averaged over anatomical subregions (instead of voxelwise values), and (ii) elastic net regression. The proposed classifiers are compared to MCI vs. CHC-classifiers from literature using different imaging modalities as features.

## 2    Methods

### 2.1    Acquisition and Pre-processing of Imaging Data

Data from two clinical centres and three different cohorts were included to increase the number of cases available for training models (Table 1). The combined data set was heterogeneous with respect to operator, MR field strength, and modalities available for each case. Three different sets of features were used.

**sMRI Features:** T1-weighted sMRI were acquired and volumes of 141 cortical and sub-cortical regions were computed by propagating anatomical labels with the geodesic information flows-algorithm [2]. These features encapsulated grey matter (GM) atrophy but did not contain direct information about CBF.

**Carotid Flow Features:** Carotid ultrasound measurements were performed in two of the cohorts. Flow velocity signals were extracted from DICOM images and used to compute the mean flow rate and flow pulsatility indices (for both ICA-L and ICA-R separately), for a total of four features. These features encapsulated the baseline total CBF, but did not contain region-specific effects.

**ASL Features:** Pseudo-continuous ASL-MRI parameters were: TR/TE, 4,000 ms/14 ms; flip angle, 40°; FOV, 240 mm × 240 mm; matrix size, 80 × 80; 17 slices; thickness, 7 mm; labelling duration, 1.65 s; post-labelling delay, 1.525 s; and labelling gap, 20 mm. The ASL-MRI CBF maps were registered against the sMRI using SPM12 and equipped with maximum probability tissue labels defined on the MNI152 atlas, provided by Neuromorphometrics, Inc. from data

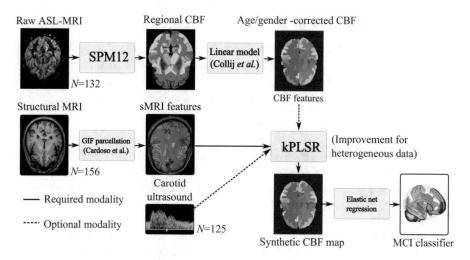

**Fig. 1.** Workflow for extracting features from heterogeneous medical imaging data and training a kPLSR model for CBF feature synthesis. An MCI classifier is trained with both synthetic and real ASL-MRI features for comparison.

collected in the OASIS project (http://www.oasis-brains.org/), and CBF was estimated in 144 anatomical regions. Regional CBF values were normalised for age and sex by using the $w$-scores method of [3] to obtain the final ASL features:

$$w_n = \frac{\mathrm{CBF}_n - (\beta_0 + \mathrm{gender}_n \cdot \beta_1 + \mathrm{age}_n \cdot \beta_2)}{\mathrm{SD\ of\ residuals}}, \tag{1}$$

where $\mathrm{CBF} = \beta_0 + \mathrm{gender} \cdot \beta_1 + \mathrm{age} \cdot \beta_2 + \varepsilon$ is a linear model trained separately for all anatomical regions on the sub-population of cognitively healthy controls.

**Table 1.** Study cohorts contributing data to this work. AD cases not used in classification were included in the kPLSR-model training to increase coverage.

| Cohort | Field strength | sMRI | Carotid | ASL | AD | MCI | Controls | Age |
|---|---|---|---|---|---|---|---|---|
| Cohort 1 | 3.0T | x | | x | 13 | 14 | 28 | $65 \pm 11$ |
| Cohort 1 | 3.0T | | | x | 13 | 14 | 13 | $66 \pm 12$ |
| Cohort 2 | 1.5T | x | x | | 0 | 53 | 48 | $74 \pm 8$ |
| Cohort 3 | 3.0T | | x | x | 0 | 10 | 15 | $75 \pm 6$ |
| Cohort 3 | 3.0T | | | x | 0 | 8 | 4 | $77 \pm 9$ |
| Total cases | | 156 | 126 | 132 | 26 | 99 | 108 | |

## 2.2 Feature Synthesis of the Regional CBF Maps

Given a matrix of inputs $X \in \mathbb{R}^{n \times p}$ and a matrix of outputs $Y \in \mathbb{R}^{n \times m}$, partial least squares-regression (PLSR) attempts to find a lower-dimensional representation of the input-output map using only $\ell \ll p$ latent variables. This is achieved by simultaneous approximate decomposition of the two matrices as:

$$X \approx TP^T, \quad Y \approx UQ^T, \tag{2}$$

where $T, U \in \mathbb{R}^{n \times \ell}$ are the loading matrices for $X$ and $Y$ respectively, and the scores $P \in \mathbb{R}^{p \times \ell}$, $Q \in \mathbb{R}^{m \times \ell}$ maximise the covariance $\mathrm{Cov}(T^T X, U^T Y)$. The feature space for synthesising ASL-MRI maps consisted of 3 demographic features (age, weight, height), 141 sMRI features, and 4 carotid flow-features. Thus the maximum number of input features was $p = 148$. A total of $n = 249$ cases were available for learning a model to predict the $m = 137$ CBF features.

As the relation between CBF and partial volumes of cortical sub-regions in the brain was likely nonlinear, the kernel version of PLSR [6] was used. In this approach, the feature samples $X$ are mapped using a nonlinear map, $\Psi(X)$, and then the standard linear PLSR is performed in the mapped feature space $(\Psi(X), Y)$. The NIPALS algorithm [8] can be formulated in such a way that only inner products of the type $\Psi(x_i)^T \Psi(x_j)$, for $i, j = 1, \ldots, n$ are required. These can then be obtained using the kernel trick as: $K(x_i, x_j) = \Psi(x_i)^T \Psi(x_j)$.

In the case of partially missing input data, we divided the features into two parts $X = [X_1 \, X_2]$, where $X_1$ contained the features that are present for all samples, and $X_2 = \emptyset$ whenever the remaining features were missing in the sample $X$. We then defined the modified kernel function:

$$[\widetilde{K}]_{i,j}(X, X) := \begin{cases} \Psi(x_{i,1})^T \Psi(x_{j,1}), & \text{if } x_i^2 = \emptyset \text{ or } x_j^2 = \emptyset \\ \Psi([x_{i,1} \, x_{i,2}])^T \Psi([x_{j,1} \, x_{j,2}]), & \text{otherwise} \end{cases}, \tag{3}$$

i.e. in the case of partially missing features the kernel function operated only on the subset of available features. Similarly, we divided the output matrix as $Y = [Y_1; Y_2]$ such that $Y_2 = \emptyset$ for all of the cases where the output data was missing, and defined the matrix $S \in \mathbb{R}^{n \times n_1}$ as having ones on the diagonal and zero otherwise. It was used to extend the outputs $Y_1$ from the restricted space to the full space, $SY_1$. The rest of the NIPALS algorithm remained the same, as shown in Algorithm 1. The kPLSR estimator $\widehat{X}_n^{\mathrm{CBF}}$ for the CBF features in the $n$th patient, learned from the demographic variables ($X_{\mathrm{train}}^{\mathrm{demo}}$), sMRI features ($X_{\mathrm{train}}^{\mathrm{sMRI}}$), and carotid flow features ($X_{\mathrm{train}}^{\mathrm{carotid}}$), was then given by the formula:

$$\widehat{X}_n^{\mathrm{CBF}} = \widetilde{K}\left(\mathcal{X}_n; \mathcal{X}_{\mathrm{train}}\right) B, \tag{4}$$

where $\mathcal{X}_{(\cdot)} = [X_{(\cdot)}^{\mathrm{demo}} X_{(\cdot)}^{\mathrm{sMRI}} X_{(\cdot)}^{\mathrm{carotid}}]$ was the combined feature vector. Figure 1 represents the workflow for extracting features, training a kPLSR model for CBF feature synthesis, and training a MCI vs. CHC binary classifier.

**Algorithm 1.** NIPALS-kPLS algorithm with heterogeneous data

---

1: $\widetilde{K}^{(0)} = \widetilde{K}(X, X)$          ▷ compute kernel matrix for mapped features
2: $Y^{(0)} = [Y_1; Y_2]$
3: **for** $\ell = 1, \ldots, \ell_{\max}$ **do**
4:      $u_0 = SY_1^{(\ell-1)}(\,:\,, 1)$       ▷ initialise loading vector in restricted output space
5:      **while** $\|\Delta u_\ell\| > tol$ **do**
6:          $t_\ell = \widetilde{K}^{(\ell-1)} u_\ell, \quad t_\ell \leftarrow t_\ell / \|t_\ell\|$        ▷ iterate in feature space
7:          $q_\ell = [Y_1^{(\ell-1)}]^T S^T t_\ell$        ▷ update output score vector
8:          $u_\ell = SY_1^{(\ell-1)} q_\ell, \quad u_\ell \leftarrow u_\ell / \|u_\ell\|$       ▷ iterate in restricted output space
9:      $\widetilde{K}^{(\ell)} = (\mathbb{I} - t_\ell t_\ell^T) \widetilde{K}^{(\ell-1)} (\mathbb{I} - t_\ell t_\ell^T)$        ▷ deflate $K$
10:     $Y^{(\ell)} = Y^{(\ell-1)} - t_\ell t_\ell^T SY_1^{(\ell-1)}$       ▷ deflate $Y$ in restricted output space
11: $T = [t_1\, t_2\, \ldots\, t_{\ell_{\max}}]$        ▷ assemble $X$ loadings
12: $U = [u_1\, u_2\, \ldots\, u_{\ell_{\max}}]$        ▷ assemble $Y$ loadings
13: $B = \widetilde{K}(X, X) U \left( T^T \widetilde{K}(X, X) U \right)^{-1} T^T SY_1$       ▷ assemble regression vector
14: **return** $(T, U, B)$

---

### 2.3 Simultaneous Classification and Feature Selection

As the amount of available training data was modest and pre-selected anatomical regions were used instead of voxelwise CBF values, standard elastic net regression (ENR) - classifier techniques were used to train three different classifier:

(i) In **Model A**, the sMRI features $X^{\mathrm{sMRI}}$ were used to train an ENR-model:

$$\min_{\beta_0, \beta} \left\{ \frac{1}{2N} \sum_{n=1}^{N} \left( Y_n - \beta_0 - X_n^{\mathrm{sMRI}} \beta \right)^2 + \lambda R(\beta; \alpha) \right\} \tag{5}$$

with the elastic net regularisation term defined as $R(\beta; \alpha) := \frac{(1-\alpha)}{2} \|\beta\|_2^2 + \alpha \|\beta\|_1$. Here $Y_n \in \mathbb{R}$ is the binary MCI diagnosis for the $n$'th patient, $X_n^{\mathrm{sMRI}}$ denotes the sMRI features, $\beta_0$ is the model intercept, and $\beta$ are the regression weights. The continuous model prediction $\widehat{Y}^A = \beta_0 + X^{\mathrm{sMRI}} \beta$ was thresholded to a binary prediction to obtain the standard ROC-curve. The hyperparameters $\lambda > 0$ and $\alpha \in (0, 1]$ were chosen to maximise the area under the ROC-curve.

(ii) In **Model B**, the synthesised CBF-features were used to train the model:

$$\min_{\beta_0, \beta} \left\{ \frac{1}{2N} \sum_{n=1}^{N} \left( Y_n - \beta_0 - \widehat{X}_n^{\mathrm{CBF}} \beta \right)^2 + \lambda R(\beta; \alpha) \right\}, \tag{6}$$

where $\widehat{X}_n^{\mathrm{CBF}}$ is the kPLSR estimator (5). In order to measure the effect of using synthetic vs. ASL-MRI-derived CBF values, Model B was trained using two different sets of data. In one case, when CBF features were missing we simply used synthetic CBF features in their place (*MRI + synthetic*). In another case, only the synthetic CBF features were used even if ASL-MRI was available (*synthetic only*). Again the continuous probability was thresholded to a binary prediction.

(iii) In **Model C**, the feature selection was performed simultaneously on both sMRI and CBF features:

$$\min_{\beta_0,\beta_1,\beta_2} \left\{ \frac{1}{2N} \sum_{n=1}^{N} \left( Y_n - \beta_0 - \widehat{X}_n^{\mathrm{CBF}}\beta_1 - X_n^{\mathrm{sMRI}}\beta_2 \right)^2 + \lambda R([\beta_1;\beta_2];\alpha) \right\}. \tag{7}$$

(iv) In **Model D**, we combined Models A and B by using the maximum probability rule, $\widehat{Y}_n^C = \max\{\widehat{Y}_n^A, \widehat{Y}_n^B\}$. The rationale for this was that a combination of two diagnostic tests with high specificity but lower sensitivity (typical for AD classifiers) may provide more sensitive diagnostic tests while avoiding the problem that simultaneous feature selection favours one set of features over the other. Models C and D were likewise trained using both synthetic CBF features alone and by combining ASL-MRI and synthetic CBF features.

## 3 Experiments

### 3.1 Synthetic CBF vs. ASL-MRI Reconstructed CBF

The Gaussian kernel, $K(x_1,x_2) = \exp(-\|x_1 - x_2\|_2^2/d)$, was used in the kPLSR model. This resulted in two model hyperparameters, the kernel width $d$ and the number of latent variables $\ell$, that had to be tuned using leave-one-out cross-validation. Only cases where the sMRI features were available ($n = 156$) were used in model training and cross-validation. Out of these, carotid ultrasound and ASL-MRI features were present in 100 and 55 cases, respectively. Hyperparameter values optimising the $R^2$-statistic were found to be $d = 35$ and $\ell = 2$. Possible bias and standard deviation of the synthesised CBF from ground truth $w$-score values were measured using a Bland-Altman - plot of $w$-scores averaged across all regions, separately for the white matter (WM) and gray matter (GM),

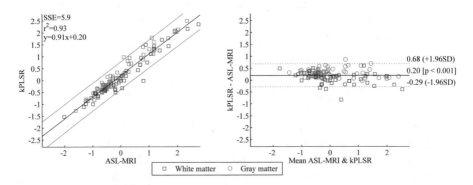

**Fig. 2.** Bland-Altman plot of kPLSR-modelled vs. ASL-MRI derived $w$-scores. Mean $w$-scores averaged over all WM/GM regions reported separately.

see Fig. 2. The mean bias was $\Delta w = 0.20$ ($p < 0.001$). The $w$-score is normalised so that its standard deviation in the normal population equals 1. The kPLS regressor slightly overestimated CBF in both the WM and GM.

**Table 2.** Performance of the classifier in the CBF imputation problem (top half), compared with studies in the literature with at least 100 cases (bottom half). ENR = Elastic Net Regression, SVM = Support Vector Machine.

| Features used for classification | Classifier | # of cases | AUC | ACC | SENS | SPEC |
|---|---|---|---|---|---|---|
| sMRI | ENR | 143 | 0.76 | 77% | 90% | 60% |
| ASL (MRI+synthetic) | ENR | 143 | 0.74 | 77% | 90% | 60% |
| ASL (synthetic only) | ENR | 143 | 0.65 | 77% | 90% | 60% |
| sMRI+ASL (joint features, MRI+synth.) | ENR | 143 | 0.77 | 88% | 100% | 70% |
| sMRI+ASL (joint features, synth. only) | ENR | 143 | 0.75 | 83% | 100% | 50% |
| sMRI+ASL (max probability, MRI+synth.) | ENR | 143 | 0.77 | 92% | 100% | 80% |
| sMRI+ASL (max probability, synth. only) | ENR | 143 | 0.72 | 92% | 100% | 80% |
| ASL (Collij et al. [3]) | SVM | 160 | 0.63 | 60% | 60% | 60% |
| sMRI (Liu et al. [4]) | SVM | 454 | - | 85% | 82% | 88% |
| PET (Ortiz et al. [5]) | SVM | 179 | 0.74 | 73% | 70% | 77% |
| sMRI+PET (Ortiz et al. [5]) | SVM | 179 | 0.91 | 86% | 90% | 82% |

## 3.2 Utility of Synthetic ASL in the CBF Imputation Problem

The MCI classifiers using Models A, B, C, and D were trained with four-fold cross-validation (4-FCV) in a training set of $n = 123$ cases. An additional randomly selected validation set of $n = 20$ cases not included in the training was used to evaluate the balanced accuracy (ACC), sensitivity (SENS), and specificity (SPEC) of each classifier using hyperparameters and cut-offs obtained in 4-FCV. A heat map of the 46 features chosen by Model B is shown in Fig. 3.

We compared our MCI classification accuracy to results reviewed in [1] with the following selection criteria: (i) the MCI vs. CHC classification problem was addressed, (ii) the feature set consisted of sMRI, ASL, or PET features, (iii) the cohort size was at least 100, and (iv) studies that used CSF biomarkers or neurocognitive test scores as features were excluded. The study with the best reported accuracy for each feature set was chosen as representative.

Results of the comparison are given in Table 2. Models A and B alone produced similar results in terms of accuracy, although Model B achieved better accuracy than was reported for ASL-MRI features in [3]. Model C improved the results slightly when MRI+synthetic CBF features were used, but the best results were obtained with Model D regardless or whether MRI+synthetic or synthetic only CBF features were used.

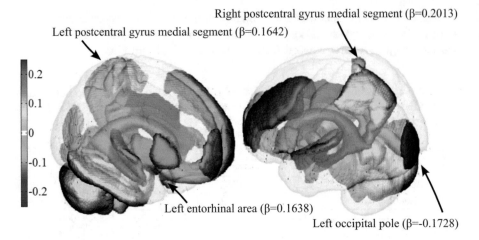

**Fig. 3.** Heat map of the coefficients $\beta$ for the CBF imputation problem. A total of 46 regions were chosen as features. Regions with largest coefficients identified.

## 4   Discussion

Kernel PLS regression on heterogeneous data was used for the robust synthesis of regional CBF values in cases where no ASL-MRI images were available. As was reported in [3], CBF features alone were not particularly informative in MCI classification, but a multi-modal classifier using synthetic CBF features outperformed pure sMRI-based classifiers in a validation test. Best classifier performance (balanced accuracy 92%, sensitivity 100%, specificity 80%) was achieved when a maximum probability-rule was used to combine classifiers using different feature sets. The benefit of our proposed method is that only basic sMRI features (partial volumes of subregions) were used and, as a result, synthetic CBF features can, therefore, be generated in large-scale brain imaging databases, such as ADNI, without the need for extensive feature computation. It is possible that more informative sMRI features, e.g. ventricular and/or hippocampal shape morphometrics, could increase the accuracy of the resultant classifiers. Provided more ASL-MRI data were available, the use of convolutional neural networks on voxelwise CBF values should also be investigated to eliminate the need for pre-selecting anatomical regions for analysis.

**Acknowledgements.** This work was funded by the FP7 project VPH-DARE@IT *"Virtual Physiological Human: DementiA Research Enabled by IT" (FP7-ICT-2011-5.2-601055).*

# References

1. Arbabshirani, M., Plis, S., Sui, J., Calhoun, V.: Single subject prediction of brain disorders in neuroimaging: promises and pitfalls. NeuroImage **145**, 137–165 (2017)
2. Cardoso, M., Modat, M., Wolz, R., Melbourne, A., Cash, D., Rueckert, D.: Geodesic information flows: spatially-variant graphs and their application to segmentation and fusion. IEEE Trans. Med. Imag. **34**(9), 1976–1988 (2015)
3. Collij, L., et al.: Application of machine learning to arterial spin labeling in mild cognitive impairment and Alzheimer disease. Radiology **281**(3), 865–875 (2016)
4. Liu, M., Zhang, D., Shen, D.: Hierarchical fusion of features and classifier decisions for Alzheimer's disease diagnosis. Hum. Brain Mapp. **35**(4), 1305–1319 (2014)
5. Ortiz, A., Munilla, J., Álvarez-Illán, I., Górriz, J., Ramírez, J.: Exploratory graphical models of functional and structural connectivity patterns for Alzheimer's disease diagnosis. Front. Comput. Neurosci. **9**, 132 (2015)
6. Rosipal, R., Trejo, L.J.: Kernel partial least squares regression in reproducing kernel Hilbert space. J. Mach. Learn. Res. **2**, 97–123 (2001)
7. van Tulder, G., de Bruijne, M.: Why does synthesized data improve multi-sequence classification? In: Navab, N., Hornegger, J., Wells, W.M., Frangi, A.F. (eds.) MICCAI 2015. LNCS, vol. 9349, pp. 531–538. Springer, Cham (2015). https://doi.org/10.1007/978-3-319-24553-9_65
8. Wold, S., Geladi, P., Esbensen, K., Öhman, J.: Multi-way principal components and PLS analysis. J. Chemom. **1**, 41–56 (1987)

# A Novel Method for Epileptic Seizure Detection Using Coupled Hidden Markov Models

Jeff Craley[1(✉)], Emily Johnson[2], and Archana Venkataraman[1]

[1] Department of Electrical and Computer Engineering, Johns Hopkins University, Baltimore, USA
jcraley@gmail.com
[2] Department of Neurology, Johns Hopkins Medical Institute, Baltimore, USA

**Abstract.** We propose a novel Coupled Hidden Markov Model to detect epileptic seizures in multichannel electroencephalography (EEG) data. Our model defines a network of seizure propagation paths to capture both the temporal and spatial evolution of epileptic activity. To address the intractability introduced by the coupled interactions, we derive a variational inference procedure to efficiently infer the seizure evolution from spectral patterns in the EEG data. We validate our model on EEG aquired under clinical conditions in the Epilepsy Monitoring Unit of the Johns Hopkins Hospital. Using 5-fold cross validation, we demonstrate that our model outperforms three baseline approaches which rely on a classical detection framework. Our model also demonstrates the potential to localize seizure onset zones in focal epilepsy.

## 1 Introduction

Epilepsy is a heterogenous neurological disorder characterized by recurring and unprovoked seizures [1]. It is estimated that 20–40% of epilepsy patients are medically refractory and do not respond to drug therapy. Alternative therapies for these patients crucially depend on being able to detect epileptic activity in the brain. The most common modality used for seizure detection is multichannel electroencephalography (EEG) acquired on the scalp. The clinical standard for seizure detection involves visual inspection of the EEG data, which is time consuming and requires extensive training. In this work, we develop an automated seizure detection procedure for clinically acquired multichannel EEG recordings.

There is a vast body of literature on epileptic seizure detection from a variety of viewpoints. The nonlinearity of EEG signals has inspired the application of techniques from chaos theory such as approximate entropy and Lyapunov exponents as in [2,3], respectively. Alternatively, wavelet and other time-frequency based features seek to capture the non-stationarity of the EEG signal as in [4]. These features are fed into standard classification algorithms to detect seizure activity. A fundamental limitation of the methods in [2,3] is that they are trained on a single channel of EEG data and fail to generalize in practice. Multichannel

© Springer Nature Switzerland AG 2018
A. F. Frangi et al. (Eds.): MICCAI 2018, LNCS 11072, pp. 482–489, 2018.
https://doi.org/10.1007/978-3-030-00931-1_55

strategies such as those in [4–7] rely heavily on prior seizure recordings to train patient specific detectors, which are often unavailable.

Unlike prior work, our approach explicitly models the spatial dynamics of a seizure through the brain over time. We build on existing work in Hidden Markov Models (HMMs) [7], adopting a Coupled HMM (CHMM) [8] to model interchannel dependencies. Specifically, the likelihood that an EEG channel will transition into a seizure state will increase if neighboring channels are in a seizure state. This coupling renders exact inference intractable. Therefore we develop a variational Expectation Maximization (EM) algorithm for our framework.

We evaluate our algorithm using 90 scalp EEG recordings from 15 epilepsy patients acquired in the Epilepsy Monitoring Unit (EMU) of the Johns Hopkins Hospital. These recordings contain up to 10 min of baseline activity before and after a seizure and have not been screened for artifacts. We compare our CHMM to classifiers evaluated on a framewise and channelwise basis. Our algorithm outperforms these baselines and demonstrates efficacy in classifying seizure intervals. Our algorithm provides localization information that could be useful for determining the seizure onset location in cases of focal epilepsy.

## 2 Generative Model of Seizure Propagation

We adopt a Bayesian framework for seizure detection. The latent variables $\mathbf{X}$ denote the seizure or non-seizure states. $\mathbf{Y}$ corresponds to observed data feature vectors computed from EEG channels as shown in Fig. 1(a). The random variable $X_i^t$ denotes the latent state of EEG channel $i$ at time $t$. We assume three possible states: pre-seizure baseline ($X_i^t = 0$), seizure propagation ($X_i^t = 1$), and post-seizure baseline ($X_i^t = 2$). The corresponding observed "emission" feature vectors $Y_i^t$ are continuous statistics computed from time window $t$ of the EEG channel $i$. For convenience, we also define the ensemble variables $\mathbf{X}^t \triangleq [X_1^t, \ldots, X_N^t]^T$ where $N$ is the number of electrodes. Given the EEG observations, our goal is to infer the latent seizure state for each chain at all times.

### 2.1 Model Formulation and Inference

Figure 1(b) shows the coupling between electrodes of the 10/20 international system [9]. We define the aunts $au(\cdot)$ of a given node as the set of electrodes connected to it in Fig. 1(b). The joint distribution of $\mathbf{X}$ and $\mathbf{Y}$ factorizes into transition priors that depend on both a channel's own previous state and those of its aunts $P(X_i^t \mid \mathbf{X}_{au(i)\,\bigcup\,i}^{t-1})$, and emission likelihoods $P(Y_i^t \mid X_i^t)$ as in Eq. (1). For simplicity we assume that all recordings begin in a non-seizure state ($X_i^0 = 0 \forall i$).

$$P(\mathbf{X}, \mathbf{Y}) = \prod_{i=1}^{N} P(Y_i^0 \mid X_i^0) \prod_{t=1}^{T} P(Y_i^t \mid X_i^t) P(X_i^t \mid \mathbf{X}_{au(i)\,\bigcup\,i}^{t-1}) \tag{1}$$

Note that the observed emissions are conditionally independent given the latent states. Below, we detail the model formulation and inference algorithm.

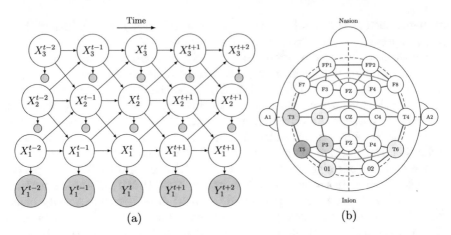

**Fig. 1.** (a) Graphical model depicting a three chain CHMM. Observed nodes are shaded gray, while latent nodes are shown in white. (b) EEG channels in the 10/20 international system [9], cross hemispheric (red) and neighboring (blue) channel connections. A seizure propagates from the red, to the orange, and finally yellow shaded electrodes.

**Coupled State Transitions.** The distribution over state vectors $\mathbf{X}^t$ forms a first order Markov chain. This distribution further factorizes into products of transition distributions of individual chains $P(\mathbf{X}^t \mid \mathbf{X}^{t-1}) = \prod_{i=1}^{N} P(X_i^t \mid \mathbf{X}_{au(i) \bigcup i}^{t-1})$. We encode these chainwise transition probabilities using time inhomogenous transition matrices as shown in Eq. (2). This structure ensures each channel begins in a non-seizure baseline state, transitions into an active seizure state, and transitions into a post-seizure state.

$$A_i^t = \begin{bmatrix} 1 - g_i^t & g_i^t & 0 \\ 0 & 1 - h_i^t & h_i^t \\ 0 & 0 & 1 \end{bmatrix} \tag{2}$$

The transition matrix $A_i^t$ is governed by neighboring and contralateral EEG channels to capture the main modes of seizure propagation, as shown in Fig. 1(b). Let $\eta_i^t$ be the number of aunts in the seizure state in the previous timestep. We model the transition probabilities into and out of the seizure state via logistic regression functions of $\eta_i^t$ as shown in Eq. (3). Parameters $\{\rho_0, \phi_0\}$ control the base onset and offset rates while $\{\rho_1, \phi_1\}$ control the effects of a channel's aunts.

$$\log\left(\frac{g_i^t}{1 - g_i^t}\right) = \rho_0 + \rho_1 \eta_i^t, \qquad \log\left(\frac{h_i^t}{1 - h_i^t}\right) = \phi_0 + \phi_1 \eta_i^t \tag{3}$$

**Emission Likelihood.** We use a Gaussian Mixture Model (GMM) to describe the emissions $Y_i^t$ of each chain. Let $C_i^t$ be the mixture from which $Y_i^t$ was generated. Let $\pi_{ij}^k$ be the prior probability of mixture component $j$ when $X_i^t = k$ for $k = 0, 1, 2$. The joint distribution over $Y_i^t$ and $C_i^t$ can be expressed as follows

$$P(Y_i^t, C_i^t = j \mid X_i^t = k) = P(Y_i^t \mid C_i^t = j)P(C_i^t = j \mid X_i^t = k)$$
$$= \pi_{ij}^k \mathcal{N}\left(Y_i^t; \mu_{ij}, \Sigma_{ij}\right) \tag{4}$$

Effectively, the emission distributions for all observed variables share the same mean parameters $\mu_{ij}$ and covariance parameters $\Sigma_{ij}$, but use different mixture weights based on the latent seizure state $k$. We tie weights for both pre- and post-seizure states, i.e. $\pi_{ij}^0 = \pi_{ij}^2$ for all channels $i$ and mixture components $j$. The data likelihood $P(Y_i^t \mid X_i^t)$ can be computed by marginalizing over $j$.

**Approximate Inference Using Variational EM.** Exact inference for the CHMM is intractable due to the coupled state transitions. Therefore we develop a structured variational algorithm [10], in which we approximate the posterior distribution over $\mathbf{X}$ as a set of $N$ independent HMM chains:

$$Q(\mathbf{X}) = \prod_{i=1}^{N} \frac{1}{Z_{Q_i}} Q_i(\mathbf{X}_i) = \prod_{i=1}^{N} \frac{1}{Z_{Q_i}} \prod_{t=1}^{T} T_i^t(X_i^t \mid X_i^{t-1}) E_i^t(X_i^t). \tag{5}$$

As seen in Eq. (5), each approximating chain includes a normalizing constant $Z_{Q_i}$, a transition term $T_i^t(X_i^t \mid X_i^{t-1})$, and an emission term $E_i^t(X_i^t)$.

The transition distribution $T_i^t(X_i^t \mid X_i^{t-1})$ is encoded by a state transition matrix $\tilde{A}_i^t$ which mimics the structure of Eq. (2). Here $\tilde{g}_i^t$ and $\tilde{h}_i^t$ are variational transition parameters analagous to the original transition parameters $g_i^t$ and $h_i^t$.

$$\tilde{A}_i^t = \begin{bmatrix} 1 - \tilde{g}_i^t & \tilde{g}_i^t & 0 \\ 0 & 1 - \tilde{h}_i^t & \tilde{h}_i^t \\ 0 & 0 & 1 \end{bmatrix} \tag{6}$$

In contrast to Eq. (6), the emission distribution $E_i^t(X_i^t)$ weighs the contribution of the observed data $Y_i^t$ through variational parameters $\tilde{l}_{i0}^t$ and $\tilde{l}_{i1}^t$. Thus $E_i^t(X_i^t = 0, 2) = \tilde{l}_{i0}^t$ and $E_i^t(X_i^t = 1) = \tilde{l}_{i1}^t$.

We learn variational parameters for each chain by minimizing the free energy of the approximation. We perform this minimization by decoupling the free energy into expectations over a single channel and expectations over the remaining channels. The index "$-i$" in Eq. (7) denotes the set of channels excluding $i$.

$$\begin{aligned} \mathcal{FE} &= -E_Q\left[\log p(\mathbf{X}, \mathbf{Y})\right] + E_Q\left[\log Q(\mathbf{X})\right] \\ &= -E_{Q_i}\left[E_{Q_{-i}}\left[\log p(\mathbf{X}_i, \mathbf{Y}_i \mid \mathbf{X}_{-i}, \mathbf{Y}_{-i})\right]\right] + E_{Q_i}\left[\log Q_i(\mathbf{X}_i)\right] \\ &\quad - E_{Q_{-i}}\left[\log p(\mathbf{X}_{-i}, \mathbf{Y}_{-i})\right] + E_{Q_{-i}}\left[\log Q_{-i}(\mathbf{X}_{-i})\right] \end{aligned} \tag{7}$$

Notice that the last line of Eq. (7) does not depend on the parameters of chain $i$, allowing a natural fixed point iteration over the parameters of a single chain while holding all other chains constant. This minimization fixes the variational parameters $\tilde{l}_i^t$ equal to the GMM likelihood of the observed data:

$$\tilde{l}_{i0}^t = p(Y_i^t \mid X_i^t = 0, 2), \qquad \tilde{l}_{i1}^t = p(Y_i^t \mid X_i^t = 1). \tag{8}$$

The updates for the variational transition parameters form logistic regressions based on the expected value of the original activations.

$$\log\left(\frac{\tilde{g}_i^t}{1-\tilde{g}_i^t}\right) = \rho_0 + \rho_1 E_{Q_{au(i)}}\left[\eta_i^t\right], \quad \log\left(\frac{\tilde{h}_i^t}{1-\tilde{h}_i^t}\right) = \phi_0 + \phi_1 E_{Q_{au(i)}}\left[\eta_i^t\right] \quad (9)$$

Once the variational parameters have been computed, the approximating distribution takes the form of an HMM where the $\tilde{l}$ parameters capture the likelihood of the data under each latent state. We can use the forward-backward algorithm [10] to compute the expected latent states $E_Q[X_i^t]$, the expected state transitions, and the expected number of aunts in the seizure state $E_Q[\eta_i^t]$.

**Learning the Model Parameters.** We use the expected values of the latent states and mixture components to update the transition parameters $\{\rho_i, \phi_i\}$ and the emission parameters $\{\mu_{ij}, \Sigma_{ij}, \pi_{ij}^k\}$. Let $\tau_i^t(j, k)$ be the expectation that channel $i$ at time $t$ is in state $k$ with mixture $j$, we can update the emission parameters according to the soft counts of the occurrence of each mixture.

$$\mu_{ij} = \frac{\sum_{k=0}^2 \sum_{t=0}^T \tau_i^t(j, k) Y_i^t}{\sum_{k=0}^2 \sum_{t=0}^T \tau_i^t(j, k)}, \quad \Sigma_{ij} = \frac{\sum_{k=0}^2 \sum_{t=0}^T \tau_i^t(j, k) \left(Y_i^t - \mu_{ij}\right)^2}{\sum_{k=0}^2 \sum_{t=0}^T \tau_i^t(j, k)} \quad (10)$$

$$\pi_{ij}^0 = \pi_{ij}^2 = \frac{\sum_{t=0}^T \tau_i^t(j, 0) + \tau_i^t(j, 2)}{\sum_{j'} \sum_{t=0}^T \tau_i^t(j', 0) + \tau_i^t(j', 2)}, \quad \pi_{ij}^1 = \frac{\sum_{t=0}^T \tau_i^t(j, 1)}{\sum_{j'} \sum_{t=0}^T \tau_i^t(j', 1)} \quad (11)$$

The update for the transition parameters $\{\rho_i, \phi_i\}$ takes the form of a weighted logistic regression. We regress the expected $\eta_i^t$ onto the expected transitions for each chain and use Newton's method to find the optimal transition parameters.

**Implementation Details.** We initialize our model by training the GMM emission distributions based on the expert annotations of seizure intervals. 3 emission mixtures resulted in a reasonable compromise between sensitivity and specificity. Transition parameters $\rho_0$, $\rho_1$, $\phi_0$, and $\phi_1$ were initialized to $-7$, $2$, $-3$, and $0$, respectively. This corresponds to expected seizures every 13 min lasting 15 s, channels turning on with a 7 fold increase per aunt node, and no cross channel influence for offset. Our EM proceedure is performed in an unsupervised fashion without further use of the labels. Model parameters are updated during the M-step of the EM algorithm. During inference, channels are updated sequentially until the scaled difference in $\mathcal{FE}$ converges to less than $10^{-4}$.

## 2.2   Baseline Comparison

We compare our model to three alternative classification schemes. The first approach is to train a logistic regression function to distinguish between baseline and seizure intervals based on a linear combination of the EEG features. The second approach uses kernel support vector machines (SVMs) to learn a possibly nonlinear decision boundary in the EEG feature space that maximally separates the baseline and seizure conditions. Here we rely on a polynomial kernel. SVM classifiers have been used extensively for seizure detection [2,5,6]. Finally we consider a GMM hypothesis testing scenario. This method trains GMMs for seizure and non-seizure states and classifies based on the ratio of the likelihoods under each GMM, roughly equating to our model with no transition prior.

## 3    Experimental Results

### 3.1    Data and Preprocessing

Our EEG data was recorded as part of routine clinical evaluation in the EMU of the Johns Hopkins Hospital. Our dataset consists of 90 seizures recordings from 15 patients with as much as 10 min of baseline before and after the seizure. Recordings were sampled at 200 Hz. We rely on expert clinical annotations denoting the seizure onset and offset to validate the performance of each method.

For preprocessing, each EEG channel was bandpass filtered through sequential application of fourth order Butterworth high and low pass filters at 1.6 Hz and 50 Hz respectively. This filtering mirrored clinical preprocessing practice for removing DC trends and high frequency components with no clinical relevance. In addition, a second order notch filter with $Q = 20$ was applied at 60 Hz to the EEG recordings to remove any remaining effect of the power supply.

We considered two emission features for analysis computed from channels in common reference: the sum of spectral coefficients in brain wave frequency bands and the log line length. Features were computed on windows of 1 s with 250 ms overlap. For spectral features, a short time Fourier transform was taken after the application of a Tukey window with shape parameter 0.25. The magnitudes of the STFT coefficients corresponding to frequencies in the theta (1–4 Hz), delta (4–8 Hz), alpha (8–13 Hz), and beta (13–30 Hz) bands were summed and the logarithm was taken, resulting in a length four feature vector. The log line length was computed as the logarithm of the sum of the absolute difference between successive samples i.e. given a signal $s$ of length $T$, $\log L = \log \left( \sum_{i=0}^{T-1} |s(i+1) - s(i)| \right)$.

### 3.2    Seizure Detection Performance

We use a 5 fold cross validation strategy for evaluation. Four folds were used to train each model and detection was evaluated on the held-out fold. Each recording was randomly assigned to a fold independently of patient. For our model, the training phase was used to learn the emission and transition parameters. Table 1 summarizes the performance for each classifier based on the average accuracy of the testing fold. The sensitivity (TPR) and specificity (TNR) denote the prediction accuracy for seizure and non-seizure frames, respectively, computed across all channels. For the probabilistic classifiers (i.e. logistic regression, GMM, and CHMM), these rates are weighted by the posterior confidence of the classifier.

The transition prior allows our CHMM to place more confidence in contiguous regions exhibiting seizure-like activity. Figure 2(a) shows an example of our classifier correctly classifying the majority of the seizure across all channels. However, this confidence comes with a reduction in specificity, as the classifier tends

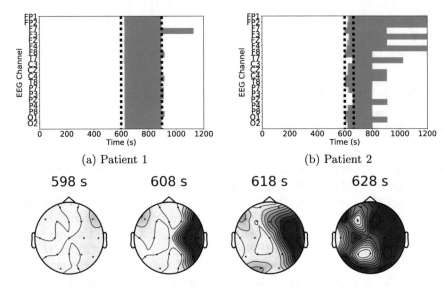

(a) Patient 1 (b) Patient 2

(c) Time lapse detail of inferred seizure onset in Patient 2

**Fig. 2.** Detection results for the CHMM. Top row: Estimated posterior probability of the latent seizure state for two epilepsy patients. White corresponds to pre- and post-seizure baseline while violet indicates seizure states. EEG channels corresponding to 10/20 system [9] channel locations are on the $y$ axis. The expert identified seizure region is denoted by the dashed black lines. (a) shows the models ability to accurately classify seizures across the whole brain and (b) shows the outward spread of a right temporal lobe seizure. Bottom row: temporal evolution of the seizure depicted in (b).

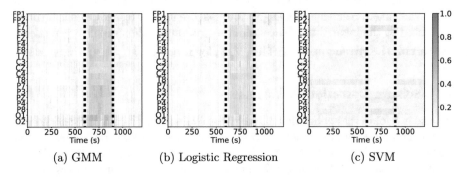

(a) GMM (b) Logistic Regression (c) SVM

**Fig. 3.** Detection results on Patient 1 for the three baseline methods. These algorithms place much lower posterior confidence in seizure intervals than the CHMM.

to associate post-seizure spectral artifacts with seizure as shown in Fig. 2(b). In future work we will investigate feature selection methods to combat this issue. Figure 2(c) shows the evolution of a focal right temporal seizure, which indicates our model's potential to localize epileptic activity on the scalp. This localization is highly relevant to clinical management of epilepsy.

Due to the heterogeneity of seizure presentations across patients, our baselines fail to perform well as shown in Fig. 3. The logistic regression and GMM correctly classify portions of seizure intervals but lack consistency. The GMM exhibits more confidence in classifying seizures than its probabilistic linear counterpart. The SVM performs poorly due to the inseperability of the EEG features in our noisy clinical dataset.

**Table 1.** Results for each method.

| Model | TPR | TNR | AUC |
|---|---|---|---|
| GMM | 21.91% | 92.39% | 0.784 |
| Logistic regression | 18.07% | **92.55%** | 0.80 |
| Kernel SVM | 10.22% | 90.27% | 0.53 |
| CHMM | **72.63%** | 79.27% | **0.839** |

## 4 Conclusion

We have presented a novel method for epileptic seizure detection based on a CHMM model. At a high level, we directly model seizure spreading by allowing the state of neighboring and symmetric EEG channels to influence the transition probabilities for a given channel. We have validated our approach on clinical EEG data from 15 unique patients. Our model outperforms three baseline approaches which perform classification on a framewise basis. By incorporating a transition prior that includes spatial and temporal contiguity to seizure regions we are able to better classify seizure intervals within EEG recordings.

**Acknowledgments.** This work was supported by a Johns Hopkins Medical Institute Synergy Award (Joint PI: Venkataraman/Johnson).

## References

1. Miller, J.W., Goodkin, H.P.: Epilepsy. Wiley, Hoboken (2014)
2. Acharya, U.R., et al.: Automated diagnosis of epileptic EEG using entropies. Biomed. Signal Process. Control. **7**(4), 401–408 (2012)
3. Güler, N.F., et al.: Recurrent neural networks employing Lyapunov exponents for EEG signals classification. Expert. Syst. Appl. **29**(3), 506–514 (2005)
4. Zandi, A.S., et al.: Automated real-time epileptic seizure detection in scalp EEG recordings using an algorithm based on wavelet packet transform. IEEE Trans. Biomed. Eng. **57**(7), 1639–1651 (2010)
5. Hunyadi, B., et al.: Incorporating structural information from the multichannel EEG improves patient-specific seizure detection. Clin. Neurophysiol. **123**(12), 2352–2361 (2012)
6. Shoeb, A.H., Guttag, J.V.: Application of machine learning to epileptic seizure detection. In: International Conference on Machine Learning, pp. 975–982 (2010)
7. Baldassano, S., et al.: A novel seizure detection algorithm informed by hidden Markov model event states. J. Neural Eng. **13**(3), 036011 (2016)
8. Brand, M., Oliver, N., Pentland, A.: Coupled hidden Markov models for complex action recognition. In: Proceedings of the 1997 IEEE Computer Society Conference on Computer vision and pattern recognition, pp. 994–999. IEEE (1997)
9. Jurcak, V., et al.: 10/20, 10/10, and 10/5 systems revisited: their validity as relative head-surface-based positioning systems. Neuroimage **34**(4), 1600–1611 (2007)
10. Murphy, K.P.: Machine Learning: A Probabilistic Perspective. The MIT Press, Cambridge (2012)

# Deep Convolutional Networks for Automated Detection of Epileptogenic Brain Malformations

Ravnoor S. Gill[1(✉)], Seok-Jun Hong[1], Fatemeh Fadaie[1],
Benoit Caldairou[1], Boris C. Bernhardt[1,2], Carmen Barba[3],
Armin Brandt[4], Vanessa C. Coelho[5], Ludovico d'Incerti[6],
Matteo Lenge[3], Mira Semmelroch[7], Fabrice Bartolomei[8],
Fernando Cendes[5], Francesco Deleo[6], Renzo Guerrini[3],
Maxime Guye[9], Graeme Jackson[7], Andreas Schulze-Bonhage[4],
Tommaso Mansi[10], Neda Bernasconi[1], and Andrea Bernasconi[1]

[1] Neuroimaging of Epilepsy Laboratory, McConnell Brain Imaging Center,
Montreal Neurological Institute, Montreal, QC, Canada
ravnoor.gill@mail.mcgill.ca
[2] Multimodal Imaging and Connectome Laboratory,
McConnell Brain Imaging Center,
Montreal Neurological Institute, Montreal, QC, Canada
[3] Children's Hospital A. Meyer-University of Florence, Florence, Italy
[4] Freiburg Epilepsy Center, University Medical Center, Freiburg, Germany
[5] University of Campinas, Campinas, Brazil
[6] Istituto Neurologico Carlo Besta, Milan, Italy
[7] The Florey Institute of Neuroscience and Mental Health
and the University of Melbourne, Melbourne, VIC, Australia
[8] Aix-Marseille Univ, INS, Marseille, France
[9] Aix-Marseille Univ, CNRS, Marseille, France
[10] Siemens Medical Solutions, Medical Imaging Technologies,
Princeton, NJ, USA

**Abstract.** Focal cortical dysplasia (FCD) is a prevalent surgically-amenable epileptogenic malformation of cortical development. On MRI, FCD typically presents with cortical thickening, hyperintensity, and blurring of the gray-white matter interface. These changes may be visible to the naked eye, or subtle and be easily overlooked. Despite advances in MRI analytics, current surface-based algorithms fail to detect FCD in 50% of cases. Moreover, arduous data pre-processing and specialized expertise preclude widespread use. Here we propose a novel algorithm that harnesses feature-learning capability of convolutional neural networks (CNNs) with minimal data pre-processing. Our classifier, trained on a patch-based augmented dataset derived from patients with histologically-validated FCD operates directly on MRI voxels to distinguish the lesion from healthy tissue. The algorithm was trained and cross-validated on multimodal MRI data from a single site (S1) and evaluated on independent data from S1 and six other sites worldwide (S2–S7; 3 scanner manufacturers and 2 field strengths) for a total of 107 subjects. The classifier showed excellent sensitivity (S1: 87%, 35/40 lesions detected; S2–S7: 91%, 61/67 lesions

© Springer Nature Switzerland AG 2018
A. F. Frangi et al. (Eds.): MICCAI 2018, LNCS 11072, pp. 490–497, 2018.
https://doi.org/10.1007/978-3-030-00931-1_56

detected) and specificity (S1: 95%, no findings in 36/38 healthy controls; 90%, no findings in 57/63 disease controls). Easy implementation, minimal pre-processing, high performance and generalizability make this classifier an ideal platform for large-scale clinical use, particularly in "MRI-negative" FCD.

**Keywords:** Magnetic resonance imaging · Clinical diagnostics Epilepsy · Deep learning · Classification

# 1  Introduction

Focal cortical dysplasia (FCD), a malformation of cortical development, is a frequent cause of drug-resistant epilepsy. This surgically-amenable lesion is characterized on histology by altered cortical laminar structure and cytological anomalies together with gliosis and demyelination, which may extend into the underlying white matter [1]. On MRI, FCD typically presents with cortical thickening, hyperintensity, and blurring of the gray-white matter interface. These changes may be visible to the naked eye on T1- and T2-weighted MRI, or subtle and easily overlooked [2].

Over the last decade, a number of automated algorithms have been developed [3]. Contemporary FCD detection methods rely on surface-based approaches [4–7], which allow to effectively model sulco-gyral morphology. While they have shown effectiveness, they have been mainly used as a proof of principle and applied to lesions previously seen on MRI, but rarely validated histologically. Despite advances in MRI analytics, current algorithms fail to detect subtle FCD [2]. Importantly, since training and validation have been performed on data from the same center and scanner, generalizability to independent cohorts remains unclear. Finally, arduous pre-processing and specialized expertise preclude their broader integration into clinical workflows.

Conventional machine-learning systems require careful engineering and considerable domain knowledge to design features from which the classifier can learn patterns. Conversely, convolutional neural networks (CNNs), a class of deep neural networks, have the capacity to extract a hierarchy of increasingly complex features from the data [8]. In biomedical imaging, CNNs have gained popularity in brain tissue classification, and segmentation of brain tumors and multiple sclerosis plaques (see Litjens et al. [9] for review). To the best of our knowledge, no study has deployed CNNs to detect cortical brain malformations.

Exploiting the complementary diagnostic power of T1- and T2-weighted contrasts, we propose a novel algorithm with minimal data pre-processing and which harnesses feature-learning proficiency of CNNs to distinguish FCD from healthy tissue directly on MRI voxels. Our algorithm was trained and tested on data from a single site (S1) and tested on independent data from S1 and six sites worldwide (S2–S7), for a total of 107 individuals. Furthermore, it was tested against a benchmark surface-based algorithm, making this study the first deep-learning approach for FCD detection with multicentric validation.

## 2  Methods

### 2.1  MRI Acquisition

At S1, multimodal MRI was acquired on a 3T Siemens TimTrio using a 32-channel head coil, including: 3D T1-weighted MPRAGE (T1w; TR = 2300 ms, TE = 2.98 ms, flip angle = 9°, FOV = 256 mm$^2$, voxel size = 1 × 1 × 1 mm$^3$), and T2-weighted 3D fluid-attenuated inversion recovery (FLAIR; TR = 5000 ms, TE = 389 ms, flip angle = 120°, FOV = 230 mm$^2$, 0.9 × 0.9 × 0.9 mm$^3$).

### 2.2  Image Pre-processing

For all datasets, T1w and FLAIR images underwent intensity non-uniformity correction [10] and normalization. T1w images were then linearly registered (affine, 9 degrees of freedom) to the age-appropriate MNI152 symmetric template (1 × 1 × 1 mm$^3$) stratified across seven age-groups [0–4.5, 4.5–8.5, 7–11, 7.5–13.5, 10–14, 13–18.5, 18.5–43 years old] [11]. Age-appropriate templates minimize the interpolation effects of linear registration, thereby limiting blurring effects that may mimic lesional tissue and manifest as false positives. FLAIR images were linearly mapped to T1w images in MNI space. Skull-stripping was performed to exclude non-brain tissue.

### 2.3  Patch-Based Input Sampling

*Balanced Inputs Based on 3D Volumetric Images.* Data imbalance is a challenging issue in FCD lesion detection where the number of healthy voxels significantly outweighs pathological voxels (<1% of total voxels). To prevent biasing the classifier towards healthy voxels, we constructed a patch-based dataset by randomly under-sampling the healthy voxels such that the feature set was composed of equal number of examples from both classes. To this end, we sub-sampled multi-contrast 3D patches from the co-registered 3D T1w and FLAIR images, with each input image modality representing a channel. The data was normalized within each input modality with zero mean and unit variance. For each normalized training image, we computed 3D patches (16 × 16 × 16) centered on the voxel of interest. The set of all computed patches were aggregated as P = {n × 2 × 16 × 16 × 16}, where n and 2 denote the number of training patches and input MRI modalities, respectively.

*Sampling Heuristics.* On a per-subject level (1.7 million patches × 32 KBytes/patch = 26.3 GB), the training is quite memory-intensive to complete within a reasonable timeframe. To circumvent this issue, we sampled only hyperintense voxels based on the FLAIR contrast by thresholding the subject-level z-normalized images and discarding the bottom 10 percentile intensities. This thresholding yielded a crude gray matter mask, which covered the hyperintense white matter as well. This approach is also biologically meaningful as FCD lesions are primarily located in the gray matter [12]; moreover, both their gray matter and white matter components are consistently hyperintense on FLAIR [13].

## 2.4 Network Architecture and Design

A typical convolutional neural network (CNN) consists of three stages: convolutions, nonlinearity, and pooling. Here, we designed two identical CNNs whose weights are optimized independently. This two-phase cascaded training procedure has been shown to allow efficient training in both CNNs [14, 15] and conventional machine learning [4, 6] paradigms when the distribution of labels is unbalanced. $CNN_1$ was trained to maximize putative lesional voxels, while $CNN_2$ reduced the number of misclassified voxels (*i.e.*, removing false positives while maintaining optimal sensitivity). Each fully convolutional network was composed of three stacks of convolution and max-pooling layers with 48, 96 and 2 filters, respectively. The rectified linear activation (ReLU) non-linearity function was applied to the first two of the three convolutional layers. Softmax non-linearity was used after the final convolution to normalize the result of the kernel convolutions into a binominal distribution over the healthy and lesional labels. See Fig. 1 for network parameters.

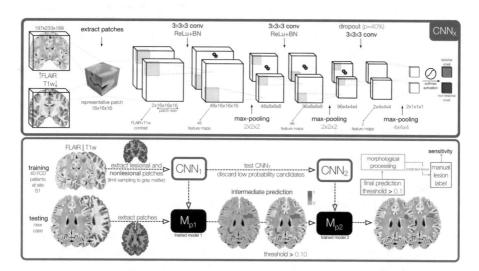

**Fig. 1.** Top panel: Convolutional network architecture ($CNN_x$) for two-label (lesional vs. non-lesional) classification. Bottom panel: Training and testing schema using two-stage CNNx cascade ($CNN_1/CNN_2$).

## 2.5 Classification Paradigm

*Training algorithm.* We used a validation set (75/25 training data split) to optimize the CNN weights. The training set is used to adjust the weights of the neural network, while the validation set measures the performance of trained CNN after each epoch and continues until the validation error plateaus. The model is randomly initialized, and network parameters learns iteratively via the adaptive learning rate method (AdaDelta) by minimizing the binary cross-entropy loss. Binary cross-entropy loss is mathematically defined as:

$crossentropy(p, q) = -(p \cdot log q + (1 - p) \cdot log(1 - q))$, where: p is the true/label distribution, and q is the model/predicted distribution.

Regularization contingencies, including batch-normalization (BN) and Dropout were implemented to prevent overfitting to the training data. At each iteration, BN regularization was implemented after the first two of the three convolutional layers and Dropout (p = 0.4) before the last layer, thereby randomly deactivating 40% of the units (or network connections).

*Inference/Testing Algorithm.* The proposed pipeline was trained on the S1 cohort of 40 consecutive patients with histologically-confirmed FCD lesions. This trained model cascade then served probabilistic predictions on unseen datasets acquired at S1-S7 sites. For each test subject, input images were first partitioned into patches with voxel sampling limited to the FLAIR mask (intra-subject Z-score >0.1). The balanced patch dataset was evaluated using $CNN_1$, which effectively discards improbable lesion candidates. The remaining voxels (threshold >10%) were re-evaluated by $CNN_2$ to obtain the final probabilistic lesion mask. Since, the cost of misclassifying the lesion as healthy tissue is severe, we applied a conservative threshold (>10%) on the probabilistic prediction masks. A simple post-processing routine involving successive morphological erosion, dilation, and extraction of connected components (>75 voxels), was executed to remove flat blobs and noise. The final segmentation masks were compared to manual expert annotations of the lesions.

## 3  Experiment and Results

### 3.1  Subjects

We studied retrospective cohorts with FCD lesions histologically-confirmed after surgery from seven tertiary epilepsy centers worldwide (n = 107). The presurgical workup included neurologic examination, assessment of seizure history, neuroimaging, and video-EEG telemetry. Since the routine MRI was initially reported as unremarkable in 56 patients (52%), the location of the seizure focus was established using intracranially-implanted electrodes; in all patients, retrospective inspection revealed a subtle FCD in the seizure onset region.

*Training Cohort.* The primary site (S1) comprised 40 patients (20 males, 35 adults; mean ± SD age = 27 ± 9 years).

*Independent Testing Cohorts.* Independent test cohorts comprised 67 histologically-confirmed FCD (37 adults and 30 children; mean ± SD age = 33 ± 11 years, 9 ± 6 years, respectively) from six sites with different scanners, and field strengths (1.5T, 3T). The control group consisted of 38 healthy individuals (age = 30 ± 7 years) and 63 disease controls with temporal lobe epilepsy (TLE) and histologically-verified hippocampal sclerosis (age = 31 ± 8), matched for age and sex to S1 cohort.

## 3.2 Performance Evaluation

*Evaluation of Classification for S1.* Two experts segmented independently 40 lesions on co-registered T1w and FLAIR images. Inter-rater dice agreement index $[D = 2|M_1 \cap M_2|/(|M_1| + |M_2|)]$ ($M_1$: 1st label, $M_2$: 2nd label; $M_1 \cap M_2$: intersection of $M_1$ and $M_2$) was $0.91 \pm 0.11$. The union of the two ground truth labels served to train the classifier. The classifier was trained using 5-fold cross validation repeated 20 times. Sensitivity was the proportion of patients in whom a detected cluster co-localized with the lesion label. Specificity was determined with respect to controls (*i.e.*, proportion of controls in whom no FCD lesion cluster was falsely identified), and disease controls with TLE. We also report the number of clusters detected in patients remote from the lesion label (*i.e.*, false positives).

*Evaluation of Classifier Generalizability.* We tested the sensitivity of the classifier trained on S1 was tested on a held-out dataset of eight FCD patients from S1 and 59 independent FCD datasets from S2–S7. For the cross-site unbiased reporting of results blinded to clinical information, the prediction maps (in stereotaxic space) were sent back to respective sites to confirm or dispute the detection of the lesion.

*Comparison with a Benchmark Surface-Based Classifier.* We analysed the S1 dataset using a previously published method [6] based on an ensemble of RUSBoosted decision trees across two classification stages, which uses a total of 30 intensity and morphology features calculated on multimodal T1-weighted and FLAIR images. The classifiers were trained using 5-fold cross validation averaged across 10 iterations.

## 3.3 Results

The 5-fold cross-validation of the CNNs resulted in a sensitivity of $87 \pm 4\%$, with an average of 35/40 lesions detected. In these cases, $2 \pm 1$ extra-lesional clusters were also detected. Specificity was 95% in healthy controls ($3 \pm 1$ clusters in 2/38) and 90% in TLE ($1 \pm 0$ cluster in 7/63).

For cross-dataset classification at seven sites, overall sensitivity was 91% (61/67 lesions detected) with $3 \pm 2$ extra-lesional clusters observed in 47/67 cases. Per-site sensitivity for S1-S7 was 100% (8/8 lesions detected, $2 \pm 2$ extra-lesional clusters), 86% (17/19, $4 \pm 2$), 89% (8/9, $2 \pm 1$), 75% (6/8, $2 \pm 1$), 100% (5/5, $5 \pm 2$), 91% (10/11, $2 \pm 3$), and 100% (7/7, $2 \pm 2$), respectively. Stratifying patients based on age, sensitivity in children (2-18.5 years old) was 90% (27/30 FCD detected, $4 \pm 3$ extra-lesional clusters) while in adults (>19 years old) it was 92% (34/37, $3 \pm 2$). Figure 2 shows test case examples.

Training and testing a surface-based classifier based on S1 dataset yielded a lower performance with a sensitivity of $83 \pm 2\%$ (33/40 lesions detected), with $4 \pm 5$ extra-lesional clusters. Specificity was 92% in healthy controls ($1 \pm 0$ cluster in 3/38).

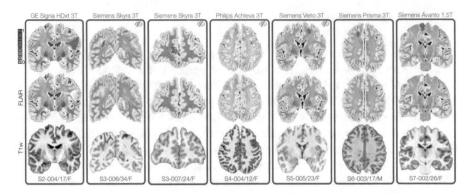

**Fig. 2.** Classification results using the cascaded CNNx trained on 40 FCD patients at site S1 (Siemens TrioTim 3T) to demonstrate generalizability for lesion detection along three axes of heterogeneity: scanner type, field strength (top labels), and age (bottom labels). The seven cases obtained using different scanners at six sites (excluding S1) are shown. The top row indicates the strength of prediction overlaid on the FLAIR, while the second/third rows show the corresponding FLAIR and T1w, respectively. The bottom labels are read as site-patient-ID/age/gender. MRI-negative cases are identified with ✑.

## 4   Discussion

We present the first deep learning method to segment FCD, with multicentric valida-tion. Operating on routine multi-contrast MRI in voxel-space, our algorithm provides the highest performance to date. Furthermore, we demonstrated generalizability of a model trained on a single-site dataset by showing robust performance across inde-pendent cohorts from various centres worldwide with different age, scanner hardware and sequence parameters. Notably, >50% of lesions were missed by conventional radiological inspection.

Operating at two consecutive levels, our classifier resulted in both high sensitivity and specificity. The number of false positive findings in healthy and disease controls were rather modest. Even though our algorithm was trained on an adult dataset, its performance was equally good in children. With respect to the latter, the use of age-appropriate templates taking into account the developmental trajectories, *i.e.,* age-varying tissue contrast, white matter myelination and cortical maturation, is likely to have contributed to the excellent performance by limiting the interpolation effects that would have occurred during registration using an adult template. Moreover, the overall high performance across cohorts strongly suggests that the network learns and opti-mizes parameters specific to FCD pathology, a fact validated by histological confir-mation in all cases.

Compared to a state-of-the-art surface-based classifier, both sensitivity and speci-ficity were higher using the current algorithm. Applying a surface-based approach to S2–S7 would have been challenging due to the large variability in image quality, which would require site-specific fine-tuning of algorithm parameters. A comprehensive comparison is part of future work. In addition, owing to the considerable time investment to manually correct brain tissue segmentation and surface extraction errors, which may have negative downstream effects on the fidelity of features extracted, the current approach is both time-effective and superior.

In conclusion, easy implementation, minimal pre-processing, significant performance gains and inference time of <6 minutes/case make this classifier an ideal platform for large-scale clinical use, particularly in "MRI-negative" FCD.

**Disclaimer.** This feature is based on research and is not commercially available. Due to regulatory reasons its future availability cannot be guaranteed.

# References

1. Blümcke, I., et al.: The clinicopathologic spectrum of focal cortical dysplasias: a consensus classification proposed by an ad hoc task force of the ILAE diagnostic methods commission. Epilepsia **52**, 158–174 (2011)
2. Bernasconi, A., Bernasconi, N., Bernhardt, B.C., Schrader, D.: Advances in MRI for "cryptogenic" epilepsies. Nat. Rev. Neurol. **7**, 99–108 (2011)
3. Kini, L.G., Gee, J.C., Litt, B.: Computational analysis in epilepsy neuroimaging: a survey of features and methods. NeuroImage: Clin. **11**, 515–529 (2016)
4. Hong, S.-J., Kim, H., Schrader, D., Bernasconi, N., Bernhardt, B.C., Bernasconi, A.: Automated detection of cortical dysplasia type II in MRI-negative epilepsy. Neurology **83**, 48–55 (2014)
5. Adler, S., et al.: Novel surface features for automated detection of focal cortical dysplasias in paediatric epilepsy. NeuroImage: Clin. **14**, 18–27 (2017)
6. Gill, R.S., et al.: Automated detection of epileptogenic cortical malformations using multimodal MRI. In: Cardoso, M.J., et al. (eds.) DLMIA/ML-CDS -2017. LNCS, vol. 10553, pp. 349–356. Springer, Cham (2017). https://doi.org/10.1007/978-3-319-67558-9_40
7. Tan, Y.-L., et al.: Quantitative surface analysis of combined MRI and PET enhances detection of focal cortical dysplasias. Neuroimage **166**, 10–18 (2017)
8. LeCun, Y., Bengio, Y., Hinton, G.: Deep learning. Nature **521**, 436–444 (2015)
9. Litjens, G., et al.: A survey on deep learning in medical image analysis. Med. Image Anal. **42**, 60–88 (2017)
10. Sled, J.G., Zijdenbos, A.P., Evans, A.C.: A nonparametric method for automatic correction of intensity nonuniformity in MRI data. IEEE Trans. Med. Imaging **17**, 87–97 (1998)
11. Fonov, V., Evans, A.C., Botteron, K., Almli, C.R., McKinstry, R.C., Collins, D.L.: Brain development cooperative group: unbiased average age-appropriate atlases for pediatric studies. Neuroimage **54**, 313–327 (2011)
12. Sisodiya, S.M., Fauser, S., Cross, J.H., Thom, M.: Focal cortical dysplasia type II: biological features and clinical perspectives. Lancet Neurol. **8**, 830–843 (2009)
13. Hong, S.-J., et al.: Multimodal MRI profiling of focal cortical dysplasia type II. Neurology **88**, 734–742 (2017)
14. Havaei, M., Davy, A., Warde-Farley, D., Biard, A., Courville, A., Bengio, Y., Pal, C., Jodoin, P.-M., Larochelle, H.: Brain tumor segmentation with deep neural networks. Med. Image Anal. **35**, 18–31 (2017)
15. Valverde, S., et al.: Improving automated multiple sclerosis lesion segmentation with a cascaded 3D convolutional neural network approach. Neuroimage **155**, 159–168 (2017)

# Binary Glioma Grading: Radiomics versus Pre-trained CNN Features

Milan Decuyper[⊠], Stijn Bonte, and Roel Van Holen

Medical Imaging and Signal Processing, Ghent University, Ghent, Belgium
milan.decuyper@ugent.be

**Abstract.** Determining the malignancy of glioma is highly important for initial therapy planning. In current clinical practice, often a biopsy is performed to verify tumour grade which involves risks and can negatively impact overall survival. To avoid biopsy, non-invasive tumour characterisation based on MRI is preferred and to improve accuracy and efficiency, the use of computer-aided diagnosis (CAD) systems is investigated. Existing radiomics CAD techniques often rely on manual segmentation and are trained and evaluated on data from one clinical centre. Therefore, there is a need for accurate and automatic CAD systems that are robust to large variations in imaging protocols between different institutions. In this study, we extract features from T1ce MRI with a pretrained CNN and compare their predictive power with hand-engineered radiomics features for binary grade prediction. Performance was evaluated on the BRATS 2017 database containing MRI and manual segmentation data of 285 patients from multiple institutions. State-of-the-art performance with an AUC of 96.4% was achieved with radiomics features extracted from manually segmented tumour volumes. Pre-trained CNN features had a strong predictive value as well and an AUC score of 93.5% could be obtained when propagating the tumour region of interest (ROI). Additionally, using a pre-trained CNN as feature extractor, we were able to design an accurate, automatic, fast and robust binary glioma grading system achieving an AUC score of 91.1% without requiring ROI annotations.

## 1 Introduction

The optimal treatment strategy of newly diagnosed glioma strongly relies on tumour malignancy. Diffuse glioma, the most common form of primary brain tumours, are divided into grades II to IV according to malignancy by the World Health Organization (WHO) [1]. Glioblastoma multiform (GBM) is the most aggressive type of primary brain tumour and has a very poor prognosis with a 5-year survival rate of only 4–5% [2]. Current standard of care for GBMs consists of early resection combined with chemotherapy and radiotherapy. Lower-grade gliomas (LGGs), on the other hand, have more favourable outcomes and possible treatment strategies include: a wait-and-scan approach, a biopsy for histopathological verification or immediate resection [3]. A recent study by Wijnenga et

© Springer Nature Switzerland AG 2018
A. F. Frangi et al. (Eds.): MICCAI 2018, LNCS 11072, pp. 498–505, 2018.
https://doi.org/10.1007/978-3-030-00931-1_57

al. [3] shows that biopsy as initial strategy negatively impacts overall survival with a reported hazard ratio of 2.69 (95% CI 1.19–6.06; p = 0.02) compared to wait-and-scan. The invasive procedure involves high risks, is subject to sampling error and the results may be subjective, depending on the neuropathologist performing the histopathological analysis [4]. Hence a biopsy to confirm diagnosis and grade of the tumour should be avoided and accurate non-invasive grading is preferred.

Conventional MR imaging with gadolinium-based contrast agents is an established technique for non-invasive brain tumour characterisation [5,6]. Through MRI, information is obtained regarding contrast enhancement, necrosis, oedema, mass effect, which are considered important predictors of tumour malignancy. Nevertheless, brain tumour grading using this diagnostic technique is not always reliable with reported sensitivities ranging between 55% and 83% [5]. For example, low-grade glioma demonstrating contrast enhancement can be misdiagnosed as high-grade or conversely 40–45% of non-enhancing lesions are found to be highly malignant gliomas after histopathological verification [6]. Moreover, the ever-increasing amount of MR image data raises the burden of accurate data analysis and dramatically increases the workload of radiologists.

Computer-aided diagnosis (CAD) may provide a way to handle this data explosion and increase diagnostic accuracy [7]. CAD systems can automatically process MR images, calculate quantitative features describing tumour characteristics and combine them to estimate tumour type and grade through the use of artificial intelligence. The time required for diagnosis can be reduced and accuracy and treatment planning enhanced while avoiding the need for biopsy. Towards computer-aided brain tumour diagnosis, the use of radiomics has been investigated [7–9]. Radiomics involves the extraction and analysis of quantitative image features and typically consists of three stages: tumour segmentation, feature extraction and finally classification or analysis of the radiomics features. Zacharaki et al. [8] investigated the classification of brain tumours into different types and grades based on conventional and perfusion MRI. In the proposed method, shape, intensity and Gabor texture features were extracted from regions of interest manually traced by expert neuroradiologists. On a dataset of 102 glioma from 98 patients, an accuracy of 87% was achieved for discriminating high-grade from low-grade glioma with a support vector machine (SVM). A system for grade identification (low- versus high-grade) of astrocytoma from T2-weighted images was designed in the work by Subashini et al. [9]. Tumours were isolated with fuzzy c-means segmentation from which shape, intensity and texture features were calculated. A learning vector quantisation classifier trained on 164 images and evaluated on 36 images achieved an accuracy of 91%. An overview of MRI based medical image analysis studies regarding brain tumour segmentation and grade classification is provided by Mohan and Subashini [7]. In current radiomics studies, often input of domain experts is required, such as manual segmentation data, making these methods not reproducible and not fully automatic. Additionally, most CAD methods are trained and evaluated on data from one clinical centre. Hence these systems are potentially not robust or applicable to data from other centres due to large variations in imaging protocols.

Our goal is to investigate the use of deep learning to develop an accurate, reproducible and fully automatic CAD system. State-of-the-art deep learning models, like convolutional neural networks (CNNs) achieve high performances in object recognition tasks [10]. We investigate the application of these techniques on medical imaging data and study their performance for brain tumour diagnosis. Deep learning has extensively been used in medical image analysis [11] and is increasingly employed in brain tumour segmentation challenges [12]. Binary brain tumour grading using a CNN trained from scratch on data from BRATS 2014 was evaluated by Pan et al. [13]. Sensitivity and specificity scores of 73% were achieved with only a limited and imbalanced dataset. Automated diagnosis with deep learning remains a challenging task as large-scale datasets of brain tumour scans comparable to ImageNet are unavailable. Therefore, in this work, we will try to overcome this lack of large training sets through the use of transfer learning. The application of pre-trained CNNs for survival prediction based on MRI has been investigated by Ahmed et al. [14]. An accuracy of 82% was achieved for differentiating long-term from short-term survival cases on a limited dataset of 22 GBM patients.

To conclude, state-of-the-art performance in binary tumour grading is currently achieved through radiomics with reported accuracies of 87% up to 91%. Only one study using deep learning for binary grade prediction was found reaching sensitivity and specificity scores of 73%. In this paper, we investigate the use of hand-engineered radiomics features and features extracted through a pre-trained CNN to achieve state-of-the-art performance in discriminating GBMs from lower-grade glioma. This allowed us to compare the predictive value of the radiomics features with pre-trained CNN features on the same heterogeneous dataset. In the radiomics approach, shape, intensity and texture features are extracted from T1ce scans manually segmented into different tumour tissues. Deep features, on the other hand, are extracted using a CNN trained on ImageNet [10].

## 2    Materials and Methods

### 2.1    Data

The data used in this work originates from the BRATS 2017 database [12,15]. It contains multi-institutional routine clinically-acquired pre-operative MRI scans of 210 glioblastoma (GBMs) and 75 lower-grade glioma (WHO grade II and III) with pathologically confirmed diagnosis. For each case a T1, T2, T1ce and FLAIR sequence is available. The MRI scans originate from multiple institutions and were acquired with different clinical protocols and scanners resulting in a very heterogeneous dataset. All subject's sequences are co-registered to the same anatomical template, interpolated to a $1 \, mm^3$ voxel size and skull-stripped. Additionally, manual segmentation labels are provided denoting the GD-enhancing, peritumoural oedema and the necrotic and non-enhancing tumour regions. In this study, only the T1ce sequence and segmentation data were used to perform binary grade prediction.

## 2.2  Feature Extraction: Radiomics

In the radiomics feature extraction approach, all scans were first bias corrected using SPM12 (version 6906, Wellcome Trust Centre for Neuroimaging, University College London) running on MATLAB R2017b (The MathWorks, Inc., Natick, MA). Next, since MRI scans are recorded in arbitrary units, the image intensities were normalised following the robust white stripe normalisation [16]. The manual segmentation labels were used to define five different tumour regions: total abnormal region, tumour core, enhancing tissue, necrosis and oedema. In every region we calculated 207 quantitative features: 14 histogram, 8 size and shape, 138 grey-level co-occurence, 22 grey-level run-length matrix, 12 neighbourhoord grey-tone difference matrix and 13 grey-level size-zone matrix features, according to the definitions in Aerts et al. [17] and Willaime et al. [18].

## 2.3  Feature Extraction: Pre-trained CNN

Instead of extracting hand-engineered features from the segmented tumour volumes, deep features were extracted using a pre-trained convolutional neural network. The VGG-11 architecture was used consisting of 8 convolutional and 3 fully connected layers [19]. The model, pre-trained on the ImageNet dataset, was loaded from the pyTorch torchvision package. Features were obtained by forward propagating an MRI slice through the network and extracting the 4096-dimensional output of the first fully connected layer. The first layer was chosen under the assumption that earlier layers learn more generally applicable features than layers deeper into the network. Before being propagated through the network, the slices were pre-processed to match the expected input of the pre-trained pytorch models. The image intensities were scaled to a range between [0,1], the slice was resized to a shape of $224 \times 224$ through bilinear interpolation and finally normalised with mean and standard deviation values provided by pyTorch. Because the model expects RGB images, the MRI slice was provided at the R channel and the B and G channels were set to zero.

Feature extraction and corresponding grading performance was evaluated for four different ways of providing the T1ce scan at the input of the network (see Fig. 1). In a first approach, the segmentation data was used to select the slice in the T1ce scan containing the largest tumour contour and crop this slice to the size of the tumour (Fig. 1: method 1). After applying the pre-processing steps explained above, the tumour patch was propagated through the network, thereby obtaining one 4096-dimensional feature vector with a corresponding label indicating LGG or GBM.

For the second method, all tumour slices were propagated through the network after being cropped to the size of the tumour (Fig. 1: method 2). Hence, multiple feature vectors are obtained for each patient and every slice or feature vector was classified into one of three classes: (1) LGG, (2) GBM where only oedema is visible, (3) GBM with contrast enhancement and necrosis. In each slice, either a LGG or a GBM is visible. Additionally, a GBM may in some slices only display oedema and no contrast enhancement and necrosis. Because these slices may have a similar appearance as LGG slices, this could be confusing for the classifier and therefore a separate class was added for GBM slices only demonstrating oedema.

In the third method, the same slice was selected as in the first approach, but now it was not cropped (Fig. 1: method 3). Hence the entire slice was propagated through the network.

To design a system able to classify a T1ce scan without requiring segmentation information, a fourth method was investigated. Here, every slice of the T1ce scan was propagated through the network (Fig. 1: method 4). One entire scan contains 155 slices, so 155 feature vectors were obtained for each patient and a fourth class, besides the three classes of the second method, was added for slices containing no tumour. Using this approach, no segmentation data is required to classify slices from a T1ce sequence of a new patient resulting in a fully automatic CAD system.

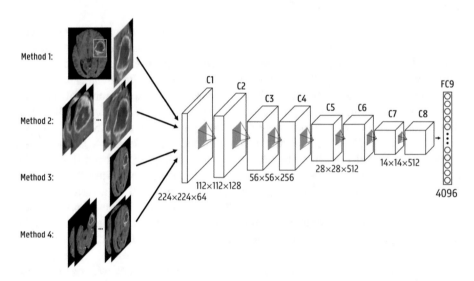

**Fig. 1.** Feature extraction with the pre-trained VGG-11 CNN. Method 1: Propagate tumour region of the slice containing the largest tumour contour. Method 2: Propagate tumour region of all tumour slices. Method 3: Propagate entire slice containing the largest tumour contour. Method 4: Propagate all slices

## 2.4 Classification

After feature extraction, classification was performed with the goal to predict whether a patient has a glioblastoma or lower-grade glioma. The feature vectors were first scaled to unit norm and features showing no variance between different samples were removed. For classification, the python scikit-learn *RandomForestClassifier* was used with 200 decision trees. All Random Forest (RF) models were trained for the binary classification task except for the second and fourth method of feature extraction with the pre-trained CNN. In those cases, the RF model was trained to classify a slice into one of 3, respectively 4 classes as explained in Sect. 2.3. For each patient, multiple slices were classified. All predictions were combined by calculating their mean probability and the sum of the probabilities of the two GBM classes was used as the final probability value of having a GBM. The performance of the classifier was evaluated on a separate test set containing 57 (20%) of the 285 glioma cases. The class ratio of 210:75 was equal in both training and test set. To enhance sensitivity and specificity of the model, the probability threshold of classifying a glioma as GBM was optimised through 5-fold cross-validation. The training and evaluation process was repeated 50 times with different random splits in train and test set to estimate average performance and variability of the model.

## 3   Results

For each of the feature extraction methods, a RF model was trained and evaluated to asses the predictive value of the resulting feature vectors. The area under the ROC curve (AUC), accuracy, sensitivity and specificity scores are reported in Table 1. The RF model trained on the radiomics features achieves the highest performance with an average AUC score of 96%. With features extracted using a pre-trained CNN, best results were obtained when zooming in on the tumour region and using all tumour slices (CNN, method 2). When using features extracted from the entire slice containing the largest tumour contour (CNN, method 3), performance is lower with an AUC of 87% compared to 92%. However, when predicting glioma grade based on all slices of the T1ce scan (CNN, method 4), performance could be improved to an AUC score of 91%. Classifying a T1ce scan was possible within 0.3 s with *CNN: method 1* and *3*, 12 s with *CNN: method 2* and 30 s with *CNN: method 4* on a Macbook Pro with 2.8 GHz Intel Core i7 CPU where propagating all slices through the CNN required most of the computation time.

**Table 1.** Mean (std) (%) area under the ROC curve, accuracy, sensitivity and specificity classification scores.

| Feature extraction method | AUC | Acc. | Sens. | Spec. |
|---|---|---|---|---|
| Radiomics | 96.4(2.6) | 89.6(3.8) | 89.9(5.4) | 88.8(8.6) |
| CNN: Method 1 | 92.2(3.9) | 83.8(4.6) | 83.3(5.2) | 85.2(9.6) |
| CNN: Method 2 | 93.5(3.0) | 86.1(4.3) | 85.4(5.4) | 88.5(8.1) |
| CNN: Method 3 | 86.8(4.6) | 79.1(4.9) | 78.6(6.4) | 80.7(9.6) |
| CNN: Method 4 | 91.1(3.6) | 82(5.3) | 81.5(7.2) | 83(9.6) |

## 4    Discussion

The results shown in Table 1 show that the best performance is achieved with the radiomics approach, matching or even outperforming state-of-the-art accuracies reported today. The achieved performance, however, was obtained when extracting radiomics features from manually segmented tumour tissues which is time-consuming and introduces subjectivity. A lot of research has been performed towards automatic segmentation algorithms and the difference in performance between using a state-of-the-art automatic segmentation algorithm or manual segmentation remains to be investigated.

Although performance is slightly lower compared to the radiomics results, accurate grading could be achieved with a pre-trained CNN as feature extractor as well. With the first method of feature extraction through a CNN, an AUC is achieved of 92% while only requiring a bounding box around the tumour which is considerably less time-consuming than accurate segmentation of the different tissues. Furthermore, when estimating grade based on all tumour slices, performance could be improved to an AUC of 93.5%. These classification scores are more than 10% higher than currently reported binary grading performance with deep learning. Moreover, an automatic segmentation algorithm could be used to define the bounding box and we expect that small variations or inaccuracies will not have a large influence on performance. Features extracted from the entire slice were less informative but by calculating an ensemble prediction from all slices, accurate grading could still be achieved reaching a performance similar to the first method. This way, a binary grading system could be designed that is fast, does not require segmentation or manual input to classify new T1ce sequences and is trained on a very heterogeneous dataset making it robust to variations in imaging protocols. These results show that a CNN, trained on an entirely different image dataset containing natural images, is able to extract informative features from MRI sequences as well. Their predictive value is lower than radiomics features extracted from manually segmented tumour volumes, but we expect that by fine-tuning the network on brain tumour MRI, results can further be improved. Future work will focus on gathering more data, allowing to specialise CNNs on brain MRI and open the path towards more accurate and automatic brain tumour characterisation.

## 5    Conclusion

In this work, we compared the predictive value of radiomics features with features extracted using a pre-trained CNN for binary brain tumour grading. Classification results showed that the best performance is achieved with shape, intensity and texture features extracted from manually segmented tumour volumes. Features from a pre-trained CNN, on the other hand, had a high predictive value as well and allowed to design an accurate, fast, automatic and robust binary grading system. These results indicate that a pre-trained CNN, with possible fine-tuning and more data, holds the potential to develop an accurate, reproducible an fully automatic CAD system.

# References

1. Louis, D.N., Perry, A., et al.: The 2016 world health organization classification of tumors of the central nervous system: a summary. Acta Neuropathol. **131**(6), 803–820 (2016)
2. Carlsson, S.K., Brothers, S.P., Wahlestedt, C.: Emerging treatment strategies for glioblastoma multiforme. EMBO Mol. Med. **6**(11), 1359–1370 (2014)
3. Wijnenga, M.M.J., Mattni, T., et al.: Does early resection of presumed low-grade glioma improve survival? A clinical perspective. J. Neuro-Oncol. **133**(1), 137–146 (2017)
4. Jackson, R.J., Fuller, G.N., et al.: Limitations of stereotactic biopsy in the initial management of gliomas. Neuro-oncology **3**(3), 193–200 (2001)
5. Law, M., Yang, S., et al.: Glioma grading: sensitivity, specificity, and predictive values of perfusion MR imaging and proton MR spectroscopic imaging compared with conventional MR imaging. AJNR **24**(10), 1989–1998 (2003)
6. Jansen, N.L., Graute, V., et al.: MRI-suspected low-grade glioma: is there a need to perform dynamic FET PET? EJNMMI **39**(6), 1021–1029 (2012)
7. Mohan, G., Subashini, M.M.: MRI based medical image analysis: survey on brain tumor grade classification. Biomed. Signal Process. Control. **39**, 139–161 (2018)
8. Zacharaki, E.I., Wang, S., et al.: MRI-based classification of brain tumor type and grade using SVM-RFE. In: Proceedings - 2009 IEEE International Symposium on Biomedical Imaging: From Nano to Macro, ISBI 2009, pp. 1035–1038. IEEE, June 2009
9. Subashini, M.M., Sahoo, S.K., et al.: A non-invasive methodology for the grade identification of astrocytoma using image processing and artificial intelligence techniques. Expert. Syst. Appl. **43**, 186–196 (2016)
10. Russakovsky, O., Deng, J., et al.: ImageNet large scale visual recognition challenge. Int. J. Comput. Vis. **115**(3), 211–252 (2015)
11. Litjens, G., Kooi, T., et al.: A survey on deep learning in medical image analysis. Med. Image Anal. **42**(December 2012), 60–88 (2017)
12. Menze, B.H., Jakab, A., et al.: The multimodal brain tumor image segmentation benchmark (BRATS). IEEE Trans. Med. Imaging **34**(10), 1993–2024 (2015)
13. Pan, Y., Huang, W., et al.: Brain tumor grading based on neural networks and convolutional neural networks. In: 2015 37th Annual International Conference of the EMBC, pp. 699–702. IEEE, August 2015
14. Ahmed, K.B., Hall, L.O., et al.: Fine-tuning convolutional deep features for MRI based brain tumor classification. In: Proceedings of SPIE 10134, Medical Imaging 2017: Computer-Aided Diagnosis, p. 101342E, March 2017
15. Bakas, S., Akbari, H., et al.: Advancing the cancer genome atlas glioma MRI collections with expert segmentation labels and radiomic features. Sci. Data **4**, 170117 (2017)
16. Shinohara, R.T., Sweeney, E.M., et al.: Statistical normalization techniques for magnetic resonance imaging. NeuroImage Clin. **6**, 9–19 (2014)
17. Aerts, H.J.W.L., Velazquez, E.R., et al.: Decoding tumour phenotype by noninvasive imaging using a quantitative radiomics approach. Nat. Commun. **5**, 4006 (2014)
18. Willaime, J.M., Turkheimer, F.E., et al.: Quantification of intra-tumour cell proliferation heterogeneity using imaging descriptors of 18F fluorothymidine-positron emission tomography. Phys. Med. Biol. **58**(2), 187–203 (2013)
19. Simonyan, K., Zisserman, A.: Very deep convolutional networks for large-scale image recognition. CoRR abs/1409.1, pp. 1–14 (2014)

# Automatic Irregular Texture Detection in Brain MRI Without Human Supervision

Muhammad Febrian Rachmadi[1,2]([✉]) [iD], Maria del C. Valdés-Hernández[2] [iD],
and Taku Komura[1]

[1] School of Informatics, University of Edinburgh, Edinburgh, UK
febrian.rachmadi@ed.ac.uk
[2] Centre for Clinical Brain Sciences, University of Edinburgh, Edinburgh, UK

**Abstract.** We propose a novel approach named one-time sampling irregularity age map (OTS-IAM) to detect any irregular texture in FLAIR brain MRI without any human supervision or interaction. In this study, we show that OTS-IAM is able to detect FLAIR's brain tissue irregularities (i.e. hyperintensities) without any manual labelling. One-time sampling (OTS) scheme is proposed in this study to speed up the computation. The proposed OTS-IAM implementation on GPU successfully speeds up IAM's computation by more than 17 times. We compared the performance of OTS-IAM with two unsupervised methods for hyperintensities' detection; the original IAM and the Lesion Growth Algorithm from public toolbox Lesion Segmentation Toolbox (LST-LGA), and two conventional supervised machine learning algorithms; support vector machine (SVM) and random forest (RF). Furthermore, we also compared OTS-IAM's performance with three supervised deep neural networks algorithms; Deep Boltzmann machine (DBM), convolutional encoder network (CEN) and 2D convolutional neural network (2D Patch-CNN). Based on our experiments, OTS-IAM outperformed LST-LGA, SVM, RF and DBM while it was on par with CEN.

**Keywords:** Irregular texture detection · MRI
Unsupervised detection · Hyperintensities detection

## 1 Introduction

Magnetic resonance imaging (MRI) aims to facilitate identifying brain pathologies, like T2-FLAIR white matter hyperintensities (WMH) that are commonly found in patients with dementia/Alzheimer's Disease (AD), stroke and multiple sclerosis. It is believed that WMH are associated with the progression of dementia [10]. However, detecting brain pathology automatically in MRI using computers is challenging as MRI appearance varies depending on studies, scanners and protocols.

© Springer Nature Switzerland AG 2018
A. F. Frangi et al. (Eds.): MICCAI 2018, LNCS 11072, pp. 506–513, 2018.
https://doi.org/10.1007/978-3-030-00931-1_58

Supervised machine learning algorithms such as support vector machine (SVM), random forest (RF) and convolutional neural networks (CNN) are usually used for automatic detection of brain pathology and have been tested in many studies [3,6]. However, manual labels of brain pathology needed for training process are not always available and usually expensive to produce as they have to be performed by an expert (*i.e.*, physician). Furthermore, the quality of the manual label itself depends and varies according to expert's skills which creates another question about reproducibility in different sets of data. These variations usually can be quantified by using intra-/inter-observer evaluations, but it does not solve the subjectivity problem.

Unsupervised machine learning algorithms which work without manual labels do not have the aforementioned dependencies. Some examples of these methods are Lesion Growth Algorithm from Lesion Segmentation Tool toolbox (LST-LGA) [8] and Lesion-TOADS [9] which were developed for unsupervised detection of hyperintensities, the main brain tissue's irregular textural characteristic in T2-FLAIR images. Unfortunately, performances of these methods are limited compared to supervised methods [6].

A newly unsupervised method named irregularity age map (IAM) was proposed in [7], and it was reported that IAM works better than LST-LGA, which is still the baseline and *state-of-the-art* method for unsupervised detection of WMH. However, the IAM has undergone only a limited evaluation, as the original study only used 20 cross-sectional MRI data, its performance was not compared against supervised machine-learning methods and it took 2.9 h to process a single MRI volume with only 35 axial slices. In this study we propose a 13.4 times faster version of IAM named OTS-IAM, which could be implemented on GPU, evaluate it on longitudinal samples and compare its performance with state-of-the-art supervised and unsupervised methods.

In summary, our main contributions in this study are:

1. Proposing one-time sampling (OTS) for IAM (*i.e.*, OTS-IAM).
2. Proposing a new post-processing step to improve OTS-IAM's performance.
3. Full evaluation of OTS-IAM on 60 MRI data from Alzheimer's Disease Neuroimaging Initiative (ADNI) database.
4. Full comparison of OTS-IAM's performance with performances of the original IAM, LST-LGA, SVM, RF, DBM, CEN and patch-based 2D CNN.

## 2  One-Time Sampling Irregularity Age Map

Like the original IAM, one-time sampling irregularity age map (OTS-IAM) is influenced by a previous work in computer graphic [1] in which a novel way of calculating age map in texture image was proposed. Age map quantifies irregular textures into a map of probability values between 0 and 1, dubbed as *age values*. The same approach of generating age map for brain MRI was then proposed and named irregularity age map (IAM) in [7] which reported that IAM works well as an unsupervised method for WMH detection. To calculate age map for MRI, four important steps need to be performed: (1) brain masks preparation,

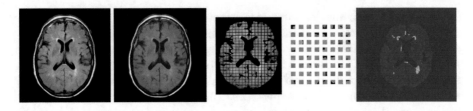

**Fig. 1. From left to right**: original T2-FLAIR MRI, original T2-FLAIR MRI over-laid by ICV (red) and CSF (blue) masks, clean T2-FLAIR MRI divided into non-overlapping grids of *source patches*, 64 examples of randomly sampled overlapping *target patches*, and final result of IAM/OTS-IAM computation. In this visualisation, dimension of both *source patches* and *target patches* are $8 \times 8$.

(2) patch generation, (3) age value calculation and (4) final age map generation. These steps are visualised in Fig. 1.

Brain masks are essential because OTS-IAM works by comparing brain tis-sues, so non-brain tissues such as skull, cerebrospinal fluid (CSF), veins and meninges need to be excluded. To exclude these non-brain tissues, intracranial volume (ICV) and CSF masks are needed. In this study, ICV mask is gener-ated by using optiBET [4] while CSF mask is generated by using an in-house method/protocol. There is no pre-processing step before computation of OTS-IAM other than the generation of these two masks. We here introduce a post-processing step to exclude the non-white matter area of the brain and improve the quality of the final result. This added step uses the normal appearing white matter (NAWM) binary mask, which was generated by using FSL-FAST [11].

Patch generation generates two sets of patches; *source patches*, generated by dividing an MRI slice into non-overlapping grid-patches, and *target patches*, gen-erated by randomly-sampling all possible overlapping patches. The assumption of this step is; if we successfully sample target patches mostly from normal brain tissues, which contain regular textures, and calculate distance values between a source patch and a set of target patches, then irregular textures located within the source patch will produce high absolute distance values. In this study, we use four different sizes of source/target patches; $1 \times 1$, $2 \times 2$, $4 \times 4$ and $8 \times 8$. Unlike previous studies [1,7] which use all possible target patches, we here use a set of randomly sampled target patches to speed up the computation.

Age value calculation is the core of the OTS-IAM where a distance value called *age value* is computed by using the function defined below. Let $\mathbf{s}$ be a source patch and $\mathbf{t}$ a target patch, the *age value* of the two patches $d$ is:

$$d = \alpha \cdot |\max(\mathbf{s} - \mathbf{t})| + (1 - \alpha) \cdot |\text{mean}(\mathbf{s} - \mathbf{t})|. \tag{1}$$

where $\alpha = 0.5$ in this study. Both maximum and mean values of the subtracted patch are used to include maximum and average differences between source and target patches in the calculation. Note that source/target patches are matrices in the size of either $1 \times 1$, $2 \times 2$, $4 \times 4$ or $8 \times 8$. Also, note that each *source patch* will be computed against a set of *target patches*, so each source patch has a set of age values. To get the final age value for one source patch, the corresponding set of age values is sorted in ascending order and, then, the mean of the first 100 age values is calculated. The rationale is simple; the mean of the first 100 age values produced by irregular source patches is still comparably higher than the mean of the first 100 age values produced by normal source patches. All final age values from all source patches are then normalised from 0 to 1 real values to create an age map of one MRI slice.

The final age map generation consists of three sub-steps, which are *blending four age maps from age value calculation, penalty* and *global normalisation*. *Blending of four age maps* is performed by using the following formulation:

$$blendedAgeMap = \alpha \cdot map_1 + \beta \cdot map_2 + \gamma \cdot map_4 + \delta \cdot mri_8 \qquad (2)$$

where $\alpha + \beta + \gamma + \delta$ is equal to 1 and $map_1$, $map_2$, $map_4$ and $map_8$ are age maps from $1 \times 1$, $2 \times 2$, $4 \times 4$ and $8 \times 8$ source/target patches. In this study, $\alpha = 0.65$, $\beta = 0.2$, $\gamma = 0.1$ and $\delta = 0.05$ as weight blending parameters. Before the blending, age maps resulted from different size of source/target patches are up-sampled to fit the original size of MRI slice and then smoothed by using Gaussian filter. The blended age map is then *penalised* using formulation below:

$$p_o = p_i \times v_i \qquad (3)$$

where $p_i$ is voxel from the blended age map, $v_i$ is voxel from the original MRI and $p_o$ is the penalised voxel. Lastly, all age maps from different slices of MRI are normalised together to produce 0 to 1 probability values.

Some important notes on OTS-IAM's computation are: (1) source and target patches are of the same size at the same time, (2) the centre of source/target patches needs to be inside the ICV and outside of the CSF masks at the same time to be included in the age value calculation, (3) if a slice does not have any source patch is skipped to accelerate the computation and (4) there is no indication of OTS-IAM decreasing performance by using only a subset of target patches (*i.e.*, randomly sampled target patches).

## 2.1   One-Time Sampling IAM vs. the Original IAM

While the original IAM has been reported to work well on WMH detection, its computation takes a considerable amount of time owed to the nature of doing one sampling for each source patch. For the sake of clarity in this study, we named this scheme *multiple-time sampling* (MTS) scheme. The original IAM has a MTS scheme as it pre-establishes the condition that every target patch should not be too close to the source patch (*i.e.*, location-based condition). The MTS scheme makes every source patch to have its own set of target patches, so extra time to do sampling for each source patch is unavoidable.

To accelerate the overall IAM's computation, we propose here the one-time sampling (OTS) scheme for IAM, where target patches are randomly sampled only once for each slice, abandoning the location-based condition of the MTS. In other words, age values of all source patches from one slice will be computed against the same set of target patches. We call this combination of OTS and IAM one-time sampling IAM (OTS-IAM).

OTS scheme also enables the possibility of GPU implementation for IAM where the number of target patches randomly sampled from a slice is limited to 2,048 samples. In comparison, the original IAM, which runs on CPU, uses an iterative sampling approach of high number of target patches (*i.e.*, 10%–75% of all possible target patches in an MRI slice), which cannot be applied on GPU because of GPU's limited memory storage and management capabilities. Fixed number of target patches in power of two allows GPU implementation, especially facilitating GPU memory allocation, for OTS-IAM computation.

## 3   MRI Data, Other Machine Learning Algorithms and Experiment Setup

A set of 60 T2-Fluid Attenuation Inversion Recovery (T2-FLAIR) MRI data from 20 subjects of the ADNI database was used for DSC evaluation where every one of them either has absence or mild presence of vascular pathology. Each T2-FLAIR MRI volume has dimension of $256 \times 256 \times 35$. Data used in this study were obtained from the ADNI [5] public database[1].

We compare performances of OTS-IAM with other machine learning algorithms that are commonly used for WMH segmentation; namely the original IAM, Lesion Growth Algorithm from Lesion Segmentation Tool (LST-LGA), SVM, RF, DBM, convolutional encoder network (CEN) and CNN. LST-LGA is the current *state-of-the-art* for unsupervised hyperintensities detection, so it is used as a direct comparison for IAM/OTS-IAM. The rests are representations of supervised conventional machine learning algorithms (*i.e.*, SVM and RF) and supervised deep learning algorithms (*i.e.*, DBM, CEN and CNN). All of them are commonly used for WMH detection/segmentation. Due to page limitation, we could not further elaborate the usage of these algorithms. However, all experiment setup for these algorithms, such as training/testing and algorithm's configurations, follow [6] for reproducibility reason.

For simplicity and reproducibility, automatic WMH detection results are compared with manual labelling of WMH by using Dice similarity coefficient (DSC) [2]. DSC measures spatial coincidence between ground truth and automatic segmentation results. Higher DSC score means better performance, and the DSC score itself can be computed as follow:

$$DSC = \frac{2 \times TP}{FP + 2 \times TP + FN} \tag{4}$$

where $TP$ is true positive, $FP$ is false positive and $FN$ is false negative.

---

[1] Database can be accessed at http://adni.loni.usc.edu. A complete listing of ADNI investigators can be found at http://adni.loni.usc.edu/wp-content/uploads/how_to_apply/ADNI_Acknowledgement_List.pdf.

**Table 1.** Algorithm's information and experiment results based on DSC metric for each tested algorithm. **Abbreviations**: "S/US" for supervised/unsupervised, "Deep Net." for deep neural networks algorithm, "Y/N" for Yes/No, "T2F/T1W" for T2-FLAIR/T1-weighted, "TRSH" for optimum threshold and "Train/Test" for training/testing time. Note that "Speed up" is used only for IAM's instances and compared to the original IAM.

| No | Method | S/US | Deep Net. | Input modality | TRSH | DSC | Train (min) | Test (s) | Test (min) | Speed up |
|----|--------|------|-----------|----------------|------|-----|-------------|----------|------------|----------|
| 1 | LST-LGA | US | N | T2F | 0.134 | 0.2936 | - | 40 | - | - |
| 2 | IAM-CPU | US | N | T2F | 0.230 | 0.3534 | - | 13067 | 218 | - |
| 3 | IAM-CPU-postprocessed | US | N | T2F | 0.179 | 0.3930 | - | 13067 | 218 | - |
| 4 | OTS-IAM-CPU-postprocessed | US | N | T2F | 0.164 | 0.4297 | - | 10410 | 174 | 1.26 |
| 5 | OTS-IAM-GPU-postprocessed | US | N | T2F | 0.159 | 0.4346 | - | 746 | 13 | 17.52 |
| 6 | SVM | S | N | T2F & T1W | 0.925 | 0.2630 | 32 | 87 | - | - |
| 7 | RF | S | N | T2F & T1W | 0.995 | 0.3633 | 40 | 43 | - | - |
| 8 | DBM | S | Y | T2F | 0.687 | 0.3235 | 1420 | 20 | - | - |
| 9 | CEN | S | Y | T2F | 0.284 | 0.4308 | 160 | 7 | - | - |
| 10 | 2D Patch-CNN | S | Y | T2F | 0.801 | 0.5225 | 525 | 30 | - | - |

# 4    Result

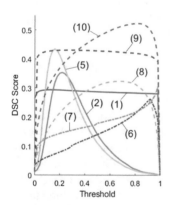

**Fig. 2.** DSC curves for some algorithms listed in Table 1. Numbers in the legend correspond to each algorithm's number.

Table 1 shows the overall results for all tested methods. From Table 1, we can see that all IAM configurations (*i.e.*, IAM-CPU, OTS-IAM-CPU and OTS-IAM-GPU) outperformed LST-LGA by large margins. GPU implementation of OTS-IAM also successfully speeded up IAM's computation by more than 17 times. Note that testing time listed in Table 1 excludes registrations and brain masks generation processes. DSC curves of some algorithms tested in this study are shown in Fig. 2. Whereas, Fig. 3 (left) shows visualisation of age map on MRI with large burden of hyperintensities while Fig. 3 (right) shows visualisation of age map generated from T1-weighted sequence. Figure 3 (right) particularly shows that OTS-IAM could be used on different sequences of MRI.

Table 1 also shows that performances of IAM and OTS-IAM not only outperformed LST-LGA but also some other supervised machine learning algorithms, which are SVM, RF and DBM. Furthermore, OTS-IAM-GPU also slightly outperformed CEN, a supervised deep neural networks algorithm. These comparisons can be seen more easily from DSC curves depicted in Fig. 2.

Based on our repetitive experiments, smaller number of randomly-sampled target patches used in OTS-IAM does not reduce the quality of the result, but rather improve it instead. The GPU implementation of OTS-IAM uses a fixed number of 2,048 target patches instead of 10%–75% of all possible target patches as per the original IAM and the CPU implementation of OTS-IAM (*i.e.*, OTS-IAM-CPU). This shows that the OTS scheme is not only able to accelerate computation time of IAM but also the quality of IAM's result.

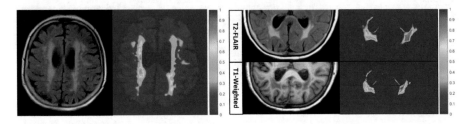

**Fig. 3.** Visualisation of OTS-IAM's results on MRI with large burden of hyperintensities (left) and on T2-FLAIR and T1-weighted of MRI after thresholding (right). Notice how each area of hyperintensities has its own level of damage. These are unlabelled MRI data from ADNI dataset.

## 5   Discussion

Through this study, we have demonstrated the use of the newly proposed method OTS-IAM to automatically detect irregular texture without any human supervision or manual labelling. In this study, OTS-IAM successfully detected irregular textures of hyperintensities in T2-FLAIR and also hypointensities in T1-weighted (Fig. 3). OTS-IAM works on texture level, not intensity level, where the size of texture patches (*i.e.*, source and target patches) can be adjusted easily as needed, including 3D patches instead of 2D patches.

Based on our experiments, incorporation of one-time sampling (OTS) scheme to IAM not only speeded up IAM's computation but also improved IAM's performance. There was no indication of performance degradation in the use of a subset of all possible target patches where uniformed random sampling was applied. The use of limited number of target patches for the GPU implementation of OTS-IAM (*i.e.*, only 2048 target patches) strengthened this argument. It would be very interesting to know the minimum number of target patches needed to further speed up IAM's computation without any performance degradation.

As an unsupervised method, OTS-IAM works independently from any subjective human expertise which usually influences supervised machine learning algorithms. The results also shows that all IAM instances outperformed LST-LGA, the *state-of-the-art* of unsupervised hyperintensities detection, by large margins. OTS-IAM also outperformed some of supervised machine learning algorithm tested; SVM, RF, DBM and CEN.

In the future, OTS-IAM might be used not only for irregular texture detection but also for other purposes. For example, OTS-IAM could be used to provide unsupervised labels of hyperintensities for pre-training step of supervised deep neural networks algorithms. By the help of OTS-IAM, the most tedious and expensive step of producing manual labels of hyperintensities could be minimised and thousands of hyperintensities age maps could be produced automatically.

**Acknowledgement.** Funds from Indonesia Endowment Fund for Education (LPDP) of Ministry of Finance, Republic of Indonesia and Row Fogo Charitable Trust (Grant No. BRO-D.FID3668413) (MCVH) are gratefully acknowledged. Data collection and sharing for this project was funded by the Alzheimer's Disease Neuroimaging Initiative (ADNI) (National Institutes of Health Grant U01 AG024904) and DOD ADNI (Department of Defense W81XWH-12-2-0012).

# References

1. Bellini, R., Kleiman, Y., Cohen-Or, D.: Time-varying weathering in texture space. ACM Trans. Graph. (TOG) **35**(4), 141 (2016). https://doi.org/10.1145/2897824. 2925891
2. Dice, L.R.: Measures of the amount of ecologic association between species. Ecology **26**(3), 297–302 (1945). https://doi.org/10.2307/1932409
3. Kamnitsas, K., et al.: Efficient multi-scale 3D CNN with fully connected CRF for accurate brain lesion segmentation. Med. Image Anal. **36**, 61–78 (2017). https://doi.org/10.1016/j.media.2016.10.004
4. Lutkenhoff, E.S., et al.: Optimized brain extraction for pathological brains (optiBET). PloS one **9**(12), e115551 (2014). https://doi.org/10.1371/journal.pone.0115551
5. Mueller, S.G., et al.: The alzheimer's disease neuroimaging initiative. Neuroimaging Clin. N. Am. **15**(4), 869–877 (2005). https://doi.org/10.1016/j.nic.2005.09.008
6. Rachmadi, M.F., Valdés-Hernández, M.C., Agan, M.L.F., Komura, T.: Deep learning vs. conventional machine learning: pilot study of wmh segmentation in brain mri with absence or mild vascular pathology. J. Imaging **3**(4), 66 (2017). https://doi.org/10.3390/jimaging3040066
7. Rachmadi, M.F., Valdés-Hernández, M.C., Komura, T.: Voxel-based irregularity age map (IAM) for brain's white matter hyperintensities in MRI. In: 2017 International Conference on Advanced Computer Science and Information Systems (ICACSIS), pp. 321–326. IEEE (2017). https://doi.org/10.1109/ICACSIS.2017.8355053
8. Schmidt, P., et al.: An automated tool for detection of FLAIR-hyperintense white-matter lesions in multiple sclerosis. Neuroimage **59**(4), 3774–3783 (2012). https://doi.org/10.1016/j.neuroimage.2011.11.032
9. Shiee, N., Bazin, P.L., Ozturk, A., Reich, D.S., Calabresi, P.A., Pham, D.L.: A topology-preserving approach to the segmentation of brain images with multiple sclerosis lesions. NeuroImage **49**(2), 1524–1535 (2010). https://doi.org/10.1016/j.neuroimage.2009.09.005
10. Wardlaw, J.M., et al.: Neuroimaging standards for research into small vessel disease and its contribution to ageing and neurodegeneration. Lancet Neurol. **12**(8), 822–838 (2013). https://doi.org/10.1016/S1474-4422(13)70124-8
11. Zhang, Y., Brady, M., Smith, S.: Segmentation of brain MR images through a hidden markov random field model and the expectation-maximization algorithm. IEEE Trans. Med. Imaging **20**(1), 45–57 (2001). https://doi.org/10.1109/42.906424

# Learning Myelin Content in Multiple Sclerosis from Multimodal MRI Through Adversarial Training

Wen Wei[1,2,3(✉)], Emilie Poirion[3], Benedetta Bodini[3], Stanley Durrleman[2,3],
Nicholas Ayache[1], Bruno Stankoff[3], and Olivier Colliot[2,3]

[1] UCA, Inria, Epione project-team, Sophia Antipolis, France
wen.wei@inria.fr
[2] Inria, Aramis project-team, Paris, France
[3] Sorbonne Université, Inserm, CNRS, Institut du cerveau et la moelle (ICM),
AP-HP-Hôpital Pitié-Salpêtrière, Boulevard de l'hôpital, Paris, France

**Abstract.** Multiple sclerosis (MS) is a demyelinating disease of the central nervous system (CNS). A reliable measure of the tissue myelin content is therefore essential to understand the physiopathology of MS, track progression and assess treatment efficacy. Positron emission tomography (PET) with $[^{11}C]$PIB has been proposed as a promising biomarker for measuring myelin content changes in-vivo in MS. However, PET imaging is expensive and invasive due to the injection of a radioactive tracer. On the contrary, magnetic resonance imaging (MRI) is a non-invasive, widely available technique, but existing MRI sequences do not provide, to date, a reliable, specific, or direct marker of either demyelination or remyelination. In this work, we therefore propose Sketcher-Refiner Generative Adversarial Networks (GANs) with specifically designed adversarial loss functions to predict the PET-derived myelin content map from a combination of MRI modalities. The prediction problem is solved by a sketch-refinement process in which the sketcher generates the preliminary anatomical and physiological information and the refiner refines and generates images reflecting the tissue myelin content in the human brain. We evaluated the ability of our method to predict myelin content at both global and voxel-wise levels. The evaluation results show that the demyelination in lesion regions and myelin content in normal-appearing white matter (NAWM) can be well predicted by our method. The method has the potential to become a useful tool for clinical management of patients with MS.

## 1 Introduction

Multiple Sclerosis (MS) is the most common neurological disability in young adults [1]. MS pathophysiology predominately involves autoimmune aggression of central nervous system (CNS) myelin sheaths, which results in inflammatory demyelinating lesions. These demyelinating lesions in CNS can cause different

A. F. Frangi et al. (Eds.): MICCAI 2018, LNCS 11072, pp. 514–522, 2018.
https://doi.org/10.1007/978-3-030-00931-1_59

symptoms such as sensory, cognitive or motor skill dysfunctions [1]. In MS, spontaneous myelin repair occurs, allowing restoration of secure and rapid conduction and protecting axons from degeneration. However, remyelination is generally insufficient to prevent irreversible disability. Therefore, a reliable measure of the tissue myelin content is essential to understand MS physiopathology, and also to quantify the effects of new promyelinating therapies.

Positron emission tomography (PET) with [$^{11}$C]PIB has been proposed as a promising biomarker for measuring myelin content changes in-vivo in MS [2]. In particular, it has been recently described that [$^{11}$C]PIB PET can be used to visualize and measure myelin loss and repair in MS lesions [3]. The reader familiar with Alzheimer's disease (AD) may note that the tracer is the same as the one used in AD. PIB has the property to bind to proteins characterized by a similar conformation, contained in both amyloid plaques and myelin. Nevertheless, note that the signal in myelin is more subtle than for amyloid plaques. However, PET is expensive and not offered in the majority of medical centers in the world. Moreover, it is invasive due to the injection of a radioactive tracer. On the contrary, MR imaging is a widely available and non-invasive technique, but existing MRI sequences do not provide, to date, a reliable, specific, or direct marker of demyelination or remyelination. Therefore, it would be of considerable interest to be able to predict the PET-derived myelin content map from multimodal MRI.

In recent years, various methods for medical image enhancement and synthesis using deep neural networks (DNN) have been proposed, such as reconstruction of 7T-like T1-w MRI from 3T T1-w MRI [4] and generation of FLAIR from T1-w MRI [5]. Several works have proposed to predict PET images from MRI or CT images [6–8]. A single 2D GAN has been proposed to generate FDG PET from CT for tumor detection in lung [6] and liver region [7]. However, they do not take into account the spatial nature of 3D images and can cause discontinuous predictions between adjacent slices. Additionally, a single GAN can be difficult to train when the inputs become complex. A two-layer DNN has been proposed to predict FDG PET from T1-w MRI [8] for AD diagnosis. However, a 2-layer DNN is not powerful enough to fully incorporate the complex information from multimodal MRI. Importantly, all these works were devoted to the prediction of FDG PET. Predicting myelin information (as defined by PIB PET) is a more difficult task because the signal is more subtle and with weaker relationship to anatomical information that could be found in T1-w MRI or CT.

In this work, we therefore propose Sketcher-Refiner GANs consisting in two conditional GANs (cGANs) with specifically designed adversarial loss to predict the PET-derived myelin content from multimodal MRI. Compared to previous works [6,7] using GANs for medical imaging, our method solves the prediction problem by a sketch-refinement process which decomposes the difficult problem into more tractable subproblems and makes the GANs training process more stable. In our method, the sketcher generates the preliminary anatomy and physiology information and the refiner refines and generates images reflecting the tissue myelin content in the human brain. In addition, the adaptive adversarial loss is designed to force the refiner to pay more attention to the prediction of the myelin content in MS lesions and normal appearing white matter (NAWM). To our knowledge, this is, to date, the first work to predict, from multi-sequence MR images, the myelin content measured usually by PET imaging.

## 2   Method

### 2.1   Sketcher-Refiner Generative Adversarial Networks (GANs)

GANs are generative models consisting of two components: a generator G and a discriminator D. Given a real image $y$, the generator G aims to learn the mapping $G(z)$ from a random noise vector $z$ to the real data. The discriminator $D(y)$ evaluates the probability that $y$ is a true image. To constrain the output of the generator, cGANs [9] were proposed in which the generator and the discriminator both receive a conditional variable $x$. More precisely, D and G play the two-player conditional minimax game with the following cross-entropy loss function:

$$\min_{G} \max_{D} \mathcal{L}(D,G) = \mathbb{E}_{x,y \sim p_{\text{data}}(x,y)}[\log D(x,y)] -$$
$$\mathbb{E}_{x \sim p_{\text{data}}(x), z \sim p_z(z)}[\log(1 - D(x,G(x,z)))] \tag{1}$$

where $p_{\text{data}}$ and $p_z$ are the distributions of real data and the input noise.

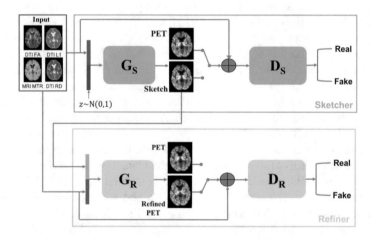

**Fig. 1.** The proposed sketcher-refiner GANs. The sketcher receives MR images and generates the preliminary anatomy and physiology information. The refiner receives MR images $I_M$ and the sketch $I_S$. Then it refines and generates PET images.

The goal is to predict the $[^{11}\text{C}]$PIB PET distribution volume ratio (DVR) parametric map $I_P$ which reflects the myelin content, from multimodal MRI $I_M$. Figure 1 is the architecture of our method consisting of two cGANs named **Sketcher** and **Refiner**. Our method decomposes the prediction problem into two steps: (1) sketching anatomy and physiology information and (2) refining and generating images reflecting the tissue myelin content in the human brain. To the end, the sketcher and the refiner have the following cross-entropy losses:

$$
\min_{G_S} \max_{D_S} \mathcal{L}(D_S, G_S) = \mathbb{E}_{I_M, I_P \sim p_{\text{data}}(I_M, I_P)}[\log D_S(I_M, I_P)] -
$$
$$
\mathbb{E}_{I_M \sim p_{\text{data}}(I_M), z \sim p_z(z)}[\log(1 - D_S(I_M, G_S(I_M, z)))] \tag{2}
$$

$$
\min_{G_R} \max_{D_R} \mathcal{L}(D_R, G_R) = \mathbb{E}_{I_M, I_P \sim p_{\text{data}}(I_M, I_P)}[\log D_R(I_M, I_P)] -
$$
$$
\mathbb{E}_{I_M \sim p_{\text{data}}(I_M), I_s \sim G_S(I_M, z)}[\log(1 - D_R(I_M, G_R(I_M, I_s)))] \tag{3}
$$

where $D_S$, $D_R$ and $G_S$, $G_R$ represent the discriminator and the generator in the sketcher and the refiner respectively.

## 2.2   Adversarial Loss with Adaptive Regularization

Previous works [10] have shown that it can be useful to combine the GAN objective function with a traditional constraint, such as L1 and L2 loss. They further suggested using L1 loss rather than L2 loss to encourage less blurring. We hence mixed the GANs' loss function with the following L1 loss for the sketcher.

$$
\mathcal{L}_{L1}(G_S) = \frac{1}{N} \sum_{i=1}^{N} |I_P^i - G_S(I_M^i, z^i)| \tag{4}
$$

where $N$ is the number of subjects and $i$ denotes the index of a subject.

In CNS, myelin constitutes most of the white matter (WM). To force the generator on MS lesions where demyelination happens, the whole image is divided into three regions of interest (ROIs): lesions, NAWM and "other". We thus defined for the refiner a weighted L1 loss in which the weights are adapted to the number of voxels in each ROI indicated as $N_{\text{Les}}$, $N_{\text{NAWM}}$ and $N_{\text{other}}$. Given the masks of the three ROIs: $R_{\text{Les}}$, $R_{\text{NAWM}}$ and $R_{\text{other}}$, the weighted L1 loss for the refiner is defined as follows:

$$
\mathcal{L}_{L1}(G_R) = \frac{1}{N \times M} \sum_{i=1}^{N} \Big( \frac{1}{N_{\text{Les}}} \sum_{j \in R_{\text{Les}}} |I_P^{i,j} - \hat{I}_P^{i,j}| +
$$
$$
\frac{1}{N_{\text{NAWM}}} \sum_{j \in R_{\text{NAWM}}} |I_P^{i,j} - \hat{I}_P^{i,j}| + \frac{1}{N_{\text{other}}} \sum_{j \in R_{\text{other}}} |I_P^{i,j} - \hat{I}_P^{i,j}| \Big) \tag{5}
$$

where $M$ is the number of voxels in a PET image, $j$ is the index of a voxel and $\hat{I}_P$ is the prediction output from the refiner.

To sum up, our overall objective functions are defined as follows:

$$
G_S^* = \arg \min_{G_S} \max_{D_S} \mathcal{L}(D_S, G_S) + \lambda_S \mathcal{L}_{L1}(G_S)
$$
$$
G_R^* = \arg \min_{G_R} \max_{D_R} \mathcal{L}(D_R, G_R) + \lambda_R \mathcal{L}_{L1}(G_R) \tag{6}
$$

where $\lambda_S$ and $\lambda_R$ are hyper-parameters which balance the contributions of two terms in the sketcher and the refiner respectively.

## 2.3   Network Architectures

Both the sketcher and the refiner have the same architectures for their generators and discriminators. For the generators, a general shape of a "U-Net" is used (see details in Fig. 2(B)). We use LeakyReLU which allows a stable training of GANs with 0.2 as slope coefficient.

For the discriminator, a traditional approach in GANs is to use a global discriminator: the discriminator is trained to globally distinguish if the input comes from the true dataset or from the generator. However, when we trained, the generator tries to over-emphasize certain image features in some regions so that it can make the global discriminator fail to differentiate a real or fake image. In our problem, each region in the PET image has its own myelin content. A key observation is that any local region in a generated image should have a myelin content that is similar to that of the real local region. Therefore, instead of using a traditional global network, we define a 3D patch discriminator trained by local patches from input images. As shown in Fig. 2(A), the input image is firstly divided into patches with size $l \times w \times h$ and then the 3D patch discriminator classifies all the patches separately. The final loss of the 3D patch discriminator is the sum of the cross-entropy losses from all the local patches. Its architecture is a traditional CNN including four stages of Conv3D-BatchNorm-LeakyReLU-Downsampling and a combination with a fully connected layer and a Softmax layer as last two layers.

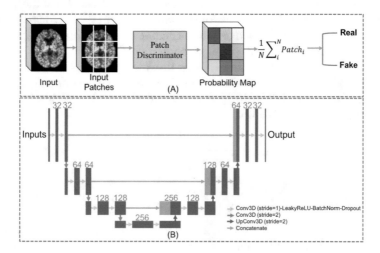

**Fig. 2.** The D and the G in our GANs. (A) The proposed 3D patch discriminator which takes all the patches and classifies them separately to output a final loss. (B) The 3D U-Net shaped generator with implementation details shown in the image.

# 3  Experiments and Evaluations

## 3.1  Overview

- **Dataset:** Our dataset includes 18 MS patients (12 women, mean age 31.4 years, sd 5.6) and 10 age- and gender-matched healthy volunteers (8 women, mean age 29.4, sd 6.3). For each subject, the MRI data includes a magnetisation transfer ratio map (MTR) and three measures derived from diffusion tensor imaging (DTI): fractional anisotropy (FA), radial diffusivity (RD), axial diffusivity (AD) while the PET data is a $[^{11}C]$PIB PET DVR parametric map. The preprocessing steps mainly consist of the intra-subject registration onto $[^{11}C]$PIB PET image space.
- **Training details:** Our sketcher-refiner GANs is implemented with the Keras library. The convolution kernel size is $3 \times 3 \times 3$ and the rate for dropout layer is 50%. The optimization is performed with the ADAM solver with $10^{-4}$, $5 \times 10^{-5}$ as initial learning rates for the sketcher and the refiner respectively. We used 3-fold cross validation with 19 subjects for training and 9 subjects for testing. Two GTX 1080 Ti GPUs are used for training.

## 3.2  Qualitative Evaluation

Figure 3 shows the qualitative comparison of our prediction results, a 2-layer DNN as in [8] and a single cGAN (corresponding to the sketcher in our approach) with corresponding input multimodal MRI and the true $[^{11}C]$PIB PET DVR parametric map. We can find that the 2-layer DNN failed to find the non-linear mapping between the multimodal MRI and the myelin content in PET. Especially, some anatomical or structural traces (that are not present in the

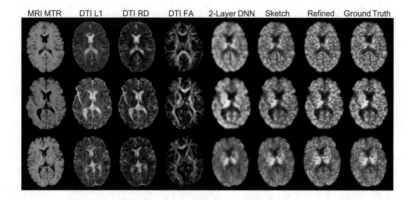

**Fig. 3.** Qualitative comparison of our method ("Refined"), a 2-layer DNN and the single cGAN (corresponding to the sketcher in our approach is denoted as "Sketch") as well as the ground truth and the input MR images.

ground truth) can still be found in the 2-layer-DNN predicted PET. This highlights that the relationship between myelin content and multimodal MRI data is complex, and only two layers are not powerful enough to encode-decode it.

It is also shown that the single-GAN predicted PET (sketch) generates a blurry output with the primitive shape and basic information. On the other hand, after the refinement process by our refiner, the output is closer to the ground truth and the myelin content is better predicted. According to this, we can also conclude that only using a single cGAN (sketcher) like in [6,7] is insufficient for our problem.

### 3.3   Global Evaluation of Myelin Prediction

To assess myelin content at a global level, three ROIs were defined: (1) WM in healthy controls (HC); (2) NAWM in MS patients; (3) lesions in MS patients. Figure 4(A) displays the comparison between the distribution of the ground truth and the predicted PET. It shows that the PET-derived overall patterns can be well reproduced by the synthetic data. Specifically, both with the gold standard and the synthetic data, there is no significant difference between NAWM in patients and WM in HC, while a statistically significant reduction of myelin content in lesions compared to NAWM can be found ($p < 0.0001$). Figure 4(B) shows the comparison at the individual participant level. The predicted DVR map is very close to the ground truth in the vast majority of participants. This demonstrates that our method can adequately predict myelin content at the individual patient level.

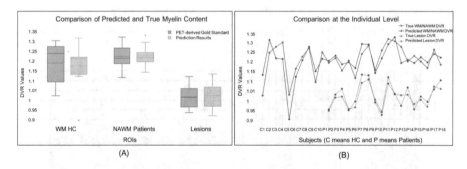

**Fig. 4.** (A) Group level evaluation. The box plots show the median (middle solid line), mean (middle dotted line) and the range of DVR for each ROI for PET-derived DVR parametric map used as gold standard (blue) and the prediction results (yellow). (B) Individual level evaluation. The red and blue colors respectively represent the mean DVR values for each subject extracted in the ground truth image and the predicted PET, respectively. Results from WM/NAWM and lesions are shown using respectively a solid line and a dotted line.

### 3.4 Voxel-Wise Evaluation of Myelin Prediction

We also evaluate the ability of our method to predict myelin content at the voxel-wise level. Within each MS lesion of each patient, each voxel was classified as a demyelinated voxel according to a procedure defined in a previous clinical study [3]. We hence measure the percentage of demyelinated voxels over total lesions load of each patient for both the ground truth and the predicted PET as shown in Fig. 5(A). Our prediction results approximate the ground truth for most of the patients. The average Dice index between the demyelination map derived from the ground truth and our predicted PET is 0.83. Figure 5(B) shows demyelinated voxels classified from both the true and the predicted PET within MS lesions. The large agreement regions demonstrate our method's strong ability to predict the demyelination in MS lesions at the voxel-wise level.

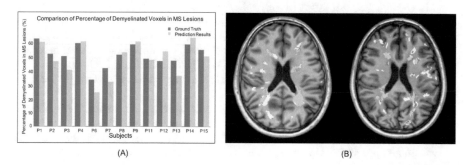

(A)                                                    (B)

**Fig. 5.** (A) Percentage of demyelinated voxels in white matter MS lesions for each patient computed from the ground truth (blue) and from our method (grey). (B) Examples of demyelinated voxels classified from the ground truth and our predicted PET within MS lesions. Agreement between methods is marked in yellow (both true and predicted PET indicated demyelination) and white (both methods did not indicate demyelination). Disagreement is marked in red (demyelination only with the true PET) and orange (only with the predicted PET).

## 4    Conclusion and Future Work

We proposed Sketcher-Refiner GANs with specific designed adversarial loss functions to predict the PET-derived myelin content from multimodal MRI. The prediction problem is solved by a sketch-refinement process in which the sketcher generates the preliminary anatomy and physiology information and the refiner refines and generates images reflecting the tissue myelin content in the human brain. Our method is evaluated for myelin content prediction at both global and voxel-wise levels. The evaluation results show that the demyelination in MS lesions, and myelin content in both patient's NAWM and control's WM can be well predicted by our method. In the future, it would be interesting to use our method on longitudinal dataset, to investigate dynamic demyelination and remyelination processes.

# References

1. Compston, A., Coles, A.: Multiple sclerosis. Lancet **372**(9648), 1502–1517 (2008)
2. Stankoff, B., et al.: Imaging central nervous system myelin by positron emission tomography in multiple sclerosis using [methyl-11c]-2-(4'-methylaminophenyl)- 6-hydroxybenzothiazole. Ann. Neurol. **69**(4), 673–680 (2011)
3. Bodini, B., et al.: Dynamic imaging of individual remyelination profiles in multiple sclerosis. Ann. Neurol. **79**(5), 726–738 (2016)
4. Bahrami, K., Shi, F., Rekik, I., Shen, D.: Convolutional neural network for reconstruction of 7T-like images from 3T MRI Using appearance and anatomical features. In: Carneiro, G., et al. (eds.) LABELS/DLMIA -2016. LNCS, vol. 10008, pp. 39–47. Springer, Cham (2016). https://doi.org/10.1007/978-3-319-46976-8_5
5. Sevetlidis, V., Giuffrida, M.V., Tsaftaris, S.A.: Whole image synthesis using a deep encoder-decoder network. In: Tsaftaris, S.A., Gooya, A., Frangi, A.F., Prince, J.L. (eds.) SASHIMI 2016. LNCS, vol. 9968, pp. 127–137. Springer, Cham (2016). https://doi.org/10.1007/978-3-319-46630-9_13
6. Bi, L., Kim, J., Kumar, A., Feng, D., Fulham, M.: Synthesis of positron emission tomography (PET) images via multi-channel generative adversarial networks (GANs). In: Cardoso, M., et al. (eds.) CMMI/SWITCH/RAMBO -2017. LNCS, vol. 10555, pp. 43–51. Springer, Cham (2017). https://doi.org/10.1007/978-3-319-67564-0_5
7. Ben-Cohen, A., Klang, E., Raskin, S.P., Amitai, M.M., Greenspan, H.: Virtual PET images from CT data using deep convolutional networks: initial results. In: Tsaftaris, S.A., Gooya, A., Frangi, A.F., Prince, J.L. (eds.) SASHIMI 2017. LNCS, vol. 10557, pp. 49–57. Springer, Cham (2017). https://doi.org/10.1007/978-3-319-68127-6_6
8. Li, R., et al.: Deep learning based imaging data completion for improved brain disease diagnosis. In: Golland, P., Hata, N., Barillot, C., Hornegger, J., Howe, R. (eds.) MICCAI 2014. LNCS, vol. 8675, pp. 305–312. Springer, Cham (2014). https://doi.org/10.1007/978-3-319-10443-0_39
9. Mirza, M., Osindero, S.: Conditional generative adversarial nets. CoRR abs/1411.1784 (2014)
10. Isola, P., Zhu, J.Y., Zhou, T., Efros, A.A.: Image-to-image translation with conditional adversarial networks. arxiv (2016)

# Deep Multi-structural Shape Analysis: Application to Neuroanatomy

Benjamín Gutiérrez-Becker[✉] and Christian Wachinger

Artificial Intelligence in Medical Imaging (AI-Med), KJP, LMU München,
Munich, Germany
benjamin.gutierrez_becker@med.uni-muenchen.de

**Abstract.** We propose a deep neural network for supervised learning on neuroanatomical shapes. The network directly operates on raw point clouds without the need for mesh processing or the identification of point correspondences, as spatial transformer networks map the data to a canonical space. Instead of relying on *hand-crafted* shape descriptors, an optimal representation is learned in the end-to-end training stage of the network. The proposed network consists of multiple branches, so that features for multiple structures are learned simultaneously. We demonstrate the performance of our method on two applications: (i) the prediction of Alzheimer's disease and mild cognitive impairment and (ii) the regression of the brain age. Finally, we visualize the important parts of the anatomy for the prediction by adapting the occlusion method to point clouds.

## 1 Introduction

Shape analysis of anatomical structures is of core importance for many tasks in medical imaging, not only as a regularization prior for segmentation tasks, but also as a powerful tool to assess differences between subjects and populations. A fundamental question when operating on shapes is to find a suitable numerical representation for a given task. Hence, many different types of parameterizations have been proposed in the past including point distribution models [2], spectral signatures [16], spherical harmonics [5], medial representations [6], and diffeomorphisms [11]. Even though these representations have proven their utility for the analysis of shapes in the medical domain, they might not be optimal for a particular task.

In recent years, deep networks have had ample success for many medical imaging tasks by learning complex, hierarchical feature representations from images. These representations have proven to outperform *hand-crafted* features in a variety of medical imaging applications [10]. One of the main reasons for the success of these methods is the use of convolutional layers, which take advantage of the shift-invariance properties of images [1]. However, the use of deep networks in medical shape analysis is still largely unexplored; mainly because typical shape representations such as point clouds and meshes do not possess an underlying Euclidean or grid-like structure.

© Springer Nature Switzerland AG 2018
A. F. Frangi et al. (Eds.): MICCAI 2018, LNCS 11072, pp. 523–531, 2018.
https://doi.org/10.1007/978-3-030-00931-1_60

In this work, we propose an alternative approach to perform supervised learning on medical shape data. Our method is based on PointNet [12], a deep neural network architecture, which operates directly on a point cloud and predicts a label in an end-to-end fashion. Point clouds present a raw and simple parameterization that avoids complexities involved with meshes and that is trivial to obtain given a segmented surface. The network does not require the alignment of point clouds, as a spatial transformer network maps the data to a canonical space before further processing. PointNet has been proposed for object classification, where the category of a single shape is predicted. For many medical applications however, not just a single anatomical structure is important for the prediction but a simultaneous view of multiple structures is required for a more comprehensive analysis of a subject's anatomy. Hence, we propose the Multi-Structure PointNet (MSPNet), which is able to simultaneously predict a label given the shape of multiple structures. We evaluate MSPNet in two neuroimaging applications, neurodegenerative disease prediction and age regression.

### 1.1   Related Work

Several shape representations have previously been used for supervised learning tasks. Spherical harmonics for approximating the hippocampal shape have been proposed in [5]. Shape information has been derived from thickness measurements of the hippocampus from a medial representation [3]. Statistical shape models to detect hippocampal shape changes were proposed by [14]. Multi-resolution shape features with non-Euclidean wavelets were employed for the analysis of cortical thickness [9]. The use of medial axis shape representations was used to compare the brain morphology of autistic and normal populations [6]. Recently, shape representation based on spectral signatures have been introduced to perform age regression and disease prediction [16,17].

All the mentioned approaches rely on computing pre-defined shape features. Alternatively, a variational auto-encoder was proposed to automatically extract features from 3D surfaces, which can in turn be used in a classification task [13]. However different to our approach, this is not an end-to-end learning since the variational encoder is not directly linked to the classification task. Consequently, the learned features capture overall variation but are not directly optimized for the given task. In addition, this approach relies on computing point correspondences between meshes and focuses on a single structure, while we simultaneously model multiple structures.

## 2   Method

We propose a method for multiple structure shape analysis that is divided into two main stages: the extraction of point clouds representing the anatomy of different structures from medical images (Sect. 2.1), and a Multi-Structure PointNet (MSPNet) (Sect. 2.2). Figure 1 illustrates the architecture of MSPNet, which is based on PointNet [12], and extends on it to allow the simultaneous processing of multiple structures.

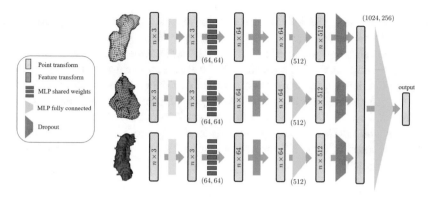

**Fig. 1.** MSPNet Architecture. The network consists of one branch per structure (illustrated for three structures), which are fused before the final multilayer perceptron (MLP). Each structure is represented by a point cloud with $n$ points that pass through transformer networks and multilayer perceptrons of the individual branch. Numbers in brackets are layer sizes.

## 2.1 Point Cloud Extraction

We extract point clouds from MRI T1-weighted images of the brain. We process the images with the FreeSurfer pipeline [4] and obtain segmentations of multiple neuroanatomical regions. From the resulting segmentations, point clouds are created by uniformly sampling the boundary of each brain structure. After this process, the anatomy of a subject is represented by a collection of $m$ point clouds $S = \{P_0, P_1, \ldots P_m\}$, where each point cloud represents a structure. A point cloud is defined as a set of $n$ points $P = [\mathbf{p}_0, \mathbf{p}_1, \ldots, \mathbf{p}_n]$, where each point is a vector of Cartesian coordinates $\mathbf{p}_i = (x, y, z)$.

## 2.2 MSPNet Architecture

We aim at finding a network architecture corresponding to a function $f : S \mapsto y$, mapping a collection of shapes described by $S$ to a prediction $y$. An overview of the network is shown in Fig. 1. MSPNet consists of multiple branches, where each branch processes the point cloud of one structure independently. This ensures that an optimal feature representation is learned per structure. At the end, the features of all branches are merged to perform a joint prediction. Each branch can be divided into the following stages: (1) point cloud alignment using a transformation network, (2) feature extraction, (3) feature alignment with a second transformation net, (4) dropout and (5) prediction. The first three stages of the architecture of each branch resemble that of a single PointNet architecture. The last two stages are particular to MSPNet.

**Point Transformation Network:** In contrast to previous approaches in deep medical shape analysis [13], MSPNet does not require point correspondences across shapes, i.e., the i-*th* points of two shapes, $\mathbf{p}_i^1$ and $\mathbf{p}_i^2$, respectively, do not need to represent the same anatomical position. We obtain the invariance to rigid transformations in MSPNet by (i) augmenting the training dataset by applying a random rigid transformation to each shape during training time and by (ii) introducing a transformation network (T-Net). This network estimates a $3 \times 3$ transformation matrix, which is applied to the input as a first step. One can think of the T-Net as a transformation into a canonical space to roughly align point clouds before any processing is done. The T-Net is shown in Fig. 2 and is composed of a multilayer perceptron (MLP), a max pooling operator and two fully connected layers.

**Feature Extractions:** The transformed points are fed into a MLP with shared weights among points. This MLP layer can be thought of as the feature extraction stage of the network. At this stage of the network, each point has access to the position of all the remaining points of the point cloud, and therefore as the output of the network, we obtain a $k$-dimensional feature vector for each point (in our case $k = 64$). Although each point is assigned a single feature vector, in practice each feature vector point contains a global signature of the input point cloud.

**Feature Transformation:** A second T-Net is applied to the computed features. This network has the same properties as the first transformation network, but its output corresponds to a $k \times k$ transformation matrix. This transformation matrix has a much higher dimension than the previous spatial transformation, which makes the optimization more challenging. To facilitate the optimization of this larger feature transformation matrix $T$, we constrain it to be close to an orthogonal matrix $C_{\mathrm{reg}} = \|I - TT^\top\|_F^2$, similar to [12]. The regularization term ensures a more stable convergence of the network. After the points are transformed they are fed to a MLP layer.

**Dropout and Prediction:** Up to this point, the architecture of each branch mirrors that of the PointNet. However the final dropout and prediction stage is particular to MSPNet. In PointNet, the last stage corresponds to a max-pooling layer performed across $n$ points, so that the output is a vector with size corresponding to the feature dimensionality. Instead of performing max-pooling, which leads to a strong shrinkage in feature space, we propose to keep the localized information per point. This leads to an increase in the network capacity, which may lend itself to overfitting. Hence, we introduce a dropout layer (keep probability $= 0.3$) for regularization. The main advantage of the new design is that more localized information is retained in the network, which we hypothesize may boost the predictive power of our network. Finally, the individual features from each branch are concatenated and fed into a last MLP to perform prediction. Batch normalization is used for all MLP layers and ReLU activations are used. The last MLP perceptron counts with intermediate dropout layers with 0.7 keep probabilities as in PointNet. To facilitate the exposition, we assumed that each structure per branch is described by the same number of $n$ points, but in practice each structure can be represented by point clouds of different dimensions.

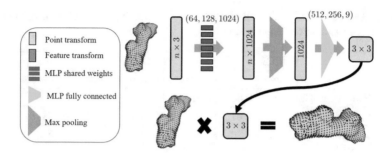

**Fig. 2.** Transformation network (T-Net) for predicting a transformation matrix to map a point cloud to canonical space before processing. A similar network is used to transform the features; the only difference is that the output corresponds to a $64 \times 64$ matrix.

## 3   Results

We evaluate the performance of MSPNet in two supervised learning tasks, classification and regression. For the classification task, we aim at using shape descriptors to discriminate between healthy controls (HC), and patients diagnosed with mild cognitive impairment (MCI) or Alzheimer's disease (AD). For the regression task, we perform age estimation of a subject based on shape information. In all our experiments, we compare to the standard PointNet architecture and spectral shape descriptors in BrainPrint [16], which achieved high performance in a competition for Alzheimer's disease classification [17]. For PointNet, the multi-structure input corresponds to a concatenation of the point clouds of all structures. We use image data from the Alzheimer's Disease Neuroimaging Initiative (ADNI) database (adni.loni.usc.edu) [8]. We work with a total of 7,974 images (2,423 HC, 978 AD, and 4,625 MCI).

### 3.1   AD and MCI Classification on Shape Data

For this experiment, we perform classification based on the shape of the left and right hippocampus and the left and right lateral ventricles, due to their key importance in Alzheimer's disease [15]. Each structure is represented by a Pointcloud of 512 points. For our experiments the dataset is split in a training, validation and test set (75%,15%,15%). Splitting is done on a per subject basis, to guarantee that the same subject does not appear in different sets. Table 1 reports the results of the classification experiment, where we report average classification precision, recall and F1-score. In both classification scenarios, PointNet shows a higher accuracy than BrainPrint, illustrating the potential of learning feature representations. Further, MSPNet showed the best performance, highlighting the benefit of individual feature learning in each branch of the network.

**Table 1.** Average precision, recall and F1-score for the mild cognitive impairment and Alzheimer's classification experiments.

| | HC-MCI | | | HC-AD | | |
|---|---|---|---|---|---|---|
| | Precision | Recall | F1-score | Precision | Recall | F1-score |
| BrainPrint | 0.57 | 0.59 | 0.57 | 0.76 | 0.77 | 0.78 |
| PointNet | 0.60 | 0.61 | 0.59 | 0.77 | 0.77 | 0.78 |
| MSPNet | 0.62 | 0.60 | **0.61** | 0.78 | 0.79 | **0.80** |

## 3.2   Age Prediction on Shape Data

For the age estimation task, we perform two different evaluations. In the first one, we perform age estimation only on the healthy controls of the ADNI database. For the second evaluation, we also include patients diagnosed with MCI and AD. The evaluations are done again on the same brain structures used for the classification task. The results of these two experiments are summarized in the mean absolute error plots of Fig. 3. For the experiment on HC MSPNet significantly outperformed BrainPrint (p-value $2.69 \times 10^{-9}$) and PointNet (p-value 0.03). In the experiment on all subjects both PointNet and MSPNet presented comparable performance, both outperforming BrainPrint (p-value 0.01).

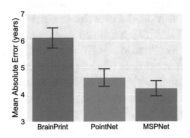 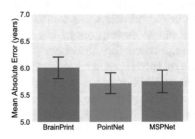

**Fig. 3.** Mean absolute error for the age prediction experiment on healthy subjects (left) and on all subjects, including MCI and AD (right)

## 3.3   Visualizing Point Importance

Of key importance for making predictions with shapes is the ability to visualize the part of the anatomy that is driving the decision. This holds in particular in the clinical context. In MSPNet, we introduce a simple yet effective method to

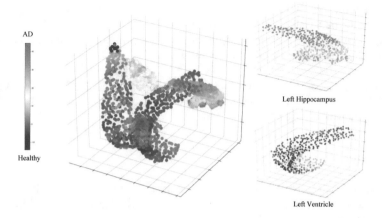

**Fig. 4.** Visualization of point importance for the HC-AD classification task for an AD subject. Figure on the left illustrates ventricles and hippocampi, while figures on the right illustrate single structures of the left hemisphere.

visualize the importance that each point has in the prediction. Our visualization is inspired by the commonly used occlusion method [7], which consists of occluding parts of a test image and observing differences in the network response. We apply a similar concept to visualize the response of MSPNet. In our case, we assess the importance of each point in the classification task by occluding this point (making the point coordinates equal to 0) together with its nearest neighbors. Then the occluded point cloud is passed through the network and the response of the output ReLU is compared to that obtained when the full point cloud is evaluated. The difference between these responses can then be assigned as the importance of this particular point. In Fig. 4, we can observe the result of using this visualization technique for one of the AD test subjects in the HC-AD classification experiment. If a point tends towards the red side of the scale, it indicates that by occluding this particular point, the network increases the activation of the AD class. This means that the region around this point is used by the network to predict AD. The exact opposite is true for points on the blue side of the scale. White points indicate that the network response was not largely affected by occluding this point. In the particular case of the example in Fig. 4, the decision of the network to give this subject a AD label is mainly driven by the left hippocampus.

## 4   Conclusion

We introduced MSPNet, a deep neural network for shape analysis on multiple brain structures. To the best of our knowledge, this is the first time that a neural network for shape analysis on point clouds is proposed in medical applications. We have shown that our method is able to achieve high accuracy in both classification and regression tasks, when compared to shape descriptors based on spectral signatures. This performance is achieved without relying on point correspondences or meshes. MSPNet learns feature representations from multiple structures simultaneously. Finally, we illustrated point-wise importance for the prediction by adapting the occlusion method.

**Acknowledgments.** This work was supported in part by SAP SE and the Bavarian State Ministry of Education, Science and the Arts in the framework of the Centre Digitisation.Bavaria (ZD.B).

# References

1. Bronstein, M.M., Bruna, J., LeCun, Y., Szlam, A., Vandergheynst, P.: Geometric deep learning: going beyond euclidean data. IEEE Signal Process. Mag. **34**(4), 18–42 (2017)
2. Cootes, T., Taylor, C., Cooper, D., Graham, J.: Active shape models-their training and application. Comput. Vis. Image Underst. **61**(1), 38–59 (1995)
3. Costafreda, S.G., et al.: Automated hippocampal shape analysis predicts the onset of dementia in mild cognitive impairment. Neuroimage **56**(1), 212–219 (2011)
4. Fischl, B., et al.: Freesurfer. Neuroimage **62**(2), 774–781 (2012)
5. Gerardin, E., et al.: Multidimensional classification of hippocampal shape features discriminates Alzheimer's disease and mild cognitive impairment from normal aging. Neuroimage **47**(4), 1476–1486 (2009)
6. Gorczowski, K., et al.: Statistical shape analysis of multi-object complexes. Comput. Vis. Pattern Recognit. **2007**, 1–8 (2007)
7. Grün, F., Rupprecht, C., Navab, N., Tombari, F.: A taxonomy and library for visualizing learned features in convolutional neural networks. arXiv preprint arXiv:1606.07757 (2016)
8. Jack, C.R., et al.: The Alzheimer's disease neuroimaging initiative (ADNI): MRI methods. J. Magn. Reson. Imaging **27**(4), 685–691 (2008)
9. Kim, W.H., Singh, V., Chung, M.K., Hinrichs, C., Pachauri, D., et al.: Multi-resolutional shape features via non-Euclidean wavelets: applications to statistical analysis of cortical thickness. NeuroImage **93**, 107–123 (2014)
10. Litjens, G., et al.: A survey on deep learning in medical image analysis. Med. Image Anal. **42**, 60–88 (2017)
11. Miller, M.I., Younes, L., Trouvé, A.: Diffeomorphometry and geodesic positioning systems for human anatomy. Technology **2**(01), 36–43 (2014)
12. Qi, C.R., Su, H., Mo, K., Guibas, L.J.: PointNet: deep learning on point sets for 3D classification and segmentation. In: CVPR (2017)
13. Shakeri, M., Lombaert, H., Tripathi, S., Kadoury, S.: Deep spectral-based shape features for Alzheimer's disease classification. In: Reuter, M., Wachinger, C., Lombaert, H. (eds.) SeSAMI 2016. LNCS, vol. 10126, pp. 15–24. Springer, Cham (2016). https://doi.org/10.1007/978-3-319-51237-2_2
14. Shen, K.K., Fripp, J., Mériaudeau, F., Chételat, G., Salvado, O., Bourgeat, P.: Detecting global and local hippocampal shape changes in Alzheimer's disease using statistical shape models. Neuroimage **59**(3), 2155–2166 (2012)
15. Thompson, P.M., Hayashi, K.M., De Zubicaray, G.I., Janke, A.L., Rose, S.E., Semple, J., et al.: Mapping hippocampal and ventricular change in Alzheimer disease. Neuroimage **22**(4), 1754–1766 (2004)

16. Wachinger, C., Golland, P., Kremen, W., Fischl, B., Reuter, M.: BrainPrint: a discriminative characterization of brain morphology. Neuroimage **109**, 232–248 (2015)
17. Wachinger, C., Reuter, M., Alzheimer's Disease Neuroimaging Initiative, et al.: Domain adaptation for Alzheimer's disease diagnostics. Neuroimage **139**, 470–479 (2016)

# Computational Modelling of Pathogenic Protein Behaviour-Governing Mechanisms in the Brain

Konstantinos Georgiadis[1(✉)], Alexandra L. Young[1], Michael Hütel[1],
Adeel Razi[3], Carla Semedo[1], Jonathan Schott[2], Sébastien Ourselin[1,2,4],
Jason D. Warren[2], and Marc Modat[1,2,4]

[1] Centre for Medical Imaging Computing, University College London, London, UK
konstantinos.georgiadis.14@ucl.ac.uk
[2] Dementia Research Centre, Institute of Neurology, University College London,
Queen Square, London WC1N 3BG, UK
[3] Wellcome Centre for Human Neuroimaging, University College London,
London, UK
[4] School of Biomedical Engineering and Imaging Sciences,
King's College London, London, UK

**Abstract.** Most neurodegenerative diseases are caused by pathogenic proteins. Pathogenic protein behaviour is governed by neurobiological mechanisms which cause them to spread and accumulate in the brain, leading to cellular death and eventually atrophy. Patient data suggests atrophy loosely follows a number of spatiotemporal patterns, with different patterns associated with each neurodegenerative disease variant. It is hypothesised that the behaviour of different pathogenic protein variants is governed by different mechanisms, which could explain the pattern variety. Machine learning approaches take advantage of the pattern predictability for differential diagnosis and prognosis, but are unable to reveal new information on the underlying mechanisms, which are still poorly understood. We propose a framework where computational models of these mechanisms were created based on neurobiological literature. Competing hypotheses regarding the mechanisms were modelled and the outcomes evaluated against empirical data of Alzheimer's disease. With this approach, we are able to characterise the impact of each mechanism on the neurodegenerative process. We also demonstrate how our framework could evaluate candidate therapies.

**Keywords:** Computational modelling · Neurodegenerative disease

## 1 Introduction

Most neurodegenerative diseases are caused by the accumulation of pathogenic proteins whose behaviour is governed by neurobiological mechanisms. The interplay among these mechanisms cause pathogenic proteins to accumulate and

© Springer Nature Switzerland AG 2018
A. F. Frangi et al. (Eds.): MICCAI 2018, LNCS 11072, pp. 532–539, 2018.
https://doi.org/10.1007/978-3-030-00931-1_61

spread in the brain network, leading to loss of function and brain atrophy. However, these mechanisms are still poorly understood and there are many hypotheses regarding what mechanisms govern pathogenic protein behaviour [14,16]. Brain imaging of patients suggests that atrophy loosely follows a number of spatiotemporal patterns, with different patterns associated with each neurodegenerative disease variant [14]. It is hypothesised that different pathogenic protein variants are governed by different mechanisms, which would explain the variety of patterns [14]. A better understanding of these mechanisms is key for the development of therapies that directly influence them and thus disease progression.

Conventional machine learning can perform differential diagnosis and prognosis [10,15] in neurodegenerative diseases. An fMRI study determined disease-specific regions as epicentres, whose functional connectivity extracted in healthy subjects correlated with atrophy progression [7,16], suggesting functional connectivity impacts disease progression. In addition, hub regions and regions with shorter functional paths to the syndrome-specific epicentre showed greater vulnerability. Despite their usefulness, such approaches and correlation studies do not provide information on the protein-behaviour governing mechanisms.

A computational modelling study of amyloid-beta and tau aggregation in AD evaluated candidate therapies [9], but on a more abstract level. Simulations and computational modelling of pathogenic protein behaviour governing mechanisms have previously been applied, but only to a small artificial neural network [5]. A model of brain network mediated trans-synaptic diffusion [10] achieved a strong correlation with atrophy of follow-up scans, however these were limited to four year follow-ups. In contrast to these works, we model many additional protein behaviour governing mechanisms, using a brain network and simulations and we predict disease progression in its entirety.

We propose that computational models of pathogenic protein behaviour governing mechanisms be created based on neurobiological literature and then used in simulations to predict atrophy. We evaluate models by comparing a simulation's atrophy prediction to the prediction of an event based model fitted to APOE4 positive AD patient data [15]. The models that best fit empirical data provide evidence in favour of the related hypotheses behind them. Thus, our framework can suggest which hypotheses warrant further investigation in neurobiological research. We also demonstrate how suggestions regarding the effectiveness of candidate therapies that directly target the mechanisms can be made by modelling their influence into the simulations.

## 2   Methodology

Structural, functional and diffusion-weighted MR images from 10 healthy subjects were used to generate a graph representation of a healthy brain network, consisting of 27 regions. We modelled multiple pathogenic protein mechanisms. After initialising the network and seeding pathogenic protein into it, we ran simulations, using our modelled mechanisms to update the network state.

## 2.1    Image Dataset and Processing

T1-weighted (T1w) MR images were parcellated into 208 regions [1]. We kept 27 symmetric grey matter regions (Fig. 1), denoted as $r \in \{1, ..., 27\}$. Each region has associated coordinates $\mathbf{c}_r$ and volume $V_r(t)$ at timestep $t$. Subject volumes were normalised by their total intracranial volume, then for each region $V_r(0)$ was set to the population averaged regional volume.

Resting-state functional scans were motion and EPI corrected and high-pass filtered (0.01 Hz). Time courses were extracted, then centered and variance-normalised. The parcellations were affinely registered from the T1w to the fMRI image space. We computed the per region and per subject synaptic signals based on the methodology of Karahanoğlu et al. [6] (activity inducing signals), which were averaged over subjects to get the population regional synaptic signals $\mathbf{Sig}_r$. The per region, per subject synaptic signals were also used to compute the synaptic activity's power spectrum, which were averaged over subjects. Using the per region power spectrums, we computed the mean frequency per region $f_r(t)$.

Diffusion data were corrected for motion, eddy-currents and EPI distortion using field maps before tensors were fitted. Tractography was then performed [13] and filtered using the approach porposed by Smith et al. [11]. We denote as $\mathbf{DTI}_{r_1,r_2}$ the connectivity matrix extracted from the tractography and defined using the brain parcellation of the T1w images.

Resting state fMRI scans were used to compute each subject's effective connectivity [2] where the structural connectivity $\mathbf{DTI}_{r_1,r_2}$ was incorporated as prior information [12] with hyperparameters $\alpha = 4, \beta = 12$. The population effective connectivity $\mathbf{EC}_{r_1,r_2}$ was calculated by performing Bayesian model reduction [3].

## 2.2    Modelling of Protein Behaviour Governing Mechanisms

We modelled the mechanisms of protein production, clearance, misfolding, extra-cellular diffusion, network-mediated diffusion, frequency-related spread and atrophy. Brain regions had associated atrophy $A_r(t)$, radius $\mathrm{Rad}_r(t)$ (we assumed spherical regions), pathogenic protein concentration $P_r(t)$ and non-pathogenic protein concentration $N_r(t)$. Pathogenic protein is hypothesised to interact with non-pathogenic protein through the mechanism of misfolding, requiring the need for a non-pathogenic protein concentration. We denote $\mathbf{D}_{r_1,r_2}$ as the Euclidean distance between the barycentric coordinates of brain regions. We calculated the correlation matrix $\mathbf{Corr}_{r_1,r_2}$ between all pairs of synaptic signals $\mathbf{Sig}_{r_1}, \mathbf{Sig}_{r_2}$. Regional atrophy was initialised to $A_r(0) = 0$ and increases as a function of the regional protein concentration (Eq. 1), with $A_t$ defining the protein concentration threshold below which no additional atrophy occurs and $A_m$ controlling the atrophy magnitude. Once $A_r(t) = 1$, then the region has fully atrophied. As atrophy increases, it linearly decreases volumes $V_r(t)$ (Eq. 2, also requiring an updating of radii), as well as the regional synaptic frequency $f_r(t)$ (Eq. 3), as we assumed that as brain regions atrophy, less synaptic activity will occur.

$$A_r(t) = A_r(t-1) + \max\left[0, A_m\left(e^{10(N_r(t)+P_r(t)-A_t)} - 1\right)\right] \tag{1}$$

$$V_r(t) = (1 - A_r(t)) V_r(0) \tag{2}$$

$$f_r(t) = (1 - A_r(t)) f_r(0) \tag{3}$$

All modelled mechanisms are applied similarly to non-pathogenic and pathogenic protein, except for misfolding. Therefore, we only display equations for pathogenic protein except for Eq. 14. Production rates of non-pathogenic $R_{\mathrm{ProdN}}$ and pathogenic $R_{\mathrm{ProdP}}$ protein model transcription and translation, the biological process through which new protein molecules are created in cells. Our model of clearance [8] summarises all the processes through which the brain removes protein molecules to be replaced by new ones. We modelled clearance $\mathrm{ClearN}_r(t), \mathrm{ClearP}_r(t)$ (Eq. 4) such that when concentrations diverge from the equilibrium of normal protein concentration levels $N_{\mathrm{Equilibrium}}, P_{\mathrm{Equilibrium}}$, where clearance rates $R_{\mathrm{ClearN}}, R_{\mathrm{ClearP}}$ are typically equal to production rates, the clearance rates adjust to compensate. We assumed pathogenic protein behaves in a "prion-like" manner [4,14], meaning that when pathogenic protein and non-pathogenic protein come in close proximity, the non-pathogenic protein misfolds and converts to the pathogenic state. Given a misfolding rate of $R_{\mathrm{Mis}}$ a concentration of non-pathogenic protein $\mathrm{Misfold}_r(t)$ misfolds to pathogenic (Eq. 5).

$$\mathrm{ClearP}_r(t) = R_{\mathrm{P}} \log\left(1 + (e-1)\frac{P_r(t)}{P_{\mathrm{Equilibrium}}}\right) \tag{4}$$

$$\mathrm{Misfold}_r(t) = R_{\mathrm{Mis}} N_r(t) P_r(t) \tag{5}$$

Many candidate mechanisms are hypothesised to spread protein [4] (e.g. diffusion, exocytosis, etc). We modelled extracellular diffusion $\mathbf{ED}_{r_1,r_2}(t)$ (Eq. 7), network-mediated diffusion $\mathbf{ND}_{r_1,r_2}(t)$ (Eq. 9) and frequency-related spread $\mathbf{FS}_{r_1,r_2}(t)$ (Eq. 10), with the elements of these matrices indicating the probability of protein spreading from region $r_2$ to $r_1$. Extracellular diffusion models Brownian motion in the extracellular space. The probability of protein spreading through extracellular diffusion to a region is given by the integral of a one dimensional (we assumed isotropic diffusion) normal distribution which is based on the region's radius, its Euclidean distance from the origin region and the standard deviation $\sigma_{\mathrm{ED}}$, which controls the extracellular diffusion speed. Network-mediated diffusion models Brownian motion adjusted to the strengths of network connections. The probability of protein spreading out of region $r_2$ is given by $\mathrm{CL}_{r_2}$ (Eq. 8) and is based on the integral of a one dimensional normal distribution, with the standard deviation $\sigma_{\mathrm{ND}}$ controlling the speed, as well as a term for the relative connection strength per unit of volume. The probability to spread into a region $r_1$ also depends on the regions' normalised connection strength $w_{r_1,r_2}$. Frequency-related spread assumes that the more frequent synaptic activity is, the more protein spreads out of a region. The probability of protein spreading from region $r_2$ to region $r_1$ is based on the general strength of frequency-related spread $R_{\mathrm{FS}}$, frequencies $f_{r_2}(t)$ and on the normalised connection strengths $w_{r_1,r_2}$ (Eq. 10).

$$\text{NED}_{r_2}(t) = \sum_{r_1} \frac{1}{\sqrt{2\pi}} \int_{\frac{\mathbf{D}_{r_1,r_2}-\text{Rad}_{r_1}(t)}{\sigma_{\text{ED}}}}^{\frac{\mathbf{D}_{r_1,r_2}+\text{Rad}_{r_1}(t)}{\sigma_{\text{ED}}}} e^{\frac{x^2}{2}} dx \tag{6}$$

$$\text{ED}_{r_1,r_2}(t) = \frac{1}{\text{NED}_{r_2}(t)\sqrt{2\pi}} \int_{\frac{\mathbf{D}_{r_1,r_2}-\text{Rad}_{r_1}(t)}{\sigma_{\text{ED}}}}^{\frac{\mathbf{D}_{r_1,r_2}+\text{Rad}_{r_1}(t)}{\sigma_{\text{ED}}}} e^{\frac{x^2}{2}} dx \tag{7}$$

$$\text{CL}_{r_2} = 2\left(\frac{1}{\sqrt{2\pi}}\int_{-\infty}^{\frac{-\text{Rad}_{r_2}(t)}{\sigma_{\text{ND}}}} e^{\frac{x^2}{2}} dx\right) \sum_{r_1} w_{r_1,r_2} \frac{\min_{r_3} V_{r_3}(0)}{V_{r_2}(0)} \tag{8}$$

$$\mathbf{ND}_{r_1,r_2}(t) = \begin{cases} \text{CL}_{r_2} w_{r_1,r_2}, & \text{if } r_1 \neq r_2 \\ \text{CL}_{r_2} w_{r_1,r_2} + (1 - \text{CL}_{r_2}), & \text{if } r_1 = r_2 \end{cases} \tag{9}$$

$$\mathbf{FS}_{r_1,r_2}(t) = \begin{cases} R_{\text{FS}}\frac{f_{r_2}(t)}{\max_{r_3} f_{r_3}(0)} w_{r_1,r_2}, & \text{if } r_1 \neq r_2 \\ R_{\text{FS}}\frac{f_{r_2}(t)}{\max_{r_3} f_{r_3}(0)} w_{r_1,r_2} + 1 - R_{\text{FS}}\frac{f_{r_2}(t)}{\max_{r_3} f_{r_3}(0)}, & \text{if } r_1 = r_2 \end{cases} \tag{10}$$

The connection strengths for network-mediated diffusion and frequency-related spread can be based on any related metric. We explored the following possibilities: connection strengths based on the correlation coefficients $w_{r_1,r_2} = |\mathbf{Corr}_{r_1,r_2}|$ between synaptic signals or based on the fibre tract connectivities $w_{r_1,r_2} = |\mathbf{DTI}_{r_1,r_2}|$ or based on the effective connectivity strengths $w_{r_1,r_2} = |\mathbf{EC}_{r_1,r_2}|$. Respectively they create the network-mediated diffusion matrices $\mathbf{ND}^C_{r_1,r_2}(t)$ with speed $\sigma_{\text{ND-C}}$, $\mathbf{ND}^D_{r_1,r_2}(t)$ with speed $\sigma_{\text{ND-D}}$ and $\mathbf{ND}^E_{r_1,r_2}(t)$ with speed $\sigma_{\text{ND-E}}$ and the frequency-related spread matrices $\mathbf{FS}^C_{r_1,r_2}(t)$ with strength $R_{\text{FS-C}}$, $\mathbf{FS}^D_{r_1,r_2}(t)$ with strength $R_{\text{FS-D}}$ and $\mathbf{FS}^E_{r_1,r_2}(t)$ with strength $R_{\text{FS-E}}$.

Each timestep, we update atrophy $A_r(t)$, volumes $V_r(t)$, radii $\text{Rad}_r(t)$ and frequencies $f_r(t)$. Production, misfolding and clearance update the concentrations, which are then transformed to quantities (Eq. 14). After spreading through the network (Eq. 15) they are transformed back to concentrations (Eq. 16).

$$\mathbf{FS}(t) = \mathbf{FS}^C(t) \times \mathbf{FS}^D(t) \times \mathbf{FS}^E(t) \tag{11}$$
$$\mathbf{ND}(t) = \mathbf{ND}^C(t) \times \mathbf{ND}^D(t) \times \mathbf{ND}^E(t) \tag{12}$$
$$\mathbf{QN}_r(t) = (N_r(t) + R_{\text{ProdN}} - \text{ClearN}_r(t) - \text{Misfold}_r(t))V_r(t) \tag{13}$$
$$\mathbf{QP}_r(t) = (P_r(t) + R_{\text{ProdP}} - \text{ClearP}_r(t) + \text{Misfold}_r(t))V_r(t) \tag{14}$$
$$\mathbf{QNP}(t) = \mathbf{FS}(t) \times \mathbf{ND}(t) \times \mathbf{ED}(t) \times \mathbf{QP}(t) \tag{15}$$
$$P_r(t+1) = \mathbf{QNP}_r(t)/V_r(t) \tag{16}$$

## 3   Results

We set: $N_r(0) = 0.01$, $R_{\text{ProdN}} = R_{\text{ClearN}} = 2e-4$, $A_t = 0.04$. We varied whether pathogenic protein was soluble ($R_{\text{ClearP}} = 2e-5$) or insoluble ($R_{\text{ClearP}} = 0$), whether there was pathogenic protein production ($P_r(0) = 0.01$, $R_{\text{ProdP}} = 2e-5$) or not ($P_r(0) = 0$, $R_{\text{ProdP}} = 0$) and whether there was pathogenic protein seeding of concentration $P_{seed}$ at the hippocampus, parahippocampal gyrus or entorhinal area (e.g. for seeding the hippocampus: $P_{r=hippocampus}(t = 0) = P_{seed}$).

We evaluated our models by comparing simulation atrophy prediction against the prediction of an event based model (EBM) [15] which computed the uncertainty matrix $\mathbf{UM}_{r,i}$ (Fig. 1), similar to Fig. 1 of Young et al. [15], but using empirical data of APOE4 positive AD patients, MCI patients and healthy controls. The EBM assumes brain regional volumes are probabilistically healthy or abnormal, fitting a normal and a uniform distribution to each brain region's volume data, which calculate the probability of health and abnormality respectively. Volume abnormality threshold $Vthres_r$ for a brain region is defined as the largest volume value such that the probability of health is equal to the probability of abnormality. We kept track of the exact timestep that each region's volume became abnormal during a simulation. We denote $OS_r$ as the event position that brain region $r$ became abnormal during a simulation (e.g. if $OS_{r=hippocampus} = 5$ then hippocampal volume became abnormal fifth, after four other brain regional volumes became abnormal). The element $\mathbf{UM}_{r,i}$ is the probability of region $r$ being $i$-th in the order of brain regional volumes becoming abnormal based on the EBM. We used the following metric to determine the goodness of fit of each model and to optimise the parameter set $\theta$ (Eq. 18):

$$\theta^\star = \min_\theta \sum_r \log(\mathbf{UM}_{r,\mathrm{OS}_r}) \tag{17}$$

$$\theta = \{R_{\mathrm{Mis}}, A_m, P_{seed}, \sigma_{\mathrm{ED}}, \sigma_{\mathrm{ND\text{-}C}}, \sigma_{\mathrm{ND\text{-}D}}, \sigma_{\mathrm{ND\text{-}E}}, R_{\mathrm{FS\text{-}C}}, R_{\mathrm{FS\text{-}D}}, R_{\mathrm{FS\text{-}E}}\} \tag{18}$$

A model of pathogenic protein without production or clearance and with hippocampus seeding best fitted the empirical data (Fig. 1) with parameters $\{R^\star_{Mis} = 0.297, A^\star_m = 0.00152, P^\star_{seed} = 0.0969, \sigma^\star_{\mathrm{ED}} = 0.000243, \sigma^\star_{\mathrm{ND\text{-}C}} = 0,$ $\sigma^\star_{\mathrm{ND\text{-}D}} = 0.0829, \sigma^\star_{\mathrm{ND\text{-}E}} = 0.00276, R^\star_{\mathrm{FS\text{-}C}} = 0, R^\star_{\mathrm{FS\text{-}D}} = 0, R^\star_{\mathrm{FS\text{-}E}} = 0\}$.

## 4   Discussion and Future Work

The simulation with the optimal parameters predicted the early stages of atrophy progression well, whereas later stages had a higher variance from the diagonal sequence given by the EBM (Fig. 1). All other parameters being equal, simulations with pathogenic protein without production or clearance better fitted the data, evidence in favour of the prion-like spread hypothesis [4,14]. Spread was primarily driven by the fibre tract connectivity, whereas functional correlations, $(\sigma^\star_{\mathrm{ND\text{-}C}} = 0)$ extracellular diffusion $(\sigma^\star_{\mathrm{ED}} = 0.000243)$ and effective connectivity $(\sigma^\star_{\mathrm{ND\text{-}E}} = 0.00276)$ only had small contributions. Frequency-related spread $(R^\star_{\mathrm{FS\text{-}C}} = 0, R^\star_{\mathrm{FS\text{-}D}} = 0, R^\star_{\mathrm{FS\text{-}E}} = 0)$ also did not contribute to the spread. This evidence suggests protein spread is driven by structural fibre tract connectivity and not by synaptic activity, in agreement with the modelling of Raj et al. [10].

If we assume that our modelling with the optimal parameters $\theta^\star$ is an accurate and sufficiently complex model of AD progression, then hypothetically our framework could easily evaluate candidate therapies by simulating therapies that

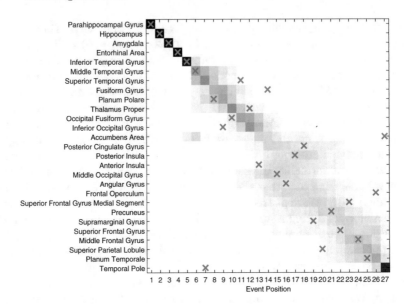

**Fig. 1.** The matrix **UM**, indicating in grey the event position uncertainty for each region, with the sequence $OS_r$ with the optimal parameters $\theta^\star$ overlayed as red crosses.

have an effect on one or more of the models. For example, altering the speed of extracellular diffusion had little effect on atrophy progression, whereas decreasing the speed of network-mediated diffusion significantly slowed down atrophy progression. Despite this example's oversimplification, it is clear how our framework can suggest potential therapy targets.

We presented a proof-of-concept of our methodology, where we aimed at inferring plausible physiological properties from empirical data with a more complicated model than proposed by Raj *et al.* [10]. Most modelling approaches rely on mathematical properties and are effective at capturing atrophy patterns (*e.g.* classification task), but are unable to elucidate the underlying mechanisms. The proposed approach, instead, aims at gaining an understanding of these mechanisms through simulation of pathogenic protein spread within the brain network. To achieve this goal, we had to make multiple assumptions, which are difficult to validate, since many neurobiological properties are still unknown.

In future work additional mechanisms (*e.g.* protein homeostasis, amyloid and tau interaction, *etc.*) will be modelled and their contribution to disease progression will be assessed for a variety of neurodegenerative diseases, under the hypothesis that different parameter values will be linked to different diseases. Adding appropriate regularisation terms to the cost function and estimating the structural and functional connectivity from a larger population of healthy controls would also be desirable. In this work, connectivity metrics were assumed to remain constant under atrophy, which is an assumption that should be relaxed in future work.

**Acknowlegments.** This work received funding from the Engineering and Physical Sciences Research Council (EP/L016478/1), the UCL Leonard Wolfson Experimental Neurology Centre (PR/ylr/18575), the Alzheimer's Society UK (AS-PG-15-025), the Australian Research Council Discovery Early Career Research Award (DE170100128), the MRC (CSUB19166), the ARUK (ARUK-Network 2012-6-ICE; ARUK-PG2014-1946; ARUK-PG2017-1946), the Brain Research Trust (UCC14191), the European Union's Horizon 2020 research and innovation programme (Grant 666992), the NIHR UCL/H Biomedical Research Centre and a Wellcome Trust Senior Clinical Fellowship [091673/Z/10/Z].

# References

1. Cardoso, M.J., et al.: Geodesic information flows: spatially-variant graphs and their application to segmentation and fusion. IEEE Trans. Med. Imaging **34**(9), 1976–1988 (2015)

2. Friston, K.J., Kahan, J., Biswal, B., Razi, A.: A DCM for resting state fMRI. Neuroimage **94**, 396–407 (2014)

3. Friston, K.J., et al.: Bayesian model reduction and empirical Bayes for group (DCM) studies. Neuroimage **128**, 413–431 (2016)

4. Frost, B., Diamond, M.I.: Prion-like mechanisms in neurodegenerative diseases. Nat. Rev. Neurosci. **11**(3), 155–159 (2010)

5. Georgiadis, K., et al.: Computational modelling of pathogenic protein spread in neurodegenerative diseases. PLoS one **13**(2), e0192518 (2018)

6. Karahanoğlu, F.I., et al.: Total activation: fMRI deconvolution through spatio-temporal regularization. Neuroimage **73**, 121–134 (2013)

7. Mandelli, M.L., et al.: Healthy brain connectivity predicts atrophy progression in non-fluent variant of primary progressive aphasia. Brain **139**(10), 2778–2791 (2016)

8. Mawuenyega, K.G.: Decreased clearance of CNS $\beta$-amyloid in Alzheimer's disease. Science **330**(6012), 1774 (2010)

9. Proctor, C.J., Boche, D., Gray, D.A., Nicoll, J.A.: Investigating interventions in Alzheimer's disease with computer simulation models. PloS one **8**(9), e73631 (2013)

10. Raj, A., LoCastro, E., Kuceyeski, A., Tosun, D., Relkin, N., Weiner, M.: Network diffusion model of progression predicts longitudinal patterns of atrophy and metabolism in Alzheimer's disease. Cell Rep. **10**(3), 359–369 (2015)

11. Smith, R.E., Tournier, J.D., Calamante, F., Connelly, A.: SIFT: spherical-deconvolution informed filtering of tractograms. NeuroImage **67**, 298–312 (2013)

12. Stephan, K.E., et al.: Tractography-based priors for dynamic causal models. Neuroimage **47**(4), 1628–1638 (2009)

13. Tournier, J.D., Calamante, F., Connelly, A.: Improved probabilistic streamlines tractography by 2nd order integration over fibre orientation distributions. Proc. ISMRM **18**, 1670 (2010)

14. Warren, J.D., et al.: Molecular nexopathies: a new paradigm of neurodegenerative disease. Trends Neurosci. **36**(10), 561–569 (2013)

15. Young, A.L., et al.: Multiple orderings of events in disease progression. In: Ourselin, S., Alexander, D.C., Westin, C.-F., Cardoso, M.J. (eds.) IPMI 2015. LNCS, vol. 9123, pp. 711–722. Springer, Cham (2015). https://doi.org/10.1007/978-3-319-19992-4_56

16. Zhou, J., et al.: Predicting regional neurodegeneration from the healthy brain functional connectome. Neuron **73**(6), 1216–1227 (2012)

# Generative Discriminative Models for Multivariate Inference and Statistical Mapping in Medical Imaging

Erdem Varol$^{(\boxtimes)}$, Aristeidis Sotiras, Ke Zeng, and Christos Davatzikos

Center for Biomedical Image Computing and Analytics, University of Pennsylvania,
Philadelphia, PA 19104, USA
erdem.varol@uphs.upenn.edu

**Abstract.** This paper presents a general framework for obtaining interpretable multivariate discriminative models that allow efficient statistical inference for neuroimage analysis. The framework, termed generative
discriminative machine (GDM), augments discriminative models with a
generative regularization term. We demonstrate that the proposed formulation can be optimized in closed form and in dual space, allowing
efficient computation for high dimensional neuroimaging datasets. Furthermore, we provide an analytic estimation of the null distribution of
the model parameters, which enables efficient statistical inference and
p-value computation without the need for permutation testing. We compared the proposed method with both purely generative and discriminative learning methods in two large structural magnetic resonance imaging
(sMRI) datasets of Alzheimer's disease (AD) (n = 415) and Schizophrenia
(n = 853). Using the AD dataset, we demonstrated the ability of GDM
to robustly handle confounding variations. Using Schizophrenia dataset,
we demonstrated the ability of GDM to handle multi-site studies. Taken
together, the results underline the potential of the proposed approach
for neuroimaging analyses.

## 1 Introduction

Voxel-based analysis [1] of imaging data has enabled the detailed mapping of
regionally specific effects, which are associated with either group differences or
continuous non-imaging variables, without the need to define *a priori* regions of
interest. This is achieved by adopting a generative model that aims to explain
signal variations as a function of categorical or continuous variables of clinical
interest. Such a model is easy to interpret. However, it does not fully exploit the
available data since it ignores correlations between different brain regions [5].

Conversely, supervised multivariate pattern analysis methods take advantage of dependencies among image elements. Such methods typically adopt a
discriminative setting to derive multivariate patterns that best distinguish the
contrasted groups. This results in improved sensitivity and numerous approaches
have been proposed to efficiently obtain meaningful multivariate brain patterns [4,6,7,10,13,14]. However, such approaches suffer from certain limitations.

© Springer Nature Switzerland AG 2018
A. F. Frangi et al. (Eds.): MICCAI 2018, LNCS 11072, pp. 540–548, 2018.
https://doi.org/10.1007/978-3-030-00931-1_62

Specifically, their high expressive power often results in overfitting due to modeling spurious distracter patterns in the data [8]. Confounding variations may thus limit the application of such models in multi-site studies [12] that are characterized by significant population or scanner differences, and at the same time hinder the interpretability of the models. This limitation is further emphasized by the lack of analytical techniques to estimate the null distribution of the model parameters, which makes statistical inference costly due to the requirement for permutation tests for most multivariate techniques.

Hybrid generative discriminative models have been proposed to improve the interpretability of discriminative models [2,11]. However, these models also do not have analytically obtainable null distribution, which makes challenging the assessment of the statistical significance of their model parameters. Last but not least, their solution is often obtained through non-convex optimization schemes, which reduces reproducibility and out-of-sample prediction performance.

To tackle the aforementioned challenges, we propose a novel framework termed *generative-discriminative machine* (GDM), which aims to obtain a multivariate model that is both accurate in prediction and whose parameters are interpretable. GDM combines ridge regression [9] and ordinary least squares (OLS) regression to obtain a model that is both discriminative, while at the same time being able to reconstruct the imaging features using a low-rank approximation that involves the group information. Importantly, the proposed model admits a closed-form solution, which can be attained in dual space, reducing computational cost. The closed form solution of GDM further enables the analytic approximation of its null distribution, which makes statistical inference and p-value computation computationally efficient.

We validated the GDM framework on two large datasets. The first consists of Alzheimer's disease (AD) patients (n = 415), while the second comprises Schizophrenia (SCZ) patients (n = 853). Using the AD dataset, we demonstrated the robustness of GDM under varying confounding scenarios. Using the SCZ dataset, we effectively demonstrated that GDM can handle multi-site data without overfitting to spurious patterns, while at the same time achieving advantageous discriminative performance.

## 2   Method

**Generative Discriminative Machine:** GDM aims to obtain a hybrid model that can both predict group differences and generate the underlying dataset. This is achieved by integrating a discriminative model (i.e., ridge regression [9]) along with a generative model (i.e., ordinary least squares regression (OLS)). Ridge and OLS are chosen because they can readily handle both classification and regression problems, while admitting a closed form solution.

Let $X \in R^{n \times d}$ denote the $n$ by $d$ matrix that contains the $d$ dimensional imaging features of $n$ independent subjects arranged row-wise. Likewise, let $Y \in R^n$ denote the vector that stores the clinical variables of the corresponding $n$ subjects. GDM aims to relate the imaging features $X$ with the clinical variables

$Y$ using the parameter vector $J \in R^d$ by optimizing the following objective:

$$\min_{J} \underbrace{\|J\|_2^2 + \lambda_1 \|Y - XJ\|_2^2}_{\text{ridge discriminator}} + \underbrace{\lambda_2 \|X^T - JY^T\|_2^2}_{\text{OLS generator}}. \tag{1}$$

If we now take into account information from k additional covariates (e.g., age, sex or other clinical markers) stored in $C \in R^{n \times k}$, we obtain the following GDM objective:

$$\min_{J, W_0, A_0} \underbrace{\|J\|_2^2 + \lambda_1 \|Y - XJ - CW_0\|_2^2}_{\text{ridge discriminator}} + \underbrace{\lambda_2 \|X^T - JY^T - A_0C^T\|_2^2}_{\text{OLS generator}}, \tag{2}$$

where $W_0 \in R^k$ contains the bias terms and $A_0 \in R^{d \times k}$ the regression coefficients pertaining to their corresponding covariates. The inclusion of the bias terms in the ridge regression term allows us to preserve the direction of the parameter vector that imaging pattern that distinguishes between the groups, while at the same time achieving accurate subject-specific classification by taking into account each sample's demographic and other information. Similarly, the inclusion of additional coefficients in the OLS term allows for reconstructing each sample by additionally taking into account its demographic or other information. Lastly, the hyperparameters $\lambda_1$ and $\lambda_2$ control the trade-off between discriminative and generative models, respectively.

**Closed Form Solution:** The formulation in Eq. 2 is optimized by the following closed form solution:

$$J = \left[ I + \lambda_1 (X^T X - X^T C(C^T C)^{-1} C^T X) + \lambda_2 (Y^T Y - Y^T C(C^T C)^{-1} C^T Y) \right]^{-1}$$
$$\times \left[ (\lambda_1 + \lambda_2)(X^T Y - X^T C(C^T C)^{-1} C^T Y) \right], \tag{3}$$

which requires a $d \times d$ matrix inversion that can be costly in neuroimaging settings. To account for that, we solve Eq. 2 in the subject space using the following dual variables $\Lambda \in R^n$:

$$\Lambda = M_{[1:n, 1:n]}^{-1} \left( I + \frac{\lambda_2 XX^T C(C^T C)^{-1} C^T - \lambda_2 XX^T}{1 + \lambda_2 (Y^T Y - Y^T C(C^T C)^{-1} C^T Y)} \right) Y, \tag{4}$$

where $M$ is the following $n + k \times n + k$ matrix:

$$M = \begin{bmatrix} -\frac{XX^T}{1 + \lambda_2(Y^T Y - Y^T C(C^T C)^{-1} C^T Y)} - I/\lambda_1 & C \\ C^T & 0 \end{bmatrix}. \tag{5}$$

The dual variables $\Lambda$ can be used to solve $J$ using the following equation:

$$J = \frac{\lambda_2 X^T Y - \lambda_2 X^T C(C^T C)^{-1} C^T Y - X^T \Lambda}{1 + \lambda_2 (Y^T Y - Y^T C(C^T C)^{-1} C^T Y)}. \tag{6}$$

**Analytic Approximation of Null Distribution:** Using the dual formulation, the GDM parameters $\boldsymbol{J}$ can be shown to be a linear combination of the group labels $\boldsymbol{Y}$ and the following matrix $\boldsymbol{Q}$:

$$\boldsymbol{Q} = \frac{\lambda_2 \boldsymbol{X}^T - \lambda_2 \boldsymbol{X}^T \boldsymbol{C}(\boldsymbol{C}^T\boldsymbol{C})^{-1}\boldsymbol{C}^T - \boldsymbol{X}^T \boldsymbol{M}^{-1}_{[1:n,1:n]}\left(\boldsymbol{I} + \frac{\lambda_2 \boldsymbol{X}\boldsymbol{X}^T \boldsymbol{C}(\boldsymbol{C}^T\boldsymbol{C})^{-1}\boldsymbol{C}^T - \lambda_2 \boldsymbol{X}\boldsymbol{X}^T}{1+\lambda_2(\boldsymbol{Y}^T\boldsymbol{Y} - \boldsymbol{Y}^T\boldsymbol{C}(\boldsymbol{C}^T\boldsymbol{C})^{-1}\boldsymbol{C}^T\boldsymbol{Y})}\right)}{1 + \lambda_2(\boldsymbol{Y}^T\boldsymbol{Y} - \boldsymbol{Y}^T\boldsymbol{C}(\boldsymbol{C}^T\boldsymbol{C})^{-1}\boldsymbol{C}^T\boldsymbol{Y})},$$

such that $\boldsymbol{J} = \boldsymbol{Q}\boldsymbol{Y}$ where $\boldsymbol{Q}$ is approximately invariant to permutation operations on $\boldsymbol{Y}$. Assuming $\boldsymbol{Y}$ is zero mean, unit variance yields that $\mathrm{E}(J_i) = 0$ and $\mathrm{Var}(J_i) = \sum_j Q_{i,j}^2$ under random permutations of $\boldsymbol{Y}$ [15,16]. Asymptotically this yields that $J_i \to \mathcal{N}(0, \sqrt{\sum_j Q_{i,j}^2})$, which allows efficient statistical inference on the parameter values of $J_i$.

## 3    Experimental Validation

We compared GDM with a purely discriminative model, namely ridge regression [9], as well as with its generative counter-part, which was obtained through the procedure outlined by Haufe et al. [8]. We chose these methods because their simple form allows the computation of their null distribution, which in turns enables the comparison of the statistical significance of their parameter maps. The covariates (i.e. $\boldsymbol{C} = [\text{age sex}]$) were linearly residualized using the training set for ridge regression and its generative counterpart.

We used two large datasets in two different settings. First, we used a subset of the ADNI study, consisting of 228 controls (CN) and 187 Alzheimer's disease (AD) patients, to evaluate out-of-sample prediction accuracy and reproducibility. Second, we used data from a multi-site Schizophrenia study, which consisted of 401 patients (SCZ) and 452 controls (CN) spanning three sites (USA n = 236, China n = 286, and Germany n = 331), to evaluate the cross-site prediction and reproducibility of each method.

For all datasets, T1-weighted MRI volumetric scans were obtained at 1.5 Tesla. The images were pre-processed through a pipeline consisting of (1) skull-stripping; (2) N3 bias correction; and (3) deformable mapping to a standardized template space. Following these steps, a low-level representation of the tissue volumes was extracted by automatically partitioning the MRI volumes of all participants into 151 volumetric regions of interest (ROI). The ROI segmentation was performed by applying a multi-atlas label fusion method. The derived ROIs were used as the input features for all methods.

**Analytical Approximation of p-Values.** To confirm that the analytical approximation of null distribution of GDM is correct, we estimated the p-values through the approximation technique as well as through permutation testing. A range of 10 to 10,000 permutations was applied to observe the error rate. This experiment was performed on the ADNI dataset. The results displayed in Fig. 1 demonstrate that the analytic approximation holds with approximately $O(1/\sqrt{\#\text{permutations}})$ error.

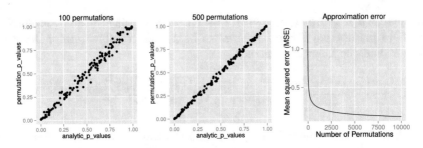

**Fig. 1.** Comparison of permutation based p-values of GDM with their analytic approximations at varying permutation levels.

**Out-of-sample Prediction and Reproducibility.** To assess the discriminative performance and reproducibility of the compared methods under varying confounding scenarios, we used the ADNI dataset. We simulated four distinct training scenarios in increasing potential for confounding effects: • Case 1: 50% AD + 50% CN subjects, mean age balanced, • Case 2: 75% CN + 25% AD, mean age balanced, • Case 3: 50% AD + 50% CN, oldest ADs, youngest CNs, • Case 4: 75% CN + 25% AD, oldest ADs, youngest CNs.

All models had their respective parameters $(\lambda_1, \lambda_2 \in \{10^{-5}, \ldots, 10^2\})$ cross-validated in an inner fold before performing out-of-sample prediction on a left out test set consisting of equal numbers of AD and CN subjects with balanced mean age. Furthermore, the inner product of training model parameters was compared between folds to assess the reproducibility of models. Training and testing folds were shuffled 100 times to yield a distribution.

The prediction accuracies and the model reproducibility for the above cases are shown in Fig. 2. The results demonstrate that while GDM is not a purely discriminative model, its predictions outperformed ridge regression in all four cases. Regarding reproducibility, the Haufe et al. [8] procedure yielded the most stable models since it yields a purely generative model. However, GDM was more reproducible than ridge regression.

**Multi-site Study.** To assess the predictive performance of the compared methods in a multi-site setting, we used the Schizophrenia dataset that comprises data from three sites. All models had their respective parameters cross-validated while training in one site before making predictions in the other two sites. Each training involved using 90% of the site samples to allow for resampling the training sets 100 times to yield a distribution. The reproducibility across the resampled sets was measured using the inner product between model parameters. The multi-site prediction and reproducibility results are visualized in Fig. 3.

In five out of six cross-site prediction settings, GDM outperformed all compared methods in terms accuracy. Also, GDM had higher reproducibility than ridge regression, while having slightly lower reproducibility than the generative procedure in Haufe et al. [8].

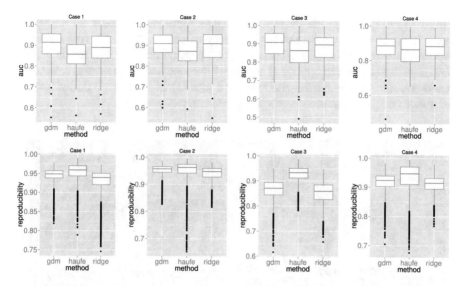

**Fig. 2.** Cross validated out-of-sample AD vs. CN prediction accuracies (top row) and normalized inner-product reproducibility of training models (bottom row) for varying training scenarios and all compared methods.

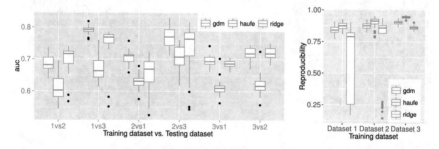

**Fig. 3.** Cross validated multi-site SCZ vs. CN prediction accuracies (left) and normalized inner-product reproducibility of training models (right) for all compared methods.

**Statistical Maps and p-Values.** To qualitatively assess and explain the predictive performance of the compared methods for the AD vs. CN scenario, we computed the model parameter maps using full resolution gray matter tissue density maps for the ADNI dataset (Fig. 4 top). Furthermore, since the null distribution of GDM, as well as ridge regression, can be estimated analytically, we computed p-values for the model parameters and displayed the regions surviving false discovery rate (FDR) correction [3] at level $q < 0.05$ (Fig. 4 bottom).

The statistical maps demonstrated that both GDM and Haufe procedure yield patterns that accurately delineate the regions associated with AD, namely the widespread atrophy present in the temporal lobe, amygdala, and hippocampus. This is in contrast with the patterns found in ridge regression that resemble a hard to interpret speckle pattern with meaningful weights only on hippocampus. This once again confirmed the tendency of purely discriminative models to capture spurious patterns. Furthermore, the p-value maps of the Haufe method and ridge regression demonstrate the wide difference between features selected by generative and discriminative methods and how GDM strikes a balance between the two to achieve superior predictive performance.

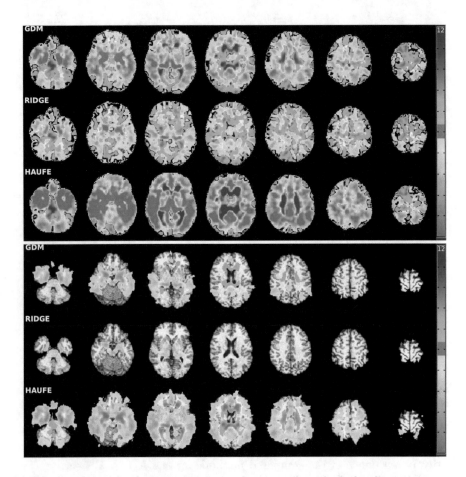

**Fig. 4.** Top: Normalized parameter maps of compared methods for discerning group differences between AD patients and controls. Bottom: Parameter $\log_{10}$ p-value maps of the compared methods for discerning group differences between AD patients and controls after FDR correction at level $q < 0.05$. Warmer colors indicate decreasing volume with AD, while colder colors indicate increasing volume with AD.

# 4    Discussion and Conclusion

The interpretable patterns captured by GDM coupled with its ability to out-perform discriminative models in terms of prediction underline its potential for neuroimaging analysis. We demonstrated that GDM may obtain highly reproducible models through generative modeling, thus avoiding overfitting that is commonly observed in neuroimaging settings. Overfitting is especially evident in multi-site situations, where discriminative models might subtly model spurious dataset effects which might compromise prediction accuracy in an out-of-site setting. Furthermore, by using a formulation that yields a closed form solution, we additionally demonstrated that is possible to efficiently assess the statistical significance of the model parameters.

While the methodology presented herein is analogous to generatively regularizing ridge regression with ordinary least squares regression, the framework proposed can be generalized to include generative regularization in other commonly used discriminative learning methods. Namely, it is possible to augment linear discriminant analysis (LDA), support vector machine (SVM), artificial neural network (ANN) objective with a similar generative term to yield an alternative generative discriminative model of learning. However, the latter two cases would not permit a closed form solution, making it impossible to analytically estimate a null distribution.

# References

1. Ashburner, J., Friston, K.J.: Voxel-based morphometry-the methods. Neuroimage **11**(6), 805–821 (2000)
2. Batmanghelich, N.K., et al.: Generative-discriminative basis learning for medical imaging. IEEE Trans. Med. Imaging **31**(1), 51–69 (2012)
3. Benjamini, Y., Hochberg, Y.: Controlling the false discovery rate: a practical and powerful approach to multiple testing. J. R. Stat. Soc. **57**, 289–300 (1995)
4. Cuingnet, R., et al.: Spatial and anatomical regularization of SVM: a general framework for neuroimaging data. IEEE PAMI **35**(3), 682–696 (2013)
5. Davatzikos, C.: Why voxel-based morphometric analysis should be used with great caution when characterizing group differences. Neuroimage **23**(1), 17–20 (2004)
6. Ganz, M., et al.: Relevant feature set estimation with a knock-out strategy and random forests. Neuroimage **122**, 131–148 (2015)
7. Grosenick, L., et al.: Interpretable whole-brain prediction analysis with GraphNet. NeuroImage **72**, 304–321 (2013)
8. Haufe, S., et al.: On the interpretation of weight vectors of linear models in multivariate neuroimaging. Neuroimage **87**, 96–110 (2014)
9. Hoerl, A.E., Kennard, R.W.: Ridge regression: biased estimation for nonorthogonal problems. Technometrics **12**(1), 55–67 (1970)
10. Kriegeskorte, N., Goebel, R., Bandettini, P.: Information-based functional brain mapping. PNAS **103**(10), 3863–3868 (2006)
11. Mairal, J., Bach, F., Ponce, J.: Task-driven dictionary learning. IEEE Trans. Pattern Anal. Mach. Intell. **34**(4), 791–804 (2012)
12. Rao, A., et al.: Predictive modelling using neuroimaging data in the presence of confounds. NeuroImage **150**, 23–49 (2017)

13. Rasmussen, P.M., et al.: Model sparsity and brain pattern interpretation of classification models in neuroimaging. Pattern Recognit. **45**(6), 2085–2100 (2012)
14. Sabuncu, M.R., Van Leemput, K.: The relevance voxel machine (RVoxM): a self-tuning bayesian model for informative image-based prediction. TMI **31**, 2290–2306 (2012)
15. Varol, E., et al.: MIDAS: regionally linear multivariate discriminative statistical mapping. NeuroImage **174**, 111–126 (2018)
16. Varol, E., et al.: Regionally discriminative multivariate statistical mapping. In: 2018 IEEE 15th International Symposium on Biomedical Imaging (ISBI 2018), pp. 1560–1563. IEEE (2018)

# Using the Anisotropic Laplace Equation to Compute Cortical Thickness

Anand A. Joshi[1]([✉]), Chitresh Bhushan[2], Ronald Salloum[1],
Jessica L. Wisnowski[1], David W. Shattuck[3], and Richard M. Leahy[1]

[1] University of Southern California, Los Angeles, CA, USA
ajoshi@usc.edu
[2] General Electric, Niskayuna, NY, USA
[3] University of California, Los Angeles, CA, USA

**Abstract.** Automatic computation of cortical thickness is a critical step when investigating neuroanatomical population differences and changes associated with normal development and aging, as well as in neuro-degenerative diseases including Alzheimer's and Parkinson's. Limited spatial resolution and partial volume effects, in which more than one tissue type is represented in each voxel, have a significant impact on the accuracy of thickness estimates, particularly if a hard intensity threshold is used to delineate cortical boundaries. We describe a novel method based on the anisotropic heat equation that explicitly accounts for the presence of partial tissue volumes to more accurately estimate cortical thickness. The anisotropic term uses gray matter fractions to incorporate partial tissue voxels into the thickness calculation, as demonstrated through simulations and experiments. We also show that the proposed method is robust to the effects of finite voxel resolution and blurring. In comparison to methods based on hard intensity thresholds, the heat equation based method yields results with in-vivo data that are more consistent with histological findings reported in the literature. We also performed a test-retest study across scanners that indicated improved consistency and robustness to scanner differences.

## 1 Introduction

Average cortical thickness can vary between 2 to 5 mm across a population of healthy subjects as well as across different brain regions in an individual [1]. Thickness of the human cerebral cortex is an important phenotypical feature and a biomarker for a range of neurological diseases and conditions, and brain development. To perform cortical thickness studies in a large population an automated approach to thickness measurement from T1-weighted MRI scans is essential. Several approaches for computation of cortical thickness are based on first estimating inner gray/white and pial surfaces and then defining the cortical thicknesses based on the distance between the two [2,3]. The Linked Distance

This work is supported by the following NIH grants: R01 NS074980, R01 NS089212.

A. F. Frangi et al. (Eds.): MICCAI 2018, LNCS 11072, pp. 549–556, 2018.
https://doi.org/10.1007/978-3-030-00931-1_63

method (LD) in BrainSuite uses distance between corresponding nodes in the two surfaces as a thickness measure. FreeSurfer's cortical thickness [4] is defined as the average of the shortest distance between the two surfaces computed in both directions. Cortical Pattern Matching (CPM) [5] finds the shortest distance between the inner and pial surfaces using the Eikonal equation. On the other hand, voxel-based methods compute thickness based on line integrals [6,7], the Laplace equation [1,8,9], or using image registration [10]. The accuracy of these approaches is impacted when individual voxels are composed of a mixture of multiple tissue types leading to partial volume effects. The convoluted geometry of the cortex together with the point spread function associated with finite resolution makes partial volume effects inevitable. Despite this, most methods use crisp definitions of cortical boundaries for surface-based calculation. In some cases partial volume effects have been accounted for by modifying the cortical surface boundaries using Eulerian or Lagrangian PDEs [9], registration of inner and pial cortical surfaces [10], closest point distances between inner and pial cortical surfaces measured in both directions [11] and using an electric field model together with a topology preserving level set approach [12]. While these methods account for partial volume effects in defining inner and pial cortical surfaces, they do not explicit account for the actual partial volume fractions that lie between the two boundaries once they are defined.

Here we describe a new thickness calculation method that explicitly models partial volume effects by modifying the Laplace equation (LE) based method of Jones et al. [8]. Rather than using the isotropic LE, we instead use the anisotropic version in which the diffusion coefficient is varied spatially in proportion to the fraction of gray matter in each voxel. Further, we use a closed formed analytic expression for thickness that can be rapidly computed without the need for computation of streamlines as in [8]. We show in Sect. 2.2, using a 1D analogy, that this closed form expression is equivalent to the result obtained using the streamline method and that the thickness measurement is robust to blurring of the image with a unit-integral kernel. Finally, we present results that compare accuracy and robustness with alternative methods for cortical thickness calculation.

## 2   Materials and Methods

### 2.1   The Anisotropic Laplace Equation (ALE) Method

We assume that the brain image has been segmented using a partial classification scheme so that each voxel is assigned a fraction of gray matter (GM), white matter (WM) and cerebrospinal fluid (CSF), with the constraint that the fractions sum to unity [13]. We model cortex as a thin sheet constrained by inner and outer cortical surfaces which are set to temperatures 0 and 1, respectively. We then model the propagation of heat between the two boundary layers of the cortex at equilibrium using the anisotropic form of Laplace's equation. In our anisotropic model, the diffusion coefficient is assumed to be inversely proportional to the fraction of gray matter in each voxel so that pure white matter and

CSF are modeled as perfect conductors. The temperature $\phi(v,t)$ as a function of spatial location $v$ and time $t$ is given by

$$\frac{\partial \phi(v,t)}{\partial t} = \text{div}\left(\frac{1}{f(v)}\nabla \phi(v,t)\right), \text{ subject to } \phi(v,t) = \begin{cases} 0 & v \in \partial\Omega_{inner} \\ 1 & v \in \partial\Omega_{pial} \end{cases} \quad (1)$$

where $\Omega$ is the domain of the computation, bounded by the closed inner surface $\partial\Omega_{inner}$ and the closed outer pial surface $\partial\Omega_{pial}$, and $f(v)$ represents the gray matter fraction at location $v$. In this formulation it is important that all partial volume voxels that contain cortical gray matter are included within the surfaces that bound $\Omega$. The equilibrium solution $\phi_\infty$ for the heat equation with anisotropic flow is given by the anisotropic Laplace equation: $\text{div}\left(\frac{1}{f(v)}\nabla\phi_\infty(v)\right) = 0$, subject to the earlier boundary conditions. Using the calculus of variations, this equation can be reduced to the harmonic energy minimization problem:

$$\phi_\infty(v) = \arg\min_\psi \int_\Omega \left\|\frac{1}{f(v)}\nabla\psi(v)\right\|^2 d\Omega. \quad (2)$$

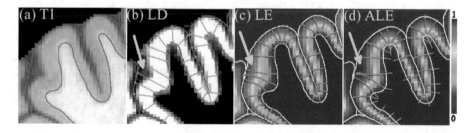

**Fig. 1.** (a) Inner and pial surface boundaries overlaid on T1-weighted image; (b) cortical thickness computation based on linked distance overlaid on the gray matter fraction image (LD); (c) the solutions of the Isotropic Laplace Equation (LE) and (d) the proposed ALE method are shown as a color coded temperature distribution with green lines depicting the streamlines. The yellow arrow depicts an example region where the three methods differ significantly.

Figure 1 illustrates the color coded temperature distribution solution to the ALE in a 2D section of cortex in comparison to the solution of the isotropic LE and the LD method within $\Omega$. We also show corresponding streamlines for the ALE using partial tissue fractions relative to those for the isotropic case. Note that the green streamlines in Fig. 1(d) extend into the white matter due to the non-zero gray matter fraction in this region. Integrals of $\phi_\infty(v)$ over these streamlines from inner to pial surface can be used to compute thickness [14]. We propose an alternative simple analytic expression by defining the cortical thickness $T(v)$ at each point on the mid-cortical surface as:

$$T(v) = f(v)\frac{1}{\|\nabla\phi_\infty(v)\|}, \text{ where } (v) \in \partial\Omega_{mid}. \quad (3)$$

Here the mid-cortical surface $(\partial\Omega)_{mid}$ is defined as the level set: $(\partial\Omega)_{mid} = \{v \in \Omega | \phi_\infty(v) = \frac{1}{2}\}$. The analytic expression in Eq. 3 is shown in Sect. 2.2 to be equivalent to the path integral for the 1-dimensional solution to the ALE, not only at $(\partial\Omega)_{mid}$ but at all points with non-zero gray matter fraction. An intuitive explanation of why this approximation works is as follows. We impose boundary conditions of temperatures 0 and 1 on the inner and pial surfaces, respectively. So for homogeneous gray matter the temperature gradient between them, $\|\nabla\phi_\infty(v)\|$, will be inversely proportion to thickness. Consequently, calculation of the reciprocal of the gradient at the midcortical surface should produce a good estimate of thickness. For the anisotropic case, we account for the increased flux in partial volume voxels that may lie on the mid-cortical surface by scaling by the gray matter fraction $f(v)$.

## 2.2    Analysis Using a 1D Model

To illustrate how ALE works, we use a 1D model. In this model we assume pure white matter is the region from $x = -\infty$ to $-L$, the cortex (containing gray matter) from $x = -L$ to $L$, and pure CSF from $x = L$ to $\infty$. We assume an arbitrary gray matter fraction distribution $f(x)$ on $(-L, L)$. We then blur this distribution with a kernel $g(x)$ with the property $\int_{-\infty}^{\infty} g(x) = 1$. Following Eq. 3, we obtain the thickness

$$T = h(x)\left(\frac{d\phi_\infty(x)}{dx}\right)^{-1} \text{ at } x : \phi_\infty(x) = 0.5. \tag{4}$$

Here $h(x) = f(x)$ for the unblurred case and $h(x) = f(x) \circledast g(x)$ for the blurred case, where $\circledast$ indicates convolution.

The 1D anisotropic heat equation is $\frac{\partial\phi(x,t)}{\partial t} = \frac{\partial}{\partial x}\left(\frac{1}{f(x)}\frac{\partial\phi(x,t)}{\partial x}\right)$ for the unblurred case, with boundary conditions $\phi(-L,t) = 0$ and $\phi(L, \text{t}) = 1$. Solving for $\phi_\infty(x)$ gives $\phi_\infty(x) = \int_{-L}^{x} f(y)dy / \int_{-L}^{L} f(y)dy$. Substituting this into Eq. 4 gives the correct value $T = \int_{-L}^{L} f(y)dy$. It is somewhat surprising that the thickness value computed in this manner does not depend on the point $x$ at which it is computed in Eq. 4, although for consistency of definition, we always compute it at the mid point between inner and outer boundary.

Now if we blur the gray matter fraction with a unit-integral kernel, solve the heat equation in steady state, and substitute in Eq. 4, the resulting solution is $T = \int_{-\infty}^{\infty}\int_{-\infty}^{\infty} f(x)g(y - x)\,dx\,dy = \int_{-\infty}^{\infty} f(x)\left(\int_{-\infty}^{\infty} g(y - x)\,dy\right)dx = \int_{-L}^{L} f(x)\,dx \int_{-\infty}^{\infty} g(y)\,dy = \int_{-L}^{L} f(x)\,dx$. In other words, the thickness calculations using ALE is unaffected by blurring with a sufficiently narrow kernel with unit integral, provided the blurred gray matter fraction of the cortex lies within the bounds defined by the inner and pial surfaces. It should be noted that in the 3D case, blurring of the image will lead to some smoothing of the computed thickness values along the surface, but this should not add bias to the thickness values.

## 3    Applications and Results

### 3.1    Average Cortical Thickness Study

We analyzed 3D structural brain MRI scans of 198 normal right handed subjects (76 male, 122 female, age range: 18–26 years) from the 1000 Functional Connectomes Project `http://fcon_1000.projects.nitrc.org`. Images were acquired using MPRAGE on a SIEMENS TRIO 3T scanner: TR = 2530 ms, TE = 3.39 ms, slice thickness = 1.33 mm, flip angle = 7°, inversion time = 1100 ms, FOV = 256 mm × 256 mm, in-plane resolution = 256 × 192, 128 slices. We used the BrainSuite software (`http://brainsuite.org`) [13] to define the partial volume fractions, as well as for extracting inner and pial cortical surfaces. All cortical surfaces were aligned to a common atlas space using BrainSuite in order to compute population based averages. Cortical thicknesses were computed as described in Sect. 2.1 where we used a finite difference method to solve the anisotropic Laplace equation.

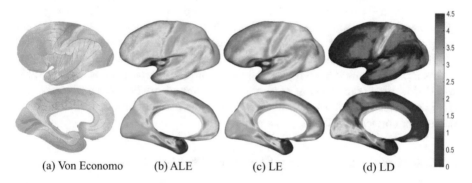

(a) Von Economo        (b) ALE        (c) LE        (d) LD

**Fig. 2.** (a) Histology based thickness map from Von Economo [15]; (b)–(d)Average cortical thickness maps of left hemisphere. Lateral (upper) and medial (lower) views from N = 198 adult subjects computed using: (b) Anisotropic Laplace Equation (ALE), (c) Isotropic Laplace Equation (LE), and (d) Linked Distance (LD).

We processed each of the 198 subject images using three methods: LD, LE and ALE, and mapped these to the atlas to compute the point-wise average cortical thickness on the surface. A surface based Laplace-Beltrami isotropic smoothing [16] of ~10 mm fwhm was applied to the thickness estimates in the original subject surface to compensate for discretization and small misregistration errors. We used a robust mean estimate in which outliers (the 5% most extreme values) were first removed for each vertex on the surface. The maps of average cortical thickness estimated are shown in Fig. 2(b)–(d). For comparison we include in Fig. 2(a) a pseudo-colored version of Von Economo's map of cortical thickness from [15] which is based on histological measurements. The patterns of thickness variation across the cortex are similar for all three methods, however the range of

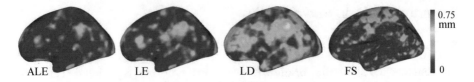

**Fig. 3.** Effect of scanner differences on cortical thickness estimates for four methods: absolute thickness difference between estimates from the Siemens Trio 3T and Siemens Avanto 1.5T scanners, averaged over 5 subjects. ALE = Anisotropic Laplace Equation, LE = Laplace Equation, LD = Linked Distance, FS = FreeSurfer.

values is quite different. The cortical thickness estimates found using ALE were more consistent with the Von Economo estimates in the parietal, occipital and temporal lobes, while they were different in the frontal lobe. It should be noted that the demographics of the subjects used in the histological study was different than the imaged population. Von Economo and Koskinas used brains of mentally healthy Caucasian subjects, 30–40 years of age [17] whereas the population in the data we used had an age range of 18–26 and were scanned in Beijing. This difference in the demographics, and especially the younger population in our study, may account for some of the increased thickness in the frontal lobe [18]. LE shows higher thickness than the Von Economo estimates everywhere on the cortex and LD showed even higher values. A similar comparison using FreeSurfer based thickness computation method was presented in [19] where it was shown that FreeSurfer also tended to overestimate the cortical thickness, although the pattern of cortical thickness was similar.

## 3.2   Test-Retest Reliability Study for Multiple Scanners

The purpose of this study was to analyze the effects of scanner differences on cortical thickness measurements in the same subject. The data for this study consisted of 5 normal subjects scanned on two different scanners. All the scans were acquired within three days at the University of Iowa. The first scan was acquired using a Siemens Trio 3T scanner: slice thickness 1 mm, TR 2530, TE 3.99 ms, inversion time 1100 ms, in-plane resolution 1 mm$^2$ and flip angle 10°. The second scan was acquired using a Siemens Avanto 1.5T scanner: slice thickness 1.5 mm, TR 26 ms, TE 7 ms, in-plane resolution 1.066 mm$^2$ and flip angle 30°. The

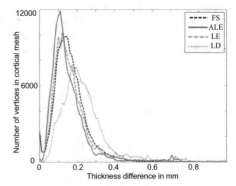

**Fig. 4.** Histogram of the average absolute differences between thickness measures for ALE, LE, LD and FS for 1.5T Siemens Avanto and 3T Siemens Trio scanners.

BrainSuite surface extraction pipeline was executed for all scans. Thickness estimates were computed using the three methods and coregistered to the atlas brain using BrainSuite. In addition, we also executed the FreeSurfer pipeline (Version 5.3.0) on these subjects. Laplace-Beltrami surface based smoothing ∼10 mm fwhm was applied and thickness differences were computed. The absolute value of thickness difference corresponding to 3T and 1.5T Siemens scanners, averaged over the five subjects was computed for all four methods as shown in Fig. 3. We also show histograms of these differences in Fig. 4. As with the simulation study, the ALE method shows the smallest absolute difference (mean 0.1575 mm, sd 0.1034 mm) between the 3T vs 1.5T scanners relative to LE (mean 0.1786 mm, sd 0.1163 mm) and LD (mean 0.2480 mm, sd 0.1520 mm), and FreeSurfer (mean 0.1679 mm, sd 0.0821 mm).

## 4    Discussion and Conclusion

Studies of cortical thickness usually compare differences in thickness in homologous areas between two groups, or changes in thickness over time during maturation, aging or disease progression. For this reason, consistency and robustness of thickness estimates is possibly as important as absolute accuracy. For this reason, we examined not only the average cortical thickness over a relatively larger population, but also the consistency of thickness estimates among subjects scanned in two different scanners. The study of 198 subjects produced average thicknesses with the ALE method that are more in line with those reported in the literature using histological studies than the alternative LE and LD methods. Differences in the frontal lobe using ALE compared to the values in the Von Economo atlas are consistent with age differences in the two different populations studied and reported changes in thickness in early adulthood in frontal cortex [18].

The consistency study in Figs. 3 and 4 shows that there is a consistent bias in all methods in the thickness estimates computed from images from the 3T scanner versus those from the 1.5T. However, the histograms of these differences confirm the reduced sensitivity to differences in resolution of the ALE method relative to FS, LE and LD.

In summary, results presented here indicate that ALE is capable of producing cortical thickness estimates that are largely consistent with those reported from histological measurements, and that these estimates are less sensitive to the effects of imaging in a different scanner for the limited range of conditions over which we have so far studied this method. Further evaluation is required, and as with all thickness estimation methods, though inter-scanner differences may be reduced, our results indicate that it is important that scanner-dependent effects be factored into any subsequent analysis.

# References

1. Hutton, C., De Vita, E., Ashburner, J., Deichmann, R., Turner, R.: Voxel-based cortical thickness measurements in MRI. NeuroImage 40(4), 1701–1710 (2008)
2. Lerch, J.P., Evans, A.C.: Cortical thickness analysis examined through power analysis and a population simulation. NeuroImage 24(1), 163–173 (2005)
3. Clarkson, M.J., et al.: A comparison of voxel and surface based cortical thickness estimation methods. NeuroImage 57(3), 856–865 (2011)
4. Fischl, B., Dale, A.M.: Measuring the thickness of the human cerebral cortex from magnetic resonance images. Proc. Natl. Acad. Sci. 97(20), 11050–11055 (2000)
5. Thompson, P.M., et al.: Detecting dynamic and genetic effects on brain structure using high-dimensional cortical pattern matching. Proc. ISB I, 473–476 (2002)
6. Aganj, I., Sapiro, G., Parikshak, N., Madsen, S.K., Thompson, P.M.: Measurement of cortical thickness from MRI by minimum line integrals on soft-classified tissue. Hum. Brain Mapp. 30(10), 3188–3199 (2009)
7. Scott, M.L.J., Bromiley, P.A., Thacker, N.A., Hutchinson, C.E., Jackson, A.: A fast, model-independent method for cerebral cortical thickness estimation using MRI. Med. Image Anal. 13(2), 269–285 (2009)
8. Jones, S.E., Buchbinder, B.R., Aharon, I.: Three-dimensional mapping of cortical thickness using Laplace's equation. Hum. Brain Mapp. 11(1), 12–32 (2000)
9. Acosta, O., Bourgeat, P., Zuluaga, M.A., Fripp, J., Salvado, O., Ourselin, S.: Automated voxel-based 3d cortical thickness measurement in a combined Lagrangian-Eulerian PDE approach using partial volume maps. Med. Image Anal. 13(5), 730–743 (2009)
10. Das, S.R., Avants, B.B., Grossman, M., Gee, J.C.: Registration based cortical thickness measurement. NeuroImage 45(3), 867–879 (2009)
11. Tustison, N.J.: Large-scale evaluation of ANTs and FreeSurfer cortical thickness measurements. NeuroImage 99, 166–179 (2014)
12. Osechinskiy, S., Kruggel, F.: Cortical surface reconstruction from high-resolution MR brain images. Int. J. Biomed. Imaging 2012, 870196 (2012)
13. Shattuck, D.W., Leahy, R.M.: BrainSuite: an automated cortical surface identification tool. Med. Image Anal. 8(2), 129–142 (2002)
14. Evans, L.C.: Partial Differential Equations. Graduate Studies in Mathematics, vol. 19. American Mathematics Society, Providence (2009)
15. von Economo, C., Koskinas, G.: The Cytoarchitectonics of the Adult Human Cortex. Springer, Vienna and Berlin (1925)
16. Joshi, A.A., Shattuck, D.W., Thompson, P.M., Leahy, R.M.: A parameterization-based numerical method for isotropic and anisotropic diffusion smoothing on non-flat surfaces. IEEE Trans. Image Process. 18(6), 1358–1365 (2009)
17. Triarhou, L.C.: The Economo-Koskinas atlas revisited: cytoarchitectonics and functional context. Stereotact. Funct. Neurosurg. 85(5), 195–203 (2007)
18. Gogtay, N., et al.: Dynamic mapping of human cortical development during childhood through early adulthood. Proc. Natl. Acad. Sci. 101(21), 8174–8179 (2004)
19. Scholtens, L.H., de Reus, M.A., van den Heuvel, M.P.: Linking contemporary high resolution magnetic resonance imaging to the von economo legacy: a study on the comparison of MRI cortical thickness and histological measurements of cortical structure. Hum. Brain Mapp. 36(8), 3038–3046 (2015)

# Dilatation of Lateral Ventricles with Brain Volumes in Infants with 3D Transfontanelle US

Marc-Antoine Boucher[1]([✉]), Sarah Lippé[2,3], Amélie Damphousse[3],
Ramy El-Jalbout[3], and Samuel Kadoury[1,3]

[1] MedICAL Laboratory, Polytechnique Montreal, Montreal, Canada
marc-antoine.boucher@polymtl.ca
[2] NED Laboratory, University of Montreal, Montreal, QC, Canada
[3] CHU Sainte-Justine Research Center, Montreal, QC, Canada

**Abstract.** Ultrasound (US) can be used to assess brain development in newborns, as MRI is challenging due to immobilization issues, and may require sedation. Dilatation of the lateral ventricles in the brain is a risk factor for poorer neurodevelopment outcomes in infants. Hence, 3D US has the ability to assess the volume of the lateral ventricles similar to clinically standard MRI, but manual segmentation is time consuming. The objective of this study is to develop an approach quantifying the ratio of lateral ventricular dilatation with respect to total brain volume using 3D US, which can assess the severity of macrocephaly. Automatic segmentation of the lateral ventricles is achieved with a multi-atlas deformable registration approach using locally linear correlation metrics for US-MRI fusion, followed by a refinement step using deformable mesh models. Total brain volume is estimated using a 3D ellipsoid modeling approach. Validation was performed on a cohort of 12 infants, ranging from 2 to 8.5 months old, where 3D US and MRI were used to compare brain volumes and segmented lateral ventricles. Automatically extracted volumes from 3D US show a high correlation and no statistically significant difference when compared to ground truth measurements. Differences in volume ratios was $6.0 \pm 4.8\%$ compared to MRI, while lateral ventricular segmentation yielded a mean Dice coefficient of $70.8 \pm 3.6\%$ and a mean absolute distance (MAD) of $0.88 \pm 0.2\,mm$, demonstrating the clinical benefit of this tool in paediatric ultrasound.

## 1 Introduction

For newborns, conditions related to cerebrospinal fluid (CSF) like ventriculomegaly (VM) are common disorders, especially for premature newborns which are frequently associated with VM, white matter injury and intraventricular

This work was supported by FRQNT (#198490), RBIQ (#5886) and CIHR grants. The authors would like to thank Inga Sophia Knoth and Caroline Dupont from Sainte-Justine Hospital for patient recruitement.

© Springer Nature Switzerland AG 2018
A. F. Frangi et al. (Eds.): MICCAI 2018, LNCS 11072, pp. 557–565, 2018.
https://doi.org/10.1007/978-3-030-00931-1_64

hemorrhage. For newborns, VM is defined as when atriums of lateral ventricles are greater than 10 mm. Mild VM is associated with neurodevelopmental disorders (learning disorders, autism and hyperactivity deficit) and arises during fetal brain development which could be detected in ultrasound (US). A previous study demonstrated that prenatal VM for full term newborns could lead to an increase in ventricle, intracranial and cortical grey matter volumes [1]. Changes in sub-cortical regions of the brain is associated with cognitive development and as such, to include diagnosis accuracy, the clinical assessment for VM should include ventricular-brain ratio.

For infants, non-invasive imaging modalities are required for macrocephaly or premature cases of newborns, as well as cases related to neurosurgery or ischemic incident. Therefore, US is often used in neonates to image the developing brain as it is cost effective and accessible. Recent 3D matrix-array transducers can acquire a volume quasi-instantly and acquisition through the fontanelle may become an alternative to MRI for some volumetric assessments, with previous studies evaluating the lateral ventricles with fairly good reliability using US [2]. Since manual segmentation is time consuming, an automatic segmentation of the lateral ventricles and brain volume in 3D US can be relevant as an objective measure to assess VM in infants with a safe and accessible imaging modalities.

A few studies focused on the segmentation of lateral ventricles in neonatal brains with 3D US. Lateral ventricles were segmented semi-automatically in 3D US with an overlap of 78.2% and mean distances of 0.65 mm [3], but require manual initialization with landmarks. The work presented in [4] showed an automatic approach that successfully segmented the ventricles on newborn cerebral 3D US images (76.7% Dice), but included patients suffering from intraventricular hemorrhage (IVH) with highly enlarged ventricles. Furthermore, the brain volume, which is essential for ventricular-brain volume ratio computation was not evaluated and there were no statistical comparison performed between 3D US and MRI volumes. This method was also applied to intraventricular hemorrhage cases of newborns, and has yet to be validated on normal and on pathological cases. To our knowledge, no study has been conducted to evaluate total brain volume or ventricular-brain volume ratio automatically in 3D US.

In this paper, we present a novel method to compute the ventricular-brain ratio for the diagnosis of VM in infants from 3D US images. Lateral ventricles are segmented with a combination of multi-atlas and deformable mesh registration approaches, from which the ventricular-brain volume ratio can be computed. Results are compared with ground truth manual segmentations on MRI data, demonstrating the clinical potential in paediatric neuroradiology to quantify ventricular enlargement. The contributions are twofold: (1) a novel optimization scheme based on a dynamic weighting factor in the fusion process, handling hyper and hypo-echoic regions within the ventricles, (2) a geometric-based brain volume estimation method, enabling volume ratios to be extracted, enabling neurodevelopment assessment.

## 2     Methods

### 2.1     Patient Data

In this study, a cohort of 12 infants aged between 2 and 8.5 months were recruited prospectively, with 3D US and T1 weighted MRI acquired within an hour apart. Ultrasound images were acquired through the fontanelle with an X6-1 matrix-array transducer (EPIQ 7 system, Philips Medical, Bothell, WA) while the MRI was acquired with a 3T MR 750 GE scanner, with a 8 channel head coil, an image resolution of $256 \times 256 \times 92$, and pixel size of $0.78 \times 0.78 \times 1.2$. 3D US was also acquired on 5 additional infants for evaluation purposes. The total brain volume from MRI, which served as ground-truth, was obtained using the cortical surface extraction sequence of Brainsuite.

### 2.2     Total Brain Volume Estimation from 3D US

In cerebral 3D US, the entire brain cannot be fully captured in a single volume even in neonates, due to the size of the transducer and limited acoustic window. Therefore, a total brain volume estimation based on an ellipsoid-fitting method was designed, which doesn't require volume stitching.

As shown in Fig. 1, when fitting a 3D ellipsoid on the skull boundary, the anterior-inferior section of the ellipsoid (shown in dashed lines) overestimates the brain volume. Therefore we estimate the brain volume as a portion of the ellipsoid volume such that: $V_{brain} = \frac{4}{3}abc\pi C_f$ where $a$, $b$ and $c$ are the semi axes of the ellipsoid and $C_f$ is a constant for all patients determined empirically by comparing the ellipsoid and ground truth brain volumes from MRI images. To apply the method on 3D US, skull stripping is first applied on the US image as illustrated in Fig. 1(c)–(d). The proposed method performs a skull detection based on intensity threshold: $V_{skull} = \{v|I(v) > I_{98} \text{ and } v \in A\}$ where $V_{skull}$ is the set of all skull voxels, $v$ a voxel in the 3D US image, $I(v)$ the intensity of this voxel, $I_{98}$ is the 98 percentile of the image intensities and $A$ an area determined from the ellipsoid geometry as follows:

$$A = \{(x,y,z)|0.8 < \frac{x^2}{a_v} + \frac{y^2}{b_v} + \frac{z^2}{c_v} < 1.3\}. \tag{1}$$

The centroid position of the brain used to estimate $A$ is constant for all patients based on empiric observations, $z_{center}$ is at 65% height level of the non-zero intensities, and $x_{center}$ and $y_{center}$ are in the middle of the non-zero intensities in the $z_{center}$ plane. Since the size of the US image is fixed, and the pixel spacing changes according to the brain size, $a_v$, $b_v$ and $c_v$ are the semi axes in fixed voxel size. Finally, the parameters of the ellipsoid shape are optimized to fit the detected boundaries from the overall appearance of the brain's shape.

Once the shape is obtained, the upper brain limit is approximated with Point 2 as the superior brain limit in 3D US and point 1 is the upper transducer position on the skull which is at the same height as point 2 (Fig. 1).

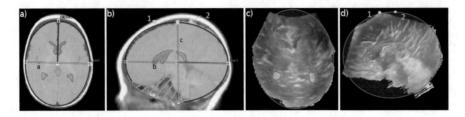

**Fig. 1.** (a) Ellipsoid fitting on the MRI axial view; (b) Ellipsoid fitting on the MRI sagittal view; (c) Ellipsoid fitted on the stripped brain in 3D US axial view; (d) Ellipsoid fitted on stripped brain in sagittal view 3D US. Lateral ventricles are outlined in red.

### 2.3  Lateral Ventricular Segmentation

The first step of the lateral ventricular segmentation method is a multi-atlas registration, where MRI atlases are registered to the infant's 3D US image. This is followed by a label fusion where the output is converted to a mesh. Finally, a deformable mesh based segmentation is applied to account for anatomical variabilities not captured by the atlases.

**Initialization.** The orientation of the US images is first corrected by rotating the volume to match the orientation observed on the MRI atlases. This is performed from a PCA on the extracted inferior skull region, to identify the principal orientation vectors of the head. Then, the brain in 3D US is extracted with the method described in 2.2 and its center position and size are calculated. Based on those measurements, a scaling and a translation are applied to the MRI atlases before the registration.

**Atlas-Based MRI/3D US Registration.** Multimodal registration between 3DUS and MRI images is performed with a locally linear correlation metric ($LC^2$) by [5] which correlates MRI intensities and gradients with US intensities. For registration purposes, several MRI atlases of infants were combined in order to take into account anatomical variability, which included a 1yr atlas Cincinnati imaging center [6], a 2–5 months atlas from the McConnel Brain Imaging Center [7] and 9 MRI volumes from the ALBERTs pediatric atlas [8].

The registration includes a rigid step with $LC^2$ and non-rigid step with $LC^2 + P$ where $P$ is a pixel weighting term. $P$ is a term created specifically to describe lateral ventricles in US by making use of the hypoechoic area (fluid cavities) and the hyperechoic area (choroid plexus). Since only the US voxels included in the MRI ventricle label are analyzed, $P$ is only added at the non-rigid registration step when the MRI labels are already roughly aligned to the US image:

$$P = \frac{C_1 \sum_{i=1}^{N} \epsilon_i max(I_L - I(v_i), 0) + (1 - \epsilon_i)max(I(v_i) - I_H, 0) + C_2}{N} \quad (2)$$

where $\epsilon_i = 1$ when $v_i$ is in the hypoechoic area and $\epsilon_i = 0$ when $v_i$ is in the hyperechoic area, $C_1, C_2$ are coefficients adjusted to the intensities and $N$ is the number of voxels in the MRI ventricle label. Moreover, $P$ is adjusted to penalize smaller labels (which statistically have higher $P$) $P_{adj}(V_k) = P(\frac{V_k}{V_M})^{\frac{1}{4}}$ where $V_k$ is the active label volume and $V_m$ the mean label volume.

The optimization of the registration process was performed using BOBYQA from [9] as proposed in [5], which does not require the metric's derivatives. Registration is repeated on the 11 MRI atlases and a selection of the top ranking ($n = 4$) exemplars is performed based on the resulting similarity metric. The fusion of registrations is accomplished with STAPLE [10] in order to create a probabilistic output of the labels on the 3D US images. A binary label for the lateral ventricles is then computed from the probabilistic map with all voxels having more than 80% probability of belonging to the lateral ventricles.

**Deformable Mesh Model.** Following the extraction of the binary labels based on the fusion process using STAPLE, morphological operations were applied to smooth the binary labels before it was converted to a surface mesh with a marching cubes algorithm. The mesh surface was sub-sampled to reduce computational complexity, by re-ordering the priority queue of mesh vertices and retriangulating the final mesh. Laplacian smoothing was performed on the mesh to smooth the surface prior to computing the normal vectors.

The mesh is deformed in an iterative fashion by minimizing the energy $E = E_I + \beta E_E$ where $E_I$ represents the internal energy of the system acting as a regularizer for the deformation and $E_E$ represents the external energy of the system which drives to deform the mesh. The internal energy is defined as:

$$E_I = \sum_{n=1}^{N_e} d(\Delta D_{1_n}, \Delta D_{2_n}) \tag{3}$$

where $N_e$ is the number of edges, $d(.,.)$ the Manhattan distance in three dimension, with $\Delta D_{1_n}$ and $\Delta D_{2_n}$ the displacement of the first and the second vertex of edge $n$ relatively to their initial position, respectively.

For every vertex, the term $P$ in Eq. (2) is computed for the transformed mesh as $P_{transform}$ and for the initial mesh as $P_{initial}$. The external energy $E_E$ is computed as follows:

$$E_E = -\sum_{i=1}^{N_v} \left\{ \begin{array}{l} \sqrt{P_{transform}} \Delta D_i \gamma \text{ if } P_{initial} >= 0.4 \vee \Delta D_i > 0 \\ \sqrt{P_{transform}} \frac{1}{|\Delta D_i|} \gamma \text{ if } P_{initial} < 0.4 \wedge \Delta D_i < 0 \end{array} \right. \tag{4}$$

where $N_v$ is the number of vertices, $\Delta D_i$ is the displacement of vertex $i$, $\gamma = 1$ if $|\Delta D_i| < l$ ($l$ threshold set according to initial mesh) and $\gamma = \frac{1}{|\Delta D_i^2|}$ otherwise.

The BFGS-limited memory version optimization algorithm is used to minimize the energy equation which is well suited for optimization problems with a high number of parameters.

## 2.4    Ventricles/Brain Volume Ratio

Once the volumes of the lateral ventricles $V_{lat.ven}$ and the total brain $V_{brain}$ are obtained, the volume ratio can be computed as follows $ratio = \frac{V_{lat.ven}}{V_{brain}}$.

# 3   Results

## 3.1   Brain Volume Comparison Between 3D US and MRI

**Parameter Selection.** Based on the comparison between the ellipsoid volume and the ground truth brain volumes in 10 MRI infant templates, optimal results were achieved when $C_f = 0.95$, meaning the brain volume represents 95% of the ellipsoid using leave-one out cross validation. A mean absolute difference of 2.7% and a maximum absolute difference of 4% was found between the estimated and ground truth brain volumes on the 10 examples used for $C_f$ determination. The 10 MRI volumes were atlases of infants all under 1 years old.

**Manual Segmentation.** To first assess the agreement between modalities, brain volumes were manually extracted in 3D US by an experienced neuro-radiologist and compared to the MRI reference brain volume (mean and standard deviation of $757^3 \pm 195 \, \text{cm}^3$). Populations were normally distributed based on Shapiro-Wilks tests. A correlation of $r = 0.988$ was found between 3D US brain volume and MRI volume. There were no statistically significant difference between both distributions based on $T$-test ($p = 0.309$) and $F$-test ($p = 0.477$). The mean absolute error was $3.12 \pm 2.65\%$.

**Automatic Segmentation.** Finally, automatic brain volume assessment was performed in 3D US on the same 12 patients. Between 3D US and MRI, the correlation was $r = 0.942$, with no statistically significant difference between both distributions ($T$: $p = 0.541$) ($F$: $p = 0.273$). The mean and standard deviation of the absolute errors was $7.73 \pm 7.52\%$ with the maximum error on a patient with abnormal brain volume due the approximation of the ellipsoid size.

## 3.2   Lateral Ventricles Volume Comparison Between 3D US and MRI

For the comparison in ventricular volumes, manual segmentation was performed in 11 out of the 12 patients, as the US image quality was poor for a patient nearing 9 months in age. The segmentations were further validated by an experienced pediatric neuro-radiologist. Compared to the reference MRI (median of $5975 \, \text{mm}^3$, with a mean volume of $11084 \, \text{mm}^3$), there was a strong correlation in lateral ventricular volumes between 3D US and MRI ($r = 0.999$), and there was no statistically significant difference between both distributions based on mean paired $T$-test and variance $F$-test ($T$: $p = 0.204$) ($F$: $p = 0.429$). The mean and standard deviation of the absolute differences was $5.8 \pm 4.92\%$. The worst individual result out of the 11 patients was due to the poor image quality linked to the infant's age (8.5 months) which is expected since the fontanelle opening is reduced due to bone maturation.

Then, automatically extracted lateral ventricular volumes in 3D US were compared to the ground truth MRI volumes on the cohort of 11 patients (mean and standard deviation MRI volumes were $5309 \pm 985 \, \text{mm}^3$). Two images had poor image quality, and one image had ventricles dilated to almost 5% of ratio and no MRI template could fit the 63 055 mm$^3$ lateral ventricle volume, showing the need to add more examples to the MRI brain template. Segmentation parameters were found empirically as $C_1 = 0.02$, $C_2 = 0.25$, $I_L = 85 + (I_{mean} - 100)$, $I_H = 115 + (I_{mean} - 100)$ where $I_{mean}$ is the mean intensity of the US image non zero voxels, $\alpha = 0.18$, $\beta = 0.82$ and $L = 2\frac{V_k}{V_M}$. For the volume comparison, a strong correlation ($r = 0.848$) and no statistically significant difference were found based on $T$-test ($p = 0.067$) and $F$-test ($p = 0.276$) although there is a small under evaluation for the volume in 3D US (mean signed error of $-6.91\%$). Absolute errors have a mean and standard deviation of $9.84\% \pm 4.61\%$.

**Table 1.** Comparison in accuracy of the lateral ventricular segmentation methods from 3D US, based on Dice coefficients, mean absolute distance and Hausdorff distance.

| Methods | DICE (%) | MAD (mm) | Hausdorff (mm) |
|---|---|---|---|
| Atlas-based with $LC^2$ [5] | $57.4 \pm 7.8$ | $1.33 \pm 0.44$ | $8.55 \pm 3.42$ |
| Atlas-based with area weights [5] | $60.4 \pm 7.5$ | $1.14 \pm 0.30$ | $7.52 \pm 2.81$ |
| Atlas-based [5] + Mesh | $65.1 \pm 4.1$ | $1.08 \pm 0.33$ | $8.46 \pm 2.98$ |
| Majority Voting (MV) | $65.0 \pm 4.0$ | $1.01 \pm 0.30$ | $7.59 \pm 3.40$ |
| STAPLE [10] | $65.5 \pm 3.8$ | $1.08 \pm 0.24$ | $7.27 \pm 3.19$ |
| Proposed method | $70.8 \pm 3.6$ | $0.88 \pm 0.20$ | $6.84 \pm 3.15$ |

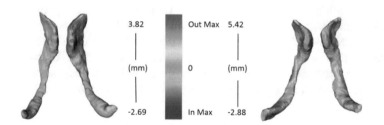

**Fig. 2.** Two examples of segmented lateral ventricles with error maps from 3D US.

### 3.3 Segmentation of Lateral Ventricles in 3D US

The segmentations were also performed on 5 additional patients with 3D US for a total of 16 infants (mean volume: $6468 \pm 320$ mm$^3$, max: $13890$ mm$^3$). The accuracy was computed using expert manual segmentations as ground truth measures and the correlation with the automatically extracted volumes was $r = 0.972$. Table 1 summarizes the results with the Dice coefficient, the mean absolute distance (MAD) and the maximal absolute distance (Hausdorff). The results demonstrate a statistically significant improvement of the proposed method to STAPLE ($p = 0.0004$ for Dice coefficient and $p = 0.0016$ for MAD measures), as well as to the Atlas-based approach with mesh modeling ($p = 0.0059$ for Dice coefficient and $p = 0.0032$ for MAD measures). Figure 2 illustrated two examples of segmented lateral ventricles lateral with color-coded error maps representing the surface distances from the ground truth.

### 3.4 Ventricular-Total Brain Volume Ratio in 3D US

For the 12 infants of the prospective cohort, the ground-truth ventricular ratios were computed from the MRI segmentations of the lateral ventricles and estimation of the brain volume. For the ratios from 3D US, the median difference was 0.00795, with a mean difference of $0.0125 \pm 0.0144$. In terms of concordance between MRI and 3D US, a correlation of $r = 0.998$ demonstrate the strong agreement, and there was no statistically significant difference based on paired $T$-test ($p = 0.672$) and $F$-test ($p = 0.437$). The absolute errors yielded a mean and standard deviation of $6.05 \pm 4.88\%$.

## 4   Conclusion

In this paper, we presented an automatic method to extract lateral ventricles as well as total brain volumes from 3D ultrasound in infant brains. This allows for an automatic assessment of the lateral ventricles dilatation with respect to total brain volumes. Compared to MRI references, the volumes yielded a high correlation and indicate no statistically significant difference between both modalities. In addition, volume ratios can be obtained with a mean ratio slightly below 0.01, which is concordant with literature. Our main contribution is the quantification of the ventricular-brain ratio in 3D US which enables a true assessment of ventricular dilatation. Future work would include adding more MRI templates and an extensive validation with additional subjects, both with higher variability in ventricular volumes, and investigate the use of convolutional neural networks.

## References

1. Gilmore, J.H., et al.: Prenatal mild ventriculomegaly predicts abnormal development of the neonatal brain. Biol. Psychiatry **64**(12), 1069–1076 (2008)
2. Gilmore, J.H., Gerig, G., Specter, B., Charles, H.C., Wilber, J.S., Hertzberg, B.S.: Infant cerebral ventricle volume: a comparison of 3-D ultrasound and magnetic resonance imaging. Ultrasound Med. Biol. **27**(8), 1143–1146 (2001)
3. Qiu, W., et al.: User-guided segmentation of preterm neonate ventricular system from 3-D ultrasound images using convex optimization. Ultrasound Med. Biol. **41**(2), 542–556 (2015)
4. Qiu, W., et al.: Automatic segmentation approach to extracting neonatal cerebral ventricles from 3D ultrasound images. Med. Image Anal. **35**, 181–191 (2017)
5. Fuerst, B., Wein, W., Müller, M., Navab, N.: Automatic ultrasound-MRI registration for neurosurgery using the 2D and 3D LC2 metric. Med. Image Anal. **18**(8), 1312–1319 (2014)

6. Altaye, M., Holland, S.K., Wilke, M., Gaser, C.: Infant brain probability templates for MRI segmentation and normalization. Neuroimage **43**(4), 721–730 (2008)
7. Fonov, V.S., Evans, A.C., McKinstry, R.C., Almli, C., Collins, D.: Unbiased nonlinear average age-appropriate brain templates from birth to adulthood. NeuroImage **47**, S102 (2009)
8. Gousias, I.S., et al.: Magnetic resonance imaging of the newborn brain: manual segmentation of labelled atlases in term-born and preterm infants. Neuroimage **62**(3), 1499–1509 (2012)
9. Powell, M.J.: The BOBYQA algorithm for bound constrained optimization without derivatives. Cambridge NA Report NA2009/06, pp. 26–46, University of Cambridge, Cambridge (2009)
10. Warfield, S.K., Zou, K.H., Wells, W.M.: Simultaneous truth and performance level estimation (STAPLE): an algorithm for the validation of image segmentation. IEEE Trans. Med. Imaging **23**(7), 903–921 (2004)

# Do Baby Brain Cortices that Look Alike at Birth Grow Alike During the First Year of Postnatal Development?

Islem Rekik[1(✉)], Gang Li[2], Weili Lin[2], and Dinggang Shen[2]

[1] BASIRA lab, CVIP group, School of Science and Engineering, Computing, University of Dundee, Dundee, UK
islem.rekik@gmail.com
[2] Department of Radiology and BRIC, University of North Carolina at Chapel Hill, Chapel Hill, NC, USA
www.basira-lab.com

**Abstract.** The neonatal brain cortex is marked with complex and high-convoluted morphology, that undergoes dramatic changes over the first year of postnatal development. A large body of existing research works investigating 'the developing brain' have focused on looking at changes in cortical morphology and charting the developmental trajectories of the cortex. However, the relationship between neonatal *cortical morphology* and its *postnatal growth trajectory* was poorly investigated. Notably, understanding the *multi-scale shape-growth relationship* may help identify early neurodevelopmental disorders that affect it. Here, we unprecedentedly explore the question: "Do cortices that look alike in shape at birth have similar kinetic growth patterns?". To this aim, we propose to analyze shape-growth relationship at three different scales. On a *global scale*, we found that neonatal cortices similar in geometric closeness are significantly correlated with their postnatal overall growth dynamics from birth till 1-year-old ($r = 0.27$). This finding was replicated when using shape similarity in morphology ($r = 0.20$). On a *local scale*, for both hemispheres, 20% of cortical regions displayed a significant high correlation ($r > 0.4$) between their similarities in morphology and dynamics. On a *connectional scale*, we identified hubs of cortical regions that were consistently similar in morphology and developed similarly across subjects including the cingulate cortex using a *novel integral shape-growth brain graph representation*.

## 1 Introduction

Little is known about how the cerebral cortex develops and works, particularly during early postnatal neurodevelopment [1]. Understanding the relationship between the neonatal cortical shape (i.e., geometry and morphology) and its postnatal kinetic behavior (i.e., dynamics[1] and velocity of growth) may produce novel diagnostic tools for better identifying neurodevelopmental disorders

---

[1] Change in space and time or 4D change.

© Springer Nature Switzerland AG 2018
A. F. Frangi et al. (Eds.): MICCAI 2018, LNCS 11072, pp. 566–574, 2018.
https://doi.org/10.1007/978-3-030-00931-1_65

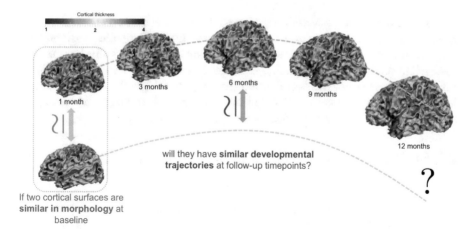

**Fig. 1.** If two cortical shapes are similar in morphology at baseline (i.e., 1 month of age), will they have similar developmental trajectories at follow-up timepoints?

at a very early stage (e.g., schizophrenia and autism) [2,3]. For instance, many brain disorders affect not only the cortex growth patterns, but also the cortex morphology [2]. However, the majority of existing studies investigate changes in cortical morphology or growth trajectories *independently*. To the best of our knowledge, the *multi-scale* relationship between the spatiotemporal dynamics and the morphology of the neonatal cerebral cortex is overlooked in the neuro-science literature.

To address this gap, we propose a multi-scale shape-growth analysis (MSGA) framework, which specifically investigates the following question: "Do cortices that look alike in shape at birth (both locally and globally) have similar kinetic growth patterns (e.g., velocity of growth) during the first year of postnatal development?" (Fig. 1). To this aim, we examine if *'similar'* shapes across individuals grow *'alike'* at three different scales: global, local and connectional. However, prior to developing our MSGA framework, one needs to define shape similarity and dynamic growth similarity metrics. Unlike several morphometric and volumetric analysis methods that cannot accurately characterize subtle cortical morphological changes in space and time, we leverage the multidirectional varifold-based cortical shape representation model introduced in [4] that quantifies the surface morphology along key directions at each vertex, and provides an informative metric that evaluates *morphological* similarity between shapes. We also use the Euclidean distance to evaluate the *geometric* closeness between shapes. As for the *kinetic*[2] *(or growth) similarity*, we use a longitudinal multi-directional varifold regression model based on [4], which allows to estimate the spatiotemporal velocity of each vertex in the baseline cortical surface within a specific time-window starting at first acquisition timepoint. Next, we propose a pairwise correlation-based similarity that quantifies the dynamic co-behavior in two different locations (or regions) of the neonatal cortex.

---

[2] Related to the velocity of shape deformation.

At a global scale, we examine the correlation between morphologically (resp. geometrically) similar neonatal cortical shapes and their kinetic developmental trajectories. At a local scale, we examine shape-velocity correlations in very distinctive cortical regions. Finally, at a connectional scale, inspired from [5], we further propose a novel integral brain 'shape-growth' graph representation to identify neonatal cortical regions that are similar in morphology, but also grow similarly in respectively the left and the right hemispheres.

## 2    Multi-scale Shape-Growth Analysis (MSGA) Framework

To investigate if two similar shapes at baseline (first acquisition timepoint) will develop similarly over time, we introduce the key ingredients of our proposed MSGA framework as follows.

**Quantification of Cortical Shape Morphology Using Multi-directional Varifold Representation.** First, we use the multidirectional varifold shape representation proposed in [4] to quantify the morphology of a cortical surface $S$ at each mesh (triangular face) center $x_i$ along two orthogonal directions: the non-oriented normal direction $\overleftrightarrow{n_i}$, and the non-oriented maximum principal curvature direction $\overleftrightarrow{\kappa_i}$. Specifically, a surface $S$ with $M$ meshes (triangles) is approximated by the sum of Dirac varifolds evaluated at the positions $x_i$ of the centers of its $M$ meshes using their corresponding non-oriented normals $\overleftrightarrow{n_i}$ and non-oriented maximum principal curvature direction $\overleftrightarrow{\kappa_i}$: $S = \sum_{i=1}^{M} \delta_{(x_i, \overleftrightarrow{n_i})} + \delta_{(x_i, \overleftrightarrow{\kappa_i})}$. More importantly, the varifold space $W^*$ is endowed with a dot-product that enables to measure the morphological similarity between two shapes $S = \sum_i \delta_{(x_i, \overleftrightarrow{n_i})} + \delta_{(x_i, \overleftrightarrow{\kappa_i})}$ and $S' = \sum_j \delta_{(x'_j, \overleftrightarrow{n'_j})} + \delta_{(x'_j, \overleftrightarrow{\kappa'_j})}$ are: $< S, S' >_{W^*} = \sum_i \sum_j K_W(x_i, x'_j)(\frac{(n_i^T n'_j)^2}{|n_i||n'_j|} + \frac{(\kappa_i^T \kappa'_j)^2}{|\kappa_i||\kappa'_j|})$, where $K_W$ is a Gaussian kernel that decays at rate $\sigma_W$. We note that $\sigma_W$ represents the scale under which morphological details of the cortical shape are overlooked.

**Quantification of Cortical Developmental Trajectories Using Multi-directional Varifold Regression Model.** Given a set of shapes $\{S^1, \ldots, S^{N_S}\}$, we nest each of these into a multidirectional varifold space $W^*$, where each shape $S^k$ (i.e., $k$–th subject in the population) is represented as a sum of two 'orthogonal' varifolds. Since each shape is measured longitudinally at different timepoints $t \in [0,1]$, we estimate its evolution trajectory through deforming the baseline multidirectional varifold $S_0^k$ onto a set of target multidirectional varifolds $\{S_1^k, \ldots, S_T^k\}$ respectively observed at different observation timepoints. To do so, we model this longitudinal shape deformation from baseline as a minimization problem [4]: $J_{W^*} = \frac{1}{2} \int_0^1 ||v_t||_V^2 dt + \gamma \sum_{j \in \{1, \ldots, T\}} ||S_{t_j}^k - \phi(S_0^k, t_j)||_{W^*}^2$, where $\gamma$ represents a trade-off between the deformation smoothness energy and the similarity between ground-truth and deformed shapes.

$\phi(\mathbf{x}, t)$, $t \in [0, 1]$ represents a smooth invertible deformation (i.e., diffeomorphism), fully defined by a set of $N_z$ control points $z_i$ and their attached initial deformation momenta $\alpha_i$. We fix the $N_z$ control points across all subjects, to compare their growth trajectories. $v_t$ denotes the estimated smooth shape deformation velocity field, which belongs to a reproducing kernel Hilbert space $V$, spanned by Gaussian kernel $K_V$ with standard deviation $\sigma_V$. Specifically, the vertex-wise spatiotemporal piece-wise continuous velocity $v$ is defined at a location $\mathbf{x}$ and timepoint $t$ as: $v(\mathbf{x}, t) = \sum_{i=1}^{N_z} K_V(\mathbf{x}, z_i(t))\alpha_i(t)$. Notably, this multidirectional varifold deformation framework allows to establish vertex-to-vertex correspondence across subjects and timepoints. Next, we leverage the multidirectional varifold for shape representation and the estimated velocity for growth quantification to compute morphological similarity and growth similarity between pairs of shapes in our cohort.

**Geometric Shape Similarity Matrix Definition.** To quantify the *geometric* concordance between two neonatal surfaces $S_i$ and $S_j$, we use the Euclidean distance as follows: $d(S_i, S_j) = \sum_{x=1}^{N} ||S_i(x) - S_j(x)||_2$, where $N$ represents the number of vertices. We then transform the pairwise distance into a pairwise similarity using a continuous mapping $(f(x) = x^2 - 2x + 1)$. Unlike non-smooth linear regression mapping, the proposed non-linear function $f$ is continuously differentiable (or smooth), which helps better preserve the potential local smoothness that may exist in the original shape similarity matrix. Next, we generate an $N_s \times N_s$ matrix $\mathcal{S}_g$, where each element $\mathcal{S}_g(i, j)$ quantifies the similarity in geometry between two shapes $S^i$ and $S^j$.

**Morphological Shape Similarity Matrix Definition.** In a similar way, we generate a morphological shape dissimilarity matrix, where each element measures the dissimilarity in morphology between two shapes $S^i$ and $S^j$ using the multidirectional varifold distance norm $||S_i - S_j||_{W^*}$. We also reverse this matrix using the $f$-mapping to finally get the morphological shape similarity matrix $\mathcal{S}_m$.

**Velocity Similarity Matrix Definition.** To measure the similarity in developmental dynamics of two baseline shapes, we first retrieve at each vertex $x_i$ in the baseline surface $S^0$ the estimated spatiotemporal velocity signal along first, second, and third axes: $\mathbf{v^1}(x_i) = [v^1(t = 0, x_i), \ldots, v^1(t = T, x_i)]$, $\mathbf{v^2}(x_i) = [v^2(t = 0, x_i), \ldots, v^2(t = T, x_i)]$, and $\mathbf{v^3}(x_i) = [v^3(t = 0, x_i), \ldots, v^3(t = T, x_i)]$. Then, we generate the coordinate-wise Pearson correlation with its corresponding vertex $x_j$ in a different baseline surface $S^j$. Finally, we compute their average and store it in an element $\mathcal{G}_v(i, j)$ of the velocity $N_s \times N_s$ similarity matrix: $\sum_{x=1}^{N} \frac{1}{N}(corr^2(\mathbf{v^1}(x_i), \mathbf{v^1}(x_j)) + corr^2(\mathbf{v^2}(x_i), \mathbf{v^2}(x_j)) + corr^2(\mathbf{v^3}(x_i), \mathbf{v^3}(x_j)))^{1/2}$.

**Initial Momenta Similarity Matrix Definition.** Here, for each pair of neonatal cortical shapes $S^i$ and $S^j$ in our cohort, we compute the average inner product between their respective initial deformation momenta, and then assign this value, which lies in the interval $[-1, 1]$, to an element $(i, j)$ of an $N_s \times N_s$ matrix $\mathcal{G}_m$ that sparsely quantifies how similar is the 4D deformation of both shapes.

**Global Shape-Growth Analysis.** At a global scale, we compute the Pearson correlation between only the halves of the geometric similarity shape matrix $\mathcal{S}_g$ and the initial momenta similarity matrix $\mathcal{G}_m$ as they are symmetric. We also compute the correlation between the morphological similarity shape matrix $\mathcal{S}_m$ and $\mathcal{G}_m$. To better examine the global dense deformation trajectories, we compute the correlation between both geometric $\mathcal{S}_g$ and morphological $\mathcal{S}_m$ matrices and the global velocity similarity matrix $\mathcal{G}_v$ computed on the whole cortex.

**Local ROI-Based Shape-Velocity co-behavior Analysis.** To investigate the variation in strength of the shape-growth correlates in different local cortical regions, we parcellate each cortical surface into $N_r$ anatomical regions of interest (ROIs). Then, we generate *for each ROI $r$* a velocity similarity matrix $\mathcal{G}_v^r$ of size $N_s \times N_s$. Supp. Fig. 1[3] illustrates this step for a representative cortical region. On a local scale, to examine if baseline cortical regions that look similar have correlated postnatal dynamic behavior, we also create for each ROI a morphological shape similarity matrix $\mathcal{S}_m^r$ to assess the regional morphological similarity between individuals and geometric shape similarity matrix $\mathcal{S}_g^r$ to assess the regional geometric concordance or closeness. Then, for each hemisphere and for each ROI, we compute the Pearson correlation coefficient between a velocity similarity matrix $\mathcal{G}_v^r$ and each shape similarity matrix ($\mathcal{S}_g$ and $\mathcal{S}_m$).

**Connectional Shape-Growth Analysis.** To identify regional cortical connections that show strong correlation between regions that look alike and grow alike *across all subjects*, we first define for each subject $s$ shape $\mathcal{S}_m^s$ and growth $\mathcal{G}_v^s$ matrices, each of size $N_r \times N_r$. Next, we fuse all morphological shape matrices to generate the morphome (Supp. Fig. 2), and all velocity-based growth matrices to generate the kinectome using network fusion method proposed in [6]. Next, for each hemisphere, we normalize both fused shape connectivity and mean velocity connectivity matrices, then we compute their absolute difference matrix. Through sparsifying the matrix by retaining only the $P_s\%$ lowest difference values, we unprecedentedly define the *morpho-kinetome*, where similar cortical regions in morphology have correlated growth trajectories and dissimilar brain regions have uncorrelated growth trajectories. We illustrate in Supp. Fig. 3 three cases where we generate morpho-kinectomes at different sparsification levels and with different scales in shape-growth co-behavior.

---

[3] Supplementary material link: http://basira-lab.com/wp-content/uploads/2017/05/Supp-Rekik-et-al.-MICCAI-2018.pdf.

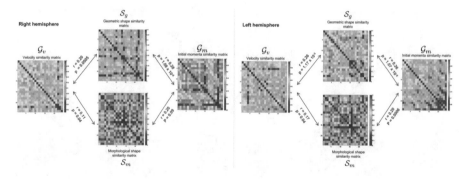

**Fig. 2.** *Global analysis: correlation between shape similarity and velocity similarity matrices.* For each hemisphere, we computed the correlation between the geometric (vs. morphological) shape similarity matrix with respectively initial momenta similarity matrix and velocity similarity matrix.

## 3   Results and Discussion

**Data and Parameter Setting.** In this study, we used 115 MR images from 23 healthy full-term born infants. Each infant was scheduled to be scanned at 5 time points (1, 3, 6, 9, and 12 months of age). At each scheduled scan, T1-, T2-, and diffusion-weighted MR images were acquired by a Siemens 3 T head-only MR scanner with a 32 channel head coil. T1-weighted images (144 sagittal slices) were acquired with the imaging parameters: $TR = 1900$ ms, $TE = 4.38$ ms, flip angle $= 7$, acquisition matrix $= 256 \times 192$, and voxel size $= 1 \times 1 \times 1$ mm³. T2-weighted images (64 axial slices) were acquired with the imaging parameters: $TR/TE = 7380/119$ ms, flip angle $= 150$, acquisition matrix $= 256 \times 128$, and voxel size $= 1.25 \times 1.25 \times 1.95$ mm³. We parcellated each cortical hemisphere into 35 ROIs using the Desikan-Killiany atlas. For the multidirectional varifold-based geodesic shape regression model, we empirically set $\gamma = 10^{-4}$, $\sigma_W = 5$, and $\sigma_V = 25$.

**Global Shape-Growth Analysis.** Our global analysis revealed a significantly positive correlation between geometrically close shapes and their growth dynamics as well as between similar shapes in morphology and their dynamic evolution. In particular, as shown in Fig. 2, the initial momenta similarity matrix correlated with the geometric shape similarity matrix in both left ($r = 0.26$, $p = 1.07 \times 10^{-5}$) and right ($r = 0.28$, $p = 0.09$) hemispheres, and slightly less correlated with the morphological shape similarity matrix ($r = 0.20$, $p < 10^{-5}$) in both hemispheres. This is quite expected since the varifold metric captures the richness and complexity of the cortical foldings that largely vary between subjects, unlike the geometric similarity that grossly approximates closeness between cortical surfaces. In addition, to not reduce the cortical growth to only a sparse set of initial momenta, we also use the velocity similarity matrix which encodes the mean correlation between all vertex-wise velocity trajectories in $x$, $y$ and $z$

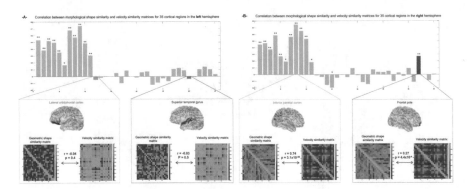

**Fig. 3.** *Local ROI-based shape-growth correlation analysis for left and right hemispheres in developing brains.* Correlation for each cortical region of interest between geometric shape similarity matrix and velocity similarity matrix. ∗∗ denotes highly significant correlations ($p \ll 10^{-3}$) and ∗ denotes significant correlation ($p < 0.05$).

spatial directions. Our findings also revealed a positive correlation between the velocity similarity matrix and the geometric shape similarity matrix in both left ($r = 0.26$, $p = 1.17 \times 10^{-5}$) and right ($r = 0.20$, $p = 5 \times 10^{-4}$) hemispheres, which similarly decreased when using the morphological shape similarity matrix for the left ($r = 0.17$, $p = 0.04$) and the right ($r = 0.11$, $p = 0.04$) hemispheres.

**Local Shape-Growth Analysis.** We extended our previous findings by exploring the shape-velocity co-behavior in 35 cortical regions. For each ROI, we computed the Pearson correlation coefficient between the velocity similarity matrix and the morphological shape similarity matrix. For both left and right hemispheres, 20% of the cortical regions had a statistically significant high correlation between their morphological closeness and growth dynamics ($r > 0.4$, $p \ll 0.001$) (Fig. 3). These included for both hemispheres the superior temporal sulcus, caudal middle frontal gyrus, fusiform gyrus, inferior parietal cortex, inferior temporal gyrus and isthmus cingulate cortex, along with the right anterior cingulate cortex –mainly belonging to the temporal, frontal and limbic lobes. We also found statistically nonsignificant negative correlations in about 10 cortical regions using the morphological similarity. For the right hemisphere, 13/35 regions had negative correlations where only the middle temporal gyrus was statistically significant ($p < 0.05$). Clearly, both cortical hemispheric developments are marked by correlated and anti-correlated shape-velocity behaviors largely consistent across cortical regions, where distinctive cortical areas exhibited highly significant correlation values. Our findings also suggest that *specific* cortical regions such as the inferior parietal cortex (IPC) in both left and right hemispheres, which is part of the default mode network and is involved in interpretation of sensory information, language and body image, exhibited a powerful positive correlation between their shape and growth similarity matrices ($r > 0.7$).

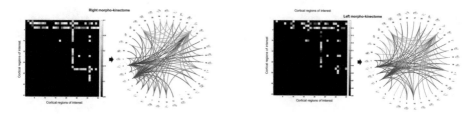

**Fig. 4.** *The morpho-kinectome.* Each node in the circular graph denotes a cortical region as in Supp. Fig. 4. Each color defines one node. The width of a circular edge represents its strength. The circular graph shows that both left and right morpho-kinectomes share two key cortical hubs: node 2 (the caudal anterior-cingulate cortex) and node 23 (posterior-cingulate cortex). We would like to note that node 4 representing the corpus callosum is overlooked as it is not part of the cortex.

**Connectional Shape-Growth Analysis Using the Proposed Morpho-Kinectome.** In the right and left morpho-kinectomes sparsified at $P_s = 15\%$, we found that the caudal anterior-cingulate cortex (ACC) and posterior-cingulate cortex (PCC) presented the key hubs (Fig. 4). Noting that developmental studies suggest that structural hubs emerge relatively early during brain development, with connectivity of posterior cingulate regions already present in the postnatal brain [7], can potentially explain the shape-growth co-behavior of neonatal cingulate regions. The cingulate cortex is involved in spatial memory, configural learning, and maintenance of discriminative avoidance learning, which are fundamental cognitive functions to infants' development and learning [7,8]. Importantly, relating the morpho-kinectome to growth connectomics and neurodevelopment [9] as well as genetic [10] and developmental [11] underpinnings would give insights into neurodevelopment and the etiology of neurodevelopmental disorders.

## 4    Conclusion

In sum, our MSGA framework demonstrated that globally similar cortical shapes similarly co-evolve. We also identified distinctive cortical regions (e.g., IPC) that grow similarly when looking similar in morphology –which may indicate that their growth is controlled by similar genetic and biological factors. We also showed that shape connectivity is a powerful predictor of growth dynamics in specific cortical regions (e.g., anterior/posterior cingulate cortices). The model presented here is a highly promising starting point, given that it can be generalized to different complex shapes and allows to examine the spatiotemporal dynamics of shapes as well as quantifying their high-dimensional (here 3D) morphology. Besides, it will be more intriguing to interpret our findings in the light of multiple covariates such as stress during development [12] and socio-economic status of recruited infants, as both are suspected to affect early brain development [13]. Eventually, our proposed MSGA framework may help better elucidate how the cortex develops and wires itself in both healthy and disordered infant brains as well as provide a comprehensive graph-based brain representation that unifies genetics, connectomics and *morpho-kinectomics*.[4]

# References

1. Li, G., et al.: Computational neuroanatomy of baby brains: a review. NeuroImage (2018)
2. Dubois, J., et al.: Primary cortical folding in the human newborn: an early marker of later functional development. Brain **131**, 2028–2041 (2008)
3. Wallace, G., Dankner, N., Kenworthy, L., Giedd, J., Martin, A.: Age-related temporal and parietal cortical thinning in autism spectrum disorders. Brain **133**, 3745–3754 (2010)
4. Rekik, I., Li, G., Lin, W., Shen, D.: Multidirectional and topography-based dynamic-scale varifold representations with application to matching developing cortical surfaces. Neuroimage **135**, 152–162 (2016)
5. Rekik, I., Li, G., Lin, W., Shen, D.: Estimation of shape and growth brain network atlases for connectomic brain mapping in developing infants. In: 2018 IEEE 15th International Symposium on Biomedical Imaging (ISBI 2018), pp. 985–989 (2018)
6. Wang, B., et al.: Similarity network fusion for aggregating data types on a genomic scale. Nat. Methods **11**, 333–337 (2014)
7. Heuvel, M.V.D., Sporns, O.: Network hubs in the human brain. Trends Cogn. Sci. **17**, 683–696 (2013)
8. Maddock, R., Garrett, A., Buonocore, M.: Remembering familiar people: the posterior cingulate cortex and autobiographical memory retrieval. Neuroscience **104**, 667–676 (2001)
9. Vertes, P., Bullmore, E.: Annual research review: growth connectomics-the organization and reorganization of brain networks during normal and abnormal development. J. Child Psychol. Psychiatry **56**, 299–320 (2015)
10. Panizzon, M., et al.: Distinct genetic influences on cortical surface area and cortical thickness. Cereb. Cortex **19**, 2728–2735 (2009)
11. Raznahan, A., et al.: How does your cortex grow? J. Neurosci. **31**, 7174–7177 (2011)
12. Hanson, J., et al.: Structural variations in prefrontal cortex mediate the relationship between early childhood stress and spatial working memory. J. Neurosci. **32**, 7917–7925 (2012)
13. Brito, N., Noble, K.: Socioeconomic status and structural brain development. Front. Neurosci. **8**, 276 (2014)

---

[4] This work was supported by NIH grants (MH100217, MH107815, MH108914, and MH110274).

# Multi-label Transduction for Identifying Disease Comorbidity Patterns

Ehsan Adeli[1(✉)], Dongjin Kwon[1,2], and Kilian M. Pohl[2]

[1] Stanford University, Stanford, USA
eadeli@stanford.edu
[2] SRI International, Menlo Park, USA

**Abstract.** Study of the untoward effects associated with the comorbidity of multiple diseases on brain morphology requires identifying differences across multiple diagnostic groupings. To identify such effects and differentiate between groups of patients and normal subjects, conventional methods often compare each patient group with healthy subjects using binary or multi-class classifiers. However, testing inferences across multiple diagnostic groupings of complex disorders commonly yield inconclusive or conflicting findings when the classifier is confined to modeling two cohorts at a time or considers class labels mutually-exclusive (as in multi-class classifiers). These shortcomings are potentially caused by the difficulties associated with modeling compounding factors of diseases with these approaches. Multi-label classifiers, on the other hand, can appropriately model disease comorbidity, as each subject can be assigned to two or more labels. In this paper, we propose a multi-label transductive (MLT) method based on low-rank matrix completion that is able not only to classify the data into multiple labels but also to identify patterns from MRI data unique to each cohort. To evaluate the method, we use a dataset containing individuals with Alcohol Use Disorder (AUD) and human immunodeficiency virus (HIV) infection (specifically 244 healthy controls, 227 AUD, 70 HIV, and 61 AUD+HIV). On this dataset, our proposed method is more accurate in correctly labeling subjects than common approaches. Furthermore, our method identifies patterns specific to each disease and AUD+HIV comorbidity that shows that the comorbidity is characterized by a compounding effect of AUD and HIV infection.

## 1 Introduction

Improvements in modern health-care together with the aging population caused populations with multiple conditions that require ongoing medical attention. The U.S. alone has approximately 75 million people living with 2 or more conditions [12] such as brain-related disorders. For brain-related conditions, disease comorbidity often leads to new cognitive impairments [8,9]. However, few studies have examined the potentially heightened burden of disease comorbidity. As

Supported in part by NIH grants U01 AA017347, R01-AA005965, R37-AA010723, K05-AA017168, F32 AA026762, and R01 MH113406.

© Springer Nature Switzerland AG 2018
A. F. Frangi et al. (Eds.): MICCAI 2018, LNCS 11072, pp. 575–583, 2018.
https://doi.org/10.1007/978-3-030-00931-1_66

an example, Alcohol Use Disorder (AUD) is common among individuals in the
United States, and its co-occurrence in individuals with human immunodeficiency virus (HIV) infection is high [8], occurring at twice the rate as occurs in
the general population [9]. Both AUD and HIV infection reduce health-related
quality of life. Adapting robust and multi-label technology can transform the
mechanistic understanding of the compounding factors of such comorbidity.

Conventional models for identifying the compounding effects of disease
comorbidity use binary (such as statistical tests [9]) or multi-class (*e.g.*, [12,15])
study designs. By nature, these methods assume samples are assigned to
mutually-exclusive labels, and hence inaccurately model the compounding factors of disease comorbidity. Thus, they often lead to inconclusive or contradicting
findings for the cases of comorbidity [14,16]. One can accurately model disease
comorbidity by allowing multiple labels to be assigned to each subject, such as
done by multi-label classification methods. However, there are two major challenges in applying multi-label methods to neuroimaging data: (1) neuroimaging
data often contain several brain morphology measures that are highly correlated;
(2) the measures are prone to noise due to inaccuracies in acquisition, preprocessing, and diagnosis of subjects. To overcome these challenges, we introduce a
multi-label transductive (MLT) classification approach based on low-rank matrix
completion (MC) [2,7] that models noise and overcomes the problem of feature
redundancy/correlation by reducing the data to low-rank subspaces. In addition,
our method identifies patterns related to each disease (label).

To gain a better understanding of MLT, consider a training data set of $N_{tr}$
samples with $l$ different labels (which lead to $c = 2^l$ number of different classes).
Let $d$ be the dimensionality of the feature space, $N = N_{tr} + N_{ts}$ the number of
total samples ($N_{ts}$ the number of testing samples). Our MLT then determines
the labels of the testing samples by first combining all features (of training and
testing) and the labels of the training data into a matrix. Here, we propose to
add columns to the matrix, in which only the labels but not the features are
defined. These allow us to compute disease specific patterns, which, together
with the testing labels are computed by MLT through matrix completion.

We apply our model to identify the impact of alcohol use and HIV infection
in the brain morphometry of individuals. We interpret the two conditions AUD
and HIV as labels that encode four classes: CTRL (samples are assigned to none
of the two labels), AUD (only assigned to the first label), HIV (only assigned to
the second label), and AUD+HIV (having both labels). Figure 1 shows the corresponding matrix whose missing values are computed by our method (see[1] for
notations).

In summary, we make two contributions: we model disease comorbidity within
a transductive multi-label setting that is robust to noise, and we identify the
disease-specific patterns by modifying the original MC algorithm [7].

---

[1] Bold capital letters denote matrices (*e.g.*, $\mathbf{D}$), and bold small letters denote vectors
(*e.g.*, $\mathbf{d}$). All non-bold letters are scalar variables. $d_{ij}$ is the scalar in row $i$ and
column $j$ of $\mathbf{D}$. $|A|$ denotes the number of elements in set $A$. $\|\mathbf{D}\|_*$ designates the
nuclear norm (sum of singular values) of $\mathbf{D}$.

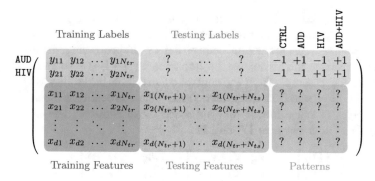

**Fig. 1.** Multi-label transduction (MLT) via matrix completion: each column represents one sample, the top rows hold the labels (two labels in this example, *i.e.*, AUD and HIV), while the rest are comprised of the features. The last columns include the representative features for each of the classes.

## 2   Multi-label Transduction via Matrix Completion

MC is the process of recovering a matrix from a sampling of its entries. Specifically, MC recovers a data matrix $\mathbf{D}$ from a matrix $\mathbf{D}^0$ for which only the subset $\Omega$ of its entries are defined (or observed). Knowing a sufficiently large number of measurements in the matrix, MC assumes that the relation between the elements of the matrix can be accurately described by a low-rank matrix [2,3,7]. Thus, MC estimates the missing entries in $\mathbf{D}$ by minimizing its rank ($\mathrm{rank}(\mathbf{D})$) while $d_{ij} = d_{ij}^0, \forall (i,j) \in \Omega$. The rank function is a non-convex, non-smooth function that can be approximated by the nuclear norm of the matrix, *i.e.*, the sum of its non-vanishing singular values, $\sigma_k$: $\|\mathbf{D}\|_* = \sum_{k=1}^{d} \sigma_k(\mathbf{D})$.

To apply MC to classification, we note that classifiers generally learn the relation between the features, $\mathbf{X}_{tr} \in \mathbb{R}^{d \times N_{tr}}$, and the labels, $\mathbf{Y}_{tr} = \mathbb{R}^{l \times N_{tr}}$, of the training samples. Learning this relation while at the same time classifying or determining the labels $\mathbf{Y}_{ts} \in \mathbb{R}^{l \times N_{ts}}$ of the $N_{ts}$ testing samples (*i.e.*, $\mathbf{X}_{ts} \in \mathbf{R}^{d \times N_{ts}}$) can again be interpreted as a Matrix Completion task [7] of a matrix that contains all training and testing data (see also Fig. 1). MC assumption about rank deficiency of the matrix is equivalent to assuming a linear relationship between the measurements and labels, which is a common assumption of classifiers. However, unlike inductive methods that learn a separate model and then apply it to testing data, MC classifies the testing samples by filling in the submatrix $\mathbf{Y}_{ts}$ of missing entries such that rank of the matrix is minimized. As discussed earlier (and illustrated in Fig. 1), we propose to extract class-specific patterns by adding columns (*i.e.*, submatrices $\mathbf{Y}_{pat} \in \mathbb{R}^{l \times c}$ and $\mathbf{X}_{pat} \in \mathbb{R}^{d \times c}$) with empty entries for the features to the matrix. As in [2,7], a row with the entries of all 1 (called '1' row) is added to the matrix modeling the bias term of linear classifiers (see Eq. (1)). In summary, $\mathbf{D}^0$ is formed by concatenating all features ($\mathbf{X}_{tr}, \mathbf{X}_{ts}, \mathbf{X}_{pat}$), labels ($\mathbf{Y}_{tr}, \mathbf{Y}_{ts}, \mathbf{Y}_{pat}$), and the '1' row.

The set of known entries in $\mathbf{D}^0$ is now defined by the 'feature' submatrix $\Omega_X$ (*i.e.*, all training $\mathbf{X}_{tr}$ and testing $\mathbf{X}_{ts}$ features) and the label submatrix $\Omega_Y$

(*i.e.*, all training labels $\mathbf{Y}_{tr}$ and the pattern-specific labels $\mathbf{Y}_{pat}$). Note, that the extracted neuroimaging measurements (used as features) may be contaminated by noise due to inaccuracies in acquisition or preprocessing. In addition, the brain of a patient might not yet be impacted by a disease so that the patient looks more like a sample from the control cohort. This variability within cohorts can also be interpreted as noise associated with the training labels. We model those sources of noise in our MC approach by introducing the matrix $\mathbf{E}$ and defining $\mathbf{D}$ as sum between $\mathbf{D}^0$ and $\mathbf{E}$:

$$\mathbf{D} = \begin{bmatrix} \mathbf{D_Y} \\ \mathbf{D_X} \\ \mathbf{D_1} \end{bmatrix} = \begin{bmatrix} \mathbf{Y}_{tr} \ \mathbf{Y}_{ts} \ \mathbf{Y}_{pat} \\ \mathbf{X}_{tr} \ \mathbf{X}_{ts} \ \mathbf{X}_{pat} \\ \mathbf{1}^\top \end{bmatrix} + \begin{bmatrix} \mathbf{E}_{\mathbf{Y}_{tr}} \ \ 0 \ \ 0 \\ \mathbf{E}_{\mathbf{X}_{tr}} \ \mathbf{E}_{\mathbf{X}_{ts}} \ 0 \\ 0^\top \end{bmatrix} = \mathbf{D}^0 + \mathbf{E}, \qquad (1)$$

where $\mathbf{D_Y}$, $\mathbf{D_X}$, and $\mathbf{D_1}$ are the label, feature and '1' rows of $\mathbf{D}$, respectively. In other words, MC determines the testing labels $\mathbf{Y}_{ts}$, representative patterns for each class $\mathbf{X}_{pat}$, and the noise $\mathbf{E}$ such that the rank of $\mathbf{D} = \mathbf{D}_0 + \mathbf{E}$, *i.e.*, $\|\mathbf{D}\|_*$, is minimized. To further constrain the optimization problem and avoid trivial solutions, we introduce the squared loss function $\mathcal{L}_x(d_{ij}, d_{ij}^0) = \frac{1}{2}(d_{ij} - d_{ij}^0)^2$ as a way to penalize large differences in features (*i.e.*, noise in features) between the two matrices and a log loss function $\mathcal{L}_y(d_{ij}, d_{ij}^0) = \frac{1}{\gamma} \log(1 + e^{-\gamma d_{ij} d_{ij}^0})$ to penalize difference in class assignment between the matrices (*i.e.*, noise in labeling). $\gamma = 1$ in our experiments. The complete minimization problem is then defined as

$$\operatorname*{argmin}_{\mathbf{D},\mathbf{E}} \ \|\mathbf{D}\|_* + \frac{\lambda_1}{|\Omega_X|} \sum_{ij \in \Omega_X} \mathcal{L}_x(d_{ij}, d_{ij}^0) + \frac{\lambda_2}{|\Omega_Y|} \sum_{ij \in \Omega_Y} \omega_{ij} \mathcal{L}_y(d_{ij}, d_{ij}^0) \qquad (2)$$

$$\text{subject to } \mathbf{D} = \mathbf{D}^0 + \mathbf{E}, \mathbf{D_1} = \mathbf{1}^\top,$$

with hyperparameters $0 \le \lambda_1, \lambda_2 \le 1$ controlling the influence of the loss functions in the minimization problem and the weight $\omega_{ij}$ of each training sample accounting for the imbalance in number of samples per cohort. This weight, computed as a processing step, weights the loss for each sample disproportional to the number of samples available in the training set with the same label $y_{ij}$ of sample $j$.

**Solving the Optimization Problem.** As in [1,2,7], Eq. (2) is a constrained, convex optimization problem that is not smooth due to the nuclear norm (first term). We estimate its solution via the fixed-point continuation (FPC) [7], one of the interior point methods that can be applied to such problems. FPC iteratively alternatives between updating the gradient of the loss terms $\mathcal{L}_x(\cdot)$ and $\mathcal{L}_y(\cdot)$, and the singular value thresholding (SVT) [3], which is used for minimizing the nuclear norm. Cabral *et al.* [2] proved the constrained optimization (similar settings as in Eq. (2)) converges to the optimal solution.

**Identifying Disease Patterns.** As a result of completing the matrix in Eq. (1), the unknown entries in $\mathbf{X}_{pat}$ are determined. Since MC minimizes the rank of the matrix, it fills the entries in $\mathbf{X}_{pat}$ such that they have maximum correlations with their respective labels in $\mathbf{Y}_{pat}$ ($i.e.$, $(-1,-1),(-1,+1),(+1,-1),(+1,+1)$). The columns of $\mathbf{X}_{pat}$ can, hence, be considered as representative patterns for each class. Since we are interested in identifying how each of these classes are different from the control group, we simply compute the difference between the representative pattern of each class and the CTRL group, $i.e.$,

$$\mathbf{p}^{\text{AUD}} = |\mathbf{x}_{pat}^{\text{CTRL}} - \mathbf{x}_{pat}^{\text{AUD}}|, \; \mathbf{p}^{\text{HIV}} = |\mathbf{x}_{pat}^{\text{CTRL}} - \mathbf{x}_{pat}^{\text{HIV}}|, \text{ and } \mathbf{p}^{\text{AUD+HIV}} = |\mathbf{x}_{pat}^{\text{CTRL}} - \mathbf{x}_{pat}^{\text{AUD+HIV}}|,$$

where $\mathbf{p}^{\text{C}}$ denotes the identified pattern for class C, and $\mathbf{x}_{pat}^{\text{C}}$ is the corresponding column in $\mathbf{X}_{pat}$ for class C. We view small differences as noise and omit them from our findings by introducing the tolerance threshold $\epsilon$ and discarding the values $\mathbf{p}^{\text{C}} < \epsilon$ ($\epsilon = 10^{-3}$ in our experiments).

# 3   Experiments

We now compare the accuracy of the proposed and alternative methods with respect to the multi-label AUD/HIV dataset. Alternative implementations include running MC separately for each label, denoted by Single Label MC (SL-MC) [7], and running our proposed method (similar to [3]) without considering the error matrix $\mathbf{E}$ (see Eq. (1); denoted as MLT-EMC). The comparison also includes the widely used multi-class SVM (MC-SVM) [4] and multi-label SVM (ML-SVM) [4,10]. For fair comparison, we ran the SVMs by weighing samples (similar to the weights $\omega_{ij}$ we used in our formulation) in the corresponding cost function according to the size of the associated class.

## 3.1   Dataset and Preprocessing

As summarized in Table 1, the dataset consists of the morphometric measurements extracted from the magnetic resonance images (MRIs) of 244 healthy controls (CTRL), 227 AUD subjects without HIV infection, 70 HIV-infected individuals that do not meet the criteria for AUD (HIV), and 61 subjects with both AUD and HIV infection (HIV+AUD). For additional details about data collection and preprocessing, please refer to [13].

**Table 1.** Details of the multi-label dataset ('svol' = supratentorial volume).

|           | Total | Sex | | Age (years) | svol ($\times 10^6$) |
|-----------|-------|-----|-----|-------------|-----------------|
|           |       | F   | M   |             |                 |
| CTRL      | 244   | 122 | 123 | $45.59 \pm 17.17$ | $1.27 \pm 0.13$ |
| AUD       | 227   | 67  | 159 | $48.49 \pm 10.04$ | $1.27 \pm 0.11$ |
| HIV       | 70    | 20  | 45  | $51.81 \pm 8.44$  | $1.27 \pm 0.15$ |
| AUD+HIV   | 61    | 23  | 43  | $50.97 \pm 8.12$  | $1.23 \pm 0.14$ |

We apply the cross-sectional approach of FreeSurfer 5.3.0 software to the skull-stripped T1w MRI of each subject in order to measure the *mean curvature (Mean-Curv)*, *surface area (SurfArea)*, *gray matter volume (GrayVol)*, and *average thickness (ThickAvg)* of 34 bilateral cortical Regions Of Interest (ROIs), the volumes of 8 bilateral subcortical ROIs (*i.e.*, thalamus, caudate, putamen, pallidum, hippocampus, amygdala, accumbens, cerebellar cortex), the volumes of 5 subregions of the corpus callosum (posterior, mid-posterior, central, mid-central and ante-rior), the volume of all white matter hypointensities, the left and right lateral and third ventricles, and the supratentorial volume (svol). In addition to svol, each subject is thus represented by the z-scores of 298 morphometric features.

**Confounding Factors.** With respect to the CTRL group, *age*, *sex*, and *svol* significantly impact ($p$-value $< 0.001$) the morphometric measurements accord-ing to the paired t-test between each demographic factor and feature. To omit their influence from the analysis, we capture the relationship between each fea-ture and the confounding factors by parameterizing a generalized linear model (GLM) [11] on the CTRL cohort of each training run within the cross-validation. After parameterizing GLM, the model is applied to the measurements of each subject to compute the residual scores that are indifferent to the confounding factors. Note, the GLM model is only trained on the data from the training folds not to involve testing data in the preprocessing stage.

**Evaluation.** The classification accuracy of each method is measured via 10-fold nested cross-validation with the hyperparameters determined via 5-fold inner cross-validation. For MLT and SL-MC, the search space of the hyperparame-ters $\lambda_1$ and $\lambda_2$ is $\{0.001, 0.01, 0.1, 0.5, 0.9, 1\}$, and for MLT-EMC we do not have those hyperparameters as there is no error terms associated. The FPC hyperpa-rameters are set according to [2]. We also rely on the literature to set the search space for the hyperparameters of the alternative approaches, each based on their respective references, MC-SVM [4] and ML-SVM [10] (*e.g.*, the search space for the hyperparameter $C$ of SVM is $\{0.01, 0.1, 1, 10, 100\}$).

We summarize the outcome of each approach through several accuracy scores. Specifically, for each class C with $P^C$ positive and $N^C$ negative samples, we com-pute for each approach the precision (Pre) and recall (Rec) based on the true positive (TP), true negative (TN), false positive (FP) and false negative (FN) of the classifier: $\text{Pre}^C = \text{TP}^c/(\text{TP}^c + \text{FP}^c)$ and $\text{Rec}^C = \text{TP}^c/(\text{TP}^c + \text{FN}^c)$. We also report the area under the ROC curve (AUC) and balanced accuracy (BAc) score for all methods: $\text{BAc}^C = 1/2 \cdot (\text{TP}^c/P^c + \text{TN}^c/N^c)$. To evaluate the performance with respect to each separate class, we also use true positive rate (TPR) metric, which shows the portion of subjects identified correctly: $\text{TPR}^C = \text{TP}^c/P^c$.

## 3.2    Results and Comparison

Table 2 summarizes the accuracy scores with respect to the two defined labels (*i.e.*, AUD and HIV). BAc scores marked with a '†' are associated with a Fisher's exact test [6] that was significantly better than chance (*i.e.*, $p$-value $< 0.001$). Our proposed method

**Table 2.** Comparison for each single label. † sign indicates a $p$-value $< 0.001$ in a Fisher exact test.

| Method | AUD | | | | HIV | | | |
|---|---|---|---|---|---|---|---|---|
| | BAc | Pre | Rec | AUC | BAc | Pre | Rec | AUC |
| Ours (MLT) | **.68**† | **.67** | **.69** | **.69** | **.78**† | **.75** | .79 | **.80** |
| SL-MC [7] | .61 | .63 | .58 | .65 | .72† | .68 | .70 | .70 |
| MLT-EMC [3] | .52 | .53 | .51 | .55 | .59 | .60 | .56 | .62 |
| MC-SVM [4] | .63 | .61 | .63 | .60 | .72† | .68 | .74 | .70 |
| ML-SVM [10] | .66† | .64 | .65 | .66 | .75† | .69 | **.80** | .74 |

obtains better results in terms of both balanced accuracy and AUC compared to all other methods. In comparison with the second best method (ML-SVM), our BAc scores are by at least 2% and our AUC scores by at least 3% better in both experiments. Also, our method obtains a better balance between precision and recall, especially for the highly imbalanced case of HIV. The results confirm our intuition that the multi-label setting is better in modeling the problem compared to the multi-class methods, as both multi-label methods involved in the comparison (MLT and ML-SVM) lead to the best results. Both SL-MC and MC-SVM perform inferior to them in terms of both balanced accuracy and AUC. Furthermore, MLT-EMC, *i.e.*, the method without the noise term, obtains the worst results.

Table 3 summarizes the TPR with respect to all four classes (CTRL, AUD, HIV, and AUD+HIV) and the overall mean across the four classes. In a four-class classification problem, a TPR higher than 0.25 is considered better than chance. The scores agree with the

**Table 3.** Class specific TPR and overall mean.

| Method | CTRL | AUD | HIV | AUD + HIV | Mean |
|---|---|---|---|---|---|
| Ours (MLT) | **.57** | **.54** | **.57** | **.59** | **.57** |
| SL-MC [7] | .38 | .35 | .32 | .33 | .35 |
| MLT-EMC [3] | .37 | .32 | .33 | .31 | .33 |
| MC-SVM [4] | .42 | .44 | .40 | .44 | .43 |
| ML-SVM [10] | .55 | .51 | .52 | .56 | .54 |

findings of Table 2 in that the multi-label methods (*i.e.*, MLT and ML-SVM) lead to the best models. Our proposed method outperforms ML-SVM by 3% with respect to the overall mean TPR.

Figure 2 shows the identified patterns by our proposed method. These patterns are composed of approximately 20% of the 298 features for AUD, 16% for HIV and 40% for AUD+HIV, with approximately 5% of the features shared among all these cohorts. These identified patterns suggest that AUD and HIV infection are associated with deficits in cortical and subcortical regions, which agrees with the HIV and Alcohol literature [5, 8, 9,

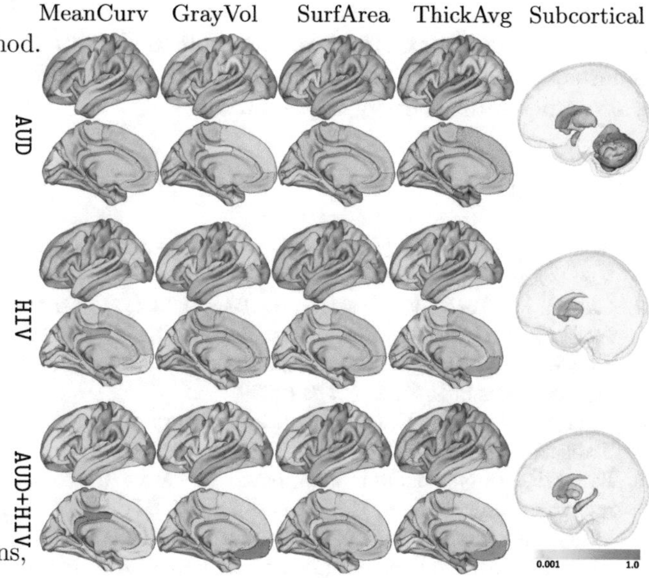

**Fig. 2.** Identified patterns for AUD, HIV, and AUD+HIV.

13]. Furthermore, several previous studies have reported cortical thickness and gray matter volume as important markers for AUD and HIV infection [9,13], which agrees with our findings. Subcortical regions, including hippocampus, thalamus and basal ganglia structures (*i.e.*, Caudate, Putamen, Palladium), are reported in the literature to be affected by HIV and AUD [5], which are also found important by our method. Specifically, the primary motor cortex region and the basal ganglia subcortical structures are shown to be more severely affected as a result of comorbidity of AUD and HIV, compared to each single one of them.

**Discussions.** The identified patterns of `AUD+HIV` record the largest number of relevant regions, which can document a compounding effect of `AUD` and `HIV`. Another interesting finding of our results is that our method avoids underestimating the impact of the disease to a small number of brain regions as commonly done by sparse classifiers. In addition, since our method spans the data matrix to a low-rank subspace, it can implicitly alleviate the redundancy and correlation among the features that are (linearly) correlated.

## 4    Conclusion

In this paper, we introduced a multi-label transductive classifier that not only classifies the data into multiple labels but also can identify the patterns specific to each disease (labels) and their comorbidity. We experimented on a large set of data with control samples and subjects with alcohol use disorder, HIV infection or both. Our model led to the superior accuracy scores in comparison to state-of-the-art methods. Our method was also able to identify disease-related (and their comorbidity) patterns, which revealed that the comorbidity was characterized by a compounding effect of the two disorders.

## References

1. Adeli, E., Fathy, M.: Non-negative matrix completion for action detection. Image Vis. Comput. **39**, 38–51 (2015)
2. Cabral, R., Torre, F., Costeira, J., Bernardino, A.: Matrix completion for weakly-supervised multi-label image classification. IEEE TPAMI **37**(1), 121–135 (2015)
3. Candès, E.J., Recht, B.: Exact matrix completion via convex optimization. Found. Comput. Math. **9**(6), 717–772 (2009)
4. Chang, C.C., Lin, C.J.: LIBSVM: a library for support vector machines. ACM Trans. IST **2**(27), 1–27 (2011)
5. Fama, R.: Impairments in component processes of executive function and episodic memory in alcoholism, HIV infection, and HIV infection with alcoholism comorbidity. Alcohol.: Clin. Exp. Res. **40**(12), 2656–2666 (2016)
6. Fisher, R.A.: The logic of inductive inference. J. R. Stat. Soc. **98**(1), 39–82 (1935)
7. Goldberg, A., Recht, B., Xu, J., Nowak, R., Zhu, X.: Transduction with matrix completion: three birds with one stone. In: NIPS, pp. 757–765 (2010)
8. Gongvatana, A.: A history of alcohol dependence augments HIV-associated neurocognitive deficits in persons aged 60 and older. Neurovirology **20**(5), 505–513 (2014)

9. Justice, A., Sullivan, L., Fiellin, D.: HIV/AIDS, comorbidity, and alcohol: can we make a difference? Alcohol Res. Health **33**(3), 258 (2010)
10. Li, X., Guo, Y.: Active learning with multi-label SVM classification. In: IJCAI (2013)
11. Madsen, H., Thyregod, P.: Introduction to General and Generalized Linear Models. CRC Press, Boca Raton (2010)
12. Parekh, A.K., Barton, M.B.: The challenge of multiple comorbidity for the us health care system. JAMA **303**(13), 1303–1304 (2010)
13. Pfefferbaum, A.: Accelerated aging of selective brain structures in human immunodeficiency virus infection. Neurobiol. Aging **35**(7), 1755–1768 (2014)
14. Tsoumakas, G., Katakis, I.: Multi-label classification: an overview. Int. J. Data Warehous. Min. **3**(3), 1–13 (2006)
15. Wang, X., Wang, F., Hu, J.: A multi-task learning framework for joint disease risk prediction and comorbidity discovery. In: ICPR, pp. 220–225. IEEE (2014)
16. Wosiak, A., Glinka, K., Zakrzewska, D.: Multi-label classification methods for improving comorbidities identification. Comput. Biol. Med. **100**, 279–288 (2017)

# Text to Brain: Predicting the Spatial Distribution of Neuroimaging Observations from Text Reports

Jérôme Dockès[1]([✉]), Demian Wassermann[1], Russell Poldrack[2],
Fabian Suchanek[3], Bertrand Thirion[1], and Gaël Varoquaux[1]

[1] INRIA, CEA, Université Paris-Saclay, Paris, France
jerome.dockes@inria.fr
[2] Stanford University, Stanford, USA
[3] Télécom ParisTech, Paris, France

**Abstract.** Despite the digital nature of magnetic resonance imaging, the resulting observations are most frequently reported and stored in text documents. There is a trove of information untapped in medical health records, case reports, and medical publications. In this paper, we propose to mine brain medical publications to learn the spatial distribution associated with anatomical terms. The problem is formulated in terms of minimization of a risk on distributions which leads to a least-deviation cost function. An efficient algorithm in the dual then learns the mapping from documents to brain structures. Empirical results using coordinates extracted from the brain-imaging literature show that (i) models must adapt to semantic variation in the terms used to describe a given anatomical structure, (ii) voxel-wise parameterization leads to higher likelihood of locations reported in unseen documents, (iii) least-deviation cost outperforms least-square. As a proof of concept for our method, we use our model of spatial distributions to predict the distribution of specific neurological conditions from text-only reports.

## 1 Introduction

Hundreds of thousands of studies, case reports, or patient records, capture observations in human neuroscience, basic or clinical. Statistical analysis of this large amount of data could provide new insights. Unfortunately, most of the spatial information that these data contain is difficult to extract *automatically*, because it is hidden in unstructured text, in sentences such as: "[...] in the anterolateral temporal cortex, especially the temporal pole and inferior and middle temporal gyri" [1].

This data cannot be processed easily by a machine, as a machine does not know where the temporal cortex is. As we will show, simply looking up such terms

**Electronic supplementary material** The online version of this chapter (https://doi.org/10.1007/978-3-030-00931-1_67) contains supplementary material, which is available to authorized users.

© Springer Nature Switzerland AG 2018
A. F. Frangi et al. (Eds.): MICCAI 2018, LNCS 11072, pp. 584–592, 2018.
https://doi.org/10.1007/978-3-030-00931-1_67

in atlases does not suffice. Indeed, even atlases disagree [2]. Furthermore, joint processing of many reports faces varying terminologies, with regions represented in different atlases that differ and overlap. Finally, not all terms in a report carry the same importance, and practitioners use terms that are not the exact labels of any atlas. Coordinate-based meta-analyses capture the spatial distribution of a term from the literature [3,4], but they also lack a model to combine terms.

Here, we propose to map case reports automatically to the brain locations that they discuss: we learn mappings of anatomical terms to brain regions from medical publications. We propose a new learning framework for translating anatomical terms to brain images – a process that we call "encoding". We learn such a mapping, quantify its performance, and compare possible choices of representation of spatial data. We then show in a proof of concept that our model can predict the brain area for textual case reports.

## 2  Methods: Formalizing Text-to-Brain-Map Translation

### 2.1  Problem Setting: From Text to Spatial Distributions

We want to predict the likelihood of the location of relevant brain structures described in a document. For this purpose, we perform supervised learning on a corpus of brain-imaging studies, each containing: (i) a text, and (ii) the locations – $i.e.$ the stereotactic coordinates – of its observations. Indeed, Functional Magnetic Resonance Imaging (fMRI) studies report the coordinates of activation peaks ($e.g.$, [5, Table 1]), and Voxel Based Morphometry (VBM) analyses report the location of differences in gray matter density ($e.g.$, [1, Table 2]). Following neuroimaging meta-analyses [3], we frame the problem in terms of spatial distributions of observations in the brain. In a document, observed locations $\mathcal{L} = \{l_a \in \mathbb{R}^3, a = 1 \dots c\}$ are sampled from a probability density function (pdf) $p$ over the brain. **Our goal is to predict this pdf $p$ from the text $\mathcal{T}$.** We denote $q$ our predicted pdf. A predicted pdf $q$ should be close to $p$, or take high values at the coordinates actually reported in the study: $\prod_{l \in \mathcal{L}} q(l)$ must be large. In a supervised learning setting, we start from a collection of studies $\mathcal{S} = (\mathcal{T}, \mathcal{L})$, with $\mathcal{T}$ the text and $\mathcal{L}$ the locations. Building the prediction engine then entails the choice of a model relating the predicted pdf $p$ to the text $\mathcal{T}$, the choice of a loss, or data-fit term, and some regularization on the model parameters. We now detail how we make each of these choices to construct a prediction.

**Model.** We start by modelling the dependency of our spatial pdf $q$ on the study text $\mathcal{T}$. This entails both choosing a representation for $q$ and writing it as a function of the text. While $q$ is defined on a subvolume of $\mathbb{R}^3$, the brain volume, we build it using a partition to work on a finite probably space: this can be either a regular grid of voxels or a set of anatomical regions (i.e. an atlas) $\mathcal{R} = \{\mathcal{R}_k, k = 0 \dots m\}$. As such a partitioning imposes on each region to be homogeneous, $q$ is then formally written on $\mathbb{R}^3$ in terms of the indicator functions of the parts[1]: $\{r_k = \frac{\mathbb{I}_k}{\|\mathbb{I}_k\|_1}, \; k = 1 \dots m\}$. Importantly, the volume of each part $\|\mathbb{I}_k\|_1$ appears as a normalization constant.

To link $q$ to the text $\mathcal{T}$ of the study, we start by building a term-frequency vector representation of $\mathcal{T}$, which we denote $\boldsymbol{x} \in \mathbb{R}^d$. $d$ is the size of our vocabulary of English words $\mathcal{W} = \{w_t\}$, and $\boldsymbol{x}_t$ is the frequency of word $w_t$ in the text. We assign to each atlas region a weight that depends linearly on $\boldsymbol{x}$:

$$q(z) = \sum_{t=1}^{d} \sum_{k=1}^{m} \boldsymbol{x}_t \boldsymbol{\beta}_{t,k} r_k(z) \quad \forall z \in \mathbb{R}^3 \tag{1}$$

where $\boldsymbol{\beta} \in \mathbb{R}^{d \times m}$ are model parameters, which we will learn.

Using an atlas is a form of regularization: constraining the prediction to be in the span of $\{r_k\}$ reduces the size of the search space. Fine partitions, *e.g.* atlases with many regions or voxel grids, yield models with more expressive power, but more likely to overfit. Choosing an atlas thus amounts to a bias-variance tradeoff.

**Label-Constrained Encoder.** A simple heuristic to turn a text into a brain map is to use atlas labels and ignore interactions between terms. The probability of a region is taken to be proportional to the frequency of its label in the text. The vocabulary is then the set of labels: $d = m$. As the word $w_k$ is the label of $\mathcal{R}_k$, $\boldsymbol{\beta}$ is diagonal. For example, for a region $\mathcal{R}_k$ in the atlas labelled "parietal lobe", the probability on $\mathcal{R}_k$ depends only on the frequency of the phrase "parietal lobe" in the text. We call this model *label-constrained encoder*.

## 2.2 Loss Function: Measuring Errors on Spatial Distributions

**Strategy.** We will fit the coefficients $\boldsymbol{\beta}$ of our model, see Eq. (1), by minimizing a risk $\mathcal{E}(p, q)$: the expectation of a distance between $p$ and $q$.

**A Plugin Estimator of $p$.** We do not have access to the true pdf, $p$; we need a plugin estimator, which we denote $\hat{p}$. By construction of our prediction $q$, the best approximation of $p$ we can hope for belongs to the span of our regions $\{r_k\}$. Hence, we build our estimator $\hat{p}$ in this space, setting the probability of a region to be proportional to the number of coordinates that fell inside it:

$$\hat{p} = \sum_{k=1}^{m} \frac{|\{a, \mathbb{I}_k(l_a) = 1\}|}{c} r_k = \sum_{k=1}^{m} \frac{1}{c} \sum_{a=1}^{c} \mathbb{I}_k(l_a) r_k \triangleq \sum_{k=1}^{m} \hat{\boldsymbol{y}}_k r_k. \tag{2}$$

When regions are voxels, there are too many regions and too few coordinates. Hence we use Gaussian Kernel Density Estimation (KDE) to smooth the estimated pdf[2]. Our supplementary material details a fast KDE implementation.

---

[1] $\mathcal{R}_0$ denotes the volume outside of the brain, or background, on which $q$ is 0.

[2] Using an atlas is also a form of KDE, with kernel $(z, z') \mapsto 1/\|\mathbb{I}_k\|_1$ if $z$ and $z'$ belong to the same region $\mathcal{R}_k, k \in \{1, \dots m\}$, 0 otherwise.

**Choice of $\mathcal{E}$.** We use two common distance functions for our loss. The first is Total Variation (TV), a common distance for distributions. Note that $p$ defines a probability measure on the finite sample space $\mathcal{R}$, $\mathcal{P}(\mathcal{R}_k) = \int_{\mathcal{R}_k} p(z)dz$, where $\mathcal{R} = \{\mathcal{R}_k, k = 1 \ldots m\}$ and $\mathcal{R}_k = \mathrm{supp}(r_k)$. $q$ defines $\mathcal{Q}$ in the same way. Then,

$$\mathrm{TV}(\mathcal{P}, \mathcal{Q}) = \sup_{\mathcal{A} \subset \mathcal{R}} |\mathcal{P}(\mathcal{A}) - \mathcal{Q}(\mathcal{A})|. \tag{3}$$

Since $\mathcal{R}$ is finite, a classical result (see [6]) shows that this supremum is attained by taking $\mathcal{A} = \{\mathcal{R}_k | \mathcal{P}(\mathcal{R}_k) > \mathcal{Q}(\mathcal{R}_k)\}$ (or its complementary) and:

$$\mathrm{TV}(\mathcal{P}, \mathcal{Q}) = \frac{1}{2} \sum_{k=1}^{m} |\mathcal{P}(\mathcal{R}_k) - \mathcal{Q}(\mathcal{R}_k)| = \frac{1}{2} \int_{\mathbb{R}^3} |p(z) - q(z)|dz. \tag{4}$$

The TV is half of the $\ell_1$ distance between the pdfs. $\|\hat{p} - q\|_1$ is therefore a natural choice for our loss. The second choice is $\|\hat{p} - q\|_2^2$, which is a popular distance and has the appeal of being differentiable everywhere.

**Factorizing the Loss.** Let us call $v_k$ the volume of $r_k$, i.e. the size of its support: $v_k \triangleq \|\mathbb{I}_k\|_1$, $k = 1 \ldots m$. Remember that $r_k = \frac{1}{v_k}\mathbb{I}_k$. Our loss can now be factorized (see supplementary material for details):

$$\int_{\mathbb{R}^3} \delta(\hat{p}(z) - q(z))dz = \sum_{k=1}^{m} v_k \delta \left( \frac{\hat{y}_k}{v_k} - \frac{\sum_{t=1}^{d} x_t \beta_{t,k}}{v_k} \right) \tag{5}$$

Here, $\delta$ is either the absolute value of the difference or the squared difference.

### 2.3 Training the Model: Efficient Minimization Approaches

To set the model parameters $\beta$, we used $n$ example studies $\{S_i = (\mathcal{T}_i, \mathcal{L}_i), i = 1 \ldots n\}$. We learn $\beta$ by minimizing the empirical risk on $\{S_i\}$ and an $\ell_2$ penalty on $\beta$. We add to the previous notations the index $i$ of each example: $p_i, q_i, \hat{y}_i, x_i$. $\hat{Y} \in \mathbb{R}^{n \times m}$ is the matrix such that $\hat{Y}_i = \hat{y}_i$, and $X \in \mathbb{R}^{n \times d}$ such that $X_i = x_i$.

**Case $\delta = \ell_2^2$.** The empirical risk is

$$\sum_{i=1}^{n} \sum_{k=1}^{m} \left( \frac{\hat{Y}_{i,k}}{\sqrt{v_k}} - \sum_{t=1}^{d} \frac{1}{\sqrt{v_k}} X_{i,t} \beta_{t,k} \right)^2. \tag{6}$$

Defining $Y'_{:,k} = \frac{\hat{Y}_{:,k}}{\sqrt{(v_k)}}$ and $\beta'_{:,k} = \frac{\beta_{:,k}}{\sqrt{(v_k)}}$, with an $\ell_2$ penalty, the problem is:

$$\operatorname*{argmin}_{\beta'} \left( \|Y' - \beta' X\|_2^2 + \lambda \|\beta'\|_2^2 \right) \tag{7}$$

where $\lambda \in \mathbb{R}_+$. This is the least-squares ridge regression predicting $\hat{p}$ expressed in the orthonormal basis of our search space $\{\frac{r_k}{\|r_k\|_2}\}$.

**Case** $\delta = \ell_1$**.** The empirical risk becomes

$$\sum_{i=1}^{n}\sum_{k=1}^{m}|\hat{Y}_{i,k} - \sum_{t=1}^{d}X_{i,t}\beta_{t,k}| \tag{8}$$

This problem is also known as a least-deviations regression, a particular case of quantile regression [7], [8]. Unlike $\ell_2$ regression, which provides an estimate of the conditional mean of the target variable, $\ell_1$ provides an estimate of the median. Quantile regression has been studied (e.g. by economists), as it is more robust to outliers and better-suited than least-squares when the noise is heteroscedastic [7]. Adding an $\ell_2$ penalty, we have the minimization problem:

$$\hat{\beta} = \underset{\beta}{\operatorname{argmin}}\left(\|\hat{Y} - X\beta\|_1 + \lambda\|\beta\|_2^2\right) \tag{9}$$

Unpenalized quantile regression is often written as a linear program and solved with the simplex algorithm [9], iteratively reweighted least squares, or interior point methods [10]. [11] uses a coordinate-descent to solve a differentiable approximation of the quantile loss (the Huber loss) with elastic-net penalty. Here, we minimize Eq. (9) via its dual formulation (c.f. supplementary material):

$$\hat{\nu} = \underset{\nu}{\operatorname{argmax}}\left(Tr(\nu^T\hat{Y} - \frac{1}{4\lambda}\nu^T XX^T\nu)\right) \quad \text{s.t. } \|\nu\|_\infty \leq 1, \tag{10}$$

where $\nu \in \mathbb{R}^{n \times m}$. The primal solution is given by $\hat{\beta} = \frac{X^T\hat{\nu}}{2\lambda}$. As the dual loss $g$ is differentiable and the constraints are *bound* constraints, we can use an efficient quasi-Newton method (L-BFGS, [12]). $g$ and its gradient are fast to compute as $X$ is sparse. $\lambda$ is set by cross-validation on the training set. We use warm-start on the regularization path (decreasing values for $\lambda$) to initialize each problem.

**Training the Label-Constrained Encoder.** The columns of $\beta$ can be fitted independently from each other. If we want $\beta$ to be diagonal, we only include one feature in each regression: we fit $m$ univariate regressions $\hat{y}_{:,k} \simeq X_{:,k}\beta_{k,k}$.

## 2.4    Evaluation: A Natural Model-Comparison Metric

Our metric is the mean log-likelihood of an article's coordinates in the predicted distribution, which diverges wherever $q = 0$. we add a uniform background to the prediction, to ensure that it is non-zero everywhere:

$$\text{the predicted pdf is written} \quad q' = \frac{1}{2}(\sum_{k=1}^{m}\frac{\mathbb{I}_k}{v_k} + q) \tag{11}$$

$$\text{the score for a study } \mathcal{S}_i = (\mathcal{T}_i, \mathcal{L}_i), \mathcal{L}_i = \{l_{i,a}\} \text{ is} \quad \frac{1}{c_i}\sum_{a=1}^{c_i}\log(q_i'(l_{i,a})) \tag{12}$$

# 3   Empirical Study

## 3.1   Data: Mining Neuroimaging Publications

We downloaded roughly 140K neuroimaging articles from online sources including Pubmed Central and commercial publishers. About 14K of these contain coordinates, which we extracted, as in [4]. We built a vocabulary of around 1000 anatomical region names by grouping the labels of several atlases and the Wikipedia page "List of regions in the human brain"[3]. So in practice, $n \approx 14 \cdot 10^3$ and $d \approx 1000$. $m$ depends on the atlas (or voxel grid) and ranges from 20 to 30K.

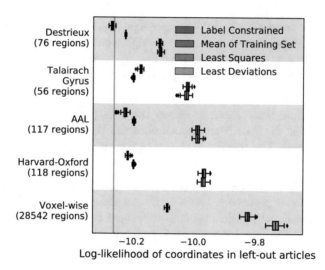

**Fig. 1. Log-Likelihood of coordinates reported by left-out articles in the predicted distribution (Eq. (12)).** The vertical line represents the test log-likelihood given a uniform distribution over the brain. Voxel-wise encoding is better than relying on any atlas. In this setting, $\ell_1$ regression significantly outperforms least squares.

## 3.2   Text-to-Brain Encoding Performance

**Comparison of Atlases and Models.** We perform 100 folds of shuffle-split cross-validation (10% in test set). As choices of $\{\mathcal{R}_k\}$, we compare several atlases and a grid of cubic 4-mm voxels. We also compare $\ell_1$ and $\ell_2$ regression, and label-constrained $\ell_2$. The label-constrained encoder is not used for the voxel grid, as it does not have labels. As a baseline, we include a prediction based on the average of the brain maps seen during training (i.e. independent of the text).

Figure 1 gives the results: for all models, voxel-wise encoding performs better than any atlas. Large atlas regions regularize too much. Despite its higher dimensionality, voxel-wise encoding learns better representations of anatomical

---

[3] https://en.wikipedia.org/wiki/List_of_regions_in_the_human_brain.

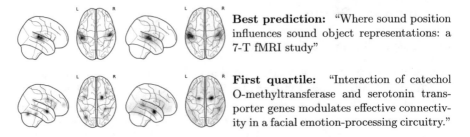

**Best prediction:** "Where sound position influences sound object representations: a 7-T fMRI study"

**First quartile:** "Interaction of catechol O-methyltransferase and serotonin transporter genes modulates effective connectivity in a facial emotion-processing circuitry."

**Fig. 2.** True map (left) and prediction (right) for best prediction and $1^{st}$ quartile

terms. The label-constrained model performs poorly, sometimes below chance, as the labels of a single atlas do not cover enough words and interactions between terms are important. For voxel-wise encoding, $\ell_1$ regression outperforms $\ell_2$. The best encoder is therefore learned using a $\ell_1$ loss and a voxel partition.

**Prediction Examples.** Figure 2 shows the true pdf (estimated with KDE) and the prediction for the articles which obtained respectively the best and the first-quartile scores. The median is shown in the supplementary material.

**Examples of Coefficients Learned by the Linear Regression.** The coefficients of the linear regression (rows of $\boldsymbol{\beta}$) are the brain maps that the model associates with each anatomical term. For frequent terms, they are close to what experts would expect (see for example Figs. 3 and 4).

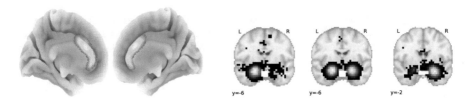

**Fig. 3.** Regression coefficient for "anterior cingulate"

**Fig. 4.** Regression coefficients for "left amygdala", "amygdala", and "right amygdala"

### 3.3   Leveraging Text Without Coordinates: Neurological Examples

Our framework can leverage unstructured spatial information contained in a large corpus of unannotated text. To showcase this, assume that we want to know which parts of the brain are associated with Huntington's disease. Our labelled corpus by itself is insufficient: only 21 documents mention the term "huntington". But we use it to learn associations between anatomical terms and locations in the brain (Sect. 2). This gives us access to the spatial information

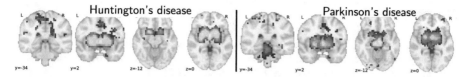

**Fig. 5. Predicted density for Huntington's and Parkinson's.** In agreement with Huntington's physiopathology [13], our method highlights the putamen, and the caudate nucleus. Also, in the case of Parkinson's [14], the brain stem, the thalamus, and the motor cortex are highlighted.

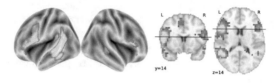

**Fig. 6. Predicted density for aphasia,** centered on Broca's and Wernicke's areas, in agreement with the literature [15].

contained in the unlabelled corpus, which was out of reach before (Sect. 3.2). We contrast the mean encoding of articles which mention "huntington" against the mean distribution (taking the difference of their *log*). Since the large corpus contains more information about Huntington's disease (over 400 articles mention it), this is sufficient to see the striatum highlighted in the resulting map (Fig. 5, left). Figure 5 (right) shows the experiment for Parkinson, and Fig. 6 for Aphasia.

## 4    Conclusion

We have introduced a theoretical framework to translate textual description of studies into spatial distributions over the brain. Such a translation enables pooling together many studies which only provide text (no images or coordinates), for statistical analysis of their results in brain space. The statistical model gives a natural metric to validate. This metric enables comparing representations, showing that voxel-wise encoding is a better approach than relying on atlases. Building prediction models tailored to our task leads to a linear regression with an $\ell_1$ loss (least absolute deviation), the total-variation distance between the true and the predicted spatial distributions. Such a model can be trained efficiently on dozens of thousands of data points and outperforms simpler approaches.

Applied to descriptions of pathologies that lack spatial information, our model synthesizes accurate brain maps that reflect the domain knowledge. Predicting spatial distributions of medical observations from text opens new alleys for clinical research from patient health records and case reports.

**Acknowledgements.** This project received funding from: the European Union's H2020 Research Programme under Grant Agreement No. 785907 (HBP SGA2), the Metacog Digiteo project, the MetaMRI associate team, and ERC NeuroLang.

# References

1. Mummery, C.J., Patterson, K., Price, C.J., et al.: A voxel-based morphometry study of semantic dementia: relationship between temporal lobe atrophy and semantic memory. Ann. Neurol. **47**(1), 36–45 (2000)
2. Bohland, J., Bokil, H., Allen, C., Mitra, P.: The brain atlas concordance problem: quantitative comparison of anatomical parcellations. PloS one **4**(9), e7200 (2009)
3. Laird, A.R., Fox, P.M., Price, C.J., et al.: ALE meta-analysis: controlling the false discovery rate and performing statistical contrasts. Hum. Brain Mapp. **25**, 155 (2005)
4. Yarkoni, T., Poldrack, R.A., Nichols, T.E., et al.: Large-scale automated synthesis of human functional neuroimaging data. Nat. Methods **8**, 665 (2011)
5. Van der Zwaag, W., Gentile, G., et al.: Where sound position influences sound object representations: a 7-T fMRI study. Neuroimage **54**(3), 1803–1811 (2011)
6. Gibbs, A.L., Su, F.E.: On choosing and bounding probability metrics. Int. Stat. Rev. **70**(3), 419–435 (2002)
7. Koenker, R., Bassett Jr., G.: Regression quantiles. Econom.: J. Econom. Soc. **46**, 33–50 (1978)
8. Chen, C., Wei, Y.: Computational issues for quantile regression. Sankhyā: Indian J. Stat. **67**, 399–417 (2005)
9. Koenker, R., d'Orey, V.: Remark AS R92: a remark on algorithm as 229: computing dual regression quantiles and regression rank scores. J. R. Stat. Soc. Ser. C **43**(2), 410–414 (1994)
10. Portnoy, S., Koenker, R., et al.: The gaussian hare and the laplacian tortoise: computability of squared-error versus absolute-error estimators. Stat. Sci. **12**(4), 279–300 (1997)
11. Yi, C., Huang, J.: Semismooth newton coordinate descent algorithm for elastic-net penalized huber loss regression and quantile regression. J. Comput. Graph. Stat. **26**, 547 (2017)
12. Byrd, R.H., Lu, P., Nocedal, J., Zhu, C.: A limited memory algorithm for bound constrained optimization. SIAM J. Sci. Comput. **16**(5), 1190–1208 (1995)
13. Walker, F.O.: Huntington's disease. Lancet **369**(9557), 218–228 (2007)
14. Davie, C.A.: A review of parkinson's disease. Br. Med. Bull. **86**, 109 (2008)
15. Damasio, A.R.: Aphasia. N. Engl. J. Med. **326**(8), 531–539 (1992)
16. Surguladze, S., et al.: Interaction of catechol O-methyltransferase and serotonin transporter genes modulates effective connectivity in a facial emotion-processing circuitry. Transl. Psychiatry **2**(1), e70 (2012)
17. Van Dam, W.O., Rueschemeyer, S.A., Bekkering, H.: How specifically are action verbs represented in the neural motor system: an fMRI study. Neuroimage **53**(4), 1318–1325 (2010)
18. Silverman, B.W.: Density Estimation for Statistics and Data Analysis, vol. 26. CRC Press, Boca Raton (1986)
19. Simonoff, J.S.: Smoothing Methods in Statistics. Springer Science & Business Media, New York (2012). https://doi.org/10.1007/978-1-4612-4026-6
20. Wand, M.: Fast computation of multivariate kernel estimators. J. Comput. Graph. Stat. **3**(4), 433–445 (1994)
21. Gramacki, A., Gramacki, J.: FFT-based fast computation of multivariate kernel density estimators with unconstrained bandwidth matrices. J. Comput. Graph. Stat. **26**, 459–462 (2016)

# Neuroimaging and Brain Segmentation Methods: Brain Segmentation Methods

# Semi-supervised Learning for Segmentation Under Semantic Constraint

Pierre-Antoine Ganaye, Michaël Sdika[(✉)], and Hugues Benoit-Cattin

Univ Lyon, INSA-Lyon, Université Claude Bernard Lyon 1, UJM-Saint Etienne,
CNRS, Inserm CREATIS UMR 5220, U1206, 69100 Lyon, France

**Abstract.** Image segmentation based on convolutional neural networks is proving to be a powerful and efficient solution for medical applications. However, the lack of annotated data, presence of artifacts and variability in appearance can still result in inconsistencies during the inference. We choose to take advantage of the invariant nature of anatomical structures, by enforcing a semantic constraint to improve the robustness of the segmentation. The proposed solution is applied on a brain structures segmentation task, where the output of the network is constrained to satisfy a known adjacency graph of the brain regions. This criteria is introduced during the training through an original penalization loss named NonAdjLoss. With the help of a new metric, we show that the proposed approach significantly reduces abnormalities produced during the segmentation. Additionally, we demonstrate that our framework can be used in a semi-supervised way, opening a path to better generalization to unseen data.

**Keywords:** Medical image segmentation
Convolutional neural network · Semi-supervised learning
Adjacency graph · Constraint

## 1 Introduction

In medical imaging, semantic segmentation is of major importance, it helps to quantify the volume and positions of anatomical structures [1,2] and lesions [3]. In the case of brain segmentation, it enables to track the volume of structures over time, providing valuable evidences to hypothesize over a possible malfunction. Multi-atlas methods [4,5] are established solutions for this problem, they are based on the registration and fusion of image atlases, which results in consistent segmentations that preserve the topology of the structures, taking into account the inter-structures relationships. In CNN based approach [1], Moeskops *et. al.* proposed a patch based segmentation architecture which leverages contextual information around the center pixel. Encoder-decoder model [6] has also been used for the same task, by first pre-training on a different dataset annotated with Freesurfer and then fine-tuning with a novel error corrective loss.

© Springer Nature Switzerland AG 2018
A. F. Frangi et al. (Eds.): MICCAI 2018, LNCS 11072, pp. 595–602, 2018.
https://doi.org/10.1007/978-3-030-00931-1_68

These pipelines are optimized to maximize the similarity between the segmentation and the ground truth, based on cost functions like the cross entropy and the dice similarity [6]. The precision and robustness of these methods are bounded by the quality and quantity of the data. At best it should represent the variability in appearance and the segmentations should bring consensus among examiners. However the lack of annotations is a common factor in medical imaging, due to the complexity of the task. Solutions have been explored to harness properties such as anatomical invariance and semantic knowledge, allowing to improve the modeling capacities of CNNs by constraining the loss. It can take the form of an implicit knowledge (contextual information, spatial position) integrated as an input of the model or a soft penalty (volume, shape) specified by an expert.

Previous works on learning under constraints [7–9] demonstrate the interest in such method, Stewart *et al.* [8] suggested to use physics laws as a domain prior and proves its applicability to object tracking. In the medical community, [10] trained a by-patch segmentation model by integrating a spacial representation of the patch, implying that the position is correlated to the label of interest, enforcing an automatically learned spatial constraint. In [9], Oktay *et al.* went further by learning a representation of the label space with an auto-encoder, extracting shape and location priors of the structures. The final model is constrained to minimize the label representation of the segmentation and the ground truth.

In this paper, we investigate how segmentation abnormalities can be reduced by introducing knowledge about the connectivity of anatomical structures. We apply the proposed method to brain structures segmentation on MR T1w images. First, an encoder-decoder model inspired from [6] is trained on a dataset. Second, a labels adjacency prior is extracted from the training set, with the objective of matching the network's output with it. A novel loss function is applied on the trained network, in a simple fine-tuning step. Finally, we take advantage of the semi-supervised nature of this constraint by applying it on an external dataset. Doing so provides better generalization, without compromising the quality.

In Sect. 2 we introduce the segmentation architecture, together with the adjacency constraint term. In Sect. 3 we present the various experiments realized to demonstrate the interest in such method. In Sect. 4 the results of the experiments are commented.

## 2   Methods

### 2.1   Encoder-Decoder Architecture

The architecture of our 2D network (Fig. 1) is directly inspired by [6], with some minor changes. This network is composed of an encoding path, followed by a decoding path where the features are upsampled with max-unpooling. During the decoding, features from the encoder are reused via skip-connections and concatenated with the upsampled path, at each resolution level. The input of the network is composed of 7 adjacent slices, while segmenting only the central slice. The other difference is that we used convolution kernels of size $3 \times 3$, proving to be more efficient in terms of parameters and with equal performance.

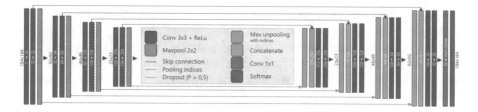

**Fig. 1.** Encoder-decoder architecture

## 2.2 Adjacency Graph of the Anatomical Structures

In this work, structural invariance of the segmentation map is assumed, the connectivity between regions should be the same from one subject to another. The adjacency graph of a segmentation map is defined as a graph where each label of the map is a vertex and where edges are weighted by the number of voxels joining two regions. Formally, the weight of the edge connecting the structures $i$ and $j$ for a given subject is defined by:

$$\mathbf{A}_{ij} = \sum_x \sum_{v \in V} \delta_{i,s(x)} \delta_{j,s(x-v)}, \tag{1}$$

where $x$ is a voxel, $s(x)$ the mapping to the corresponding label, $\delta$ the Kronecker delta and $V$ defines a neighborhood which does not include 0.

The matrix $\mathbf{A}$ ultimately describes how many contours are shared between pairs of structures in the 3D volume. Although, this matrix can vary from one subject to another, it is assumed in this work that its binary version $\tilde{\mathbf{A}} = (\mathbf{A} > 0)$ does not change. One can consequently define the set of forbidden transitions between structures: $F = \{(i,j)| \ \tilde{\mathbf{A}}_{ij} = 0\}$.

## 2.3 Loss Functions

*Constraint Training.* Knowing which regions interact together allows to determine which adjacencies should be considered as abnormal and finally not present in the output of an automatic segmentation system. We propose in this work to train the network $\phi$ such that its weights $\mathbf{w}$ minimize inconsistencies in the output segmentation by solving the constrained optimization problem:

$$\min_{G(\mathbf{w})=0} \frac{1}{|D_S|} \sum_{(\mathbf{x},y) \in D_S} L(\phi(\mathbf{x}, \mathbf{w}), y)) \tag{2}$$

where $L$ is a segmentation loss (Dice or cross-entropy for example) and

$$G(\mathbf{w}) = \sum_{\mathbf{x} \in D_{NA}} \sum_{(i,j) \in F} a_{ij}(\phi(\mathbf{x}, \mathbf{w})). \tag{3}$$

The $a_{ij}$ function measures the adjacency between labels $i$ and $j$ from the network probability output and will be discussed below. $D_S$ and $D_{NA}$ are respectively the training dataset for the segmentation and the Non-Adjacency loss.

*Computing the Adjacency Loss.* The $a_{ij}$ function takes as input a map $p(.)$ of probability vectors (the output of the network) and should return, as Eq. 1, a measure of connectivity between the $i$ and $j$ labels. A simple idea to build $a_{ij}$ would be to define an approximation $f$ of the $\delta_{.,s(x)}$ function from the $p$ map and plug it into Eq. 1. If $f$ is such an approximation, $a_{ij}$ can be computed as follows:

$$a_{ij}(f) = \sum_x \sum_{v \in V} f_i(x) f_j(x - v), \tag{4}$$

$$= \sum_x f_i(x) \sum_{v \in V} f_j(x - v), \tag{5}$$

$$= \sum_x f_i(x) \tilde{f}_j(x) \tag{6}$$

where $\tilde{f} = f * \mathbb{1}_V$ is the convolution of $f$ and the indicatrix of the neighborhood (a kernel with all one values in $V$, $0 \notin V$).

Note that if $f_i(x) = \delta_{i, \arg\max_k p_k(x)}$, we obtain exactly $a_{ij}(f) = \mathbf{A}_{ij}$. However, as the derivative of the argmax function is 0 a.e., gradient descent algorithms such as SGD would not be possible for the training. We will investigate two families of smooth approximation:

$$f^P(p) = p^\beta \quad \text{and} \quad f^{\mathrm{norm}}(p) = \left( \frac{p}{\max_k p_k} \right)^\beta, \tag{7}$$

where the exponent is meant component-wise.

*Semi-supervised Learning.* Evaluating the loss of the semantic constraint does not require the ground truth segmentation, thus the framework of this method is extended to semi-supervised learning, where the model is simultaneously optimized to minimize the classical segmentation loss on the annotated dataset $D_S$ and constrained by the connectivity term on an external non-annotated dataset $D_{NA}$ complemented by the ground truth dataset $D_S$.

## 2.4   Numerical Resolution

*Constrained Optimization.* The constrained learning is performed by fine-tuning a model trained to solve the original task (without any form of penalization). During this second step, $G(w)$ is weighted by a coefficient $\lambda$ and added as a penalization to the segmentation loss. The $\lambda$ coefficient is linearly increased as a function of the iteration index until it reaches a predefined $\lambda_{\max}$.

*Multi-objective Model Selection.* A standard way to select the best network is to choose the iteration with minimal loss on the validation dataset. In our case, selecting the best model involves balancing between the quality of the segmentation and its fidelity to anatomical properties. To solve this multi-objective problem, we opt to pick the model that maximizes the average graph loss among the five best average dice.

# 3  Experiments

## 3.1  Data

The proposed method was evaluated on brain-region segmentation from T1-weighted MR images, using the MICCAI 2012 multi-atlas challenge and IBSRv2 datasets. Each dataset was split into training/validation/test subsets as presented in Table 1. The OASIS dataset [11] was used as the source of unlabelled training data for the semi-supervised experiments, excluding the subjects who also appear in MICCAI 2012. In IBSRv2, 6 of the 39 labels were removed from the segmentation problem (such as Lesions, Blood vessel or Unknown).

**Table 1.** The three brain MRI datasets used for the experiments.

|          | Nb subjects | Nb labels | Nb train | Nb validation | Nb test |
|----------|-------------|-----------|----------|---------------|---------|
| MICCAI12 | 35          | 135       | 10       | 5             | 20      |
| IBSRv2   | 18          | 33        | 10       | 3             | 5       |
| OASIS    | 406         |           | 284      | 122           |         |

All the images were affine registered to a reference atlas in the MNI space. Bias field correction was applied with N4ITK. The mean and standard deviation were estimated on each of the datasets and respectively centered and reduced.

## 3.2  Implementation Details

The cross entropy and the average dice similarity [6] are the loss functions. The numerical optimization is performed with SGD, with a batch size of 8, the initial learning rate is set to 0.01 and updated following the poly rate policy [12], for 300 epochs. While applying the penalization, $\lambda_{max}$ (see 2.5) is set to 0.01, the learning rate is lowered to $1e-3$, for 50 epochs.

Due to the number of classes and important volume variations between structures, we noticed that class imbalance was causing issues during the optimization of the cross entropy. Following the work of Roy *et al.* in [6], median frequency weighting was applied with success. The dice loss is left unaffected by this problem. We used elastic deformation [13] as the main data augmentation method. The code and the models' parameters will be made available publicly[1].

## 3.3  Evaluation

To quantify how many abnormalities based on the adjacency prior are present, we introduce two new metrics, $CA^{unique}$ and $CA^{volume}$:

$$CA^{unique}(a) = 100\frac{|O \cap H|}{|H|} \quad \text{and} \quad CA^{volume}(a) = \frac{\sum\limits_{(i,j)\in(O \cap H)} a_{ij}}{vol_{contour}}$$

---

[1] https://github.com/trypag/NonAdjLoss.

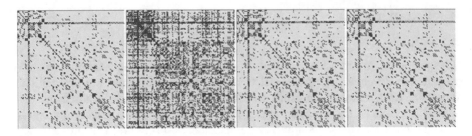

**Fig. 2.** Binary adjacency graphs matrices. Blue shows correct connections, red shows impossible adjacencies. From left to right: ground truth $\tilde{A}$ for the MICCAI12 dataset, after training without constraint, after training with NonAdjLoss, after semi-supervised training of NonAdjLoss with 100 images.

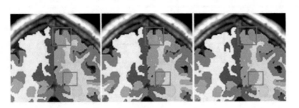

**Fig. 3.** Segmentation maps of one patient for: ground truth, model without loss, model with NonAdjLoss (from left to right). Red boxes highlight areas where inconsistencies were corrected.

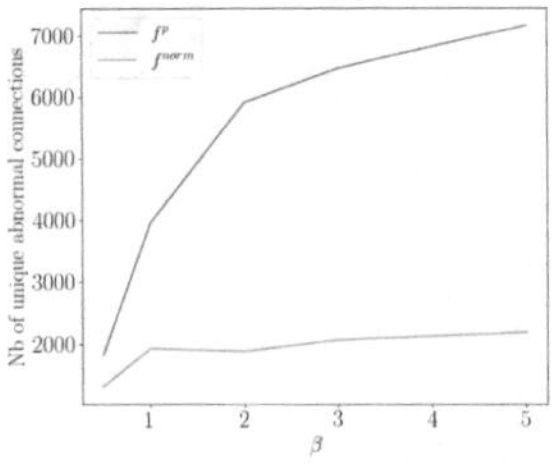

**Fig. 4.** Number of unique abnormal connections as a function of $f^p$, $f^{norm}$ and $\beta$.

where $a$ is the adjacency graph of a segmentation map, $O = \{(i,j) \mid a_{ij} > 0\}$, $H = \{(i,j) \mid \tilde{A}_{ij} = 0\}$ and $vol_{contour}$ the volume of contours voxels in the inferred segmentation (Fig. 3).

## 4    Results

To select the best approximation of $f$, we evaluated several values of $\beta$ for $f^p$ and $f^{norm}$. For these models optimized with NonAdjLoss on the train dataset of MICCAI12, we measured the total number of unique abnormal connections on part of the validation set (20 images from OASIS). Figure 4 shows that $f^{norm}$ is significantly better at reducing the number of forbidden transitions than $f^p$. The optimal value of $\beta$ lies at 0.5, we opt to use this configuration for the rest of the experiments. The effect of training with NonAdjLoss is demonstrated in Fig. 2, where the adjacency graphs obtained from the segmentation of the MIC-CAI12 test dataset are presented. After optimizing with the penalty term, the number of unique abnormalities (red dots) considerably decreases, while correct connections (blue dots) are preserved. We further improve the results by performing semi-supervised training on 100 unannotated images from OASIS,

**Table 2.** Distance, similarity and connectivity metrics for each models. MSD is the mean surface distance. All the metrics are averaged over the test dataset. Our loss $(n) = \text{NonAdjLoss}(n)$, where $n$ is the number of images used for the semi-supervised training. (average $\pm$ standard deviation).

| MICCAI12 | Dice | Hausdorff | MSD | $CA^{unique}$ | $CA^{volume}$ |
|---|---|---|---|---|---|
| Baseline | 0.735 ± 0.11 | 21.19 ± 9.82 | 1.25 ± 0.43 | 2.57 ± 5.7e−2 | 7.6e−3 ± 2.5e−2 |
| Our loss (0) | 0.730 ± 0.10 | 13.20 ± 4.70 | 1.13 ± 0.35 | 0.21 ± 6.1e−3 | 2.7e−4 ± 1.0e−3 |
| Our loss (20) | 0.733 ± 0.10 | 11.42 ± 4.21 | 1.08 ± 0.33 | 0.02 ± 3.9e−4 | 7.3e−6 ± 2.4e−5 |
| Our loss (100) | 0.739 ± 0.10 | 11.20 ± 4.13 | 1.05 ± 0.34 | 0.01 ± 1.5e−4 | 1.5e−6 ± 2.4e−6 |
| IBSRv2 | | | | | |
| Baseline | 0.825 ± 0.11 | 20.86 ± 21.14 | 0.82 ± 0.36 | 5.44 ± 1.8e−2 | 5.4e−4 ± 1.7e−4 |
| Our loss (0) | 0.833 ± 0.10 | 15.45 ± 20.32 | 0.78 ± 0.31 | 0 ± 0 | 0 ± 0 |
| Our loss (20) | 0.833 ± 0.10 | 14.54 ± 19.57 | 0.77 ± 0.31 | 0 ± 0 | 0 ± 0 |
| Our loss (50) | 0.835 ± 0.10 | 15.16 ± 19.26 | 0.77 ± 0.30 | 0.12 ± 1.5e−3 | 2.2e−6 ± 4.2e−6 |

effectively showing that the true objective of minimizing inconsistent predictions is achieved. It also demonstrates that semi-supervision has the ability to strengthen the generalization of the constraint, by learning from unseen cases.

In Table 2, we evaluate classical metrics to measure the quality of segmentation and quantify abnormalities. For the MICCAI12 dataset, we can see that the proposed methodology does not harm the dice similarity, keeping a steady level, while considerably lowering the Hausdorff distance (significantly better than the baseline for all the proposed models, with 95% confidence). This means that training with the NonAdjLoss enables to correct segmentation errors that are spatially far away from their ground truth, thus reducing the level of inconsistency. The $CA^{unique}$ metric provides a percentage of unique abnormal connections in the segmentations, for both datasets we prove to gradually decrease it by applying the proposed loss, sometimes even resulting in no abnormality. The only exception is the model trained with semi-supervision on 50 images of IBSRv2, we suggest that it is due to an optimization problem. $CA^{volume}$ indicates the overall volume of inconsistent segmentations, quantifying how many abnormalities were observed. Again we notice the same pattern as before, gradually diminishing the errors.

## 5   Conclusion

To our knowledge, this is the first time in the literature that a loss constraint based on a label connectivity prior is proposed. It can be applied to any image segmentation problem where invariance in the label space is ensured, without needing to modify the network's architecture. Furthermore, while no segmentation quality measure was impaired, the Hausdorff and MSD were clearly improved. The higher the number of labels, the more constrained the problem is, which leads to a potentially better efficiency of the method. Not requiring the ground truth annotation is also a serious advantage to extend to semi-supervised training, enforcing the generalization of the new loss on larger datasets.

**Acknowledgments.** This work was funded by the CNRS PEPS "APOCS" and was performed within the framework of the LABEX PRIMES (ANR-11-LABX-0063) of Université de Lyon, within the program "Investissements d'Avenir" (ANR-11-IDEX-0007) operated by the French National Research Agency (ANR). We gratefully acknowledge the support of NVIDIA Corporation with the donation of the Titan X Pascal GPU used for this research. Also we would like to thank the IN2P3 computing center for sharing their resources.

# References

1. Moeskops, P., Viergever, M.A., Mendrik, A.M., de Vries, L.S., Benders, M.J.N.L., Išgum, I.: Automatic segmentation of MR brain images with a convolutional neural network. IEEE TMI **35**(5), 1252–1261 (2016)
2. Roth, H.R., et al.: DeepOrgan: multi-level deep convolutional networks for automated pancreas segmentation. In: Navab, N., Hornegger, J., Wells, W.M., Frangi, A.F. (eds.) MICCAI 2015. LNCS, vol. 9349, pp. 556–564. Springer, Cham (2015). https://doi.org/10.1007/978-3-319-24553-9_68
3. Havaei, M., et al.: Brain tumor segmentation with deep neural networks. Med. Image Anal. **35**, 18–31 (2017)
4. Heckemann, R.A., Hajnal, J.V., Aljabar, P., Rueckert, D., Hammers, A.: Automatic anatomical brain MRI segmentation combining label propagation and decision fusion. NeuroImage **33**(1), 115–126 (2006)
5. Wang, H., Yushkevich, P.A.: Multi-atlas segmentation with joint label fusion and corrective learning—an open source implementation. Front. Neuroinformatics **7**, 27 (2013)
6. Roy, A.G., Conjeti, S., Sheet, D., Katouzian, A., Navab, N., Wachinger, C.: Error corrective boosting for learning fully convolutional networks with limited data. In: Descoteaux, M., Maier-Hein, L., Franz, A., Jannin, P., Collins, D.L., Duchesne, S. (eds.) MICCAI 2017. LNCS, vol. 10435, pp. 231–239. Springer, Cham (2017). https://doi.org/10.1007/978-3-319-66179-7_27
7. Xu, J., Zhang, Z., Friedman, T., Liang, Y., den Broeck, G.V.: A semantic loss function for deep learning with symbolic knowledge (2018)
8. Stewart, R., Ermon, S.: Label-free supervision of neural networks with physics and domain knowledge (2017)
9. Oktay, O., et al.: Anatomically constrained neural networks (ACNNs): application to cardiac image enhancement and segmentation. IEEE TMI **37**(2), 384–395 (2018)
10. de Brebisson, A., Montana, G.: Deep Neural Networks for Anatomical Brain Segmentation. arXiv:1502.02445 [cs, stat], February 2015
11. Marcus, D.S., Fotenos, A.F., Csernansky, J.G., Morris, J.C., Buckner, R.L.: Open access series of imaging studies: longitudinal MRI data in nondemented and demented older adults. J. Cogn. Neurosci. **22**(12), 2677–2684 (2010)
12. Chen, L., Papandreou, G., Kokkinos, I., Murphy, K., Yuille, A.L.: Deeplab: semantic image segmentation with deep convolutional nets, atrous convolution, and fully connected CRFs. CoRR abs/1606.00915 (2016)
13. Simard, P.Y., Steinkraus, D., Platt, J.C.: Best practices for convolutional neural networks applied to visual document analysis. In: Proceedings of the Seventh International Conference on Document Analysis and Recognition, pp. 958–963, August 2003

# Autofocus Layer for Semantic Segmentation

Yao Qin[1,2(✉)], Konstantinos Kamnitsas[1,3], Siddharth Ancha[1,4], Jay Nanavati[1], Garrison Cottrell[2], Antonio Criminisi[1], and Aditya Nori[1]

[1] Microsoft Research, Cambridge, UK
[2] University of California, San Diego, USA
[3] Imperial College London, London, UK
[4] Carnegie Mellon University, Pittsburgh, USA
yaq007@eng.ucsd.edu

**Abstract.** We propose the autofocus convolutional layer for semantic segmentation with the objective of enhancing the capabilities of neural networks for multi-scale processing. Autofocus layers adaptively change the size of the effective receptive field based on the processed context to generate more powerful features. This is achieved by parallelising multiple convolutional layers with different dilation rates, combined by an attention mechanism that learns to focus on the optimal scales driven by context. By sharing the weights of the parallel convolutions we make the network scale-invariant, with only a modest increase in the number of parameters. The proposed autofocus layer can be easily integrated into existing networks to improve a model's representational power. We evaluate our mod els on the challenging tasks of multi-organ segmentation in pelvic CT and brain tumor segmentation in MRI and achieve very promising performance.

## 1 Introduction

Semantic segmentation is a fundamental problem in medical image analysis. Automatic segmentation systems can improve clinical pipelines, facilitating quantitative assessment of pathology, treatment planning and monitoring of disease progression. They can also facilitate large-scale research studies, by extracting measurements from magnetic resonance images (MRI) or computational tomography (CT) scans of large populations in an efficient and reproducible manner.

For high performance, segmentation algorithms are required to use multi-scale context [6], while still aiming for pixel-level accuracy. Multi-scale processing provides detailed cues, such as texture information of a structure, combined with contextual information, such as a structure's surroundings, which can facilitate decisions that are ambiguous when based only on local context. Note that such

---

Y. Qin, K. Kamnitsas and S. Ancha—Most of this work was performed while the authors were interns at MSR, Cambridge.

© Springer Nature Switzerland AG 2018
A. F. Frangi et al. (Eds.): MICCAI 2018, LNCS 11072, pp. 603–611, 2018.
https://doi.org/10.1007/978-3-030-00931-1_69

a mechanism is also part of the human visual system, via foveal and peripheral vision.

A large volume of research has sought algorithms for effective multi-scale processing. An overview of traditional approaches can be found in [6]. Contemporary segmentation systems are often powered by convolutional neural networks (CNNs). The various network architectures proposed to effectively capture image context can be broadly grouped into three categories. The first type creates an image pyramid at multiple scales. The image is down-sampled and processed at different resolutions. Farabet *et al.* trained the same filters to perform on all such versions of an image to achieve scale invariance [5]. In contrast, DeepMedic [9] proposed learning dedicated pathways for several scales, to enable 3D CNNs to extract more patterns from a larger context in a computationally efficient manner. The second type uses an encoder that gradually down-samples to capture more context, followed by a decoder that learns to upsample the segmentations, combining multi-scale context using skip connections [11]. Later extensions include U-net [15], which used a larger decoder to learn upsampling features instead of segmentations as in [11]. Learning to upsample with a decoder, however, increases model complexity and computational requirements, when downsampling may not be even necessary. Finally, driven by this idea, [3,16] proposed dilated convolutions to process greater context without ever downsampling the feature maps. Taking it further, DeepLab [3] introduced the module Atrous Spatial Pyramid Pooling (ASPP), where dilated convolutions with varying rates are applied in parallel to capture multi-scale information. The activations from all scales are naively fused via summation or concatenation.

We propose the *autofocus layer*, a novel module that enhances the multi-scale processing of CNNs by learning to select the 'appropriate' scale for identifying different objects in an image. Our work on autofocus shares similarities with ASPP in that we also use parallel dilated convolutional filters to capture both local and more global context. The crucial difference is that instead of naively aggregating features from all scales, the autofocus layer adaptively chooses the optimal scale to focus on in a data-driven, learned manner. In particular, our autofocus module uses an attention mechanism [1] to indicate the importance of each scale when processing different locations of an image (Fig. 1). The computed attention maps, one per scale, serve as filters for the patterns extracted at that scale. Autofocus also enhances interpretability of a network as the attention maps reveal how it locally 'zooms in or out' to segment different context. Compared to the use of attention in [4], our solution is modular and independent of architecture.

We extensively evaluate and compare our method with strong baselines on two tasks: multi-organ segmentation in pelvic CT and brain tumor segmentation in MRI. We show that thanks to its adaptive nature, the autofocus layer copes well with biological variability in the two tasks, improving performance of a well-established model. Despite its simplicity, our system is competitive with more elaborate pipelines, demonstrating the potential of the autofocus mechanism.

Additionally, autofocus can be easily incorporated into existing architectures by replacing a standard convolutional layer.

## 2 Method

### 2.1 Dilated Convolution

As they are fundamental to our work, we first present the basics of dilated convolutions [3,16] while introducing notation. The standard 3D dilated convolutional layer at depth $l$ with dilation rate $r$ can be represented as a mapping $\mathbf{Conv}_l^r : \mathbf{F}_{l-1} \rightarrow \mathbf{F}_l^r$, where $\mathbf{F}_{l-1} \in \mathbb{R}^{W' \times H' \times D' \times C'}$ and $\mathbf{F}_l^r \in \mathbb{R}^{W \times H \times D \times C}$ are input and output tensors with $C$ channels (feature maps) of size $(W \times H \times D)$. For neurons in $\mathbf{F}_l^r$, the size $\phi_l^{\{x,y,z\}} \in \mathbb{N}^3$ of their receptive field on the input image can be controlled via $r$. For dilated convolution layers with kernel size $\boldsymbol{\theta}_l^{\{x,y,z\}} \in \mathbb{N}^3$, $\phi_l^{\{x,y,z\}}$ can be derived recursively as follows:

$$\phi_l^{\{x,y,z\}} = \phi_{l-1}^{\{x,y,z\}} + r_l(\boldsymbol{\theta}_l^{\{x,y,z\}} - \mathbf{1})\boldsymbol{\eta}_l^{\{x,y,z\}}, \tag{1}$$

Here $\boldsymbol{\eta}_l^{\{x,y,z\}} \in \mathbb{N}^3$ denotes the stride of the receptive field at layer $l$, which is a product of the strides of kernels in preceding layers. It can be observed from Eq. (1) that greater context can be captured by increasing dilation $r_l$ but in less detail as the input signal is probed more sparsely. Thus greater $r_l$ leads to a 'zoom out' behavior. Usually, the dilation rate $r$ is a hyperparameter that is manually set and fixed for each layer. Standard convolution is a special case when $r = 1$. Below we describe the autofocus mechanism that adaptively chooses the optimal dilation rate for different areas of the input.

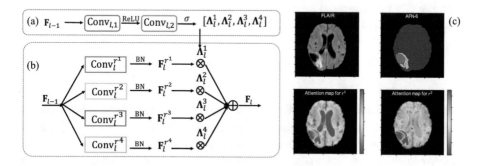

**Fig. 1.** An autofocus convolutional layer with the number of candidate dilation rates $K = 4$. (a) The attention model. (b) A weighted summation of activations from parallel dilated convolutions. (c) An example of attention maps for a small ($r^1$) and larger ($r^2$) dilation rate. The first row is the input and the segmentation result of AFN-6 (described in Sect. 2.3). The second row shows how the module 'zooms out' for more context when processing large or ambiguous structures.

## 2.2   Autofocus Convolutional Layer

Unambiguously classifying different objects in an image is likely to require different combinations of local and global information. For example, large structures may be better segmented by processing a large receptive field $\phi_l$ at the expense of fine details, while small objects may require focusing on high resolution local information. Consequently, architectures that statically define multi-scale processing may be suboptimal. Our adaptive solution, the autofocus module, is summarized in Fig. 1 and formalized in the following.

Given activations of the previous layer $\mathbf{F}_{l-1}$, we capture multi-scale information by processing it in parallel via $K$ convolutional layers with different dilation rates $r^k$. They produce $K$ tensors $\mathbf{F}_l^{r^k}$ (Fig. 1(b)), each set to have same number of channels $C$. They detect patterns at $K$ different scales which we merge in a data-driven manner by introducing a soft attention mechanism [1].

Within the module we construct a small attention network (Fig. 1(a)) that processes $\mathbf{F}_{l-1}$. In this work it consists of two convolutional layers. The first, **Conv**$_{l,1}$, applies $3 \times 3 \times 3$ kernels, produces half the number of channels than those in $\mathbf{F}_{l-1}$ (empirically chosen) and is followed by a ReLU activation function $f$. The second, **Conv**$_{l,2}$, applies $1 \times 1 \times 1$ filters and produces a tensor with $K$ channels, one per scale. It is followed by an element-wise softmax $\sigma$ that normalizes the $K$ activations for each voxel to add up to one. Let this normalized output be $\mathbf{\Lambda}_l = [\mathbf{\Lambda}_l^1, \mathbf{\Lambda}_l^2, \cdots, \mathbf{\Lambda}_l^K] \in \mathbb{R}^{W \times H \times D \times K}$. Formally:

$$\mathbf{\Lambda}_l = \sigma(\text{Conv}_{l,2}(f(\text{Conv}_{l,1}(\mathbf{F}_{l-1})))) \tag{2}$$

In the above, $\mathbf{\Lambda}_l^k \in \mathbb{R}^{W \times H \times D}$ is an attention map that corresponds to the $k$-th scale. For any specific spatial location (voxel), the corresponding $K$ values from the $K$ attention maps $\mathbf{\Lambda}_l^k$ can be interpreted as how much focus to put on each scale. Thus the final output of the autofocus layer is computed by fusing the outputs from the parallel dillated convolutions as follows:

$$\mathbf{F}_l = \sum_{k=1}^{K} \mathbf{\Lambda}_l^k \cdot \mathbf{F}_l^{r^k} \tag{3}$$

where $\cdot$ is an element-wise multiplication. Note that the attention weights $\mathbf{\Lambda}_l^k$ are shared across all channels of tensor $\mathbf{F}_l^{r^k}$ for scale $k$. Since the attention maps are predicted by a fully convolutional network, different attention is predicted for each voxel, driven by the image context for the optimal choice of scale (Fig. 1(c)).

The increase in representational power offered by each autofocus layer naturally comes with computational requirements as the module is based in parallelism of $K$ dilated convolutional layers. Therefore an appropriate balance should be sought, which we investigate in Sect. 3 with very promising results.

**Scale Invariance:** The size of some anatomical structures such as bones and organs may vary, while the overall appearance is rather similar. For others, size

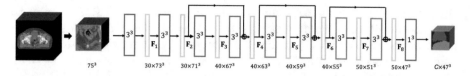

**Fig. 2.** The AFNET-4 model. Layers 1–2 are standard convolutions and 3–4 are dilated with rate 2. Layers 4–8 are autofocus layers, denoted with red. All layers except the classification layer use $3^3$ kernels. Yellow rectangles represent ReLU layers. Residual connections are used. Number and size of feature maps shown as (number × size).

may correlate with appearance. For instance, the texture of large developed tumors differs from early-stage small tumors. This suggests that scale invariance could be leveraged to regularize learning but must be done appropriately. We make the parallel filters in an autofocus layer share parameters. This makes the number of trainable parameters independent of $K$, with only the attention module adding parameters over a standard convolution. As a result, each parallel filter seeks patterns with similar appearance but of different sizes. Hence, the network is *adaptively* scale-invariant – the attention mechanism chooses the scale in a data-driven manner, unlike Farabet *et al.* [5], whose network learns shared filters between different scales but naively concatenates all their responses.

### 2.3   Autofocus Neural Networks

The proposed autofocus layer can be integrated into existing architectures to improve their multi-scale processing capabilities by replacing standard or dilated convolutions. To demonstrate this, we chose DeepMedic (DM) [9] with residual connections [8] as a starting point. DM uses different pathways with high and low resolution inputs for multi-scale processing. Instead, we keep only its high-resolution pathway and seek to empower it with our method. First, we enhance it with standard dilated convolutions with rate 2 in its last 6 hidden layers to enlarge its receptive field, arriving at the BASIC model that serves as another baseline. We now define a family of AFNETs by converting the last $n$ hidden layers of BASIC to autofocus layers—denoted as "AFN-$n$", where $n \in \{1, \ldots, 6\}$. Figure 2 shows AFNET-4. The proposed AFNETs are trained end-to-end.

## 3   Evaluation

We extensively evaluate AFNETs on the tasks of multi-organ and brain tumor segmentation. Specifically, on both tasks we perform: (1) a study where we successively add autofocus to more layers of the BASIC network to explore its impact, and (2) comparison of AFNETs with baselines. Finally, (3) we evaluate on the public benchmark BRATS'15 and show that our method competes with state-of-the-art pipelines regardless its simplicity, showing its potential.

**Baselines:** We compare AFNETs with the previously defined BASIC model to show the contribution of autofocus layer over standard dilated convolutions. Similarly, we compare with DeepMedic [9], denoted as DM, to compare our adaptive multi-scale processing with the static multi-scale pathways. Finally, we place an ASPP module [3] on top of BASIC, comparison of which against AFN-1 shows contribution of the attention mechanism. ASPP-C and ASPP-S represent fusion of ASPP activations via concatenation and summation respectively. Source codes and pretrained models in PyTorch framework are online available at: https:// github.com/yaq007/Autofocus-Layer.

### 3.1 ADD and UW Datasets of Pelvic CT Scans

**Material:** We use two databases of pelvic CT scans, collected from patients diagnosed with prostate cancer in different clinical centers. The first, referred to as ADD, contains 86 scans with varying number of $512 \times 512$ slices and $3\,mm$ inter-slice spacing. UW consists of 34 scans of $512 \times 512$ slices with $1\,mm$ inter-slice spacing. Expert oncologists manually delineated in all images the following structures: prostate gland, seminal vesicles (SV), bladder, rectum, left femur and right femur. Each scan is normalized so that its intensities have zero mean and unit variance. We also re-sample UW to the spacing of ADD. To produce a stringent test of the models' generalization, we train them for this multi-class problem using the ADD data and then evaluate them on UW data.

**Configuration Details:** BASIC, ASPP and AFN models were trained with the ADAM optimizer for 300 epochs to minimize the soft dice loss [13]. Each batch consists of 7 segments of size $75^3$. The learning rate starts at 0.001 and is reduced to 0.0001 after 200 epochs. We use dilation rates 2, 6, 10 and 14 ($K = 4$) for both ASPP and the autofocus modules. It takes around 20 hours to train an AFNET with 2 NVIDIA TITAN X GPUs. Performance of DeepMedic was obtained by training the public software [9] with default parameters, but without augmentation and by sampling each class equally, similar to other methods.

### 3.2 Brain Tumor Segmentation Data (BRATS 2015)

**Material:** The training database of BRATS'15 [12] consists of multi-modal MR scans of 274 cases, along with corresponding annotations of the tumors. We normalize each scan so that intensities belonging to the brain have zero mean and unit variance. For our ablation study, we train all models on the same 193 subjects and evaluate their performance on 54 subjects. The subsets were chosen randomly, including both high and low grade gliomas. Results on the remaining 23 cases aren't reported as they were used for configuration during development. Following standard protocol, we report performance for segmenting the *whole* tumor, *core* and *enhancing* tumor. Finally, to compare with other methods, we train AFNET-6 on all 274 images, segment the 110 test cases of BRATS'15 (no annotations publicly available) and submit predictions for online evaluation.

**Table 1.** Performance on multi-organ segmentation problem of baseline models and AFN on Uw database, after being trained on ADD. Absolute dice scores are shown.

|  | BASIC | ASPP-S | ASPP-C | DM | AFN-1 | AFN-2 | AFN-3 | AFN-4 | AFN-5 | AFN-6 |
|---|---|---|---|---|---|---|---|---|---|---|
| Prostate | 50.94 | 55.83 | 58.67 | 69.66 | 63.36 | 75.43 | **76.66** | 75.30 | 73.81 | 76.15 |
| Bladder | 72.83 | 81.53 | 80.43 | 93.54 | 88.76 | 92.38 | 93.56 | **94.49** | 94.28 | 93.32 |
| Rectum | 64.39 | 67.30 | 67.04 | 70.74 | 71.46 | 76.20 | 78.45 | **79.80** | 78.96 | 78.82 |
| SV | 53.97 | 50.11 | 59.37 | 56.75 | 61.68 | 65.03 | **65.12** | 64.83 | 60.87 | 63.24 |
| LFemur | 91.60 | 94.13 | 93.81 | 94.68 | 93.24 | **95.18** | 93.52 | 94.59 | 93.42 | 95.16 |
| RFemur | 91.65 | 94.34 | 92.78 | 94.63 | 91.93 | 94.38 | 91.61 | 94.60 | 94.16 | **95.75** |
| Mean Dice | 70.90 | 74.98 | 75.35 | 78.89 | 78.41 | 83.10 | 83.15 | **83.94** | 82.58 | 83.74 |

**Table 2.** Ablation study on BRATS'15 training database via cross-validation on 54 random held-out cases. Dice scores shown in format mean(standard deviation).

|  | BASIC | ASPP-S | ASPP-C | DM | AFN-1 | AFN-2 | AFN-3 | AFN-4 | AFN-5 | AFN-6 |
|---|---|---|---|---|---|---|---|---|---|---|
| Whole | 87.90 | 87.83 | 87.90 | 88.93 | 88.03 | 88.63 | 88.42 | 88.88 | 89.19 | **89.30** |
|  | (8.57) | (8.36) | (8.08) | (7.05) | (8.29) | (8.02) | (9.28) | (8.31) | (7.87) | **(8.00)** |
| Core | 72.61 | 74.08 | 73.58 | 71.42 | 73.93 | 73.43 | 73.79 | 73.91 | **74.32** | 74.11 |
|  | (26.39) | (24.16) | (24.57) | (27.48) | (25.04) | (24.90) | (26.00) | (25.49) | **(24.80)** | (24.19) |
| Enh | 74.37 | 73.06 | 73.47 | **76.08** | 73.17 | 73.94 | 74.21 | 74.08 | 74.48 | 75.62 |
|  | (25.23) | (26.93) | (25.36) | **(25.12)** | (26.98) | (25.83) | (27.58) | (25.70) | (26.56) | (25.02) |

**Configuration Details:** Settings are similar to Kamnitsas *et al.* [9] for a fair comparison. For each method in Table 2 we report the average of three runs with different seeds.

## 3.3 Results

**Ablation Study:** Results from the ablation study on the cervical CT database and the BRATS database are summarized in Tables 1 and 2 respectively. We observe the following: (a) Building AFN-1 by converting the last layer of BASIC to autofocus improves performance, while (b) the gains surpass those by the popular ASPP for most classes of the tasks. It is important to note that ASPP adds multiple parallel convolutional layers without sharing weights between them. This incurs a large increase in the number of parameters, and is therefore partly the reason for improvements of ASPP over BASIC (see Table 3). (c) Converting more layers of the BASIC baseline to autofocus layers tends to improve performance. An exception is AFN-4 vs. AFN-5/6 on the Uw dataset. We speculate that

**Table 3.** Number of trainable parameters in convolutional kernels of different models.

| Models | BASIC | ASPP-S | ASPP-C | DM | AFN-1 | AFN-6 |
|---|---|---|---|---|---|---|
| Params | 315,725 | 967,330 | 478,435 | 662,555 | 349,904 | 450,209 |

**Table 4.** Dice scores achieved by state-of-the-art methods on BRATS'15 test database. $^\dagger$ are semi-automatic. $^*$ used CNN ensembles and more extensive augmentation.

| Models | AFN-6 | peres1$^{\dagger*}$[14] | bakas1$^\dagger$[2] | kamnk1/DM [9] | kayab1$^*$ [10] | isenf1$^*$ [7] |
|---|---|---|---|---|---|---|
| Whole | 84% | 83% | 81% | 84% | 85% | 85% |
| Core | 69% | 72% | 63% | 67% | 72% | 74% |
| Enh | 63% | 60% | 58% | 63% | 61% | 64% |
| Cases | 110/110 | 53/110 | 53/110 | 110/110 | 110/110 | 110/110 |

this is due to randomness in training and suboptimal optimization. (d) Empowering the high-resolution pathway of DeepMedic with adaptive autofocus quickly outperforms the gains from the static second pathway on pelvic scan and brain tumor segmentation except for the enhancing tumor. We speculate that gains are more profound in the former task due to the greater variation in the size of structures, where the adaptive nature of autofocus shines. Finally we note that by sharing weights across scales, AFNETs have small number of trainable parameters, shown in Table 3, which could enable rapid learning from little data, which is however left for future work. On the downside, the multiple scales on each autofocus layer increase memory and computation requirements.

**Comparison with State-of-the-Art on BRATS'15:** Performance on test data of BRATS'15 obtained via the online evaluation platform is shown on Table 4, along with other top published methods. AFN-6 compares favorably to the semi-automatic methods that topped the BRATS'15 challenge [2, 14], as well as DeepMedic with the second static lower-resolution pathway. Note that in [14] high and low grade gliomas were separated by visual inspection and then passed to an appropriately specialized CNN, giving them an advantage over other methods. Our model is only surpassed by the pipelines of [7, 10], who both used ensembles of CNNs with deep supervision and more aggressive data augmentation. The promising performance obtained by our simple method indicates the potential of the autofocus layer, which can be adopted in more elaborate systems.

## 4   Conclusion

We proposed an autofocus convolutional layer for segmentation of biomedical images. An autofocus layer can adapt the network's receptive field at different spatial locations in a data-driven manner. Our extensive evaluation of AFNETs shows that they cope well with biological variability in different tasks and generalize well on both MR and CT images. We have shown that the autofocus convolutional layer can be integrated into existing network architectures to substantially increase their representational power with only a small increase in model parameters. In addition, the interpretability of autofocus layers can leverage understanding of deep learning systems. Investigating the potential of autofocus modules in regression problems would be interesting future work.

**Acknowledgments.** G.W.C. and Y.Q. were partially supported by Guangzhou Science and Technology Planning Project (Grant No. 201704030051).

# References

1. Bahdanau, D., Cho, K., Bengio, Y.: Neural machine translation by jointly learning to align and translate. In: ICLR (2015)
2. Bakas, S., et al.: GLISTRboost: combining multimodal MRI segmentation, registration, and biophysical tumor growth modeling with gradient boosting machines for glioma segmentation. In: Crimi, A., Menze, B., Maier, O., Reyes, M., Handels, H. (eds.) BrainLes 2015. LNCS, vol. 9556, pp. 144–155. Springer, Cham (2016). https://doi.org/10.1007/978-3-319-30858-6_13
3. Chen, L.C., Papandreou, G., Kokkinos, I., Murphy, K., Yuille, A.L.: Deeplab: semantic image segmentation with deep convolutional nets, atrous convolution, and fully connected CRFs. arXiv preprint arXiv:1606.00915 (2016)
4. Chen, L.C., Yang, Y., Wang, J., Xu, W., Yuille, A.L.: Attention to scale: Scale-aware semantic image segmentation. In: CVPR (2016)
5. Farabet, C., Couprie, C., Najman, L., LeCun, Y.: Learning hierarchical features for scene labeling. IEEE Trans. PAMI **35**(8), 1915–1929 (2013)
6. Galleguillos, C., Belongie, S.: Context based object categorization: a critical survey. Comput. Vis. Image Underst. **114**(6), 712–722 (2010)
7. Isensee, F., Kickingereder, P., Wick, W., Bendszus, M., Maier-Hein, K.H.: Brain tumor segmentation and radiomics survival prediction: contribution to the BRATS 2017 challenge. In: Crimi, A., Bakas, S., Kuijf, H., Menze, B., Reyes, M. (eds.) BrainLes 2017. LNCS, vol. 10670, pp. 287–297. Springer, Cham (2018). https://doi.org/10.1007/978-3-319-75238-9_25
8. Kamnitsas, K., et al.: Deepmedic for brain tumor segmentation. In: Crimi, A., Menze, B., Maier, O., Reyes, M., Winzeck, S., Handels, H. (eds.) BrainLes 2016. LNCS, vol. 10154, pp. 138–149. Springer, Cham (2016). https://doi.org/10.1007/978-3-319-55524-9_14
9. Kamnitsas, K., et al.: Efficient multi-scale 3d CNN with fully connected CRF for accurate brain lesion segmentation. MedIA **36**, 61–78 (2017)
10. Kayalibay, B., Jensen, G., van der Smagt, P.: CNN-based segmentation of medical imaging data. arXiv preprint arXiv:1701.03056 (2017)
11. Long, J., Shelhamer, E., Darrell, T.: Fully convolutional networks for semantic segmentation. In: CVPR, pp. 3431–3440 (2015)
12. Menze, B.H.: The multimodal brain tumor image segmentation benchmark (BRATS). IEEE Trans. Med. Imaging **34**(10), 1993–2024 (2015)
13. Milletari, F., Navab, N., Ahmadi, S.A.: V-net: fully convolutional neural networks for volumetric medical image segmentation. In: 2016 Fourth International Conference on 3D Vision (3DV), pp. 565–571. IEEE (2016)
14. Pereira, S., Pinto, A., Alves, V., Silva, C.A.: Brain tumor segmentation using convolutional neural networks in MRI images. IEEE TMI **35**, 1240–1251 (2016)
15. Ronneberger, O., Fischer, P., Brox, T.: U-net: convolutional networks for biomedical image segmentation. In: Navab, N., Hornegger, J., Wells, W.M., Frangi, A.F. (eds.) MICCAI 2015. LNCS, vol. 9351, pp. 234–241. Springer, Cham (2015). https://doi.org/10.1007/978-3-319-24574-4_28
16. Yu, F., Koltun, V.: Multi-scale context aggregation by dilated convolutions. In: ICLR (2016)

# 3D Segmentation with Exponential Logarithmic Loss for Highly Unbalanced Object Sizes

Ken C. L. Wong, Mehdi Moradi[(✉)], Hui Tang, and Tanveer Syeda-Mahmood

IBM Research – Almaden Research Center, San Jose, CA, USA
mmoradi@us.ibm.com

**Abstract.** With the introduction of fully convolutional neural networks, deep learning has raised the benchmark for medical image segmentation on both speed and accuracy, and different networks have been proposed for 2D and 3D segmentation with promising results. Nevertheless, most networks only handle relatively small numbers of labels ($<10$), and there are very limited works on handling highly unbalanced object sizes especially in 3D segmentation. In this paper, we propose a network architecture and the corresponding loss function which improve segmentation of very small structures. By combining skip connections and deep supervision with respect to the computational feasibility of 3D segmentation, we propose a fast converging and computationally efficient network architecture for accurate segmentation. Furthermore, inspired by the concept of focal loss, we propose an exponential logarithmic loss which balances the labels not only by their relative sizes but also by their segmentation difficulties. We achieve an average Dice coefficient of 82% on brain segmentation with 20 labels, with the ratio of the smallest to largest object sizes as 0.14%. Less than 100 epochs are required to reach such accuracy, and segmenting a $128 \times 128 \times 128$ volume only takes around 0.4 s.

## 1 Introduction

With the introduction of fully convolutional neural networks (CNNs), deep learning has raised the benchmark for medical image segmentation on both speed and accuracy [7]. Different 2D [5,8] and 3D [1,2,6,10] networks were proposed to segment various anatomies such as the heart, brain, liver, and prostate from medical images. Regardless of the promising results of these networks, 3D CNN image segmentation is still challenging. Most networks were applied on datasets with small numbers of labels ($<10$) especially in 3D segmentation. When more detailed segmentation is required with much more anatomical structures, previously unseen issues, such as computational feasibility and highly unbalanced object sizes, need to be addressed by new network architectures and algorithms.

There are only a few frameworks proposed for highly unbalanced labels. In [8], a 2D network architecture was proposed to segment all slices of a 3D brain volume. Error corrective boosting was introduced to compute label weights

A. F. Frangi et al. (Eds.): MICCAI 2018, LNCS 11072, pp. 612–619, 2018.
https://doi.org/10.1007/978-3-030-00931-1_70

that emphasize parameter updates on classes with lower validation accuracy. Although the results were promising, the label weights were only applied to the weighted cross-entropy but not the Dice loss, and the stacking of 2D results for 3D segmentation may result in inconsistency among consecutive slices.

In [9], the generalized Dice loss was used as the loss function. Instead of computing the Dice loss of each label, the weighted sum of the products over the weighted sum of the sums between the ground-truth and predicted probabilities was computed for the generalized Dice loss, with the weights inversely proportional to the label frequencies. In fact, the Dice coefficient is unfavorable to small structures as a few pixels of misclassification can lead to a large decrease of the coefficient, and this sensitivity is irrelevant to the relative sizes among structures. Therefore, balancing by label frequencies is nonoptimal for Dice losses.

To address the issues of highly unbalanced object sizes and computational efficiency in 3D segmentation, we have two key contributions in this paper. **(I)** We propose the exponential logarithmic loss function. In [4], to handle the highly unbalanced dataset of a two-class image classification problem, a modulating factor computed solely from the softmax probability of the network output is multiplied by the weighted cross-entropy to focus on the less accurate class. Inspired by this concept of balancing classification difficulties, we propose a loss function comprising the logarithmic Dice loss which intrinsically focuses more on less accurately segmented structures. The nonlinearities of the logarithmic Dice loss and the weighted cross-entropy can be further controlled by the proposed exponential parameters. In this manner, the network can achieve accurate segmentation on both small and large structures. **(II)** We propose a fast converging and computationally efficient network architecture by combining the advantages of skip connections and deep supervision, which has only about 1/14 of the parameters of, and is twice as fast as, the V-Net [6]. Experiments were performed on brain magnetic resonance (MR) images with 20 highly unbalanced labels. Combining these two innovations achieved an average Dice coefficient of 82% with the average segmentation time as 0.4 s.

## 2   Methodology

### 2.1   Proposed Network Architecture

3D segmentation networks require much more computational resources than 2D networks. Therefore, we propose a network architecture which aims at accurate segmentation and fast convergence with respect to limited resources (Fig. 1). Similar to most segmentation networks, our network comprises the encoding and decoding paths. The network is composed of convolutional blocks, each comprises $k$ cascading $3 \times 3 \times 3$ convolutional layers of $n$ channels associated with batch normalization (BN) and rectified linear units (ReLU). For better convergence, a skip connection with a $1 \times 1 \times 1$ convolutional layer is used in each block. Instead of concatenation, we add the two branches together for less memory consumption, so the block allows efficient multi-scale processing and deeper networks can be trained. The number of channels ($n$) is doubled after

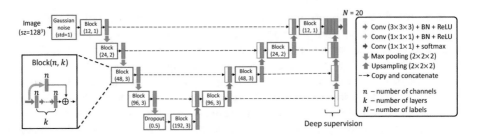

**Fig. 1.** Proposed network architecture optimized for 3D segmentation. Blue and white boxes indicate operation outputs and copied data, respectively.

each max pooling and is halved after each upsampling. More layers ($k$) are used with tensors of smaller sizes so that more abstract knowledge can be learned with feasible memory use. Feature channels from the encoding path are concatenated with the corresponding tensors in the decoding path for better convergence. We also include a Gaussian noise layer and a dropout layer to avoid overfitting.

Similar to [5], we utilize deep supervision which allows more direct back-propagation to the hidden layers for faster convergence and better accuracy [3]. Although deep supervision significantly improves convergence, it is memory expensive especially in 3D networks. Therefore, we omit the tensor from the block with the most channels (Block(192, 3)) so that training can be performed on a GPU with 12 GB of memory. A final layer of $1 \times 1 \times 1$ convolution with the softmax function provides the segmentation probabilities.

## 2.2   Exponential Logarithmic Loss

We propose a loss function which improves segmentation on small structures:

$$L_{\text{Exp}} = w_{\text{Dice}} L_{\text{Dice}} + w_{\text{Cross}} L_{\text{Cross}} \tag{1}$$

with $w_{\text{Dice}}$ and $w_{\text{Cross}}$ the respective weights of the exponential logarithmic Dice loss ($L_{\text{Dice}}$) and the weighted exponential cross-entropy ($L_{\text{Cross}}$):

$$L_{\text{Dice}} = \mathbf{E}\left[(-\ln(\text{Dice}_i))^{\gamma_{\text{Dice}}}\right] \text{ with } \text{Dice}_i = \frac{2\left(\sum_{\mathbf{x}} \delta_{il}(\mathbf{x}) \ p_i(\mathbf{x})\right) + \epsilon}{\left(\sum_{\mathbf{x}} \delta_{il}(\mathbf{x}) + p_i(\mathbf{x})\right) + \epsilon} \tag{2}$$

$$L_{\text{Cross}} = \mathbf{E}\left[w_l \left(-\ln(p_l(\mathbf{x}))\right)^{\gamma_{\text{Cross}}}\right] \tag{3}$$

with $\mathbf{x}$ the pixel position and $i$ the label. $l$ is the ground-truth label at $\mathbf{x}$. $\mathbf{E}[\bullet]$ is the mean value with respect to $i$ and $\mathbf{x}$ in $L_{\text{Dice}}$ and $L_{\text{Cross}}$, respectively. $\delta_{il}(\mathbf{x})$ is the Kronecker delta which is 1 when $i = l$ and 0 otherwise. $p_i(\mathbf{x})$ is the softmax probability which acts as the portion of pixel $\mathbf{x}$ owned by label $i$ when computing $\text{Dice}_i$. $\epsilon = 1$ is the pseudocount for additive smoothing to handle missing labels in training samples. $w_l = \left((\sum_k f_k)/f_l\right)^{0.5}$, with $f_k$ the frequency of label $k$, is the label weight for reducing the influences of more frequently seen labels. $\gamma_{\text{Dice}}$ and $\gamma_{\text{Cross}}$ further control the nonlinearities of the loss functions, and we use $\gamma_{\text{Dice}} = \gamma_{\text{Cross}} = \gamma$ here for simplicity.

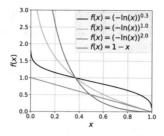

**Fig. 2.** Loss functions with different nonlinearities, where $x$ can be Dice$_i$ or $p_l(\mathbf{x})$.

The use of the Dice loss in CNN was proposed in [6]. The Dice coefficient is unfavorable to small structures as misclassifying a few pixels can lead to a large decrease of the coefficient. The use of label weights cannot alleviate such sensitivity as it is irrelevant to the relative object sizes, and the Dice coefficient is already a normalized metric. Therefore, instead of size differences, we use the logarithmic Dice loss which focuses more on less accurate labels. Figure 2 shows a comparison between the linear ($\mathbf{E}\left[1 - \text{Dice}_i\right]$) and logarithmic Dice loss.

We provide further control on the nonlinearities of the losses by introducing the exponents $\gamma_{\text{Dice}}$ and $\gamma_{\text{Cross}}$. In [4], a modulating factor, $(1-p_l)^\gamma$, is multiplied by the weighted cross-entropy to become $w_l(1-p_l)^\gamma(-\ln(p_l))$ for two-class image classification. Apart from balancing the label frequencies using the label weights $w_l$, this focal loss also balances between easy and hard samples. Our exponential loss achieves a similar goal. With $\gamma > 1$, the loss focuses more on less accurate labels than the logarithmic loss (Fig. 2). Although the focal loss works well for the two-class image classification in [4], we got worse results when applying to our segmentation problem with 20 labels. This may be caused by the over suppression of the loss function when the label accuracy becomes high. In contrast, we could get better results with $0 < \gamma < 1$. Figure 2 shows that when $\gamma = 0.3$, there is an inflection point around $x = 0.5$, where $x$ can be Dice$_i$ or $p_l(\mathbf{x})$. For $x < 0.5$, this loss behaves similarly to the losses with $\gamma \geq 1$ with decreasing gradient magnitude as $x$ increases. This trend reverses for $x > 0.5$ with increasing gradient magnitude. In consequence, this loss encourages improvements at both low and high prediction accuracy. This characteristics is the reason of using the proposed exponential form instead of the one in [4].

## 2.3   Training Strategy

Image augmentation is used to learn invariant features and avoid overfitting. As realistic nonrigid deformation is difficult to implement and computationally expensive, we limit the augmentation to rigid transformations including rotation (axial, $\pm 30°$), shifting ($\pm 20\%$), and scaling ($[0.8, 1.2]$). Each image has an 80% chance to be transformed in training, thus the number of augmented images is proportional to the number of epochs. The optimizer Adam is used with the

Nesterov momentum for fast convergence, with the learning rate as $10^{-3}$, batch size as one, and 100 epochs. A TITAN X GPU with 12 GB of memory is used.

## 3   Experiments

### 3.1   Data and Experimental Setups

A dataset of 43 3D brain MR images from different patients was neuroanatomically labeled to provide the training and validation samples. The images were produced by the T1-weighted MP-RAGE pulse sequence which provides high tissue contrast. They were manually segmented by highly trained experts with the results reviewed by a consulting neuroanatomist. Each segmentation had 19 semantic labels of brain structures, thus 20 labels with the background included (Table 1(a)). As there were various image sizes (128 to 337) and spacings (0.9 to 1.5 mm), each image was resampled to isotropic spacing using the minimum spacing, zero padded on the shorter sides, and resized to $128 \times 128 \times 128$.

Table 1(a) shows that the labels were highly unbalanced. The background occupied 93.5% of an image on average. Without the background, the relative sizes of the smallest and largest structures were 0.07% and 50.24%, respectively, thus a ratio of 0.14%.

We studied six loss functions using the proposed network, and applied the best one to the V-Net architecture [6], thus a total of seven cases were studied. For $L_{\mathrm{Exp}}$, we set $w_{\mathrm{Dice}} = 0.8$ and $w_{\mathrm{Cross}} = 0.2$ as they provided the best results. Five sets of data were generated by shuffling and splitting the dataset, with 70% for training and 30% for validation in each set. Experiments were performed on all five sets of data for each case studied for more statistically sound results. The actual Dice coefficients, not the $\mathrm{Dice}_i$ in (2), were computed for each validation image. Identical setup and training strategy were used in all experiments.

### 3.2   Results and Discussion

Table 1(b) shows the Dice coefficients averaged from the five experiments. The linear Dice loss ($\mathbf{E}\left[1 - \mathrm{Dice}_i\right]$) had the worst performance. It performed well with the relatively large structures such as the gray and white matters, but the performance decreased with the sizes of the structures. The very small structures, such as the nucleus accumbens and amygdala, were missed in all experiments. In contrast, the logarithmic Dice loss ($L_{\mathrm{Dice}}(\gamma = 1)$) provided much better results, though the large standard deviation of label 2 indicates that there were misses. We also performed experiments with the weighted cross-entropy ($L_{\mathrm{Cross}}(\gamma = 1)$), whose performance was better than the linear Dice loss but worse than the logarithmic Dice loss. The weighted sum of the logarithmic Dice loss and weighted cross-entropy ($L_{\mathrm{Exp}}(\gamma = 1)$) outperformed the individual losses, and it provided the second best results among the tested cases. As discussed in Sect. 2.2, $L_{\mathrm{Exp}}(\gamma = 2)$ was ineffective even on larger structures. This is consistent with our observation in Fig. 2 that the loss function is over suppressed when the accuracy

**Table 1.** Semantic brain segmentation. (a) Semantic labels and their relative sizes on average (%) without the background. CVL represents cerebellar vermal lobules. The background occupied 93.5% of an image on average. (b) Dice coefficients between prediction and ground truth averaged from five experiments (format: mean $\pm$ std%). The best results are highlighted in blue. $w_{\text{Dice}} = 0.8$ and $w_{\text{Cross}} = 0.2$ for all $L_{\text{Exp}}$.

(a) Semantic labels and their relative sizes on average (%).

| | | | | | | | |
|---|---|---|---|---|---|---|---|
| 1. Cerebral grey | (50.24) | 2. 3rd ventricle | (0.09) | 3. 4th ventricle | (0.15) | 4. Brainstem | (1.46) |
| 5. CVL I-V | (0.39) | 6. CVL VI-VII | (0.19) | 7. CVL VIII-X | (0.26) | 8. Accumbens | (0.07) |
| 9. Amygdala | (0.21) | 10. Caudate | (0.54) | 11. Cerebellar grey | (8.19) | 12. Cerebellar white | (2.06) |
| 13. Cerebral white | (31.23) | 14. Hippocampus | (0.58) | 15. Inf. lateral vent. | (0.09) | 16. Lateral ventricle | (2.11) |
| 17. Pallidum | (0.25) | 18. Putamen | (0.73) | 19. Thalamus | (1.19) | | |

(b) Average Dice coefficients (mean$\pm$std%) with respective to the ground truth.

Proposed network with linear Dice loss, logarithmic Dice loss, and weighted cross-entropy

| | | | | |
|---|---|---|---|---|
| $\mathbf{E}\,[1 - \text{Dice}_i]$ (2) | 1. 87$\pm$1   2. 47$\pm$38   3. 32$\pm$40   4. 72$\pm$36   5. 50$\pm$41   6. 30$\pm$37   7. 31$\pm$38 <br> 8. 0$\pm$0   9. 0$\pm$0   10. 34$\pm$42   11. 88$\pm$1   12. 86$\pm$1   13. 88$\pm$1   14. 32$\pm$39 <br> 15. 0$\pm$0   16. 54$\pm$44   17. 0$\pm$0   18. 51$\pm$42   19. 35$\pm$43   **Average: 43$\pm$11** | | | |
| $L_{\text{Dice}}(\gamma = 1)$ (2) | 1. 84$\pm$1   2. 61$\pm$30   3. 83$\pm$2   4. 90$\pm$1   5. 81$\pm$2   6. 73$\pm$2   7. 78$\pm$2 <br> 8. 68$\pm$2   9. 74$\pm$2   10. 85$\pm$1   11. 87$\pm$1   12. 85$\pm$1   13. 88$\pm$1   14. 79$\pm$2 <br> 15. 59$\pm$3   16. 89$\pm$1   17. 79$\pm$1   18. 86$\pm$2   19. 88$\pm$1   **Average: 80$\pm$2** | | | |
| $L_{\text{Cross}}(\gamma = 1)$ (3) | 1. 87$\pm$1   2. 56$\pm$5   3. 79$\pm$3   4. 86$\pm$2   5. 76$\pm$3   6. 67$\pm$2   7. 73$\pm$6 <br> 8. 59$\pm$4   9. 65$\pm$4   10. 83$\pm$2   11. 87$\pm$2   12. 85$\pm$1   13. 89$\pm$1   14. 75$\pm$3 <br> 15. 54$\pm$6   16. 89$\pm$1   17. 76$\pm$3   18. 84$\pm$1   19. 86$\pm$1   **Average: 77$\pm$2** | | | |

Proposed network with $L_{\text{Exp}}$ at different values of $\gamma$

| | | | | |
|---|---|---|---|---|
| $L_{\text{Exp}}(\gamma = 1)$ (1) | 1. 87$\pm$2   2. 78$\pm$3   3. 84$\pm$1   4. 90$\pm$1   5. 82$\pm$1   6. 74$\pm$2   7. 78$\pm$3 <br> 8. 68$\pm$3   9. 75$\pm$1   10. 83$\pm$3   11. 87$\pm$1   12. 86$\pm$0   13. 89$\pm$1   14. 80$\pm$1 <br> 15. 64$\pm$1   16. 90$\pm$1   17. 80$\pm$2   18. 86$\pm$2   19. 88$\pm$1   **Average: 81$\pm$1** | | | |
| $L_{\text{Exp}}(\gamma = 2)$ (1) | 1. 79$\pm$7   2. 61$\pm$15   3. 74$\pm$6   4. 75$\pm$10   5. 67$\pm$12   6. 62$\pm$8   7. 66$\pm$10 <br> 8. 52$\pm$17   9. 56$\pm$15   10. 64$\pm$12   11. 78$\pm$8   12. 78$\pm$7   13. 84$\pm$4   14. 64$\pm$11 <br> 15. 46$\pm$10   16. 77$\pm$10   17. 60$\pm$16   18. 67$\pm$15   19. 67$\pm$15   **Average: 67$\pm$11** | | | |
| $L_{\text{Exp}}(\gamma = 0.3)$ (1) | 1. 88$\pm$1   2. 77$\pm$2   3. 84$\pm$1   4. 91$\pm$1   5. 82$\pm$1   6. 74$\pm$1   7. 78$\pm$2 <br> 8. 69$\pm$2   9. 75$\pm$2   10. 86$\pm$1   11. 89$\pm$1   12. 86$\pm$1   13. 89$\pm$0   14. 81$\pm$1 <br> 15. 62$\pm$5   16. 91$\pm$1   17. 80$\pm$1   18. 87$\pm$1   19. 89$\pm$1   Average: 82$\pm$1 | | | |

V-Net with the best $L_{\text{Exp}}$ at $\gamma = 0.3$

| | | | | |
|---|---|---|---|---|
| V-Net <br> $L_{\text{Exp}}(\gamma = 0.3)$ (1) | 1. 84$\pm$2   2. 67$\pm$7   3. 80$\pm$4   4. 87$\pm$4   5. 78$\pm$3   6. 67$\pm$5   7. 73$\pm$6 <br> 8. 59$\pm$7   9. 65$\pm$5   10. 72$\pm$5   11. 85$\pm$2   12. 82$\pm$4   13. 86$\pm$2   14. 72$\pm$7 <br> 15. 48$\pm$8   16. 82$\pm$6   17. 70$\pm$7   18. 75$\pm$6   19. 78$\pm$5   **Average: 74$\pm$4** | | | |

is getting higher. In contrast, $L_{\text{Exp}}(\gamma = 0.3)$ gave the best results. Although it only performed slightly better than $L_{\text{Exp}}(\gamma = 1)$ in terms of the means, the smaller standard deviations indicate that it was also more precise.

When applying the best loss function to the V-Net, its performance was only better than the linear Dice loss and $L_{\text{Exp}}(\gamma = 2)$. This shows that our proposed network architecture performed better than the V-Net on this problem.

Figure 3 shows the validation Dice coefficients vs. epoch, averaged from the five experiments. Instead of the losses, we show the Dice coefficients as their magnitudes were consistent among cases. Similar to Table 1(b), the logarithmic Dice loss, $L_{\text{Exp}}(\gamma = 1)$, and $L_{\text{Exp}}(\gamma = 0.3)$ had good convergence and performance, with $L_{\text{Exp}}(\gamma = 0.3)$ performed slightly better. These three cases converged at about 80 epochs. The weighted cross-entropy and $L_{\text{Exp}}(\gamma = 2)$ were more fluctuating. The linear Dice loss also converged at about 80 epochs but with a much

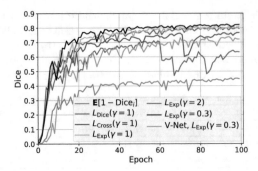

**Fig. 3.** Validation Dice coefficients vs. epoch, averaged from five experiments.

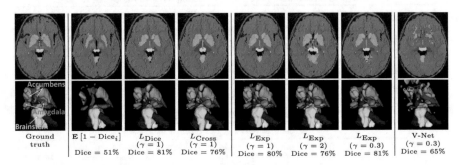

**Fig. 4.** Visualization of an example. Top: axial view. Bottom: 3D view with the cerebral grey, cerebral white, and cerebellar grey matters hidden for better illustration.

smaller Dice coefficient. Comparing between the V-Net and the proposed network with $L_{\mathrm{Exp}}(\gamma = 0.3)$, the V-Net had worse convergence especially at the earlier epochs. This shows that the proposed network had better convergence.

Figure 4 shows the visualization of an example. There are two obvious observations. First of all, consistent with Table 1(b), the linear Dice loss missed some small structures such as the nucleus accumbens and amygdala, though it performed well on large structures. Secondly, the segmentation of the V-Net deviated a lot from the ground truth. The logarithmic Dice loss, $L_{\mathrm{Exp}}(\gamma = 1)$, and $L_{\mathrm{Exp}}(\gamma = 0.3)$ had the best segmentations and average Dice coefficients. The weighted cross-entropy had the same average Dice coefficient as $L_{\mathrm{Exp}}(\gamma = 2)$, though the weighted cross-entropy over-segmented some structures such as the brainstem, and $L_{\mathrm{Exp}}(\gamma = 2)$ had a noisier segmentation.

Comparing the efficiencies between the proposed network and the V-Net, the proposed network had around 5 million parameters while the V-Net had around 71 million parameters, a 14-fold difference. Furthermore, the proposed network only took about 0.4 s on average to segment a $128 \times 128 \times 128$ volume, while the V-Net took about 0.9 s. Therefore, the proposed network was more efficient.

# 4   Conclusion

In this paper, we propose a network architecture optimized for 3D image segmentation, and a loss function for segmenting very small structures. The proposed network architecture has only about 1/14 of the parameters of, and is twice as fast as, the V-Net. For the loss function, the logarithmic Dice loss outperforms the linear Dice loss, and the weighted sum of the logarithmic Dice loss and the weighted cross-entropy outperforms the individual losses. With the introduction of the exponential form, the nonlinearities of the loss functions can be further controlled to improve the accuracy and precision of segmentation.

# References

1. Çiçek, Ö., Abdulkadir, A., Lienkamp, S.S., Brox, T., Ronneberger, O.: 3D u-net: learning dense volumetric segmentation from sparse annotation. In: Ourselin, S., Joskowicz, L., Sabuncu, M.R., Unal, G., Wells, W. (eds.) MICCAI 2016. LNCS, vol. 9901, pp. 424–432. Springer, Cham (2016). https://doi.org/10.1007/978-3-319-46723-8_49
2. Dou, Q., Yu, L., Chen, H., Jin, Y., Yang, X., Qin, J., Heng, P.A.: 3D deeply supervised network for automated segmentation of volumetric medical images. Med. Image Anal. **41**, 40–54 (2017)
3. Lee, C.Y., Xie, S., Gallagher, P.W., Zhang, Z., Tu, Z.: Deeply-supervised nets. In: International Conference on Artificial Intelligence and Statistics, pp. 562–570 (2015)
4. Lin, T.Y., Goyal, P., Girshick, R., He, K., Dollár, P.: Focal loss for dense object detection. arXiv:1708.02002 [cs.CV] (2017)
5. Mehta, R., Sivaswamy, J.: M-net: a convolutional neural network for deep brain structure segmentation. In: IEEE International Symposium on Biomedical Imaging, pp. 437–440 (2017)
6. Milletari, F., Navab, N., Ahmadi, S.A.: V-net: fully convolutional neural networks for volumetric medical image segmentation. In: IEEE International Conference on 3D Vision, pp. 565–571 (2016)
7. Ronneberger, O., Fischer, P., Brox, T.: U-net: convolutional networks for biomedical image segmentation. In: Navab, N., Hornegger, J., Wells, W.M., Frangi, A.F. (eds.) MICCAI 2015. LNCS, vol. 9351, pp. 234–241. Springer, Cham (2015). https://doi.org/10.1007/978-3-319-24574-4_28
8. Roy, A.G., Conjeti, S., Sheet, D., Katouzian, A., Navab, N., Wachinger, C.: Error corrective boosting for learning fully convolutional networks with limited data. In: Descoteaux, M., Maier-Hein, L., Franz, A., Jannin, P., Collins, D.L., Duchesne, S. (eds.) MICCAI 2017. LNCS, vol. 10435, pp. 231–239. Springer, Cham (2017). https://doi.org/10.1007/978-3-319-66179-7_27
9. Sudre, C.H., Li, W., Vercauteren, T., Ourselin, S., Jorge Cardoso, M.: Generalised dice overlap as a deep learning loss function for highly unbalanced segmentations. In: Cardoso, M.J., et al. (eds.) DLMIA/ML-CDS -2017. LNCS, vol. 10553, pp. 240–248. Springer, Cham (2017). https://doi.org/10.1007/978-3-319-67558-9_28
10. Tang, H., et al.: Segmentation of anatomical structures in cardiac CTA using multi-label V-Net. In: Medical Imaging 2018: Image Processing, vol. 10574, p. 1057407 (2018)

# Revealing Regional Associations of Cortical Folding Alterations with In Utero Ventricular Dilation Using Joint Spectral Embedding

Oualid M. Benkarim[1]([✉]), Gerard Sanroma[5], Gemma Piella[1], Islem Rekik[2],
Nadine Hahner[3], Elisenda Eixarch[3], Miguel Angel González Ballester[1,6],
Dinggang Shen[4], and Gang Li[4]

[1] BCN Medtech, Universitat Pompeu Fabra, Barcelona, Spain
oualid.benkarim@upf.edu
[2] BASIRA Lab, CVIP Group, School of Science and Engineering, Computing,
University of Dundee, Dundee, UK
[3] BCNatal, Hospital Clínic and Hospital Sant Joan de Déu, Barcelona, Spain
[4] Department of Radiology and BRIC, University of North Carolina at Chapel Hill,
Chapel Hill, NC, USA
[5] Deutsche Zentrum für Neurodegenerative Erkrankungen (DZNE), Bonn, Germany
[6] ICREA, Barcelona, Spain

**Abstract.** Fetal ventriculomegaly (VM) is a condition with dilation of one or both lateral ventricles, and is diagnosed as an atrial diameter larger than 10 mm. Evidence of altered cortical folding associated with VM has been shown in the literature. However, existing studies use a holistic approach (i.e., ventricle as a whole) based on diagnosis or ventricular volume, thus failing to reveal the spatially-heterogeneous association patterns between cortex and ventricle. To address this issue, we develop a novel method to identify spatially fine-scaled association maps between cortical development and VM by leveraging vertex-wise correlations between the growth patterns of both ventricular and cortical surfaces in terms of area expansion and curvature information. Our approach comprises multiple steps. In the first step, we define a joint graph Laplacian matrix using cortex-to-ventricle correlations. Next, we propose a spectral embedding of the cortex-to-ventricle graph into a common underlying space where their joint growth patterns are projected. More importantly, in the joint ventricle-cortex space, the vertices of associated regions from both cortical and ventricular surfaces would lie close to each other. In the final step, we perform clustering in the joint embedded space to identify associated sub-regions between cortex and ventricle. Using a dataset of 25 healthy fetuses and 23 fetuses with isolated non-severe VM within the age range of 26–29 gestational weeks, our results show that the proposed approach is able to reveal clinically relevant and meaningful regional associations.

**Keywords:** Joint spectral embedding · Ventriculomegaly · Fetal Cortical folding

© Springer Nature Switzerland AG 2018
A. F. Frangi et al. (Eds.): MICCAI 2018, LNCS 11072, pp. 620–627, 2018.
https://doi.org/10.1007/978-3-030-00931-1_71

# 1   Introduction

During the intrauterine life, the fetal brain undergoes drastic maturational changes. Cortical folding is one of the major processes that occurs during this period, and any deviation from its normal developmental course might lead to adverse postnatal outcome [3]. In prenatal ultrasound examination, ventriculomegaly (VM) is the most frequent abnormal finding in the fetal brain. VM is a condition with dilation of one or both lateral ventricles, as shown in Fig. 1A. It is diagnosed as an atrial diameter larger than 10 mm at any gestational age [4]. Evidence of altered cortical folding associated with *in utero* VM has been shown by studies in the literature. Among their findings, cortical gray matter was significantly enlarged in fetuses with isolated VM [5]. Using curvature-based analysis, studies also found reduced cortical folding in the insula, the occipital lobe and the posterior part of the temporal lobe [2,9].

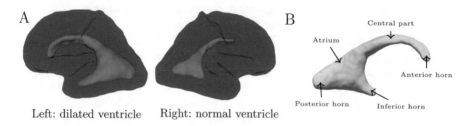

Left: dilated ventricle    Right: normal ventricle

**Fig. 1.** A: Cortical and ventricular surfaces of a 28 gestational weeks fetus with left VM. B: Regions of the lateral ventricle.

To study the association between VM and the morphology of cortical folding, existing works either use diagnosis or ventricular volume to characterize this condition. Although ventricular volume captures the extent of enlargement and is more distinctive than the dichotomous information offered by diagnosis, a single scalar value might not be sufficient to provide all the information related to ventricular enlargement (e.g., spatial information about the dilated ventricular regions). In this work, we aim to find associations between ventricular regions (see Fig. 1B) and cortical folding by incorporating into our analysis the ventricular surfaces. For this purpose, we propose a novel approach to jointly analyze the cortical and ventricular shapes based on their growth patterns. The motivation of using growth patterns is their ability to reflect the underlying micro-structural brain changes. The main idea of our approach is to find a common underlying representation of the vertex-wise growth patterns of both cortical and ventricular surfaces such that vertices with associated patterns from both anatomical surfaces can lie close to each other. In this way, regional associations can be conveniently found by identifying clusters containing vertices from both surfaces in the new latent space. The contributions of our work are threefold:

- We propose a novel approach for joint analysis of different anatomical shapes based on their growth patterns.

– We identify, for the first time, spatially fine-scaled associations related to *in utero* VM between ventricular surfaces and alterations in cortical folding.
– We use fusion of similarity matrices to capture associations based on multiple cortical features.

## 2    Method

Given $P$ subjects and their corresponding cortical and ventricular surfaces with $N_c$ and $N_v$ vertices respectively, for each subject, the growth patterns $\mathbf{x}_i$ for each vertex are represented by:

$$\mathbf{x}_i = [x_i^1, x_i^2, \cdots, x_i^P], \tag{1}$$

where $x_i^k$ is the feature (e.g., local surface area) of the $k$-th subject at vertex $i$. In this study, growth patterns were built using a cross-sectional dataset. Although it is preferable to use longitudinal data, repeated *in utero* imaging is difficult due to ethical and practical issues.

Cortical and ventricular growth patterns are not necessarily to be represented using the same feature (e.g., we can use area for ventricles while curvature for cortices). We assume that there exists a common underlying representation for these heterogeneous growth patterns, $\mathbf{x}_i$, such that vertices of associated regions from both surfaces can lie close to each other and, most likely, form dense clusters. Thus, we propose to find a shared representation of cortical and ventricular growth patterns using joint projection onto a common space:

$$Y = \underset{Y}{\operatorname{argmin}} \sum_{i,j} \|Y_i^c - Y_j^c\|^2 S_c(i,j) + \sum_{i,j} \|Y_i^v - Y_j^v\|^2 S_v(i,j)$$

$$+\mu \sum_{i,j} \|Y_i^c - Y_j^v\|^2 S_{cv}(i,j), \tag{2}$$

where $Y = [Y^c, Y^v]^T$ is the common latent representation with $N = (N_c + N_v)$ rows such that the first $N_c$ rows correspond to the embedded cortical growth patterns (i.e., $Y^c$) and the remaining $N_v$ rows belong to the ventricle (i.e., $Y^v$), $T$ stands for matrix transpose, $S_c$ and $S_v$ are the intra-structure similarity matrices, $S_{cv}$ is the similarity matrix between cortical and ventricular growth patterns, and $\mu$ is a tradeoff parameter. Given two similar (i.e., high $S_{cv}(i,j)$) growth patterns, $\mathbf{x}_i^c$ and $\mathbf{x}_j^v$ from cortex and ventricle respectively, the third term in Eq. (2) enforces their projections (i.e., $Y_i^c$ and $Y_j^v$) to fall close to each other. This also occurs for similar growth patterns from the same surface (enforced by the first and second terms).

Since we are interested in identifying associations between the growth patterns of both structures, similarity between the growth patterns is defined in terms of correlation. First, we build the inter-structure similarity matrix based on the correlations between the growth patterns of both surfaces as follows:

$$S_{cv}(i,j) = \frac{1 + \rho(\mathbf{x}_i^c, \mathbf{x}_j^v)}{2}, \tag{3}$$

where $\rho$ is Pearson's correlation coefficient. Similarly, intra-structure similarity matrices ($S_c$ and $S_v$) are built to capture within surface correlations. The joint similarity matrix is constructed by filling its block-diagonal with the intra-structure matrices and the off-diagonal with the inter-structure matrix:

$$S = \begin{pmatrix} S_c & \mu S_{cv} \\ \mu S_{cv}^T & S_v \end{pmatrix}. \tag{4}$$

Then, we compute the normalized Laplacian of the joint similarity matrix:

$$L = I - D^{-1/2} S D^{-1/2}, \tag{5}$$

where $D$ is the degree matrix of $S$ (i.e., a diagonal matrix such that $D(i,i) = \sum_j S(i,j)$), and $I$ is the identity matrix. Laplacian eigenmaps [1] can then be used to solve Eq. (2) based on the joint Laplacian and find the common underlying space $Y$.

To discover the regional relationships induced by ventricular enlargement, we cluster the embedded growth patterns using hierarchical clustering. Associated regions are identified by clusters containing vertices from both shapes.

## Features for Cortical Growth Patterns

The area of ventricular surfaces increases dramatically with the enlargement and can be considered reliable in capturing the ventricular dilation. However, alterations in cortical folding can be characterized by multiple distinct features. Therefore, we extend our approach to include both area and curvedness (derived from curvature) as features for cortical surfaces by fusing the similarity matrices created for each of them with ventricular area: $S_1$ built using area for both structures, and $S_2$ using curvedness for cortices. For each similarity matrix, $S_m$, $m \in \{1, 2\}$, we derive two matrices [10]:

$$P_m(i,j) = \begin{cases} \frac{S_m(i,j)}{2 \sum_{k \neq i} S_m(i,k)} & i \neq j \\ 1/2 & \text{otherwise.} \end{cases} \tag{6}$$

$$W_m(i,j) = \begin{cases} \frac{S_m(i,j)}{2 \sum_{k \in \mathcal{N}_i} S_m(i,k)} & j \in \mathcal{N}_i \\ 0 & \text{otherwise,} \end{cases} \tag{7}$$

where $\mathcal{N}_i$ denotes the neighborhood of the $i$-th vertex in terms of the vertices with most correlated growth patterns. Fusion is then iteratively conducted:

$$P_1^{t+1} = W_1 P_2^t W_1^T, \qquad P_2^{t+1} = W_2 P_1^t W_2^T, \tag{8}$$

where $P_m$ and $W_m$ are the dense and sparse similarity matrices derived from $S_m$ (i.e., $S_1$ and $S_2$). In this way, the reliable similarity information is diffused across matrices. Finally, the dense matrices are averaged to obtain the fused matrix:

$$P_f = (P_1 + P_2)/2. \tag{9}$$

The fused similarity matrix, $P_f$, is able to capture common and complementary associations, and remove spurious and isolated correlations. We use $P_f$ (rather than $S$) to compute the joint Laplacian and project the growth patterns.

## 3    Experiments

### 3.1    Data and Preprocessing

In our experiments, we used a fetal brain MRI dataset of 25 controls and 23 subjects with isolated non-severe ventriculomegaly (INSVM) between 26 and 29 gestational weeks. The INSVM cohort was composed of fetuses with unilateral or bilateral ventricular width between 10–14.9 mm, with no abnormal karyotype, infections or malformations with risk of abnormal neurodevelopment. T2-weighted MR images were acquired on a 1.5T scanner (SIEMENS MAGNETOM Aera) with an 8-channel body coil. For each subject, multiple orthogonal 2D scans (i.e., 4 axial, 2 coronal, and 2 sagittal stacks) were collected, from which a high-resolution 3D image was reconstructed using the method in [7].

Tissue segmentation was performed on the reconstructed images using a learning-based method [8]. Then, cortical and ventricular surfaces were extracted for each hemisphere. In order to establish vertex-wise correspondences, for each structure, surfaces were co-registered and resampled to the same number of vertices [11].

### 3.2    Experimental Setup

Ventricular growth patterns were built with area information from each vertex, which was computed as one third of the total area of adjacent triangles [6]. For cortices, we used both area and curvedness. Thus, we conducted 3 different experiments, using: (1) correlations between ventricular area and cortical area, (2) ventricular area and cortical curvedness, and (3) fusing both similarity matrices (i.e., using both area and curvedness from the cortices and area for the ventricles). For clustering, we used 2 to 25 clusters to illustrate the number of correlated regions identified with different clusters. The optimal associations between ventricles and cortices were determined by finding the most appropriate number of cluster using the silhouette coefficient: $s(i) = (b(i) - a(i))/max(a(i), b(i))$, where $i$ indexes vertices in the embedded space, $a(i)$ is the mean distance between the $i$-th vertex to the rest of vertices in its cluster, and $b(i)$ is the minimum average distance computed with the vertices in the rest of clusters.

### 3.3    Results

Although there may exist contralateral associations and unilateral ventricular enlargement may be associated with alterations in the opposite hemisphere, for this work, associations were only studied for each hemisphere independently. Figure 2 shows the associations identified by our approach in the left hemisphere between ventricular dilation and cortical folding. Surfaces are displayed such that cortical and ventricular regions found to be associated are depicted with the same color code. From these results, regardless of using area or curvedness to characterize the cortical growth patterns, we can observe that, with 3 clusters, the posterior part of the ventricular surface and the posterior part of the cortical surface fall into the same cluster (blue for area and pink for curvedness). This pattern is replicated for the anterior part (green for area and cyan for curvedness) and further preserved with increasing clusters, as clusters emerge in the posterior/anterior parts of both surfaces. As the number of clusters increases, we obtain more localized and fine-grained associations (i.e., shared clusters), which emphasizes the strength of the maintained associations.

Comparing the associations found when using area expansion and curvature information for the cortex, we can see that, with 8 clusters, the anterior horn and part of the ventricular body are associated with a region nearby the anterior cingulate gyrus. This association is captured with cortical area (green) for a larger number of clusters than curvedness (cyan). The most important association found by curvedness is between the posterior (i.e., occipital) horn and the occipital lobe (pink). With 8 clusters, the association includes the calcarine and the parieto-occipital fissures, although only a small part of the latter fissure is preserved with 20 clusters. This association is also found by cortical area with 8 (blue) and 15 clusters (yellow).

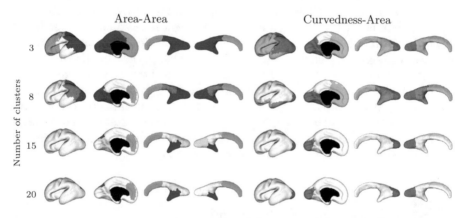

**Fig. 2.** Associations identified in left hemisphere using correlations between: (a) cortical area and ventricular area, and (b) cortical curvedness and ventricular area. Cortex-ventricle associations are color-coded, with white depicting no associations.

Still, using a single feature to describe the growth patterns might not be able to capture all putative associations or give rise to spurious ones. Results using the fused similarity matrix for different number of clusters are shown for both hemispheres in Fig. 3. Noteworthy is that associations found in both hemispheres are in large overlap, with the only difference being the correlation between the ventricular body and the anterior horn with the anterior cingulate gyrus (red) in the left cortex. Nonetheless, in both hemispheres, the posterior horn and the occipital lobe belonged in the same cluster (green in left hemisphere), and the inferior horn and the atrium (blue and cyan in left and right hemispheres, respectively) showed to be correlated with the superior part of the parietal lobe. Associations corresponding to the best clustering in terms of silhouette score are

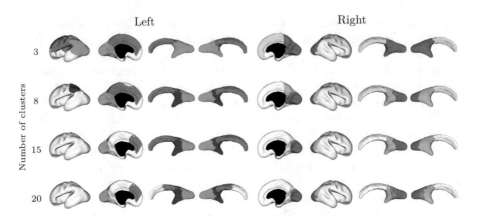

**Fig. 3.** Associations found using fused similarity matrix for each hemisphere separately. Cortex-ventricle associations are color-coded, with white depicting no associations.

shown in Fig. 4. The highest values of silhouette coefficient were found with 17 and 14 clusters for the left and right hemispheres, respectively. The atrium was identified bilaterally (blue and pink for left and right hemispheres, respectively). Since the atrium is the ventricular region used in clinical practice to diagnose VM, this highlights the clinic relevance of our results. In the cortex, the occipital lobe was found to be associated with the posterior horn (green and pink) in both hemispheres, which is consistent with findings in the literature [2,9]. In association with VM, our approach is able to identify meaningful cortical and ventricular regions. Furthermore, it provides the means to establish relationships between these regions and gain more insight into the fine-grained associations between ventricular enlargement and cortical development.

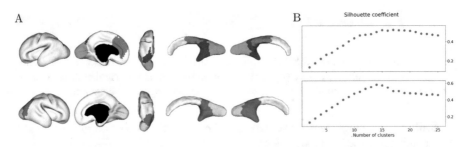

**Fig. 4.** A: Regional associations identified using fused similarity matrix with the optimal number of clusters in terms of silhouette coefficient for left (top) and right (bottom) hemispheres. B: Silhouette scores for different number of clusters for each hemisphere.

## 4    Conclusions

In this work, we have presented a novel approach to identify spatially fine-grained correlations between different shapes based on their growth patterns. This is the first work that approaches the study of associations between fetal VM and cortical folding alterations by jointly analyzing cortical and ventricular shapes. Our results demonstrate that the proposed approach is able to identify clinically relevant regions (e.g., atrium in the ventricle and occipital lobe in the cortex) and further establish relationships between these regions. For future work, instead of fusing similarity matrices from different features prior to performing the embedding, multi-view approaches can be explored. Also, additional features (e.g., local gyrification index) can be used to identify other correlated regions.

**Acknowledgments.** This research was partially funded by the "Fundació La Marató de TV3" (no. 20154031) and supported in part by National Institutes of Health grants (MH100217, MH107815, MH108914, and MH116225). It has also been funded by Instituto de Salud Carlos III (PI16/00861) integrados en el Plan Nacional de I+D+I y cofinanciados por el ISCIII-Subdirección General de Evaluacin y el Fondo Europeo de Desarrollo Regional (FEDER) "Una manera de hacer Europa", "la Caixa" Foundation, and The Cerebra Foundation for the Brain-Injured Child, Carmarthen, Wales.

# References

1. Belkin, M., Niyogi, P.: Laplacian eigenmaps for dimensionality reduction and data representation. Neural Comput. **15**(6), 1373–1396 (2003)
2. Benkarim, O.M., Hahner, N., Piella, G., et al.: Cortical folding alterations in fetuses with isolated non-severe ventriculomegaly. NeuroImage: Clin. **18**, 103–114 (2018)
3. Benkarim, O.M., Sanroma, G., Zimmer, V.A., et al.: Toward the automatic quantification of in utero brain development in 3D structural MRI: a review. Hum. Brain Mapp. **38**(5), 2772–2787 (2017)
4. Cardoza, J., Goldstein, R., Filly, R.: Exclusion of fetal ventriculomegaly with a single measurement: the width of the lateral ventricular atrium. Radiology **169**(3), 711–714 (1988)
5. Kyriakopoulou, V., Vatansever, D., Elkommos, S., et al.: Cortical overgrowth in fetuses with isolated ventriculomegaly. Cereb Cortex **24**(8), 2141–2150 (2014)
6. Li, G., Nie, J., Wang, L., et al.: Mapping region-specific longitudinal cortical surface expansion from birth to 2 years of age. Cereb Cortex **23**(11), 2724–2733 (2013)
7. Murgasova, M., Quaghebeur, G., Rutherford, M.: Reconstruction of fetal brain MRI with intensity matching and complete outlier removal. Med. Image Anal. **16**(8), 1550–1564 (2012)
8. Sanroma, G., Benkarim, O.M., Piella, G., Ballester, M.Á.G.: Building an ensemble of complementary segmentation methods by exploiting probabilistic estimates. In: Wang, L., Adeli, E., Wang, Q., Shi, Y., Suk, H.-I. (eds.) MLMI 2016. LNCS, vol. 10019, pp. 27–35. Springer, Cham (2016). https://doi.org/10.1007/978-3-319-47157-0_4
9. Scott, J., Habas, P., Rajagopalan, V., et al.: Volumetric and surface-based 3D MRI analyses of fetal isolated mild ventriculomegaly. Brain Struct. Funct. **218**(3), 645–655 (2013)
10. Wang, B., Mezlini, A.M., Demir, F., et al.: Similarity network fusion for aggregating data types on a genomic scale. Nat. Methods **11**, 333–337 (2014)
11. Xia, J., Zhang, C., Wang, F., et al.: Fetal cortical parcellation based on growth patterns. In: IEEE International Symposium on Biomedical Imaging (ISBI), pp. 696–699 (2018)

# CompNet: Complementary Segmentation Network for Brain MRI Extraction

Raunak Dey[(✉)] and Yi Hong

Computer Science Department, University of Georgia, Athens, USA
rd31879@uga.edu

**Abstract.** Brain extraction is a fundamental step for most brain imaging studies. In this paper, we investigate the problem of skull stripping and propose complementary segmentation networks (CompNets) to accurately extract the brain from T1-weighted MRI scans, for both normal and pathological brain images. The proposed networks are designed in the framework of encoder-decoder networks and have two pathways to learn features from both the brain tissue and its complementary part located outside of the brain. The complementary pathway extracts the features in the non-brain region and leads to a robust solution to brain extraction from MRIs with pathologies, which do not exist in our training dataset. We demonstrate the effectiveness of our networks by evaluating them on the OASIS dataset, resulting in the state of the art performance under the two-fold cross-validation setting. Moreover, the robustness of our networks is verified by testing on images with introduced pathologies and by showing its invariance to unseen brain pathologies. In addition, our complementary network design is general and can be extended to address other image segmentation problems with better generalization.

## 1 Introduction

Image segmentation aims to locate and extract objects of interest from an image, which is one of the fundamental problems in medical research. Take the brain extraction problem as an example. To study the brain, magnetic resonance imaging (MRI) is the most popular modality choice. However, before the quantitative analysis of brain MRIs, e.g., measuring normal brain development and degeneration, uncovering brain disorders such as Alzheimer's disease, or diagnosing brain tumors or lesions, skull stripping is typically a preliminary but essential step, and many approaches have been proposed to tackle this problem.

In literature, the approaches developed for brain MRI extraction can be divided into two categories: traditional methods (manual, intensity or shape model based, hybrid, and PCA-based methods [3,11]) and deep learning methods [7,10]. Deep neural networks have demonstrated the improved quality of the predicted brain mask, compared to traditional methods. However, these deep networks focus on learning image features mainly for brain tissue from a training dataset, which is typically a collection of normal (or apparently normal) brain MRIs, because these images are more commonly available than brain scans with

© Springer Nature Switzerland AG 2018
A. F. Frangi et al. (Eds.): MICCAI 2018, LNCS 11072, pp. 628–636, 2018.
https://doi.org/10.1007/978-3-030-00931-1_72

pathologies. Thus, their model performance is sensitive to unseen pathological tissues.

In this paper, we propose a novel deep neural network architecture for skull stripping from brain MRIs, which improves the performance of existing methods on brain extraction and more importantly is invariant to brain pathologies by only training on publicly available regular brain scans. In our new design, a network learns features for both brain tissue and non-brain structures, that is, we consider the complementary information of an object that is outside of the region of interest in an image. For instance, the structures outside of the brain, e.g., the skull, are highly similar and consistent among the normal and pathological brain images. Leveraging such complementary information about the brain can help increase the robustness of a brain extraction method and enable it to handle images with unseen structures in the brain.

We explore multiple complementary segmentation networks (CompNets). In general, these networks have two pathways in common: one to learn what is the brain tissue and to generate a brain mask; the other to learn what is outside of the brain and to help the other branch generate a better brain mask. There are three variants, i.e., the probability, plain, and optimal CompNets. In particular, the probability CompNet needs an extra step to generate the ground truth for the complementary part such as the skull, while the plain and optimal CompNets do not need this additional input. The optimal CompNet is built upon the plain one and introduces dense blocks (a series of convolutional layers fully connected to each other [4]) and multiple intermediate outputs [1], as shown in Fig. 1. This optimal CompNet has an end-to-end design, fewer number of parameters to estimate, and the best performance among the three CompNets on both normal and pathological images from the OASIS dataset. In addition, this network is generic and can be applied in image segmentation if the complementary part of an object contributes to the understanding of the object in the image.

## 2   CompNets: Complementary Segmentation Networks

An encoder-decoder network, like U-Net [9], is often used in image segmentation. Current segmentation networks mainly focus on objects of interest, which may lead to the difficulty in its generalization to unseen image data. In this section, we introduce our novel complementary segmentation networks (short for CompNet), which increase the segmentation robustness by incorporating the learning process of the object of interest with the learning of its complementary part in the image.

The architecture of our optimal CompNet is depicted in Fig. 1. This network has three components. The first component is a segmentation branch for the region of interest (ROI) such as the brain, which follows the U-Net architecture and generates the brain mask. Given normal brain scans, a network with only this branch focuses on extracting features of standard brain tissue, resulting in its difficulty of handling brain scans with unseen pathologies. To tackle this problem, we augment the segmentation branch by adding a complementary one to learn structures in the non-brain region, because they are relatively consistent among

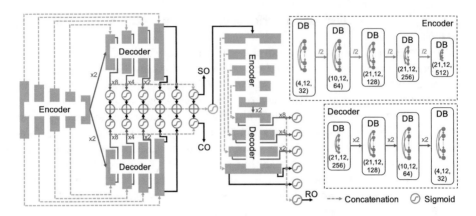

**Fig. 1.** Architecture of our complementary segmentation network, the optimal Comp-Net. The dense blocks (DB), corresponding to the gray bars, are used in each encoder and decoder. The triple $(x, y, z)$ in each dense block indicates that it has $x$ convolutional layers with a kernel size $3 \times 3$; each layer has $y$ filters, except for the last one that has $z$ filters. SO: segmentation output for the brain mask; CO: complementary segmentation output for the non-brain mask; RO: reconstruction output for the input image. These three outputs produced by the Sigmoid function are the final predictions; while all other Sigmoids produce intermediate outputs, except for the green one that generates input for the image reconstruction sub-network. Best viewed in color.

normal and pathological images. However, due to the lack of true masks for the complementary part, we adopt a sub-encoder-decoder network to reconstruct the input brain scan based on the outputs of the previous two branches. This third component guides the learning process of the complementary branch, similar to the unsupervised W-Net [13]. It provides direct feedback to the segmentation and complementary branches and expects reasonable predictions from them as input to reconstruct the original input image. The complementary branch indirectly affects the segmentation branch since they share the encoder to extract features.

The optimal CompNet includes dense blocks and multiple intermediate outputs, which help reduce the number of parameters to estimate and make the network easier to optimize. For readability and a better understanding, we start with a discussion of the plain version in detail.

**Plain CompNet.** The plain network is a simplified version of the network shown in Fig. 1. Similar to U-Net, the encoder and decoder blocks (the gray bars in Fig. 1) of the segmentation and reconstruction sub-networks have two convolutional layers in each, with a kernel of size $3 \times 3$. The number of convolutional filters in the encoder starts from 32, followed by 64, 128, 256, and 512, while the number in the decoder starting from 256, followed by 128, 64, and 32. Each convolutional layer is followed by batch normalization [5] and dropout [12]. After each gray bar in the encoder, the feature maps are downsampled by 2 using max pooling; while for the decoder the feature maps are upsampled by 2 using deconvolutional layers. Each segmentation branch of this plain network has only one final output from the last layer of the decoder, after applying the Sigmoid function. The two outputs of the segmentation branches are combined through addition and passed as input to the reconstruction sub-network. In this sub-network, the output from the last layer of the decoder is the reconstructed image. Like U-Net, we have the concatenation of feature maps from an encoder to its decoder at the same resolution level, shown by the gray arrows in Fig. 1.

We use the Dice coefficient $(Dice(A, B) = 2|A \cap B|/(|A| + |B|)$ [2]) in the objective function to measure the goodness of segmentation predictions and the mean squared error (MSE) to measure the goodness of the reconstruction. In particular, the learning goal of this network is to maximize the Dice coefficient between the predicted mask for the ROI $(\hat{Y}_S)$ and its ground truth $(Y_S)$, minimize the Dice coefficient between the predicted mask for the non-ROI $(\hat{Y}_C)$ and the ROI true mask, and minimize the MSE between the reconstructed image $(\hat{X}_R)$ and the input image $(X)$. We formulate the loss function for one sample as

$$Loss(Y_S, \hat{Y}_S, \hat{Y}_C, X, \hat{X}_R) = -Dice(Y_S, \hat{Y}_S) + Dice(Y_S, \hat{Y}_C) + MSE(X, \hat{X}_R). \quad (1)$$

Here, the reconstruction loss ensures that the complementary output is not an empty image and that the summation of segmentation and complementary outputs is not an entirely white image, because such inputs without a whole brain and skull structure map will result in a substantial reconstruction error.

**Optimal CompNet.** The plain CompNet has nearly 18 million parameters. Introducing dense connections among convolutional layers can considerably reduce the number of parameters of a network and mitigate the vanishing gradient problem in a deep neural network. Therefore, we replace each gray block in the plain CompNet with a dense block, as shown in Fig. 1. Each dense block has different numbers of convolutional layers and filters. Specifically, the dense blocks in each encoder have 4, 10, 21, 21, and 21 convolutional layers, respectively, and the ones in each decoder have 21, 21, 10, and 4 layers, respectively. All these convolutional layers use the same kernel size $3 \times 3$ and the number of convolutional filters is 12 in each layer, except for the last one, changing from 32, to 64, to 128, to 256, and to 512 in the five dense blocks of the encoder while changing from 256, to 128, to 64, and to 32 in the four dense blocks of the decoder. This design aims to increase the amount of information that is transferred from one dense block to its next one by using more feature maps. In addition, we place dropout at the transition between dense blocks. Through adopting these dense blocks, our optimal CompNet becomes much deeper while having fewer parameters (15.3 million) to optimize, compared to the plain one.

Another change made to the plain CompNet is introducing multiple intermediate outputs [1]. These early outputs can mitigate the vanishing gradient problem in a deep neural network by shorting the distance from the input to the output. As shown in Fig. 1, each decoder in the segmentation and reconstruction sub-networks has six outputs, one after each Sigmoid function. The first five outputs are intermediate outputs, which are generated from the original and upsampled feature maps of the first convolutional layer in each dense block of the decoder and the feature maps of the last convolutional layer in the last dense block. We observe that having an intermediate output at the beginning of each dense-block provides better performance than having it at the end. An extra one at the end of the last dense-block allows collecting features learned by this block. The concatenation of all feature maps used for the intermediate outputs generates the sixth output, which is the final output of that branch for prediction. Furthermore, we use addition operations to integrate each pair of intermediate or final outputs from the two segmentation branches and then use the concatenation operation to collect all of them, resulting in the input for the reconstruction sub-network via a Sigmoid function (the green one in Fig. 1). Each Sigmoid layer produces a feature map with one channel using a $1 \times 1 \times 1$ convolutional filter, which normalizes its response value within $[0, 1]$.

**Probability CompNet.** The reconstruction sub-network is to guide the learning process of the complementary pathway. One might replace it by providing the ground truth of the complementary part for training, e.g., generating the skull mask. This strategy is our first attempt, and it could be non-trivial for images with a noisy background. After having the true masks for both brain and non-brain regions, we can build a network containing only the segmentation and complementary branches in Fig. 1, by removing the reconstruction component. To leverage the complementary information, we build connections between the convolutional layers of the two branches at the same resolution level. In particular, the feature maps of a block from one segmentation branch are converted to a probability map, which is inverted and multiplied to the feature maps at the same resolution level of the other branch. We perform the same operations on the other branch. Essentially, one branch informs the other to focus on learning features of its complementary part. This network can also handle brain extraction from pathological images; however, both brain and skull masks are needed for training, and the image background noise will influence the result. Although we can set an intensity threshold to denoise the background, this hyper-parameter may vary among images collected from different brain MRI scanners.

## 3   Experiments

**Datasets.** We evaluate CompNets on the OASIS dataset [8], which consists of a collection of T1-weighted brain MRI scans of 416 subjects aged 18 to 96 and 100 of them clinically diagnosed mild to moderate Alzheimer's disease. We use a subset with 406 subjects that have both brain images and masks available, with image dimension of $256 \times 256 \times 256$. These subjects are randomly shuffled and equally divided into two chunks for training and testing with two-fold cross-validation, similar to [7] for comparison on (apparently) normal brain images.

To further evaluate the robustness of our networks, in one chunk of the OASIS subset we introduce brain pathologies, such as synthetic 3D brain tumors and lesions with different intensity distributions, into the images at different locations and with different sizes, as well as damaged skulls and non-brain tissue membranes, as shown in the first column of Fig. 2. We train networks on the other chunk of unchanged images and test them on this chunk of noisy images.

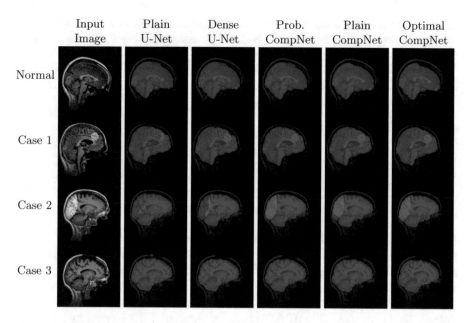

**Fig. 2.** Qualitative comparison among five networks, plain and dense U-Nets, probability, plain, and optimal CompNets, on four image samples: a normal one, one with pathology inside of the brain (case 1), one with pathology on the boundary of the brain (case 2), and one with a damaged skull (case 3). The true (red) and predicted (blue) masks are superimposed over the original images. The purple color indicates a perfect overlap between the ground truth and the prediction. Best viewed in color.

**Experimental Settings.** Apart from dropout with a rate of 0.3, we also use $L_2$ regularizer to penalize network weights with large magnitudes and its control hyperparameter $\lambda$ is set to 2e−4. For training, we use the Adam optimizer [6] with a learning rate 1e−3. All networks run up to 10 epochs.

**Experimental Results.** We compare our CompNets with a 3D deep network proposed in [7], a plain U-Net (the backbone of the plain CompNet), and a dense U-Net (the backbone of the optimal CompNet). These networks are tested on (apparently) normal images (with two-fold cross-validation) and on pathological images (with being trained on the other fold with clean images). Given a 3D brain MRI scan of a subject, our networks accept 2D slices and predict brain masks slice by slice, which are stacked back to a 3D mask without any post-processing.

Figure 2 demonstrates the qualitative comparison among predicted brain masks. For (apparently) normal brain scans, all networks produce visually acceptable brain masks. However, the plain and dense U-Nets have difficulties in handling images with pathologies, especially the pathological tissue on or near the boundary of the brain. Part of the pathological tissue in the brain is considered as non-brain tissue. The plain U-Net even oversegments part of the skull as the brain when the skull intensity changes, as shown in the case 3.

**Table 1.** Quantitative comparison (mean and standard deviation in percentage) among different networks on (apparently) normal and pathological images. *This paper is not directly comparable to our networks, because it was evaluated on mixed data samples, including 77 images (57%) from OASIS data set. (Prob.: Probability; Opti.: Optimal)

| | (Apparently) normal images | | | Pathological images | | |
|---|---|---|---|---|---|---|
| | Dice | Sensitivity | Specificity | Dice | Sensitivity | Specificity |
| Kleesiek et al. [7]* | 95.77 ± 0.01 | 94.25 ± 0.03 | 99.36 ± 0.003 | – | – | – |
| Plain U-Net | 92.30 ± 6.20 | 95.60 ± 1.48 | 96.20 ± 0.09 | 79.90 ± 8.10 | 93.80 ± 5.10 | 95.20 ± 2.15 |
| Dense U-Net | 96.40 ± 4.10 | 97.50 ± 0.70 | 96.90 ± 0.01 | 85.43 ± 5.80 | 96.13 ± 3.20 | 97.10 ± 1.27 |
| Prob. CompNet | 95.10 ± 0.19 | 96.73 ± 0.90 | 96.03 ± 0.02 | 92.10 ± 5.23 | 96.32 ± 1.90 | 98.86 ± 0.50 |
| Plain CompNet | 96.70 ± 0.22 | 97.93 ± 0.62 | 98.57 ± 0.06 | 95.21 ± 3.75 | 96.32 ± 1.03 | 99.21 ± 0.10 |
| Opti. CompNet | **98.27 ± 0.30** | **98.26 ± 0.58** | **99.80 ± 0.05** | **97.62 ± 2.21** | **97.84 ± 0.80** | **99.76 ± 0.12** |

In contrast, our CompNets can correctly recognize the brain, and the optimal CompNet presents the best visual results for all four cases. We then use Dice score, sensitivity, and specificity to quantify the segmentation performance of each network, as reported in Table 1. The optimal CompNet consistently performs the best among all networks for either normal (averaged Dice of 98.27%) or pathological (averaged Dice 97.62%) images, although its performance on images with pathologies is slightly downgraded by <0.7% on average and <2.6% in the worst case.

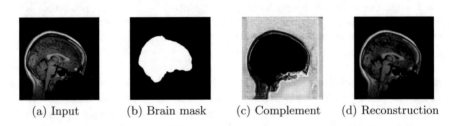

(a) Input          (b) Brain mask          (c) Complement          (d) Reconstruction

**Fig. 3.** Three outputs (b–d) of our optimal CompNet for an input brain scan (a).

Figure 3 shows the three outputs from the optimal CompNet: the masks for the brain and its complement and the reconstructed image. According to the combination of the brain mask and its complementary one, we can identify different parts in the brain image. This confirms that the brain branch works as expected; more importantly, the complementary branch has learned features for separating the non-brain region from the brain tissue. This enables the network to handle unseen brain tissue and be insensitive to pathologies in brain scans.

## 4    Discussion and Conclusions

In this paper, we proposed a complementary network architecture to segment the brain from an MRI scan. We observed that the complimentary segmentation branch of the optimal CompNet learns a mask outside of the brain and can help recognize different structures in the non-brain region. The complimentary design makes our network insensitive to pathologies in brain scans and helps segment the brain correctly. We used synthetic pathological images due to the lack of publicly available brain scans with both pathologies and skulls. Our source code is publicly available for in-house testing on real pathological images[1].

Furthermore, our current networks accept 2D slices from a 3D brain image, but the design can be extended to 3D networks for directly handling 3D images. Implementing 3D CompNets will be one of our future work plans. In addition, our complementary network design is not specific to the brain extraction problem but can be generalized to other image segmentation problems if the complementary part helps learn and understand the objects of interest. Another future work is the analysis of the theoretical and geometric implications of our CompNets.

# References

1. Dey, R., Lu, Z., Hong, Y.: Diagnostic classification of lung nodules using 3D neural networks. In: International Symposium on Biomedical Imaging, pp. 774–778 (2018)
2. Dice, L.R.: Measures of the amount of ecologic association between species. Ecology **26**(3), 297–302 (1945)
3. Han, X., et al.: Brain extraction from normal and pathological images: a joint PCA/image-reconstruction approach. arXiv:1711.05702 (2017)
4. Huang, G., Liu, Z., Weinberger, K.Q., van der Maaten, L.: Densely connected convolutional networks. In: CVPR, vol. 1, p. 3 (2017)
5. Ioffe, S., Szegedy, C.: Batch normalization: accelerating deep network training by reducing internal covariate shift. In: ICML, pp. 448–456 (2015)
6. Kingma, D.P., Ba, J.: Adam: a method for stochastic optimization. arXiv:1412.6980 (2014)
7. Kleesiek, J., Urban, G., Hubert, A., Schwarz, D., Maier-Hein, K., Bendszus, M., Biller, A.: Deep MRI brain extraction: a 3D convolutional neural network for skull stripping. NeuroImage **129**, 460–469 (2016)
8. Marcus, D.S., Wang, T.H., Parker, J., Csernansky, J.G., Morris, J.C., Buckner, R.L.: Open access series of imaging studies (OASIS): cross-sectional MRI data in young, middle aged, nondemented, and demented older adults. J. Cogn. Neurosci. **19**(9), 1498–1507 (2007)
9. Ronneberger, O., Fischer, P., Brox, T.: U-Net: convolutional networks for biomedical image segmentation. In: Navab, N., Hornegger, J., Wells, W.M., Frangi, A.F. (eds.) MICCAI 2015. LNCS, vol. 9351, pp. 234–241. Springer, Cham (2015). https://doi.org/10.1007/978-3-319-24574-4_28
10. Salehi, S.S.M., Erdogmus, D., Gholipour, A.: Auto-context convolutional neural network for geometry-independent brain extraction in magnetic resonance imaging. arXiv:1703.02083 (2017)
11. Souza, R., Lucena, O., Garrafa, J., Gobbi, D., Saluzzi, M., Appenzeller, S., et al.: An open, multi-vendor, multi-field-strength brain MR dataset and analysis of publicly available skull stripping methods agreement. NeuroImage **170**, 482–494 (2017)

---

[1] https://github.com/raun1/Complementary_Segmentation_Network.git.

12. Srivastava, N., Hinton, G., Krizhevsky, A., Sutskever, I., Salakhutdinov, R.: Dropout: a simple way to prevent neural networks from overfitting. J. Mach. Learn. Res. **15**(1), 1929–1958 (2014)
13. Xia, X., Kulis, B.: W-net: a deep model for fully unsupervised image segmentation. arXiv:1711.08506 (2017)

# One-Pass Multi-task Convolutional Neural Networks for Efficient Brain Tumor Segmentation

Chenhong Zhou[1], Changxing Ding[1(✉)], Zhentai Lu[2], Xinchao Wang[3], and Dacheng Tao[4]

[1] School of Electronic and Information Engineering,
South China University of Technology, Guangzhou, China
`chxding@scut.edu.cn`
[2] Guangdong Provincial Key Laboratory of Medical Image Processing,
School of Biomedical Engineering, Southern Medical University, Guangzhou, China
[3] Department of Computer Science, Stevens Institute of Technology, Hoboken, USA
[4] UBTECH Sydney AI Centre, SIT, FEIT, University of Sydney, Sydney, Australia

**Abstract.** The model cascade strategy that runs a series of deep models sequentially for coarse-to-fine medical image segmentation is becoming increasingly popular, as it effectively relieves the class imbalance problem. This strategy has achieved state-of-the-art performance in many segmentation applications but results in undesired system complexity and ignores correlation among deep models. In this paper, we propose a light and clean deep model that conducts brain tumor segmentation in a single-pass and solves the class imbalance problem better than model cascade. First, we decompose brain tumor segmentation into three different but related tasks and propose a multi-task deep model that trains them together to exploit their underlying correlation. Second, we design a curriculum learning-based training strategy that trains the above multi-task model more effectively. Third, we introduce a simple yet effective post-processing method that can further improve the segmentation performance significantly. The proposed methods are extensively evaluated on BRATS 2017 and BRATS 2015 datasets, ranking first on the BRATS 2015 test set and showing top performance among 60+ competing teams on the BRATS 2017 validation set.

## 1 Introduction

Brain tumors are one of the most fatal cancers worldwide [1]. Timely diagnosis of brain tumors from multimodal Magnetic Resonance Imaging (MRI) is of critical importance for treatment planning [2]. Automatic segmentation methods are highly desired in terms of efficiency and objectivity. However, automatic

**Electronic supplementary material** The online version of this chapter (https:// doi.org/10.1007/978-3-030-00931-1_73) contains supplementary material, which is available to authorized users.

A. F. Frangi et al. (Eds.): MICCAI 2018, LNCS 11072, pp. 637–645, 2018.
https://doi.org/10.1007/978-3-030-00931-1_73

brain tumor segmentation is still a challenging task due to large diversity of tumor shape, size, and location. Besides, there are four intra-tumoral classes, i.e., edema, necrosis, non-enhancing, and enhancing tumor. They are grouped into three overlapped regions which are required to be segmented for quantitative evaluation, i.e., complete tumor (all four classes), tumor core (all four classes except edema), and enhancing tumor (the enhancing tumor class only).

In recent years, Convolutional Neural Networks (CNNs) have been widely adopted for MRI-based brain tumor segmentation. CNN model architectures [3–6] have rapidly evolved from single-label prediction (predicting the label of a single voxel of the input patch) to dense-prediction (making predictions for voxels within the input patch simultaneously). To relieve the class imbalance problem, many recent works adopt the Model Cascade (MC) strategy for medical image segmentation [7,8]. For example, Wang et al. [8] decomposed multi-class brain tumor segmentation into a sequence of three successive binary segmentation tasks, each of which is solved by an independent network. MC relieves the class imbalance problem effectively by coarse-to-fine segmentation; therefore, its results are very encouraging. However, it comes with a price of system complexity and ignores the correlation among tasks.

Here we approach the above problems of MC via multi-task learning. We observe that multi-class brain tumor segmentation can be decomposed into three different but related tasks. Instead of training them individually like MC, we propose a One-pass Multi-task Network (OM-Net) that integrates the three tasks in a single model, which not only exploits their correlation in training but also simplifies the inference process by one-pass computation. Moreover, we design an effective training scheme based on curriculum learning, which is helpful to improve the convergence quality of OM-Net. Besides, to further improve performance, we propose a simple yet effective post-processing method to refine the segmentation results of OM-Net. Finally, the proposed approach obtains the first position on BRATS 2015 test set and achieves very competitive performance on BRATS 2017 validation set, respectively.

## 2   Methods

### 2.1   Model Cascade: A Strong Baseline

In this section, we first present an MC-based segmentation framework, as a strong baseline for OM-Net. We split multi-class brain tumor segmentation into three different but related tasks and each of them is implemented by an independent network. The three tasks are described as follows.

(1) *Coarse segmentation to detect complete tumor.* A network is trained to locate the complete tumor as a Region of Interest (ROI). Training patches are sampled randomly within the brain. To reduce overfitting, we train the network as a more difficult five-class segmentation problem. In testing, we still employ it as a binary segmentation task by merging the probability of four intra-tumoral classes. (2) *Refined segmentation for complete tumor and its intra-tumoral classes.* The coarse tumor mask obtained above is dilated by 5 voxels

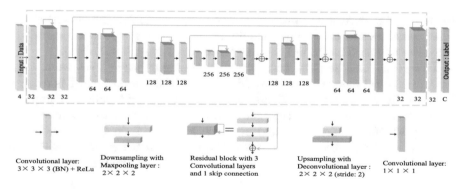

**Fig. 1.** Network architecture used in each task. The building blocks are represented by colored cubes with numbers below being the number of feature maps. C equals to 5, 5, and 2 for the first, second, and third task, respectively. (Best viewed in color)

to reduce false negatives. Then, the second network predicts labels of all voxels within the dilated region. Training patches are sampled randomly within the dilated ground-truth area of complete tumor. (3) *Precise segmentation for enhancing tumor.* Enhancing tumor is hard to segment due to the very unbalanced training data. We train the third network specially to segment enhancing tumor. Training patches for this network are sampled randomly within the ground-truth area of tumor core which covers all enhancing tumor voxels.

Network architecture for each task is identical except for the final convolutional classification layer. We use a 3D variant of the FusionNet [9], as illustrated in Fig. 1. Size of input patches for the network is $32 \times 32 \times 16 \times 4$, where the number 4 indicates the four MRI modalities. In testing, MC needs to run the three networks successively because the ROI of one network is determined by all its preceding networks. More specifically, the first network produces a coarse mask for complete tumor. The second network classifies all voxels in the dilated mask and obtains the precise region of complete tumor. Finally, we determine the precise enhancing tumor region by scanning all voxels in the complete tumor region using the third network. The tumor core region is meanwhile determined by merging results of the last two networks. Therefore, the entire inference process of MC requires alternate GPU-CPU computations for three times.

## 2.2   One-Pass Multi-task Network (OM-Net)

The above MC baseline can already achieve promising performance. However, it suffers from system complexity and ignores the correlation among the three tasks. We observe that the networks used for the three tasks are almost identical and their essential difference lies in training data. Inspired by this fact, we propose to transform the MC baseline into a single multi-task learning model. This model includes three tasks with their respective training data being the same as those in MC. Each task owns an independent convolutional layer, one classification layer, and one loss layer. All the other model parameters are shared to utilize the underlying correlation among the tasks. In this model, predictions of the three classifiers can be obtained simultaneously in a single-pass. Therefore, we name the proposed model as One-pass Multi-task Network (OM-Net).

Observing that the three tasks are of increasing difficulty level, we propose to train OM-Net more effectively based on curriculum learning, which is realized by gradually increasing the difficulty of training tasks and is proved to improve the convergence quality of deep models [10]. Model architecture and training strategy of OM-Net are illustrated in Fig. 2. First, we train OM-Net with the first task only until the loss curve tends to flatten, which enables OM-Net to learn the basic knowledge of differentiating tumor and normal tissues.

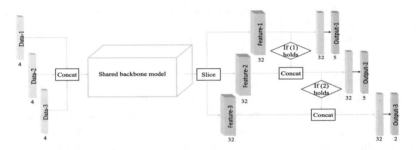

**Fig. 2.** Architecture of OM-Net. Data-$i$, Feature-$i$, and Output-$i$ denote training data, feature, and classification layer for the $i$-th task, respectively. The shared backbone model refers to the network layers outlined by the yellow dashed line in Fig. 1.

Then, we add the second task to OM-Net. As shown in Fig. 2, Data-1 and Data-2 are concatenated along the batch dimension as the input for OM-Net. Features produced by the shared backbone model are sliced at the same position on the batch dimension to obtain task-specific features and are then used to train task-specific parameters. Moreover, we argue that not only knowledge (model parameters) but also learning material (training data), can be transferred from the easier course (task) to the more difficult course (task) in curriculum learning. Therefore, training patches in Data-1 that conform to the following sampling strategy can be reused in the second task:

$$\frac{\sum_{i=1}^{n} \mathbf{1}\left\{l_i \in C_{complete}\right\}}{n} \geq 0.4, \tag{1}$$

where $l_i$ is the label of the $i$-th voxel in the patch, $n$ is the total number of voxels in the patch, and $C_{complete}$ refers to the all tumor classes. We concatenate the features of patches in Data-1 that satisfy the above sampling condition to Feature-2 and then calculate the loss for the second task. Training process in this step continues until the loss curve of the second task tends to flatten.

Finally, we introduce the third task and its training data to OM-Net. The concatenation and slicing operations are similar to those in the second step.

Training patches from Data-1 and Data-2 that conform to the following sampling strategy can be reused in the third task:

$$\frac{\sum_{i=1}^{n} \mathbf{1}\left\{l_i \in C_{core}\right\}}{n} \geq 0.5, \tag{2}$$

where $C_{core}$ refers to the tumor classes that belong to tumor core. The three tasks in OM-Net are trained together until convergence.

During inference, OM-Net obtains the predictions of the three tasks simultaneously. The way that OM-Net utilizes these results for final segmentation is exactly the same as that in MC. It is worth noting that OM-Net is essentially different from one existing multi-task model for brain tumor segmentation [11]. The model in [11] aims to provide more diverse supervision signals for the same training data. In comparison, OM-Net integrates tasks that have respective training data and aims to accomplish coarse-to-fine segmentation by a single model.

### 2.3  Post-processing

We further propose a novel post-processing method to refine the segmentation results of OM-Net. Our proposed method is mainly inspired by [6], but is more robust and easier to use in practice. It consists of two steps. First, isolated small clusters whose volume is less than one-tenth of the maximum 3D connected tumor area are removed. This step is identical to step 3 in [6]. Second, it is observed that when the volume of predicted enhancing tumor is less than five percent of the volume of the complete tumor, non-enhancing voxels tend to be falsely predicted as edema [6]. We find that this problem also happens in OM-Net and propose a fully-automatic method to relieve this problem. Specifically, we employ the K-means clustering algorithm to cluster the predicted edema voxels into two groups according their intensity values in MRI images. For each group, we compute the average probability of all its voxels belonging to the non-enhancing class, according to the prediction results of OM-Net. Labels of voxels in the group with the higher averaged probability are converted to non-enhancing, while those in the other group remain unchanged.

Compared with the approach in [6] that depends on manually determined threshold, our proposed approach is automatic and flexible. In the experiment section, we find it promotes the performance of OM-Net significantly.

## 3  Experiments

We evaluate the performance of the proposed methods on BRATS 2017 and BRATS 2015 datasets, respectively. The brain of each patient is scanned with four modalities, i.e., Flair, T1, T1c, and T2. All the images have been skull-striped and co-registered. For pre-processing, voxel intensities inside the brain are normalized to have zero mean and unit variance for each modality image. We sample around 400,000, 400,000, and 200,000 patches for the first, second, and third task, respectively. All networks are implemented based on the $C3D^1$ package, a modified version of *Caffe*[12]. We adopt SoftmaxWithLoss as the loss function and use stochastic gradient descent to train all networks. The initial learning rate of all networks is 0.001 and then divided by 2 after every 4 epochs. Each network in MC is trained for 20 epochs. OM-Net is trained for 1 epoch, 1 epoch, and 18 epochs for each of its three steps, respectively.

### 3.1   Results on BRATS 2017 Dataset

The training set of BRATS 2017 [2, 13–15] contains 285 MRI images. The validation set of BRATS 2017 contains 46 MRI images with hidden ground-truth and evaluation on this set is conducted online. For more convenient evaluation, we randomly divide the training set into two subsets, i.e., a training subset including 260 MRI images and a local validation subset including 25 MRI images.

We first carry out a number of experiments on the local validation subset. Quantitative comparison results are tabulated in Table 1[2]. Here MC1, MC2, and MC3 indicate the one-model, two-model, and three-model cascades, respectively. In order to justify the effectiveness of the curriculum learning-based training strategy, we further test OM-Net$^0$ (a naive multi-task learning model without training data transfer or step-wise training) and OM-Net$^d$ (a multi-task learning model with training data transfer but no step-wise training). OM-Net$^{p^1}$ and OM-Net$^p$ denote OM-Net with the first post-processing step and both post-processing steps, respectively. In addition, we also provide qualitative comparisons between MC3, OM-Net, and OM-Net$^p$ in the *supplementary materials*.

**Table 1.** Performance on the local validation subset of BRATS 2017 (%)

| Method | Parameters | Dice | | | Positive predictive value | | | Sensitivity | | |
|---|---|---|---|---|---|---|---|---|---|---|
| | | Complete | Core | Enhancing | Complete | Core | Enhancing | Complete | Core | Enhancing |
| MC1 | 13.81 M | 90.41 | 78.48 | 72.91 | 95.09 | 87.23 | 75.72 | 87.05 | 77.68 | 83.44 |
| MC2 | 27.62 M | 91.08 | 79.11 | 75.14 | 91.19 | 86.67 | 81.84 | 91.55 | 78.03 | 80.26 |
| MC3 | 41.43 M | 91.08 | 79.58 | 79.95 | 91.19 | 85.98 | 84.96 | 91.55 | 79.56 | 81.61 |
| OM-Net | 13.86 M | 91.10 | 79.87 | **80.87** | 92.42 | 87.60 | 85.44 | 90.23 | 78.31 | 82.80 |
| OM-Net$^0$ | 13.86 M | 90.40 | 79.42 | 79.96 | 91.25 | 87.57 | 82.68 | 90.52 | 76.87 | 83.91 |
| OM-Net$^d$ | 13.86 M | 91.11 | 79.92 | 80.24 | 92.11 | 87.78 | 84.19 | 90.73 | 78.09 | 83.28 |
| OM-Net$^{p^1}$ | 13.86 M | 91.28 | 79.88 | 80.84 | 93.38 | 87.71 | 85.44 | 89.87 | 78.26 | 82.75 |
| OM-Net$^p$ | 13.86 M | **91.28** | **82.50** | 80.84 | 93.38 | 83.60 | 85.44 | 89.87 | 83.60 | 82.75 |

First, Table 1 shows the Dice scores are steadily improved with the increase of model number in MC, which justifies the effectiveness of each model in MC. However, larger number of models leads to system complexity and more storage consumption. Second, with only one-third of the parameters of MC3, OM-Net

---

[1]   https://github.com/facebook/C3D.

[2]   Dice score is the overall evaluation index, identical to F measure. Therefore, we only highlight the best Dice scores in bold in Tables 1 and 2.

achieves better Dice scores consistently, especially for tumor core and enhancing tumor. Third, OM-Net outperforms both OM-Net$^0$ and OM-Net$^d$, demonstrating the effectiveness of the proposed training strategy. Fourth, the first post-processing step slightly improves the Dice score for complete tumor as it removes part of false positives. The proposed second step significantly improves the Dice score of tumor core by as much as 2.62%. The above results justify the effectiveness of the proposed approaches.

Additionally, we evaluate the performance of OM-Net$^p$ on BRATS 2017 validation set and compare it with the other 60+ participants. OM-Net$^p$ achieves Dice scores of 77.841%, 90.386%, and 82.792% for enhanced tumor (ET), whole tumor (WT), and tumor core (TC), respectively, and ranks second on the online leaderboard in terms of the averaged Dice score. The approach proposed in [8] currently ranks first, outperforming OM-Net$^p$ by 0.74%, 0.11%, and 0.99% on the Dice scores for ET, WT, and TC, respectively. However, the approach in [8] is a complicated ensemble system that includes as many as 9 models. In comparison, there is only a single model in our approach.

### 3.2   Results on BRATS 2015 Dataset

The BRATS 2015 dataset consists of 274 MRI images for training and 110 MRI images for testing. We use all training data to train OM-Net and MC3. Evaluation is conducted on the test set. The results are tabulated in Table 2.

**Table 2.** Performance on BRATS 2015 test set (%)

| Method | Dice | | | Positive predictive value | | | Sensitivity | | |
|---|---|---|---|---|---|---|---|---|---|
| | Complete | Core | Enhancing | Complete | Core | Enhancing | Complete | Core | Enhancing |
| MC3 | 86 | 70 | 63 | 86 | 82 | 60 | 88 | 67 | 72 |
| OM-Net | 86 | 71 | 64 | 86 | 83 | 61 | 88 | 68 | 72 |
| OM-Net$^{p^1}$ | 87 | 71 | 64 | 87 | 83 | 61 | 89 | 68 | 72 |
| OM-Net$^p$ | **87** | **75** | **64** | 87 | 81 | 61 | 89 | 75 | 72 |
| Isensee et al. [16] | 85 | 74 | 64 | 83 | 80 | 63 | 91 | 73 | 72 |
| Zhao et al. [6] | 84 | 73 | 62 | 89 | 76 | 63 | 82 | 76 | 67 |
| Kamnitsas et al. [5] | 85 | 67 | 63 | 85 | 86 | 63 | 88 | 60 | 67 |

First, we compare the results of MC3, OM-Net, OM-Net$^{p^1}$, and OM-Net$^p$. Table 2 shows that OM-Net consistently outperforms MC3, with 1% higher Dice scores on both tumor core and enhancing tumor. Besides, the first post-processing step improves the Dice score of OM-Net by 1% on the complete tumor region; the proposed second post-processing step significantly improves the Dice score of tumor core by 4%. The above results are consistent with the conclusions on the BRATS 2017 data. Second, we compare the performance of OM-Net$^p$ with the other leading approaches on the BRATS 2015 test set. It is observed in Table 2 that OM-Net$^p$ beats the state-of-the-art approaches on Dice scores and ranks first currently on the online leaderboard.

# 4    Conclusion

We propose the OM-Net model trained with the curriculum learning-based strategy to relieve the class imbalance problem in brain tumor segmentation. Unlike the popular MC framework, OM-Net integrates multiple networks in MC into a single deep model and conducts coarse-to-fine segmentation in a single pass. Therefore, it substantially saves model parameters and reduces system complexity. OM-Net is also advantageous as it effectively utilizes the correlation between the tasks. With a single and light model, the proposed approach ranks first on BRATS 2015 test set and achieves top performance on BRATS 2017 dataset.

**Acknowledgments.** Changxing Ding is the corresponding author and he was supported in part by the National Natural Science Foundation of China (No. 61702193), Guangzhou Key Lab of Body Data Science (No. 201605030011), and the Program for Guangdong Introducing Innovative and Entrepreneurial Teams (Grant No.: 2017ZT07X183). Zhentai Lu was supported by Guangdong Natural Science Foundation (No. 2016A030313574). Dacheng Tao was supported by Australian Research Council Projects (FL-170100117, DP-180103424 and LP-150100671).

# References

1. Bauer, S., Wiest, R., Nolte, L.P., Reyes, M.: A survey of MRI-based medical image analysis for brain tumor studies. Phys. Med. Biol. **58**, R97 (2013)
2. Menze, B.H., Jakab, A., Bauer, S., et al.: The multimodal brain tumor image segmentation benchmark (BRATS). IEEE TMI **34**(10), 1993–2024 (2015)
3. Pereira, S., et al.: Brain tumor segmentation using convolutional neural networks in MRI images. IEEE Trans. Med. Imag. **35**(5), 1240–1251 (2016)
4. Havaei, M., Davy, A., Warde-Farley, D., et al.: Brain tumor segmentation with deep neural networks. Med. Image Anal. **35**, 18–31 (2017)
5. Kamnitsas, K., et al.: Efficient multi-scale 3D CNN with fully connected CRF for accurate brain lesion segmentation. Med. Image Anal. **36**, 61–78 (2017)
6. Zhao, X., Wu, Y., Song, G., et al.: A deep learning model integrating FCNNs and CRFs for brain tumor segmentation. Med. Image Anal. **43**, 98–111 (2018)
7. Christ, P.F., et al.: Automatic Liver and lesion segmentation in CT using cascaded fully convolutional neural networks and 3D conditional random fields. In: Ourselin, S., Joskowicz, L., Sabuncu, M.R., Unal, G., Wells, W. (eds.) MICCAI 2016. LNCS, vol. 9901, pp. 415–423. Springer, Cham (2016). https://doi.org/10.1007/978-3-319-46723-8_48
8. Wang, G., et al.: Automatic brain tumor segmentation using cascaded anisotropic convolutional neural networks. arXiv preprint arXiv:1709.00382 (2017)
9. Quan, T.M., et al.: Fusionnet: a deep fully residual convolutional neural network for image segmentation in connectomics. arXiv preprint arXiv:1612.05360 (2016)
10. Bengio, Y., et al.: Curriculum learning. In: ICML, pp. 41–48. ACM (2009)
11. Shen, H., Wang, R., Zhang, J., McKenna, S.: Multi-task fully convolutional network for brain tumour segmentation. In: Valdés Hernández, M., González-Castro, V. (eds.) MIUA 2017. CCIS, vol. 723, pp. 239–248. Springer, Cham (2017). https://doi.org/10.1007/978-3-319-60964-5_21
12. Jia, Y., Shelhamer, E., Donahue, J., et al.: Caffe: convolutional architecture for fast feature embedding. In: 22nd ACM MM, pp. 675–678 (2014)

13. Bakas, S.: Advancing the cancer genome atlas glioma MRI collections with expert segmentation labels and radiomic features. Nat. Sci. Data **4**, 170117 (2017)
14. Bakas, S., et al.: Segmentation labels and radiomic features for the pre-operative scans of the TCGA-LGG collection. Cancer Imaging Arch. (2017)
15. Bakas, S., et al.: Segmentation labels and radiomic features for the pre-operative scans of the TCGA-GBM collection. Cancer Imaging Arch. (2017)
16. Isensee, F., Kickingereder, P., Wick, W., Bendszus, M., Maier-Hein, K.H.: Brain tumor segmentation and radiomics survival prediction: contribution to the BRATS 2017 challenge. In: Crimi, A., Bakas, S., Kuijf, H., Menze, B., Reyes, M. (eds.) BrainLes 2017. LNCS, vol. 10670, pp. 287–297. Springer, Cham (2018). https://doi.org/10.1007/978-3-319-75238-9_25

# Deep Recurrent Level Set for Segmenting Brain Tumors

T. Hoang Ngan Le[✉], Raajitha Gummadi[✉], and Marios Savvides[✉]

Carnegie Mellon University, Pittsburgh, USA
{thihoanl,rgummadi}@andrew.cmu.edu, msavvid@ri.cmu.edu

**Abstract.** Variational Level Set (VLS) has been a widely used method in medical segmentation. However, segmentation accuracy in the VLS method dramatically decreases when dealing with intervening factors such as lighting, shadows, colors, etc. Additionally, results are quite sensitive to initial settings and are highly dependent on the number of iterations. In order to address these limitations, the proposed method incorporates VLS into deep learning by defining a novel end-to-end trainable model called as Deep Recurrent Level Set (DRLS). The proposed DRLS consists of three layers, i.e., Convolutional layers, Deconvolutional layers with skip connections and LevelSet layers. Brain tumor segmentation is taken as an instant to illustrate the performance of the proposed DRLS. Convolutional layer learns visual representation of brain tumor at different scales. Brain tumors occupy a small portion of the image, thus, deconvolutional layers are designed with skip connections to obtain a high quality feature map. Level-Set Layer drives the contour towards the brain tumor. In each step, the Convolutional Layer is fed with the LevelSet map to obtain a brain tumor feature map. This in turn serves as input for the LevelSet layer in the next step. The experimental results have been obtained on BRATS2013, BRATS2015 and BRATS2017 datasets. The proposed DRLS model improves both computational time and segmentation accuracy when compared to the classic VLS-based method. Additionally, a fully end-to-end system DRLS achieves state-of-the-art segmentation on brain tumors.

## 1 Introduction

According to CBTRUS (Central Brain Tumor Registry of the United States), an estimated 78,980 new cases of primary malignant and non-malignant brain and other CNS tumors are expected to be diagnosed in the United States in 2018. Moreover, 16,616 deaths are likely to be attributed to primary malignant brain and other CNS tumors in the US in 2018. Magnetic resonance imaging (MRI) and computed tomography (CT) scans provide high-resolution images of the brain. Based on the degree of excitation and repetition times, different modalities of MRI images maybe obtained, i.e. FLAIR, T1, T2, T1c.

These modalities prove to be highly useful in detecting different subregions of the brain tumor, namely: edema (Whole tumor), non-enhancing solid core

A. F. Frangi et al. (Eds.): MICCAI 2018, LNCS 11072, pp. 646–653, 2018.
https://doi.org/10.1007/978-3-030-00931-1_74

(tumor core), necrotic/cystic core and enhancing core. Manual detection, segmentation of the brain tumors for cancer diagnosis, from large amounts of MRI images generated during clinical routine, is a difficult and time consuming task. Thus, there is substantial importance for automatic brain tumor image segmentation from MRI for diagnosis and radiotherapy.

A fundamental difficulty with segmenting brain tumors automatically is that they can appear anywhere in the brain, and vary in their shape, size and structure. Additionally, brain tumors along with their surrounding edema are often diffused, poorly contrasted, and have extended tentacle-like structures. VLS method with Active Contour (AC) is widely applied in image segmentation [1] due to its ability to automatically handle such various topological changes. Some of the remarkable works [2–4] on brain tumor segmentation that utilized VLS have shown the potential of VLS in achieving highly accurate brain tumor segmentation. However, the segmentation accuracy in the VLS based methods dramatically reduces when dealing with numerous intervening factors such as lighting, shadows, colors and backgrounds with large variety or complexity. MRI images are modalities that contain such factors.

The limitations of the VLS approaches can be summarized as follows: (1) they are largely handicapped in capturing variations of real-world objects. (2) they fail to memorize and to fully infer target objects. (3) they are limited in segmenting multiple objects with semantic information. (4) the segmentations generated by them are quite sensitive to numerous predefined parameters such as initial contour and number of iterations. To overcome the limitation of VLS in solving the problem of brain tumor segmentation, our motivation is to answer the following questions: (1) How to incorporate LS into deep learning to inherit the merits of both LS algorithm and deep learning? (2) Is it possible to replace the softmax function by a LS energy minimization to get a better outcome on MRI images? If so, how is curve evolution performed with forward and backward processes of the deep learning framework. (3) How are VLS iteration processes performed in the deep framework?

To address these issues and boost the classic VLS methods to learn-able deep network approaches, we propose a new formulation of VLS integrated in a deep framework, called the **Deep Recurrent Level Set (DRLS)** which combines the advantages of both fully convolutional network (FCN)[5] and LS method [6]. For MRI images, the local, global, and contextual information is important to obtain a high quality feature map. To achieve this goal, the proposed DRLS is designed by incorporating LevelSet Layers into VGGNet-16 [7] with three types of layers: Convolutional layers, deconvolutional layers and LevelSet layers. The proposed DRLS contains two main parts corresponding to visual representation and curve evolution. The first part extracts features using a Fully Convolutional Network (FCN) while incorporating skip connection to up-sample the feature map. The second part is composed of a level set layer that drives the contour such that the energy function attains a minima as shown in Fig. 1. Notably, our target is to show that it is completely possible to promote VLS to the higher level of learnable framework.

## 2   Literature Review

Over the years, discriminative, generative and deep learning methods have been used to segment brain tumors from MRI images. The following is brief description of such methodologies.

**Classical Segmentation Methods.** Anitha et al. [8], proposed segmentation using adaptive pillar K-means followed by extracting crucial features from the segmented image using discrete wavelet transforms. The features are put through two-tier classifiers namely, k-Nearest Neighbor Classifier(k-NN) and self-organizing maps(SOM). Dimah et al. [9] proposed a level set based approach for tumor segmentation by using histogram based clustering. The method also provides a local statistical characterization of the image by integrating the probabilistic non-negative matrix factorization (PNMF) framework into level set formulation. Tustison et al. [10] used asymmetry and first order statistical features to train concatenated Random Forests (RF) by introducing the output of the first RF as an input to the another.

**Level-Set Methods.** Some of the initial works that utilized VLS for brain tumor segmentations are [2,3]. [2] combined VLS evolution and global smoothness with the flexibility of topology changes followed by mathematical morphology. Thus, achieving significant advantages over conventional statistical classification. [3] introduced a threshold-based scheme that uses level sets for 3D tumor segmentation (TLS). A global threshold is used to design the speed function which is defined based on confidence interval and is iteratively updated throughout the evolution process that require different degrees of user involvement. Thapaliya, et al. [4] introduced a new SPF that can efficiently stop the contours at weak or blurred edges.

**Deep Learning Methods.** In the year 2015, the top finisher of BRATS 2015 challenge was the first to apply CNN to brain tumor segmentation [11]. The proposed CNN architecture exploits both local features as well as more global contextual features simultaneously and was 30 times faster than the then state of art solutions. Additionally, the architecture uses convolutional implementation of a fully connected layer thereby allowing a 40 fold speed up. Urban, et al. [12] proposed a 3D CNN architecture which extracts 3D voxel patches from different brain MRI modalities. The tissue label of the center voxel is predicted by feeding 3D voxels into a 4-layered CNN architecture. In order to avoid high computations of 3D voxels, Zikic et al. [13] transformed the 4D data into 2D data such that standard 2D-CNN architectures can be used to solve the brain tumor segmentation task.

Recently, [14] evaluated a 11-layered CNN architecture on BRATS dataset by implementing small $3 \times 3$ sized filters in the convolutional layers and reported comparative dice scores. In order to improve the performance and overcome the limitation of training data, CNNs is designed in a another fashion which combines with other classification methods or clustering methods [15]. One of the state of the art deep learning - based approach for segmenting brain tumor was developed called DMRes which is an improvement of Deep Medic [16].

# 3  Proposed Network

The pipeline of the proposed network is as illustrated in Fig. 1.

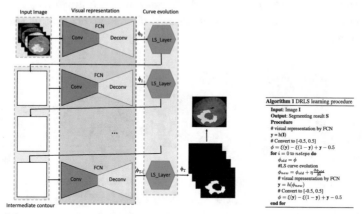

(a) The proposed DRLS with 2 main parts: visual representation and curve evolution

(b) Psedo code of DRLS

**Fig. 1.** The proposed DRLS network and algorithm

## 3.1  Formulation of Level Sets

Consider a binary image segmentation problem in 2D space, $\Omega$. The boundary $C$ of an open set $\omega \in \Omega$ is defined as: $C = \partial \omega$. In VLS framework, the boundary $C$ can be represented by the zero level set $\phi$ as follows:

$$\forall (x,y) \in \Omega \begin{cases} C = & \{(x,y) : \phi(x,y) = 0\} \\ inside(C) & \{(x,y) : \phi(x,y) > 0\} \\ output(C) & \{(x,y) : \phi(x,y) < 0\} \end{cases} \tag{1}$$

For image segmentation, $\Omega$ denotes the entire domain of an image **I**. The zero LS function $\phi$ divides the region $\Omega$ into two regions: region inside $\omega$ (foreground), denoted as inside(C) and region outside $\omega$ (background) denoted as outside(C). The length of the contour C is defined as: $Length(C) = \int_\Omega |\nabla H(\phi(x,y))| dxdy = \int_\Omega \delta(\phi(x,y))|\nabla\phi(x,y)| dxdy$ and the area inside the contour C is defined as $Area(C) = \int_\Omega H(\phi(x,y)) dxdy$

## 3.2  Recurrent Fully Convolutional Neural Network (RFCN)

[17] extended the classic CNNs to infer and learn from arbitrary-sized inputs. Later, [5] proposed a FCN model which adapts and extends deep classification architectures to learn efficiently from whole input and whole ground truth images. By casting fully connected layers into convolutional neural network with kernels that cover their entire input regions, FCN allows to take input of any

size and generate spatial outputs in one forward pass. To map the coarse feature map into a pixel-wise prediction of the input image, FCN up-samples the coarse feature map by a stack of deconvolution layers. Figures 2(a-1, a-2) show the comparison between classic CNN and FCN.

The RFCN is an extension of FCN architecture and given in Fig. 2(a-3). In the proposed RFCN, the output feature map of the current step is the input to the next step.

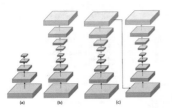
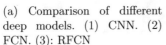
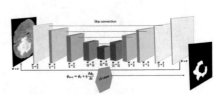

(a) Comparison of different deep models. (1) CNN. (2) FCN. (3): RFCN

(b) VGG-16 with 5 conv. layer and 5 deconv. layers on which DRLS is built

**Fig. 2.** The details of the proposed DRLS

## 3.3 Deep Recurrent Level Set (DRLS) - Proposed

Our proposed DRLS network is built based on VGG-16 with three different layers: convolutional layer, deconvolutional layer and LevelSet layer as shown in Fig. 1.

**Convolutional Layer:** The output feature map is computed by convolving the input feature map with convolution kernels $\mathbf{Y}^{(s,\theta)} = f^s(\mathbf{X}, \theta) = \mathbf{X} * \mathbf{W}^s + \mathbf{b}$ where $\mathbf{X}$ is the input feature map. $\mathbf{W}, \mathbf{b}$ are convolution kernel and bias. $\mathbf{W}^s$ indicates convolution at a stride $s$. $\mathbf{Y}^{(s,\theta)}$ denotes the output feature map generated by the convolutional layers with total stride of $s$ and parameterized by $\theta$. Because of the stride of conv and pooling layers, the final output feature maps $\mathbf{Y}^{(s,\theta)}$ is downsampled by a factor of the total stride of $s$ compared to the input feature map.

**Deconvolutional Layer:** The deconv layer is used to upsample the input feature maps using the stored max-pooling indices from the corresponding conv feature map. Here, a skip connection is introduced to concatenate the output of deconv feature map with the corresponding convolutional feature map. Figure 2(b) illustrates the network's architecture.

Let $g^s(; \tau)$ denote a deconv layer parameterized by $\tau$ that up-samples the input by a factor of $s$. The output is then concatenated with the corresponding convolutional layer $\mathbf{Y}^{(s-1,\theta)}$ via skip connection as $\hat{\mathbf{Y}}^{s,\tau} = concat\left[g^s(\mathbf{Y}^{(s,\theta)}; \tau), \mathbf{Y}^{(s-1,\theta)}\right]$.

**LevelSet Layer.** The network is trained to minimize the following energy function:

$$E(c_1, c_2, \phi) = \int_\Omega \mu H(\phi) + \nu\delta(\phi)|\nabla\phi|$$
$$+ \alpha(H(\phi) - GT)^2 + \lambda_1|H(\phi) - c_1|^2 H(\phi) + \lambda_2|H(\phi) - c_2|^2(1 - H(\phi))dxdy \tag{2}$$

In Eq. 2, the first term defines the area inside the contour C whereas the second term defined the length of the contour $C$. In the third term, $GT$ is the groundtruth. Minimizing this term with $\alpha > 0$ supervises the network to learn where a brain tumor occurs in the MRI images. The last two terms correspond to energy inside and outside of the contour $C$. To optimize the energy function, the calculus of variations is used. The derivative of energy function $E$ w.r.t $\phi$ is,

$$\frac{\partial E}{\partial \phi} = \delta(\varphi)\left[\mu - \nu div\frac{\nabla\phi}{|\nabla\phi|} + 2\alpha(H(\phi) - GT) + \lambda_1(H(\phi) - c_1)^2\right.$$
$$\left. + 2\lambda_1(H(\phi) - c_1)H(\phi) - \lambda_2(H(\phi) - c_2)^2 + 2\lambda_2(H(\phi) - c_2)(1 - H(\phi))\right] \tag{3}$$

The derivatives of energy function $E$ w.r.t $c_1$ and $c_2$ are,

$$\frac{\partial E}{\partial c_1} = -2\lambda_1(H(\phi) - c_1)H(\phi) \quad \frac{\partial E}{\partial c_2} = -2\lambda_2(H(\phi) - c_2)(1 - H(\phi)) \tag{4}$$

Fix $\phi$ and minimize the energy function w.r.t $c_1$, $c_2$:

$$c_1 = \frac{\int_\Omega H(\phi)(x,y)H(\phi)dxdy}{\int_\Omega H(\phi)dxdy} \ , \ c_2 = \frac{\int_\Omega H(\phi)(x,y)(1 - H(\phi))dxdy}{\int_\Omega(1 - H(\phi))dxdy} \tag{5}$$

Fix $c_1$ and $c_2$, and minimize the energy function w.r.t $\phi$

$$\frac{\partial\phi}{\partial t} = \delta_\epsilon(\varphi)\left[-\mu + \nu div\frac{\nabla\varphi}{|\nabla\varphi|} - \lambda_1(H(\phi) - c_1)^2 - 2\lambda_1(H(\phi) - c_1)H(\phi)\right.$$
$$\left. + \lambda_2(H(\phi) - c_2)^2 - 2\lambda_2(H(\phi) - c_2)(1 - H(\phi))\right] \tag{6}$$

## 4    Experimental Results

**Dataset and Measurements:** The proposed DRLS method is evaluated on BRATS17. Additionally, BRATS13 and BRATS15 datasets were used for comparing its performance with the state of the art techniques. The dataset is divided into training (80%) and testing (20%) datasets. The network is first trained on 168 HGG and 60 LGG training set of BRATS17 and then is fine tuned on the training sets of BRATS15 and BRATS13, respectively. To evaluate the performance of the proposed method, standard metrics are used as suggested in BRATS challenge [18] namely, Dice, Sensitivity(Sens) and Specificity(Spec). Besides metric scores, time consumption is also a key factor.

**Results:** The algorithm was tested on 42 HGG and 15 LGG from BRATS17, 44 HGG and 11 LGG from BRATS15 and 9 HGG and 7 LGG from BRATS13 for comparison with other methods.

**Table 1.** Performance of DRLS in comparison with other methods on BRATS13

| Methodology | Dice score | | | Sensitivity | | | Specificity | | |
|---|---|---|---|---|---|---|---|---|---|
| | WT | CT | ET | WT | CT | ET | WT | CT | ET |
| Havei et al. [11] | 0.88 | 0.79 | 0.73 | 0.87 | 0.79 | 0.80 | 0.89 | 0.79 | 0.68 |
| Urban et al. [12] | 0.87 | 0.77 | 0.73 | 0.92 | 0.79 | 0.70 | - | - | - |
| Zikic et al. [13] | 0.837 | 0.736 | 0.69 | - | - | - | - | - | - |
| Pereira et al. [14] | 0.88 | 0.83 | 0.77 | 0.89 | 0.83 | 0.81 | - | - | - |
| **Proposed DRLS** | 0.89 | 0.79 | 0.74 | 0.90 | 0.89 | 0.93 | 0.91 | 0.82 | 0.73 |

**Table 2.** Performance of DRLS in comparison with other methods on BRATS15

| Methodology | Dice score | | | Sensitivity | | | Specificity | | |
|---|---|---|---|---|---|---|---|---|---|
| | WT | CT | ET | WT | CT | ET | WT | CT | ET |
| Pereira et al. [14] | 0.78 | 0.65 | 0.7 | - | - | - | - | - | - |
| Pavel et al. [15] | 0.83 | 0.75 | 0.77 | - | - | - | - | - | - |
| Chang et al. [19] | 0.87 | 0.81 | 0.72 | - | - | - | - | - | - |
| Deep Medic [16] | 0.896 | 0.754 | 0.718 | 0.903 | 0.73 | 0.73 | - | - | - |
| DMRes [16] | 0.896 | 0.763 | 0.724 | 0.922 | 0.754 | 0.763 | - | - | - |
| **Proposed DRLS** | 0.88 | 0.82 | 0.73 | 0.91 | 0.76 | 0.78 | 0.90 | 0.81 | 0.71 |

Tables 1 and 2 have shown that the proposed algorithm outperforms other methods in terms of Dice scores and Sensitivity.

Besides metric scores, time consumption is also a key factor. Certain methods such as Tustison et al. [10] take 100 min to compute predictions per brain. However, when run on 4 GPUs the proposed DRLS shows a run time of just 55 seconds per patient. The algorithm is robust to outliers, runs fast and consistently shows improved core tumor segmentation.

# References

1. Osher, S., Sethian, J.A.: Fronts propagating with curvature-dependent speed: algorithms based on Hamilton-Jacobi formulations. J. Comput. Phys. **79**(1), 12–49 (1988)
2. Ho, S., Bullitt, E., Gerig, G.: Level-set evolution with region competition: automatic 3-D segmentation of brain tumors. In: ICPR (2002)
3. Taheri, S., Ong, S.H., Chong, V.F.H.: Level-set segmentation of brain tumors using a threshold-based speed function. Image Vis. Comput. **28**(1), 26–37 (2010)
4. Thapaliya, K., Pyun, J.-Y., Park, C.-S., Kwon, G.-R.: Level set method with automatic selective local statistics for brain tumor segmentation in MR images. Computer. Med. Imaging Graph. **37**(7), 522–537 (2013)
5. Long, J., Shelhamer, E., Darrell, T.: Fully convolutional networks for semantic segmentation. In: Proceedings of the IEEE Conference on Computer Vision and Pattern Recognition, pp. 3431–3440 (2015)

6. Chan, T.F., Vese, L.A.: Active contours without edges. IEEE Trans. Image Process. (TIP) **10**(2), 266–277 (2001)
7. Simonyan, K., Zisserman, A.: Very deep convolutional networks for large-scale image recognition. arXiv preprint arXiv:1409.1556 (2014)
8. Anitha, V., Murugavalli, S.: Brain tumor classification using two-tier classifier with adaptive segmentation technique. IET Comput. Vis. **10**(1), 9–17 (2016)
9. Dera, D., Bouaynaya, N., Fathallah-Shaykh, H.M.: Assessing the non-negative matrix factorization level set segmentation on the brats benchmark. In: Proceedings MICCAI-BRATS Workshop 2016 (2016)
10. Tustison, N.J.: Optimal symmetric multimodal templates and concatenated random forests for supervised brain tumor segmentation (simplified) with ANTsR. Neuroinformatics **13**, 209–225 (2015)
11. Havaei, M., et al.: Brain tumor segmentation with deep neural networks. Med. Image Anal. **35**, 18–31 (2017)
12. Urban, G., Bendszus, M., Hamprecht, F., Kleesiek, J.: Multi-modal brain tumor segmentation using deep convolutional neural networks. In: MICCAI BraTS (Brain Tumor Segmentation) Challenge, Proceedings, Winning Contribution, pp. 31–35 (2014)
13. Zikic, D., Ioannou, Y., Brown, M., Criminisi, A.: Segmentation of brain tumor tissues with convolutional neural networks. In: Proceedings MICCAI-BRATS, pp. 36–39 (2014)
14. Pereira, S., Pinto, A., Alves, V., Silva, C.A.: Brain tumor segmentation using convolutional neural networks in MRI images. IEEE Trans. Med. Imaging **35**(5), 1240–1251 (2016)
15. Dvořák, P., Menze, B.: Local structure prediction with convolutional neural networks for multimodal brain tumor segmentation. In: Menze, B., et al. (eds.) MCV 2015. LNCS, vol. 9601, pp. 59–71. Springer, Cham (2016). https://doi.org/10.1007/978-3-319-42016-5_6
16. Kamnitsas, K.: Efficient multi-scale 3D CNN with fully connected CRF for accurate brain lesion segmentation. Med. Image Anal. **36**, 61–78 (2017)
17. Matan, O., Burges, C.J., LeCun, Y., Denker, J.S.: Multi-digit recognition using a space displacement neural network. In: Advances in Neural Information Processing Systems, pp. 488–495 (1992)
18. Menze, B.H., Jakab, A., Bauer, S., et al.: The multimodal brain tumor image segmentation benchmark (BRATS). IEEE Trans. Med. Imaging **34**(10), 1993–2024 (2015)
19. Chang, P.D.: Fully convolutional neural networks with hyperlocal features for brain tumor segmentation. In: Proceedings MICCAI-BRATS Workshop, pp. 4–9 (2016)

# Pulse Sequence Resilient Fast Brain Segmentation

Amod Jog$^{(\boxtimes)}$ and Bruce Fischl

Athinoula A. Martinos Center for Biomedical Imaging,
Massachusetts General Hospital and Harvard Medical School, Boston, MA, USA
ajog@mgh.harvard.edu

**Abstract.** Accurate automatic segmentation of brain anatomy from $T_1$-weighted ($T_1$-w) magnetic resonance images (MRI) has been a computationally intensive bottleneck in neuroimaging pipelines, with state-of-the-art results obtained by unsupervised intensity modeling-based methods and multi-atlas registration and label fusion. With the advent of powerful supervised convolutional neural networks (CNN)-based learning algorithms, it is now possible to produce a high quality brain segmentation within seconds. However, the very supervised nature of these methods makes it difficult to generalize them on data different from what they have been trained on. Modern neuroimaging studies are necessarily multi-center initiatives with a wide variety of acquisition protocols. Despite stringent protocol harmonization practices, it is not possible to standardize the whole gamut of MRI imaging parameters across scanners, field strengths, receive coils etc., that affect image contrast. In this paper we propose a CNN-based segmentation algorithm that, in addition to being highly accurate and fast, is also resilient to variation in the input $T_1$-w acquisition. Our approach relies on building approximate forward models of $T_1$-w pulse sequences that produce a typical test image. We use the forward models to augment the training data with test data specific training examples. These augmented data can be used to update and/or build a more robust segmentation model that is more attuned to the test data imaging properties. Our method generates highly accurate, state-of-the-art segmentation results (overall Dice overlap = 0.94), within seconds and is consistent across a wide-range of protocols.

**Keywords:** CNN · Pulse sequence model · Segmentation · Brain
MRI

## 1 Introduction

Whole brain segmentation is one of the most important tasks in a neuroimage processing pipeline. A segmentation output consists of labels for white matter,

**Electronic supplementary material** The online version of this chapter (https://doi.org/10.1007/978-3-030-00931-1_75) contains supplementary material, which is available to authorized users.

© Springer Nature Switzerland AG 2018
A. F. Frangi et al. (Eds.): MICCAI 2018, LNCS 11072, pp. 654–662, 2018.
https://doi.org/10.1007/978-3-030-00931-1_75

cortex, and subcortical structures such as the thalamus, hippocampus, amygdala, and others. Structure volumes and shape statistics that rely on volumetric segmentation are regularly used to quantify differences between healthy and diseased populations [3]. An ideal segmentation algorithm of course needs to be highly accurate, but also, critically, it needs to be robust to variations in input data. Large modern MRI datasets are necessarily multi-center initiatives to gain access to a larger pool of subjects. It is very difficult to achieve perfect acquisition harmonization across different sites due to variations in scanner manufacturers, field strengths, receive coils, pulse sequences, available contrast, and resolution. Variations in site-specific parameters introduce bias and increase variation in downstream image processing including segmentation [5,7]. Low computational load is a yet another desirable property of a segmentation algorithm. A fast, memory-light segmentation algorithm enables quicker processing of large datasets and wider adoption.

Existing segmentation algorithms can be broadly classified into three types: (a) unsupervised, (b) multi-atlas registration-based, and (c) supervised. Unsupervised algorithms [3,12] fit the observed intensities of the input image to an underlying atlas-based model and perform maximum a posteriori labeling. They assume a functional form (e.g. Gaussian) of intensity distributions and results can degrade if the input distribution differs from this assumption. Efforts have been made to develop a hybrid approach [10] that is robust to input sequences and also leverages manually labeled training data. Unsupervised methods are usually computationally intensive, taking 0.5–4 h to run. Multi-atlas registration and label fusion (MALF) algorithms achieve state-of-the-art [1,14] segmentation accuracy. However, they require multiple computationally expensive registrations followed by label fusion. Registration quality can also suffer if the test image contrast properties are significantly different from the atlas images. Recently, with the success of deep learning methods in medical imaging, supervised segmentation approaches built on 3D CNNs have produced accurate segmentations with a runtime of a few seconds to minutes [8,13]. Despite the powerful local and global context that these models provide, they are vulnerable to subtle contrast differences that are inevitably present in multi-scanner MRI studies. However, with appropriate training using test-data specific augmentation, as we will show in Sect. 3, these differences can be essentially removed.

We present PSACNN–Pulse Sequence Augmented Convolutional Neural Network; a CNN-based multi-label segmentation approach that employs an augmentation scheme of generating training image patches on-the-fly that appear as if they have been imaged using the pulse sequence of the test data. PSACNN training consists of three major steps: (1) Estimating test data pulse sequence parameters, (2) applying test data pulse sequence forward models to training data nuclear magnetic resonance (NMR) parameters to create test data specific training features, (3) training a deep CNN to predict the segmentation using the augmentation. We will describe each of these steps in detail in Sect. 2.

## 2   Method

Let $\mathcal{A} = \{A_{(1)}, A_{(2)}, \ldots, A_{(M)}\}$ be a collection of $M$ $T_1$-w images with a corresponding expert manually labeled image set $\mathcal{Y} = \{Y_{(1)}, Y_{(2)}, \ldots Y_{(M)}\}$. The paired collection $\{\mathcal{A}, \mathcal{Y}\}$ is referred to as the atlas image set or the training image set. We assume that $\mathcal{A}$ is acquired using the pulse sequence $\Gamma_A$, where $\Gamma_A$ can be MPRAGE (magnetization prepared gradient echo) [9] or FLASH (fast low angle shot), SPGR (spoiled gradient echo), and others. In addition, let $\mathcal{B} = \{\boldsymbol{\beta}_{(1)}, \boldsymbol{\beta}_{(2)}, \ldots, \boldsymbol{\beta}_{(M)}\}$ be the corresponding NMR parameter maps. For each $i \in \{1, \ldots, M\}$ we have $\boldsymbol{\beta}_{(i)} = [\rho_{(i)}, T_{1(i)}, T_{2(i)}]$, where $\rho_{(i)}$ is a map of proton densities, and $T_{1(i)}$ and $T_{2(i)}$ store the longitudinal $(T_1)$, and transverse $(T_2)$ relaxation time maps respectively. Most atlas datasets do not acquire or generate $\boldsymbol{\beta}$ maps. We generate them using image synthesis for our atlas dataset using a previously acquired dataset [4] that estimated $\rho$ and $T_1$ maps from multi-echo FLASH images. We describe how to complement existing atlas sets with these maps in the Supplementary Material.

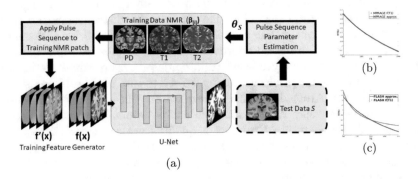

**Fig. 1.** (a) Workflow of test-data specific augmentation for training, (b) fit of $T_1$ component in the MPRAGE equation (blue) and our approximation (red), (c) fit of $T_1$ component (blue) with our approximation (red) for FLASH.

***Step 1: Estimating Test Data Acquisition Parameters:*** Let $S$ be the test image we want to obtain the segmentation for, acquired via pulse sequence $\Gamma_S$. We assume that we have access to the test image $S$ (or test dataset) prior to training so that we can design test data-specific augmentations. We would like to estimate the pulse sequence parameters of $\Gamma_S$ directly from the image $S$. The intensity $S(\mathbf{x})$ at voxel $\mathbf{x}$ is given by the imaging equation of $\Gamma_S$. Equation (1) shows the imaging equation for the FLASH sequence. $S(\mathbf{x})$ is a function of the acquisition parameters repeat time $(TR)$, echo time $(TE)$, flip angle $(\alpha)$, gain $(G)$, and tissue NMR parameters $\boldsymbol{\beta}(\mathbf{x}) = [\rho(\mathbf{x}), T_1(\mathbf{x}), T_2^*(\mathbf{x})]$.

$$S_{\text{FLASH}}(\mathbf{x}) = \Gamma_{\text{FLASH}}(\boldsymbol{\beta}(\mathbf{x}), [TR, TE, \alpha]) = G\rho \sin\alpha \frac{(1 - e^{-\frac{TR}{T_1}})}{1 - \cos\alpha e^{-\frac{TR}{T_1}}} e^{-\frac{TE}{T_2^*}} \quad (1)$$

The MPRAGE sequence is more complex to model [15]. In general, it is difficult to derive an imaging equation for most pulse sequences, and given $\boldsymbol{\beta}(\mathbf{x})$ and the acquisition parameter set, it is necessary to run a full Bloch equation simulation to obtain voxel intensities. Additionally, the number of scanner parameters that can affect signal intensity is very large on modern MRI scanners and for many datasets detailed parameter sets may not be available. Therefore, we would like to estimate these directly from the image intensities. However, it is generally not possible to robustly estimate all pulse sequence parameters purely from the image intensities themselves. Even when the equation is well-understood, the highly nonlinear dependence of intensities on imaging parameters makes their estimation unstable. Therefore, we have formulated approximations of the MPRAGE and FLASH equations with a smaller number of parameters that can be estimated directly and robustly from the image intensities. Our approximation for the FLASH sequence is shown in Eqs. (2)–(3).

$$\log(S_{\text{FLASH}}) = \log(G\sin\alpha) + \log(\rho) + \log(\frac{(1 - e^{-\frac{TR}{T_1}})}{1 - \cos\alpha e^{-\frac{TR}{T_1}}}) - \frac{TE}{T_2^*} \quad (2)$$

$$\approx \theta_0 + \log(\rho) + \frac{\theta_1}{T_1} + \frac{\theta_2}{T_2}, \quad (3)$$

where $\boldsymbol{\theta}_{\text{FLASH}} = \{\theta_0, \theta_1, \theta_2\}$ forms our parameter set. For the range of values of $T_1$ in the human brain $(500, 3000)$ ms at 1.5 T, our approximation fits the signal dependence on $T_1$ well (see Figs. 1(b) and 1(c)). Equation (4) is our approximation for the MPRAGE imaging equation provided by Wang et al. [15].

$$\log(S_{\text{MPRAGE}}) \approx \theta_0 + \log(\rho) + \theta_1 T_1 + \theta_2 T_1^2. \quad (4)$$

Given test image $S$ we estimate $\boldsymbol{\theta}_S$ by a strategy that is similar to Jog et al. [6]. Let $S_c$, $S_g$, $S_w$ be the mean intensities of cerebrospinal fluid (CSF), gray matter (GM), and white matter (WM), respectively. These can be obtained by a three-class classification scheme by fitting a Gaussian mixture model to the intensity distribution. Let $\boldsymbol{\beta}_c$, $\boldsymbol{\beta}_g$, $\boldsymbol{\beta}_w$ be the mean NMR values for CSF, GM, and WM classes, respectively, that are obtained from previously acquired data [4]. Thus, we have a system of three linear equations $S_c = \Gamma_S(\boldsymbol{\beta}_c; \boldsymbol{\theta}_S); S_g = \Gamma_S(\boldsymbol{\beta}_g; \boldsymbol{\theta}_S); S_w = \Gamma_S(\boldsymbol{\beta}_w; \boldsymbol{\theta}_S)$ and three unknown parameters in $\boldsymbol{\theta}_S$ that can be solved to obtain the estimate $\hat{\boldsymbol{\theta}}_S$ (see Fig. 1(a)).

***Step 2: Feature Extraction and Augmentation:*** We use the U-Net CNN architecture [11] as our predictor. GPU memory constraints prevent us from using the $256 \times 256 \times 256$ brain image as an input. For each $A_{(i)}$ in $\mathcal{A}$, we sample a $32 \times 32 \times 32$-sized patch $\mathbf{p}(\mathbf{x})$ at voxel $\mathbf{x}$. We observed that training a purely patch-based U-Net leads to segmentation errors due to incorrect localization. To provide more global context, we use information from the nonlinear warp produced by the FreeSurfer pipeline [4] that aligns the FreeSurfer atlas to the test image $S$. For each voxel $\mathbf{x}$ in $S$, we extract the FreeSurfer atlas 3D coordinates $\mathbf{w}(\mathbf{x})$, that warp to $\mathbf{x}$, and produce $\mathbf{w}_p(\mathbf{x})$–a $32 \times 32 \times 32 \times 3$-sized patch of warped FreeSurfer atlas coordinates. The final feature vector $\mathbf{f}(\mathbf{x}) = [\mathbf{p}(\mathbf{x}), \mathbf{w}_p(\mathbf{x})]$, is a concatenation of the intensity patch and the coordinates (see Fig. 1(a) 2D patches shown for illustration). Test data-specific augmented patches are generated by extracting NMR parameter patches from the $i^{\text{th}}$ NMR map for each $i \in \{1, \ldots, M\}$, $\boldsymbol{\beta}_{(i)} \in \mathcal{B}$ where the NMR patch at voxel $\mathbf{x}$ is denoted by $[\boldsymbol{\rho}_{(i)}(\mathbf{x}), \mathbf{T}_{1(i)}(\mathbf{x}), \mathbf{T}_{2(i)}(\mathbf{x})]$. The augmented patches are given by:

$$\mathbf{p}'(\mathbf{x}) = \Gamma_S([\boldsymbol{\rho}_{(i)}(\mathbf{x}), \mathbf{T}_{1(i)}(\mathbf{x}), \mathbf{T}_{2(i)}(\mathbf{x})]; \hat{\boldsymbol{\theta}}_S) \qquad (5)$$

In addition, we also sample the 3D parameter space of $\boldsymbol{\theta}_S$ and generate non-test-data specific augmented patches to increase the breadth of available training. The augmented patches are also concatenated with the warped FreeSurfer atlas coordinate patches $\mathbf{w}_p(\mathbf{x})$ to generate the augmented features $\mathbf{f}'(\mathbf{x}) = [\mathbf{p}'(\mathbf{x}), \mathbf{w}_p(\mathbf{x})]$ and added to the training (Fig. 1(a)).

Original and augmented features with corresponding manually labeled image patches are used to train the U-Net, which has three pooling and corresponding upsampling stages (see Fig. 1(a)). Each orange block consists of two 3D convolutional layers followed by a batch normalization layer. The first block has 32 $3 \times 3 \times 3$-sized filters and the subsequent encoding blocks have twice the number of filters as the previous block. The red arrows signify skip connections where data from the encoding layers is concatenated as input to the decoding layers. All activations are ReLu except for the last layer that has a softmax activation. The output is $32 \times 32 \times 32 \times L$ where $L = 44$ is the number of labels. We use the Adam optimization algorithm to minimize soft Dice-based loss calculated over all labels. The batch size is 32 and is divided equally between augmented and original training features. The total size of unaugmented training was $\approx 10^6$ patches from 16 subjects in $\mathcal{A}$. Validation data was generated from three subjects. During prediction, we extract features from test image $S$ and apply the trained U-Net to generate overlapping patches of label probabilities. These are combined together by averaging in the overlapping regions to produce the final probability map. The label with the highest probability is selected as the label in the hard segmentation.

## 3   Experiments and Results

*3.1: Same Scanner Dataset.* In this experiment we compare the performance of PSACNN against unaugmented CNN (CNN), FreeSurfer segmentation (ASEG) [3], SAMSEG [10], and MALF [14], on test data with the same acquisition protocol as the training data. The complete dataset consists of 39 subjects with 1.5 T Siemens Vision MPRAGE acquisitions (TR=9.7 ms, TE=4 ms TI=20 ms, voxel-size $= 1 \times 1 \times 1.5$ mm$^3$) with expert manual labels done as per

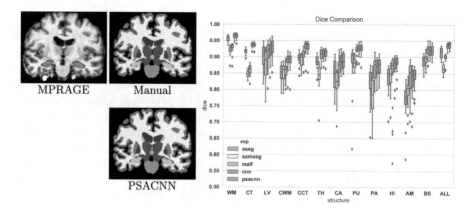

**Fig. 2.** Acronyms: white matter (WM), cortex (CT), lateral ventricle (LV), cerebellar white matter (CWM), cerebellar cortex (CCT), thalamus (TH), caudate (CA), putamen (PU), pallidum (PA), hippocampus (HI), amygdala (AM), brain-stem (BS), overlap for all structures (ALL).

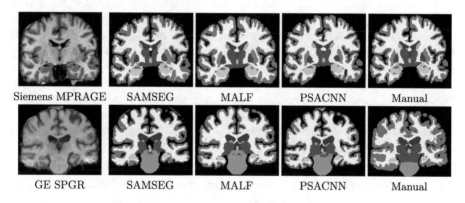

**Fig. 3.** Input Siemens and GE acquisitions with segmentation results from SAMSEG, MALF, and PSACNN, along with manual segmentations.

the protocol described in [2]. We chose a subset of 16 subjects as training data for CNN, PSACNN, and as the atlas set for MALF. We used 3 subjects for validation for the CNNs, and 20 for testing. Figure 2 shows the results for all the algorithms and an example segmentation produced by PSACNN. CNN (red) and PSACNN (blue) have significantly higher Dice overlap (ALL Dice $= 0.9376$) than the next best method as tested using a paired t-test ($p < 0.01$) for most structures.

**3.2: Different Scanner Datasets.** In this experiment we compare the accuracy of PSACNN against other methods on two datasets; (a) Siemens dataset: 14 subject MPRAGE scans acquired on a 1.5 T Siemens SONATA scanner with the same pulse sequence as the training data, and (b) GE dataset: 13 subject SPGR (spoiled gradient recalled) scans acquired on a 1.5 T GE

Signa (TR $= 35$ ms, TE $= 5$ ms, $\alpha = 45°$, voxel size $= 0.9375 \times 0.9375 \times 1.5$ mm$^3$) scanner. Both datasets have expert manual segmentations generated with the same protocol as the training data. Figure 4 shows comparison of Dice coefficients for all the methods. On the Siemens dataset CNN and PSACNN are significantly better than the rest (Fig. 4(a)), as is expected due to its similarity with the training data. However, on the GE scans, which present a noticeably different tissue contrast (Fig. 3), all methods show a reduced overlap (Fig. 4(b)), but CNN has the worst performance of all as it is unable to generalize. PSACNN (ALL Dice $= 0.7636$) on the other hand, is robust to the contrast change due to pulse sequence-based augmentation, and produces segmentations that are comparable to the state-of-the-art algorithms such as SAMSEG (ALL Dice $= 0.7708$) and MALF (ALL Dice $= 0.7804$) in accuracy, with an order of magnitude lower processing time.

**3.3: Multi-scanner Consistency.** In this experiment we tested the consistency of the segmentation produced by the various methods on four datasets acquired from 15 subjects; *MEF*: multi-echo FLASH acquired on Siemens 3 T TRIO scanner, *TRIO*: MPRAGE acquired on Siemens TRIO, *GE*: MPRAGE on 1.5 T GE Signa, and *Siemens*: MPRAGE on 1.5 T Siemens SONATA scanner. *Siemens* scan parameters are closest to that of the training dataset and we calculated absolute difference in structure volumes obtained using different segmentation methods on datasets *MEF*, *TRIO*, *GE* with respect to *Siemens*. In Table 1 we report only WM volume differences due to lack of space and also because total white matter volumes are a good indicator of accumulated effects of pulse sequence variation. MALF shows the highest consistency for *GE* and is second best for the rest. PSACNN provides the best performance for *TRIO* and second best for the *GE* dataset.

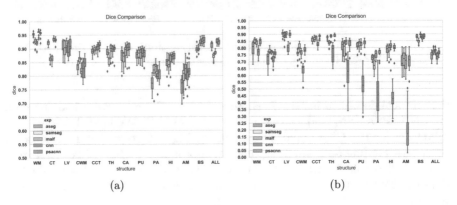

(a)                                    (b)

**Fig. 4.** Dice evaluations on (a) Siemens dataset, (b) GE dataset. For acronyms refer to Fig. 2 caption.

**Table 1.** Mean and standard deviation (Std. Dev.) of absolute WM volume differences between datasets *MEF, TRIO, GE,* with *Siemens* for different algorithms.

| | MEF | | TRIO | | GE | |
|---|---|---|---|---|---|---|
| | Mean | (Std) | Mean | (Std.) | Mean | (Std) |
| ASEG | 21.38 | (19.16) | 2.48 | (2.72) | 9.13 | (15.64) |
| SAMSEG | 2.22 | (1.26)* | 2.12 | (2.78) | 7.49 | (11.53) |
| MALF | 4.14 | (1.64) | 2.18 | (3.00) | 4.33 | (4.71)* |
| CNN | 87.13 | (12.14) | 2.18 | (2.61) | 9.02 | (17.70) |
| PSACNN | 7.13 | (4.40) | 1.65 | (2.65)* | 6.56 | (10.68) |

* Statistically significantly better than the next best method ($p < 0.01$) using a paired t-test.

## 4    Discussion and Conclusions

We have described PSACNN, a CNN-based segmentation algorithm that can adapt itself to the test data by generating test-data specific augmentation using an approximate forward model of MRI image formation. The augmentation can be used to robustly train or fine-tune any underlying predictor. We show state-of-the-art segmentation performance on diverse datasets. The prediction is fast and takes less than a minute to run on a single process. In the future we intend to use more accurate imaging equations and simulate other pulse sequences to increase the range of application.

## References

1. Asman, A.J., Landman, B.A.: Formulating spatially varying performance in the statistical fusion framework. IEEE TMI **31**(6), 1326–1336 (2012)
2. Buckner, R., et al.: A unified approach for morphometric and functional data analysis in young, old, and demented adults using automated atlas-based head size normalization: reliability and validation against manual measurement of total intracranial volume. NeuroImage **23**(2), 724–738 (2004)
3. Fischl, B., Salat, D.H., et al.: Whole brain segmentation: automated labeling of neuroanatomical structures in the human brain. Neuron **33**(3), 341–355 (2002)
4. Fischl, B., Salat, D.H., van der Kouwe, A.J., et al.: Sequence-independent segmentation of magnetic resonance images. NeuroImage **23**(S1), S69–S84 (2004)
5. Han, X., Jovicich, J., Salat, D., et al.: Reliability of MRI-derived measurements of human cerebral cortical thickness: the effects of field strength, scanner upgrade and manufacturer. NeuroImage **32**(1), 180–194 (2006)
6. Jog, A., Carass, A., Roy, S., Pham, D.L., Prince, J.L.: MR image synthesis by contrast learning on neighborhood ensembles. MedIA **24**(1), 63–76 (2015)
7. Jovicich, J., et al.: Brain morphometry reproducibility in multi-center 3T MRI studies: a comparison of cross-sectional and longitudinal segmentations. NeuroImage **83**, 472–484 (2013)

8. Li, W., Wang, G., Fidon, L., Ourselin, S., Cardoso, M.J., Vercauteren, T.: On the compactness, efficiency, and representation of 3D convolutional networks: brain parcellation as a pretext task. In: Niethammer, M., et al. (eds.) IPMI 2017. LNCS, vol. 10265, pp. 348–360. Springer, Cham (2017). https://doi.org/10.1007/978-3-319-59050-9_28

9. Mugler, J.P., Brookeman, J.R.: Three-dimensional magnetization-prepared rapid gradient-echo imaging (3D MPRAGE). Mag. Reson. Med. **15**(1), 152–157 (1990)

10. Puonti, O., Iglesias, J.E., Leemput, K.V.: Fast and sequence-adaptive whole-brain segmentation using parametric Bayesian modeling. NeuroImage **143**, 235–249 (2016)

11. Ronneberger, O., Fischer, P., Brox, T.: U-Net: convolutional networks for biomedical image segmentation. In: Navab, N., Hornegger, J., Wells, W.M., Frangi, A.F. (eds.) MICCAI 2015. LNCS, vol. 9351, pp. 234–241. Springer, Cham (2015). https://doi.org/10.1007/978-3-319-24574-4_28

12. Shiee, N., et al.: A topology-preserving approach to the segmentation of brain images with multiple sclerosis lesions. NeuroImage **49**(2), 1524–1535 (2010)

13. Wachinger, C., Reuter, M., Klein, T.: DeepNAT: deep convolutional neural network for segmenting neuroanatomy. NeuroImage **170**, 435–445 (2017)

14. Wang, H., Suh, J.W., Das, S.R., Pluta, J.B., Craige, C., Yushkevich, P.A.: Multi-atlas segmentation with joint label fusion. IEEE Trans. Pattern Anal. Mach. Intell. **35**(3), 611–623 (2013)

15. Wang, J., He, L., Zheng, H., Lu, Z.L.: Optimizing the magnetization-prepared rapid gradient-echo (MP-RAGE) sequence. PloS one **9**(5), e96899 (2014)

# Improving Cytoarchitectonic Segmentation of Human Brain Areas with Self-supervised Siamese Networks

Hannah Spitzer[1]([⊠]), Kai Kiwitz[2], Katrin Amunts[1,2], Stefan Harmeling[3], and Timo Dickscheid[1]

[1] Institute of Neuroscience and Medicine INM-1, Forschungszentrum Jülich, Jülich, Germany
h.spitzer@fz-juelich.de
[2] C. and O. Vogt Institute of Brain Research, Heinrich-Heine University Düsseldorf, Düsseldorf, Germany
[3] Institute of Computer Science, Heinrich-Heine University Düsseldorf, Düsseldorf, Germany

**Abstract.** Cytoarchitectonic parcellations of the human brain serve as anatomical references in multimodal atlas frameworks. They are based on analysis of cell-body stained histological sections and the identification of borders between brain areas. The de-facto standard involves a semi-automatic, reproducible border detection, but does not scale with high-throughput imaging in large series of sections at microscopical resolution. Automatic parcellation, however, is extremely challenging due to high variation in the data, and the need for a large field of view at microscopic resolution. The performance of a recently proposed Convolutional Neural Network model that addresses this problem especially suffers from the naturally limited amount of expert annotations for training. To circumvent this limitation, we propose to pre-train neural networks on a self-supervised auxiliary task, predicting the 3D distance between two patches sampled from the same brain. Compared to a random initialization, fine-tuning from these networks results in significantly better segmentations. We show that the self-supervised model has implicitly learned to distinguish several cortical brain areas – a strong indicator that the proposed auxiliary task is appropriate for cytoarchitectonic mapping.

**Keywords:** Self-supervised learning · Deep learning
Brain parcellation · Human brain · Histology

## 1 Introduction

Analysis of cytoarchitectonic cortical areas in high-resolution histological images of brain sections is essential for identifying the segregation of the human brain into cortical areas and nuclei [2]. Areas can be distinguished based on their specific architecture, the presence of cell clusters and specific cell types according to

© Springer Nature Switzerland AG 2018
A. F. Frangi et al. (Eds.): MICCAI 2018, LNCS 11072, pp. 663–671, 2018.
https://doi.org/10.1007/978-3-030-00931-1_76

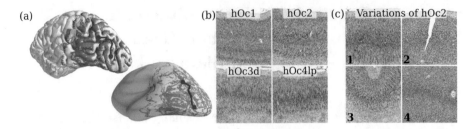

**Fig. 1.** (a) Sampling locations on pial and inflated left surface (red dots). The approximate geodesic distance between two points is shown in blue. (b) Example patches (1019 × 1019 px) extracted from 2 μm resolution histological sections showing areas hOc1–hOc4lp. Small variations in the laminar pattern distinguish the areas. (c) Examples of hOc2. (1) Intra-area variability, (2) artifacts, (3) high curvature, and (4) oblique histological cuts make identification of areas a challenging task.

their morphology, the visibility of columns, and other features. Borders between cortical areas can be identified in a reproducible manner by a well-accepted method, which relies on image analysis and multivariate statistical tools to capture maximal changes in the distribution of cell bodies from the cortical surface to the white matter border [8]. However, a completely automatic method that allows area segmentation in a large series of human brain sections is still missing.

Automatic identification of areas in cell-body stained whole brain sections is an extremely challenging segmentation task considering staining and sectioning artifacts, different relative orientations of the sectioning plane wrt. the brain surface, and high inter-subject variability (Fig. 1b, c). To reliably identify differences in the distribution of cell bodies in the cortex, automatic methods need to rely on high resolutions (1–2 μm) and a large field of view (approx. 2 mm) at the same time. Since expert annotation of brain regions is very labor-intensive, the amount of training data available for automatic algorithms is limited.

Previously, we have shown that despite these limitations, it is possible to employ Convolutional Neural Networks (CNNs) for segmentation of cytoarchitectonic areas [10]. However, the performance of this model is not yet accurate enough for fully automatic segmentation.

In this work, we introduce a way to bypass the limitation of labeled training data by exploiting unlabeled high resolution cytoarchitectonic sections. We formulate a self-supervised auxiliary task based on the estimation of spatial distances between image patches sampled from the same brain, using a Siamese network architecture. In particular, we determine the approximate geodesic distance between two image patches by exploiting the inherent 3D structure of a whole brain 3D reconstruction (Fig. 1a), as provided, e.g., by the *BigBrain* [1].

We make the following contributions: (1) By applying transfer learning, we significantly improve the accuracy of area classification in the visual cortex. (2) Carefully examining the training objective, we show that the Siamese network gains significant performance when predicting absolute 3D coordinates in addition to the pairwise distances. (3) We show that the self-supervised model learns to identify anatomically plausible cytoarchitectonic borders, although it was never trained to develop a concept of brain areas.

## 2    Related Work

**Automatic Area Parcellation.** Recently, CNNs were successfully employed
for parcellation of MRT volumes [3,6]. In contrast to these works, we aim to
improve cytoarchitectonic mapping in microscopic scans of human whole brain
cell-body stained tissue sections. A CNN method for automatically classifying
cortical areas in such data was introduced by [10]. Building on a U-Net architecture [7], their key insight is to include prior knowledge in form of probabilistic
atlas information to deal with the difficult variations in texture and limited training data. In this work, we improve upon their results by leveraging unlabeled
data in a self-supervised approach.

**Transfer Learning and Self-supervised Learning.** In recent years, several
studies have successfully used transfer learning to apply well performing models
trained on the ImageNet dataset to new tasks [13]. For the task at hand however,
we require an unusually large receptive field and address a very specific data
domain. This forces us to train a custom CNN architecture from scratch.

Siamese networks have been employed for learning highly nonlinear image
features for keypoint matching, image retrieval, or object pose estimation [5,9].
Leveraging spatial dependencies in input images or exploiting motion information contained in video, such features can be learned in a self-supervised manner
[4,11]. We take up this idea and leverage the 3D relationship between individual brain sections. In [5] a Siamese regression network for pose estimation is
presented, combining targets for pose regression from a single image with prediction of distance in pose between two input images. Extending this approach,
we include prediction of the 3D coordinates of input patches as an additional
objective in our model, and explain the benefits gained by this modification.

## 3    Method

Our aim is to improve the accuracy of the supervised area segmentation in
[10]. Since classical cytoarchitectonic mapping is an extremely time consuming
expert task, we cannot easily overcome the problem of limited training data. We

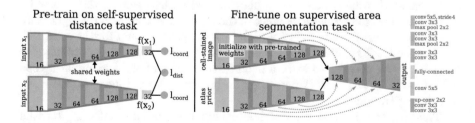

**Fig. 2.** Siamese network architecture for the auxiliary distance task (left) and extended
U-net architecture for the area segmentation task (right). The network branches marked
in red share the same architecture. Based on [10].

therefore propose to exploit unlabeled brain sections from a 3D reconstructed whole-brain volume for automatic brain mapping, of which much larger amounts can be acquired in reasonable time. Our method consists of two consecutive steps (Fig. 2): (1) Pre-train weights on a self-supervised task using a Siamese network. (2) Fine-tune from these weights on the area segmentation task using a small dataset with brain region labels.

### 3.1   Self-supervised Siamese Network on Auxiliary Distance Task

Considering a dataset of unlabeled brain sections from one human brain [1], we formulate the self-supervised feature learning task: Given two input patches sampled randomly from the cortex in arbitrary sections, learn to predict the geodesic distance along the surface of the brain between these two patches (Fig. 1a).

We use a Siamese network that computes a regression function based on two input patches $(x_1, x_2)$ (Fig. 2). The network consists of two branches with identical CNN architecture and shared weights, computing features $(f(x_1), f(x_2))$. The branch architecture corresponds to the texture filtering branch of the extended U-Net architecture of [10] with a 32-channel dense layer added on top of the last convolutional layer. We define the predicted distance as the squared Euclidean distance between the feature vectors, and the distance loss as:

$$l_{\text{dist}} = \|\|f(x_1) - f(x_2)\|_2^2 - y_{\text{dist}}\|_1. \tag{1}$$

The groundtruth distance $y_{\text{dist}}$ is computed by finding the closest points of the inputs on the brain surface and calculating their shortest distance along this surface. With this formulation of $l_{\text{dist}}$, the model learns a Euclidean feature embedding $f(x)$ of inputs $x$ wrt. the geodesic distance along the brain surface.

We have successfully trained models using loss (1); however, our experiments have shown that convergence is faster and a higher accuracy regarding the predicted distances is reached, when we include the prediction of the 3D location of the inputs as an additional task in the training. To this end we add an additional dense layer $d$ calculating the predicted coordinate for each input $x$ based on $f(x)$ and formulate the coordinate loss $l_{\text{coord}}$ as follows:

$$l_{\text{coord}} = \|d(f(x)) - y_{\text{coord}}\|_1, \tag{2}$$

with $y_{\text{coord}}$ the 3D location of input $x$ on the inflated surface. By defining $y_{\text{coord}}$ on the inflated surface, we ensure a high correlation between the distance of the coordinates and the geodesic distance of the inputs. For points on the right hemisphere, we reverse the left-right coordinate of $y_{\text{coord}}$ to account for the essentially mirror symmetric topology of areas on the two hemispheres. The coordinate loss $l_{\text{coord}}$ helps the network to learn a good feature embedding, agglomerating spatially close samples, even though they do not necessarily appear together as a pair during training.

The final training loss is a weighted combination of $l_{\text{dist}}$ and $l_{\text{coord}}$ together with a L2 weight regularization that regularizes all weights and biases, except those in the final dense layers:

$$L = l_{\text{dist}}(x_1, x_2) + \alpha(l_{\text{coord}}(x_1) + l_{\text{coord}}(x_2)) + \lambda\|w\|_2. \tag{3}$$

**Implementation Details.** We generate our dataset from the *BigBrain* [1], a dataset of 7400 consecutive histological cell-body stained sections that were registered to a 3D volume at 20 μm resolution. A surface mesh is available at 200 μm resolution. We sample 200k 1019 × 1019 px patches at 2 μm resolution from sections 0–3000 (occipital and parietal lobe, encompasses visual cortex), leaving out 1/12th of the sections for testing (cf. Fig. 1a for the sampling locations, Fig. 1b for example patches). To ensure that the laminar structure of the cortex is clearly visible, we only sample from the center of non-oblique cortex, i.e., where the cutting plane was 90° ± 45° degrees to the brain surface. From these samples we build 200k pairs in such a way that each patch occurs at least once, pairs always lie on the same hemisphere, and the resulting degree distribution of connections between pairs follows a power law. It would also be possible to include pairs across the hemispheres and calculate their distance by mapping points from the right hemisphere to the surface of the left hemisphere. We choose to stick to intrahemispheric distances due to interhemispheric differences in tissue size and area spread. We set the coordinate loss weight $\alpha$ to 10 and the weight decay factor $\lambda$ to 0.001. Networks are trained for 16 epochs with SGD using an initial learning rate of 0.01, decaying by factor 2 every 3 epochs.

## 3.2 Fine-Tuning the Extended U-Net on the Area Segmentation Task

For the area segmentation we use the extended U-net architecture proposed in [10]. This model combines local image features extracted from high resolution image patches with a topological prior given by a cytoarchitectonic probabilistic atlas (http://jubrain.fz-juelich.de). For the two input types, the model has two separate downsampling branches that are joined before the upsampling branch. We use the same dataset as described in [10], comprising 111 cell-body stained sections from 4 different brains, partially annotated with 13 visual areas using the observer-independent method [8]. For training, 2025 × 2025 px patches with 2 μm resolution were randomly extracted from the dataset.

We initialize the texture filtering branch with the weights from the Siamese distance regression network. Compared to [10], we reduce the overall learning rate, but train the atlas data branch with a higher learning rate to account for the different initializations of the branches. In detail, we first train for 8k iterations without the atlas data followed by additional 10k iterations including the atlas information (batch size 40). Initial learning rates for these phases were 0.05 and 0.025 (0.25 for the atlas data branch), with learning rate decay at iterations 3k, 5k, and 6k by factor 2. Choosing a good learning rate was essential for the good performance of the fine-tuning.

## 4 Experiments

We investigate the benefit of transfer learning from the self-supervised network on the original task of classifying brain areas. In particular, we show the influence

| Experiment (#train, loss) | Aux. $err_{dist}$ | Area Segm. Dice | $err_{seg}$ |
|---|---|---|---|
| Baseline [10] | - | 0.72 | 21.2 |
| 20k, $l_{dist}$ | 5.88 | 0.75 | 17.2 |
| 20k, $l_{coord}$ | 8.13 | 0.79 | 16.8 |
| 20k, $L$ | 4.54 | 0.79 | 15.2 |
| **200k, $L$** | **3.73** | **0.80** | **14.4** |

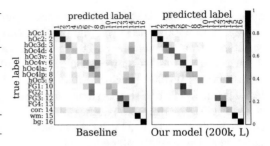

Baseline        Our model (200k, L)

**Fig. 3.** Quantitative results. Left: Column 2 evaluates the models on the auxiliary distance regression task, and columns 3-4 show the performance of the fine-tuned models on the area segmentation task. Right: Confusion matrices for the baseline model and our fine-tuned model on the area segmentation task. Model were trained on visual areas hOc1–FG4, non-annotated other cortex, white matter, and background.

of the different loss components. Furthermore, we demonstrate that the self-supervised network can distinguish several cytoarchitectonic areas without being explicitly trained on brain area classification. As evaluation metrics for the area segmentation task we report both the Dice score (harmonic mean of precision and recall) and the pixel distance error $err_{seg}$ that assigns to each misclassified pixel a penalty depending on the distance to the nearest pixel that is of the misclassified class [10, 12]. For the self-supervised distance task the mean difference between the predicted and the groundtruth distances $err_{dist}$ is reported.

**Siamese Network Loss.** The loss function $L$ defined in Eq. (3) for the self-supervised network combines a distance loss $l_{dist}$ with a coordinate loss $l_{coord}$. In order to evaluate the influence of each loss component, we trained a self-supervised model on 10% of the training set (20k samples) for 50 epochs. The model performs best when combining $l_{dist}$ with $l_{coord}$ (Fig. 3, rows 2-4 of the table). The inclusion of $l_{coord}$ doubles the performance on the distance task, showing that $l_{coord}$ has the expected effect of guiding the model towards a more representative feature embedding. When training only on $l_{coord}$ the performance of the fine-tuned network is almost as good as training on the combined loss. However, the pixel distance error $err_{seg}$ is then larger. A possible intuitive explanation is that $l_{dist}$ allows the model to see more realistic relationships between samples of the cortex than $l_{coord}$, where distances between coordinates only approximately represent the geodesic distance. Thus the combined loss enables the model to better allocate individual samples in the feature embedding and make less errors on the area segmentation task. Training on the full dataset moderately increases performance on the area segmentation task.

**Fine-Tuned Area Segmentation Model.** Compared to the randomly initialized network in [10], the Dice score increases to 0.80, while $err_{seg}$ drops from 21.2 to 14.4. The drastic reduction of $err_{seg}$ indicates that due to the pre-training, the model can locate patches more accurately in the brain and less often confuses spatially distant areas. Thus, the effect of pre-training on the distance regression

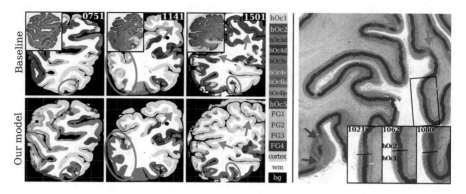

**Fig. 4.** Qualitative results. Left: Results on the area segmentation task with partial manual annotations in upper left corner. Compared to the baseline [10], our method predicts several areas significantly more accurate (circles) and has learned to deal much better with the "other cortex" class (arrows). Right: Squared Euclidean distances between averaged feature vectors of neighboring image patches, visualized by colored points along the cortical ribbon. Blue values indicate lower distances than yellowish values. Large distances occur at the border between hOc1/hOc2 over consecutive sections (green, enlarged boxes), at regions of high curvature (red) and at oblique regions (blue). This shows that the model has learned to recognize changes in cytoarchitecture.

task is similar to that of including the topological atlas prior in the supervised network [10]. The confusion matrices (Fig. 3) and example segmentations (Fig. 4) reveal that the fine-tuned model predicts more areas reliably, and overall exhibits less noise in the segmentation.

**Self-supervised Learning of Primary Visual Cortex.** To better understand the feature embedding that the self-supervised network learns, we average blocks of nine neighboring feature vectors (each $200\,\mu m$ apart) and plot the squared Euclidean distances between neighboring averaged feature vectors. This way we can appreciate the differences that the model predicts between neighboring regions. There are three main factors that cause the model to see large differences between neighboring parts of the cortex: (1) Oblique parts of the cortex, (2) regions with high curvature, and (3) borders between cortical brain areas. The latter is particularly exciting: It shows us that the model actually discovered relevant properties of some cytoarchitectonic regions to solve the distance regression. In Fig. 4 we show that the network has correctly identified the border between hOc1/hOc2 and tracks it through several sections.

## 5  Discussion and Conclusion

Exploiting prior knowledge and the inherent structure of the data is beneficial for tasks with limited training data. Our experiments show that the self-supervised distance task is a suitable auxiliary task for classifying cortical brain areas.

It significantly increases the Dice score which is a measure for the quality of the segmentation. In our evaluation, we have shown the importance of both components of our loss function to learn a good feature embedding. Additionally, we have demonstrated that our self-supervised model, trained with only the distances between samples as training signal, learns to identify several areal borders.

Inspired by this success, we plan to evaluate more auxiliary tasks based on inherent and relevant structures of 3D human brain reconstructions, such as local curvature or the relative orientation of the sectioning plane wrt. the brain surface and further evaluate the unsupervised features and their applicability towards identifying areal borders.

**Acknowledgements.** This work was partially supported by the Helmholtz Association through the Helmholtz Portfolio Theme "Supercomputing and Modeling for the Human Brain", and by the European Union's Horizon 2020 Framework Research and Innovation under Grant Agreement No. 7202070 (Human Brain Project SGA1) and 785907 (Human Brain Project SGA2). Computing time was granted by the John von Neumann Institute for Computing (NIC) and provided on the supercomputer JURECA at Jülich Supercomputing Centre (JSC).

# References

1. Amunts, K., et al.: BigBrain: an ultrahigh-resolution 3D human brain model. Science **340**(6139), 1472–1475 (2013)
2. Amunts, K., Zilles, K.: Architectonic mapping of the human brain beyond Brodmann. Neuron **88**(6), 1086–1107 (2015)
3. de Brébisson, A., Montana, G.: Deep neural networks for anatomical brain segmentation. In: CVPRW, pp. 20–28 (2015)
4. Doersch, C., Gupta, A., Efros, A. A.: Unsupervised visual representation learning by context prediction. In: ICCV, pp. 1422–1430 (2015)
5. Doumanoglou, A., Balntas, V., Kouskouridas R., Kim, T.: Siamese regression networks with efficient mid-level feature extraction for 3D object pose estimation. CoRR, abs/1607.02257 (2016)
6. Glasser, M.F., et al.: A multi-modal parcellation of human cerebral cortex. Nature **539**(7615), 171–178 (2016)
7. Ronneberger, O., Fischer, P., Brox, T.: U-Net: convolutional networks for biomedical image segmentation. In: Navab, N., Hornegger, J., Wells, W.M., Frangi, A.F. (eds.) MICCAI 2015. LNCS, vol. 9351, pp. 234–241. Springer, Cham (2015). https://doi.org/10.1007/978-3-319-24574-4_28
8. Schleicher, A., Amunts, K., Geyer, S., Morosan, P., Zilles, K.: Observer-independent method for microstructural parcellation of cerebral cortex: a quantitative approach to cytoarchitectonics. Neuroimage **9**(1), 165–177 (1999)
9. Simo-Serra, E., Trulls, E., Ferraz, L., Kokkinos, I., Fua, P., Moreno-Noguer, F.: Discriminative learning of deep convolutional feature point descriptors. In: ICCV, pp. 118–126 (2015)
10. Spitzer, H., Amunts, K., Harmeling, S., Dickscheid, T.: Parcellation of visual cortex on high-resolution histological brain sections using convolutional neural networks. In: ISBI, pp. 920–923 (2017)

11. Wang, X., Gupta, A.: Unsupervised learning of visual representations using videos. In: ICCV (2015)
12. Yasnoff, W., Mui, J., Bacus, J.: Error measures for scene segmentation. Pattern Recognit. **9**(4), 217–231 (1977)
13. Yosinski, J., Clune, J., Bengio, Y., Lipson, H.: How transferable are features in deep neural networks? In: NIPS, pp. 3320–3328 (2014)

# Registration-Free Infant Cortical Surface Parcellation Using Deep Convolutional Neural Networks

Zhengwang Wu[1], Gang Li[1], Li Wang[1], Feng Shi[2], Weili Lin[1],
John H. Gilmore[1], and Dinggang Shen[1(✉)]

[1] Department of Radiology, University of North Carolina at Chapel Hill, Chapel Hill,
NC, USA
dgshen@med.unc.edu
[2] Shanghai United Imaging Intelligence Co., Ltd., Shanghai, China

**Abstract.** Automatic parcellation of infant cortical surfaces into anatomical regions of interest (ROIs) is of great importance in brain structural and functional analysis. Conventional cortical surface parcellation methods suffer from two main issues: *(1)* Cortical surface registration is needed for establishing the atlas-to-individual correspondences; *(2)* The mapping from cortical shape to the parcellation labels requires designing of specific hand-crafted features. To address these issues, in this paper, we propose a novel cortical surface parcellation method, which is free of *surface registration and designing of hand-crafted features*, based on deep convolutional neural network (DCNN). Our main idea is to formulate surface parcellation as a patch-wise classification problem. Briefly, we use DCNN to train a classifier, whose inputs are the local cortical surface patches with multi-channel cortical shape descriptors such as mean curvature, sulcal depth, and average convexity; while the outputs are the parcellation label probabilities of cortical vertices. To enable effective convolutional operation on the surface data, we project each spherical surface patch onto its intrinsic tangent plane by a geodesic-distance-preserving mapping. Then, after classification, we further adopt the graph cuts method to improve spatial consistency of the parcellation. We have validated our method based on 90 neonatal cortical surfaces with manual parcellations, showing superior accuracy and efficiency of our proposed method.

## 1 Introduction

The highly folded human cerebral cortex can be parcellated into many neurobiologically meaningful regions [1], which are the foundation of many brain structural and functional studies. However, manual parcellation is extremely time-consuming, difficult to reproduce, and expertise-dependent. Therefore, automatic methods for cortical surface parcellation is highly desired. Previously,

This work is partially supported by NIH grants: MH100217, MH107815, MH108914, MH109773, MH110274, MH116225 and MH117943.

© Springer Nature Switzerland AG 2018
A. F. Frangi et al. (Eds.): MICCAI 2018, LNCS 11072, pp. 672–680, 2018.
https://doi.org/10.1007/978-3-030-00931-1_77

several methods have been proposed [1–6]. However, these methods have two main issues: (1) Cortical surface registration is needed to establish the atlas-to-individual cortical correspondences. (2) The mapping from cortical shape to the parcellation labels requires designing of specific hand-crafted features. However, the infant brain undergoes a rapid expansion, which is also highly heterogeneous across subjects; This raises challenges for *surface registration and designing of hand-crafted features.*

To address these issues, for the first time, we explore to build a highly non-linear mapping from the cortical shape domain to the parcellation label domain, by leveraging the strong representation ability of the deep convolutional neural network (DCNN). To cope with the heterogeneous expansion patterns across subjects, as well as to reduce the amount of required data for effective training, we use a patch-wise training strategy. To extend the convolution operation from the Euclidean space to the cortical surface mesh in Riemannian space, we further project the spherical cortical surface patch to its tangent plane by a geodesic-distance-preserving mapping, which can be done efficiently by leveraging the 2D topological nature of the cerebral cortex. Benefited from the strong representation power of DCNN, our method is *free of registration and free of designing hand-crafted features.* After patch-wise classification, we further use the graph cuts method to improve the spatial consistency of the parcellation. To validate our proposed method, we manually parcellated 90 term-born neonatal cortical surfaces by following the FreeSurfer parcellation protocols [2], using an in-house developed toolkit. Comparisons between our automatic parcellations and the manual parcellations showed the effectiveness and efficiency of our method. The main contribution of this paper lies in twofold: (1) We propose an automatic infant cortical surface parcellation method based on DCNN, working on the surface mesh. (2) Our method requires no surface registration and designing of hand-crafted features.

## 2    Method

### 2.1    Materials and Image Processing

A total of 90 term-born neonates were recruited. Their T2-weighted brain MR images were collected using a Siemens head-only 3 T scanner. All images were processed using a standard pipeline [7], which includes: (a) intensity inhomogeneity correction; (b) skull stripping; (c) cerebellum and brain stem removal; (d) tissue segmentation; (e) masking and filling non-cortical structures; (f) separation of the left and right hemispheres. Then, the topologically correct and geometrically accurate inner and outer cortical surfaces were reconstructed using a topology-preserving deformable surface method based on tissue segmentation [8]. Each cortical surface was represented by a surface mesh. Each vertex on the mesh was coded with 3 shape descriptors, i.e., the mean curvature, sulcal depth, and average convexity. These three shape descriptors reflect local geometric shape of the cortical surface, which have been used for cortical surface parcellation [2]. After that, we followed the FreeSurfer parcellation protocols [2] to manually label each cortical surface by an in-house toolkit. The rationality of using FreeSurfer folding-based parcellation protocols lies in the fact that all major gyral and sulcal patterns of the cortex are established at term birth and are stable during postnatal brain development [9].

## 2.2   Convolution on Cortical Surface

Our main idea is to train a DCNN model to build the mapping from the cortical shape domain to the parcellation label domain. However, unlike in the Euclidean space, there is no straightforward convolution operation for surface mesh, which sits in the Riemannian space and has no consistent neighborhood definition. Literally, two strategies have been proposed to extend the convolution operation to the surface mesh [10], including (a) convolution in other domains, e.g., the spectral domain obtained by the graph Laplacian [11]; (b) projecting the original surface to a certain intrinsic space, e.g., the tangent space (which is an Euclidean space with consistent neighborhood definition [12]). In this paper, we adopted the later strategy, i.e., projecting each surface patch to its local tangent plane by preserving the geodesic distance.

However, there are two challenges: (a) the geodesic distance is scale dependent, while the infant brain size varies dramatically across individuals; and (b) the geodesic distance computation for each vertex on cortical surface is time consuming. To address these issues, we leverage the 2D topological nature of the cerebral cortex. First, the inner cortical surface was inflated to a standard sphere by minimizing geometric distortion between original cortical surface and its spherical representation [13], which also normalizes the size variation. Then, on the spherical surface, the geodesic distance is homogenous everywhere. Therefore, the positions of two equal-size (measured in geodesic distance) spherical patches only differ each other in a simple rotation, indicating that we can obtain consistent spherical patches at all vertices by simple rotations.

The projection rule is illustrated in Fig. 1(a). For a vertex $V$, given a point $P_i$ on the tangent plane, we need to find the corresponding point $P_i'$ on the spherical surface. Given $P_i$, we can obtain: (a) the distance $r$ from $V$ to $P_i$; (b) the angle $\theta$ between the vector $\overrightarrow{VP_i}$ and the positive $x$-axis. Then, on the spherical surface, we can locate $P_i'$ by preserving: (a) the geodesic distance $r$ (i.e., the arc length on sphere) from $V$ to $P_i'$; (b) the angle $\theta$ between two great circles. One great circle is inscribed to the vector $\overrightarrow{VP_i}$, while the other is inscribed to the $x$-axis, as illustrated in Fig. 1(a).

With this projection rule, for each vertex V on the spherical surface, we can sample an Euclidean local patch $\{P_i(V), i = 1, \ldots, K^2\}$ on the tangent plane using the Cartesian coordinates, with $i$ indicating the point index and $K^2$ as the point number for a $K \times K$ patch. Of note, the sampled Euclidean patch in Cartesian coordinates can allow direct application of the conventional convolution definition. Therefore, we can define the convolution operation for the spherical patch $\{P_i'(V), i = 1, \ldots, K^2\}$ on the spherical surface based on the one-to-one correspondence between $P_i(V)$ and $P_i'(V)$, using the sampled Euclidean

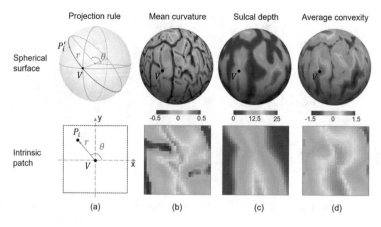

**Fig. 1.** Projection of a spherical surface patch to its tangent plane by preserving the geodesic distance and angle. (a) Illustration of projection rule. $P_i$ and $P_i'$ are the corresponding points on the local tangent plane of $V$ and spherical surface. They have the same geodesic distance $r$ to $V$ and the same angle $\theta$. In the tangent plane, $r$ is the distance from $V$ to $P_i$, and $\theta$ is the angle between $\overrightarrow{VP_i}$ and the positive $x$-axis. On the spherical surface, $r$ is the geodesic distance from $V$ to $P_i'$, and $\theta$ is the angle between two great circles, who are inscribed to the vector $\overrightarrow{VP_i}$ and the positive $x$-axis, respectively. (b)-(d) The spherical surfaces color-coded by the mean curvature, sulcal depth, and average convexity (top row), respectively; and their corresponding intrinsic patches at vertex $V$ (bottom row).

patch as a bridge. Herein, we name the spherical patch $\{P_i'(V), i = 1, \ldots, K^2\}$ as the intrinsic patch at vertex $V$ for better specificity and clarity. The top row of Fig. 1(b)–(d) shows the spherical cortical surfaces color-coded by the mean curvature, sulcal depth, and average convexity, respectively. While the bottom row of Fig. 1(b)–(d) shows the corresponding intrinsic patches at vertex $V$.

Once we located the intrinsic patch positions for a vertex $V_1$, i.e., $\{P_i'(V_1), i = 1, \ldots, K^2\}$, then for any other vertex on the spherical surface, e.g., $V_2$, we can simply rotate $\{P_i'(V_1)\}$ to $V_2$ to obtain the positions $\{P_i'(V_2)\}$ of the intrinsic patch at $V_2$. As mentioned before, this is because the geodesic distance on the sphere is homogeneous. Thus, we only need to construct the intrinsic patch for one time, which dramatically improves the efficiency of our method.

It is worth noting that different subjects may have different sphere tessellation. Therefore, we use the same tessellation to resample each spherical cortical surface. Through this strategy, we have the consistent intrinsic patches across different subjects. Note that our projection is different from the way in [12] in twofold: (a) unlike the arbitrary angular direction at different vertices, our local intrinsic patches at different vertices share the consistent angular direction, since they are obtained based on the rotation; (b) the projection only needs to conduct one time, which is much more efficient.

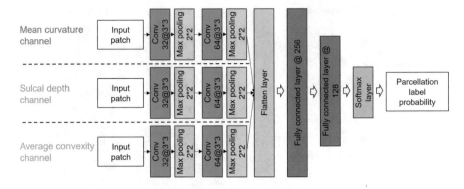

**Fig. 2.** The 3-channel DCNN architecture for cortical surface parcellation.

### 2.3   DCNN Architecture

After extending the convolution operation on the cortical surface, we use the DCNN to train the parcellation classifier. For each cortical surface, we have 3 shape descriptors, which reflect the folding patterns of the cortical surface in different views. Basically, the mean curvature measures the cortical folding in a fine view, and the average convexity measures the cortical folding in a coarse view. The sulcal depth measures the cortical folding by combining both the coarse and fine views. They thus provide complementary shape information. We include all of them into a 3-channel DCNN, with its architecture shown in Fig. 2. Each channel first independently performs the convolution and max-pooling operations. Then, all three channels are flattened and connected to the fully connected layer. Finally, the softmax layer outputs probability of each vertex belonging to each parcellation label.

### 2.4   Improving Spatial Consistency with Graph Cuts

Once the DCNN classifier is trained, for each vertex on a testing surface, we can extract the intrinsic patch and then apply the trained classifier for parcellation. However, since each vertex is classified independently without considering the spatial consistency, thus possibly producing spatially inconsistent labels. To improve the parcellation, we further use the graph cuts method to explicitly impose the spatial consistency. Specifically, based on the manual parcellation protocol, we know that the manual parcellations tend to split two cortical regions at the highly bent sulcal fundi [2]. Therefore, we explicitly formulate parcellation as a cost minimization procedure, i.e., $E = E_d + \lambda E_s$. Here, $E_d$ is the data fitting term, $E_s$ is the smoothness term, and $\lambda$ is a weight used to balance them. The data fitting term is defined as: $E_d = -\sum_V \log p_V(l_V)$, where $p_V(l_V)$ is the probability of assigning vertex $V$ as label $l_V$, which is obtained from the DCNN output. And $l_V = 1, \ldots, 36$ corresponds to 36 labels defined in the parcellation protocols [2]. The smoothness term is defined as: $E_s = \sum_{V^* \in \mathcal{N}_V} C_{V,V^*}(l_V, l_{V^*})$, where vertex $V^*$ is the direct neighbor of $V$, and $C_{V,V^*}(l_V, l_{V^*})$ is the cost to

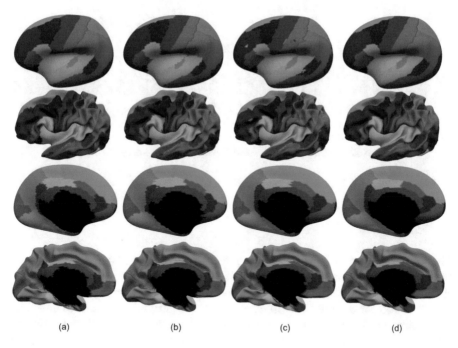

**Fig. 3.** Visual comparison of cortical parcellation results using different methods. (a) Manual parcellation; (b) Multi-atlas method with majority voting; (c) DCNN without graph cuts; (d) DCNN with graph cuts.

label vertex $V$ as $l_V$ and also label vertex $V^*$ as $l_{V^*}$. This cost can be defined as:

$$C_{V,V^*}(l_V, l_{V^*}) = \frac{1 + \overrightarrow{n_V} \cdot \overrightarrow{n_{V^*}}}{2} \times \frac{e^{-|H_V|} + e^{-|H_{V^*}|}}{2} \times (1 - \delta(l_V, l_{V^*})) \qquad (1)$$

Herein, $\overrightarrow{n}$ denotes the unit normal direction of a vertex, and $H$ denotes the mean curvature of a vertex. $\delta(l_V, l_{V^*})$ is the Dirac delta function: if $l_V = l_{V^*}$, $\delta(l_V, l_{V^*}) = 1$; otherwise, $\delta(l_V, l_{V^*}) = 0$. This cost definition adaptively encourages the label smoothness based on local cortical geometry. At the highly-bent cortical area (e.g., the sulcal fundi), two vertices $V$ and $V^*$ having the different labels generally have quite different normal directions and also large curvature magnitudes, therefore both the first and second terms in Eq. (1) is are small. On the other hand, if $V$ and $V^*$ have the same label, they generally have similar normal directions and large curvature magnitudes, therefore only the second term in Eq. (1) is small. However, if two vertices are on the flat cortical area, i.e., their normal directions are generally similar and their curvature magnitudes are close to 0, then both the first and second terms in Eq. (1) are close to 1. The minimization of the above cost function can be efficiently solved using the graph cuts method [14].

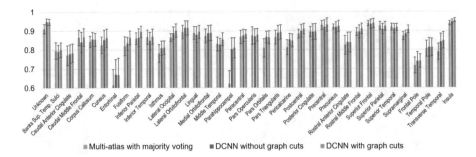

**Fig. 4.** Quantitative comparison of different parcellation methods for each ROI.

## 3     Experiments

To evaluate our method, we use a 3-fold cross-validation on 90 neonatal subjects. Specifically, the dataset is equally partitioned into 3 groups in a random manner. At each fold, two groups of data are used to train the network, while the other group is used for testing. The process is repeated until all groups have been used for testing. During the training, we augment the data by randomly rotating each training patch to improve the model generalization ability. We empirically set the patch size as $35 \times 35$. In the graph cuts method, we set the parameter $\lambda$ as 1. To assess the performance, we use Dice ratio to measure the overlap of the manual parcellation and the automatic parcellation.

For comparison, we adopt a popular multi-atlas based parcellation method (using majority voting). Specifically, we treat the training surfaces with manual parcellations as multiple atlases and use surface registration to propagate the parcellations of all atlases onto each testing subject. Then, we use majority voting to determine the final parcellation on testing subject. Herein, we use spherical demons [15] for registration, which has been well validated in cortical surfaces registration.

Figure 3 shows parcellation results using different methods. Specifically, Fig. 3(a) shows the manual parcellations. Figure 3(b)–(d) show the parcellation results using (b) multi-atlas with majority voting, (c) DCNN without graph cuts, and (d) DCNN with graph cuts, respectively. From this figure, we can see that the parcellation results of DCNN with graph cuts are highly consistent with manual parcellations.

Figure 4 further shows the quantitative comparison of parcellation results using different methods. From this figure, we can see that DCNN with graph cuts achieves the best performance. On average, the multi-atlas method with majority voting achieves average Dice ratio $84.54 \pm 0.08$ %. DCNN without graph cuts achieves the average Dice ratio $86.18 \pm 0.06$ %. And, DCNN with graph cuts achieves the average Dice ratio $87.06 \pm 0.06$ %. Compared to the multi-atlas method with majority voting, DCNN with graph cuts achieves better results in 34 out of 36 ROIs.

Of note, our method is registration free, thus much more efficient than the multi-atlas method. Specifically, for DCNN based parcellation, once the classifier is trained, the parcellation can be accomplished in less than a minute in a general PC. While, for the multi-atlas method, the computation time is much longer. For example, for the parcellation with 60 atlases, the computation time is around 2–3 h, since all 60 atlases need to be registered onto the testing subject one-by-one.

## 4    Conclusion

In this paper, we propose a *registration free* method for infant cortical surface parcellation, by using DCNN to learn the mapping from cortical shape domain to the parcellation label domain in a patch-wise manner. Our proposed method has been validated on 90 neonatal cortical surfaces. Both visual and quantitative comparisons show the effectiveness and efficiency of our method. In future, we will test our proposed method on more datasets and also make it publically available.

## References

1. Glasser, M.F., et al.: A multi-modal parcellation of human cerebral cortex. Nature **536**, 171–178 (2016)
2. Desikan, R.S., et al.: An automated labeling system for subdividing the human cerebral cortex on MRI scans into gyral based regions of interest. Neuroimage **31**, 968–980 (2006)
3. Li, G., et al.: Automatic cortical sulcal parcellation based on surface principal direction flow field tracking. NeuroImage **46**, 923–937 (2009)
4. Li, G., Shen, D.: Consistent sulcal parcellation of longitudinal cortical surfaces. NeuroImage **57**, 76–88 (2011)
5. Wu, Z., Li, G., Meng, Y., Wang, L., Lin, W., Shen, D.: 4D infant cortical surface atlas construction using spherical patch-based sparse representation. In: Descoteaux, M., Maier-Hein, L., Franz, A., Jannin, P., Collins, D.L., Duchesne, S. (eds.) MICCAI 2017. LNCS, vol. 10433, pp. 57–65. Springer, Cham (2017). https://doi.org/10.1007/978-3-319-66182-7_7
6. Li, G., et al.: Computational neuroanatomy of baby brains: a review. NeuroImage (2018, in press). https://doi.org/10.1016/j.neuroimage.2018.03.042
7. Li, G., et al.: Construction of 4D high-definition cortical surface atlases of infants: methods and applications. Med. Image Anal. **25**, 22–36 (2015)
8. Li, G., Nie, J., et al.: Consistent reconstruction of cortical surfaces from longitudinal brain MR images. Neuroimage **59**, 3805–3820 (2012)
9. Hill, J., et al.: A surface-based analysis of hemispheric asymmetries and folding of cerebral cortex in term-born human infants. J. Neurosci. **30**, 2268–2276 (2010)
10. Bronstein, M.M., et al.: Geometric deep learning: going beyond euclidean data. IEEE Signal Process. Mag. **34**, 18–42 (2017)
11. Bruna, J., et al.: Spectral networks and locally connected networks on graphs. arXiv preprint arXiv:1312.6203 (2013)

12. Masci, J., et al.: Geodesic convolutional neural networks on Riemannian manifolds. In: Proceedings of ICCV workshops, pp. 37–45 (2015)
13. Fischl, B., et al.: Cortical surface-based analysis: II: inflation, flattening, and a surface-based coordinate system. Neuroimage **9**, 195–207 (1999)
14. Boykov, Y., Kolmogorov, V.: An experimental comparison of min-cut/max-flow algorithms for energy minimization in vision. IEEE Trans. Pattern Anal. Mach. Intell. **26**, 1124–1137 (2004)
15. Yeo, B.T., et al.: Spherical demons: fast diffeomorphic landmark-free surface registration. IEEE Trans. Med. Imag. **29**, 650–668 (2010)

# Joint Segmentation of Intracerebral Hemorrhage and Infarct from Non-Contrast CT Images of Post-treatment Acute Ischemic Stroke Patients

Hulin Kuang, Mohamed Najm, Bijoy K. Menon, and Wu Qiu[✉]

Department of Clinical Neuroscience, University of Calgary, Calgary, Canada
qiu.wu.ch@gmail.com

**Abstract.** Cerebral infarct volume observed from follow-up non contrast CT (NCCT) scans is an important radiologic outcome measurement of the effectiveness of acute ischemic patient treatment. Post-treatment infarct typically includes ischemic infarct only. But in around 10% of ischemic stroke patients, intracerebral hemorrhage is present along with ischemic infarction. Post-treatment infarct is currently segmented manually, making it tedious and error prone. In order to measure post-treatment infarct volume more efficiently from follow-up NCCT images, a novel joint segmentation approach is proposed for segmenting ischemic infarct and hemorrhage simultaneously, which makes use of multi-region time-implicit contour evolution combined with random forest (RF) learned semantic information. The quantitative evaluation using 16 patient images shows that the proposed segmentation approach is delivering favorable results compared to the gold standard of manual segmentation in terms of dice similarity coefficient (DSC) and the mean (MAD) and maximum (MAXD) absolute surface distance. To the best of our knowledge, this paper reports the first attempt of simultaneously segmenting ischemic infarct and hemorrhage from follow-up NCCT images of post-treatment acute stroke patients.

**Keywords:** Intracerebral hemorrhage · Ischemic infarct
Multi-region image segmentation · Convex optimization
Random forest · Non contrast CT

## 1 Introduction

Recent randomized clinical trials revolutionized stroke care by confirming the benefit of modern endovascular therapy (EVT) in patients with acute ischemic stroke (AIS) due to proximal anterior circulation occlusion [1]. It is important for the EVT therapy being tested within the clinical trials to show the capability of saving brain tissue by measuring post-treatment infarct volume in addition to showing benefit across clinical scales, such as the modified Rankin Scale

© Springer Nature Switzerland AG 2018
A. F. Frangi et al. (Eds.): MICCAI 2018, LNCS 11072, pp. 681–688, 2018.
https://doi.org/10.1007/978-3-030-00931-1_78

(mRS) and the National Institute of Health Stroke Scale (NIHSS). A recent study shows that the beneficial effect of EVT on functional outcome could be explained by a reduction in post-treatment infarct volume measured manually [2]. Post-treatment infarct of AIS patients typically includes ischemic infarct only. But in around 10% of AIS patients [3] hemorrhage occurs after endovascular interventions, including intracerebral hemorrhage (ICH) (Fig. 1) and subarachnoid hemorrhage (SAH). These are feared complications. In those patients, post-treatment ischemic infarct is accompanied by hemorrhage, leading to two components within and around the final infarct (Fig. 1(a)). Non-contrast CT scanning is widely available and is the most commonly used diagnostic tool for AIS patients. Manually contouring ischemic infarct and hemorrhage from the follow up NCCT images is tedious, time consuming, and observer dependent. Therefore, an auto- or semi-automated infarct segmentation from clinically used follow-up NCCT images would be desired.

However, it is challenging to segment ischemic infarct and hemorrhage from NCCT images, as shown in Fig. 1(a), due to: (1) low signal to noise ratio of clinically used NCCT images compared to MR images, (2) low contrast of brain soft tissue making anatomical structures less differentiable, (3) anisotropy of NCCT images with big thickness (5 mm) while having high in-plane resolution $(0.625 > 0.625 \, \text{mm}^2)$, (4) interference from atrophy, leukoaraiosis, and partial volume effect at the area around cerebrospinal fluid (CSF) that has similar intensity distributions to ischemic infarct, and intra-ventricular calcification and artefact showing the same appearance as hemorrhage. While previous studies mostly focused on segmenting either infarct [4] or hemorrhage [5] individually, these techniques cannot be directly applied on brain scans of the patients with these infarcts and hemorrhage occurring at the same time. Zimmerman et al. [6] manually measured intracerebral hemorrhage and infarct. Bardera et al. [7] and Gillebert et al. [4] developed two different models to segment hemorrhage and infarct sequentially.

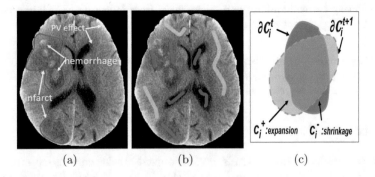

(a)                    (b)                    (c)

**Fig. 1.** A post-treatment NCCT image of an AIS patient with manual segmentations of hemorrhage and ischemic infarct super-imposed in (a); (b): initialization to the algorithm, red: infarct, green: hemorrhage, blue: ventricle, yellow: background; (c) scheme of the proposed contour evolution for a single contour.

**Contributions:** In this work, segmenting hemorrhage and ischemic infarct is integrated into the same framework, for which, a novel joint segmentation approach is proposed for extracting these two different lesions simultaneously from the follow up NCCT images. The proposed approach makes use of multi-region time-implicit contour evolution combined with random forest (RF) learned semantic information. We show that the evolution of multiple contours can be globally optimized through convex optimization technique, leading to an efficient and parallelized algorithm.

## 2    Method

We propose a segmentation approach to simultaneously partition the image into 4 disjoint regions $C_i$, $i = 1 \ldots 4$, denoting ischemic infarct, hemorrhage, CSF, and other tissues, using multi-region time-implicit level-set evolution. It should be noted that the CSF was segmented as an additional region in order to remove the false positives caused by the partial volume effect of CSF.

### 2.1    Preprocessing

The input non-contrast CT images are first skull-stripped using an in-house software. Then, some voxels are sampled manually for each region, as shown in Fig. 1(b), to initialize the subsequent contour evolution. The user-sampled voxels are not only used to estimate prior intensity probability functions (PDFs) $f_i(x)$ for each region, but also hard coded in the algorithm serving as constraints supervising the optimization scheme.

### 2.2    Multi-region Time-Implicit Level-Set Evolution

A fully time-implicit level-set scheme using global optimization [8,9] has demonstrated great advantages on both implementation and computation compared to classical level-set methods [10]. In this work, we employ the evolution of 4 time-implicit level-sets $C_i$, $i = 1 \ldots n$ to simultaneously partition the image into 4 disjoint regions, under the constraint:

$$\Omega = \cup_{i=1}^{n} C_i, \quad C_k \cap C_l = \emptyset, \quad \forall k \neq l. \tag{1}$$

where $\Omega$ means image domain, and $n = 4$ and $i = 1 \ldots 4$. Let $C_i^t$, $i = 1 \ldots n$, be the $i$-th region at the current time frame $t$, which moves to position $C_i^{t+1}$ at the next time frame $t + 1$. For each region $C_i^t$, $i = 1 \ldots n$, we define $C_i^+$ and $C_i^-$ indicating expansion and shrinkage of $C_i^t$ at time $t$ with respect two costs of $c_i^+(x)$ and $c_i^-(x)$ (Fig. 1(c)).

With these definitions, the evolution of each region is intended to minimize the discrepancy between shrinkage and expansion over the discrete time frame while minimizing contour length:

$$\min_{C_i} \sum_{i=1}^{n} \left\{ \int_{C_i^-} c_i^-(x)\, dx + \int_{C_i^+} c_i^+(x)\, dx \right\} + \sum_{i=1}^{n} \int_{\partial C_i} g(s)\, ds \tag{2}$$

subject to (1), where $g(s)$ is the weighting function along the contour boundaries. In general, the level-set $\mathcal{C}$ evolution is driven by distance functions and image features for image segmentation, for which, the cost functions $c_i^-(x)$ and $c_i^+(x)$, w.r.t. region changes are typically defined as:

$$c_i^-(x) = c_i^+(x) = \mathrm{dist}(x, \partial\mathcal{C}_i^t)/h + F_i(x) \tag{3}$$

where $h > 0$ is constant, $\mathrm{dist}(.)$ is the distance defining distances from $x$ to the contour $\partial\mathcal{C}_i^t$, and the image feature $F_i(x)$ is derived according to the specified intensity distribution [11]. In this study, we use geodesic distance function $gdist(x, \partial\mathcal{C})$ rather than traditional Euclidean distance function $dist(.)$, as geodesic distance defines the distances from $x$ to the contour $\partial\mathcal{C}^t$ using the shortest path along the image intensities, which could provide more adaptive spatial contexts [12].

In particular, in addition to image features derived from images, an additional cost $r_i(x)$ for each voxel is incorporated into the contour evolution, which includes semantic information derived from a random forest trained model [4]. The semantic information used for the random forest training and testing includes intensity information, and the mean and standard deviation of intensity in local regions, as well as the difference of those features between the current voxel and quasi-symmetry one from its contralateral side. An infarct occurrence spatial probability map obtained from training images served as an additional feature.

The cost functions are accordingly defined as follows:

$$c_i^+(x) = c_i^-(x) = -\omega_1 \log f_i(x) - \omega_2 \log r_i(x)$$
$$+ \omega_3 \frac{1}{h} \mathrm{gdist}(x, \partial\mathcal{C}_i^t) \quad \forall x \in \mathcal{C}_i^t \tag{4}$$

where the weighting parameters $\omega_1, \omega_2, \omega_3 > 0$, $\omega_1 + \omega_2 + \omega_3 = 1$ weight the contributions from the intensity PDFs $f_i(x)$, random forest learned probability $r_i(x)$, and geodesic distance for each voxel, respectively. The parameters $\omega_i$ were empirically defined and kept as constants through evaluation stage. Specifically, we have implemented two independent random forest models trained with datasets from another study for classifying hemorrhage and ischemic infarct voxels, sequentially. There were no random forest models specially derived for CSF and other tissues that means only intensity and distance priors were used for segmenting these two regions.

## 2.3 Spatially Continuous Potts Model

The introduced variational problem (2) can be equally reformulated as a Potts problem [13]. For this purpose, we define two cost functions $D_i^s(x)$ and $D_i^q(x)$ w.r.t. the current contour $\mathcal{C}_i^t$, $i = 1...n$, at time $t$:

$$D_i^s := \begin{cases} c_i^-(x), & \text{where } x \in \mathcal{C}_i^t \\ 0, & \text{otherwise} \end{cases} \tag{5}$$

$$D_i^q := \begin{cases} c_i^+(x)\,, & \text{where } x \notin C_i^t \\ 0\,, & \text{otherwise} \end{cases}. \tag{6}$$

If $u_i(x) \in \{0,1\}$, $i = 1 \ldots n$, is defined as the indicator function of the region $C_i$, the disjoint constraint in (1) can be represented by

$$\sum_{i=1}^{n} u_i(x) = 1; \quad u_i(x) \in \{0,1\} \quad \forall x \in \Omega \tag{7}$$

Via the cost functions (5) and (6). It can be proven that the variational formulation (2) can be represented as the Potts model:

$$\min_{u_i(x)\in\{0,1\}} \sum_{i=1}^{n} \langle u_i, D_i^q - D_i^s \rangle + \sum_{i=1}^{n} \int_{\Omega} g(x) \, |\nabla u_i| \, dx \tag{8}$$

subject to the contour disjointness constraint (7), where the weighted length term in (2) is encoded as the weighted total-variation function (the second term in (8)).

We introduce convex relaxation technique to optimize the energy function (8) by solving a continuous min-cut/max-flow problem [9,14], which implies that the position of the contour $C_i^{t+1}$ at the next time step $t+1$ is globally optimal, and the given contour $C_i^t$ at the current time frame is always moving to its globally optimal position. In addition, the new contour position at each evolution step is computed in a time-implicit manner, which allows a large time-step and parallelization of the algorithm, substantially speeding up calculations.

## 3   Experiments and Results

**Image Acquistion:** The size of 3D NCCT images is $512 \times 512 \times (18 - 37)$ with a voxel spacing from $0.37 \times 0.37 \times 3\,\text{mm}^3$ to $0.49 \times 0.49 \times 5\,\text{mm}^3$. Sixteen stroke patients with hemorrhage and infarct after treatment from a clinical trial [1] were used for this study. Hemorrhage and infarct in each image were manually segmented slice by slice on parallel axial view using ITK-SNAP by a trained observer and verified by an experienced clinician. Manual segmentations were used to evaluate the proposed segmentation approach. In addition, another 30 patients (different form the used 16 patients) were used to train the random forest model and derive the lesion probability $r(x)$.

**Evaluation Metrics:** Our segmentation method was evaluated by comparing the algorithm segmented results with manual segmentation in terms of the overlap of hemorrhage, infarct, and those two regions as a whole, respectively, using *volume-based metrics*: dice similarity coefficient (DSC); and *distance-based metrics*: the mean absolute surface distance (MAD) and maximum absolute surface distance (MAXD) [8,9,14,15]. In order to show the efficacy of the proposed approach, two semi-automated segmentation methods: graph cut (GC) [16] and multi-region segmentation (MLS) [17], and a fullly automated random forest method followed by the optimization of Markov random field (RF+MRF) [5] were compared. For fair comparison, the initializations to our proposed method were also used as initializations to MLS and GC. It should be noted that the major part of the proposed method was implemented in Matlab, while the derivation of lesion probability map using random forest was implemented in Python. The methods used for comparison above were re-implemented based on the papers and publicly available code.

**Results:** The quantitative results in Table 1 show that the proposed approach yielded mean DSCs of 76.4% for ischemic infarct, 56.5% for hemorrhage, and 85.0% for the two regions with favorable values of MAD and MAXD. A statistical analysis, using one-way ANOVA followed by Tukey-Kramer correction regarding DSC, MAD, and MAXD,shows that the proposed method significantly outperformed the fully automated method RF+MRF and two other semi-automated methods :MLS and GC ($p < 0.001$ for all three comparisons using all three metrics). The Spearman correlation analysis between manually and algorithm segmented infarct volume shows that the Spearman correlation coefficients are 0.954 ($p < 0.0001$) for ischemic infarct, 0.964 ($p < 0.0001$) for hemorrhage, and 0.986 ($p < 0.0001$) for the two regions, respectively (Fig. 2).

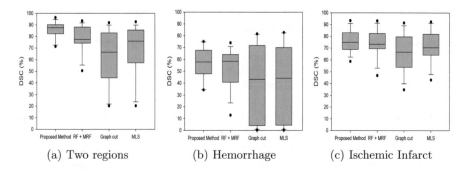

(a) Two regions      (b) Hemorrhage      (c) Ischemic Infarct

**Fig. 2.** Segmentation accuracy in terms of DSC in different regions using 16 patients.

**Table 1.** Overall performance results using 16 patient images in terms of DSC (%), MAD (mm), and MAXD (mm).

| | Ischemic infarct | | | Hemorrhage | | | Two regions | | |
|---|---|---|---|---|---|---|---|---|---|
| | DSC | MAD | MAXD | DSC | MAD | MAXD | DSC | MAD | MAXD |
| Graph cut [16] | 65.7 ± 16.5 | 4.2 ± 4.0 | 17.5 ± 12.0 | 39.8 ± 32.2 | 14.3 ± 19.6 | 29.6 ± 32.7 | 62.3 ± 22.8 | 9.6 ± 10.6 | 27.2 ± 19.4 |
| MLS [17] | 71.3 ± 13.9 | 4.4 ± 4.3 | 18.4 ± 12.4 | 40.1 ± 30.8 | 20.4 ± 16.6 | 49.1 ± 28.7 | 67.6 ± 22.6 | 9.9 ± 10.6 | 31.3 ± 18.3 |
| RF+MRF [5] | 73.5 ± 12.4 | 4.4 ± 3.6 | 18.9 ± 10.3 | 51.7 ± 16.8 | 7.0 ± 3.5 | 21.7 ± 7.6 | 78.2 ± 11.7 | 5.0 ± 3.9 | 20.9 ± 10.1 |
| Proposed | **76.4 ± 9.7** | **2.7 ± 1.0** | **13.5 ± 4.5** | **56.5±13.2** | **5.5 ± 2.7** | **17.7 ± 6.7** | **85.0 ± 7.4** | **3.2 ± 1.1** | **15.5 ± 4.3** |

# 4 Discussion and Conclusion

A novel semi-automated segmentation approach is proposed, which is capable of extracting hemorrhage and ischemic infarct simultaneously from the follow-up NCCT images of post-treatment AIS patients. The quantitative results using 16 patients demonstrate its efficacy in terms of DSC, MAD, and MAXD. Compared to other state-of-the-art segmentation methods, the proposed approach improves segmentation accuracy greatly for ischemic infarct, hemorrhage, and the two regions as a whole.

There are few methods segmenting post-treatment ischemic infarct accompanied by hemorrhage from follow-up NCCT images. Most of studies segmented these two distinct pathologies individually. Compared to other ischemic infarct segmentation methods, the mean DSC of 76.4% obtained in this study is higher than a median of 74% (42–90%) [18] and 58–62% [4] in previous studies. Compared to other hemorrhage segmentation methods, the DSC of $56.5 \pm 13.2\%$ for the hemorrhage generated in this study is higher than $55 \pm 24\%$ reported in [19], but lower than 62–78% [4] and 89.9% [5] in previous studies. However, in addition to the volume biased DSC value, the Spearman correlation coefficient of $0.964$ ($p < 0.0001$) demonstrated a strong correlation between the algorithm and manually segmented hemorrhage volume. Additionally, the proposed approach generated favorable DSC, MAD, and MAXD values provided both infarct and hemorrhage as a whole.

One limitation of this study is that the proposed approach was only validated on a small number of images. Extensive evaluation on a larger dataset is required to make it clinically applicable. In addition, given that the fully automated RF+MRF yielded poorer performance, the proposed method required user sampled voxels in 3–5 slices out of 28–44 slices in a CT image as initialization to improve performance even though it required additional <1 min. The introduced observer variability should be assessed in the future.

To summarize, we have proposed a segmentation approach to quantifying infarct volume of post-treatment AIS patients who present with ischemic infarct and hemorrhage simultaneously. The quantitative evaluations showed excellent accuracy and strong correlation with the gold standard of manual segmentation, suggesting that the proposed technique has the potential of being used to assess imaging outcomes in AIS patients after treatment.

# References

1. Goyal, M., et al.: Endovascular thrombectomy after large-vessel ischaemic stroke: a meta-analysis of individual patient data from five randomised trials. Lancet **387**(10029), 1723–1731 (2016)
2. Al-Ajlan, F.S., et al.: Intra-arterial therapy and post-treatment infarct volumes. Stroke **47**(3), 777–781 (2016)
3. Yoon, W., Jung, M.Y., Jung, S.H., Park, M.S., Kim, J.T., Kang, H.K.: Subarachnoid hemorrhage in a multimodal approach heavily weighted toward mechanical thrombectomy with solitaire stent in acute stroke. Stroke **44**(2), 414–419 (2013)

4. Gillebert, C.R., Humphreys, G.W., Mantini, D.: Automated delineation of stroke lesions using brain CT images. NeuroImage: Clin. **4**, 540–548 (2014)
5. Muschelli, J., Sweeney, E.M., Ullman, N.L., Vespa, P., Hanley, D.F., Crainiceanu, C.M.: Pitchperfect: primary intracranial hemorrhage probability estimation using random forests on CT. NeuroImage: Clin. **14**, 379–390 (2017)
6. Zimmerman, R., Maldjian, J., Brun, N., Horvath, B., Skolnick, B.: Radiologic estimation of hematoma volume in intracerebral hemorrhage trial by CT scan. Am. J. Neuroradiol. **27**(3), 666–670 (2006)
7. Bardera, A., et al.: Semi-automated method for brain hematoma and edema quantification using computed tomography. Comput. Med. Imaging Graph. **33**(4), 304–311 (2009)
8. Qiu, W., et al.: 3d MR ventricle segmentation in pre-term infants with post-hemorrhagic ventricle dilatation (PHVD) using multi-phase geodesic level-sets. NeuroImage **118**, 13–25 (2015)
9. Qiu, W., Yuan, J., Ukwatta, E., Sun, Y., Rajchl, M., Fenster, A.: Efficient 3D multi-region prostate MRI segmentation using dual optimization. In: Gee, J.C., Joshi, S., Pohl, K.M., Wells, W.M., Zöllei, L. (eds.) IPMI 2013. LNCS, vol. 7917, pp. 304–315. Springer, Heidelberg (2013). https://doi.org/10.1007/978-3-642-38868-2_26
10. Chan, T.F., Vese, L.A.: Active contours without edges. IEEE Trans Image Process. **10**(2), 266–277 (2001)
11. Boykov, Y., Kolmogorov, V., Cremers, D., Delong, A.: An integral solution to surface evolution PDEs via geo-cuts. In: Leonardis, A., Bischof, H., Pinz, A. (eds.) ECCV 2006. LNCS, vol. 3953, pp. 409–422. Springer, Heidelberg (2006). https://doi.org/10.1007/11744078_32
12. Criminisi, A., Sharp, T., Blake, A.: GeoS: geodesic image segmentation. In: Forsyth, D., Torr, P., Zisserman, A. (eds.) ECCV 2008. LNCS, vol. 5302, pp. 99–112. Springer, Heidelberg (2008). https://doi.org/10.1007/978-3-540-88682-2_9
13. Yuan, J., Bae, E., Tai, X.-C., Boykov, Y.: A continuous max-flow approach to potts model. In: Daniilidis, K., Maragos, P., Paragios, N. (eds.) ECCV 2010. LNCS, vol. 6316, pp. 379–392. Springer, Heidelberg (2010). https://doi.org/10.1007/978-3-642-15567-3_28
14. Qiu, W., et al.: 3D prostate TRUS segmentation using globally optimized volume-preserving prior. In: Golland, P., Hata, N., Barillot, C., Hornegger, J., Howe, R. (eds.) MICCAI 2014. LNCS, vol. 8673, pp. 796–803. Springer, Cham (2014). https://doi.org/10.1007/978-3-319-10404-1_99
15. Qiu, W., Yuan, J., Ukwatta, E., Fenster, A.: Rotationally resliced 3D prostate TRUS segmentation using convex optimization with shape priors. Med. Phys. **42**(2), 877–891 (2015)
16. Boykov, Y., Veksler, O., Zabih, R.: Fast approximate energy minimization via graph cuts. IEEE Trans. Pattern Anal. Mach. Intell. **23**(11), 1222–1239 (2001)
17. Vese, L.A., Chan, T.F.: A multiphase level set framework for image segmentation using the Mumford and Shah model. Int. J. Comput. Vis. **50**(3), 271–293 (2002)
18. Boers, A.M., et al.: Automated cerebral infarct volume measurement in follow-up noncontrast CT scans of patients with acute ischemic stroke. Am. J. Neuroradiol. **34**(8), 1522–1527 (2013)
19. Boers, A., et al.: Automatic quantification of subarachnoid hemorrhage on non-contrast CT. Am. J. Neuroradiol. **35**(12), 2279–2286 (2014)

# Patch-Based Mapping of Transentorhinal Cortex with a Distributed Atlas

Jin Kyu Gahm[1(✉)], Yuchun Tang[1,2], and Yonggang Shi[1]

[1] Laboratory of Neuro Imaging, USC Stevens Neuroimaging and Informatics Institute, Keck School of Medicine, University of Southern California, Los Angeles, USA
jkgahm@loni.usc.edu
[2] Research Center for Sectional and Imaging Anatomy, Shandong University Cheeloo College of Medicine, Jinan, Shandong, China

**Abstract.** The significance of the transentorhinal (TE) cortex has been well known for the early diagnosis of Alzheimer's disease (AD). However, precise mapping of the TE cortex for the detection of local changes in the region was not well established mostly due to significant geometric variations around TE. In this paper, we propose a novel framework for automated patch generation of the TE cortex, patch-based mapping, and construction of an atlas with a distributed network. We locate the TE cortex and extract a small patch surrounding the TE cortex from a cortical surface using a coarse map by FreeSurfer. We apply a recently developed intrinsic surface mapping algorithm based on Riemannian metric optimization on surfaces (RMOS) in the Laplace-Beltrami embedding space to compute fine maps between the small patches. We also develop a distributed atlas of the TE cortex, formed by a shortest path tree whose nodes are atlas subjects, to reduce anatomical misalignments by mapping only between similar patches. In our experimental results, we construct the distributed atlas of the TE cortex using 50 subjects from the Human Connectome Project (HCP), and show that detailed correspondences within the distributed network are established. Using a large-scale dataset of 380 subjects from the Alzheimer's Disease Neuroimaging Initiative (ADNI), we demonstrate that our patch-based mapping with the distribute atlas outperforms the conventional centralized mapping (direct mapping to a single atlas) for detecting atrophy of the TE cortex in the early stage of AD.

## 1 Introduction

According to the Braak staging of Alzheimer's disease [1], the neurodegenerative process of AD begins in the transentorhinal (TE) cortex, and spreads to the hippocampus, the temporal and insular cortices, and eventually to the entire cortex. Therefore, the *in vivo* detection of localized changes in the TE cortex from MRI

This work was in part supported by the National Institute of Health (NIH) under Grant R01EB022744, P41EB015922, U01EY025864, U01AG051218, P50AG05142.

A. F. Frangi et al. (Eds.): MICCAI 2018, LNCS 11072, pp. 689–697, 2018.
https://doi.org/10.1007/978-3-030-00931-1_79

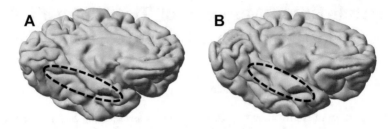

**Fig. 1.** (A) Continuous and (B) discontinuous collateral sulcus, highlighted by the dashed ellipsoids, on left cortical surfaces reconstructed from *in vivo* T1-weighted MRI. The transentorhinal (TE) cortex is also shown in red color.

is critical for the early diagnosis of AD, so accurate mapping of the TE cortex is needed. Spherical registration-based mapping of entire cortical surfaces has been most widely used [2] but has a limitation with the accurate mapping of small and variable cortical structures such as the TE cortex (see Fig. 1). Therefore, a focused mapping of small patches including the target TE and neighboring structures and construction of a TE atlas on the patches have the potential of more precisely mapping this critical region. Recently, we have developed a novel approach of intrinsic surface mapping in the Laplace-Beltrami (LB) embedding space based on Riemannian metric optimization on surfaces (RMOS) [3,4]. In our previous work, RMOS has been applied only to the genus-zero structures of the thalamus and striatum, but is applicable to general structures with various topology including patches whose Euler number is 1. In this work, we propose a novel framework for automatically generating the small patches from entire cortical surfaces using FreeSurfer's coarse maps, and producing fine maps between the small patches using RMOS.

One of the main challenges in studying the TE cortex and neighboring structures in the medial temporal lobe (MTL) is the existences of significant variations in the sulcal patterns. In particular, the continuous/discontinuous collateral sulcus (CS), and the no/shallow/deep rhinal sulcus (RS) were shown in different subjects from autopsy brains [5]. The continuous/discontinuous CS is also clearly observed on cortical surfaces from *in vivo* MRI as shown in Fig. 1. For a large-scale study of the TE cortex, therefore, mapping of the TE cortex from a single atlas with specific sulcal patterns to subjects with the various sulcal patterns, which we call a *centralized atlas*, is not appropriate. Instead, we construct an atlas network of multiple TE patches with various sulcal patterns whose connections are made only between similar patches, which we call a *distributed atlas*. Recently, such a distributed network (graph) of images was developed for image registration to reduce registration error by registering only similar images nearby in the graph [6].

Our distributed atlas of the TE cortex with RMOS maps for its connections is reusable and generally applicable to any study with different group/project subjects. In our experiments, we build the distributed atlas using a dataset of 50 subjects from the Human Connectome Project (HCP) [7]. Using a large-

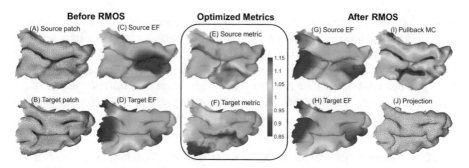

**Fig. 2.** Illustration of RMOS mapping between (A) source and (B) target patches colored with their mean curvature (MC). (C, D) The 6th eigenfunctions (EFs) of the source and target patches before RMOS; and (G, H) the EFs after RMOS, with (E, F) the optimized Riemannian metrics. (I, J) The pullback of MC from the source to the target patch, and projection of the source onto the target patch by the RMOS map.

scale study dataset of 380 subjects from the Alzheimers Disease Neuroimaging Initiative (ADNI) [8], we demonstrate that our novel framework of automated patch generation and patch-based RMOS mapping with the distributed atlas achieves significantly improved sensitivity in the detection of TE atrophy in the early stage of AD, compared to RMOS and FreeSurfer mapping with the centralized atlas.

## 2    Methods

In this section, we describe the details of our novel framework for precise mapping of the TE cortex with a distributed atlas as well as present the theoretical backgrounds for patch-based RMOS.

**Automated Patch Generation.** To extract small patches of the TE cortex from entire cortical surfaces, we first perform FreeSurfer reconstruction of T1-weighted MRI that generates surface representation and parcellation of the cortex [9,10]. We randomly choose one subject as a reference, and manually delineate on the reference cortical surface a small but extensive patch region including the TE cortex and neighboring structures around the MTL to prevent the pattern of primary sulcus such as the CS near the TE cortex from being discontinued at the boundaries of the patch. More specifically, using the FreeSurfer's cortical parcellation results (*aparc+aseg*), we include the entire parahippocampal (PH) and entorhinal cortices (EC) with extension to the lateral side; the anterior parts of the fusiform and the inferior temporal cortex; and a half of the temporal pole adjacent to EC. Then we compute coarse maps of cortical surfaces from other subjects to the reference using FreeSurfer (*mris_register*) using the surfaces (*lh.white*) and features (*lh.sulc* and *lh.inflated.H*) in the FreeSurfer reconstruction results. Then using the FreeSurfer maps, we pull back the ROI labels from the reference surface onto the cortical surface (*lh.pial*) of each subject, smooth by mesh evolution and decimate the surfaces with the ROI labels to 25 K vertices. We extract by the labels the patches with about 2 K vertices from the cortical surfaces, and remove unconnected parts and holes ensuring that the final patches have the Euler number of 1. This process is fully automatic for a large-scale study with different subjects once we provide the reference cortical surface with the labels of the patch as a part of our atlas of the TE cortex.

**RMOS Mapping of Patches.** After we extract the patches using FreeSurfer's coarse maps of large cortical surfaces, we compute fine maps between the small patches using RMOS [3,4]. Let $\mathcal{P}_i(i = 1, 2)$ denote the triangular mesh representation of two patches, and $W_i$ denote their Riemannian metrics, *i.e.*, the edge weights. Using the Riemannian metrics, the LB embeddings of $\mathcal{P}_i$ can be computed as:

$$I_{\mathcal{M}_i}^{\Phi_i}(x) = \left( \frac{f_1^i(x)}{\sqrt{\lambda_1^i}}, \frac{f_2^i(x)}{\sqrt{\lambda_2^i}} \cdots, \frac{f_n^i(x)}{\sqrt{\lambda_n^i}}, \cdots \right) \qquad \forall x \in \mathcal{P}_i , \tag{1}$$

where $\lambda_n^i$ and $f_n^i$ are the $n$-th eigenvalue and eigenfunction of $\mathcal{P}_i$. Given $L$ features $\xi_1^j$ and $\xi_2^j$ ($j = 1, \cdots, L$) defined on the patches, we compute the data term with the patch maps $u_1 : \mathcal{P}_1 \to \mathcal{P}_2$ and $u_2 : \mathcal{P}_2 \to \mathcal{P}_1$ as:

$$E_F(\mathcal{P}_1, \mathcal{P}_2) = \sum_{j=1}^{L} \left[ \int_{\mathcal{P}_1} (\xi_1^j - \xi_2^j \circ u_1)^2 d\mathcal{P}_1 + \int_{\mathcal{P}_2} (\xi_2^j - \xi_1^j \circ u_2)^2 d\mathcal{P}_2 \right] . \tag{2}$$

We iteratively alternate the energy minimization of the data term $E_F$ with a regularization term, and matching of the embeddings, *i.e.*, the eigenfunctions by updating the Riemannian metrics $W_i$ in a gradient descent way. For the data term, in this work, we use three geometric features extracted from - FreeSurfer's sulcal depth, and curvature on the inflated surface; and the mean curvature (MC) computed from the normalized cortical surface to the average surface volume. Figure 2 illustrates RMOS mapping between two non-isometric patches (A, B) whose eigenfunctions, especially the 6th eigenfunctions (C, D), are largely different. With the optimized metrics plotted in Fig. 2(E, F), the eigenfunctions become well matched (G, H). From the optimized LB embeddings, we compute the point-wise maps between the patches, and visualize the map from the source patch to the target patch by projecting the source patch onto the target patch (Fig. 2(J)), and pulling back the MC from the target patch to the source patch (I), where the MC features of the target patch are well aligned to the source patch (A).

**Distributed Atlas Construction.** Given an atlas dataset of $N$ subjects, we construct a distributed atlas that is a tree-structured graph whose nodes are the $N$ atlas subjects. Edges represent geometric dissimilarities of the patches between any pair of two nodes $\mathcal{P}_i$ and $\mathcal{P}_j$ with weights defined by the data term $E_F$ in Eq. 3 using the initial nearest point map in the LB embedding space:

$$D_{i,j} = \exp\{E_F(\mathcal{P}_i, \mathcal{P}_j)/(\alpha\sigma)\} \tag{3}$$

where $\sigma$ is the Gaussian kernel width set as the average data term, and $\alpha$ is the parameter that decides the tree height. We determine the root node, *i.e.*, the common space node, which is the shortest to all other nodes by the sum of the dissimilarity measures between their patches. Then we find the shortest path from the root node to every node that forms a tree-structured graph as the distributed atlas. We empirically adjust the parameter $\alpha$ that forms the tree of height $> 1$. The tree of height 1 is not desirable since direct mapping from the root to other nodes with significantly different sulcal patterns in the distributed atlas network may fail to establish anatomically correct correspondences. For each connection in the graph (total $N$-1 edges), we run RMOS mapping between similar patches of the edge nodes. In the RMOS maps, each point in one patch is discretized as a linear combination of vertex positions in the other patch. For any node of depth $> 1$, therefore, the composition of the maps along the path is done by combining the (point-wise) linear transformations and projecting the position of the final composite points onto the target patch.

**TE Mapping with the Distributed Atlas.** For a large-scale study with different subjects, once the reference patch label, and the distributed atlas and its connection maps are provided, mapping of the TE cortex for the new subjects is fully automatic and straightforward by following the steps: (1) Run FreeSurfer reconstruction; (2) Compute coarse maps to the reference using FreeSurfer, and extract patches with the features by the maps; (3) Compute the dissimilarities of the patches between every subject and the atlas (every node); (4) Connect each subject to the shortest node in the distributed atlas; (5) Compute fine RMOS maps of the patches for the new connections, and composite maps along the path between each leaf (subject) and the root. The distributed atlas is generally applicable to any study so it is not required to rebuild it with new subjects.

## 3   Results

In this section, we present experimental results of our novel framework for precise mapping of the TE cortex with a distributed atals that significantly improved sensitivity in the early detection of TE atrophy by AD.

**Generation of TE Patches.** We first demonstrate automated generation of patches for construction of the distributed atlas from 50 subjects of the Q1-Q3 release of HCP. Note that our experiments in this paper only focused on the left hemisphere. As shown in Fig. 3(A), we delineated the ROI including TE on the reference cortical surface of a randomly chosen subject. The pullback ROI of the reference to two other subjects by the FreeSurfer cortical maps are shown in

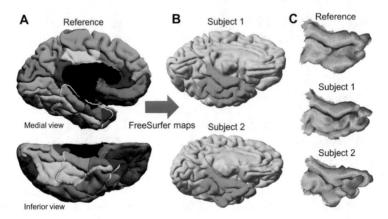

**Fig. 3.** Automated patch generation of the left TE cortex with the neighboring structures. (A) Delineation of ROI (yellow curve) around the medial temporal cortex on a reference cortical surface guided by the FreeSurfer parcellation results. (B) Pullback labels (green) of the reference ROI onto subject's cortical surfaces by the FreeSurfer's cortical maps. (C) Extracted patches by the labels from the cortical surfaces, and colored with their mean curvature that ranges $[-0.4, 0.4]$ mm$^{-1}$.

Fig. 3(B), and moderately well matched the actual anatomical ROI we defined on the reference cortical surface. We collected the sulcal depth and curvature maps on the cortical surfaces from the FreeSurfer reconstruction results, and computed the normalized MC of the original cortical surfaces. We decimated the cortical surfaces with the three feature maps to 25K, and extracted the patches and features as shown in Fig. 3(C).

**TE Cortex Delineation.** To evaluate the performance of patch-based mapping of the TE cortex in the distributed atlas, we delineated the TE cortex on each patch of the 50 HCP subjects. The TE cortex locates in the medial bank of the CS, and is the medial portion of the perirhinal cortex, which is bordered caudally by the PH cortex and ventrally and medially by the EC [11]. We used the FreeSurfer parcellation results to pinpoint the neighboring structures, and determined the TE boundaries by the CS and EC. Our delineation of the TE cortex for 3 subjects are shown in Fig. 4(A) (colored in red).

**Patch Mapping.** Fig. 4(A-C) demonstrates RMOS mapping of patches from an atlas to subjects. We set the RMOS parameters - the eigenorder 6, the regularization coefficient of 0.1, and the maximum 200 iterations. Direct mapping from the atlas to Subject b introduced large distortions in the projection of the atlas patch onto the Subject b patch, and the pullback TE labels of Subject b onto the atlas did not match well with the atlas TE labels as shown in Fig. 4(B). Instead, mapping between more similar patches, determined from the dissimilarities measures between the patches in Eq. 3, from the atlas to Subject a, and

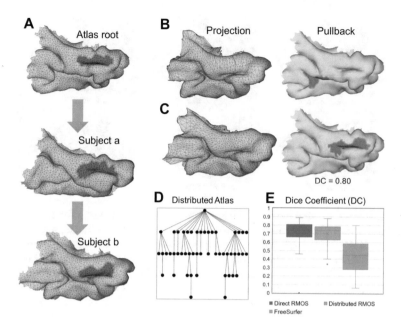

**Fig. 4.** Patch-based mapping of the TE cortex, and distributed atlas construction. (A) Demonstration of RMOS mapping of patches from the atlas to Subject b through Subject a where TE were manually delineated (with red color). (B) Projection of the atlas patch onto Subject b patch, and pullback TE labels of Subject b onto the atlas patch by direct mapping from the atlas to Subject b; (C) by composition of the maps between the atlas and Subject a, and between Subject a and b. (D) Tree-structured distributed atlas constructed using an atlas dataset of 50 subjects from the HCP. (E) Box plots of overlap measures between the atlas root and pullback TE labels of other 49 atlas subjects by the direct, distributed RMOS maps, and direct FreeSurfer cortical maps.

from Subject a to b (Fig. 4(A)), and composition of the two maps produced a high quality TE map with much less distortions as shown in Fig. 4(C).

**Distributed Atlas Construction.** We computed all the dissimilarities of the patches in Eq. 3 between every pair of the 50 subjects. We found the root node shortest to every other node from which we constructed the tree-structured distributed atlas of height 4 by setting the parameter $\alpha = 0.2$ in Eq. 3, as shown in Fig. 4(D). We evaluated the quality of the TE maps in the distributed network between the 49 subjects and one root (atlas) using the dice coefficient (DC), in comparison with direct maps from each subject to the atlas root by RMOS and FreeSurfer as shown in Fig. 4(E). Obviously, FreeSurfer's surface mapping of the entire cortex produced poor performance. Both direct and distributed mapping of the small patches using RMOS achieved much better alignment of TE. However, the direct maps completely misaligned the TE regions for the two subjects

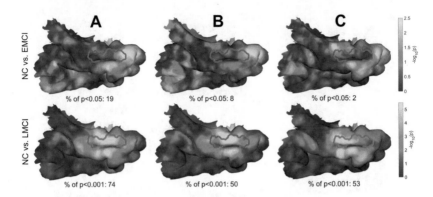

**Fig. 5.** Log-scale $p$-value maps of thickness on the atlas root patch for the 105 normal control (NC) vs. 132 early mild cognitive impairment (EMCI) subjects, and the 105 NC vs. 92 late MCI (LMCI) subjects (A) using the distributed atlas and (B) the centralized atlas with RMOS, and (C) the centralized atlas with the FreeSurfer cortical maps. The boundary of the TE cortex we delineated on the atlas root patch is drawn by a red curve. The percentage of points in the TE cortex at which the $p$-value is less than 0.05 or 0.001 is shown below each $p$-value map.

(extreme outliers in the box plot). There was no misalignment for any subject by the distributed maps.

**TE Atrophy by Early AD.** In the last experiment, we examined localized thickness changes of the TE cortex on patches during the development of AD using 380 subjects from ADNI 2 that consist of 105 normal control (NC), 132 early and 92 late mild cognitive impairment (EMCI, LMCI), and 51 AD. Using the distributed atlas constructed from the HCP data, we extracted patches, established new connections to the distributed atlas, and computed their RMOS maps. We also computed RMOS patch maps and FreeSurfer cortical maps with the centralized atlas, $i.e.$, direct mapping to the root atlas. Using the three different maps, we pulled back the cortical thickness of each subject extracted from the FreeSurfer reconstruction results (and normalized by total intracranial volume) to the root atlas, and applied vertex-wise $t$-test of the pullback thickness between the different groups. Figure 5 shows the $p$-value maps on the root atlas patch especially for thickness changes between NC vs. early AD (EMCI, LMCI). Clearly, our mapping approach with the distributed atlas found more regions of significant thickness change within the TE cortex between NC and other groups, particularly detected about 20% of the TE cortex at the posterior parts significant in atrophy in the very early stage of AD (EMCI).

## 4    Conclusion

In this paper, we developed a novel framework for automated patch generation and patch-based mapping of the TE cortex with a distributed atlas. The patch-based approach achieved higher accuracy of TE mapping than FreeSurfer, and mapping thorough the distributed atlas produced better anatomical matches than the centralized atlas mapping. The distributed atlas built with the HCP data worked well for the different ADNI data, and our novel framework of TE mapping achieved higher sensitivity in the early detection of AD. For future work, we will complete the experiments of TE atrophy for both hemispheres, and apply our distributed atlas approach for mapping of the subcortical structures or other cortical regions with a large amount of geometric variations.

# References

1. Braak, H., Braak, E.: Neuropathological stageing of Alzheimer-related changes. Acta Neuropathol. **82**(4), 239–259 (1991)
2. Fischl, B., Sereno, M.I., Dale, A.M.: Cortical surface-based analysis II: inflation, flattening, and a surface-based coordinate system. NeuroImage **9**(2), 195–207 (1999)
3. Gahm, J.K., Shi, Y.: Riemannian metric optimization for connectivity-driven surface mapping. In: Proceedings of Medical Image Computing and Computer-Assisted Intervention, Part I, pp. 228–236 (2016)
4. Gahm, J.K., Shi, Y.: Holistic mapping of striatum surfaces in the Laplace-Beltrami embedding space. In: Proceedings of Medical Image Computing and Computer-Assisted Intervention, Part I, pp. 21–30 (2017)
5. Ding, S.L., Van Hoesen, G.W.: Borders, extent, and topography of human perirhinal cortex as revealed using multiple modern neuroanatomical and pathological markers. Hum. Brain Mapp. **31**(9), 1359–1379 (2010)
6. Ying, S., Wu, G., Wang, Q., Shen, D.: Hierarchical unbiased graph shrinkage (HUGS): a novel groupwise registration for large data set. NeuroImage **84**, 626–638 (2014)
7. Essen, D.C.V., Smith, S.M., Barch, D.M., Behrens, T.E., Yacoub, E., Ugurbil, K.: The WU-Minn human connectome project: an overview. NeuroImage **80**, 62–79 (2013)
8. Mueller, S., et al.: The Alzheimer's disease neuroimaging initiative. Clin. North Am. **15**, 869–877 (2005)
9. Dale, A.M., Fischl, B., Sereno, M.I.: Cortical surface-based analysis I: segmentation and surface reconstruction. NeuroImage **9**, 179–194 (1999)
10. Fischl, B., et al.: Automatically parcellating the human cerebral cortex. Cereb. Cortex **14**(1), 11–22 (2004)
11. Taylor, K.I., Probst, A.: Anatomic localization of the transentorhinal region of the perirhinal cortex. Neurobiol. Aging **29**(10), 1591–1596 (2008)

# Spatially Localized Atlas Network Tiles Enables 3D Whole Brain Segmentation from Limited Data

Yuankai Huo[1(✉)], Zhoubing Xu[1], Katherine Aboud[1], Prasanna Parvathaneni[1], Shunxing Bao[1], Camilo Bermudez[1], Susan M. Resnick[2], Laurie E. Cutting[1], and Bennett A. Landman[1]

[1] Vanderbilt University, Nashville, TN, USA
yuankai.huo@vanderbilt.edu
[2] Laboratory of Behavioral Neuroscience, National Institute on Aging, Baltimore, MD, USA

**Abstract.** Whole brain segmentation on a structural magnetic resonance imaging (MRI) is essential in non-invasive investigation for neuroanatomy. Historically, multi-atlas segmentation (MAS) has been regarded as the *de facto* standard method for whole brain segmentation. Recently, deep neural network approaches have been applied to whole brain segmentation by learning random patches or 2D slices. Yet, few previous efforts have been made on detailed whole brain segmentation using 3D networks due to the following challenges: (1) fitting entire whole brain volume into 3D networks is restricted by the current GPU memory, and (2) the large number of targeting labels (e.g., >100 labels) with limited number of training 3D volumes (e.g., <50 scans). In this paper, we propose the spatially localized atlas network tiles (SLANT) method to distribute multiple independent 3D fully convolutional networks to cover overlapped sub-spaces in a standard atlas space. This strategy simplifies the whole brain learning task to localized sub-tasks, which was enabled by combing canonical registration and label fusion techniques with deep learning. To address the second challenge, auxiliary labels on 5111 initially unlabeled scans were created by MAS for pre-training. From empirical validation, the state-of-the-art MAS method achieved mean Dice value of 0.76, 0.71, and 0.68, while the proposed method achieved 0.78, 0.73, and 0.71 on three validation cohorts. Moreover, the computational time reduced from >30 h using MAS to ≈15 min using the proposed method. The source code is available online (https://github.com/MASILab/SLANTbrainSeg).

## 1 Introduction

Historically, multi-atlas segmentation (MAS) has been regarded as the de facto standard method on detailed whole brain segmentation (>100 anatomical regions) due to its high accuracy. Moreover, MAS only demands a small number of manually labeled examples (atlases) [1]. Recently, deep convolutional neural

© Springer Nature Switzerland AG 2018
A. F. Frangi et al. (Eds.): MICCAI 2018, LNCS 11072, pp. 698–705, 2018.
https://doi.org/10.1007/978-3-030-00931-1_80

networks (DCNN) have been applied to whole brain segmentation. To address the challenges of training a network on a small number of manually traced brains, patch-based DCNN methods have been proposed. de Brébisson et al. [2] proposed to learn 2D and 3D patches as well as spatial information, which was extended to include 2.5D patches by BrainSegNet [9]. Recently, DeepNAT [12] was proposed to perform hierarchical multi-task learning on 3D patches. Li et al. [7] introduced the 3D patch-based HC Net for high resolution segmentation. From another perspective, Roy et al. [11] proposed to use 2D fully convolutional network (FCN) to learn slice-wise image features by using auxiliary labels on initially unlabeled data. Although detailed cortical parcellations were not performed, Roy et al. revealed a promising direction on how to use initially unlabeled data to leverage training. With a large number of auxiliary labels, it is appealing to perform 3D FCN (e.g., 3D U-Net [3]) on whole brain segmentation since it typically yields higher spatial consistency than 2D or patch-based methods. However, directly applying 3D FCN to whole brain segmentation (e.g., 1 mm isotropic resolution) is restricted by the current graphics processing unit (GPU) memory. A common solution is to down sample the inputs, yet, the accuracy can be sacrificed.

In this paper, we propose the spatially localized atlas network tiles (SLANT) method for detailed whole brain segmentation (133 labels under BrainCOLOR protocol [5]) by combining canonical medical image processing techniques with deep learning. SLANT distributes a set of independent 3D networks (network tiles) to cover overlapped sub-spaces in a standard MNI atlas space.

Then, majority vote label fusion was used to obtain final whole brain segmentation from the overlapped sub-spaces. To leverage learning performance on 133 labels with only 45 manually traced training data, auxiliary labels on 5111 initially unlabeled scans were created from non-local spatial STAPLE (NLSS) MAS [1] for pre-training inspired by [11].

## 2   Methods

**Registration and Intensity Harmonization:** Affine registration [10] was employed to register all training and testing scans to MNI 305 space (Fig. 1). Then, N4 bias field correction was deployed to reduce bias. To further harmonize the intensities on large-scale MRI, we introduced a regression-based intensity normalization method. First, we defined a gray-scale MRI volume (with N voxels) as a vector $I \in \mathbb{R}^{N \times 1}$. $I$ was demeaned and normalized by standard deviation (std) to $I'$. The intensities were harmonized by a pre-trained linear regression model on sorted intensity. The sorted intensity vector $V_s$ was calculated from $V_s = \text{sort}(I'(mask > 0))$, where "sort" rearrange intensities from largest to smallest, and "$mask$" was a prior mask learned from a union operation from all atlases. To train the linear regression, mean sorted intensity vector $\overline{V_s}$ was obtained by averaging $V_s$ from all atlases. The coefficients were fitted between $V'_s$ (from $I'$) and $\overline{V_s}$, and intensity normalized image $\widehat{I'}$ is obtained from fitted $\beta_1$ and $\beta_0$: $\overline{V_s} = \beta_1 \cdot V'_s + \beta_0$, and $\widehat{I'} = \beta_1 \cdot I' + \beta_0$.

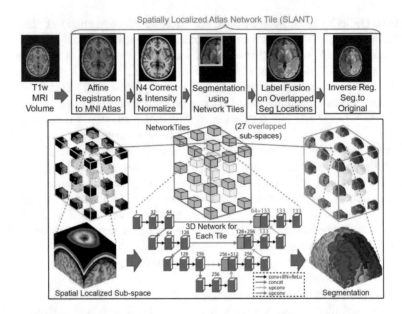

**Fig. 1.** The proposed SLANT-27 (27 network tiles) method is presented, which combines canonical medical image processing methods (registration, harmonization, label fusion) with 3D network tiles. 3D U-Net is used as each tile, whose deconvolutional channel numbers are modified to 133. The tiles are spatially overlapped in MNI space.

**Network Tiles:** After affine registration, all training brains were mapped to the same MNI atlas space ($172 \times 220 \times 156$ voxels with $1\,\mathrm{mm}$ isotropic resolution). We employed $k$ 3D U-Net as a network tiles to cover entire MNI space with/without overlaps (Fig. 2). To be compatible with 133 labels, the number of channels of deconvolutional layers in each 3D U-Net were defined as 133 (Fig. 1). Each sub-space $\psi_n$ was presented by one coordinate $(x_n, y_n, z_n)$ and sub-space size $(d_x, d_y, d_z)$, $n \in \{1, 2, \ldots, k\}$ as

$$\psi_n = [x_n : (x_n + d_x), y_n : (y_n + d_y), z_n : (z_n + d_z)] \tag{1}$$

As showed in Fig. 2, SLANT-8 covered the MNI space using eight U-Nets by covering $k = 2 \times 2 \times 2 = 8$ non-overlapped sub-spaces. To improve spatial consistency at boundaries, SLANT-27 covered $k = 3 \times 3 \times 3 = 27$ overlapped sub-spaces.

**Label Fusion:** For SLANT-27, whose sub-spaces were overlapped, the label fusion method were employed to get a single segmentation from overlapped sub-spaces. Briefly, the k segmentations $\{S_1, S_2, \ldots, S_k\}$ from network tiles were fused to achieve final segmentation $S_{MNI}$ in MNI space by performing majority vote:

$$S_{MNI}(i) = \underset{l \in \{0, 1, \ldots, L-1\}}{\arg\min} \frac{1}{k} \sum_{m=1}^{k} p(l | S_m, i) \tag{2}$$

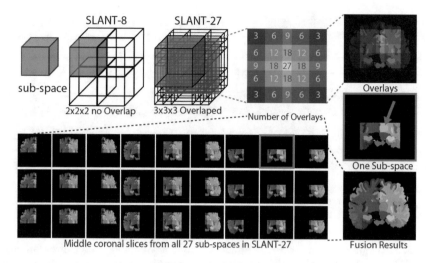

**Fig. 2.** SLANT-8 covered eight non-overlapped sub-spaces in MNI, while SLANT-27 covered 27 overlapped sub-spaces in MNI. Middle coronal slices from all 27 sub-spaces were visualized (lower left panel). The number of overlays, as well as sub-spaces overlays, were showed (upper right panel). The incorrect labels (red arrow) in one sub-space were corrected in final segmentation by performing majority vote label fusion.

where $\{0, 1, \ldots, L-1\}$ represents $L$ possible labels for a given voxel $i \in \{1, 2, \ldots, N\}$. $p(l|S_m, i) = 1$ if $S_m(i) = l$, while $p(l|S_m, i) = 0$, otherwise. Then, the $S_{MNI}$ was registered to the original space by conducting another affine registration [10]. When using SLANT-8, whose sub-spaces were not overlapped, the native concatenation was applied rather than performing the label fusion.

**Boost Learning on Unlabeled Data:** Similar to [11], the auxiliary labels on large-scale initially unlabeled MRI scans were obtained by performing existing segmentation tools. Briefly, MAS using hierarchical non-local spatial staple (NLSS) label fusion [1] was performed on 5111 multi-site scans. Next, the large-scale auxiliary labels were used for pre-training. Then, the small-scale manually labeled training data were used for fine-tuning the network.

## 3    Experiments

*Training Cohort:* 45 T1-weighted (T1w) MRI scans from Open Access Series on Imaging Studies (OASIS) dataset [8] were manually labeled to 133 labels according to BrainCOLOR protocol [5]. 5111 multi-site T1w MRI scans for auxiliary labels were achieved from night different projects (described in [5]). *Testing Cohort 1:* Five withheld T1w MRI scans from OASIS dataset with manual segmentation (BrainCOLOR protocol) were used for validation, which evaluates the performance of different methods on the same site testing data. *Testing*

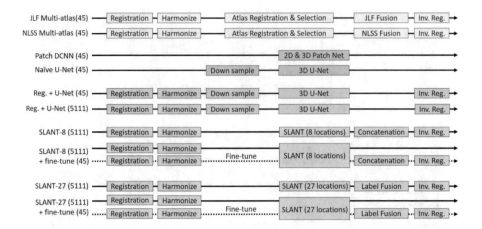

**Fig. 3.** This figure demonstrates the major components of different segmentation methods. (45) indicated the 45 OASIS manually traced images were used in training, while (5111) indicated the 5111 auxiliary label images were used in training.

*Cohort 2:* One T1 MRI scan from colin27 cohort [4] with manual segmentation (BrainCOLOR protocol) was used for testing. This cohort evaluates the performance of different methods on a widely used standard template. *Testing Cohort 3:* 13 T1 MRI scans from Child and Adolescent Neuro Development Initiative (CANDI) [6] were used for testing. This cohort evaluates the performance of different methods on an independent population, whose age range (5–15 yrs.) was not covered by OASIS training cohort (18–96 yrs.).

**Experimental Design:** The experimental design is presented in Fig. 3. First, two state-of-the-art multi-atlas label fusion methods, joint label fusion (JLF) [13] and non-local spatial staple (NLSS) [1], were used as baseline methods. The parameters were set the same as the papers, which were optimized for whole brain segmentation. Next, patch-based network [2] and naive 3D U-Net [3] methods were used using their open-source implementations. By using affine registration as preprocessing, Reg. + U-Net was trained using 45 manually labeled scans and 5111 auxiliary labeled scans. Then, the proposed SLANT methods were evaluated on covering eight non-overlapped sub-spaces (SLANT-8) and 27 overlapped sub-spaces (SLANT-27), trained by 5111 auxiliary labeled scans. Last, 45 manually labeled scans were used to fine-tune the SLANT networks.

For all 3D U-Net in baseline methods and SLANT networks, we used the same parameters with 3D batch size = 1, input resolution = $96 \times 128 \times 88$, input channel = 1, output channel = 133, optimizer = Adam, learning rate = 0.0001. The deep networks can fit into an NVIDIA Titan GPU with 12 GB memory. For all the training using 5111 scans, 6 epochs were trained ($\approx$24 training hours); while for all the training using 45 scans, 1000 epochs were trained to ensure the similar training batches as 5111 scans. For the fine-tuning using 45 scans, 30 epochs were trained.

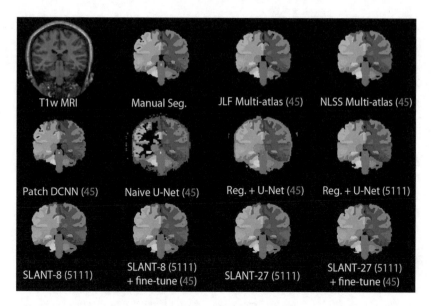

**Fig. 4.** Qualitative results of manual segmentation, MAS methods, patch-based DCNN method, U-Net approaches and proposed SLANT methods.

Results reported in this paper were from the epoch with best overall performance on OASIS cohort for each method, so that colin27 and CANDI were independent testing cohorts as external validation.

## 4   Result

The qualitative and quantitative results have been shown in Figs. 4 and 5. In Fig. 5, "45" indicated the 45 OASIS manually traced images were used in training, while "5111" indicated the 5111 auxiliary label images were used in training. The mean Dice similarity coefficient (DSC) values on 132 anatomical labels (excluding background) between automatic segmentation methods with manual segmentation in original image space were showed as boxplots in Fig. 5. From the results, the affine registration (Reg. + U-Net) significantly leveraged the U-Net performance (compared with Naive U-Net). For the same network (Reg. + U-Net), results using 5111 auxiliary labeled scans achieved better performance than using 45 manual labeled scans. From Table 1, the proposed SLANT-27 method with fine-tuning achieved superior performance on mean DSC across testing cohorts. All claims of statistical significance in this paper have been calculated using the Wilcoxon signed rank test for $p < 0.05$.

## 5   Conclusion and Discussion

In this study, we developed the SLANT method to combine the canonical medical image processing approaches with localized 3D FCN networks in MNI space.

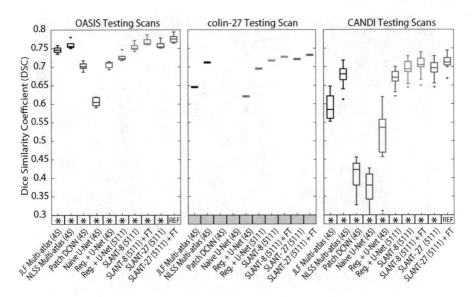

**Fig. 5.** From quantitative results, The propsed SLANT-27 using 5111 auxiliary labels for pretraining and fine-tuned (FT) by 45 manual labels achieved highest median Dice similarity coefficient (DSC) values. The SLANT-27 was used as the reference method (REF) in statistical analysis. The significant difference to REF was marked with *.

**Table 1.** Mean, std and median DSC values on three validation cohorts

| Methods | Training scan # | OASIS dataset | | Colin27 | CANDI dataset | |
|---|---|---|---|---|---|---|
| | | Mean ± Std | Median | DSC | Mean ± Std | Median |
| JLF [13] | 45 | 0.746 ± 0.009 | 0.746 | 0.646 | 0.590 ± 0.033 | 0.585 |
| NLSS [1] | 45 | 0.760 ± 0.012 | 0.756 | 0.712 | 0.677 ± 0.029 | 0.680 |
| Patch DCNN [2] | 45 | 0.702 ± 0.011 | 0.701 | 0.012 | 0.409 ± 0.038 | 0.422 |
| Naive U-Net [3] | 45 | 0.606 ± 0.012 | 0.605 | 0.000 | 0.375 ± 0.043 | 0.380 |
| Reg. + U-Net | 45 | 0.706 ± 0.009 | 0.711 | 0.621 | 0.514 ± 0.081 | 0.536 |
| Reg, + U-Net | 5111 | 0.726 ± 0.012 | 0.722 | 0.695 | 0.669 ± 0.023 | 0.671 |
| SLANT-8 | 5111 | 0.753 ± 0.011 | 0.750 | 0.717 | 0.694 ± 0.024 | 0.694 |
| SLANT-8 + FT | 5111 + 45 | 0.768 ± 0.011 | 0.763 | 0.726 | 0.704 ± 0.025 | 0.705 |
| SLANT-27 | 5111 | 0.759 ± 0.011 | 0.754 | 0.721 | 0.694 ± 0.024 | 0.697 |
| SLANT-27 + FT | 5111 + 45 | **0.776 ± 0.012** | **0.775** | **0.732** | **0.711 ± 0.023** | **0.712** |

For the same network (Reg. + U-Net), results from 5111 auxiliary labeled scans achieved better performance than the results from 45 manual labeled scans. From Figs. 4 and 5, and Table 1, we demonstrate that our proposed strategy successfully takes advantages of historical efforts (registration, harmonization, and label-fusion) and consistently yields superior performance. Moreover, the proposed method requires ≈15 min, compared with >30 h by MAS. Note that the 3D U-Net in the proposed SLANT can be replaced by other 3D segmentation networks, which might yield better performance.

**Acknowledgments.** This research was supported by NSF CAREER 1452485, NIH R01EB017230, R21EY024036, R21NS064534, R01EB006136, R03EB012461, R01NS095291, Intramural Research Program, National Institute on Aging, NIH.

# References

1. Asman, A.J., Landman, B.A.: Hierarchical performance estimation in the statistical label fusion framework. Med. Image Anal. **18**(7), 1070–1081 (2014)
2. de Brébisson, A., Montana, G.: Deep neural networks for anatomical brain segmentation. arXiv preprint arXiv:1502.02445 (2015)
3. Çiçek, Ö., Abdulkadir, A., Lienkamp, S.S., Brox, T., Ronneberger, O.: 3D U-net: learning dense volumetric segmentation from sparse annotation. In: Ourselin, S., Joskowicz, L., Sabuncu, M.R., Unal, G., Wells, W. (eds.) MICCAI 2016. LNCS, vol. 9901, pp. 424–432. Springer, Cham (2016). https://doi.org/10.1007/978-3-319-46723-8_49
4. Collins, D.L., et al.: Design and construction of a realistic digital brain phantom. Trans. Med. Imaging **17**(3), 463–468 (1998)
5. Huo, Y., Aboud, K., Kang, H., Cutting, L.E., Landman, B.A.: Mapping lifetime brain volumetry with covariate-adjusted restricted cubic spline regression from cross-sectional multi-site MRI. In: Ourselin, S., Joskowicz, L., Sabuncu, M.R., Unal, G., Wells, W. (eds.) MICCAI 2016. LNCS, vol. 9900, pp. 81–88. Springer, Cham (2016). https://doi.org/10.1007/978-3-319-46720-7_10
6. Kennedy, D.N., Haselgrove, C., Hodge, S.M., Rane, P.S., Makris, N., Frazier, J.A.: Candishare: a resource for pediatric neuroimaging data. Neuroinformatics **10**, 319–322 (2012)
7. Li, W., Wang, G., Fidon, L., Ourselin, S., Cardoso, M.J., Vercauteren, T.: On the compactness, efficiency, and representation of 3D convolutional networks: brain parcellation as a pretext task. In: Niethammer, M., Styner, M., Aylward, S., Zhu, H., Oguz, I., Yap, P.-T., Shen, D. (eds.) IPMI 2017. LNCS, vol. 10265, pp. 348–360. Springer, Cham (2017). https://doi.org/10.1007/978-3-319-59050-9_28
8. Marcus, D.S., Wang, T.H., Parker, J., Csernansky, J.G., Morris, J.C., Buckner, R.L.: Open access series of imaging studies (oasis): cross-sectional mri data in young, middle aged, nondemented, and demented older adults. J. Cogn. Neurosci. **19**(9), 1498–1507 (2007)
9. Mehta, R., Majumdar, A., Sivaswamy, J.: Brainsegnet: a convolutional neural network architecture for automated segmentation of human brain structures. J. Med. Imaging **4**(2), 024003 (2017)
10. Ourselin, S., Roche, A., Subsol, G., Pennec, X., Ayache, N.: Reconstructing a 3D structure from serial histological sections. Image Vis. Comput. **19**(1–2), 25–31 (2001)
11. Roy, A.G., Conjeti, S., Sheet, D., Katouzian, A., Navab, N., Wachinger, C.: Error corrective boosting for learning fully convolutional networks with limited data. In: Descoteaux, M., Maier-Hein, L., Franz, A., Jannin, P., Collins, D.L., Duchesne, S. (eds.) MICCAI 2017. LNCS, vol. 10435, pp. 231–239. Springer, Cham (2017). https://doi.org/10.1007/978-3-319-66179-7_27
12. Wachinger, C., Reuter, M., Klein, T.: Deepnat: deep convolutional neural network for segmenting neuroanatomy. NeuroImage **170**, 434–445 (2017)
13. Wang, H., Yushkevich, P.: Multi-atlas segmentation with joint label fusion and corrective learning-an open source implementation. Front. Neuroinformatics **7**, 27 (2013)

# Adaptive Feature Recombination and Recalibration for Semantic Segmentation: Application to Brain Tumor Segmentation in MRI

Sérgio Pereira[1,2(✉)], Victor Alves[2], and Carlos A. Silva[1]

[1] CMEMS-UMinho Research Unit, University of Minho, Guimarães, Portugal
id5692@alunos.uminho.pt, csilva@dei.uminho.pt
[2] Centro Algoritmi, University of Minho, Braga, Portugal

**Abstract.** Convolutional neural networks (CNNs) have been success-fully used for brain tumor segmentation, specifically, fully convolutional networks (FCNs). FCNs can segment a set of voxels at once, having a direct spatial correspondence between units in feature maps (FMs) at a given location and the corresponding classified voxels. In convolutional layers, FMs are merged to create new FMs, so, channel combination is crucial. However, not all FMs have the same relevance for a given class. Recently, in classification problems, Squeeze-and-Excitation (SE) blocks have been proposed to re-calibrate FMs as a whole, and suppress the less informative ones. However, this is not optimal in FCN due to the spa-tial correspondence between units and voxels. In this article, we propose feature recombination through linear expansion and compression to cre-ate more complex features for semantic segmentation. Additionally, we propose a segmentation SE (SegSE) block for feature recalibration that collects contextual information, while maintaining the spatial meaning. Finally, we evaluate the proposed methods in brain tumor segmentation, using publicly available data.

## 1 Introduction

Brain tumor segmentation plays an important role during treatment planning and follow-up evaluation. But, it is time-consuming and prone to inter- and intra-rater variability. Therefore, automatic and reliable methods are desirable. However, brain tumor segmentation is a challenging task, due to their irregular shape, appearance, and location [9,15]. Recently, CNN-based approaches have achieved state of the art results [6,9,15]. With enough data, CNNs can learn complex patterns, such as brain tumor attributes, which are, otherwise, difficult to capture by feature engineering.

Despite being used in many applications, many CNN developments are first evaluated in image classification. In VGGNet [12] it was shown that replacing a layer with large kernels by blocks of several layers with $3 \times 3$ kernels result in deeper and more powerful CNNs. Later, He et al. [2] proposed residual learning

© Springer Nature Switzerland AG 2018
A. F. Frangi et al. (Eds.): MICCAI 2018, LNCS 11072, pp. 706–714, 2018.
https://doi.org/10.1007/978-3-030-00931-1_81

using identity-based skip connections that allow better gradient flows and training of very deep CNNs. Other studies explored the recombination of FMs, either by compression with convolutional layers with $1 \times 1$ kernels [7], or by dividing a stack of FMs into smaller groups with grouped convolutions [13]. More recently, Hu et al. [3] proposed FM recalibration with the SE block. This block is inspired by the intuition that not all FMs are informative for all classes. Therefore, the SE block learns how to adaptively suppress the least discriminative FMs (recalibration). They showed that the simple addition of this block to state-of-the-art CNNs increased their representational power.

Semantic segmentation is one of the domains where CNNs have been pervasively used. Although one can use conventional CNNs with fully connected layers [9], FCNs [10] are arguably one of the most important advancements regarding CNNs for semantic segmentation. In these architectures, fully-connected layers are replaced by convolutional layers, usually, with $1 \times 1$ kernels. In this way, a set of voxels from an image patch can be efficiently classified in just one forward pass. In FCNs, there is a direct spatial correspondence between units in the FMs and the classified voxels. Most of the advancements in CNN design can be easily incorporated into FCN. For instance, the principles of VGGNet [12] and residual learning [2] were incorporated in [6,9], respectively. However, although the SE block has the attractive property of re-calibrating FMs, it was conceived to weight whole FMs, which is not optimal for FCN. Since there is a spatial correspondence between units in FMs and the voxels, it is desirable to emphasize or suppress certain regions of the FMs, instead of the whole FM.

In this paper, we explore the recombination and recalibration of FMs. In recombination, instead of reducing the FMs number only, we employ linear expansion followed by compression for mixing the information. Additionally, we study how to incorporate recalibration into FCN. In SE block, global average pooling captures the whole contextual information in a FM. Instead, we argue that dilated convolution [14] is better suited for the recalibration block in FCN. Hence, the contribution of this paper is threefold. First, we propose recombination of FMs by linear expansion and compression. Second, we explore FMs recalibration in the context of FCN. We observe that the original SE block is not optimal for FCN, and we propose a better-suited alternative. Third, we evaluate our proposal on brain tumor segmentation, using publicly available data.

## 2   Methods

We follow a hierarchical FCN-based brain tumor segmentation approach. Thus, we start by roughly segmenting the whole tumor with a binary FCN (WT-FCN). Using this segmentation, we define a cuboid region of interest (ROI) around the tumor with a margin of 10 extra voxels in each side. Finally, a second multi-class FCN (MC-FCN) is responsible for segmenting the multiple tumor structures inside the ROI. The proposed FCNs are inspired by an encoder-decoder architecture with long skip connections [10]. The input for the FCNs are image patches extracted from all the available MRI sequences. In this section, we first define the baseline FCNs; then, we present the proposed recombination and recalibration (RR) block. This block is evaluated in the more challenging multi-class segmentation problem. The WT-FCN is fixed across all experiments to isolate and make it easier to compare improvements introduced by the RR block.

### 2.1 Baseline Segmentation Approach

The architecture of the 3D WT-FCN is depicted in Fig. 1. We used both regular blocks of convolutional layers, and blocks with residual connections and pre-activation [2]. This network segments 3D patches, with the three pooling layers providing a large field of view. These two characteristics contribute for reducing the number of voxels with false positive tumor detections. The baseline MC-FCN architecture can be perceived from Fig. 2(a), by not considering the RR block. We design the MC-FCN as a 2D network, as a proof of concept to evaluate the proposed component, which makes it computationally cheaper than the WT-FCN. Hence, 2D image patches are extracted in the axial plane. Additionally, we observed no benefits from using residual connections, or from being as deep as the WT-FCN.

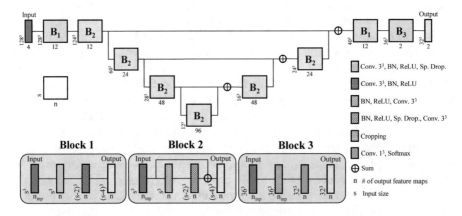

**Fig. 1.** Architecture of the WT-FCN. Downsampling is obtained by max-pooling. We use nearest neighbor upsampling to increase the FMs size, and $1 \times 1 \times 1$ conv. layers to adjust the number of FMs before addition. BN stands for batch normalization, and Sp. Drop. for spatial dropout.

### 2.2 Recombination and Recalibration

We propose recombination and recalibration of FMs as complementary operations. Recombination consists in mixing the information across FMs channels to create new combined features. In the past, convolutional layers with $1 \times 1$ kernels were proposed as cross channel parametric pooling [7] to decrease (compress) the number of FMs. Also, bottleneck blocks in ResNet compress the channels, process

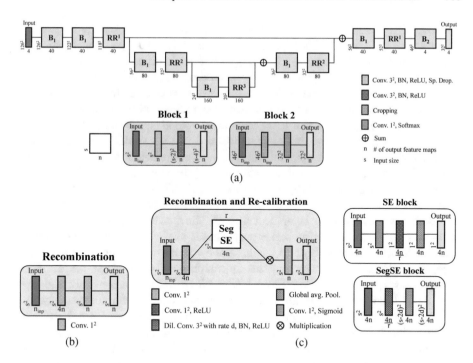

**Fig. 2.** Architecture of the MC-FCN. (a) Architecture overview with the RR block. Input sizes correspond to the RR block with SegSE. Downsampling is obtained by max-pooling. We use nearest neighbor upsampling to increase the FMs size, and $1 \times 1$ conv. layers to adjust the number of FMs, before addition. (b) Recombination block. (c) RR block, and the SE and SegSE blocks.

them, and expand to more channels. Instead, we propose linear recombination of FMs to increase the number of FMs (expansion), followed by compression to the original number. This operation is done by convolutional layers with $1 \times 1$ kernels (Fig. 2(b)). Experimentally, we found an expansion factor of 4 to work well. Linear recombination of units in a given spatial location of the FMs results from the weighted sum of the units in the same location of all FMs. Hence, expansion combines features into a higher dimension, while compression learns how to compress features and suppress the least discriminative ones.

We propose the RR block (Fig. 2(c)) that combines both FM recombination and recalibration. Feature map recalibration with the SE block, as proposed in [3], is shown in Fig. 2(c) – SE block. First, global average pooling summarizes each FM into its average value to capture contextual information. Then, two fully connected (FC) layers[1] capture cross-channels relations. The first one is a compression layer with a factor of $r$, followed by ReLU activation. The second FC layer restores the original dimension, and is followed by the sigmoid activation function. Finally, this vector is channel-wise multiplied with the input FMs,

---

[1] Equivalently implemented as convolutional layers with $1 \times 1$ kernels.

i.e., each FM is multiplied by a corresponding scalar value, resulting in each FM being scaled as a whole. Ideally, less discriminative FMs are suppressed. This approach was shown to improve learning in image classification. In this problem, a FM may have a strong response for a given class or subset of classes. However, in semantic segmentation with FCN, a patch of voxels is segmented at once. So, there is a correspondence between segmented pixels and units in FMs. In this scenario, some regions of the FMs may have strong activations that are relevant for the structure that is being segmented in that spatial location. Hence, the SE block may be not optimal for semantic segmentation, since it collapses the whole FM into a single value, regardless of the regions. Thus, the proposed RR block includes a segmentation adapted SegSE block. A straightforward approach for adapting the SE block for semantic segmentation is by simply removing the global average pooling layer. In this way, the spatial correspondence among units and voxels is maintained. However, this is also not optimal, since contextual information is important to evaluate the spatial importance of a given feature. In preliminary experiments, this approach resulted in worse performance. Therefore, we propose our SegSE block (Fig. 2(c)) that uses a convolutional layer with $3 \times 3$ kernels with dilation $d$ for context aggregation. Simultaneously, this layer is responsible for the compression stage. Experimentally, we found that the best dilation rates depend on the resolution of the FMs. This is due to the fact that deeper layers already have a larger field of view. Hence, we set the rate in $\{RR^1, RR^2, RR^3\}$ (c.f. Fig. 2(a)) to $\{3, 2, 1\}$. In preliminary experiments, we evaluated using spatial average pooling followed by convolutional and transposed convolutional layers, but, we obtained worse performance. The reason is probably due to the checkerboard artifacts that appear with this combination of layers, but not with dilated convolution.

## 3   Experimental Setup

We evaluate the proposed blocks in the Brain Tumor Segmentation Challenge (BRATS) 2017 and 2013 databases [1,8]. BRATS 2017 has two publicly available datasets: Training (285 subjects) and Leaderboard (46 subjects). In BRATS 2013 we use Training (30 subjects) and Challenge (10 subjects). For each subject, there are four MRI sequences available: T1, post-contrast T1 (T1c), T2, and FLAIR. All images are already interpolated to $1\,mm$ isotropic resolution, skull stripped, and aligned. Only the Training sets contain manual segmentations. In BRATS 2017 it distinguishes three tumor regions: edema, necrotic/non-enhancing tumor core, and enhancing tumor. In BRATS 2013 the manual segmentation have necrosis and non-enhancing tumor separately, although we fuse these labels to be similar to BRATS 2017. Evaluation is performed for the whole tumor (all regions combined), tumor core (all, excluding edema), and enhancing tumor. Since annotations are not publicly available for 2017 Learderboard and 2013 Challenge, the evaluation is computed by the CBICA IPP and SMIR online platforms[2]. The development of the RR block was conducted in the larger

---

[2] https://ipp.cbica.upenn.edu/ and https://www.smir.ch/BRATS/Start2013.

BRATS 2017 Training set, which was randomly divided into training (60%), validation (20%), and test (20%)[3]. However, networks tested in BRATS 2013 Challenge were trained in the 2013 Training set.

Image pre-processing included bias field correction, and standardization of the intensity histograms of each MRI sequence, as in [9]. During training, we use the crossentropy loss, the Adam optimizer with learning rate of $5 \times 10^{-5}$, weight decay of $1 \times 10^{-6}$, and spatial dropout probability of 0.05. Since we used convolution without padding, during multiplication or sum of FMs with different sizes, we cropped the center part of the biggest one. The compression factor $r$ in the SE and SegSE blocks was set to 10. Data augmentation included sagittal flipping and random rotations of $90°$. For training the binary whole tumor FCN, all tumor regions in manual segmentations were fused into a single label. All the hyperparameters were found using the validation set, before evaluation in the test set. The FCNs were implemented using Keras and Theano.

Metrics provided by the online evaluation platforms differ in BRATS 2017 and 2013. Hence, in BRATS 2017 we use Dice and the $95^{\text{th}}$ percentile of the Hausdorff Distance ($HD_{95}$). In BRATS 2013, the online platform computes Dice, Sensitivity, and Positive Predictive Value (PPV).

## 4    Results and Discussion

We evaluate the effect of recombination and recalibration of FMs using the SegSE block in the test set (20% of BRATS 2017 Training), and compare it with the baseline and RR with SE block. Quantitative results are presented in Table 1, and segmentation examples in Fig. 3. When we include the recombination by expansion followed by compression stage to the baseline FCN, we observe that Dice improves in all tumor regions. Although this improvement is negligible in the enhancing region, it is substantial in the whole tumor and in the core. In fact, it achieves the highest Dice for tumor core of all the blocks. However, the $HD_{95}$ is higher in all classes, when compared with the baseline.

In Table 1 we can find results for the recalibration stage, when joined with recombination to form the RR block. We observe that the SE block, proposed in [3], leads to worse Dice, when compared to the baseline with recombination layers. Actually, the Dice of the enhancing tumor is even lower than the baseline. This may be due to enhancing region usually being a smaller part of the whole tumor volume. Additionally, finer details are needed to define this region, hence its contribution to a whole feature map response may be less strong than the other tumor regions and end up being suppressed by the SE block. Therefore, we conclude that the SE block, acting as whole FM recalibration is not optimal for segmentation. Finally, it is possible to observe from Table 1 that the RR block with the proposed SegSE recalibration achieves the best scores, both in Dice (excepting core by a small margin) and $HD_{95}$. We note that the SegSE block is the only approach that substantially improves the Dice of enhancing

---

[3] Subjects id in each set are available online: https://github.com/sergiormpereira/rr_segse.

**Table 1.** Results (average) obtained in the test set (20% of BRATS 2017 Training). We evaluate recombination (Recomb.) of FMs, and RR using both SE and SegSE blocks. Bold results show the best score for each tumor region.

| Method | Dice | | | HD95 | | |
|---|---|---|---|---|---|---|
| | Whole | Core | Enh. | Whole | Core | Enh. |
| Baseline | 0.857 | 0.739 | 0.682 | 8.645 | 10.761 | 6.672 |
| Baseline + Recomb. | 0.865 | **0.769** | 0.687 | 9.720 | 11.453 | 7.790 |
| Baseline + RR SE | 0.859 | 0.756 | 0.672 | 8.939 | 13.306 | 7.319 |
| Baseline + RR SegSE | **0.866** | 0.766 | **0.698** | **8.475** | **10.513** | **6.131** |

**Table 2.** Results (average) obtained in BRATS 2017 Leaderboard set. Bold results show the best score for each tumor region. Underlined scores are the best among single-model approaches (excluding Kamnitsas).

| Method | Dice | | | HD95 | | |
|---|---|---|---|---|---|---|
| | Whole | Core | Enh. | Whole | Core | Enh. |
| Islam [4] | 0.876 | 0.761 | 0.689 | 9.820 | 12.361 | 12.938 |
| Jesson [5] | <u>0.899</u> | 0.751 | 0.713 | **<u>4.160</u>** | **<u>8.650</u>** | 6.980 |
| Kamnitsas [6] | **0.901** | **0.797** | **0.738** | 4.230 | **6.560** | **4.500** |
| Baseline | 0.878 | 0.760 | 0.692 | 6.597 | 11.915 | <u>5.978</u> |
| Baseline + RR SegSE | 0.884 | <u>0.771</u> | <u>0.719</u> | 6.202 | 10.215 | 6.702 |

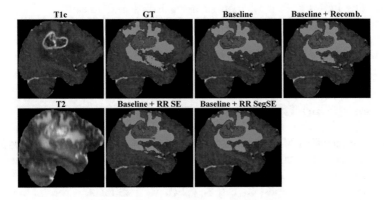

**Fig. 3.** Examples of the segmentation obtained with each of the evaluated RR blocks. The colors in segmentations mean: green – edema, blue - tumor core, and red – enhancing tumor.

tumor over the baseline. Moreover, the HD95 suggests that besides achieving good overlap scores, it also obtains the best contour definition. The SegSE stage comes at the cost of more parameters. In order to evaluate if its performance is due to these extra capacity, we proportionally increased the width of the baseline, such that its parameters number becomes similar to the network with RR + SegSE. The results obtained with this larger network in terms of Dice/HD95 were 0.852/9.049, 0.751/10.647, and 0.678/7.065 for the complete, core, and enhancing regions, respectively. So, the RR SegSE block improvements are due to better learning, and not directly to the higher capacity.

We compare with the state of the art in BRATS 2017 Leaderboard and BRATS 2013 Challenge in Tables 2 and 3, respectively. In BRATS 2017, most of the top performing methods are ensembles of FCN. In principle, taking a CNN or FCN and building an ensemble will certainly lead to better results. Since we are evaluating the effect of the SegSE block, we need to assess it in a single model. So, we compare our results with other single CNN approaches, such as

Islam [4] and Jesson [5], for the sake of fairness. Nevertheless, we present results obtained by the multi-model and multi-training settings ensemble proposed by Kamnitsas et al. [6], the winner of BRATS 2017 Challenge. In the single network approaches, Islam [4] employed a hypercolumns-inspired CNN. Jesson [5] used a FCN with multiple prediction layers and loss functions in different scales. Additionally, the authors employed a learning curriculum to deal with class imbalance. From Table 2, we observe that the baseline achieves competitive results, when compared with the single-model approaches. The Dice is comparable with Islam, while the $HD_{95}$ scores are smaller. Regarding the RR SegSE block, we confirm that it improves the baseline performance. Indeed, the results are competitive with Jesson, with better Dice for core and enhancing regions, and $HD_{95}$ in the enhancing region. In BRATS 2013 Challenge (Table 3), the proposed FCN with the RR SegSE block improves over the baseline, again. The other compared methods are all recent and top performing CNN-based approaches. Pereira [9] uses a plain CNN with fully-connected layers. Shen [11] uses a FCN enhanced by input symmetry maps and a boundary-aware loss function. Zhao [15] also proposes a FCN followed by a conditional random field trained as recurrent neural network and a sophisticated post-processing stage. We note that the proposed method achieves the highest Dice and Sensitivity scores. In fact, the baseline with the RR SegSE block is ranked 1st by the online evaluation platform.

**Table 3.** Results (average) obtained in BRATS 2013 Challenge set. Bold results show the best score for each tumor region.

| Method | Dice | | | PPV | | | Sensitivity | | |
|---|---|---|---|---|---|---|---|---|---|
| | Whole | Core | Enh. | Whole | Core | Enh. | Whole | Core | Enh. |
| Pereira [9] | 0.88 | 0.83 | 0.77 | 0.88 | **0.87** | 0.74 | 0.89 | 0.83 | 0.81 |
| Shen [11] | 0.88 | 0.83 | 0.76 | 0.87 | **0.87** | 0.73 | 0.9 | 0.81 | 0.81 |
| Zhao [15] | 0.88 | **0.84** | 0.77 | **0.9** | **0.87** | **0.76** | 0.86 | 0.82 | 0.8 |
| Baseline | 0.87 | 0.83 | 0.77 | 0.81 | 0.81 | 0.71 | **0.94** | 0.88 | 0.87 |
| Baseline + RR SegSE | **0.89** | **0.84** | **0.78** | 0.86 | 0.83 | 0.71 | 0.93 | **0.89** | **0.88** |

## 5   Conclusion

Recalibration of FMs has the power to adaptively emphasize discriminative FMs and suppress the uninformative ones. However, this is not optimal in the context of FCN for segmentation. In this work, we propose recombination and recalibration of FMs for semantic segmentation. The former employs linear expansion followed by compression of FMs for mixing features, while the later adaptively recalibrates regions of the FMs. We show that both recombination and recalibration improve over a competitive baseline. Although we opted for a simple U-net inspired network, the proposed block can be used in other more complex FCN. Still, our FCN with the RR SegSE block achieves competitive results in BRATS 2017 Leaderboard, when compared with other single-model approaches, and superior results in BRATS 2013 Challenge.

**Acknowledgments.** Sérgio Pereira was supported by a scholarship from the Fundação para a Ciência e Tecnologia (FCT), Portugal (scholarship number PD/BD/ 105803/2014). This work has been supported by COMPETE: POCI-01-0145-FEDER-007043 and FCT - Fundação para a Ciência e Tecnologia within the Project Scope: UID/CEC/00319/2013.

# References

1. Bakas, S., et al.: Advancing the cancer genome atlas glioma MRI collections with expert segmentation labels and radiomic features. Sci. Data **4**, 170117 (2017)
2. He, K., Zhang, X., Ren, S., Sun, J.: Identity mappings in deep residual networks. In: Leibe, B., Matas, J., Sebe, N., Welling, M. (eds.) ECCV 2016. LNCS, vol. 9908, pp. 630–645. Springer, Cham (2016). https://doi.org/10.1007/978-3-319-46493-0_38
3. Hu, J., et al.: Squeeze-and-excitation networks. arXiv:1709.01507 (2017)
4. Islam, M., Ren, H.: Multi-modal pixelnet for brain tumor segmentation. In: Crimi, A., Bakas, S., Kuijf, H., Menze, B., Reyes, M. (eds.) BrainLes 2017. LNCS, vol. 10670, pp. 298–308. Springer, Cham (2018). https://doi.org/10.1007/978-3-319-75238-9_26
5. Jesson, A., Arbel, T.: Brain Tumor segmentation using a 3D FCN with multi-scale loss. In: Crimi, A., Bakas, S., Kuijf, H., Menze, B., Reyes, M. (eds.) BrainLes 2017. LNCS, vol. 10670, pp. 392–402. Springer, Cham (2018). https://doi.org/10.1007/978-3-319-75238-9_34
6. Kamnitsas, K., et al.: Ensembles of multiple models and architectures for robust brain tumour segmentation. In: Crimi, A., Bakas, S., Kuijf, H., Menze, B., Reyes, M. (eds.) BrainLes 2017. LNCS, vol. 10670, pp. 450–462. Springer, Cham (2018). https://doi.org/10.1007/978-3-319-75238-9_38
7. Lin, M., et al.: Network in network. arXiv preprint arXiv:1312.4400 (2013)
8. Menze, B.H., et al.: The multimodal brain tumor image segmentation benchmark (BRATS). IEEE Trans. Med. Imaging **34**(10), 1993 (2015)
9. Pereira, S., et al.: Brain tumor segmentation using convolutional neural networks in MRI images. IEEE Trans. Med. Imaging **35**(5), 1240–1251 (2016)
10. Ronneberger, O., Fischer, P., Brox, T.: U-Net: convolutional networks for biomedical image segmentation. In: Navab, N., Hornegger, J., Wells, W.M., Frangi, A.F. (eds.) MICCAI 2015. LNCS, vol. 9351, pp. 234–241. Springer, Cham (2015). https://doi.org/10.1007/978-3-319-24574-4_28
11. Shen, H., Wang, R., Zhang, J., McKenna, S.J.: Boundary-aware fully convolutional network for brain tumor segmentation. In: Descoteaux, M., Maier-Hein, L., Franz, A., Jannin, P., Collins, D.L., Duchesne, S. (eds.) MICCAI 2017. LNCS, vol. 10434, pp. 433–441. Springer, Cham (2017). https://doi.org/10.1007/978-3-319-66185-8_49
12. Simonyan, K., Zisserman, A.: Very deep convolutional networks for large-scale image recognition. arXiv:1409.1556 (2014)
13. Xie, S., et al.: Aggregated residual transformations for deep neural networks. In: CVPR, pp. 5987–5995 (2017)
14. Yu, F., Koltun, V.: Multi-scale context aggregation by dilated convolutions. In: ICLR (2016)
15. Zhao, X.: A deep learning model integrating FCNNs and CRFs for brain tumor segmentation. Med. Image Anal. **43**, 98–111 (2018)

# Cost-Sensitive Active Learning
# for Intracranial Hemorrhage Detection

Weicheng Kuo[1(✉)], Christian Häne[1], Esther Yuh[2], Pratik Mukherjee[2],
and Jitendra Malik[1]

[1] University of California Berkeley, Berkeley, CA 94720, USA
wckuo@berkeley.edu
[2] University of California San Francisco School of Medicine,
San Francisco, CA 94143, USA

**Abstract.** Deep learning for clinical applications is subject to stringent
performance requirements, which raises a need for large labeled datasets.
However, the enormous cost of labeling medical data makes this challenging. In this paper, we build a cost-sensitive active learning system for the
problem of intracranial hemorrhage detection and segmentation on head
computed tomography (CT). We show that our ensemble method compares favorably with the state-of-the-art, while running faster and using
less memory. Moreover, our experiments are done using a substantially
larger dataset than earlier papers on this topic. Since the labeling time
could vary tremendously across examples, we model the labeling time
and optimize the return on investment. We validate this idea by core-set
selection on our large labeled dataset and by growing it with data from
the wild.

**Keywords:** Artificial intelligence · Computer aided diagnosis
Segmentation

## 1 Introduction

Clinical applications set very high bars for machine learning algorithms, because
any misdiagnosis could impact treatment plans and gravely harm the patient. To
reach the required performance, supervised learning is the leading technique, and
its success is well established. However, a challenge in supervised learning is that
it requires a large amount of labeled data, especially when deep neural networks
are used. Unfortunately, expert labeling of medical images requires enormous
time and cost. The problem is exacerbated when accurate pixelwise labeling
is required. Accordingly, medical segmentation datasets tend to be relatively
small [1,2].

Active learning (AL) aims to address the paucity of labeled data by reasoned
choice of which available unlabeled examples to annotate [3–7]. A limitation of
many prior studies of AL is that they validated AL only in a core-set selection
setting, [8] rather than demonstrating its utility in growing the labeled data,

© Springer Nature Switzerland AG 2018
A. F. Frangi et al. (Eds.): MICCAI 2018, LNCS 11072, pp. 715–723, 2018.
https://doi.org/10.1007/978-3-030-00931-1_82

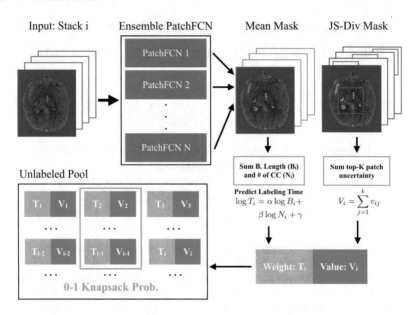

**Fig. 1.** Overview. First, the stack runs through the ensemble PatchFCNs trained on the seed set $S$, which produces the mean hemorrhage heatmap and the Jensen-Shannon (JS) divergence uncertainty heatmap. From the mean hemorrhage heatmap, we apply multiple thresholds to compute the mean boundary length $B_i$ and number of connected components $N_i$. Our log-regression model then takes $B_i$ and $N_i$ to predict the stack labeling time $T_i$. The sum of uncertainty of the top-K uncertain patches is defined to be the stack uncertainty $V_i$. Given any fixed labeling budget(time) $Q$, we treat each stack in the unlabeled pool as an item of weight $T_i$ and value $V_i$. The optimal set of items for annotation is obtained by solving a 0-1 Knapsack problem with dynamic programming.

and also did not attempt to model the cost of labeling [3,4,7]. However, the potential value/use of AL is not in achieving comparable performance with less data, but in improving the model while also minimizing labeling costs. On other problems it has been shown that labeling costs vary greatly from one example to another [3,9,10]. In the case of intracranial hemorrhage, we observe that times needed for pixelwise labeling vary up to 3 orders of magnitude for different cases (See Fig. 3). Most AL studies to date select examples without addressing this wide variation in labeling time [4–8].

In this paper, we propose a cost-sensitive AL system by combining the query-by-committee [5] approach with labeling time prediction for each example. Our uniform-cost AL system compares favorably with the state of the art [4], while the cost-sensitive system gives a further boost under labeling time constraints. All experiments are conducted on our pixelwise-labeled dataset (29095 frames), which is about two orders of magnitude larger than standard MICCAI segmentation datasets [1,2]. Moreover, our system is simpler, faster, and uses less memory than earlier works [4,8]. Through the example of intracranial hemorrhage detection, we demonstrate the potential of cost-sensitive active learning to scale up medical datasets efficiently.

## 2    Supervised Learning System

As a machine learning system we use a convolutional neural network (CNN). More specifically we use a fully convolutional neural network (FCN). FCNs are able to make pixelwise predictions. The standard approach for using an FCN is to input the entire image into the FCN and obtain pixelwise predictions with a single forward pass [11,12]. We instead use an FCN which uses a patch as input and makes predictions for presence of hemorrhage for each pixel within a specific patch at a time, which we call PatchFCN. This architecture has the advantage that the network has to make its predictions based on the local morphology and hence is less prone to overfit into the global context, which results in better test time accuracy than standard FCNs. At test time we apply the PatchFCN in a sliding window fashion (see Fig. 2). We extensively tested this network architecture in a separate technical report [13] and established that it outperforms whole image baselines for various underlying FCN architectures. We use the 38 layer dilated residual net (DRN) as specific FCN architecture. It uses dilated convolutions to preserve spatial resolution together with residual connections [14]. We also group the pixelwise predictions into regions using connected component analysis and aggregate the pixelwise predictions into frame and stack classification scores. This facilitates hemorrhage detection at the pixel, region, frame and stack level.

## 3    Cost-Sensitive Active Learning

Let us define our active learning problem as follows: given a labeled seed set $S$ and an unlabeled pool set $U$, find a small subset $P$ from $U$ for labeling that maximizes a suitable test set metric. Our system which is depicted in Fig. 1 estimates an uncertainty score for each example (see Sect. 3.1) and the labeling time (see Sect. 3.2). The goal is to select the set of examples such that the sum of their uncertainty is maximized under the constraint that the total estimated labeling time stays within a given budget. The optimal selection of items reduces to the well-known 0-1 Knapsack problem, which can be solved with dynamic programming.

### 3.1    Uncertainty Measure

Uncertainty (or informativeness) is at the core of active learning techniques. It can be estimated by single model outputs [6] or a committee of models [5]. The idea of query-by-committee (QBC) is to run multiple models on the same example and use their disagreement to estimate uncertainty. Experimentally, we found that QBC consistently works better than single-model uncertainty. Within

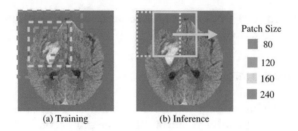

(a) Training                (b) Inference

Patch Size
■ 80
■ 120
■ 160
■ 240

**Fig. 2.** PatchFCN system. We train the network on patches and test it in a sliding window fashion. The optimal crop size is found to be $160 \times 160$ for our task.

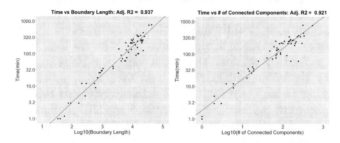

**Fig. 3.** Left: Time vs Log (Boundary Length). Right: Time vs Log (Number of Connected Components). Both plots show the goodness of our linear fit and the normality of residuals after the log transform. Note that the y-axis is actually displayed in log-scale.

the QBC framework, we have tried various uncertainty measures and found the Jensen-Shannon (JS) divergence to work best. Concretely, let's assume we have $N$ models in the committee and the output distribution of model $i$ is $P_i$. The JS divergence is then defined as:

$$JS(P_1, P_2, \ldots, P_N) = H\left(\frac{1}{N} \sum_i^N P_i\right) - \frac{1}{N} \sum_i^N H(P_i) \qquad (1)$$

where H is the entropy function.

We average all pixelwise uncertainties within each patch to obtain the uncertainty of a patch. The stack uncertainty is obtained by averaging the top $K$ uncertain patches within the stack. The choice of $K$ is a balance between taking the max ($K = 1$) or the mean ($K = \infty$) of the whole stack. In all AL experiments in this paper, we set $K = 200$ and number of models $N = 4$. We have tried larger $N$ but didn't gain any performance. Visualization of such uncertainty can be found in Fig. 6.

## 3.2    Labeling Time Prediction

First, we need to ask what is the optimal unit of labeling – patch, frame or stack? Employing our neuro-radiology expertise, we settled on labeling stacks.

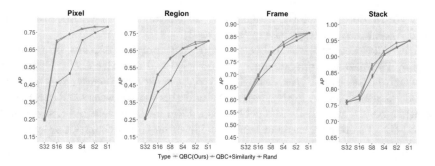

**Fig. 4.** Core-set selection curves. Our system (QBC) starts to outperform [4] (QBC + Similarity) on region, frame and stack level as the dataset grows beyond one fourth of the whole set. Both QBC algorithms maintain a large gap with random baselines on pixel and region APs. For the frame and stack APs, our system still maintains a healthy margin above the random baseline for all data sizes. The region AP is computed following the definition of [13].

While labeling patches/frames may seem more effective from a machine learning perspective, it comes with a severe overhead, i.e. the whole stacks need to be retrieved and examined by radiologists anyway. Therefore, it is less efficient than labeling the stacks.

To apply active learning in practice, we need to ensure it actually saves labeling cost or efforts. This is crucial as per-stack labeling times in our data span 3 orders of magnitude. We utilize linear regression to predict the log labeling time $\log t$ based on two features: (1) mask boundary length $B$, and (2) number of connected components $M$ under the log-transform.

$$\log t = \alpha \log B + \beta \log M + \gamma \tag{2}$$

Figure 3 shows the effectiveness of our log-transform and the goodness of fit on both features. 61 data points were used to fit the linear model, which we found to be sufficient. In order to compute the features at test time we use the pixelwise predictions of our network. We also tried using deep FCN features from an intermediate layer directly but found the prediction to be less stable.

## 4    Data Collection

Our pixelwise labeled dataset contains 1247 clinical head CT scans (29095 valid frames) performed from 2010–2017 on 64-detector-row CT scanners (GE, Siemens) at our affiliated hospitals. Each scan is a stack of 27-38 frames with in-plane resolution close to 0.5mm and z-axis resolution of 5mm. Scans were anonymized by removing all protected health information as well as skull, scalp and face. A board-certified neuroradiologist with specialization in traumatic brain injury (TBI) identified areas of acute intracranial hemorrhage at the pixel level. We randomly split the dataset into a trainval/test set of 934/313 stacks, called $S_{trainval}$, $S_{test}$ respectively (S for seed).

The unlabeled set was collected using key phrase searches of radiology reports. We searched independently for positive and negative cases. The search for positive cases over 1 year yielded 1755 cases. A separate search over a shorter period identified 640 negative cases. We call this set of cases set U (for unlabeled) to be distinguished from set S. Also, 120 randomly selected cases from U (called $U_{test}$) were annotated at stack level in order to benchmark our system in this domain.

## 5    Experiments

### 5.1    Core-Set Active Learning

A core-set is a subset of the training set where the empirical loss of a model is similar to that on the entire training set. In this experiment, we grow the core-set iteratively and study how the performance improves [4,8]. For fair comparison, we strip away the cost prediction and Knapsack-solving part of our full system (See Fig. 1), and select examples based on their uncertainty scores alone.

We use the average precision (AP) metric to compare algorithms. Figure 4 shows the performance of our query-by-committee system (QBC), suggestive annotation system (QBC + Similarity) [4], and random baseline. In this comparison, we improve [4] by using the patch-based approach for QBC + Similarity baseline, because PatchFCN [13] gives better uncertainty and similarity measures than vanilla FCN. Without it, we observed a significant performance drop. Following [4], we tried diversifying the ensemble with bootstrapping, but did not see benefit.

The experiment began with a seed set 1/32 of the training set, and doubled it by either random sampling or active learning. In the next round, this doubled set becomes the new seed set and the process repeats. In each round, we trained an ensemble for all methods in order to compute QBC uncertainty. Figure 4 shows that our system's performance at half the dataset (S2) closely matches the performance of using the whole dataset (S1) for every AP, similar to [4,8]. However, here we use a dataset that is two orders of magnitude larger and much harder to overfit on.

Our experiment indicates that on a large dataset, QBC uncertainty alone could be sufficient to yield competitive performance, if not state-of-the-art. Without bootstrapping or pairwise similarity, our system beats the random baseline by a good margin and compares favorably with [4] in performance and time complexity. The time complexity of core-set approaches [4,8] are dominated by the pairwise similarity computation, which is quadratic and can be expensive in practice when the seed set is too large to be grown by brute-force labeling. In contrast, our system has linear time complexity because it computes everything on-the-fly.

## 5.2  Cost-Sensitive Active Learning

After validating the core-set AL, we model the cost with the full system described in Fig. 1. We randomly select half of our labeled training set as the seed set to mimic the scenario where the seed set is large enough to render naive labeling impractical for growing the data. Yet at the same time we want the pool to be at least as large as the seed. In each iteration, we increment the data by allocating additional *time* to add labeled examples by solving the Knapsack problem. For the random baseline, we randomly select examples to add until no example can fit in the given time anymore. Figure 5 shows the superiority of our system (QBC) over both uniform-cost AL (UAL) and the random baseline in such setting. The result supports Fig. 6 where UAL is biased toward examples with large bleeds and long labeling times. In fact, UAL selected 8/11 stacks in the first/second rounds, whereas cost-sensitive AL (CAL) selected 94/107 stacks. Due to lack of stack diversity, UAL performs worse than CAL at the stack level.

The strong gain of CAL at (+10%) not carrying over to (+20%) is explained by the ratio of unlabeled pool to the labeled training set. When the ratio is small, the data is insufficient for AL system to choose from. In Fig. 4, the ratio starts with 3100% and stops with 100% at S2. In Fig. 5, the ratio started with 100%. After (+10%) round, the ratio is 66% for CAL and 80% for Rand. The leveling off of CAL performance shows that most of the informative examples were already selected in the (+10%) round.

## 5.3  Active Learning in the Wild

Finally, we apply our system on the unlabeled pool described in Sect. 4. First, we train an ensemble on the entire labeled set. Then we select examples from the

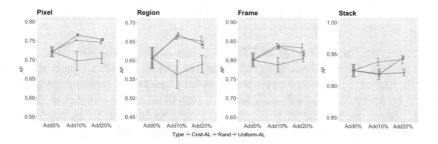

**Fig. 5.** Cost-sensitive active learning. At the first iteration, the system achieves much better performance than the random baseline for all metrics. The random baseline does not improve over the seed set. In the next round, the random baseline improves the stack AP while the ALs remain the same. The error bars of AL come from the network initialization and the stochastic gradient (SGD) training. The error bars of random baseline mostly come from the random addition of data, plus the same sources of AL randomness. The time increment is 10% of the total labeling time of the pool, which simulates the situation where our budget is only a small fraction of the total labeling cost.

**Table 1.** Left: Performance on $S_{test}$. Compared to Ensemble $S_{trainval}$, Ensemble $(S \cup U)_{train}$ performs just as well on the pixel level and slightly outperform on the stack level. Right: Performance on $U_{test}$. Ensemble $(S \cup U)_{train}$ beats Ensemble $S_{trainval}$ by a good margin on the pool set.

| $S_{test}$ | Pixel AP | Stack AP | $U_{test}$ | Stack AP |
|---|---|---|---|---|
| Ens. $(S \cup U)_{train}$ | $77.9 \pm 0.3\%$ | $\mathbf{95.6 \pm 0.9\%}$ | Ens. $(S \cup U)_{train}$ | $\mathbf{90.1 \pm 1.7\%}$ |
| Ens. $S_{trainval}$ | $\mathbf{78.2 \pm 0.2\%}$ | $95.0 \pm 0.1\%$ | Ens. $S_{trainval}$ | $85.1 \pm 0.3\%$ |

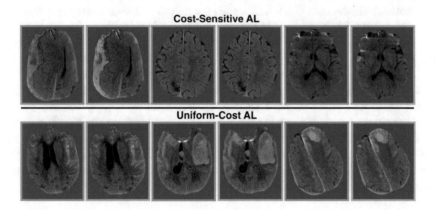

**Fig. 6.** Examples selected by cost-sensitive and uniform-cost AL systems. Blue boxes are the original images, while orange boxes are the images overlaid with Jensen-Shannon divergence. The brightness of the green color indicates uncertainty. The examples selected by uniform-cost system mostly contain massive bleeds and are substantially more time-consuming for annotation, whereas examples by the cost-sensitive system are diverse and meaningful, maximizing the return on investment.

unlabeled pool under a budget of 100 h. A neuroradiologist examined the selected cases and determined there were 115 negatives and 64 positives. There were also 51 subacute or postsurgical cases we excluded. The actual labeling time turned out to be within 10% of our estimate. We call these newly annotated examples $U_{train}$, to be distinguished from $S_{trainval}$ defined in Sect. 4. To qualitatively assess the impact of cost modeling, we show examples mined by both uniform-cost and cost-sensitive AL in Fig. 6.

For quantitative benchmarking, we trained an ensemble of 4 PatchFCNs from scratch with the newly augmented data (Ensemble $S_{trainval}+U_{train}$) and compared them with the ensemble trained on the original data (Ensemble $S_{trainval}$). The results on $S_{test}$ and $U_{test}$ are shown in Table. 1. We benchmark on two test sets here because we care about the performance on both seed $S$ and pool $U$ domains, which in practice are often not exactly the same. The gain on $S_{test}$ shows that our method works despite the domain shift, and the strong gain on $U_{test}$ demonstrates how a model trained on large data can be improved by collecting a little more data judiciously.

# 6    Conclusion

In this paper, we proposed a cost-sensitive, query-by-committee active learning system for intracranial hemorrhage detection. We validated it on a substantially larger pixelwise labeled dataset than earlier works and applied it to improve the model by annotating new data from the wild. Our study demonstrates the potential of growing large medical datasets to the next level with cost-sensitive active learning.

**Acknowledgments.** This work was supported in part by California Initiative to Advance Precision Medicine. Christian Häne received funding from the Swiss National Science foundation (165245). Amazon Web Services provided part of the compute time.

# References

1. Sirinukunwattana, K., et al.: Gland segmentation in colon histology images: the glas challenge contest. Med. Image Anal. **35**, 489–502 (2017)
2. Zhang, Y., Ying, M.T., Yang, L., Ahuja, A.T., Chen, D.Z.: Coarse-to-fine stacked fully convolutional nets for lymph node segmentation in ultrasound images. In: BIBM (2016)
3. Settles, B., Craven, M., Friedland, L.: Active learning with real annotation costs. In: NIPS Workshop on Cost-sensitive Learning (2008)
4. Yang, L., Zhang, Y., Chen, J., Zhang, S., Chen, D.Z.: Suggestive annotation: a deep active learning framework for biomedical image segmentation. In: Descoteaux, M., Maier-Hein, L., Franz, A., Jannin, P., Collins, D.L., Duchesne, S. (eds.) MICCAI 2017. LNCS, vol. 10435, pp. 399–407. Springer, Cham (2017). https://doi.org/10. 1007/978-3-319-66179-7_46
5. Seung, H.S., Opper, M., Sompolinsky, H.: Query by committee. In: Workshop on Computational Learning Theory (1992)
6. Lewis, D.D., Gale, W.A.: A sequential algorithm for training text classifiers. In: SIGIR (1994)
7. Mahapatra, D., Schüffler, P.J., Tielbeek, J.A.W., Vos, F.M., Buhmann, J.M.: Semi-supervised and active learning for automatic segmentation of crohn's disease. In: Mori, K., Sakuma, I., Sato, Y., Barillot, C., Navab, N. (eds.) MICCAI 2013. LNCS, vol. 8150, pp. 214–221. Springer, Heidelberg (2013). https://doi.org/10.1007/978-3-642-40763-5_27
8. Sener, O., Savarese, S.: Active learning for convolutional neural networks: a core-set approach. In: ICLR (2018)
9. Settles, B.: Active learning. In: Lectures on AI and ML (2012)
10. Tomanek, K.: Resource-aware annotation through active learning (2010)
11. Long, J., Shelhamer, E., Darrell, T.: Fully convolutional networks for semantic segmentation. In: CVPR (2015)
12. Yuh, E., Mukherjee, P., Manley, G.: Interpretation and quantification of emergency features on head computed tomography. Provisional Application no. 62/269,778 (2015)
13. Kuo, W.C., Häne, C., Yuh, E., Mukherjee, P., Malik, J.: PatchFCN for intracranial hemorrhage detection. In: arXiv preprint arXiv:1806.03265 (2018)
14. Yu, F., Koltun, V., Funkhouser, T.: Dilated residual networks. In: CVPR (2017)

# Author Index

Printed in the United States
By Bookmasters